AF776243

Hemoglobin and Its Diseases

A subject collection from *Cold Spring Harbor Perspectives in Medicine*

OTHER SUBJECT COLLECTIONS FROM *COLD SPRING HARBOR PERSPECTIVES IN MEDICINE*

Addiction
Parkinson's Disease
Type 1 Diabetes
Angiogenesis: Biology and Pathology
HIV: From Biology to Prevention and Treatment
The Biology of Alzheimer Disease

SUBJECT COLLECTIONS FROM *COLD SPRING HARBOR PERSPECTIVES IN BIOLOGY*

DNA Replication
The Endoplasmic Reticulum
Wnt Signaling
Protein Synthesis and Translational Control
The Synapse
Extracellular Matrix Biology
Protein Homeostasis
Calcium Signaling
The Golgi
Germ Cells
The Mammary Gland as an Experimental Model
The Biology of Lipids: Trafficking, Regulation, and Function
Auxin Signaling: From Synthesis to Systems Biology
The Nucleus
Neuronal Guidance: The Biology of Brain Wiring
Cell Biology of Bacteria
Cell–Cell Junctions
Generation and Interpretation of Morphogen Gradients
Immunoreceptor Signaling
NF-κB: A Network Hub Controlling Immunity, Inflammation, and Cancer
Symmetry Breaking in Biology
The Origins of Life
The p53 Family

Hemoglobin and Its Diseases

A subject collection from *Cold Spring Harbor Perspectives in Medicine*

EDITED BY

David J. Weatherall
Weatherall Institute of Molecular Medicine

Alan N. Schechter
National Institutes of Health

David G. Nathan
Dana-Farber Cancer Institute

COLD SPRING HARBOR LABORATORY PRESS
Cold Spring Harbor, New York • www.cshlpress.org

Hemoglobin and Its Diseases

A Subject Collection from *Cold Spring Harbor Perspectives in Medicine*

Articles online at www.perspectivesinmedicine.org

All rights reserved
© 2013 by Cold Spring Harbor Laboratory Press, Cold Spring Harbor, New York
Printed in the United States of America

Executive Editor	Richard Sever
Managing Editor	Maria Smit
Project Manager	Barbara Acosta
Permissions Administrator	Carol Brown
Production Editor	Diane Schubach
Production Manager/Cover Designer	Denise Weiss
Publisher	John Inglis

Front cover artwork: The image shows the phenomenon of red cell rosetting. This is how normal red cells aggregate around malaria parasite–infected cells and are a marker of the severity of malaria. It is significantly diminished in α-thalassemic red cells that are infected. Image supplied by D.J.P. Ferguson, Oxford University.

Library of Congress Cataloging-in-Publication Data

Hemoglobin and its diseases/edited by David J. Weatherall, Alan N. Schechter, David G. Nathan.
p. ; cm
"A subject collection from Cold Spring Harbor perspectives in medicine."
Includes bibliographical references and index.
ISBN 978-1-936113-45-3 (hardcover : alk. paper)
I. Weatherall, D. J. II. Schechter, Alan N. III. Nathan, David G., 1929- IV. Cold Spring Harbor perspectives in medicine.
[DNLM: 1. Hematologic Diseases–Collected Works. 2. Hemoglobins–Collected Works. WH 120]

612.1′111–dc23

2012027357

10 9 8 7 6 5 4 3 2 1

All World Wide Web addresses are accurate to the best of our knowledge at the time of printing.

Authorization to photocopy items for internal or personal use, or the internal or personal use of specific clients, is granted by Cold Spring Harbor Laboratory Press, provided that the appropriate fee is paid directly to the Copyright Clearance Center (CCC). Write or call CCC at 222 Rosewood Drive, Danvers, MA 01923 (978-750-8400) for information about fees and regulations. Prior to photocopying items for educational classroom use, contact CCC at the above address. Additional information on CCC can be obtained at CCC Online at www.copyright.com.

For a complete catalog of all Cold Spring Harbor Laboratory Press publications, visit our website at www.cshlpress.org.

Contents

Preface

THE INHERITED DISORDERS OF HEMOGLOBIN are by far the commonest Mendelian diseases. It is estimated that between 300,000 and 400,000 babies are born with these conditions each year. They were the first genetic diseases to be characterized at the molecular level; hence, early work in this field led to the era of molecular medicine.

It is a surprising fact that, although work on the inherited disorders of hemoglobin opened up the whole field of human molecular genetics between 1950 and 1970, more recently it has become sidetracked from the mainstream of human genetics. Indeed, there has been a steady decline in its place in the literature of human genetics over the last 40 years. This is particularly surprising because the human hemoglobin field has disclosed many hundreds of the different mutations that are associated with monogenic disease, taught us more about the genetic basis for the phenotypic heterogeneity of inherited diseases than any other condition, and has provided increasing insight into the evolutionary and population biology of these conditions.

Why have the hemoglobin disorders drifted away from the mainstream of human molecular genetics over recent years? Although it is not absolutely clear why this has happened it probably reflects the way in which human genetics in general, and medical genetics in particular, have evolved over the last half-century. Medical genetics only became a clinical specialty in a few key centers in the United States and Europe in the 1950s and 1960s, and it was natural that they should focus on chromosomal conditions and monogenic diseases, which were common in their particular populations. Although there were increasing numbers of immigrant communities in these regions in which inherited hemoglobin disorders were not uncommon, they tended to be seen in hematology rather than genetics centers. In this way the two fields gradually drifted apart. This tendency is unfortunate because the remarkable technical developments in genomics have so much to offer each field.

The inherited disorders of hemoglobin occur at their highest frequency in the developing countries of the tropical belt, although they are being also seen with increasing frequency in emigrants from this region to the richer countries. As the poorer countries go through an epidemiological transition characterized by improved hygiene, nutrition, and public health facilities and better medical care in general, babies with hemoglobin disorders who would have died in the first few years of life are now surviving to present for diagnosis and treatment. This phenomenon occurred in many Mediterranean populations after World War II and will undoubtedly increase the global burden of hemoglobin disorders in many developing countries in the future. For example, it has been estimated recently that if early deaths due to sickle cell anemia are controlled in sub-Saharan Africa, the population of patients with this disorder could reach five to six million in the foreseeable future.

For these reasons, in this book we have attempted to give a broad coverage of the hemoglobin disorders, ranging from their basic biology and molecular pathology, through their clinical manifestations and management, to the various aspects of their population biology and the reasons for their high and increasing frequency. As well as being of value to workers in the hemoglobin field, we hope that this broad view will also be of interest to workers in medical genetics in general because of the many lessons that have been learned from the study of the abnormal hemoglobins, which are relevant to other monogenic diseases and, at least to some extent, to polygenic disease as well.

We acknowledge the great help that we have had from our colleagues who have given a great deal of time to produce their individual chapters. We are also grateful to the considerable help that we have received from the staff of Cold Spring Harbor Laboratory Press, particularly Barbara Acosta, for their continued help through the long generation of this book. Finally, we thank Liz Rose for her invaluable help in every aspect of the production of this book at each stage of its development.

DAVID J. WEATHERALL
ALAN N. SCHECHTER
DAVID G. NATHAN

Hemoglobin and Its Disorders: 150 Years of Study

David J. Weatherall[1], Alan N. Schechter[2], and David G. Nathan[3]

[1]Weatherall Institute of Molecular Medicine, University of Oxford, John Radcliffe Hospital, Oxford OX3 9DS, United Kingdom

[2]Molecular Medicine Branch, NIDDK, National Institutes of Health, Bethesda, Maryland 20892

[3]Dana-Farber Cancer Institute, Children's Hospital, Boston, Massachusetts 02215

Correspondence: aschechter@helix.nih.gov

HEMOGLOBIN AND THE EVOLUTION OF NATURAL PHILOSOPHY INTO MOLECULAR MEDICINE

One hundred and fifty years ago, Felix Hoppe gave the name hemoglobin to the red pigment in blood whose light absorption spectrum changed on addition of oxygen. This publication in 1862 and related work by this German physician, who was later known as Hoppe-Seyler and who in 1877 founded the long published *Zeitschrift fur Physiologische Chemie*, clearly marks the creation of the field we now know as physiological chemistry or biochemistry. In the seventeenth and eighteenth centuries, the study of air and other gases by natural philosophers—who we would now identify as physicists or chemists, such as Boyle, Lower, Priestly, Lavoisier, and Davy—had led to the identification and recognition of the importance of oxygen as a component of air and the slow recognition that this gas interacted with blood via the pulmonary circulation in mammals. Harvey had missed this fundamental aspect of the circulation of the blood, presumably because without the knowledge of oxygen no function could be ascribed to it.

Hoppe-Seyler's work was very quickly followed by that of the mathematician George Stokes in England who used various chemical reductants and spectroscopic measurements to show the reversibility of oxygen binding to hemoglobin. At approximately the same time, hemoglobin was crystallized from the blood of many species of animals with significant differences noted in the formation of these crystals. Acid treatment of these hemoglobins had identified an iron-containing species, later known as hemin, which was a constituent of the colorless protein species that remained after this separation. Precise chemical analyses showed that in the various hemoglobin preparations the iron content was always close to 0.335%, suggesting a surprisingly large minimum molecular mass of 16,670 Da, assuming one iron atom per protein molecule. Thus, by the end of the nineteenth century, the basic structure and function of hemoglobin inside the red blood cells, which themselves had long puzzled microscopists because of their lack of a nucleus, had been ascertained. Perhaps most importantly, the concepts of physical and organic chemistry had been combined with that of physiology in the study of this ubiquitous and abundant protein with vital functions in much of the animal kingdom. Further, these techniques and concepts strongly shaped the development of biochemistry in the first half of the last century, a pattern of influence that has persisted as biochemistry itself

Copyright © 2013 Cold Spring Harbor Laboratory Press; all rights reserved
Cite this article as *Cold Spring Harb Perspect Med* doi: 10.1101/cshperspect.a013052

evolved into molecular biology and, later, molecular medicine.

In the 1920s, Gilbert Adair in England used osmotic pressure measurements to infer that the functional molecular mass of hemoglobin was about 67,000; this value was quickly confirmed by The Svedbeg in Sweden using his newly created method of ultracentrifugation. These results were astounding to many, as the concept of macromolecules was still poorly formed at a time when nucleic acids were commonly thought of as being made of distinct units of four nucleotides. However, in view of the fact that the results had come from several means of study, they were gradually accepted and led to the realization that the functional unit of hemoglobin was likely to be a tetramer of four heme-containing protein species. In the late 1950s, this model was further refined by early results from the rapidly developing field of peptide chemistry. The concept that the tetramer was composed, even in a single species or in a single individual, of pairs of several varieties of identical dimeric subunits and that the dimers in turn could be combinations of each of several different monomeric species, i.e., α-like and β-like heme proteins, became apparent. Full protein sequencing by Braunitzer and others was soon accomplished and the chemical bases of these heterogeneities became clear. These chemical results also gave the necessary information to rationalize the mathematical models that had been used to explain the studies of function, i.e., binding of oxygen and other ligands, that had also started at the beginning of the twentieth century (see below).

In 1904, Christian Bohr in Denmark discovered both the unexpected sigmoidal shape of curves of percentage oxygen saturation as a function of partial pressure of oxygen and, with colleagues such as August Krogh, the sensitivity of these curves to carbon dioxide concentration, later realized to be primarily a pH effect (Fig. 1) (Bohr et al. 1904). In later work by these investigators and others, including Hasselbalch, Barcroft, Hill, Haldane, Roughton, and Henderson, the basic physiology of gas binding to hemoglobin, both inside the red cell and free in solution, were established. Perhaps the last major surprising experimental discovery of this story

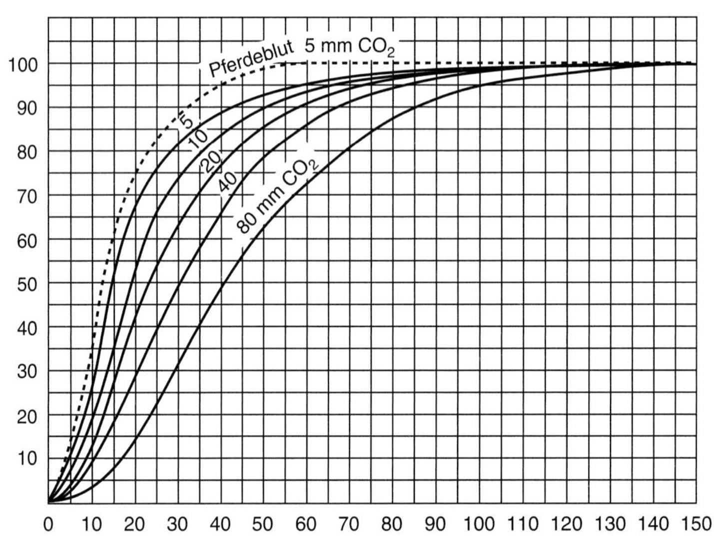

Figure 1. Experiment of Bohr et al. (1904) showing the effect of carbon dioxide tension on the binding of O_2 to hemoglobin. The ordinate shows the percentage of saturation of hemoglobin with O_2. The abscissa shows oxygen tension in mmHg. Note that the experimental points show sigmoid binding. (This figure was taken directly from the original paper of Bohr and associates; reproduced by Bunn and Forget 1986, and reused here, with permission.)

was that of Ruth and Rheinhold Benesch, who in the late 1960s, showed that 2,3-diphosphoglyceric acid (also called 2,3-bisphosphoglyceric acid), which is of unusually high concentration in the erythrocyte, was a strong modulator of oxygen affinity, as is pH in the "Bohr" effect.

Again these studies, although crucial for understanding the role of hemoglobin in transporting oxygen (and carbon dioxide and carbon monoxide and, later, nitric oxide—the four "biological" gases), had broader implications for our understanding of protein function in general. In 1935, Linus Pauling, interested in the nature of the chemical bonds linking oxygen to the heme iron atoms, realized that the cooperativity of oxygen binding to hemoglobin implied by the sigmoidal (rather than typical hyperbolic) saturation curves indicated chemical interactions among the oxygen-binding sites on each tetramer. This concept and the even more general one of thermodynamic linkage in the energetics of all ligand binding to hemoglobin, developed initially by L.J. Henderson, were the forerunners of the concept of allosteric interactions, developed after the Second World War by Jeffries Wyman, Jacques Monod, and others and in a different form by Daniel Koshland.

By the 1960s and 1970s, hemoglobin studies were poised to follow the reconceptualization of biochemistry as molecular biology, with both genetic and structural developments. The development of recombinant DNA technology, including the sequencing and synthesizing of DNA fragments and the detection and cloning of genes were applied to the globin-gene clusters in many species. Perhaps the landmark study in a rush of developments was that of Tom Maniatis in cloning the rabbit β-globin gene, showing that even in the complex eukaryotic genome individual genes could be identified and isolated. This opened the way for Efstratiadis, Weissman, Forget, and others to clone the genes for the two (α-like and β-like) human globin clusters, to work out their genomic arrangements, to identify the corresponding messenger RNA molecules, and begin the study of transcriptional control, which is still a poorly understood field. Again, the abundance and ease of study of the hemoglobin "system" made it an attractive and tractable goal for the application of the methods of what was then the new frontier of biochemistry.

The structural advances trace to the decision of Max Perutz (Fig. 2) working in Cambridge, United Kingdom in the late 1930s to attempt to apply the techniques of X-ray crystallography structural determination to the beautiful crystals of hemoglobin that had first been prepared in the mid-nineteenth century. At that time, these methods had not been successful with macromolecules and it took many decades to develop ways to allow their application. Working with many others, especially John Kendrew who focused on the simpler myoglobin molecule, Perutz was able to determine a fairly high-resolution image of hemoglobin molecules by the early 1960s, with detailed functional implications obvious by his classic papers in 1970. These results opened up the floodgates for the determination of macromolecular structure by these methods, so that as of this writing several thousand proteins and other molecules and often their interactions are now available. Thus, hemoglobin studies were at the forefront of both the genetic and structural components of molecular biology. From the point of hemoglobin studies, the

Figure 2. From Max Perutz (1914–2002); reprinted, with permission, from The Medical Research Council, United Kingdom.

work of Perutz—culminating slightly more than a century after Hoppe's paper—has led to the belief that the physiological aspects of the field are quite "mature" and it has been suggested that there are few major findings left in its study. Recent work concerning the interaction of nitric oxide with hemoglobin may challenge this assumption. These early advances in an understanding of the structure and function of hemoglobin are discussed in more detail by Edsall (1972) and Bunn and Forget (1986).

However, medical advances during the last 60 years also gave immense impetus to another aspect of the study of hemoglobin. The elucidation of sickle cell anemia as a "molecular disease" because of changes in the chemical structure of the protein and of the thalassemia syndromes as largely defects of transcriptional control in mutated genes, allowed the study of hemoglobin to largely define the paradigm for what has recently been called "molecular medicine." This paradigm, which even more recently has been in some apposition to "genomic medicine," is the basis for much of the theme of the remainder of this introduction and indeed of this entire collection in the Cold Spring Harbor monograph series on Perspectives in Medicine.

THE BIRTH OF MEDICAL GENETICS AND THE DISCOVERY OF INHERITED DISORDERS OF HEMOGLOBIN

Although hints that some diseases might have a familial tendency go back as far as the ancient Greeks, genetics only became established as a field of medical research at the beginning of the twentieth century. In 1899 and 1901, the British clinician Archibald Garrod published accounts of a condition called alkaptonuria, characterized by joint disease and the passage of dark pigment in the urine (Bearn 1993). In his second paper, he emphasized the frequency of consanguinity in their parents. Although he did not appreciate the significance of these findings, William Bateson, a botanist and one of the leading protagonists of Mendel's work at the time, became aware of them and pointed out to Garrod that they were just what might be expected in a disease that had a Mendelian recessive form of inheritance, or, in other words, that it was passed on by each of the symptomless parents to the affected child. Garrod studied several more inherited disorders of this type and, in 1908, presented this work in his Croonian Lectures to the Royal College of Physicians of London. But although these lectures, which were published later entitled *Inborn Errors of Metabolism* (Garrod 1909), marked the beginning of the discipline of medical genetics, little attention was paid to this work for many years.

Indeed, it was not until after the Second World War that medical genetics finally was established as an important specialty. However, during the first half of the twentieth century, a number of important genetic diseases were first described, although in each case it took some time before the form of genetic transmission was established. In 1910, James Herrick, a Chicago physician, presented a paper to the Association of American Physicians in which he described a student from Grenada, West Indies, with chronic anemia and curiously elongated and sickle-shaped red blood cells (Herrick 1910). Although over subsequent years further cases were described, and it became clear that this disease was largely although not entirely restricted to those of African origin, it took many years for its pattern of inheritance as an autosomal recessive to be determined; in the late 1940s, studies in both the United States and Africa confirmed that this was the case.

In 1925, a second form of inherited anemia was discovered by Thomas Cooley, a pediatrician from Detroit. Cooley's first account of what became known as thalassemia appeared in a single-page abstract in *Transactions of the American Pediatric Society* (Cooley and Lee 1925). It describes five young children with anemia and enlargement of the spleen and liver, together with discoloration of the skin and a curious Mongoloid appearance of the skull. Cooley was also impressed by the remarkable radiological changes in the skull and long bones of these children and by the abnormal morphology of their red cells, which were pale and varied in shape and size. A similar although milder condition was described quite independently in the same year by three

Italian workers, Rietti, Greppi, and Micheli, and for many years was referred to in the Italian literature as "La Malattia di Rietti-Greppi-Micheli" (Weatherall and Clegg 2001). This disease received its name "thalassemia" in a classical paper on the first autopsies of this condition (Whipple and Bradford 1932). All the cases seen in the United States up to that time had a Mediterranean origin and, hence, Whipple and Bradford invented the word thalassemia, which mixes the Greek roots for the "sea" and "blood." Like sickle cell anemia, it was many years before it became clear that thalassemia is also a disorder inherited as a Mendelian recessive (Weatherall and Clegg 2001). During subsequent years, reports of patients with a disease resembling thalassemia gradually appeared from many different parts of the world and it was soon clear that this condition is not restricted to the Mediterranean region.

EARLY DISCOVERIES ON THE NATURE OF THE INHERITED DISORDERS OF HEMOGLOBIN

During the 1930s and 1940s, several independent streams of work finally came together to provide evidence regarding the precise nature of the inherited disorders of hemoglobin and, incidentally, began the era of "molecular medicine."

First, as discussed earlier, Max Perutz, a refugee from Germany who was working in Cambridge, England, after many fruitless years of research, was able to describe the fine structure of hemoglobin as evidenced by X-ray crystallography. He showed that it was a complex molecule consisting of four peptide chains: two identical α-chains and two β-chains each attached to a heme moiety. He also began to describe the allosteric properties of this molecule required as an effective oxygen transporter.

In 1945, the distinguished American protein chemist Linus Pauling was traveling on an overnight train with William Castle, the Boston hematologist, who told him that he had noticed that, when the red cells of patients with sickle cell disease were deprived of oxygen and formed a sickle shape they showed unusual properties when examined under polarized light (Conley 1980). To Pauling, this meant that some form of molecular reorganization must have occurred and that this might indicate that the hemoglobin of patients with sickle cell disease could be in some way abnormal. At about this time, the Swedish chemist Arne Tiselius had developed the technique of electrophoresis, which made it possible to separate proteins of different structure and charge in an electric field. Pauling, together with his postdoctoral student, Harvey Itano, set about building a Tiselius apparatus and examined the hemoglobin of patients with sickle cell disease and their relatives. They found that their hemoglobin moved in an electric field at a different rate from the hemoglobin of normal individuals and, even more importantly, the red cells of unaffected carriers of the sickle cell gene contained both normal and abnormal hemoglobins. This work was published in *Science* in November 1949 under the title "Sickle Cell Anemia, a Molecular Disease" (Pauling et al. 1949). Although Pauling briefly retracted this paper shortly after it was published because of his inability to find differences in the amino acid composition between sickle cell and normal hemoglobin, in 1956, Ingram, working in Cambridge, United Kingdom, was able to show that sickle cell hemoglobin differs from adult hemoglobin by a single amino acid substitution in its β-chain; glutamic acid in normal hemoglobin is replaced by valine in sickle cell hemoglobin (Ingram 1956).

Moving boundary electrophoresis was too complicated to be established in routine laboratories but methods for detecting abnormal hemoglobins using much simpler approaches by filter paper electrophoresis were soon devised and a widespread search for different hemoglobin variants began. Furthermore, a remarkable family was discovered in Baltimore in which there were two hemoglobin variants, one affecting the α-chain and the other the β-chain of hemoglobin (Smith and Torbert 1958). From the way that these variants segregated within this family it became quite clear that there are different sets of genes involved in α- and β-globin-chain synthesis and that these are almost certainly on different chromosomes.

Work in many different laboratories at about this time showed that human hemoglobin changes its pattern during development, with two different embryonic hemoglobins later followed by a fetal hemoglobin and then adult hemoglobin. The different peptide chains of these hemoglobins were sequenced and a detailed picture of the genetic control of hemoglobin gradually emerged.

These studies of abnormal human hemoglobins began to have a major influence on the better understanding of thalassemia. Indeed, as early as 1952, it was suggested that this disease might result from a defect in adult hemoglobin production with persistent production of fetal hemoglobin; elevated levels of the latter had already been observed in many children with a severe form of thalassemia. In 1955, a minor adult hemoglobin was discovered and named hemoglobin A_2 (Kunkel and Wallenius 1955). It was found that the relative level of this component was elevated in the parents of patients with some although not all forms of thalassemia. Another important discovery about this time was the finding of a rapidly migrating hemoglobin variant in a Chinese family, again associated with the clinical picture of thalassemia; this was subsequently found to contain no α-chains and to consist entirely of β-chains, with the structure of β_4 (Jones and Schroeder 1963). Because new hemoglobins, as they were discovered, were designated a letter of the alphabet, this new variant was called hemoglobin H. In 1957, a fetal version of this variant, consisting of four γ-chains, or γ_4, was found in the umbilical cord of an infant whose parents showed evidence of thalassemia (Ager and Lehmann 1958).

But perhaps the most important observation of all, leading to the concept that thalassemias result from reduced globin-chain production, came from studies of the hemoglobin patterns of patients who had inherited the sickle cell gene from one parent and a thalassemia gene from the other, a condition called sickle cell thalassemia. This condition was first described in Italy in the mid-1940s (Silvestroni and Bianco 1944–1945) and, in 1952, the associated hemoglobin pattern was first analyzed. It was found that the red cells of patients who had received a thalassemia gene from one parent and the sickle cell gene from the other contained more hemoglobin S than A, that is, the reverse of what is found in individuals with the sickle cell trait, who have considerably more hemoglobin A than S in their red cells (Sturgeon et al. 1952). Because the sickle cell mutation was known to affect the β-globin chain, this suggested that the action of the thalassemia gene was to preferentially suppress β-globin synthesis. Later, patients were found in whom this interaction did not occur, suggesting that they might have defective α-chain synthesis, which had also been proposed as the basis for patients with a thalassemic blood picture who had hemoglobins H or Bart's in their blood.

These observations were put together in a theoretical paper suggesting that the thalassemias might result from different defects in the production of the α-and β-chains of hemoglobin (Ingram and Stretton 1959). This hypothesis became amenable to investigation in the 1960s as better methods for analyzing hemoglobin production were developed.

HEMOGLOBIN SYNTHESIS AND THE BEGINNINGS OF ITS ANALYSIS AT THE MOLECULAR LEVEL

During the early 1960s, methods were developed for studying hemoglobin synthesis in vitro (Heywood et al. 1964; Marks and Burka 1964) and it soon became apparent that the thalassemias are not so much disorders of hemoglobin production as imbalanced globin-chain synthesis (Weatherall et al. 1965). In other words, in β-thalassemia, excess α-chains are produced that are not capable of forming a viable hemoglobin and precipitate to form inclusion bodies in the bone marrow, which had first been observed by Fessas (1963). In the α-thalassemias, excess β-chains in adult life form β_4 tetramers or hemoglobin H, whereas excess γ-chains in fetal life produce γ_4 tetramers, or hemoglobin Bart's. Several studies about this time suggested that there were different forms of β-thalassemia, some associated with no β-globin-chain production, others with the production of normal β-globin chains but at a reduced rate. Detailed studies of the rates of initiation, elongation, and termination of the β-globin chains in the latter form of thalassemia showed no abnormalities suggesting that this disorder might result from a reduced output of normal β-globin messenger RNA (Clegg et al. 1968).

By the early 1970s, methods were becoming available for studying the function of messenger RNA isolated from human reticulocytes and it was found that some forms of β-thalassemia are associated with reduced amounts of β-globin messenger RNA (Bank and Marks 1971; Benz and Forget 1971; Nathan et al. 1971; Nienhuis and Anderson 1971). In 1974, two groups independently showed, using complimentary DNA (cDNA) prepared from α-globin messenger RNA, that the α-globin genes were deleted from stillborn babies with a severe form of thalassemia that had been designated the hemoglobin Bart's hydrops syndrome (Ottolenghi et al. 1974; Taylor et al. 1974). This was the first time that a disease had been clearly defined at the DNA level. As we shall see in the sections that follow, over the next 20 years considerable progress was made toward an understanding of the molecular pathology and pathophysiology of the hemoglobinopathies, leading to new approaches to their prevention and management.

FURTHER DEVELOPMENTS IN THE UNDERSTANDING OF THE MOLECULAR PATHOLOGY AND PATHOPHYSIOLOGY OF THE HEMOGLOBIN DISORDERS: CLINICAL APPLICATIONS

Thalassemia

From the late 1970s onward, a highly productive period began with the identification of various thalassemia mutations and deletions (Sankaran and Nathan 2010). Numerous clinical scientists, including Kan, Forget, Weatherall, Orkin, Higgs, and Kazazian, were empowered by the tools and insight provided by basic scientists such as Baltimore, Sharp, Maniatis, Leder, Lodish, and the late and much mourned Daniel Nathans. The work began with the use of Southern blotting to elucidate deletions resulting in thalassemia. Although this was highly successful in the α-thalassemias (Embury et al. 1979; Orkin et al. 1979a), it was of limited use in β-thalassemia (Orkin et al. 1979b), and only in specific instances. This problem was solved with the use of gene cloning and DNA sequencing to identify point mutations that result in the thalassemias. In most instances, recurrent common mutations were found in many patients, making routine sequencing laborious and limited in its ability to identify new mutations. However, the finding that different β-globin-gene mutations exist on haplotype blocks, at least at the β-globin locus, at first by Kan in sickle cell anemia and then by Antonarakis, Kazazian, and Orkin in β-thalassemia, helped to surmount this limitation (Kan and Dozy 1978; Antonarakis et al. 1982; Orkin et al. 1982).

The discovery of hundreds of mutations that cause the thalassemia syndromes then followed. Not only was a large set of mutations identified that result in the thalassemias, but this also led to the elucidation of many mechanisms that can impair mRNA production or function in the cell. Insights were gained into processes as diverse as transcription, mRNA modification, splicing, and translation. These findings presaged the period that would follow with the identification of mutations in various genes that cause a diverse group of Mendelian disorders. Even in the past few years, new insights into the molecular lesions that can cause human disease continue to be learned from the thalassemias, as exemplified by a Melanesian form of α-thalassemia that is attributable to the creation of a new promoter, which competes during transcription with the endogenous α-globin promoters (De Gobbi et al. 2006).

Even before the insights from molecular biology began to be used to elucidate the pathophysiology of the thalassemia syndromes, an effort was initiated to use measurements of globin-chain imbalance for prenatal diagnosis. The methods pioneered by several scientists, including Clegg, Weatherall, Huehns, Stamatoyannopoulos, and Kazazian, led directly to the first successful prenatal diagnosis by Kan and colleagues (Kan et al. 1975a,b). Advances in these techniques were made by Chang and colleagues (1974), leading to more feasible diagnostic methods. Cao, Loukopoulos, and others in the Mediterranean region helped to apply these methods in their respective populations, resulting in a large reduction in the incidence of new patients with thalassemia in those populations (Aleporou-Marinou et al. 1980; Cao et al. 1991). With the discovery of the molecular basis for many of the common thalassemia mutations, DNA-based prenatal genetic diagnostic methods were soon applied and supplanted the older methods (Orkin et al. 1978; Old et al. 1982; Boehm et al. 1983). This allowed earlier detection in the first trimester and, as a result, these methods were more widely adopted in a variety of countries. However, there is still controversy surrounding these approaches, limiting their use to populations in which prenatal genetic diagnosis or premarital genetic counseling is accepted (Colah et al. 2008).

As mentioned earlier, the finding of imbalanced globin-chain synthesis was a key development toward an understanding of the complex pathophysiology of thalassemia (Nathan and Gunn 1966; Gabuzda et al. 1967). It is now firmly established that disease severity in the thalassemias is determined by the degree of imbalance in α-versus non-α-globin synthesis. β-Thalassemia syndromes are more severe than most α-thalassemia syndromes (other than complete absence of α-chain production) because free α-chains are much less stable than free β-chains and precipitate in developing erythroblasts where they cause extensive mechanical and metabolic damage resulting in ineffective erythropoiesis (Rund and Rachmilewitz 2005). Evidence of the importance of imbalanced rather than deficient globin synthesis derives as well from clinical observations. Patients with high levels of γ-chain synthesis or (less commonly) those with concomitant α-thalassemia, both of which reduce imbalance, have considerably milder disease as do some patients with β-thalassemia in which some β-chains are produced as opposed to the forms in which there is no β-chain synthesis.

α-Chains are normally synthesized slightly in excess of non-α-chains (Lodish and Nathan 1972) but the excess is far greater in β-thalassemia. The ratio of α- to non-α-chain synthesis is 2:1 in β-thalassemia heterozygotes, a gap well tolerated by the protective systems in erythroblasts. But homozygotes or compound heterozygotes may have ratios as high as 50:1. Only a complex process of protein quality control including an important chaperone, the ubiquitin-proteasome system and certain lysosome-autophagy pathways recently examined in some detail by Weiss and his colleagues (Khandros and Weiss 2004; Kong et al. 2004; E Khandros and C Thom, unpubl.) spares the developing erythroblasts from total annihilation as the highly unstable α-chains find no non-α-chain with which to mate. α-Chain precipitation also triggers reactive oxygen species for which the antioxidant defenses of the cell must be brought to the battle (Rund and Rachmilewitz 2001). Even with those systems operating at maximal capacity, only a few mature red cells emerge from the marrow, and they are usually salvaged by dint of their capacity to synthesize the γ-chains of fetal hemoglobin. Such patients may have virtually 100% fetal hemoglobin but with a total circulating hemoglobin concentration of 1 g/dL.

In contrast to the dire consequences of poorly tolerated α-chain excess, the threefold reduction of α-chain synthesis observed in the form of α-thalassemia known as hemoglobin H disease is relatively benign. This is owing to the far better solubility of free β-chains that form a tetramer (hemoglobin H) that slowly precipitates over the life span of the red cells and is selectively removed in the spleen (Nathan et al. 1969). Complete loss of α-chain production is lethal. Fetuses with this defect do not survive without heroic intervention.

Thus, the central pathophysiology of the β-thalassemias is ineffective erythropoiesis, which leads to anatomic abnormalities including a greatly expanded marrow space, highly fragile bones, distorted facies, ectopic erythroid masses, and hepatosplenomegaly, together with hyperbilirubinemia with gallstones, excessive iron absorption, and all of the consequences of iron overload, particularly diabetes mellitus, pituitary failure and consequent hypogonadism, and fatal heart failure and arrhythmias.

All of these physical manifestations, to say nothing of psychological and sociological suffering, are worsened by red cell transfusions and concomitant intravenous iron overload. Before the development of effective approaches to iron chelation therapy, patients with thalassemia died in early childhood in the less well-developed world and before the age of 20 in the developed world.

The past 50 years has seen far better treatment emerge from universities and pharmaceutical companies, but costs remain a huge problem, certainly in the underdeveloped world and in the wealthier societies as well. It must be recognized that thalassemia is a completely preventable disease. Societies that reject prenatal diagnosis and selective abortion of affected fetuses must seriously consider the consequences of that decision.

Sickle Cell Anemia

Described initially as a curiosity by James Herrick in 1910, sickle cell anemia is a massive public health problem across the malaria zone worldwide and especially in equatorial Africa where it has arisen several times and is a leading cause of infant mortality, disability, and shortened life span. Hahn and Gillespie (1927) showed that the sickle shape of affected red cells is because of an abnormality of deoxyhemoglobin. Later, Harris (1950) showed that deoxyhemoglobin S forms "tactoids," bundles of hemoglobin fibers that run lengthwise in the sickle erythrocyte and distort the erythrocyte membrane. After the classic work of Pauling, Itano, and Ingram, described earlier, had revealed the single amino acid substitution that causes the disease, the basis of the pathophysiology became much clearer. The fine structure of the fibers, the role of contact points in their stabilization, and hence the influence of various hemoglobins, particularly hemoglobin F, on the rate and extent of sickling, reviewed in great detail by Edelstein (1981) and Ferrone et al. (1985), has shown how pH, temperature, ionic strength, and particularly hemoglobin concentration all strongly enhance the rate of nucleation of hemoglobin molecules until a "critical polymer" of about 30 tetramers is reached and the growth of the polymeric fiber feeds on itself. But the actual number of morphologically sickled cells may not be determinative with respect to the symptoms of sickle cell disease. Noguchi et al. (1983) have shown that polymer formation and attendant lack of deformability may be the controlling factors that influence clinical manifestations of this protean illness.

For very good reasons, close attention has been paid to the control of the rate of polymer formation in sickle cell disease. Down-regulation of the rate of polymer formation by means of increasing fetal hemoglobin to sufficient levels in a significant fraction of the red cells would surely be the best therapeutic approach to the problem. But the distortion of the erythrocyte membrane by the fibers also plays a major pathophysiological role. Disruption of the spectrin-actin lattice was shown by Lux et al. (1976). Later, Hoover et al. (1979) followed by Hebbel et al. (1980) and by Mohandas and Evans (1984) showed that the membrane disruption is associated with tight adherence of the sickle red cells to vascular endothelium, which in turn excites the endothelial cells to unleash a storm of cytokines that contribute to a state of ischemic reperfusion injury. Various therapeutic attacks on these inflammatory aspects of the disease are now in progress (Field et al. 2011).

These early observations on the pathophysiology of sickle cell anemia laid the way for a far better understanding of the disease and better possibilities for its management, work that is described in several of the articles that follow.

CONCLUSIONS

Although much of the research in this field that was performed in the second half of the twentieth century and afterward focused mainly on the thalassemias and sickle cell anemia, many other rarer forms of hemoglobinopathy were discovered that involved either the stability or oxygen-carrying properties of hemoglobin and, incidentally, told us a great deal about structure/function relationships in this complex molecule. Because somewhere between 300 and 400,000 babies are born each year with an inherited disorder of hemoglobin, as well as providing the story of the early development of molecular medicine, the further exploration of this field has extremely important clinical implications, particularly for the poorer countries of the world.

REFERENCES

Ager JAM, Lehmann H. 1958. Observations on some "fast" haemoglobins: K, J, N and Bart's. *Brit Med J* **1:** 929–931.

Aleporou-Marinou V, Sakarelou-Papapetrou N, Antsaklis A, Fessas P, Loukopoulos D. 1980. Prenatal diagnosis of thalassemia major in Greece: Evaluation of the first large series of attempts. *Ann NY Acad Sci* **344:** 181–188.

Antonarakis SE, Boehm CD, Giardina PVJ, Kazazian HH. 1982. Nonrandom association of polymorphic restriction sites in the β-globin gene complex. *Proc Natl Acad Sci* **79:** 137–141.

Bank A, Marks PA. 1971. Hemoglobin synthesis in thalassemia. *Ser Haematol* **4:** 97–115.

Bearn AG. 1993. *Archibald Garrod and the individuality of man.* Clarendon, Oxford.

Benz EJ, Forget BG. 1971. Defect in messenger RNA for human hemoglobin synthesis in β thalassemia. *J Clin Invest* **50:** 2755–2760.

Boehm CD, Antonarakis SE, Phillips JA III, Stetten G, Kazazian HH Jr. 1983. Prenatal diagnosis using DNA polymorphisms. Report on 95 pregnancies at risk for sickle-cell disease or β-thalassemia. *N Engl J Med* **308:** 1054–1058.

Bohr C, Hasselbalch L, Krogh A. 1904. Ueber einen in biologischer Bezichung wichtigen Einfluss, den die Kohlensaurespannung des Blutes auf dessen Sauerstoffbendung ubt. *Skand Arch Physiol* **16:** 402–412.

Bunn HF, Forget BG. 1986. Historical Background. In *Hemoglobin: Molecular, genetic and clinical aspects* (ed. Bunn HF, Forget BG), pp. 1–12. W.B. Saunders, Philadelphia.

Cao A, Rosatelli C, Pirastu M, Galanello R. 1991. Thalassemias in Sardinia: Molecular pathology, phenotype-genotype correlation and prevention. *Am J Pediatr Hematol Oncol* **13:** 179–188.

Chang H, Hobbins JC, Cividalli G, Frigoletto FD, Mahoney MJ, Kan YW, Nathan DG. 1974. In utero diagnosis of: Hemoglobinopathies. Hemoglobin synthesis in fetal red cells. *N Engl J Med* **290:** 1067–1068.

Clegg JB, Weatherall DJ, Na-Nakorn S, Wasi P. 1968. Haemoglobin synthesis in β-thalassaemia. *Nature* **220:** 664–668.

Colah R, Surve R, Wadia M, Solanki P, Mayekar P, Thomas M, Gorakshakar A, Dastur A, Mohanty D. 2008. Carrier screening for β-thalassemia during pregnancy in India: A 7-year evaluation. *Genet Test* **12:** 181–185.

Conley CL. 1980. Sickle-cell anemia—The first molecular disease. In *Blood, pure and eloquent* (ed. Wintrobe MM), pp. 319–317. McGraw-Hill, New York.

Cooley TB, Lee P. 1925. A series of cases of splenomegaly in children with anemia and peculiar bone changes. *Trans Am Pediatr Soc* **37:** 29.

De Gobbi M, Viprakasit V, Hughes JR, Fisher C, Buckle VJ, Ayyub H, Gibbons RJ, Vernimmen D, Yoshinaga Y, de Jong P, et al. 2006. A regulatory SNP causes a human genetic disease by creating a new transcriptional promoter. *Science* **312:** 1215–1217.

Edelstein SJ. 1981. Molecular topology in crystals and fibers of hemoglobin S. *J Mol Biol* **150:** 557–575.

Edsall JT. 1972. Blood and hemoglobion: The evolution of knowledge of functional adaptation in a biochemical system. *J Hist Biol* **5:** 205.

Embury SH, Lebo RV, Dozy AM, Kan YW. 1979. Organization of the α-globin genes in the Chinese α-thalassemia syndromes. *J Clin Invest* **63:** 1307–1310.

Ferrone FA, Hofrichter J, Eaton WA. 1985. Kinetics of sickle hemoglobin polymerization. II. A double nucleation mechanism. *J Mol Biol* **183:** 611–631.

Fessas P. 1963. Inclusions of hemoglobin in erythroblasts and erythrocytes of thalassemia. *Blood* **21:** 21–32.

Field JJ, Nathan DG, Linden J. 2011. Targeting iNKT cells for the treatment of sickle cell disease. *Clin Immunol* **140:** 177–183.

Gabuzda TG, Nathan DG, Gardner FH, Kreimer-Birnbaum M, Bannerman RM. 1967. Hemoglobin F and β-thalassemia. *Science* **157:** 1079.

Garrod AE. 1909. Inborn errors of metabolism. In *The Croonian lecture delivered before the Royal College of Physicians of London in June 1908.* Frowde, Hodder & Stoughton, Oxford University Press, London.

Hahn E, Gillespie EB. 1927. Sickle cell anemia: Report of a case greatly improved by splenectomy. *Arch Intern Med* **39:** 233.

Harris JW. 1950. Studies on the destruction of red blood cells. VIII. Molecular orientation in sickle cell hemoglobin solutions. *Proc Soc Exp Biol Med* **75:** 197–201.

Hebbel RP, Yamada O, Moldow CF, Jacob HS, White JG, Eaton JW. 1980. Abnormal adherence of sickle erythrocytes to cultured vascular endothelium: Possible mechanism for microvascular occlusion in sickle cell disease. *J Clin Invest* **65:** 154–160.

Herrick JB. 1910. Peculiar elongated and sickle-shaped red blood corpuscles in a case of severe anemia. *Arch Intern Med* **6:** 517–521.

Heywood JD, Karon M, Weissman S. 1964. Amino acids: Incorporation into α- and β-chains of hemoglobin by normal and thalassemic reticulocytes. *Science* **146:** 530–531.

Hoover R, Rubin R, Wise G, Warren R. 1979. Adhesion of normal and sickle erythrocytes to endothelial monolayer cultures. *Blood* **54:** 872–876.

Ingram VM. 1956. Specific chemical difference between the globins of normal human and sickle-cell anaemia haemoglobin. *Nature* **178:** 792–794.

Ingram VM, Stretton AOW. 1959. Genetic basis of the thalassemia diseases. *Nature* **184:** 1903–1909.

Jones RT, Schroeder WA. 1963. Chemical characterization and subunit hybridization of human hemoglobin H and associated compounds. *Biochemistry* **2:** 1357–1367.

Kan YW, Dozy AM. 1978. Polymorphisms of DNA sequence adjacent to human β-globin structural gene: Relation to sickle mutation. *Proc Natl Acad Sci* **75:** 5631.

Kan YW, Golbus MS, Klein P, Dozy AM. 1975a. Successful application of prenatal diagnosis in a pregnancy at risk for homozygous β-thalassemia. *N Eng J Med* **292:** 1096.

Kan YW, Golbus MS, Trecartin R. 1975b. Prenatal diagnosis of homozygous β-thalassaemia. *Lancet* **2:** 790–791.

Khandros E, Weiss MJ. 2004. Protein quality control during erythropoiesis and hemoglobin synthesis. *Hematol Oncol Clin North Am* **24:** 1071–1088.

Kong Y, Zhou S, Kihm AJ, Katein AM, Yu X, Gell DA, Mackay JP, Adachi K, Foster-Brown L, Louden CS, et al. 2004. Loss of α-hemoglobin-stabilizing protein impairs erythropoiesis and exacerbates β-thalassemia. *J Clin Invest* **114:** 1457–1466.

Kunkel HG, Wallenius G. 1955. New hemoglobin in normal adult blood. *Science* **122:** 288.

Lodish HF, Nathan DG. 1972. Regulation of hemoglobin synthesis. Preferential inhibition of α and β globin synthesis. *J Biol Chem* **247:** 7822–7829.

Lux SE, John KM, Karnovsky MJ. 1976. Irreversible deformation of the spectrin-actin lattice in irreversibly sickled cells. *J Clin Invest* **58:** 955–963.

Marks PA, Burka ER. 1964. Hemoglobin synthesis in human reticulocytes: A defect in globin formation in thalassemia major. *Ann NY Acad Sci* **119:** 513.

Mohandas N, Evans E. 1984. Adherence of sickle erythrocytes to vascular endothelial cells: Requirement for both cell membrane changes and plasma factors. *Blood* **64:** 282–287.

Nathan DG, Gunn RB. 1966. Thalassemia: The consequences of unbalanced hemoglobin synthesis. *Am J Med* **41:** 815–830.

Nathan DG, Stossel TB, Gunn RB, Zarkowsky HS, Laforet MT. 1969. Influence of hemoglobin precipitation on erythrocyte metabolism in α and β thalassemia. *J Clin Invest* **48:** 33–41.

Nathan DG, Lodish HF, Kan YW, Housman D. 1971. β Thalassemia and the translation of globin messenger RNA. *Proc Natl Acad Sci* **68:** 2514.

Nienhuis AW, Anderson WF. 1971. Isolation and translation of hemoglobin messenger RNA from thalassemia, sickle cell anemia, and normal human reticulocytes. *J Clin Invest* **50:** 2458–2462.

Noguchi CT, Torchia DA, Schechter AN. 1983. Intracellular polymerization of sickle hemoglobin. Effects of cell heterogeneity. *J Clin Invest* **72:** 846–852.

Old JM, Ward RHT, Petrou M, Karagozlu F, Modell B, Weatherall DJ. 1982. First-trimester fetal diagnosis for haemoglobinopathies: Three cases. *Lancet* **2:** 1413–1416.

Orkin SH, Alter BP, Altay C, Mahoney MJ, Lazarus H, Hobbins JC, Nathan DG. 1978. Application of endonuclease mapping to the analysis and prenatal diagnosis of thalassemias caused by globin-gene deletion. *N Eng J Med* **299:** 166–172.

Orkin SH, Old J, Lazarus H, Altay C, Gurgey A, Weatherall DJ, Nathan DG. 1979a. The molecular basis of α-thalassemias: Frequent occurrence of dysfunctional α loci among non-Asians with Hb H disease. *Cell* **17:** 33–43.

Orkin SH, Old JM, Weatherall DJ, Nathan DG. 1979b. Partial deletion of β-globin gene DNA in certain patients with β^0-thalassemia. *Proc Natl Acad Sci* **76:** 2400–2404.

Orkin SH, Kazazian HH Jr, Antonarakis SE, Goff SC, Boehm CD, Sexton JP, Waber PG, Giardina PJV. 1982. Linkage of β-thalassaemia mutations and β-globin gene polymorphisms with DNA polymorphisms in human β-globin gene cluster. *Nature* **296:** 627–631.

Ottolenghi S, Lanyon WG, Paul J, Williamson R, Weatherall DJ, Clegg JB, Pritchard J, Pootrakul S, Wong HB. 1974. The severe form of α thalassaemia is caused by a haemoglobin gene deletion. *Nature* **251:** 389–392.

Pauling L, Itano HA, Singer SJ, Wells IG. 1949. Sickle-cell anemia, a molecular disease. *Science* **110:** 543–548.

Rund D, Rachmilewitz E. 2001. Pathophysiology of α- and β-thalassemia: Therapeutic implications. *Sem Hematol* **38:** 343–349.

Rund D, Rachmilewitz E. 2005. β-Thalassemia. *N Engl J Med* **353:** 1135–1146.

Sankaran VG, Nathan DG. 2010. Thalassemia: An overview of 50 years of clinical research. *Hematol Oncol Clin North Am* **24:** 1005–1020.

Silvestroni E, Bianco I. 1944–1945. Microdrepanocitoanemia, in un sogetto di razza bianca. *Boll Atti Acad Med* **70:** 347.

Smith EW, Torbert JV. 1958. Study of two abnormal hemoglobins with evidence for a new genetic locus for hemoglobin formation. *Bull Johns Hopkins Hosp* **102:** 38–45.

Sturgeon P, Itano HA, Valentine WN. 1952. Chronic hemolytic anemia associated with thalassemia and sickling trait. *Blood* **7:** 350–357.

Taylor JM, Dozy A, Kan YW, Varmus HE, Lie-Injo LE, Ganeson J, Todd D. 1974. Genetic lesion in homozygous α-thalassaemia (hydrops foetalis). *Nature* **251:** 392–393.

Weatherall DJ, Clegg JB. 2001. *The thalassaemia syndromes.* Blackwell Science, Oxford.

Weatherall DJ, Clegg JB, Naughton MA. 1965. Globin synthesis in thalassemia: An in vitro study. *Nature* **208:** 1061–1065.

Whipple GH, Bradford WL. 1932. Racial or familial anemia of children. Associated with fundamental disturbances of bone and pigment metabolism (Cooley-Von Jaksch). *Am J Dis Child* **44:** 336–365.

Erythropoiesis: Development and Differentiation

Elaine Dzierzak and Sjaak Philipsen

Erasmus MC, Erasmus Stem Cell Institute, Department of Cell Biology, 3000 CA Rotterdam, The Netherlands

Correspondence: e.dzierzak@erasmusmc.nl

Through their oxygen delivery function, red blood cells are pivotal to the healthy existence of all vertebrate organisms. These cells are required during all stages of life—embryonic, fetal, neonatal, adolescent, and adult. In the adult, red blood cells are the terminally differentiated end-product cells of a complex hierarchy of hematopoietic progenitors that become progressively restricted to the erythroid lineage. During this stepwise differentiation process, erythroid progenitors undergo enormous expansion, so as to fulfill the daily requirement of $\sim 2 \times 10^{11}$ new erythrocytes. How the erythroid lineage is made has been a topic of intense research over the last decades. Developmental studies show that there are two types of red blood cells—embryonic and adult. They develop from distinct hemogenic/hematopoietic progenitors in different anatomical sites and show distinct genetic programs. This article highlights the developmental and differentiation events necessary in the production of hemoglobin-producing red blood cells.

ONTOGENY OF THE HEMATOPOIETIC SYSTEM

Microscopic observations identified the first blood cells in the conceptus as those of the erythroid lineage. The early production of erythroid lineage cells occurs in the yolk sac, a transient extraembryonic tissue. These erythroid cells are nucleated and short-lived. They are derived from mesodermal cells that are formed from epiblast cells ingressing through the primitive streak (Lawson et al. 1991; Kinder et al. 1999). The newly formed mesodermal cells migrate posteriorly, enter the yolk sac, and come in close contact with endoderm cells (Fig. 1A). It is this interaction between the two cell layers that is required for the initiation of erythropoiesis (Belaoussoff et al. 1998). Interestingly, the mesodermal cells that migrate into the yolk sac form blood islands containing not only red blood cells, but also endothelial cells. The overlapping ontogenic appearance of both erythroid and endothelial cells indicates a common mesodermal precursor—the hemangioblast—with bilineage potential (Sabin 1920; Murray 1932). This is further supported by the overlap in genetic programs for the two lineages (i.e., expression of Flk-1 [KDR], Scl [Tal1], and CD34) and the lack of both lineages in embryos deficient for some of these genes (Shalaby et al. 1995; Tavian et al. 1999; Park et al. 2005). Surprisingly, hemangioblasts in vivo are localized not in the yolk sac but in the posterior primitive streak (Huber et al. 2004). As they migrate to the yolk sac, they begin their commitment to endothelial and hematopoietic progenitors, with several of these cells contributing to the formation of each blood island (Ueno and Weissman

Copyright © 2013 Cold Spring Harbor Laboratory Press; all rights reserved
Cite this article as *Cold Spring Harb Perspect Med* doi: 10.1101/cshperspect.a011601

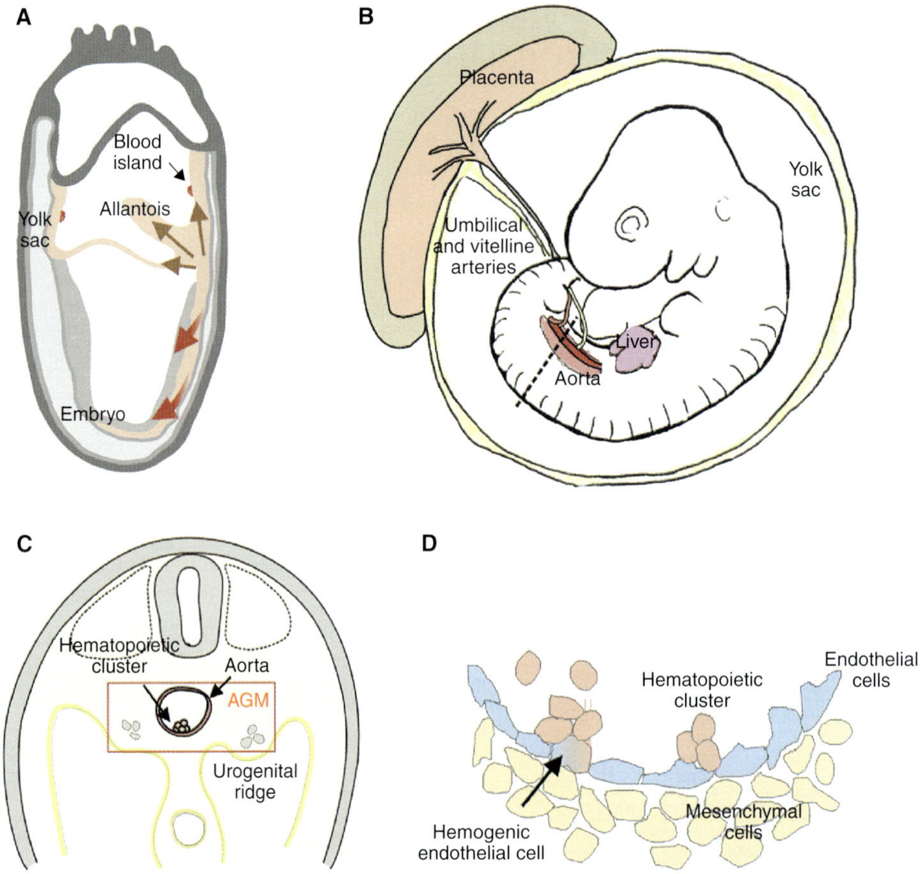

Figure 1. Ontogeny of the mouse hematopoietic system. (*A*) During gastrulation of the mouse conceptus, emerging mesodermal cells (brown arrows) migrate to the extraembryonic compartments (yolk sac and allantois). Blood islands begin to form in the yolk sac upon interactions of the mesoderm (orange) with the endoderm (gray). (Red arrows) Mesoderm migrating into the embryo proper. (*B*) Embryonic day 10.5 (E10.5) mouse embryo. Hematopoietic sites include the aorta (AGM region), yolk sac, umbilical and vitelline arteries, placenta, and liver. (Dotted line) The transverse section in panel *C*. (*C*) Transverse section showing the AGM region of an E10.5 mouse embryo. Urogenital ridges are lateral to the aorta. A hematopoietic cluster on the ventral wall of the aorta is shown. (*D*) Close-up of the aorta showing a hemogenic endothelial cell transitioning to a hematopoietic cluster cell. Nonhemogenic endothelial cells (blue) and mesenchymal cell (yellow) are shown.

2006). Yolk sac blood islands containing primitive erythrocytes are detectable in the mouse conceptus at embryonic day 7.5 (E7.5) (Russell and Bernstein 1966) and in the human conceptus at ~16–20 d of gestation (Tavian and Peault 2005).

In contrast to the rapid commitment and differentiation of primitive erythroid cells from migrating mesodermal cells, the adult erythroid lineage takes its origin from a more complex cell lineage differentiation hierarchy, with hematopoietic stem cells (HSCs) as the founder cells for this hierarchy. A cohort of rare HSCs is harbored in the bone marrow and continuously replenishes the blood system throughout adult life. To produce adult red blood cells and the many other functionally distinct lineages of differentiated blood cells, such as lymphocytes, macrophages, granulocytes, and megakaryocytes, HSCs undergo a relatively long and extensive process of cellular differentiation and proliferation. In the mouse embryos, HSCs appear 3 d later in ontogeny than yolk sac blood islands (Muller et al. 1994; Medvinsky and Dzierzak 1996). This late appearance suggests an independent mesodermal origin for HSCs. How the adult HSC-based hierarchy develops in the embryo, and from what precursors HSCs are derived, is an important focus of developmental studies.

Again, the first clues on HSC development came from microscopic observations of a variety of developing vertebrate embryos (amphibian, avian, and mammalian). Clusters of hematopoietic cells were found within the embryo in the aorta–gonad–mesonephros (AGM) region, closely associated with the ventral lumenal aspect of the aorta (Fig. 1B–D) (for review, see Dzierzak and Speck 2008). The results of avian embryo grafting experiments performed in the mid-1970s (Dieterlen-Lievre 1975; Martin et al. 1978; Beaupain et al. 1979; Dieterlen-Lievre and Martin 1981) suggested that the development of the adult hematopoietic system begins with these clusters. Grafts of quail embryo bodies onto chick yolk sacs before vascularization revealed that the adult blood system of the chimera was quail-derived. Moreover, the embryonic aorta was found to contain hematopoietic progenitors (Cormier and Dieterlen-Lievre 1988), suggesting that this vessel is responsible for the generation of adult hematopoietic cells. Similarly, lineage-tracing experiments, in which individual blastomeres of the 32-cell stage *Xenopus* embryo were marked, showed that the blastomeres contributing to the formation of the ventral blood islands (yolk sac) are distinct from the blastomeres that contribute to the formation of the aorta, hematopoietic clusters, and pronephros (Ciau-Uitz et al. 2000). Thus, the specific localization of intrabody hematopoiesis in frog embryos has been found to be associated with the dorsal aorta and pronephros region (Turpen and Knudson 1982). These nonmammalian vertebrate embryo studies concluded that the first emergence of hematopoietic cells is in the extraembryonic yolk sac; the subsequent emergence of hematopoietic cells occurs both extraembryonically and intraembryonically; and the permanent contributors to the adult hematopoietic system are intraembryonically derived hematopoietic cells, with most extraembryonic hematopoietic cells becoming extinct. Importantly, the close association of hematopoietic cell clusters and endothelial cells on the ventral aspect of the aorta (Dieterlen-Lievre 1975) led to the hypothesis that the late-emerging, permanent adult hematopoietic system is derived from specialized "hemogenic" endothelial cells (Fig. 1D).

In contrast to the nonmammalian vertebrate species, the origin of the first HSCs in mammalian embryos was uncertain and, until the 1990s, thought to be the yolk sac (Moore and Metcalf 1970). Cell tracing studies of early stage mammalian embryos required experimental approaches different from the grafting approaches (which rely on ex utero development and large embryo size) taken in avian and amphibian embryos. One approach includes the stringent functional test for HSCs in different regions of the early/midgestation mouse conceptus—the in vivo transplantation of such cells into adult irradiated recipients. Much like adult bone marrow cell transplantations, long-term high-level self-renewing hematopoietic repopulation of adult recipients reveals the presence of HSCs in the injected cell population. Transplantation of cells from several embryonic hematopoietic tissues (yolk sac, intraembryonic AGM region, chorio-allantoic placenta, liver) showed that the first tissue to contain HSCs was the AGM (Fig. 1B,C) (Muller et al. 1994; Gekas et al. 2005; Ottersbach and Dzierzak 2005). HSCs were further localized to the aorta, vitelline, and umbilical arteries at E10.5 (de Bruijn et al. 2000). Moreover, organ explant cultures showed that the AGM region de novo produces the first adult type HSCs (Medvinsky and Dzierzak 1996). One to 1.5 d later, the mouse yolk sac, placenta, and liver also contain HSCs. Because the vascular connection between the embryo body and the extraembryonic hematopoietic sites occurs at E8.25 (Downs 1998) and the circulation carries cells between these sites, the origin of the HSCs found in these other tissues remains unknown.

The presence of hematopoietic clusters in the mouse aorta and vitelline/umbilical arteries at the time of HSC generation is intriguing (Fig. 1D). Functional studies have shown that

AGM HSCs express the same molecules (Wood et al. 1997; de Bruijn et al. 2002; North et al. 2002) as aortic hematopoietic cluster cells and, in some cases, aortic endothelial cells. Genetic (conditional knockout) studies of the Runx1 transcription factor (Chen et al. 2009) show an endothelial cell requirement for this factor in the generation of HSCs and vascular clusters, suggesting that HSCs are generated from functioning vascular endothelial cells that retain hematopoietic potential. Recent live imaging of the midgestation mouse aorta has confirmed this, clearly showing in real time the endothelial-to-hematopoietic cell transition (Boisset et al. 2010). The emerging cells are rare, localized to the ventral aspect of the aorta, and begin to express all of the expected HSC markers as they transit from a flat endothelial to a round hematopoietic morphology (Boisset et al. 2010; Yokomizo et al. 2012). Thus, HSCs are derived from endothelial cells.

With the focus on hemoglobin-producing erythrocytes—primitive erythrocytes generated early in the yolk sac and adult erythrocytes from vascular endothelial-derived HSCs—there are intermediate types of hematopoietic progenitors also generated in the embryo (Fig. 2). Before HSCs are generated at E10.5, definitive erythro-myeloid progenitors are found in the E8.5/E9 yolk sac and placenta (Alvarez-Silva et al. 2003; McGrath and Palis 2005), and multipotent (erythroid–myeloid–lymphoid) hematopoietic progenitors (Godin et al. 1995) are found in the E9 AGM, most likely in hematopoietic clusters. Tissue explant culture before progenitor assay reveals that already by E8 the

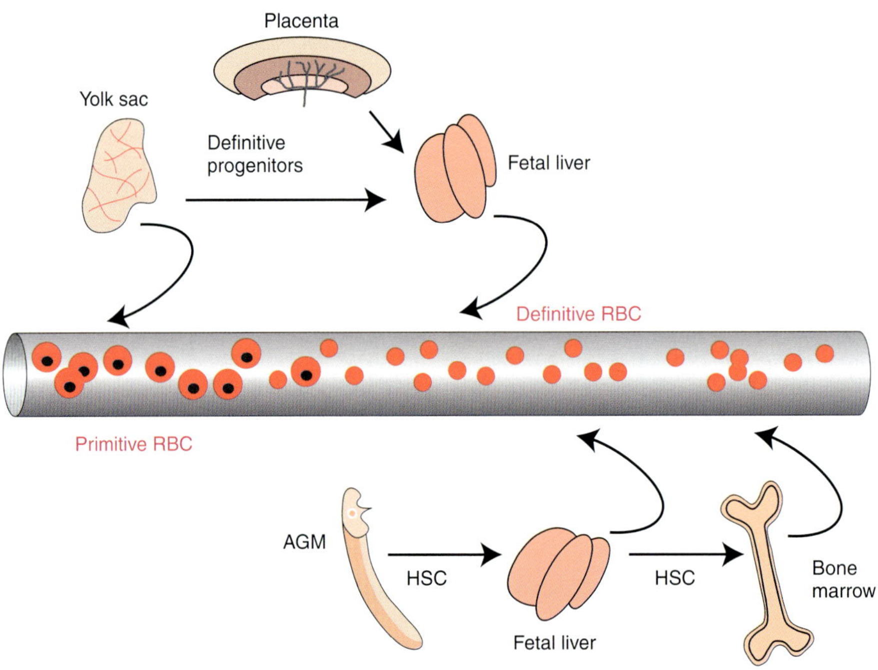

Figure 2. Ontogeny of erythroid lineage cells in the circulation. Embryonic erythrocytes (primitive red blood cells [RBCs]) are made by the yolk sac at E7.5 and are found in the circulation until ~E11/12. At E9, the yolk sac and placenta generate definitive progenitors that migrate to the fetal liver, where they differentiate to definitive RBCs (expressing fetal/adult globin) and enter the circulation. At E10.5, the AGM generates the first HSCs that migrate to the fetal liver and differentiate to the erythroid lineage (among other lineages), and these definitive RBCs enter the circulation. Fetal liver HSCs migrate and colonize the bone marrow at birth, where they provide lifelong production of definitive RBCs for the circulation. The spleen also is a site of differentiation for erythroid cells (not shown).

 Cite this article as *Cold Spring Harb Perspect Med* doi: 10.1101/cshperspect.a011601

yolk sac, AGM, and allantois (Cumano et al. 1996; Zeigler et al. 2006; Corbel et al. 2007) contain cells with hemogenic potential. Interestingly, in the absence of circulation in *Ncx1*$^{-/-}$ embryos (heartbeat defect), the numbers of erythro-myeloid progenitors in the yolk sac were found to be equivalent to the cumulative number of progenitors in wild-type embryos in all anatomic sites (Lux et al. 2007). No progenitors were found in the *Ncx1*$^{-/-}$ AGM, suggesting that the yolk sac normally generates all of these progenitors and distributes them to other tissues. As the cells colonize the liver, the newly emigrated progenitors expand and differentiate toward the definitive erythroid lineage (Ema and Nakauchi 2000).

ERYTHROID LINEAGE DIFFERENTIATION DURING ONTOGENY

Primitive erythrocytes, emerging in the extraembryonic yolk sac, are relatively large cells. The primitive cells are characterized by the expression of embryonic globins (εy, βh1, and ζ in the mouse; ε, γ, and ζ in man), which form a variety of distinguishable hemoglobin tetramers in man ($\zeta_2\varepsilon_2$ [Gower1], $\alpha_2\varepsilon_2$ [Gower2], $\zeta_2\gamma_2$ [Portland1], $\zeta_2\beta_2$ [Portland2]). In mammals, these are the only erythroid cells that retain their nucleus when they enter the circulation. Once released in the bloodstream, the macrocytic primitive erythrocytes retain proliferative capacity, and mitotic figures are observed in the circulating blood of early mammalian embryos (Lin et al. 1996). Intravascular erythropoiesis is not normally observed at any other developmental stage. The long-held view that primitive erythrocytes remain nucleated and that they disappear from the circulation very quickly during the embryonic-to-fetal transition period has been recently revised. In fact, primitive cells enucleate very efficiently between E12.5 and E14.5, resulting in macrocytic, enucleated erythrocytes (Fraser et al. 2007). Beginning at E11.5, newly emigrating erythro-myeloid progenitors (and subsequently HSCs) differentiate in the fetal liver to produce the first definitive erythrocytes (Fig. 2). There is an immediate switch to fetal/adult globins when definitive erythropoiesis starts in the fetal liver. Expression of a specific fetal β-like globin (γ-globin) is a feature of anthropoid primates. Hemoglobin tetramers consisting of α- and γ-globin chains ($\alpha_2\gamma_2$) are known as HbF in humans. These hemoglobins allow the developing fetus to extract oxygen more efficiently from the maternal blood. Mammalian definitive erythrocytes expel their nucleus before they enter the circulation and are smaller in size than primitive erythrocytes.

Around the time of birth, the site of erythropoiesis switches to the bone marrow and the spleen. Humans rely mainly on the bone marrow for steady-state adult erythropoiesis, but in mice the spleen remains an important erythropoietic organ during adult life. Under erythroid stress conditions, for example, low oxygen pressure or anemia, the spleen of both mouse and man is used to expand the erythropoietic capacity (Socolovsky 2007).

Fetal globin expression is silenced in adult erythropoiesis. Hemoglobin tetramers composed of α- and β-globin ($\alpha_2\beta_2$, HbA1) account for ~97% of all hemoglobin in adult erythrocytes. HbA2 ($\alpha_2\delta_2$) and HbF account, respectively, for ~2% and ~1% of total hemoglobin in most adults. High HbF is restricted to a few cells, termed F cells, in normal adults (Boyer et al. 1975). Clonal analysis has shown that F cells are derived from the same progenitor cells as the cells containing adult globin, ruling out that fetal and adult stem cell lineages coexist (Papayannopoulou et al. 1977). Some individuals maintain higher levels of HbF throughout adult life. This condition, known as hereditary persistence of fetal hemoglobin (HPFH), is caused in the majority of cases by deletions in the β-globin locus, but also by point mutations in the γ-globin gene promoters (Stamatoyannopoulos 2005). In rare cases, the HPFH phenotype does not segregate with the β-globin locus, suggesting a *trans*-regulatory mechanism (see also Sankaran and Orkin 2013). The HPFH condition is not clinically manifest, but it is a factor ameliorating the effects of β-thalassemia and sickle cell disease. Reactivation of γ-globin expression in adults with these disorders is therefore a very attractive therapeutic approach, because the large majority of patients will have normal γ-globin genes that they have switched off after birth. Despite the fact that mice do not have a fetal β-like globin gene, the γ-globin genes in human β-globin locus transgenes are expressed in the early fetal liver and silenced subsequently in development between E13.5 and E14.5 (Dillon and Grosveld 1991; Strouboulis et al. 1992). Furthermore, γ-globin transgenes with HPFH mutations in the promoter phenocopy the HPFH condition in mice (Berry et al. 1992; Peterson et al. 1995).

ERYTHROID DIFFERENTIATION WITHIN THE ADULT HEMATOPOIETIC HIERARCHY

Erythroid cells are derived from HSCs, but the differentiation path they follow is still a matter of debate (Fig. 3) (Buza-Vidas et al. 2007). Immature progenitors are capable of giving rise to multilineage colonies, but whether there is a fixed order to cell fate decisions (i.e., at each stage a cell has limited choices) or whether there is greater plasticity to a cell's potential fates remains to be resolved. The classical model, put forward by the Weissman group, posits a decision between two fates at each step in differentiation, thereby gradually diminishing lineage potential (Fig. 3). More recently, an alternative scheme proposed by Adolfsson et al. (2005) identified a highly proliferative lymphoid-primed multipotent progenitor (LMPP) with GM potential but devoid of the ability to adopt erythroid or megakaryocytic lineage fates (Fig. 3, dotted arrows).

Cell populations corresponding to these progenitors can be purified by flow cytometric sorting based on the expression of various cell surface markers. Lineage output of these cells is determined by the mature differentiated cells emerging in colony assays in response to growth factors. Immature erythroid-restricted progenitors are identified as the burst forming unit-erythroid (BFU-E), so-called because its earliest progeny are motile, giving rise to a multi-subunit colony (or burst). These large colonies containing up to several thousand hemoglobinized cells appear after 5–8 d (mouse) or 10–14 d (human) in methylcellulose cultures. Their growth is dependent on several factors, the most important of which are stem cell factor (SCF), thrombopoietin (TPO), interleukin 3 (IL3), IL11, and FLT3-ligand. BFU-E occurs at a frequency of $40–120/10^5$ bone marrow cells and also circulates in the peripheral blood at a frequency of $10–40/10^5$ light-density mononuclear cells (Migliaccio et al. 2001).

More mature erythroid progenitors, colony-forming units-erythroid (CFU-E), consist of small colonies of 16–125 cells that appear after 2–3 d (mouse) or 5–8 d (human) in methylcellulose culture. They are fivefold to eightfold more abundant than BFU-E in bone marrow and under normal circumstances do not appear in the circulation. As progenitors undergo the differentiation process, their numbers increase, with their proliferative potential simultaneously decreasing.

Several techniques have been described for the production of erythroblasts in liquid cultures (Fibach et al. 1989; Dolznig et al. 2001; Migliaccio et al. 2002). Although culture conditions differ in each protocol, SCF, erythropoietin (EPO), dexamethasone (Dex), and transferrin are commonly present, usually supplemented by insulin or IGF-1. In adult peripheral blood, the majority of the in vitro erythroid expansion potential resides in $CD34^-$ cells (van den Akker et al. 2010). These liquid cultures are advantageous for producing large numbers of erythroblasts from peripheral blood samples, thus enabling functional analyses of normal or abnormal erythropoiesis without the need for bone marrow sampling. In addition, erythroid cultures can also be obtained using mouse or human embryonic stem cells as the starting material (Carotta et al. 2004; Pilat et al. 2005; Olivier et al. 2006), and large-scale production of fully matured human erythrocytes has been reported (Neildez-Nguyen et al. 2002; Giarratana et al. 2005). The advent of induced pluripotent stem (iPS) cells raises prospects for production of patient-specific erythrocytes for transfusion purposes. However, at present, it is prohibitively expensive and a daunting task to generate the vast numbers of cells required for clinical practice (Douay and Andreu 2007; see also Arora and Daley 2012).

 Cite this article as *Cold Spring Harb Perspect Med* doi: 10.1101/cshperspect.a011601

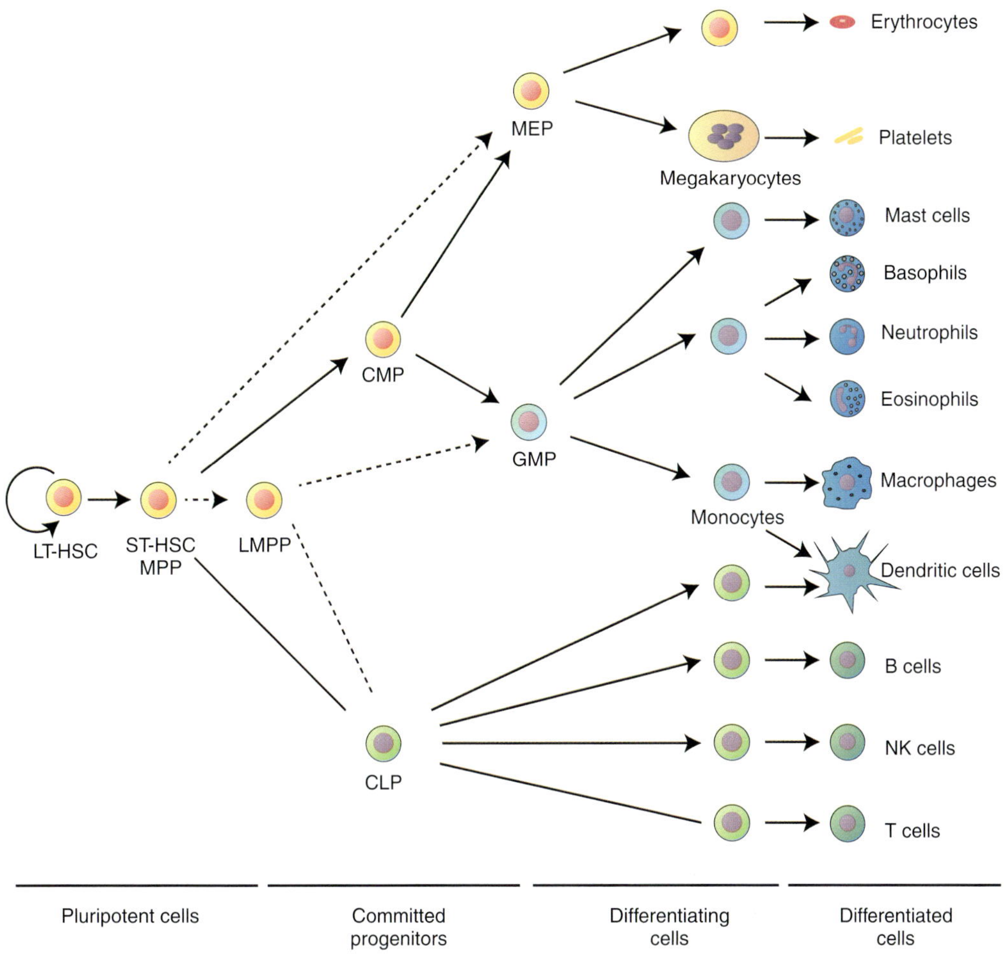

Figure 3. Proposed models for the hematopoietic hierarchy. In the model proposed by the Weissman group (Kondo et al. 1997; Akashi et al. 2000; Manz et al. 2002) (solid arrows), multipotential progenitors (MPPs or short-term HSCs [ST-HSCs]) give rise to either a common lymphocyte progenitor (CLP) or a common myeloid progenitor (CMP), which, in turn, gives rise to either a granulocyte-macrophage progenitor (GMP, equivalent to CFU-GM) or a megakaryocyte-erythroid progenitor (MEP). The alternate model suggested by the Jacobson group (Adolfsson et al. 2005) (dotted arrows) involves the generation of MEPs directly from the MPPs/ST-HSCs, whereas a lymphoid-primed multipotential progenitor (LMPP) has the potential to generate both CLPs and GMPs. LT-HSC, Long-term hematopoietic stem cell; NK cell, natural killer cell. (This figure is modified from data by Ferreira et al. 2005; reprinted, with permission, from the author as defined by the American Society for Microbiology.)

ERYTHROPOIESIS: REGULATION AND NICHE

Erythrocytes represent the most common cell type in adult blood. Human blood contains $\sim 5 \times 10^6$ erythrocytes per microliter (normal range 4.7×10^6 to 6.1×10^6 for males and 4.2×10^6 to 5.4×10^6 for females); these cells have an average life span of 120 d. New erythrocytes are constantly produced in the bone marrow, which provides a niche consisting of endothelial cells of the vascular system, osteoblasts, stromal cells, hematopoietic cells, and the extracellular matrix. This complex niche supports direct cell–cell contact and exposure of the developing hematopoietic cells to cell adhesion molecules, growth factors, and cytokines. The earliest erythroid progenitors are responsive to cytokines including TPO, GM-CSF, IL3, and IL11, and in particular to SCF. SCF binds to its receptor KIT, a tyrosine kinase that signals through several pathways including PI-3 kinase, Src kinases, and PLC-γ. At subsequent stages, SCF acts synergistically with EPO in the proliferation and expansion of the developing erythroid progenitors and may play a role in phosphorylating the EPO receptor itself (Wu et al. 1995; see Bunn 2013). Erythroid cells at the terminal stages of differentiation have shed their nucleus, endoplasmic reticulum, and mitochondria, and, consequently, they are no longer able to proliferate. To maintain the red blood cell count in the ~5 L of blood of an adult individual, $\sim 2.4 \times 10^6$ new erythrocytes have to be produced each second. The new cells enter the circulation as reticulocytes that are still engaged in protein translation. In humans, it takes ~1 wk for reticulocytes to complete the maturation process. In the bone marrow, a series of intermediate erythroid precursors can be recognized that progressively gains erythroid characteristics (Fig. 4). Development from the pro-erythroblast to the reticulocyte involves four to five rapid cell divisions, resulting in a progressive reduction in cell size. Mature erythrocytes have a diameter of only 6–8 μm. Their small size and biconcave shape creates a large surface area for gas exchange and allows the cells to enter the microcapillaries in the tissues.

Under steady-state conditions, ~1% of the erythrocytes are cleared every day and replaced by new cells. Remarkably, the rate of erythropoiesis can increase significantly from this baseline level in response to hypoxic stress, which occurs when adequate oxygen supply to all tissues is compromised by insufficient numbers of functional erythrocytes. Increasing red cell production is the primary response to counteract hypoxic stress. EPO production in the kidneys is directly regulated by tissue oxygen tension through the activity of the hypoxia-inducible transcription factor complex. Although EPO is the main regulator of red cell production (see Bunn 2013), other factors have an auxiliary role in the expansion of the erythroid progenitor compartment through their support of progenitor self-renewal. In particular, bone morphogenetic protein 4 (BMP4)/SMAD5 (Lenox et al. 2005), STAT5 (Socolovsky et al. 2001), SCF/KIT (Menon et al. 2006), and the glucocorticoid receptor (GR) (Bauer et al. 1999) are known to be involved.

A decreased apoptotic rate of erythroid progenitors may also contribute to the increase in red cell production that occurs under erythropoietic stress. It is believed that under steady-state conditions up to 60% of pro-erythroblasts succumb to apoptosis in the mouse spleen (Liu

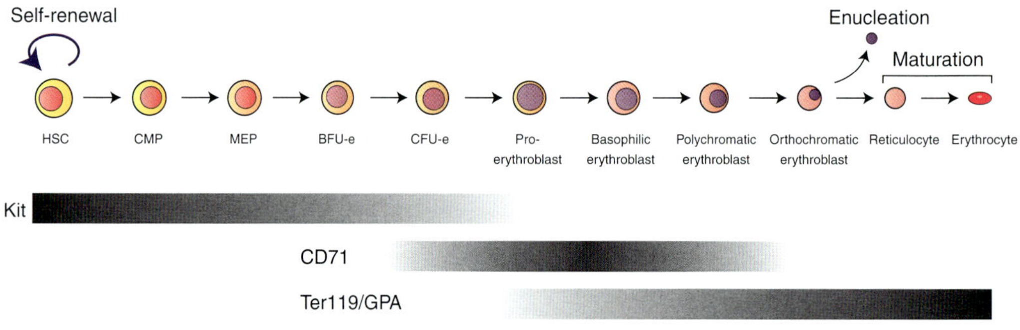

Figure 4. Erythroid differentiation in the mouse. The expression of the most commonly used cell surface markers to identify the various stages is indicated by the bars. Cells at the CFU-e and pro-erythroblast stage are the most sensitive to, and dependent on, the presence of EPO. Gray, low expression; black, high expression; HSC, hematopoietic stem cell; CMP, common myeloid progenitor; MEP, megakaryocyte-erythroid progenitor; BFU-e, burst-forming unit, erythroid; CFU-e, colony-forming unit, erythroid.

 Cite this article as *Cold Spring Harb Perspect Med* doi: 10.1101/cshperspect.a011601

et al. 2006). Mice deficient for the STAT5 transcription factor, the major downstream signaling target of EPOR, have a blunted stress erythropoietic response. The erythroid progenitors have an increased rate of apoptosis attributed to their failure to up-regulate expression of the antiapopotic BCL-X_L protein (Socolovsky et al. 1999, 2001). Furthermore, the FAS/FASL death receptor/death ligand system has been proposed to regulate erythroid homeostasis (De Maria et al. 1999). Under stress conditions, expression of FAS and FASL is reduced in erythroid cells and could contribute to the enhanced survival of these cells, further indicating that the survival function of the EPOR/STAT5 pathway is an important modulator of the erythropoietic rate (Liu et al. 2006).

ERYTHROBLASTIC ISLANDS ARE STRUCTURAL UNITS OF TERMINAL ERYTHROID DIFFERENTIATION

The erythroblastic island is a structural unit in the bone marrow where terminal erythroid differentiation takes place (Bessis 1958). It consists of a central macrophage (also known as the nurse cell) surrounded by differentiating erythroid progenitors (Fig. 5). Erythroblastic islands are also found in the mouse fetal liver (Fig. 5), but a role for an analogous structure has not been identified in the yolk sac. The fact that yolk sac cells enter the circulation without undergoing enucleation, a process in which the macrophage plays an important role, supports the notion that primitive erythropoiesis does not require the formation of islands. However, macrophages are formed during primitive hematopoiesis, and further investigations are necessary to determine the role of these cells in primitive erythropoiesis.

Several proteins on the surface of the macrophage and the erythroblasts mediate interactions between the macrophage and erythroblasts, and between the erythroblasts themselves. The erythroblast-macrophage protein (EMP or macrophage-erythroblast attacher MAEA) is expressed on both cell types and mediates adhesion between the cells. MAEA knockout mice survive until birth, displaying defects in terminal erythroid maturation (Soni et al. 2006). Formation of erythroblastic islands is impaired and MAEA-null macrophages appear to lack the extensive cytoplasmatic extensions typical of mature macrophages. MAEA-null macrophages are unable to interact with wild-type erythroblasts. In contrast, MAEA-null erythroblasts can interact with wild-type macrophages, but this does not rescue the enucleation defect of these erythroblasts. In enucleating wild-type erythroblasts, MAEA colocalizes with F-actin aggregates present at the constriction between the extruding nucleus and the reticulocyte. The actin cytoskeleton in MAEA-null erythroblasts is predominantly localized to the cell membrane,

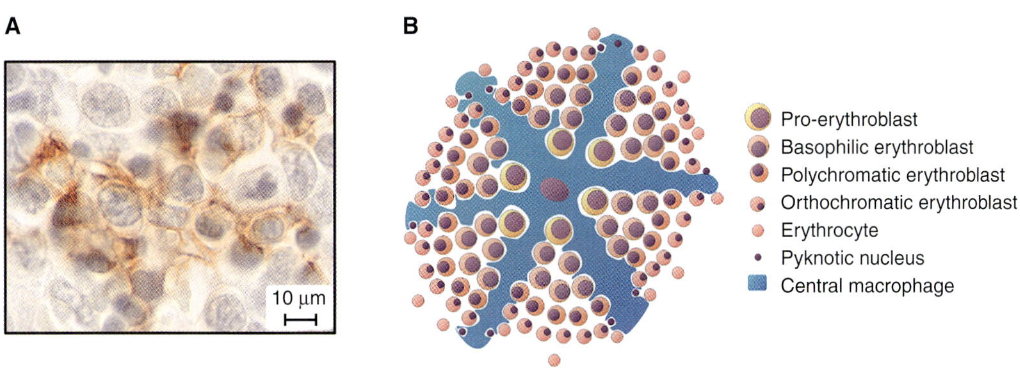

Figure 5. The erythroblastic island. (*A*) Erythroblastic island in E13.5 fetal liver. The cytoplasmic extensions of the central macrophage (stained with the F4/80 antibody) (brown) are surrounding erythroid cells at various stages of differentiation. (*B*) Schematic drawing of an erythroblastic island.

and very little cytoplasmic actin is observed. Collectively, it can be concluded that MAEA is required for the organization of the erythroblastic island and for efficient enucleation of erythroblasts.

The role of other cell adhesion molecules and interactions with extracellular matrix proteins such as laminin and fibronectin is less clearly defined. Erythroblasts express α_4 (Itga4)/CD29/β_1 integrin (Itgb1), and the CD242 (Icam4, LW blood group) counter-receptor. Macrophages express CD51 (α_v integrin, Itgav) and the vascular cell adhesion molecule 1 (Vcam1) counter-receptor. Perturbation of the integrin system can adversely affect erythroblastic island formation (Lee et al. 2006) and stress erythropoiesis (Scott et al. 2003), but the generic role of cell adhesion in tissue integrity and the considerable potential for redundancy preclude the assignment of specific functions to individual factors.

A role for signaling between erythroid cells has been most convincingly shown for death ligand/death receptor interactions. Erythroid cells express the CD95 (FAS) death receptor throughout differentiation. Only mature cells express the ligand CD95L (FASL), but these cells are insensitive to FAS signaling. In contrast, immature cells activate proapoptotic proteases in response to FASL exposure. The levels of the essential transcription factors TAL1 and GATA1 are reduced through cleavage by activated caspases, most notably by caspase 3 (CASP3) (De Maria et al. 1999; Zeuner et al. 2003; Ribeil et al. 2007). This will stop progression along the erythroid differentiation pathway and may induce cell death. A decreased survival rate of erythroid progenitor cells is a hallmark of many acquired and hereditary anemias. Patients with rheumatoid arthritis develop anemia in $\sim$50% of cases, and treatment with a blocking antibody directed against the death receptor ligand TNFα improved the anemia and decreased the elevated numbers of apoptotic erythroid progenitors in the bone marrow (Papadaki et al. 2002). An increased rate of apoptosis is also observed in β-thalassemia patients. This results in a phenomenon known as ineffective erythropoiesis (Centis et al. 2000; Mathias et al. 2000; Pootrakul et al. 2000; Papadaki et al. 2002). Up to 80% of erythroid progenitors may be lost because of apoptosis occurring at the polychromatic erythroblast stage (Mathias et al. 2000). The apoptotic erythroblasts are cleared very rapidly by macrophages. These cells are activated by increased interferon-γ (IFNG) levels in approximately one-third of the patients (Wanachiwanawin et al. 1999), and this may contribute to the quantitative differences in ineffective erythropoiesis between different patients. In addition, many inflammatory diseases are accompanied by anemia, and in a mouse model, IFNG has been implicated in anemia of chronic disease by activating macrophages, reducing erythrocyte life span and inhibiting erythropoiesis (Libregts et al. 2011).

In normal erythropoiesis, expelled nuclei are phagocytosed very rapidly by the central macrophages. Once released, the nuclei contain very low levels of ATP and start to expose phosphatidylserine on their surface, because it requires energy to maintain phosphatidylserine exclusively localized to the inner leaflet of the cell membrane. Surface exposure of phosphatidylserine is an early event in apoptosis and serves as an "eat me" signal for apoptotic cells. This signal is also used for the engulfment of expelled nuclei by the macrophages (Yoshida et al. 2005).

TRANSCRIPTIONAL REGULATION OF ERYTHROPOIESIS AND THE STEM CELL/EARLY PROGENITOR COMPARTMENT

A wide variety of transcription factors is involved in the establishment of hematopoietic cell lineages. For example, TAL1, a basic helix–loop–helix transcription factor, and LMO2, an LIM-domain transcription factor, are critical for the onset of hematopoiesis because null embryos show complete absence of primitive and definitive hematopoiesis (Robb et al. 1995, 1996; Shivdasani et al. 1995; Porcher et al. 1996). The zinc-finger transcription factor GATA2 is also essential for the early stages of hematopoiesis. Primitive and definitive hematopoiesis is abrogated when the GATA2 gene is deleted, and it appears to play a role in the proliferation of the early precursors rather than in

 Cite this article as *Cold Spring Harb Perspect Med* doi: 10.1101/cshperspect.a011601

their differentiation (Tsai et al. 1994; Tsai and Orkin 1997). RUNX1 is crucial for the early stages of definitive hematopoiesis. In RUNX1-null embryos, yolk sac hematopoiesis is normal but fetal liver hematopoiesis is absent (Okuda et al. 1996).

Once multipotent progenitors arise from the HSC and proliferate, the process of lineage restriction starts. The first step in this process involves restriction to either the myeloid or the lymphoid lineage. Transcription factors that are preferentially expressed in cells destined to one of those fates have been identified, but few have been extensively studied. PU.1, a member of the Ets family of transcription factors, is the best-studied transcription factor known to be involved in lineage restriction at this stage. PU.1's effect on lineage commitment is dependent on its expression level. A high level of expression leads to commitment to the myeloid lineage, whereas lower levels of expression lead to commitment to the lymphoid lineage (Scott et al. 1994; Nerlov and Graf 1998; Huang et al. 2007). Thus, owing to their role in the stem cell/progenitor compartment, inactivation of these factors also affects erythropoiesis.

Because the number of transcription factors with identifiable functions specifically during erythropoiesis is too vast to allow a thorough discussion of each, we will focus on KLF1. This transcription factor is required for terminal erythroid differentiation and provides one of the best-studied examples of an erythroid transcriptional regulator. Expression of the KLF1 (formerly known as erythroid Krüppel-like factor [EKLF]) is largely restricted to the erythroid cell lineage (Southwood et al. 1996). KLF1 specifically binds the sequence 5′-CCACACCCT-3′ (Miller and Bieker 1993). β-Thalassemia patients that carry mutations in the KLF1 binding site in the β-globin promoter display strongly reduced expression of β-globin (Orkin et al. 1984; Feng et al. 1994; Faustino et al. 1996). The role of KLF1 has been studied extensively in mice (Nuez et al. 1995; Perkins et al. 1995). Mice heterozygous for the KLF1 gene appear completely healthy. In the absence of KLF1, however, the fetuses develop a fatal anemia and die around E14. The primitive erythroid cells function sufficiently for normal survival of the KLF1 knockout embryos up to ~E12. Embryonic globin gene expression is not influenced by the absence of KLF1, even though KLF1 normally binds these genes in vivo (Zhou et al. 2006). When the embryos switch to definitive erythropoiesis in the fetal liver, KLF1-null mutants rapidly develop anemia because of a deficit in β-globin expression. Remarkably, the number of CFU-E in the KLF1-null fetal livers is similar to those found in wild-type fetal livers, showing that KLF1 deficiency causes a very late erythroid defect (Nuez et al. 1995; Perkins et al. 1995). Furthermore, fetal liver-derived erythroid cells of KLF1$^{-/-}$ fetuses have an abnormal morphology, and most of the cells retain a nucleus, indicating that the adult β-globin genes are not the only target genes of KLF1. Indeed, KLF1$^{-/-}$ mice could not be rescued by expression of exogenous human γ-globin (Perkins et al. 2000). Genome-wide expression analyses of KLF1$^{-/-}$ erythroid cells showed that KLF1 activates many other erythroid-specific genes, including genes encoding proteins associated with the erythroid cell membrane, such as band 4.9/dematin. These proteins are important for the function and stability of the erythrocytes. In addition, KLF1 has a role in cell cycle regulation linked to the proliferation arrest required for terminal differentiation (Tallack et al. 2007). Thus, KLF1 is an activator of essential erythroid genes that are up-regulated during terminal erythroid differentiation.

The first reported loss-of-function mutations in human KLF1 were linked to the rare blood group In(Lu) (inhibitor of Lutheran antigen expression) and always occurred in the presence of a normal KLF1 allele. Gene expression profiling identified more than 650 putative KLF1 target genes, including the blood group antigens BCAM and CD44, which are suppressed in the In(Lu) individuals (Singleton et al. 2008). A study of a large Maltese pedigree concluded that haploinsufficiency for KLF1 causes HPFH. All carriers displayed high HbF levels, although with considerable variation (mean 8.4%; range 3.3%–19.5%) (Borg et al. 2010). Part of this variability could be explained by SNP haplotypes at the *BCL11A* locus (Menzel et al. 2007), which encodes a repressor of γ-globin expression (Sankaran et al. 2008). Importantly, KLF1 was shown to be a direct activator of BCL11A expression (Borg et al. 2010; Zhou et al. 2010). Therefore, attenuation of KLF1 activity could be a fruitful approach to raise HbF levels in patients with β-type hemoglobinopathies. Supporting this notion, a sickle cell anemia patient with KLF1 haploinsufficiency was reported recently (Gallienne et al. 2012). This patient displayed a high HbF level (20.3%) and had a particularly mild disease phenotype. In contrast, missense mutations of critical residues in the DNA binding domain of KLF1 can have a dominant phenotype with severe consequences. Two unrelated patients with congenital dyserythropoietic anemia (CDA) carried a missense mutation (p.E325K) in the DNA-binding domain of KLF1. Although the patients displayed very high levels of HbF (31.6% and 44%), this mutation had a dominant effect that resulted in severe hemolytic anemia (Arnaud et al. 2010). Given the broad impact of KLF1 on erythroid-specific gene expression (Drissen et al. 2005; Pilon et al. 2008; Singleton et al. 2008; Borg et al. 2010), the array of human erythroid phenotypes associated with KLF1 mutations will likely be expanded as more cases are described in detail. This is illustrated by two recent studies reporting that zinc protoporphyria (Perseu et al. 2011) and increased HbA2 levels (Perseu et al. 2011) are linked to KLF1 mutations.

CONCLUDING REMARKS

In conclusion, much progress has been made in understanding the ontogeny, expansion, and terminal differentiation of erythroid cells, and future work will contribute to the ultimate goal of fully comprehending the molecular control of erythropoiesis at the systems level.

ACKNOWLEDGMENTS

We acknowledge support from the Landsteiner Society for Blood Research, Netherlands Institute for Regenerative Medicine FES subsidy, National Institutes of Health R37 DK54077, Netherlands Organization for Scientific Research (NWO), and Netherlands Genomics Initiative (NGI).

REFERENCES

Reference is also in this collection.

Adolfsson J, Mansson R, Buza-Vidas N, Hultquist A, Liuba K, Jensen CT, Bryder D, Yang L, Borge OJ, Thoren LA, et al. 2005. Identification of $Flt3^+$ lympho-myeloid stem cells lacking erythro-megakaryocytic potential. *Cell* **121:** 295–306.

Akashi K, Traver D, Miyamoto T, Weissman IL. 2000. A clonogenic common myeloid progenitor that gives rise to all myeloid lineages. *Nature* **404:** 193–197.

Alvarez-Silva M, Belo-Diabangouaya P, Salaun J, Dieterlen-Lievre F. 2003. Mouse placenta is a major hematopoietic organ. *Development* **130:** 5437–5444.

Arnaud L, Saison C, Helias V, Lucien N, Steschenko D, Giarratana MC, Prehu C, Foliguet B, Montout L, de Brevern AG, et al. 2010. A dominant mutation in the gene encoding the erythroid transcription factor KLF1 causes a congenital dyserythropoietic anemia. *Am J Hum Genet* **87:** 721–727.

* Arora N, Daley GQ. 2012. Pluripotent stem cells in research and treatment of hemoglobinopathies. *Cold Spring Harb Perspect Med* **2:** a011841.

Bauer A, Tronche F, Wessely O, Kellendonk C, Reichardt HM, Steinlein P, Schutz G, Beug H. 1999. The glucocorticoid receptor is required for stress erythropoiesis. *Genes Dev* **13:** 2996–3002.

Beaupain D, Martin C, Dieterlen-Lievre F. 1979. Are developmental hemoglobin changes related to the origin of stem cells and site of erythropoiesis? *Blood* **53:** 212–225.

Belaoussoff M, Farrington SM, Baron MH. 1998. Hematopoietic induction and respecification of A-P identity by visceral endoderm signaling in the mouse embryo. *Development* **125:** 5009–5018.

Berry M, Grosveld F, Dillon N. 1992. A single point mutation is the cause of the Greek form of hereditary persistence of fetal haemoglobin. *Nature* **358:** 499–502.

Bessis M. 1958. L'ilot erythroblastique, unite fonctionnelle de la moelle osseuse. *Rev Hematol* **13:** 8–11.

Boisset JC, van Cappellen W, Andrieu-Soler C, Galjart N, Dzierzak E, Robin C. 2010. In vivo imaging of haematopoietic cells emerging from the mouse aortic endothelium. *Nature* **464:** 116–120.

Borg J, Papadopoulos P, Georgitsi M, Gutierrez L, Grech G, Fanis P, Phylactides M, Verkerk AJ, van der Spek PJ, Scerri CA, et al. 2010. Haploinsufficiency for the erythroid transcription factor KLF1 causes hereditary persistence of fetal hemoglobin. *Nat Genet* **42:** 801–805.

Boyer SH, Belding TK, Margolet L, Noyes AN. 1975. Fetal hemoglobin restriction to a few erythrocytes (F cells) in normal human adults. *Science* **188:** 361–363.

* Bunn HF. 2013. Erythropoietin. *Cold Spring Harb Perspect Med* **3:** a011619.

Buza-Vidas N, Luc S, Jacobsen SE. 2007. Delineation of the earliest lineage commitment steps of haematopoietic stem cells: New developments, controversies and major challenges. *Curr Opin Hematol* **14:** 315–321.

Carotta S, Pilat S, Mairhofer A, Schmidt U, Dolznig H, Steinlein P, Beug H. 2004. Directed differentiation and mass cultivation of pure erythroid progenitors from mouse embryonic stem cells. *Blood* **104:** 1873–1880.

Centis F, Tabellini L, Lucarelli G, Buffi O, Tonucci P, Persini B, Annibali M, Emiliani R, Iliescu A, Rapa S, et al. 2000. The importance of erythroid expansion in determining the extent of apoptosis in erythroid precursors in patients with β-thalassemia major. *Blood* **96:** 3624–3629.

Chen MJ, Yokomizo T, Zeigler BM, Dzierzak E, Speck NA. 2009. Runx1 is required for the endothelial to haematopoietic cell transition but not thereafter. *Nature* **457:** 887–891.

Ciau-Uitz A, Walmsley M, Patient R. 2000. Distinct origins of adult and embryonic blood in *Xenopus*. *Cell* **102:** 787–796.

Corbel C, Salaun J, Belo-Diabangouaya P, Dieterlen-Lievre F. 2007. Hematopoietic potential of the pre-fusion allantois. *Dev Biol* **301:** 478–488.

Cormier F, Dieterlen-Lievre F. 1988. The wall of the chick embryo aorta harbours M-CFC, G-CFC, GM-CFC and BFU-E. *Development* **102:** 279–285.

Cumano A, Dieterlen-Lievre F, Godin I. 1996. Lymphoid potential, probed before circulation in mouse, is restricted to caudal intraembryonic splanchnopleura. *Cell* **86:** 907–916.

de Bruijn MF, Speck NA, Peeters MC, Dzierzak E. 2000. Definitive hematopoietic stem cells first develop within the major arterial regions of the mouse embryo. *EMBO J* **19:** 2465–2474.

de Bruijn MF, Ma X, Robin C, Ottersbach K, Sanchez MJ, Dzierzak E. 2002. Hematopoietic stem cells localize to the endothelial cell layer in the midgestation mouse aorta. *Immunity* **16:** 673–683.

De Maria R, Zeuner A, Eramo A, Domenichelli C, Bonci D, Grignani F, Srinivasula SM, Alnemri ES, Testa U, Peschle C. 1999. Negative regulation of erythropoiesis by caspase-mediated cleavage of GATA-1. *Nature* **401:** 489–493.

Dieterlen-Lievre F. 1975. On the origin of haemopoietic stem cells in the avian embryo: An experimental approach. *J Embryol Exp Morphol* **33:** 607–619.

Dieterlen-Lievre F, Martin C. 1981. Diffuse intraembryonic hemopoiesis in normal and chimeric avian development. *Dev Biol* **88:** 180–191.

Dillon N, Grosveld F. 1991. Human γ-globin genes silenced independently of other genes in the β-globin locus. *Nature* **350:** 252–254.

Dolznig H, Boulme F, Stangl K, Deiner EM, Mikulits W, Beug H, Mullner EW. 2001. Establishment of normal, terminally differentiating mouse erythroid progenitors: Molecular characterization by cDNA arrays. *FASEB J* **15:** 1442–1444.

Douay L, Andreu G. 2007. Ex vivo production of human red blood cells from hematopoietic stem cells: What is the future in transfusion? *Transfus Med Rev* **21:** 91–100.

Downs KM. 1998. The murine allantois. *Curr Top Dev Biol* **39:** 1–33.

Drissen R, von Lindern M, Kolbus A, Driegen S, Steinlein P, Beug H, Grosveld F, Philipsen S. 2005. The erythroid phenotype of EKLF-null mice: Defects in hemoglobin metabolism and membrane stability. *Mol Cell Biol* **25:** 5205–5214.

Dzierzak E, Speck NA. 2008. Of lineage and legacy: The development of mammalian hematopoietic stem cells. *Nat Immunol* **9:** 129–136.

Ema H, Nakauchi H. 2000. Expansion of hematopoietic stem cells in the developing liver of a mouse embryo. *Blood* **95:** 2284–2288.

Faustino P, Lavinha J, Marini MG, Moi P. 1996. β-Thalassemia mutation at −90C → T impairs the interaction of the proximal CACCC box with both erythroid and nonerythroid factors. *Blood* **88:** 3248–3249.

Feng WC, Southwood CM, Bieker JJ. 1994. Analyses of β-thalassemia mutant DNA interactions with erythroid Kruppel-like factor (EKLF), an erythroid cell-specific transcription factor. *J Biol Chem* **269:** 1493–1500.

Ferreira R, Ohneda K, Yamamoto M, Philipsen S. 2005. GATA1 function, a paradigm for transcription factors in hematopoiesis. *Mol Cell Biol* **25:** 1215–1227.

Fibach E, Manor D, Oppenheim A, Rachmilewitz EA. 1989. Proliferation and maturation of human erythroid progenitors in liquid culture. *Blood* **73:** 100–103.

Fraser ST, Isern J, Baron MH. 2007. Maturation and enucleation of primitive erythroblasts during mouse embryogenesis is accompanied by changes in cell-surface antigen expression. *Blood* **109:** 343–352.

Gallienne AE, Dreau HM, Schuh A, Old J, Henderson S. 2012. Ten novel mutations in the erythroid transcription factor KLF1 gene associated with increased fetal hemoglobin levels in adults. *Haematologica* **97:** 340–343.

Gekas C, Dieterlen-Lievre F, Orkin SH, Mikkola HK. 2005. The placenta is a niche for hematopoietic stem cells. *Dev Cell* **8:** 365–375.

Giarratana MC, Kobari L, Lapillonne H, Chalmers D, Kiger L, Cynober T, Marden MC, Wajcman H, Douay L. 2005. Ex vivo generation of fully mature human red blood cells from hematopoietic stem cells. *Nat Biotechnol* **23:** 69–74.

Godin I, Dieterlen-Lievre F, Cumano A. 1995. Emergence of multipotent hemopoietic cells in the yolk sac and para-aortic splanchnopleura in mouse embryos, beginning at 8.5 days postcoitus. *Proc Natl Acad Sci* **92:** 773–777.

Huang G, Zhang P, Hirai H, Elf S, Yan X, Chen Z, Koschmieder S, Okuno Y, Dayaram T, Growney JD, et al. 2007. PU.1 is a major downstream target of AML1 (RUNX1) in adult mouse hematopoiesis. *Nat Genet* **40:** 51–60.

Huber TL, Kouskoff V, Fehling HJ, Palis J, Keller G. 2004. Haemangioblast commitment is initiated in the primitive streak of the mouse embryo. *Nature* **432:** 625–630.

Kinder SJ, Tsang TE, Quinlan GA, Hadjantonakis AK, Nagy A, Tam PP. 1999. The orderly allocation of mesodermal cells to the extraembryonic structures and the anteroposterior axis during gastrulation of the mouse embryo. *Development* **126:** 4691–4701.

Kondo M, Weissman IL, Akashi K. 1997. Identification of clonogenic common lymphoid progenitors in mouse bone marrow. *Cell* **91:** 661–672.

Lawson KA, Meneses JJ, Pedersen RA. 1991. Clonal analysis of epiblast fate during germ layer formation in the mouse embryo. *Development* **113:** 891–911.

Lee G, Lo A, Short SA, Mankelow TJ, Spring F, Parsons SF, Yazdanbakhsh K, Mohandas N, Anstee DJ, Chasis JA. 2006. Targeted gene deletion demonstrates that the cell adhesion molecule ICAM-4 is critical for erythroblastic island formation. *Blood* **108:** 2064–2071.

Lenox LE, Perry JM, Paulson RF. 2005. BMP4 and Madh5 regulate the erythroid response to acute anemia. *Blood* **105:** 2741–2748.

Libregts SF, Gutierrez L, de Bruin AM, Wensveen FM, Papadopoulos P, van Ijcken W, Ozgur Z, Philipsen S, Nolte MA. 2011. Chronic IFN-γ production in mice induces anemia by reducing erythrocyte life span and inhibiting erythropoiesis through an IRF-1/PU.1 axis. *Blood* **118:** 2578–2588.

Lin CS, Lim SK, D'Agati V, Costantini F. 1996. Differential effects of an erythropoietin receptor gene disruption on primitive and definitive erythropoiesis. *Genes Dev* **10:** 154–164.

Liu Y, Pop R, Sadegh C, Brugnara C, Haase VH, Socolovsky M. 2006. Suppression of Fas–FasL coexpression by erythropoietin mediates erythroblast expansion during the erythropoietic stress response in vivo. *Blood* **108:** 123–133.

Lux CT, Yoshimoto M, McGrath K, Conway SJ, Palis J, Yoder MC. 2007. All primitive and definitive hematopoietic progenitor cells emerging prior to E10 in the mouse embryo are products of the yolk sac. *Blood* **111:** 3435–3438.

Manz MG, Miyamoto T, Akashi K, Weissman IL. 2002. Prospective isolation of human clonogenic common myeloid progenitors. *Proc Natl Acad Sci* **99:** 11872–11877.

Martin C, Beaupain D, Dieterlen-Lievre F. 1978. Developmental relationships between vitelline and intra-embryonic haemopoiesis studied in avian "yolk sac chimaeras." *Cell Differ* **7:** 115–130.

Mathias LA, Fisher TC, Zeng L, Meiselman HJ, Weinberg KI, Hiti AL, Malik P. 2000. Ineffective erythropoiesis in β-thalassemia major is due to apoptosis at the polychromatophilic normoblast stage. *Exp Hematol* **28:** 1343–1353.

McGrath KE, Palis J. 2005. Hematopoiesis in the yolk sac: More than meets the eye. *Exp Hematol* **33:** 1021–1028.

Medvinsky A, Dzierzak E. 1996. Definitive hematopoiesis is autonomously initiated by the AGM region. *Cell* **86:** 897–906.

Menon MP, Karur V, Bogacheva O, Bogachev O, Cuetara B, Wojchowski DM. 2006. Signals for stress erythropoiesis are integrated via an erythropoietin receptor–phosphotyrosine-343–Stat5 axis. *J Clin Invest* **116:** 683–694.

Menzel S, Garner C, Gut I, Matsuda F, Yamaguchi M, Heath S, Foglio M, Zelenika D, Boland A, Rooks H, et al. 2007. A QTL influencing F cell production maps to a gene encoding a zinc-finger protein on chromosome 2p15. *Nat Genet* **39:** 1197–1199.

Migliaccio AR, Campisi S, Migliaccio G. 2001. Standardization of progenitor cell assay for cord blood banking. *Ann Ist Super Sanita* **37:** 595–600.

Migliaccio G, Di Pietro R, di Giacomo V, Di Baldassarre A, Migliaccio AR, Maccioni L, Galanello R, Papayannopoulou T. 2002. In vitro mass production of human erythroid cells from the blood of normal donors and of thalassemic patients. *Blood Cells Mol Dis* **28:** 169–180.

Miller IJ, Bieker JJ. 1993. A novel, erythroid cell-specific murine transcription factor that binds to the CACCC element and is related to the Kruppel family of nuclear proteins. *Mol Cell Biol* **13:** 2776–2786.

Moore MA, Metcalf D. 1970. Ontogeny of the haemopoietic system: Yolk sac origin of in vivo and in vitro colony forming cells in the developing mouse embryo. *Br J Haematol* **18:** 279–296.

Muller AM, Medvinsky A, Strouboulis J, Grosveld F, Dzierzak E. 1994. Development of hematopoietic stem cell activity in the mouse embryo. *Immunity* **1:** 291–301.

Murray P. 1932. The development in vitro of the blood of the early chick embryo. *Proc Roy Soc London* **11:** 497–521.

Neildez-Nguyen TM, Wajcman H, Marden MC, Bensidhoum M, Moncollin V, Giarratana MC, Kobari L, Thierry D, Douay L. 2002. Human erythroid cells produced ex vivo at large scale differentiate into red blood cells in vivo. *Nat Biotechnol* **20:** 467–472.

Nerlov C, Graf T. 1998. PU.1 induces myeloid lineage commitment in multipotent hematopoietic progenitors. *Genes Dev* **12:** 2403–2412.

North TE, de Bruijn MF, Stacy T, Talebian L, Lind E, Robin C, Binder M, Dzierzak E, Speck NA. 2002. Runx1 expression marks long-term repopulating hematopoietic stem cells in the midgestation mouse embryo. *Immunity* **16:** 661–672.

Nuez B, Michalovich D, Bygrave A, Ploemacher R, Grosveld F. 1995. Defective haematopoiesis in fetal liver resulting from inactivation of the EKLF gene. *Nature* **375:** 316–318.

Okuda T, van Deursen J, Hiebert SW, Grosveld G, Downing JR. 1996. AML1, the target of multiple chromosomal translocations in human leukemia, is essential for normal fetal liver hematopoiesis. *Cell* **84:** 321–330.

Olivier EN, Qiu C, Velho M, Hirsch RE, Bouhassira EE. 2006. Large-scale production of embryonic red blood cells from human embryonic stem cells. *Exp Hematol* **34:** 1635–1642.

Orkin SH, Antonarakis SE, Kazazian HH Jr. 1984. Base substitution at position −88 in a β-thalassemic globin gene. Further evidence for the role of distal promoter element ACACCC. *J Biol Chem* **259:** 8679–8681.

Ottersbach K, Dzierzak E. 2005. The murine placenta contains hematopoietic stem cells within the vascular labyrinth region. *Dev Cell* **8:** 377–387.

Papadaki HA, Kritikos HD, Valatas V, Boumpas DT, Eliopoulos GD. 2002. Anemia of chronic disease in rheumatoid arthritis is associated with increased apoptosis of bone marrow erythroid cells: Improvement following anti-tumor necrosis factor-α antibody therapy. *Blood* **100:** 474–482.

Papayannopoulou T, Brice M, Stamatoyannopoulos G. 1977. Hemoglobin F synthesis in vitro: Evidence for control at the level of primitive erythroid stem cells. *Proc Natl Acad Sci* **74:** 2923–2927.

Park C, Ma YD, Choi K. 2005. Evidence for the hemangioblast. *Exp Hematol* **33:** 965–970.

Perkins AC, Sharpe AH, Orkin SH. 1995. Lethal β-thalassaemia in mice lacking the erythroid CACCC-transcription factor EKLF. *Nature* **375:** 318–322.

Perkins AC, Peterson KR, Stamatoyannopoulos G, Witkowska HE, Orkin SH. 2000. Fetal expression of a human Aγ globin transgene rescues globin chain imbalance but not hemolysis in EKLF null mouse embryos. *Blood* **95:** 1827–1833.

Perseu L, Satta S, Moi P, Demartis FR, Manunza L, Sollaino MC, Barella S, Cao A, Galanello R. 2011. KLF1 gene mutations cause borderline HbA(2). *Blood* **118:** 4454–4458.

Peterson KR, Li QL, Clegg CH, Furukawa T, Navas PA, Norton EJ, Kimbrough TG, Stamatoyannopoulos G. 1995. Use of yeast artificial chromosomes (YACs) in studies of mammalian development: Production of β-globin locus YAC mice carrying human globin developmental mutants. *Proc Natl Acad Sci* **92:** 5655–5659.

Pilat S, Carotta S, Schiedlmeier B, Kamino K, Mairhofer A, Will E, Modlich U, Steinlein P, Ostertag W, Baum C, et al. 2005. HOXB4 enforces equivalent fates of ES-cell-derived and adult hematopoietic cells. *Proc Natl Acad Sci* **102:** 12101–12106.

Pilon AM, Arcasoy MO, Dressman HK, Vayda SE, Maksimova YD, Sangerman JI, Gallagher PG, Bodine DM. 2008. Failure of terminal erythroid differentiation in EKLF-deficient mice is associated with cell cycle perturbation and reduced expression of E2F2. *Mol Cell Biol* **28:** 7394–7401.

Pootrakul P, Sirankapracha P, Hemsorach S, Moungsub W, Kumbunlue R, Piangitjagum A, Wasi P, Ma L, Schrier SL. 2000. A correlation of erythrokinetics, ineffective erythropoiesis, and erythroid precursor apoptosis in Thai patients with thalassemia. *Blood* **96:** 2606–2612.

Porcher C, Swat W, Rockwell K, Fujiwara Y, Alt FW, Orkin SH. 1996. The T cell leukemia oncoprotein SCL/tal-1 is essential for development of all hematopoietic lineages. *Cell* **86:** 47–57.

Ribeil JA, Zermati Y, Vandekerckhove J, Cathelin S, Kersual J, Dussiot M, Coulon S, Moura IC, Zeuner A, Kirkegaard-Sorensen T, et al. 2007. Hsp70 regulates erythropoiesis by preventing caspase-3-mediated cleavage of GATA-1. *Nature* **445:** 102–105.

Robb L, Lyons I, Li R, Hartley L, Kontgen F, Harvey RP, Metcalf D, Begley CG. 1995. Absence of yolk sac hematopoiesis from mice with a targeted disruption of the *scl* gene. *Proc Natl Acad Sci* **92:** 7075–7079.

Robb L, Elwood NJ, Elefanty AG, Kontgen F, Li R, Barnett LD, Begley CG. 1996. The *scl* gene product is required for the generation of all hematopoietic lineages in the adult mouse. *EMBO J* **15:** 4123–4129.

Russell ES, Bernstein SE. 1966. Blood and blood formation. In *Biology of the laboratory mouse* (ed. Green EL), pp. 351–372. McGraw-Hill, New York.

Sabin F. 1920. *Studies on the origin of blood vessels and of red blood corpuscles as seen in the living blastoderm of chicks during the second day of incubation*, Vol. 9, p. 214. Carnegie Institute of Washington, Washington, DC.

* Sankaran VG, Orkin SH. 2013. The switch from fetal to adult hemoglobin. *Cold Spring Harb Perspect Med* **3:** a011643.

Sankaran VG, Menne TF, Xu J, Akie TE, Lettre G, Van Handel B, Mikkola HK, Hirschhorn JN, Cantor AB, Orkin SH. 2008. Human fetal hemoglobin expression is regulated by the developmental stage-specific repressor BCL11A. *Science* **322:** 1839–1842.

Scott EW, Simon MC, Anastasi J, Singh H. 1994. Requirement of transcription factor PU.1 in the development of multiple hematopoietic lineages. *Science* **265:** 1573–1577.

Scott LM, Priestley GV, Papayannopoulou T. 2003. Deletion of α4 integrins from adult hematopoietic cells reveals roles in homeostasis, regeneration, and homing. *Mol Cell Biol* **23:** 9349–9360.

Shalaby F, Rossant J, Yamaguchi TP, Gertsenstein M, Wu XF, Breitman ML, Schuh AC. 1995. Failure of blood-island formation and vasculogenesis in Flk-1-deficient mice. *Nature* **376:** 62–66.

Shivdasani RA, Mayer EL, Orkin SH. 1995. Absence of blood formation in mice lacking the T-cell leukaemia oncoprotein tal-1/SCL. *Nature* **373:** 432–434.

Singleton BK, Burton NM, Green C, Brady RL, Anstee DJ. 2008. Mutations in EKLF/KLF1 form the molecular basis of the rare blood group In(Lu) phenotype. *Blood* **112:** 2081–2088.

Socolovsky M. 2007. Molecular insights into stress erythropoiesis. *Curr Opin Hematol* **14:** 215–224.

Socolovsky M, Fallon AE, Wang S, Brugnara C, Lodish HF. 1999. Fetal anemia and apoptosis of red cell progenitors in Stat5a$^{-/-}$5b$^{-/-}$ mice: A direct role for Stat5 in Bcl-X(L) induction. *Cell* **98:** 181–191.

Socolovsky M, Nam H, Fleming MD, Haase VH, Brugnara C, Lodish HF. 2001. Ineffective erythropoiesis in Stat5a$^{-/-}$b$^{-/-}$ mice due to decreased survival of early erythroblasts. *Blood* **98:** 3261–3273.

Soni S, Bala S, Gwynn B, Sahr KE, Peters LL, Hanspal M. 2006. Absence of erythroblast macrophage protein (Emp) leads to failure of erythroblast nuclear extrusion. *J Biol Chem* **281:** 20181–20189.

Southwood CM, Downs KM, Bieker JJ. 1996. Erythroid Kruppel-like factor exhibits an early and sequentially localized pattern of expression during mammalian erythroid ontogeny. *Dev Dyn* **206:** 248–259.

Stamatoyannopoulos G. 2005. Control of globin gene expression during development and erythroid differentiation. *Exp Hematol* **33:** 259–271.

Strouboulis J, Dillon N, Grosveld F. 1992. Developmental regulation of a complete 70-kb human β-globin locus in transgenic mice. *Genes Dev* **6:** 1857–1864.

Tallack MR, Keys JR, Perkins AC. 2007. Erythroid Kruppel-like factor regulates the G1 cyclin dependent kinase inhibitor p18^{INK4c}. *J Mol Biol* **369:** 313–321.

Tavian M, Peault B. 2005. Embryonic development of the human hematopoietic system. *Int J Dev Biol* **49:** 243–250.

Tavian M, Hallais MF, Peault B. 1999. Emergence of intraembryonic hematopoietic precursors in the pre-liver human embryo. *Development* **126:** 793–803.

Tsai FY, Orkin SH. 1997. Transcription factor GATA-2 is required for proliferation/survival of early hematopoietic cells and mast cell formation, but not for erythroid and myeloid terminal differentiation. *Blood* **89:** 3636–3643.

Tsai FY, Keller G, Kuo FC, Weiss M, Chen J, Rosenblatt M, Alt FW, Orkin SH. 1994. An early haematopoietic defect in mice lacking the transcription factor GATA-2. *Nature* **371:** 221–226.

Turpen JB, Knudson CM. 1982. Ontogeny of hematopoietic cells in *Rana pipiens*: Precursor cell migration during embryogenesis. *Dev Biol* **89:** 138–151.

Ueno H, Weissman IL. 2006. Clonal analysis of mouse development reveals a polyclonal origin for yolk sac blood islands. *Dev Cell* **11:** 519–533.

van den Akker E, Satchwell TJ, Pellegrin S, Daniels G, Toye AM. 2010. The majority of the in vitro erythroid expansion potential resides in $CD34^-$ cells, outweighing the contribution of $CD34^+$ cells and significantly increasing the erythroblast yield from peripheral blood samples. *Haematologica* **95:** 1594–1598.

Wanachiwanawin W, Wiener E, Siripanyaphinyo U, Chinprasertsuk S, Mawas F, Fucharoen S, Wickramasinghe SN, Pootrakul P, Visudhiphan S. 1999. Serum levels of tumor necrosis factor-α, interleukin-1, and interferon-γ in β^0-thalassemia/HbE and their clinical significance. *J Interferon Cytokine Res* **19:** 105–111.

Wood HB, May G, Healy L, Enver T, Morriss-Kay GM. 1997. CD34 expression patterns during early mouse development are related to modes of blood vessel formation and reveal additional sites of hematopoiesis. *Blood* **90:** 2300–2311.

Wu H, Klingmuller U, Besmer P, Lodish HF. 1995. Interaction of the erythropoietin and stem-cell-factor receptors. *Nature* **377:** 242–246.

Yokomizo T, Yamada-Inagawa T, Yzaguirre AD, Chen MJ, Speck NA, Dzierzak E. 2012. Whole-mount three-dimensional imaging of internally localized immunostained cells within mouse embryos. *Nat Protoc* **7:** 421–431.

Yoshida H, Kawane K, Koike M, Mori Y, Uchiyama Y, Nagata S. 2005. Phosphatidylserine-dependent engulfment by macrophages of nuclei from erythroid precursor cells. *Nature* **437:** 754–758.

Zeigler BM, Sugiyama D, Chen M, Guo Y, Downs KM, Speck NA. 2006. The allantois and chorion, when isolated before circulation or chorio-allantoic fusion, have hematopoietic potential. *Development* **133:** 4183–4192.

Zeuner A, Eramo A, Testa U, Felli N, Pelosi E, Mariani G, Srinivasula SM, Alnemri ES, Condorelli G, Peschle C, et al. 2003. Control of erythroid cell production via caspase-mediated cleavage of transcription factor SCL/Tal-1. *Cell Death Differ* **10:** 905–913.

Zhou D, Pawlik KM, Ren J, Sun CW, Townes TM. 2006. Differential binding of erythroid Krupple-like factor to embryonic/fetal globin gene promoters during development. *J Biol Chem* **281:** 16052–16057.

Zhou D, Liu K, Sun CW, Pawlik KM, Townes TM. 2010. KLF1 regulates BCL11A expression and γ- to β-globin gene switching. *Nat Genet* **42:** 742–744.

Erythropoietin

H. Franklin Bunn

Hematology Division, Department of Medicine, Brigham and Women's Hospital, Harvard Medical School, Boston, Massachusetts 02115

Correspondence: hfbunn@rics.bwh.harvard.edu

During the past century, few proteins have matched erythropoietin (Epo) in capturing the imagination of physiologists, molecular biologists, and, more recently, physicians and patients. Its appeal rests on its commanding role as the premier erythroid cytokine, the elegant mechanism underlying the regulation of its gene, and its remarkable impact as a therapeutic agent, arguably the most successful drug spawned by the revolution in recombinant DNA technology. This concise review will begin with a synopsis of the colorful history of this protein, culminating in its purification and molecular cloning. It then covers in more detail the contemporary understanding of Epo's physiology as well as its structure and interaction with its receptor. A major part of this article focuses on the regulation of the *Epo* gene and the discovery of HIF, a transcription factor that plays a cardinal role in molecular adaptation to hypoxia. In the concluding section, a synopsis of Epo's role in disorders of red blood cell production will be followed by an assessment of the remarkable impact of Epo therapy in the treatment of anemias, as well as concerns that provide a strong impetus for the development of even safer and more effective treatment.

In 1890, Viault (1890) observed that 2 weeks after traveling in Peru from sea level (Lima) to the mountain area of Morococha (4200 m) his red blood cell count went from 5.0 million to 7.1 million/mm^3. Values of five other sojourners in his party ranged from 7.1 to 8.0 million. These simple observations provided the first convincing demonstration of the robust burst of erythropoiesis in man soon after exposure to high altitude hypoxia. During the transition into the next millennium the mechanism underlying this phenomenon became a topic of heated debate. Friedrich Miescher (1893), well known for his discovery of DNA, proposed that a decrease in oxygen tension within the bone marrow provided a direct stimulus to erythroid cells. A half-century elapsed before this theory was disproven by carefully executed measurements of oxygen saturation in bone marrow specimens of patients with erythrocytosis, both primary (Berk et al. 1948) and secondary (Stohlman et al. 1954).

In 1906, Carnot and Deflandre (1906) proposed an alternate mechanism for hypoxic induction of erythropoiesis. They observed an increase in red blood cell counts following infusion of normal rabbits with serum from anemic animals and concluded that erythropoiesis is regulated by a humoral "factor" in the plasma. Attempts to reproduce this experiment over the ensuing decades yielded equivocal or negative results, thus casting doubt on this hypothesis.

Copyright © 2013 Cold Spring Harbor Laboratory Press; all rights reserved
Cite this article as *Cold Spring Harb Perspect Med* doi: 10.1101/cshperspect.a011619

However, in the middle of the twentieth century, Krumdieck (1943) and Erslev (1953) modified the experimental design of Carnot and Deflandre by the addition of accurate measurements of reticulocytes and convincingly showed in rabbits the induction of new red cell production within 3–6 days following injection of anemic serum injected.

The notion that hypoxic stimulation of erythropoiesis involved an indirect humoral mechanism was strongly buttressed by experiments of Reissmann (1950) and Ruhenstroth-Bauer (1950). They used parabiotic pairs of rats whose circulations were connected at the capillary level by overlapping flaps of skin and elegantly showed that when one rat was exposed to low oxygen tension, whereas the other breathed room air and remained normoxic, both animals developed a surge of new red cell production and erythrocytosis.

Taken together, these studies led to the conception of a circulating erythroid-stimulating hormone, "erythropoietin" (Epo). Organ ablation studies in rats (Jacobson et al. 1957) and man (Nathan et al. 1964) firmly established that the kidney was the major site but not the sole site of Epo production. These findings led Eugene Goldwasser and his colleagues to undertake an intense and prolonged effort to isolate Epo. Initial attempts to obtain Epo from kidneys were unsuccessful owing to the release of proteolytic enzymes during tissue homogenization. In the search for a more tractable source of Epo, Goldwasser first turned to plasma of anemia sheep, then to urine from Argentinians with severe iron deficiency owing to hookworm infestation, and finally to urine from Japanese patients with aplastic anemia. This 15-year endeavor was greatly facilitated by the development of a sensitive and specific assay using incorporation of radio-labeled iron into newly produced red cells (Fried et al. 1956). By 1977, Goldwasser and his team were able to prepare 8 mg of highly purified human Epo (Miyake et al. 1977). Amino-terminal amino acid sequencing of this preparation enabled the synthesis of semidegenerate oligonucleotide probes that could then be used for the molecular cloning of the *Epo* gene (Jacobs et al. 1985; Lin et al. 1985). This advance opened up a new era in the exploration of the physiology and molecular biology of Epo and was exploited in the development of recombinant human Epo as a therapeutic agent for patients with various types of anemia.

For more detailed information on the history of Epo, see reviews by Grant and Root (1952), Erslev (1980, 1993), and Goldwasser (1996).

PHYSIOLOGY[1]

In humans and other mammals, erythropoiesis normally proceeds at a low basal rate, replacing senescent red blood cells with young reticulocytes. In humans, red cell production can be enhanced as much as eightfold the baseline rate in a variety of clinical settings including hemorrhage, hemolysis, and other types of stress that impair the oxygenation of arterial blood or the delivery of oxygen to the tissues. Epo is the primary, and probably the sole mediator of hypoxic induction of erythropoiesis. During fetal development, Epo is produced mainly in the liver. However, in keeping with the above-mentioned report of Jacobson et al. (1957), following birth, the kidney accounts for ~80% of Epo production. Realizing that renal cortical blood flow closely matches oxygen consumption, Erslev et al. (1985) proposed that the proximal tubule is the ideal location for Epo production. In situ hybridization studies have indeed shown that *Epo* mRNA expression in the kidney is localized to a subset of peritubular fibroblasts in the cortex close to the boundary with the medulla (Koury et al. 1988, 1989; Lacombe et al. 1988; Bachmann et al. 1993; Maxwell et al. 1993a). In the liver, Epo is produced both in hepatocytes and in interstitial cells (Koury et al. 1991; Schuster et al. 1992). The latter have been shown to be Ito presinusoidal cells (Maxwell et al. 1994). In both Ito cells and the above-mentioned renal fibroblasts, in situ expression of Epo correlates with that of ecto-

[1]For more information on all aspects of erythropoietin, see the recent detailed and comprehensive review by Wenger and Kurtz (2011).

5′-nucleotidase, probably because this enzyme is also controlled by local oxygen tension (Schmid et al. 1994).

The notion that Epo production is markedly up-regulated by hypoxia and that it stimulates erythropoiesis in a dose-dependent manner led to the now well accepted paradigm of a negative feedback loop, similar to those underlying the regulation of blood glucose by insulin and of peripheral endocrine hormones by pituitary tropins. As shown in Figure 1A, hypoxia induces an increase in Epo hormone production in the

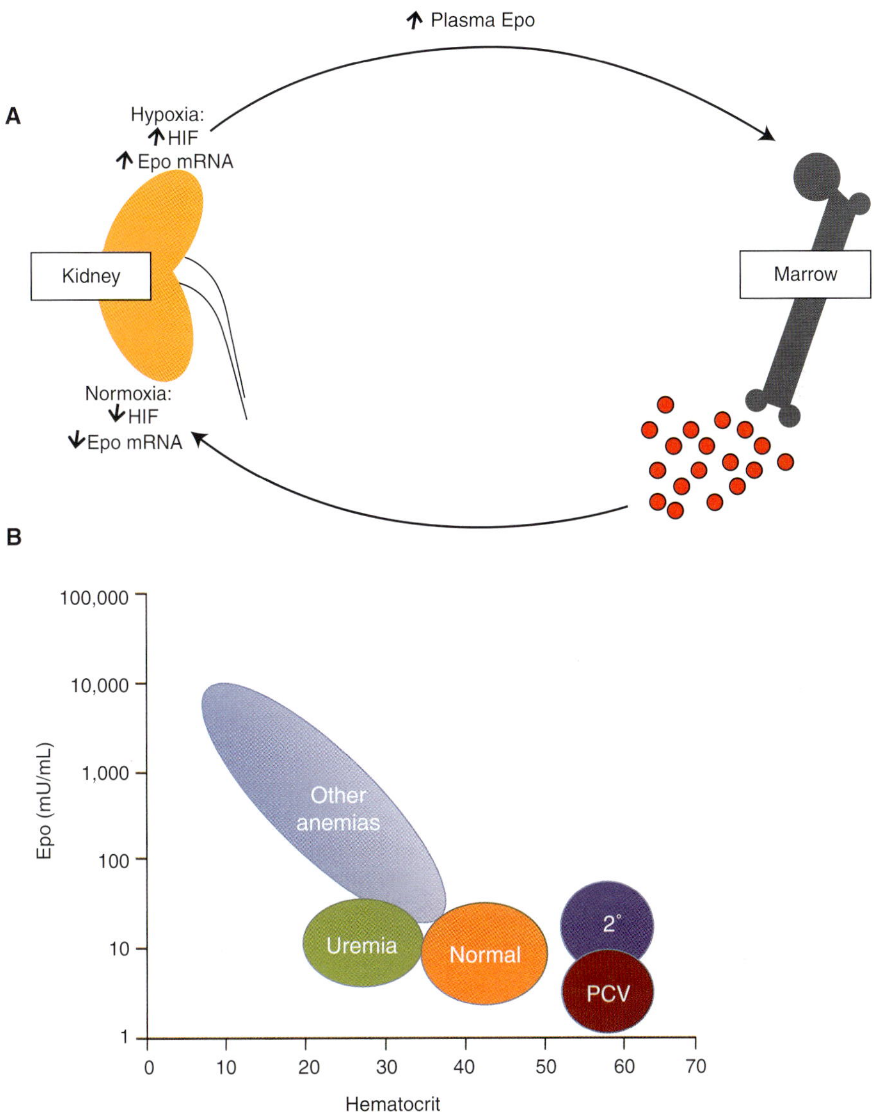

Figure 1. Regulation of red cell production by Epo. (*A*) Decreased oxygen delivery to specialized cells in the kidney results in increased expression and secretion of Epo, which circulates in the plasma and stimulates marrow progenitors, thereby increasing red cell production. If the increase in red cell mass relieves the hypoxic signal, Epo expression is down-regulated. (*B*) Plasma Epo levels (milliunits/mL) in patients with different types of and degrees of anemia and in those with primary erythrocytosis and secondary erythrocytosis. HIF, hypoxia inducible factor; PCV, polycythemia vera.

kidney, which then circulates in the plasma and binds to receptors abundantly expressed on erythroid progenitor cells, thereby promoting the viability, proliferation, and terminal differentiation of erythroid precursors, and causing an increase in red blood cell mass. The oxygen-carrying capacity of the blood is thereby enhanced, increasing tissue oxygen tension, thus completing the feedback loop and suppressing further expression of Epo.

Further insight into the physiology of Epo production has come from measurements of plasma levels in normal individuals as well as in those with different types of anemia and erythrocytosis (Erslev et al. 1987). These studies became more conclusive with the development of accurate immune assays. Figure 1B depicts plasma Epo levels in normal individuals as well as in those with different types of anemia of varying degree. Note that the plasma Epo levels are plotted on a logarithmic scale. In patients with very severe anemia, Epo may be increased up to 1000-fold. The striking inverse relationship between plasma Epo levels and hematocrit or hemoglobin is seen in many different types of anemia, including blood loss, impaired red cell production, and hemolysis.

BIOCHEMISTRY AND CELL BIOLOGY[2]

Structure

The molecular cloning of the *Epo* gene led straightaway to high-level production of recombinant human erythropoietin (rhEpo) in cell culture in sufficient purity and quantity for both definitive scientific pursuits and the development of therapy (discussed at the end of this article). Human *Epo* messenger RNA (mRNA) encodes a 193-residue polypeptide. Following cleavage of a canonical leader sequence in the endoplasmic reticulum and glycosylation in the Golgi, a 166-residue polypeptide is released (Jacobs et al. 1985; Lin et al. 1985). The primary structure of rhEpo was shown to be identical to that of the endogenous hormone except for the in vivo posttranslational cleavage of an arginine at the carboxyl terminus. Endogenous Epo and rhEpo have about 40% carbohydrate (w/w) of very similar structure, consisting of three *N*-linked polysaccharide groups and one *O*-linked group. Glycosylation of Epo slightly impedes its biological activity but is essential for ensuring prolonged circulation in the plasma (Goldwasser et al. 1974).

From its primary structure it was predicted that Epo, like other hematopoietic cytokines, folds into a globular three-dimensional structure consisting of a bundle of four amphipathic α helices connected by loops lacking secondary structure and stabilized by a critical disulphide bridge between the amino-terminal and carboxy-terminal helices (Bazan 1990; Boissel et al. 1993). This prediction was borne out by nuclear magnetic resonance spectroscopy of Epo in solution (Cheetham et al. 1998) as well as by X-ray crystallographic analysis of Epo in a complex with two extracellular domains of the Epo receptor (Syed et al. 1998).

Interaction with EpoR

Epo circulates in plasma with a plasma half-life of ~7–8 h and binds to high-affinity (~100 pM) receptors present in relatively small numbers (~1000/cell) on the surface of erythroid progenitor cells (CFUe) in the bone marrow. The receptor-binding domains on the surface of Epo that were predicted from the properties of site-directed mutants (Wen et al. 1994; Elliott et al. 1996, 1997) have been confirmed and extended by structural analysis (Syed et al. 1998; Remy et al. 1999). EpoR is present on the surface of erythroid progenitors as a homodimer, even in the absence of ligand (Livnah et al. 1999). As shown in Figure 2, on binding to Epo, the receptor undergoes a conformational change that brings its intracellular domains into close apposition (Livnah et al. 1999; Remy et al. 1999), enabling cross phosphorylation via the binding of Jak2 kinase and the initiation of the signal transduction cascade.

The growth of CFUe in vitro was shown to be Epo dependent (McLeod et al. 1974). Ligand

[2]For more detailed information on Epo structure, interaction with receptor, and signal transduction, see review by Jelkmann and Wagner (2004).

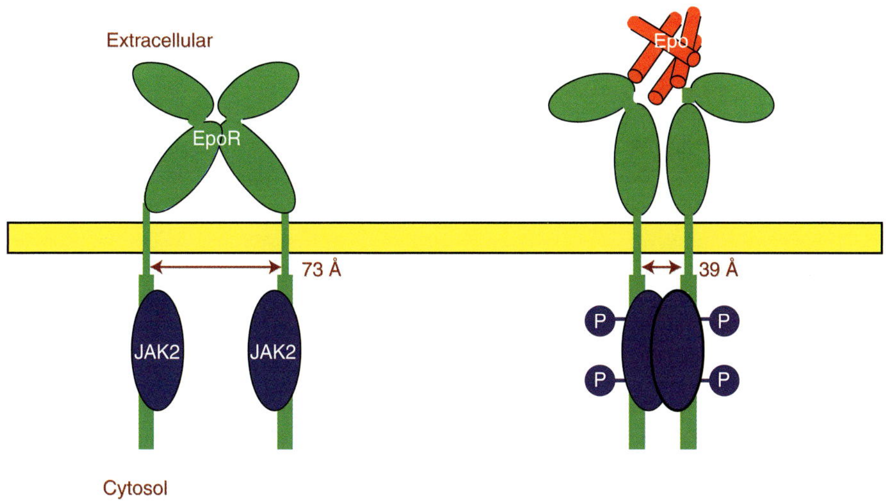

Figure 2. Epo-dependent signaling. When Epo binds to its dimeric receptor (EpoR) on erythroid progenitor cells, the two receptor monomers are pulled together allowing phosphorylation of JAK2 kinase, which initiates the signal transduction cascade.

binding results in salvaging CFUe from apoptosis (Koury and Bondurant 1990) thereby enabling proliferation of erythroid cells as well as induction of terminal differentiation. The finding that both *Epo* mRNA and Epo protein are expressed in erythroid progenitors (Hermine et al. 1991; Stopka et al. 1998) has raised the intriguing possibility that tonic low-level erythropoiesis may be supported by autocrine stimulation, whereas circulating (hormonal) Epo provides a more robust stimulus to erythropoiesis during hypoxic stress.

In man, Epo's hematopoietic role appears to be restricted to the erythron, whereas in rodents, Epo stimulates megakaryocyte proliferation and maturation as well (Clark and Dessypris 1986; McDonald et al. 1987; Berridge et al. 1988).

NONHEMATOPOIETIC EFFECTS OF Epo[3]

As stressed in the sections above, by far the most abundant sites of Epo production are the kidney and liver, whereas expression of EpoR is vastly more robust in marrow erythroid progenitors than in other cells. However, it has become increasingly evident that Epo has significant biological effects apart from regulation of red cell production. Attention has focused primarily on the central nervous system along with the heart, kidney, liver, and vasculature.

Within the brain, *Epo* and *Epo*-receptor mRNA are expressed widely (Tan et al. 1991; Masuda et al. 1994; Digicaylioglu et al. 1995; Sakanaka et al. 1998) and throughout development (Dame et al. 2000) in neurons, astrocytes, and endothelial cells (Yamaji et al. 1996). In contrast, demonstration of Epo and EpoR protein expression in situ has been considerably more challenging and less convincing. Oxygen-regulated expression of *Epo* has been observed in astrocytes both in vitro (Masuda et al. 1994; Marti et al. 1996) and in vivo (Tan et al. 1991; Marti et al. 1996), suggesting the possibility of a paracrine function for Epo in neural tissue.

The most compelling evidence of the importance of Epo signaling in the central nervous system comes from two independent sources. Targeted knockout of EpoR expression in the mouse brain has shown reduced neural cell proliferation and impaired poststroke neurogenesis (Tsai et al. 2006). Although global deletion of EpoR is lethal to mouse embryos, rescue by

[3]For more information on this topic, see review by Noguchi et al. (2008).

expression of erythroid-specific EpoR results in survival to adulthood (Suzuki et al. 2002) but impaired neural cell proliferation and viability (Chen et al. 2007). In addition, administration of Epo protects against experimental brain injury in vivo (Sakanaka et al. 1998; Brines et al. 2000). Moreover, structural derivatives of Epo that lack erythroid stimulating activity have been shown to confer neuroprotection in animal models of stroke, spinal chord compression, diabetic neuropathy, and autoimmume encephalomyelitis (Leist et al. 2004). In a preliminary trial in patients with acute stroke, recombinant human erythropoietin (rhEpo) therapy was well tolerated and readily penetrated the blood–brain barrier (Ehrenreich et al. 2002). The authors' cbservations of reduction in infarct size and clinical improvement have not been confirmed in a more extensive clinical trial (Ehrenreich et al. 2009).[4]

High doses of rhEpo (Parsa et al. 2004; Nishiya et al. 2006) and the above-mentioned Epo derivatives (Fiordaliso et al. 2005; Brines et al. 2008) confer protection in animal models of cardiac ischemia/reperfusion and myocardial infarction. Moreover rhEpo is also effective in protecting the kidney and liver from ischemia/reperfusion injury (Yang et al. 2003; Patel et al. 2004; Sharples et al. 2004; Sepodes et al. 2006; Schmeding et al. 2009) as well as other experimental threats to kidney (Vaziri et al. 1994; Bahlmann et al. 2004) or liver (Le Minh et al. 2007; Schmeding et al. 2008, 2009) integrity.

The salutary effects of rhEpo and derivatives thereof on experimental models of ischemia and other types of injury may be due in part to their effect on the vascular endothelium. In vitro rhEpo stimulates the proliferation, mobilization, and differentiation of endothelial progenitor and precursor cells (Bahlmann et al. 2003; Heeschen et al. 2003) and also enhances endothelial cell viability (Chen et al. 2008) and survival by blocking apoptosis (Bahlmann et al. 2004). Thus, exogenous Epo can significantly enhance neovascularization induced by inflammation and ischemia (Heeschen et al. 2003).

[4]For more information on Epo and the brain, see reviews by Jelkmann (2005) and Noguchi et al. (2007).

Perhaps the most puzzling aspect of nonerythroid effects of Epo is the nature of the receptor. In erythroid cells, EpoR mRNA is expressed at sufficiently high levels for abundant display of EpoR protein on cell surface and dimerization that is required for high-affinity Epo binding. In contrast, the mRNA expression in nonerythroid cells is so low that convincing demonstration of surface EpoR protein by immunological means has been problematic at best. It is very likely that such low-level expression would not be sufficient for the mass action needed for EpoR dimerization to occur. Accordingly, the low-affinity binding site for Epo may be a heterodimer composed of EpoR and another partner, perhaps CD131, the β common cytokine receptor (Brines et al. 2008; Brines 2010). This may explain the requirement of high levels of exogenous Epo for biological effects in nonerythroid cells and possibly how an Epo derivative could have activity only in nonerythroid cells.

REGULATION OF THE *Epo* GENE[5]

Initial investigation of the molecular basis of *Epo* gene regulation was greatly facilitated by the discovery that the human hepatoma cell lines Hep3B and HepG2 produce significant amounts of Epo constitutively, with marked induction in response to hypoxia (Goldberg et al. 1987). The magnitude and time course of the induction of *Epo* mRNA paralleled Epo protein production. More recently, a human renal cell line has also been shown to display oxygen-dependent regulation of Epo production (Frede et al. 2011).

Tissue-specific expression of the *Epo* gene and its induction by hypoxia are dependent on far upstream *cis* elements and an enhancer element downstream from the polyadenylation signal. Transgenic mice experiments indicate the presence of element(s) between 0.4 and 6 kb 5′ of the promoter, which suppress promiscuous expression, and element(s) 9.5 and 14 kb upstream, which are necessary for kidney-specific

[5]For more information on this topic, see review by Fandrey (2004).

expression (Semenza et al. 1989, 1990, 1991a; Madan et al. 1995).

The *Epo* gene has a weak promoter, owing in part to suppression of transcription by a GATA motif (Imagawa et al. 1994, 1997). Mutation of this response element in a 180-kb *Epo* transgene resulted in constitutive in vivo expression in renal peritubular cells (Obara et al. 2008). Enhanced binding of GATA-2 to this site may contribute to the inhibition of Epo production by inflammatory cytokines (La Ferla et al. 2002). GATA-4 is expressed exclusively in hepatocytes and binds strongly to this GATA motif, enhancing Epo expression (Dame et al. 2004). The marked reduction in expression of hepatic GATA-4 following birth is likely an important contributor to the switch in Epo production from liver to kidney (Dame et al. 2004).

A crucial 3′ enhancer (Beck et al. 1991; Pugh et al. 1991; Semenza et al. 1991b; Semenza and Wang 1992) binds to two transcription factors, hypoxia-inducible factor 1 (HIF) and the nuclear receptor HNF-4. As shown in Figure 3, these two DNA-binding proteins interact with the transcriptional coactivator p300/CBP, triggering transcriptional activation.

HIF

Hypoxic induction of Epo depends in large part on the transcription factor HIF. HIF is activated in virtually all cells (Maxwell et al. 1993b; Wang and Semenza 1993b) by exposure to hypoxia, as well as cobalt or iron chelators. The activation of HIF enables it to bind to a consensus sequence (5′-TACGTGCT-3′) first identified in the Epo 3′ enhancer (Fig. 3) (Semenza et al. 1991b). Subsequently, this hypoxia response element has been identified in more than a hundred genes and shown to be required for hypoxic induction of transcription. In addition to Epo, an impressive number of other physiologically relevant genes are regulated by HIF (Wenger et al. 2005) including those involved in angiogenesis, intermediary metabolism, and, of particular interest to readers of this collection, iron homeostasis (Peyssonnaux et al. 2007).

Wang and Semenza (1995) purified HIF by its affinity for the *Epo* 3′ enhancer and showed that it is a heterodimer composed of a 120 kD α subunit and a 91–94 kD β subunit. Molecular cloning revealed that both subunits are basic helix-loop-helix proteins in the PAS family of transcription factors (Wang et al. 1995a). HIF-α, or HIF-1α, is a novel protein, whereas HIF-1β is the previously cloned and characterized aryl hydrocarbon receptor nuclear translocator (ARNT) (Hoffman et al. 1991).

The HIF-α subunit is widespread among animals, including not only mammals and fish but also invertebrates such as flies and worms. In man and other mammals, three homologous HIF-α genes have been identified: HIF-1α, HIF-2α (Tian et al. 1997), and HIF-3α (Gu et al. 1998). HIF-1α is the most abundant and present in most organs and tissues. In contrast, HIF-2α has more restricted expression and was originally thought to be limited to the vascular endothelium (Tian et al. 1997). Knockout of HIF-1α in mice is embryonic lethal (Iyer et al. 1998; Ryan et al. 1998), whereas HIF-2α null mice survive gestation but have impaired survival owing to phenotypic abnormalities that are strain-dependent. Most genes that are

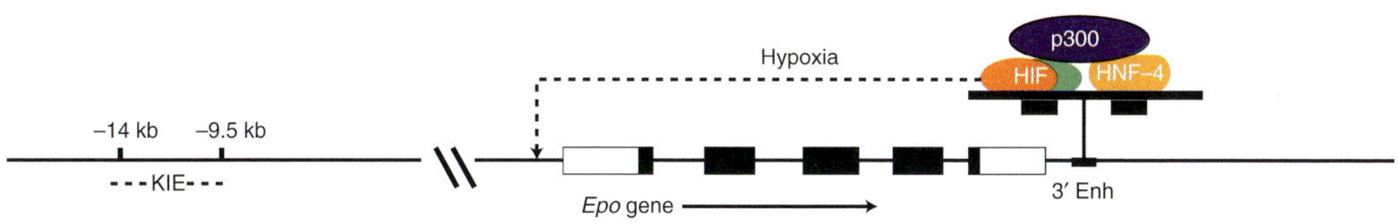

Figure 3. Diagram of the *Epo* gene. The five exons of the *Epo* gene are shown as rectangles with the coding regions in black. Far upstream of the *Epo* promoter is a kidney-inducible element (KIE) that is required for high-level up-regulation of *Epo* mRNA in the kidney. Just downstream from exon 5 is a critical enhancer that binds to HNF-4 and also to HIF when the Epo-producing cell is hypoxic. These two transcription factors bind to the transcriptional activator p300. This enhanceosome is a powerful inducer of *Epo* transcription.

induced by hypoxia in an HIF-dependent manner require HIF-1α (Warnecke et al. 2004). In contrast, Epo production and its up-regulation by hypoxia require HIF-2α (Gruber et al. 2007) in the kidney (Scortegagna et al. 2005; Kapitsinou et al. 2010; Paliege et al. 2010) and an Epo-producing renal cell line (Frede et al. 2011), as well as in the fetal and adult liver (Rankin et al. 2007).

Oxygen Sensing and HIF Activation

To understand the mechanism by which HIF-1 is activated, it is necessary to determine how hypoxia impacts on the expression of its two subunits. The steady-state levels of HIF-α mRNA and ARNT mRNA are not significantly affected by oxygen tension (Gradin et al. 1996; Huang et al. 1996; Wood et al. 1996). At the protein level, the ARNT subunit remains abundant, irrespective of the oxygenation of the cell. In contrast, the HIF-α subunit cannot be detected in oxygenated cells (Wang et al. 1995b; Huang et al. 1996). HIF-α protein can only be detected in deoxygenated cells or in those exposed to iron chelators or certain transition-metal ions, most notably cobalt, all of which induce HIF-dependent gene transcription (Semenza and Wang 1992; Wang and Semenza 1993a). These observations suggest that the activation of HIF-1 depends on an increase in the amount of HIF-α protein in deoxygenated cells.

In cells exposed to 21% O_2, HIF-α is remarkably unstable (Huang et al. 1996). It forms a complex with von Hippel–Lindau protein (vHL) (Maxwell et al. 1999), enabling it to undergo ubiquitination and subsequent degradation in proteasomes (Salceda and Caro 1997; Huang et al. 1998). Low oxygen tension abrogates this process, thereby allowing the HIF-α subunit to accumulate so that it can enter the nucleus and form a stable heterodimer that can participate in transcriptional regulation. Oxygen-dependent degradation of HIF-α depends on a large (~200-residue) interior segment (Huang et al. 1998). When this domain is deleted, the protein is stable and capable of *trans*-activation, even in oxygenated cells. When this domain is fused to a foreign protein, it confers rapid oxygen-dependent degradation.

The oxygen-dependent degradation domain of HIF-1α contains two highly conserved LXXLAP sequences, which are recognition sites for an oxygen- and iron-dependent prolyl hydroxylase that converts the proline into a hydroxyproline (Ivan et al. 2001; Jaakkola et al. 2001; Masson et al. 2001; Yu et al. 2001). This posttranslational modification is necessary and sufficient for the binding of vHL to HIF-α. As shown in Figure 4, this interaction enables the ubiquitin E3 ligase to dock at this complex, thereby triggering the covalent addition of a number of small ubiquitin molecules (Cockman et al. 2000; Ohh et al. 2000). The polyubiquitinated HIF-α is then recognized by the proteasome where it is rapidly degraded. Ratcliffe and his colleagues (Epstein et al. 2001) showed that the hydroxylation of these two specific proline residues is mediated by a dioxygenase that is homologous to classic collagen prolyl hydroxylase. Shortly thereafter, a human homolog of HIF prolyl hydroxylase was biochemically purified and characterized (Ivan et al. 2002). These enzymes have an active center where an iron atom and a molecule of α-ketoglutarate interact with a conserved core of two histidines, an aspartate and an arginine residue. The two LXXLAP motifs within HIF-α each bind at this site, along with oxygen, catalyzing hydroxylation of the proline residue. This oxidative modification is inhibited by iron chelators as well as by cobaltous ion, which presumably displaces iron from the active site. The effect of the iron chelation and cobalt on the activity of HIF prolyl hydroxylases provides a satisfying explanation for how these agents mimic hypoxia in activating HIF. These enzymes are the proximate oxygen sensor for regulating HIF. Among the three HIF-α prolyl hyderoxylases, PHD2 plays the predominant role at most sites (Berra et al. 2003) including renal production of Epo (Minamishima et al. 2008; Takeda et al. 2008).

Hemoglobin is encapsulated in circulating erythrocytes not only in man and other mammals but also in all other vertebrates. Oxygen homeostasis in these complex organisms is

Cite this article as *Cold Spring Harb Perspect Med* doi: 10.1101/cshperspect.a011619

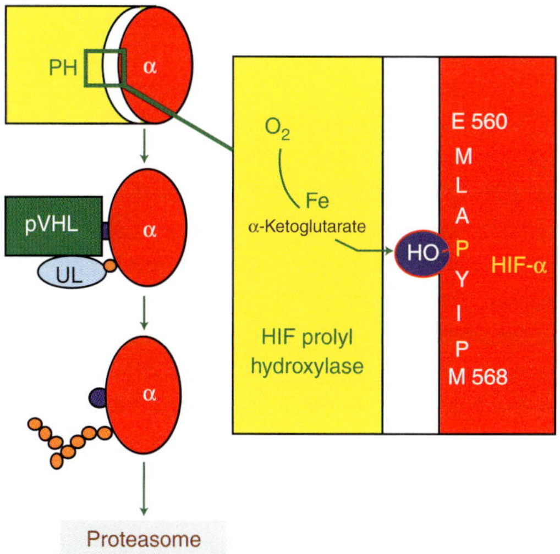

Figure 4. Pathway by which the hypoxia-inducible transcription factor is up-regulated by low intracellular oxygen tension. In normally oxygenated cells, the α subunit (HIF-α) undergoes hydroxylation of two proline residues, one of which is shown here. In the presence of iron (Fe) and α-ketoglutarate, this oxygen-dependent posttranslational modification is catalyzed by an HIF-α-specific prolyl hydroxylase (PH). The von Hippel–Lindau protein (pVHL) binds to hydroxylated HIF-α. Subsequent docking of a ubiquitin ligase (UL) enables HIF-α to be polyubiquitinated.

well served by HIF-dependent regulation of a cytokine that is required for erythropoiesis. In contrast, in many invertebrate phyla, oxygen transport depends on the presence of free soluble polymeric hemoglobins in circulating hemolymph, and in some, hemoglobin production is enhanced during hypoxic stress. One of the most dramatic observations in marine biology is the change in the fresh-water crustacean *Daphnia magna* from pale gray to bright red soon after transfer from normoxic to hypoxic water (Fox 1947, 1949). Two of their four tandem globin genes have been shown to have functional HIF response elements within their promoters (Gorr et al. 2004). Thus, in keeping with the relative simplicity of this organism, the regulation of oxygen-carrying capacity of the blood is direct, rather than mediated indirectly via a cytokine.

HNF-4

Downstream from the HIF-1 binding site, the *Epo* 3′ enhancer contains two tandem consensus steroid hormone response elements separated by 2 bp (Fig. 1). Mutations at these sites either abolish or markedly inhibit hypoxic induction of reporter genes (Blanchard et al. 1992; Semenza and Wang 1992; Pugh et al. 1994; Hu et al. 1997). These elements in the *Epo* enhancer interact with an orphan nuclear receptor, i.e., a DNA-binding protein that shares structural homology with hormone-binding nuclear receptors but lacks a known ligand (Blanchard et al. 1992). The orphan receptor HNF-4 plays a critical role in *Epo* gene regulation (Galson et al. 1995; Makita et al. 2001). The expression of HNF-4 is limited to renal cortex and liver, sites of Epo production, and also intestine. Thus the binding of HNF-4 to the *Epo* enhancer is likely to contribute to both the high-level induction of the *Epo* gene as well as its tissue specificity.

p300

As shown in Figure 1, the carboxy-terminal portion of HIF-1α binds specifically to p300 (Arany et al. 1996), a general transcriptional activator that participates in a number of biological functions such as induction of tissue-specific expression, regulation of cell cycle, and stimulation of differentiation pathways. This very large protein, which is closely homologous to CREB-binding protein (CBP), does not bind to DNA but does interact with a number of other proteins including HNF-4 and other nuclear receptors. Thus, as depicted in Figure 1, the HIF-1 heterodimer, activated by hypoxia, participates in a macromolecular assembly with p300 (or a related family member) and with HNF-4 to transduce a signal to the Epo promoter, enabling activation of transcription. It is very likely that such a combinatorial process applies not only to *Epo* but also to other genes that are induced by hypoxia (Ebert and Bunn 1998). Most, if not all, cells contain a protein factor called FIH (factor-inhibiting HIF) (Mahon et al. 2001) that hydroxylates an asparagine at the carboxyl terminus of HIF-α in an oxygen- and iron-dependent manner, thereby blocking binding to p300 and inhibiting transcription (Lando et al. 2002a,b).

IMPACT OF Epo PRODUCTION ON CLINICAL DISORDERS

In a variety of clinical settings, anemia can be caused by underproduction of Epo, whereas erythrocytosis can result from overproduction.

Underproduction of Epo

Patients with renal failure generally develop severe anemia, owing in part to suppression of erythropoiesis from a buildup of metabolic wastes and in part to a moderate reduction in red cell life span. However, the most important contributor to the anemia of uremia is insufficient Epo production (Fig. 1B) (Caro et al. 1979; Cotes 1989). The degree of anemia correlates roughly with the extent of renal functional impairment. In renal failure owing to a wide variety of etiologies, insufficient Epo production results either from direct damage to Epo-producing cells in the kidney or from the suppression of Epo production by inflammatory cytokines.

Patients with inflammatory disorders have a less marked impairment of Epo production. Examples include rheumatoid arthritis (Means 1994), cancer (Miller et al. 1990; Spivak 1994), and AIDS (Fischl et al. 1990). It is likely that inflammatory cytokines suppress *Epo* gene expression in these disorders.

Exposure to certain metals may result in disordered structure and function of the renal proximal tubule, resulting in suppression of Epo production, out of proportion to impairment of global renal function. Such a pathogenetic process is likely the basis for the marked anemia often encountered in cancer patients who have been treated with *cis*-platinum or individuals suffering from chronic cadmium intoxication (Horiguchi et al. 2000). Epo production is also inhibited in patients with increased plasma viscosity owing to monoclonal dysproteinemias (Singh et al. 1993).

Overproduction of Epo

Measurement of plasma Epo is useful in distinguishing primary from secondary erythrocytosis (Erslev and Caro 1984; Messinezy et al. 2002). Individuals with autonomous red cell overproduction due to polycythemia vera (or rarely to mutations at the carboxyl terminus of EpoR) have low plasma Epo levels, whereas individuals with various forms of chronic arterial hypoxemia have erythrocytosis owing to increased plasma Epo levels. The highest documented hematocrit levels have been observed in patients with right to left cardiac shunts. The erythropoietic response in patients with hypoxemia owing to chronic obstructive pulmonary disease is variable, depending in part on whether there is coexisting infection, which, as explained above, can suppress erythropoietin expression.

Increased oxygen affinity of red cells causes impaired delivery of oxygen to tissues and therefore hypoxia at the cellular level, which triggers increased Epo expression and therefore secondary erythrocytosis (Prchal and Prchal 1999). Such a "shift to the left" in the oxyhemoglobin dissociation curve can be caused by mutations in α- or β-globin subunits or in two red cell enzymes. Deficiency in cytochrome b5 reductase causes congenital mehemoglobinemia and mild erythrocytosis. A more pronounced increase in red cell mass has been reported in very rare families with deficiency of bisphosphoglycerate mutase that causes a deficit in red cell 2,3-BPG, a critical modulator of intracellular hemoglobin function (Rosa et al. 1978). These individuals with left-shifted O_2-binding curves are generally asymptomatic because impaired oxygen delivery is balanced by increased oxygen-carrying capacity.

Occasional individuals have chronic elevation of plasma Epo that is inappropriate and autonomous, i.e., not due to arterial hypoxemia or cellular hypoxia. These individuals usually have either sporadic or germline mutation in genes responsible for oxygen sensing and HIF regulation.

Erythrocytosis due to inappropriate overproduction of Epo can be encountered in various neoplasms, particularly those arising in kidney, liver, and cerebellum, organs that are physiologic sites of erythropoietin expression. In particular, secondary erythrocytosis occurs in patients with renal carcinomas, Wilms tumor, hepatomas, and cerebellar hemangioblastomas. In some cases, the tumor cells secrete Epo (Jelkmann 1992). In other cases, the surrounding normal kidney or liver tissue secretes Epo, presumably because of local ischemia.

Highly vascular renal and central nervous system tumors can arise in families and sporadically owing to mutations in the von Hippel–Lindau (VHL) gene. As explained earlier in this article, this protein is required for the oxygen-dependent degradation of HIF-α. Inactivating mutations lead to constitutive activation of HIF-α and therefore overexpression of HIF responsive genes (Gnarra et al. 1996; Iliopoulos et al. 1996), particularly vascular endothelial growth factor (VEGF), which causes the tumors to have enhanced vascularity. In a subset of affected individuals, overexpression of Epo leads to erythrocytosis.

A number of families with congenital erythrocytosis have been encountered in Chuvashia, a circumscribed region in the Russian Federation in which there may have been inbreeding and a high likelihood of a founder effect (Sergeyeva et al. 1997; Gordeuk and Prchal 2006). The erythrocytosis (mean hemoglobin levels of 23 g/dL and hematocrit of 67%) follows an autosomal recessive pattern of inheritance. Affected individuals have elevated erythropoietin levels, and increased incidence of thrombotic and hemorrhagic complications. These individuals are homozygous for a missense mutation at the carboxyl terminus of the VHL gene (R200W) leading to HIF activation (Ang et al. 2002). The clinical phenotype is similar in a R200W "knock-in" mouse model (Hickey et al. 2007). Isolated families with congenital erythrocytosis have been encountered elsewhere that harbor the R200W mutation (Pastore et al. 2003; Cario et al. 2005; Perrotta et al. 2006). Affected individuals are usually homozygotes but some are compound heterozygotes, having R200W along with another VHL mutation (Pastore et al. 2003; Cario et al. 2005).

Families as well as isolated individuals have been encountered in which erythrocytosis owing to elevated plasma Epo levels is explained by heterozygosity for an activating mutation in HIF-2α (Gale et al. 2008; Martini et al. 2008; Percy et al. 2008a,b, 2012; Furlow et al. 2009; van Wijk et al. 2010). These mutations generally impair the ability of HIF-2α to bind to VHL or PHD2. In other families, affected individuals are heterozygous for loss-of-function mutations in PHD2 (Percy et al. 2006, 2007; Ladroue et al. 2008). Genome-wide association studies comparing lowland Tibetans to highlanders with erythrocytosis have shown highly significant polymorphisms in the vicinity of HIF-2α (Beall et al. 2010; Yi et al. 2010) and PHD2 (Simonson 2010). Both these selected "experiments of nature" and the broader population studies, attest to the primacy of HIF-2α and PHD2 in regulating the *Epo* gene.

THERAPY WITH rhEpo AND ERYTHROID-STIMULATING AGENTS[6]

By far the most common use of rhEpo has been in patients with chronic renal failure. Over one million patients worldwide have responded effectively to either rhEpo or a derivative, darbepoietin alfa (Aranesp), which has an enhanced

[6]For more information on this subject, see reviews by Bunn (2007), Elliott (2008), and Jelkmann (2008).

plasma half-life (~22 h) owing to the introduction of two additional *N*-linked glycosylation sites.

The importance of erythropoietin in the pathogenesis of the anemia of renal failure has been affirmed by the dramatic efficacy of rhEpo therapy. Figure 5 shows one of the first cases treated with rhEpo. Within a few days after initiation of rhEpo therapy, the hematocrit approached normal, necessitating a reduction in dose. The marked increase in red cell mass following treatment was accompanied by enhanced utilization of iron stores, as reflected in a decline in serum iron and serum ferritin. Other patients who have normal or low iron stores before rhEpo therapy need concomitant administration of iron to achieve an optimal erythropoietic response.

The other common use of rhEpo has been in treating anemia in cancer patients with chemotherapy-associated anemia. In addition, rhEpo is effective in AIDs patients who have developed severe anemia as a result of antiviral therapy. A minority of patients with myelodysplastic anemia also responds to rhEpo. In some clinics, rhEpo is used in the management of the anemia of prematurity as well in surgical patients, in both the preoperative and postoperative periods.

The impressive efficacy and safety of rhEpo therapy notwithstanding, a number of problems warrant careful consideration. The most evident are the high cost of treatment and the need for parenteral administration. Of equal concern are outcome studies on large numbers of renal dialysis patients showing that doses of rhEpo and darbepoietin alfa that raise the hemoglobin level >12 g/dL are associated with thrombosis of arteriovenous access, impaired control of hypertension, and a small but sig-

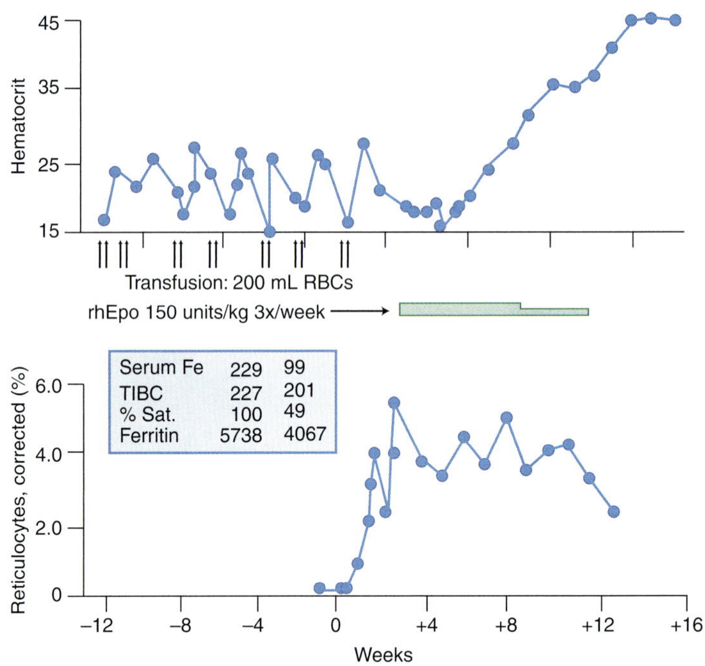

Figure 5. Response of an anephric patient to recombinant human erythropoietin (rhEpo) therapy. Note that, before therapy, the patient was severely anemic and transfusion dependent. Treatment with rhEpo resulted in a reticulocytosis followed by a progressive increase in hemoglobin. The dose of rhEpo had to be lowered to prevent the hemoglobin from rising too high. Before rhEpo therapy, the patient was severely iron overloaded. The marked increase in red cell mass following therapy was accompanied by a significant reduction in iron stores. RBC, red blood cells; Fe, iron; TIBC, total iron binding capacity; Sat., saturation (Data from Eschbach et al. 1987.)

 Cite this article as *Cold Spring Harb Perspect Med* doi: 10.1101/cshperspect.a011619

nificant increase in mortality (Besarab et al. 1998; Singh et al. 2006; Phrommintikul et al. 2007). Cancer patients are inherently at higher risk to develop thrombosis, but this risk is significantly enhanced in those receiving rhEpo therapy (Bohlius et al. 2006; Bennett et al. 2008).

It is likely that the prothrombotic effects of rhEpo dose are due to the high doses given rather than to the increase in hemoglobin level. As mentioned earlier in this article, nonerythropoietic effects are widespread and complex. Early reports suggest that rhEpo may also promote enhanced tumor growth in vivo (Henke et al. 2003; Leyland-Jones 2003) and in vitro (Mohyeldin et al. 2005). However, EpoR expression is low in the great majority of tumors and though expressed at the mRNA level, it cannot be detected on the cell surface (Sinclair et al. 2008). Moreover, more recent meta-analyses mentioned above do not support the notion that rhEpo therapy promotes tumor progression in vivo (Bohlius et al. 2006; Bennett et al. 2008).

RECENT DEVELOPMENTS

In view of the above-mentioned concerns regarding long-term rhEpo therapy, there has been increasing interest in developing erythroid-stimulating agents that offer advantages over rhEpo and darbepoietin alfa. One approach entailed the production of a large phage library expressing short (~15-residue) random peptides that are then screened for sequences that bind with relatively high affinity and specificity to the Epo receptor. This strategy led to the identification of peptides that lack any homology with Epo but are active both in vitro and in vivo in stimulating erythropoiesis through the same signaling pathway that is triggered by Epo (Livnah et al. 1996; Wrighton et al. 1996). However, in contrast to recent progress with thrombopoietin mimetic agents, little progress has been made to date in developing these Epo mimetic peptides into safe and effective drugs.

Another approach involves structural modification of Epo in the interest of prolonging its longevity in the circulation. Continuous erythropoietin receptor activator (CERA) is a derivative in which a 30 kDa methoxy-polyethylene glycol polymer has been covalently linked to Epo. It has a half-life in the circulation of ~135 h, about sixfold greater than that of darbepoietin alfa (Macdougall 2005; Macdougall et al. 2006; Provenzano et al. 2007). This attribute greatly reduces the frequency of parenteral administration. CERA has been through extensive clinical trials and has passed muster in terms of both safety and efficacy. As a result, it received approval from the European Commission and the American Food and Drug Administration in 2008. However, it has not yet gained wide use in clinical practice.

The most novel strategy for pharmacologic stimulation of erythropoiesis involves the activation of HIF by inhibition of HIF-α prolylhydroxylases. As shown in Figure 4, small molecule analogs of α-ketoglutarate could prevent HIF-α hydroxylation, even in well-oxygenated cells and thereby induce expression not only of endogenous Epo but also of other HIF-dependent genes that are required for robust erythropoiesis. Several companies have developed inhibitors that induce Epo and erythropoiesis in mice (Safran et al. 2006; Yan et al. 2010) and in humans (Yan et al. 2010). Because these agents suppress expression of hepcidin (Volke et al. 2009), they may be particularly effective in the treatment of anemia of chronic inflammation. HIF prolyl hydroxylase inhibitors offer for the first time the possibility of inexpensive oral therapy for stimulation of erythropoiesis. Of course, a major concern is specificity because such a large number of genes are regulated by HIF. The lead compounds that are currently undergoing clinical trials appear to inhibit HIF-2α specifically. If so, they may also enhance duodenal iron absorption because the divalent metal transporter 1 (DMT1), like Epo, is up-regulated by HIF-2α.

REFERENCES

Ang SO, Chen H, Hirota K, Gordeuk VR, Jelinek J, Guan Y, Liu E, Sergueeva AI, Miasnikova GY, Mole D, et al. 2002. Disruption of oxygen homeostasis underlies congenital Chuvash polycythemia. *Nat Genet* **32:** 614–621.

Arany Z, Huang LE, Eckner R, Bhattacharya S, Jiang C, Goldberg MA, Bunn HF, Livingston DM. 1996. Participation by the p300/CBP family of proteins in the cellular response to hypoxia. *Proc Natl Acad Sci* **93:** 12969–12973.

Bachmann S, Le Hir M, Eckardt K. 1993. Co-localization of erythropoietin mRNA and ecto-5′-nucleotidase immunoreactivity in peritubular cells of rat renal cortex indicates that fibroblasts produce erythropoietin. *J Histochem Cytochem* **41:** 335–341.

Bahlmann FH, DeGroot K, Duckert T, Niemczyk E, Bahlmann E, Boehm SM, Haller H, Fliser D. 2003. Endothelial progenitor cell proliferation and differentiation is regulated by erythropoietin. *Kidney Int* **64:** 1648–1652.

Bahlmann FH, Song R, Boehm SM, Mengel M, von Wasielewski R, Lindschau C, Kirsch T, de Groot K, Laudeley R, Niemczyk E, et al. 2004. Low-dose therapy with the long-acting erythropoietin analogue darbepoetin α persistently activates endothelial Akt and attenuates progressive organ failure. *Circulation* **110:** 1006–1012.

Bazan JF. 1990. Haemopoietic receptors and helical cytokines. *Immunol Today* **11:** 350–354.

Beall CM, Cavalleri GL, Deng L, Elston RC, Gao Y, Knight J, Li C, Li JC, Liang Y, McCormack M, et al. 2010. Natural selection on EPAS1 (HIF2α) associated with low hemoglobin concentration in Tibetan highlanders. *Proc Natl Acad Sci* **107:** 11459–11464.

Beck I, Ramirez S, Weinmann R, Caro J. 1991. Enhancer element at the 3′-flanking region controls transcriptional response to hypoxia in the human erythropoietin gene. *J Biol Chem* **266:** 15563–15566.

Bennett CL, Silver SM, Djulbegovic B, Samaras AT, Blau CA, Gleason KJ, Barnato SE, Elverman KM, Courtney DM, McKoy JM, et al. 2008. Venous thromboembolism and mortality associated with recombinant erythropoietin and darbepoetin administration for the treatment of cancer-associated anemia. *JAMA* **299:** 914–924.

Berk L, Burchenal J, Wood TM, Castle W. 1948. Oxygen saturation of sternal marrow blood with special reference to pathogenesis of polycythemia vera. *Proc Soc Exp Biol Med* **69:** 316–320.

Berra E, Benizri E, Ginouves A, Volmat V, Roux D, Pouyssegur J. 2003. HIF prolyl-hydroxylase 2 is the key oxygen sensor setting low steady-state levels of HIF-1α in normoxia. *EMBO J* **22:** 4082–4090.

Berridge MV, Fraser JK, Carter JM, Lin FK. 1988. Effects of recombinant human erythropoietin on megakaryocytes and on platelet production in the rat. *Blood* **72:** 970–977.

Besarab A, Bolton WK, Browne JK, Egrie JC, Nissenson AR, Okamoto DM, Schwab SJ, Goodkin DA. 1998. The effects of normal as compared with low hematocrit values in patients with cardiac disease who are receiving hemodialysis and epoetin. *N Engl J Med* **339:** 584–590.

Blanchard KL, Acquaviva AM, Galson DL, Bunn HF. 1992. Hypoxic induction of the human erythropoietin gene: Cooperation between the promoter and enhancer, each of which contains steroid receptor response elements. *Mol Cell Biol* **12:** 5373–5385.

Bohlius J, Wilson J, Seidenfeld J, Piper M, Schwarzer G, Sandercock J, Trelle S, Weingart O, Bayliss S, Djulbegovic B, et al. 2006. Recombinant human erythropoietins and cancer patients: Updated meta-analysis of 57 studies including 9353 patients. *J Natl Cancer Inst* **98:** 708–714.

Boissel J-P, Lee W-R, Presnell SR, Cohen FE, Bunn HF. 1993. Erythropoietin structure-function relationships: Mutant proteins that test a model of tertiary structure. *J Biol Chem* **268:** 15983–15993.

Brines M. 2010. The therapeutic potential of erythropoiesis-stimulating agents for tissue protection: A tale of two receptors. *Blood Purificat* **29:** 86–92.

Brines ML, Ghezzi P, Keenan S, Agnello D, de Lanerolle NC, Cerami C, Itri LM, Cerami A. 2000. Erythropoietin crosses the blood-brain barrier to protect against experimental brain injury. *Proc Natl Acad Sci* **97:** 10526–10531.

Brines M, Patel NS, Villa P, Brines C, Mennini T, De Paola M, Erbayraktar Z, Erbayraktar S, Sepodes B, Thiemermann C, et al. 2008. Nonerythropoietic, tissue-protective peptides derived from the tertiary structure of erythropoietin. *Proc Natl Acad Sci* **105:** 10925–10930.

Bunn HF. 2007. New agents that stimulate erythropoiesis. *Blood* **109:** 868–873.

Cario H, Schwarz K, Jorch N, Kyank U, Petrides PE, Schneider DT, Uhle R, Debatin KM, Kohne E. 2005. Mutations in the von Hippel-Lindau (VHL) tumor suppressor gene and VHL-haplotype analysis in patients with presumable congenital erythrocytosis. *Haematologica* **90:** 19–24.

Carnot P, Deflandre C. 1906. Sur l'activite hemopoietique des differents organeau au cours de la regeneration du sang. *CR Searces Acad Sci* **143:** 432–435.

Caro J, Brown S, Miller O, Murray T, Erslev AJ. 1979. Erythropoietin levels in uremic nephric and anephric patients. *J Lab Clin Med* **93:** 449–458.

Cheetham JC, Smith DM, Aoki KH, Stevenson JL, Hoeffel TJ, Syed RS, Egrie J, Harvey TS. 1998. NMR structure of human erythropoietin and a comparison with its receptor bound conformation. *Nat Struct Biol* **5:** 861–866.

Chen ZY, Asavaritikrai P, Prchal JT, Noguchi CT. 2007. Endogenous erythropoietin signaling is required for normal neural progenitor cell proliferation. *J Biol Chem* **282:** 25875–25883.

Chen J, Connor KM, Aderman CM, Smith LE. 2008. Erythropoietin deficiency decreases vascular stability in mice. *J Clin Invest* **118:** 526–533.

Clark DA, Dessypris EN. 1986. Effects of recombinant erythropoietin on murine megakaryocytic colony formation in vitro. *J Lab Clin Med* **108:** 423–429.

Cockman ME, Masson N, Mole DR, Jaakkola P, Chang GW, Clifford SC, Maher ER, Pugh CW, Ratcliffe PJ, Maxwell PH. 2000. Hypoxia inducible factor-α binding and ubiquitylation by the von Hippel-Lindau tumor suppressor protein. *J Biol Chem* **275:** 25733–25741.

Cotes PM. 1989. Physiological studies of erythropoietin in plasma. In *Erythropoietin* (ed. Jelkmann W, Gross AJ), pp. 57–79. Springer, Berlin.

Dame C, Bartmann P, Wolber E, Fahnenstich H, Hofmann D, Fandrey J. 2000. Erythropoietin gene expression in different areas of the developing human central nervous system. *Brain Res Dev Brain Res* **125:** 69–74.

Dame C, Sola MC, Lim KC, Leach KM, Fandrey J, Ma Y, Knopfle G, Engel JD, Bungert J. 2004. Hepatic erythropoietin gene regulation by GATA-4. *J Biol Chem* **279:** 2955–2961.

Digicaylioglu M, Bichet S, Marti HH, Wenger RH, Rivas LA, Bauer C, Gassmann M. 1995. Localization of specific erythropoietin binding sites in defined areas of the mouse brain. *Proc Natl Acad Sci* **92:** 3717–3720.

Ebert BL, Bunn HF. 1998. Regulation of transcription by hypoxia requires a multiprotein complex that includes hypoxia-inducible factor 1, an adjacent transcription factor, and pp300/CREB binding protein. *Mol Cell Biol* **18:** 4089–4096.

Ehrenreich H, Hasselblatt M, Dembowski C, Cepek L, Lewczuk P, Stiefel M, Rustenbeck HH, Breiter N, Jacob S, Knerlich F, et al. 2002. Erythropoietin therapy for acute stroke is both safe and beneficial. *Mol Med* **8:** 495–505.

Ehrenreich H, Weissenborn K, Prange H, Schneider D, Weimar C, Wartenberg K, Schellinger PD, Bohn M, Becker H, Wegrzyn M, et al. 2009. Recombinant human erythropoietin in the treatment of acute ischemic stroke. *Stroke* **40:** e647–e656.

Elliott S. 2008. Erythropoiesis-stimulating agents and other methods to enhance oxygen transport. *Brit J Pharmacol* **154:** 529–541.

Elliott S, Lorenzini T, Chang D, Barzilay J, Delorme E, Giffin J, Hesterberg L. 1996. Fine-structure epitope mapping of anti-erythropoietin monoclonal antibodies reveals a model of recombinant human erythropoietin protein structure. *Blood* **87:** 2702–2713.

Elliott S, Lorenzini T, Chang D, Barzilay J, Delorme E. 1997. Mapping of the active site of recombinant human erythropoietin. *Blood* **89:** 493–502.

Epstein AC, Gleadle JM, McNeill LA, Hewitson KS, O'Rourke J, Mole DR, Mukherji M, Metzen E, Wilson MI, Dhanda A, et al. 2001. *C. elegans* EGL-9 and mammalian homologs define a family of dioxygenases that regulate HIF by prolyl hydroxylation. *Cell* **107:** 43–54.

Erslev A. 1953. Humoral regulation of red cell production. *Blood* **8:** 349–357.

Erslev A. 1980. Blood and mountains. In *Blood, pure and eloquent* (ed. Wintrobe M), pp. 256–280. McGraw Hill, New York.

Erslev AJ. 1993. The discovery of erythropoietin. *ASAIO J* **39:** 89–92.

Erslev AJ, Caro J. 1984. Pure erythrocytosis classified according to erythropoietin titers. *Am J Med* **76:** 57–61.

Erslev AJ, Caro J, Besarab A. 1985. Why the kidney? *Nephron* **41:** 213–216.

Erslev AJ, Wilson J, Caro J. 1987. Erythropoietin titers in anemic, nonuremic patients. *J Lab Clin Med* **109:** 429–433.

Eschbach JW, Ergie JC, Downing MR, Browne JK, Adamson JW. 1987. Correction of the anemia of endstage renal disease with recombinant human erythropoietin. *N Engl J Med* **316:** 73–78.

Fandrey J. 2004. Oxygen-dependent and tissue-specific regulation of erythropoietin gene expression. *Am J Physiol Regul Integr Comp Physiol* **286:** R977–R988.

Fiordaliso F, Chimenti S, Staszewsky L, Bai A, Carlo E, Cuccovillo I, Doni M, Mengozzi M, Tonelli R, Ghezzi P, et al. 2005. A nonerythropoietic derivative of erythropoietin protects the myocardium from ischemia-reperfusion injury. *Proc Natl Acad Sci* **102:** 2046–2051.

Fischl M, Galpin JE, Levine JD, Groopman JE, Henry DH, Kennedy P, Miles S, Robbins W, Starrett B, Zalusky R, et al. 1990. Recombinant human erythropoietin for patients with AIDS treated with zidovudine. *N Engl J Med* **322:** 1488–1493.

Fox HM. 1947. Daphnia haemoglobin. *Nature* **160:** 431.

Fox HM. 1949. Haemoglobin in Crustacea. *Nature* **164:** 59.

Frede S, Freitag P, Geuting L, Konietzny R, Fandrey J. 2011. Oxygen-regulated expression of the erythropoietin gene in the human renal cell line REPC. *Blood* **117:** 4905–4914.

Fried W, Plzak L, Jacobson L, Goldwasser E. 1956. Studies on erythropoiesis II. Assay of erythropoietin in hypophysectomized rats. *Proc Soc Exp Biol Med* **92:** 203–207.

Furlow PW, Percy MJ, Sutherland S, Bierl C, McMullin MF, Master SR, Lappin TR, Lee FS. 2009. Erythrocytosis-associated HIF-2α mutations demonstrate a critical role for residues C-terminal to the hydroxylacceptor proline. *J Biol Chem* **284:** 9050–9058.

Gale DP, Harten SK, Reid CD, Tuddenham EG, Maxwell PH. 2008. Autosomal dominant erythrocytosis and pulmonary arterial hypertension associated with an activating HIF2 α mutation. *Blood* **112:** 919–921.

Galson DL, Tsuchiya T, Tendler DS, Huang LE, Ren Y, Ogura T, Bunn HF. 1995. The orphan receptor hepatic nuclear factor 4 functions as a transcriptional activator for tissue-specific and hypoxia-specific erythropoetin gene expression and is antagonized by EAR3/COUP-TF1. *Mol Cell Biol* **15:** 2135–2144.

Gnarra JR, Zhou S, Merrill MJ, Wagner JR, Krumm A, Papavassiliou E, Oldfield EH, Klausner RD, Linehan WM. 1996. Post-transcriptional regulation of vascular endothelial growth factor mRNA by the product of the VHL tumor suppressor gene. *Proc Natl Acad Sci* **93:** 10589–10594.

Goldberg MA, Glass GA, Cunningham JM, Bunn HF. 1987. The regulated expression of erythropoietin by two human hepatoma cell lines. *Proc Natl Acad Sci* **84:** 7972–7976.

Goldwasser E. 1996. Erythropoietin, a somewhat personal history. *Perspect Biol Med* **40:** 18–32.

Goldwasser E, Kung CK-H, Eliason J. 1974. On the mechanism of erythropoietin-induced differentiation. *J Biol Chem* **249:** 4202–4206.

Gordeuk VR, Prchal JT. 2006. Vascular complications in Chuvash polycythemia. *Semin Thromb Hemost* **32:** 289–294.

Gorr TA, Cahn JD, Yamagata H, Bunn HF. 2004. Hypoxia-induced synthesis of hemoglobin in the crustacean Daphnia magna is hypoxia-inducible factor-dependent. *J Biol Chem* **279:** 36038–36047.

Gradin K, McGuire J, Wenger RH, Kvietikova I, Whitelaw M, Toftgard R, Tora L, Gassman M, Poellinger L. 1996. Functional interference between hypoxia and dioxin signal transduction pathways: Competition for recruitment of the Arnt transcription factor. *Mol Cell Biol* **16:** 5221–5231.

Grant WC, Root WS. 1952. Fundamental stimulus for erythropoiesis. *Physiol Rev* **32:** 449–498.

Gruber M, Hu CJ, Johnson RS, Brown EJ, Keith B, Simon MC. 2007. Acute postnatal ablation of Hif-2α results in anemia. *Proc Natl Acad Sci* **104:** 2301–2306.

Gu YZ, Moran SM, Hogenesch JB, Wartman L, Bradfield CA. 1998. Molecular characterization and chromosomal localization of a third α-class hypoxia inducible factor subunit, HIF3α. *Gene Expr* **7:** 205–213.

Heeschen C, Aicher A, Lehmann R, Fichtlscherer S, Vasa M, Urbich C, Mildner-Rihm C, Martin H, Zeiher AM, Dimmeler S. 2003. Erythropoietin is a potent physiologic stimulus for endothelial progenitor cell mobilization. *Blood* **102:** 1340–1346.

Henke M, Laszig R, Rube C, Schafer U, Haase KD, Schilcher B, Mose S, Beer KT, Burger U, Dougherty C, et al. 2003. Erythropoietin to treat head and neck cancer patients with anaemia undergoing radiotherapy: Randomised, double-blind, placebo-controlled trial. *Lancet* **362:** 1255–1260.

Hermine O, Beru N, Pech N, Goldwasser E. 1991. An autocrine role for erythropoietin in mouse hematopoietic cell differentiation. *Blood* **78:** 2253.

Hickey MM, Lam JC, Bezman NA, Rathmell WK, Simon MC. 2007. von Hippel-Lindau mutation in mice recapitulates Chuvash polycythemia via hypoxia-inducible factor-2α signaling and splenic erythropoiesis. *J Clin Invest* **117:** 3879–3889.

Hoffman EC, Reyes H, Chu F-F, Sander F, Conley LH, Brooks BA, Hankinson O. 1991. Cloning of a factor required for activity of the Ah (dioxin) receptor. *Science* **252:** 954–958.

Horiguchi H, Kayama F, Oguma E, Willmore WG, Hradecky P, Bunn HF. 2000. Cadmium and platinum suppression of erythropoietin production in cell culture. *Blood* **96:** 3743–3747.

Hu B, Wright E, Campbell L, Blanchard KL. 1997. In vivo analysis of DNA-protein interactions on the human erythropoietin enhancer. *Mol Cell Biol* **17:** 851–856.

Huang LE, Arany Z, Livingston DM, Bunn HF. 1996. Activation of hypoxia-inducible transcription factor depends primarily upon redox-sensitive stabilization of its a subunit. *J Biol Chem* **271:** 32253–32259.

Huang LE, Gu J, Schau M, Bunn HF. 1998. Regulation of hypoxia-inducible factor 1α is mediated by an oxygen-dependent degradation domain via the ubiquitin-proteasome pathway. *Proc Natl Acad Sci* **95:** 7987–7992.

Iliopoulos O, Levy AP, Jiang CWG, Kaelin J, Goldberg MA. 1996. Negative regulation of hypoxia-inducible genes by the von Hippel-Lindau protein. *Proc Natl Acad Sci* **93:** 10595–10599.

Imagawa S, Izumi T, Miura Y. 1994. Positive and negative regulation of the erythropoietin gene. *J Biol Chem* **269:** 9038–9044.

Imagawa S, Yamamoto M, Miura Y. 1997. Negative regulation of the erythropoietin gene expression by the GATA transcription factors. *Blood* **89:** 1430–1439.

Ivan M, Kondo K, Yang H, Kim W, Valiando J, Ohh M, Salic A, Asara JM, Lane WS, Kaelin WG Jr. 2001. HIFα targeted for VHL-mediated destruction by proline hydroxylation: Implications for O_2 sensing. *Science* **292:** 464–468.

Ivan M, Haberberger T, Gervasi DC, Michelson KS, Gunzler V, Kondo K, Yang H, Sorokina I, Conaway RC, Conaway JW, et al. 2002. Biochemical purification and pharmacological inhibition of a mammalian prolyl hydroxylase acting on hypoxia-inducible factor. *Proc Natl Acad Sci* **99:** 13459–13464.

Iyer NV, Kotch LE, Agani F, Leung SW, Laughner E, Wenger RH, Gassmann M, Gearhart JD, Lawler AM, Yu AY, et al. 1998. Cellular and developmental control of O_2 homeostasis by hypoxia-inducible factor 1α. *Genes Dev* **12:** 149–162.

Jaakkola P, Mole DR, Tian YM, Wilson MI, Gielbert J, Gaskell SJ, Kriegsheim A, Hebestreit HF, Mukherji M, Schofield CJ, et al. 2001. Targeting of HIF-α to the von Hippel-Lindau ubiquitylation complex by O2-regulated prolyl hydroxylation. *Science* **292:** 468–472.

Jacobs K, Shoemaker C, Rudersdorf R, Neill SD, Kaufman RJ, Mufson A, Seehra J, Jones SS, Hewick R, Fritsch EF, et al. 1985. Isolation and characterization of genomic and cDNA clones of human erythropoietin. *Nature* **313:** 806–810.

Jacobson LO, Marks E, Gaston E, Goldwasser E. 1957. Role of the kidney in erythropoiesis. *Nature* **179:** 633–634.

Jelkmann W. 1992. Erythropietin: Structure, control of production, and function. *Physiol Rev* **72:** 449–489.

Jelkmann W. 2005. Effects of erythropoietin on brain function. *Curr Pharm Biotechnol* **6:** 65–79.

Jelkmann W. 2008. Developments in the therapeutic use of erythropoiesis stimulating agents. *Brit J Haematol* **141:** 287–297.

Jelkmann W, Wagner K. 2004. Beneficial and ominous aspects of the pleiotropic action of erythropoietin. *Ann Hematol* **83:** 673–686.

Kapitsinou PP, Liu Q, Unger TL, Rha J, Davidoff O, Keith B, Epstein JA, Moores SL, Erickson-Miller CL, Haase VH. 2010. Hepatic HIF-2 regulates erythropoietic responses to hypoxia in renal anemia. *Blood* **116:** 3039–3048.

Koury MJ, Bondurant MC. 1990. Erythropoietin retards DNA breakdown and prevents programmed death in erythroid progenitor cells. *Science* **248:** 378–381.

Koury ST, Bondurant MC, Koury MJ. 1988. Localization of erythropoietin synthesizing cells in murine kidneys by in situ hybridization. *Blood* **71:** 524–527.

Koury ST, Koury MJ, Bondurant MC, Caro J, Graber SE. 1989. Quantitation of erythropoietin-producing cells in kidneys of mice by in situ hybridization: Correlation with hematocrit, renal erythopoietin mRNA, and serum erythropoietin concentration. *Blood* **74:** 645–651.

Koury ST, Bondurant MC, Koury MJ, Semenza GL. 1991. Localization of cells producing erythropoietin in murine liver by in situ hybridization. *Blood* **77:** 2497–2503.

Krumdieck N. 1943. Erythropoietic substance in the serum of anemic animals. *Proc Soc Exp Biol Med* **54:** 14–17.

Lacombe C, Silva J-LD, Bruneval P, Fournier J-G, Wendling F, Casadevall N, Camilleri J-P, Bariety J, Varet B, Tambourin P. 1988. Pertubular cells are the site of erythropoietin synthesis in the murine hypoxic kidney. *J Clin Invest* **81:** 620–623.

Ladroue C, Carcenac R, Leporrier M, Gad S, Le Hello C, Galateau-Salle F, Feunteun J, Pouyssegur J, Richard S, Gardie B. 2008. PHD2 mutation and congenital erythrocytosis with paraganglioma. *N Engl J Med* **359:** 2685–2692.

La Ferla K, Reimann C, Jelkmann W, Hellwig-Burgel T. 2002. Inhibition of erythropoietin gene expression signaling involves the transcription factors GATA-2 and NF-κB. *FASEB J* **16:** 1811–1813.

Lando D, Peet DJ, Gorman JJ, Whelan DA, Whitelaw ML, Bruick RK. 2002a. FIH-1 is an asparaginyl hydroxylase enzyme that regulates the transcriptional activity of hypoxia-inducible factor. *Genes Dev* **16:** 1466–1471.

Lando D, Peet DJ, Whelan DA, Gorman JJ, Whitelaw ML. 2002b. Asparagine hydroxylation of the HIF transactivation domain a hypoxic switch. *Science* **295:** 858–861.

Leist M, Ghezzi P, Grasso G, Bianchi R, Villa P, Fratelli M, Savino C, Bianchi M, Nielsen J, Gerwien J, et al. 2004. Derivatives of erythropoietin that are tissue protective but not erythropoietic. *Science* **305:** 239–242.

Le Minh K, Klemm K, Abshagen K, Eipel C, Menger MD, Vollmar B. 2007. Attenuation of inflammation and apoptosis by pre- and posttreatment of darbepoetin-α in acute liver failure of mice. *Am J Pathol* **170:** 1954–1963.

Leyland-Jones B. 2003. Breast cancer trial with erythropoietin terminated unexpectedly. *Lancet Oncol* **4:** 459–460.

Lin F-K, Suggs S, Lin C-H, Browne JK, Smalling R, Egrie JC, Chen KK, Fox GM, Martin F, Stabinsky Z, et al. 1985. Cloning and expression of the human erythropoietin gene. *Proc Natl Acad Sci* **82:** 7580–7584.

Livnah O, Stura EA, Johnson DL, Middleton SA, Mulcahy LS, Wrighton NC, Dower WJ, Jolliffe LK, Wilson IA. 1996. Functional mimicry of a protein hormone by a peptide agonist: The EPO receptor complex at 2.8 A. *Science* **273:** 464–471.

Livnah O, Stura EA, Middleton SA, Johnson DL, Jolliffe LK, Wilson IA. 1999. Crystallographic evidence for preformed dimers of erythropoietin receptor before ligand activation. *Science* **283:** 987–990.

Macdougall IC. 2005. CERA (Continuous erythropoietin receptor activator): A new erythropoiesis-stimulating agent for the treatment of anemia. *Curr Hematol Rep* **4:** 436–440.

Macdougall IC, Robson R, Opatrna S, Liogier X, Pannier A, Jordan P, Dougherty FC, Reigner B. 2006. Pharmacokinetics and pharmacodynamics of intravenous and subcutaneous continuous erythropoietin receptor activator (C.E.R.A.) in patients with chronic kidney disease. *Clin J Am Soc Nephrol* **1:** 1211–1215.

Madan A, Lin C, Hatch SL, Curtin PT. 1995. Regulated basal, inducible, and tissue-specific human erythropoietin gene expression in transgenic mice requires multiple *cis* DNA sequences. *Blood* **85:** 2735–2741.

Mahon PC, Hirota K, Semenza GL. 2001. FIH-1: A novel protein that interacts with HIF-1α and VHL to mediate repression of HIF-1 transcriptional activity. *Genes Dev* **15:** 2675–2686.

Makita T, Hernandez-Hoyos G, Chen TH, Wu H, Rothenberg EV, Sucov HM. 2001. A developmental transition in definitive erythropoiesis: Erythropoietin expression is sequentially regulated by retinoic acid receptors and HNF4. *Genes Dev* **15:** 889–901.

Marti HH, Wenger RH, Rivas LA, Straumann U, Digicaylioglu M, Henn V, Yonekawa Y, Bauer C, Gassmann M. 1996. Erythropoietin gene expression in human, monkey and murine brain. *Eur J Neurosci* **8:** 666–676.

Martini M, Teofili L, Cenci T, Giona F, Torti L, Rea M, Foa R, Leone G, Larocca LM. 2008. A novel heterozygous HIF2AM535I mutation reinforces the role of oxygen sensing pathway disturbances in the pathogenesis of familial erythrocytosis. *Haematologica* **93:** 1068–1071.

Masson N, Willam C, Maxwell PH, Pugh CW, Ratcliffe PJ. 2001. Independent function of two destruction domains in hypoxia-inducible factor-α chains activated by prolyl hydroxylation. *EMBO J* **20:** 5197–5206.

Masuda S, Okano M, Yamagishi K, Nagao M, Ueda M, Sasaki R. 1994. A novel site of erythropoietin production: Oxygen-dependent production in cultured rat astrocytes. *J Biol Chem* **269:** 19488–19493.

Maxwell PH, Osmond MK, Pugh CW, Heryet A, Nicholls LG, Tan CC, Doe BG, Ferguson DJP, Johnson MH, Ratcliffe PJ. 1993a. Identification of the renal erythropoietin-producing cells using transgenic mice. *Kidney Int* **44:** 1149–1162.

Maxwell PH, Pugh CW, Ratcliffe PJ. 1993b. Inducible operation of the erythropoietin 3′ enhancer in multiple cell lines: Evidence for a widespread oxygen-sensing mechanism. *Proc Natl Acad Sci* **90:** 2423–2427.

Maxwell PH, Ferguson DJP, Osmond MK, Pugh CW, Heryet A, Doe BG, Johnson MH, Ratcliffe PJ. 1994. Expression of a homologously recombined erythopoietin-SV40 T antigen fusion gene in mouse liver: Evidence for erythropoietin production by Ito cells. *Blood* **84:** 1823–1830.

Maxwell PH, Wiesener MS, Chang G-W, Clifford SC, Vaux EC, Cockman ME, Wykoff CC, Pugh CW, Maher ER, Ratcliffe PJ. 1999. The tumor suppressor protein VHL targets hypoxia-inducible factors for oxygen-dependent proteolysis. *Nature* **399:** 271–275.

McDonald TP, Cottrell MB, Clift RE, Cullen WC, Lin FK. 1987. High doses of recombinant erythropoietin stimulate platelet production in mice. *Exp Hematol* **15:** 719–721.

McLeod DL, Shreeve MM, Axelrad AA. 1974. Improved plasma culture system for production of erythrocytic colonies in vitro: Quantitative assay method for CFU-E. *Blood* **44:** 517–534.

Means RT Jr. 1994. Clinical application of recombinant erythropoietin in the anemia of chronic disease. *Hematol Oncol Clin North Am* **8:** 933–944.

Messinezy M, Westwood NB, El-Hemaidi I, Marsden JT, Sherwood RS, Pearson TC. 2002. Serum erythropoietin values in erythrocytoses and in primary thrombocythaemia. *Brit J Haematol* **117:** 47–53.

Miescher F. 1893. Über die Beziehungen zwischen Meereshöhe und Beschaffenheit des Blutes. *Corresp Bl Schweiz Aerzte* **23:** 809–830.

Miller CB, Jones RJ, Piantadosi S, Abeloff MD, Spivak JL. 1990. Decreased erythropoietin response in patients with the anemia of cancer. *N Engl J Med* **322:** 1689–1692.

Minamishima YA, Moslehi J, Bardeesy N, Cullen D, Bronson RT, Kaelin WG Jr. 2008. Somatic inactivation of the PHD2 prolyl hydroxylase causes polycythemia and congestive heart failure. *Blood* **111:** 3236–3244.

Miyake T, Hung CKH, Goldwasser E. 1977. Purification of human erythropoietin. *J Biol Chem* **252:** 5558–5564.

Mohyeldin A, Lu H, Dalgard C, Lai SY, Cohen N, Acs G, Verma A. 2005. Erythropoietin signaling promotes invasiveness of human head and neck squamous cell carcinoma. *Neoplasia* **7:** 537–543.

Nathan DG, Schupak E, Stohlman F Jr, Merrill JP. 1964. Erythropoiesis in anephric man. *J Clin Invest* **43:** 2158–2165.

Nishiya D, Omura T, Shimada K, Matsumoto R, Kusuyama T, Enomoto S, Iwao H, Takeuchi K, Yoshikawa J, Yoshiyama M. 2006. Effects of erythropoietin on cardiac remodeling after myocardial infarction. *J Pharm Sci* **101:** 31–39.

Noguchi CT, Asavaritikrai P, Teng R, Jia Y. 2007. Role of erythropoietin in the brain. *Crit Rev Oncol Hemat* **64:** 159–171.

Noguchi CT, Wang L, Rogers HM, Teng R, Jia Y. 2008. Survival and proliferative roles of erythropoietin beyond the erythroid lineage. *Expert Rev Mol Med* **10:** e36.

Obara N, Suzuki N, Kim K, Nagasawa T, Imagawa S, Yamamoto M. 2008. Repression via the GATA box is essential for tissue-specific erythropoietin gene expression. *Blood* **111:** 5223–5232.

Ohh M, Park CW, Ivan M, Hoffman MA, Kim TY, Huang LE, Pavletich N, Chau V, Kaelin WG. 2000. Ubiquitination of hypoxia-inducible factor requires direct binding to the β-domain of the von Hippel-Lindau protein. *Nat Cell Biol* **2:** 423–427.

Paliege A, Rosenberger C, Bondke A, Sciesielski L, Shina A, Heyman SN, Flippin LA, Arend M, Klaus SJ, Bachmann S. 2010. Hypoxia-inducible factor-2α-expressing interstitial fibroblasts are the only renal cells that express erythropoietin under hypoxia-inducible factor stabilization. *Kidney Int* **77:** 312–318.

Parsa CJ, Kim J, Riel RU, Pascal LS, Thompson RB, Petrofski JA, Matsumoto A, Stamler JS, Koch WJ. 2004. Cardioprotective effects of erythropoietin in the reperfused ischemic heart: A potential role for cardiac fibroblasts. *J Biol Chem* **279:** 20655–20662.

Pastore YD, Jelinek J, Ang S, Guan Y, Liu E, Jedlickova K, Krishnamurti L, Prchal JT. 2003. Mutations in the VHL gene in sporadic apparently congenital polycythemia. *Blood* **101:** 1591–1595.

Patel NS, Sharples EJ, Cuzzocrea S, Chatterjee PK, Britti D, Yaqoob MM, Thiemermann C. 2004. Pretreatment with EPO reduces the injury and dysfunction caused by ischemia/reperfusion in the mouse kidney in vivo. *Kidney Int* **66:** 983–989.

Percy MJ, Zhao Q, Flores A, Harrison C, Lappin TR, Maxwell PH, McMullin MF, Lee FS. 2006. A family with erythrocytosis establishes a role for prolyl hydroxylase domain protein 2 in oxygen homeostasis. *Proc Natl Acad Sci* **103:** 654–659.

Percy MJ, Furlow PW, Beer PA, Lappin TR, McMullin MF, Lee FS. 2007. A novel erythrocytosis-associated PHD2 mutation suggests the location of a HIF binding groove. *Blood* **110:** 2193–2196.

Percy MJ, Beer PA, Campbell G, Dekker AW, Green AR, Oscier D, Rainey MG, van Wijk R, Wood M, Lappin TR, et al. 2008a. Novel exon 12 mutations in the HIF2A gene associated with erythrocytosis. *Blood* **111:** 5400–5402.

Percy MJ, Furlow PW, Lucas GS, Li X, Lappin TR, McMullin MF, Lee FS. 2008b. A gain-of-function mutation in the HIF2A gene in familial erythrocytosis. *N Engl J Med* **358:** 162–168.

Percy MJ, Chung YJ, Harrison C, Mercieca J, Hoffbrand VJ, Dinardo CL, Dantos PCJL, Fonseca GHH, Gualalndro SFM, Pereira AC, et al. 2012. Two new mutations in the HIF2A gene associated with erythrocytosis. *Am J Hematol* **87:** 439–442.

Perrotta S, Nobili B, Ferraro M, Migliaccio C, Borriello A, Cucciolla V, Martinelli V, Rossi F, Punzo F, Cirillo P, et al. 2006. Von Hippel-Lindau-dependent polycythemia is endemic on the island of Ischia: Identification of a novel cluster. *Blood* **107:** 514–519.

Peyssonnaux C, Zinkernagel AS, Schuepbach RA, Rankin E, Vaulont S, Haase VH, Nizet V, Johnson RS. 2007. Regulation of iron homeostasis by the hypoxia-inducible transcription factors (HIFs). *J Clin Invest* **117:** 1926–1932.

Phrommintikul A, Haas SJ, Elsik M, Krum H. 2007. Mortality and target haemoglobin concentrations in anaemic patients with chronic kidney disease treated with erythropoietin: A meta-analysis. *Lancet* **369:** 381–388.

Prchal JF, Prchal JT. 1999. Molecular basis for polycythemia. *Curr Opin Hematol* **6:** 100–109.

Provenzano R, Besarab A, Macdougall IC, Ellison DH, Maxwell AP, Sulowicz W, Klinger M, Rutkowski B, Correa-Rotter R, Dougherty FC. 2007. The continuous erythropoietin receptor activator (C.E.R.A.) corrects anemia at extended administration intervals in patients with chronic kidney disease not on dialysis: Results of a phase II study. *Clin Nephrol* **67:** 306–317.

Pugh CW, Tan CC, Jones RW, Ratcliffe PJ. 1991. Functional analysis of an oxygen-regulated transcriptional enhancer lying 3′ to the mouse erythropoietin gene. *Proc Natl Acad Sci* **88:** 10553–10557.

Pugh CW, Ebert BL, Ebrahim O, Ratcliffe PJ. 1994. Characterisation of functional domains within the mouse erythropoietin 3′ enhancer conveying oxygen-regulated responses in different cell lines. *Biochem Biophys Acta* **1217:** 297–306.

Rankin EB, Biju MP, Liu Q, Unger TL, Rha J, Johnson RS, Simon MC, Keith B, Haase VH. 2007. Hypoxia-inducible factor-2 (HIF-2) regulates hepatic erythropoietin in vivo. *J Clin Invest* **117:** 1068–1077.

Reissmann KR. 1950. Studies on the mechanism of erythropoietic stimulation in parabiotic rats during hypoxia. *Blood* **5:** 372–380.

Remy I, Wilson IA, Michnick SW. 1999. Erythropoietin receptor activation by a ligand-induced conformation change. *Science* **283:** 990–993.

Rosa R, Prehu M-O, Beuzard Y, Rosa J. 1978. The first case of a complete deficiency of diphosphoglycerate mutase in human erythrocytes. *J Clin Invest* **62:** 907.

Ruhenstroth-Bauer G. 1950. Versuche zum Nachweis eines spezifischen erythropetischen Hormons. *Arch Exp Pathol Pharmakol* **211:** 32–56.

Ryan HE, Lo J, Johnson RS. 1998. HIF-1α is required for solid tumor formation and embryonic vascularization. *EMBO J* **17:** 3005–3015.

Safran M, Kim WY, O'Connell F, Flippin L, Gunzler V, Horner JW, Depinho RA, Kaelin WG Jr. 2006. Mouse model for noninvasive imaging of HIF prolyl hydroxylase activity: Assessment of an oral agent that stimulates erythropoietin production. *Proc Natl Acad Sci* **103:** 105–110.

Sakanaka M, Wen TC, Matsuda S, Masuda S, Morishita E, Nagao M, Sasaki R. 1998. In vivo evidence that erythropoietin protects neurons from ischemic damage. *Proc Natl Acad Sci* **95:** 4635–4640.

Salceda S, Caro J. 1997. Hypoxia-inducible factor 1α (HIF-1α) protein is rapidly degraded by the ubiquitin-proteasome system under normoxic conditions. Its stabilization by hypoxia depends on redox-induced changes. *J Biol Chem* **272:** 22642–22647.

Schmeding M, Boas-Knoop S, Lippert S, Ruehl M, Somasundaram R, Dagdelen T, Neuhaus P, Neumann UP. 2008. Erythropoietin promotes hepatic regeneration after extended liver resection in rats. *J Gastroenterol Hepatol* **23:** 1125–1131.

Schmeding M, Hunold G, Ariyakhagorn V, Rademacher S, Boas-Knoop S, Lippert S, Neuhaus P, Neumann UP. 2009. Erythropoietin reduces ischemia-reperfusion injury after liver transplantation in rats. *Transpl Int* **22:** 738–746.

Schmid TC, Loffing J, Le Hir M, Kaissling B. 1994. Distribution of ecto-5′-nucleotidase in the rat liver: Effect of anaemia. *Histochemistry* **101:** 439–447.

Schuster SJ, Koury ST, Bohrer M, Salceda S, Caro J. 1992. Cellular sites of extrarenal and renal erythropoietin production in anemic rats. *Brit J Hematol* **81:** 153–159.

Scortegagna M, Ding K, Zhang Q, Oktay Y, Bennett MJ, Bennett M, Shelton JM, Richardson JA, Moe O, Garcia JA. 2005. HIF-2α regulates murine hematopoietic development in an erythropoietin-dependent manner. *Blood* **105:** 3133–3140.

Semenza GL, Wang GL. 1992. A nuclear factor induced by hypoxia via de novo protein synthesis binds to the human erythropoietin gene enhancer at a site required for transcriptional activation. *Mol Cell Biol* **12:** 5447–5454.

Semenza GL, Trystman M, Gearhart JD, Antonarakis S. 1989. Polycythemia in transgenic mice expressing the human erythropoietin gene. *Proc Natl Acad Sci* **86:** 2301–2305.

Semenza GL, Dureza RC, Traystman MD, Gearhart JD, Antonarakis SE. 1990. Human erythropoietin gene expression in transgenic mice: Multiple transcription initiation sites and *cis*-acting regulatory elements. *Mol Cell Biol* **10:** 930–938.

Semenza GL, Koury ST, Nejfelt MK, Gearhart JD, Antonarakis SE. 1991a. Cell-type-specific and hypoxia-inducible expression of the human erythropoietin gene in transgenic mice. *Proc Natl Acad Sci* **88:** 8725–8729.

Semenza GL, Nejfelt MK, Chi SM, Antonarakis SE. 1991b. Hypoxia-inducible nuclear factors bind to an enhancer element located 3′ to the human erythropoietin gene. *Proc Natl Acad Sci* **88:** 5680–5684.

Sepodes B, Maio R, Pinto R, Sharples E, Oliveira P, McDonald M, Yaqoob M, Thiemermann C, Mota-Filipe H. 2006. Recombinant human erythropoietin protects the liver from hepatic ischemia-reperfusion injury in the rat. *Transpl Int* **19:** 919–926.

Sergeyeva A, Gordeuk VR, Tokarev YN, Sokol L, Prchal JH, Prchal JT. 1997. Congenital polycythemia in Chuvashia. *Blood* **89:** 2148–2154.

Sharples EJ, Patel N, Brown P, Stewart K, Mota-Philipe H, Sheaff M, Kieswich J, Allen D, Harwood S, Raftery M, et al. 2004. Erythropoietin protects the kidney against the injury and dysfunction caused by ischemia-reperfusion. *J Am Soc Nephrol* **15:** 2115–2124.

Simonson TS, Yang Y, Huff CD, Yun H, Qin G, Witherspoon DJ, Bai Z, Lorenzo FR, Xing J, Jorde LB, et al. 2010. Genetic evidence for high-altitude adaptation in Tibet. *Science* **329:** 72–75.

Sinclair AM, Rogers N, Busse L, Archibeque I, Brown W, Kassner PD, Watson JE, Arnold GE, Nguyen KC, Powers S, et al. 2008. Erythropoietin receptor transcription is neither elevated nor predictive of surface expression in human tumour cells. *Brit J Cancer* **98:** 1059–1067.

Singh A, Eckardt KU, Zimmermann A, Gotz KH, Hamann M, Ratcliffe PJ, Kurtz A, Reinhart WH. 1993. Increased plasma viscosity as a reason for inappropriate erythropoietin formation. *J Clin Invest* **91:** 251–256.

Singh AK, Szczech L, Tang KL, Barnhart H, Sapp S, Wolfson M, Reddan D. 2006. Correction of anemia with epoetin alfa in chronic kidney disease. *N Engl J Med* **355:** 2085–2098.

Spivak JL. 1994. Recombinant human erythropoietin and the anemia of cancer [editorial]. *Blood* **84:** 997–1004.

Stohlman FJ, Rath C, Rose J. 1954. Evidence for a humoral regulation of erythropoiesis. *Blood* **9:** 721–733.

Stopka T, Zivny JH, Stopkova P, Prchal JF, Prchal JT. 1998. Human hematopoietic progenitors express erythropoietin. *Blood* **91:** 3766–3772.

Suzuki N, Ohneda O, Takahashi S, Higuchi M, Mukai HY, Nakahata T, Imagawa S, Yamamoto M. 2002. Erythroid-specific expression of the erythropoietin receptor rescued its null mutant mice from lethality. *Blood* **100:** 2279–2288.

Syed RS, Reid SW, Li C, Cheetham JC, Aoki KH, Liu B, Zhan H, Osslund TD, Chirino AJ, Zhang J, et al. 1998. Efficiency of signalling through cytokine receptors depends critically on receptor orientation. *Nature* **395:** 511–516.

Takeda K, Aguila HL, Parikh NS, Li X, Lamothe K, Duan LJ, Takeda H, Lee FS, Fong GH. 2008. Regulation of adult erythropoiesis by prolyl hydroxylase domain proteins. *Blood* **111:** 3229–3235.

Tan CC, Eckardt K-U, Ratcliffe PJ. 1991. Organ distribution of erythopoietin messenger RNA in normal and uremic rats. *Kidney Int* **40:** 69–76.

Tian H, McKnight SL, Russell DW. 1997. Endothelial PAS domain protein 1 (EPAS1), a transcription factor selectively expressed in endothelial cells. *Genes Dev* **11:** 72–82.

Tsai PT, Ohab JJ, Kertesz N, Groszer M, Matter C, Gao J, Liu X, Wu H, Carmichael ST. 2006. A critical role of erythropoietin receptor in neurogenesis and post-stroke recovery. *J Neurosci* **26:** 1269–1274.

van Wijk R, Sutherland S, Van Wesel AC, Huizinga EG, Percy MJ, Bierings M, Lee FS. 2010. Erythrocytosis associated with a novel missense mutation in the HIF2A gene. *Haematologica* **95:** 829–832.

Vaziri ND, Zhou XJ, Liao SY. 1994. Erythropoietin enhances recovery from cisplatin-induced acute renal failure. *Am J Physiol* **266:** F360–F366.

Viault F. 1890. Sur l'augmentation considerable du nombre des globules rouges dans le sang chez les habitants des hautes plateaux de l'Amerique du sud. *CR Acad Sci Paris* **111:** 918–919.

Volke M, Gale DP, Maegdefrau U, Schley G, Klanke B, Bosserhoff AK, Maxwell PH, Eckardt KU, Warnecke C. 2009. Evidence for a lack of a direct transcriptional suppression of the iron regulatory peptide hepcidin by hypoxia-inducible factors. *PloS ONE* **4:** e7875.

Wang GL, Semenza GL. 1993a. Desferrioxamine induces erythropoietin gene expression and hypoxia-inducible factor 1 DNA-binding activity: Implications for models of hypoxia signal transduction. *Blood* **82:** 3610–3615.

Wang GL, Semenza GL. 1993b. General involvement of hypoxia-inducible factor 1 in transcriptional response to hypoxia. *Proc Natl Acad Sci* **90:** 4304–4308.

Wang GL, Semenza GL. 1995. Purification and characterization of hypoxia-inducible factor-1. *J Biol Chem* **270:** 1230–1237.

Wang GL, Jiang B-H, Rue EA, Semenza GL. 1995a. Hypoxia-inducible factor 1 is a basic helix-loop-helix-PAS heterodimer regulated by cellular O_2 tension. *Proc Natl Acad Sci* **92:** 5510–5514.

Wang GL, Jiang BH, Rue EA, Semenza GL. 1995b. Hypoxia-inducible factor 1 is a basic-helix-loop-helix-PAS heterodimer regulated by cellular O2 tension. *Proc Natl Acad Sci* **92:** 5510–5514.

Warnecke C, Zaborowska Z, Kurreck J, Erdmann VA, Frei U, Wiesener M, Eckardt KU. 2004. Differentiating the functional role of hypoxia-inducible factor (HIF)-1α and HIF-2α (EPAS-1) by the use of RNA interference: Erythropoietin is a HIF-2α target gene in Hep3B and Kelly cells. *FASEB J* **18:** 1462–1464.

Wen D, Boissel J-P, Showers M, Ruch BC, Bunn HF. 1994. Erythropoietin structure-function relationships: Identification of functionally important domains. *J Biol Chem* **269:** 22839–22846.

Wenger RH, Kurtz A. 2011. Erythropoietin. *Compr Physiol* **1:** 1759–1794.

Wenger RH, Stiehl DP, Camenisch G. 2005. Integration of oxygen signaling at the consensus HRE. *Sci STKE* **2005:** re12.

Wood SM, Gleadle JM, Pugh CW, Hankinson O, Ratcliffe PJ. 1996. The role of the aryl hydrocarbon receptor nuclear translocator (ARNT) in hypoxic induction of gene expression. *J Biol Chem* **269:** 15117–15123.

Wrighton NC, Farrell FX, Chang R, Kashyap AK, Barbone FP, Mulcahy LS, Johnson DL, Barrett RW, Jolliffe LK, Dower WJ. 1996. Small peptides as potent mimetics of the protein hormone erythropoietin. *Science* **273:** 458–463.

Yamaji R, Okada T, Moriya M, Naito M, Tsuruo T, Miyatake K, Nakano Y. 1996. Brain capillary endothelial cells express two forms of erythropoietin receptor mRNA. *Eur J Biochem* **239:** 494–500.

Yan L, Colandrea VJ, Hale JJ. 2010. Prolyl hydroxylase domain-containing protein inhibitors as stabilizers of hypoxia-inducible factor: Small molecule-based therapeutics for anemia. *Expert Opin Ther Pat* **20:** 1219–1245.

Yang CW, Li C, Jung JY, Shin SJ, Choi BS, Lim SW, Sun BK, Kim YS, Kim J, Chang YS, et al. 2003. Preconditioning with erythropoietin protects against subsequent ischemia-reperfusion injury in rat kidney. *FASEB J* **17:** 1754–1755.

Yi X, Liang Y, Huerta-Sanchez E, Jin X, Cuo ZX, Pool JE, Xu X, Jiang H, Vinckenbosch N, Korneliussen TS, et al. 2010. Sequencing of 50 human exomes reveals adaptation to high altitude. *Science* **329:** 75–78.

Yu F, White SB, Zhao Q, Lee FS. 2001. HIF-1α binding to VHL is regulated by stimulus-sensitive proline hydroxylation. *Proc Natl Acad Sci* **98:** 9630–9635.

Evolution of Hemoglobin and Its Genes

Ross C. Hardison

Center for Comparative Genomics and Bioinformatics, Huck Institute of Genome Sciences, Department of Biochemistry and Molecular Biology, The Pennsylvania State University, University Park, Pennsylvania 16802

Correspondence: rch8@psu.edu

Insights into the evolution of hemoglobins and their genes are an abundant source of ideas regarding hemoglobin function and regulation of globin gene expression. This article presents the multiple genes and gene families encoding human globins, summarizes major events in the evolution of the hemoglobin gene clusters, and discusses how these studies provide insights into regulation of globin genes. Although the genes in and around the α-like globin gene complex are relatively stable, the β-like globin gene clusters are more dynamic, showing evidence of transposition to a new locus and frequent lineage-specific expansions and deletions. The *cis*-regulatory modules controlling levels and timing of gene expression are a mix of conserved and lineage-specific DNA, perhaps reflecting evolutionary constraint on core regulatory functions shared broadly in mammals and adaptive fine-tuning in different orders of mammals.

A wide range of animals, vertebrate and invertebrate, use hemoglobins to transport oxygen, carrying it from lungs, gills, or other respiratory organs to peripheral tissues that need the oxygen for efficient metabolism. Hence it is natural to compare the structure and function of hemoglobin proteins between species both to explore adaptation and to discover aspects of biochemistry and physiology that are conserved. Comparative studies also have been conducted on the genes and gene clusters that encode the hemoglobins, revealing a rich history of gene duplications and losses as well as translocations. One motivation for comparative studies has been to use the insights from the evolutionary analyses to better understand mechanisms of gene regulation. Many human hemoglobinopathies result from inadequate expression of globin genes, and attempts to modulate globin gene expression are a fundamental approach to seek novel avenues to therapy. This article summarizes broad aspects of the evolution of hemoglobins and related globins, starting with the five basic types of globins found in humans, and then progressively focuses more specifically on hemoglobin gene clusters in vertebrates and eutherian mammals. Finally, the impact of these and related studies on hemoglobin gene regulation is discussed.

HEMOGLOBINS WITHIN THE FAMILY OF GLOBIN HEME PROTEINS

Hemoglobins were originally discovered as abundant proteins in red blood cells of mammals and other jawed vertebrates (gnathostomes) that

Copyright © 2013 Cold Spring Harbor Laboratory Press; all rights reserved
Cite this article as *Cold Spring Harb Perspect Med* doi: 10.1101/cshperspect.a011627

bind and release oxygen reversibly. The major hemoglobin in adult humans, hemoglobin A, is a heterotetramer composed of two α-globin and two β-globin polypeptides, each with an associated heme group. These are encoded by the duplicated *HBA1* and *HBA2* genes and by the *HBB* gene, respectively (Fig. 1). Hemoglobins are produced only in erythroid cells, where they are the major protein. The multi-subunit hemoglobin binds oxygen cooperatively in the lungs, transports the oxygen through the blood, and releases it in peripheral tissues to support oxidative metabolism. The heme molecule contains an iron atom (Fe) in the reduced state (ferrous or +2 oxidation state), and five of the six coordination sites of the Fe are occupied, four by the porphyrin ring and one by the "proximal" histidine (His) of the surrounding globin polypeptide. The sixth coordination site is bound reversibly by oxygen. Oxygen is loaded onto hemoglobin, transported, and unloaded with no chemical (covalent) change to either the oxygen or the heme groups, and the iron in the heme group stays in the reduced state. This contrasts with the familiar role of many heme proteins, such as cytochromes, which catalyze redox reactions involving changes in the oxidation state of the heme iron.

Myoglobin (encoded by the *MB* gene) is a related, monomeric heme-bound globin protein found predominantly in skeletal and heart muscle. The Fe in the heme is pentacoordinate. It has long been described as an oxygen storage protein, and it facilitates diffusion of oxygen to the mitochondria (Wittenberg and Wittenberg 1987).

Additional heme-containing globins were discovered by mining the wealth of information in the sequence of the human genome and transcripts produced from it. Cytoglobin, encoded by the *CYGB* gene (Fig. 1), is found in many tissues (Burmester et al. 2002; Trent and Hargrove 2002), in sharp contrast to the stringently tissue-specific expression pattern of hemoglobin and myoglobin genes. The most distantly related globin found in the human genome is neuroglobin, encoded by *NGB* (Burmester et al. 2000). Its mRNA is abundant in brain tissue but also is present in many other tissues. It is related to invertebrate nerve globins, indicating that an

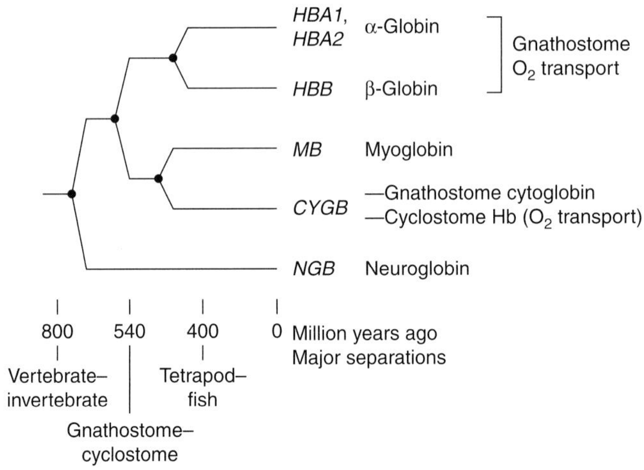

Figure 1. Model for evolution of vertebrate globin genes. The deduced times of duplication and divergence are shown along the horizontal axis, and contemporary human globin genes are shown at the *right*. Major duplication events in globin gene evolution are noted along the tree, and time of origin of some major animal groups is indicated along the horizontal axis. For simplicity, only the gene tree is shown. For more information on the inferred times of these duplications relative to speciation events, refer to figures in the primary references (Burmester et al. 2002; Opazo et al. 2008a,b; Hoffmann et al. 2010). Times of major separations used in creation of the figure are taken from TimeTree (Hedges et al. 2006).

ancestral gene was present before the divergence of vertebrates and invertebrates more than 800 million years ago (Fig. 1). In contrast to the pentacoordinate heme complex in hemoglobins and MB, heme forms a hexacoordinate complex with both NGB and CYGB, having two His residues, termed proximal and distal, coordinated with the Fe. Ligands such as oxygen and nitric oxide compete with the distal His for binding, but despite this, NGB and CYBG still have high affinity for the ligands. Both these hexacoordinate heme globins have been implicated in nitric oxide metabolism, with CYGB showing nitric oxide dioxygenase activity, converting nitric oxide to nitrate (Oleksiewicz et al. 2011), and NGB showing nitrite reductase activity to form nitric oxide (Tiso et al. 2011). The latter activity also has been shown for myoglobin (Hendgen-Cotta et al. 2008) and deoxy-hemoglobin (Gladwin and Kim-Shapiro 2008). Physiologically, the nitrite reductase activity could provide a means to produce nitric oxide under hypoxic conditions, signaling from which could regulate mitochondrial respiration and protect tissues (nerves by NGB, heart muscle by MB) from damage under ischemic conditions (Dietz 2011). A role for CYGB in oxygen-requiring reactions, such as hydroxylation, has not been ruled out (Fago et al. 2004). These proposed enzymatic roles in nitric oxide and other metabolism may harken back to functions performed by ancestral hemoglobins in primordial life (Hardison 1998, 1999; Tiso et al. 2011).

The five types of globin genes listed in Figure 1 are located on five different chromosomes: *HBA1* and *HBA2* at chromosomal position 16p13.3, *HBB* at 11p15.4, *MB* at 22q12.3, *CYGB* at 17q25.1, and *NGB* at 14q24.3. *MB*, *CYGB*, and *NGB* are present as single-copy genes, whereas *HBB* and *HBA*s are in clusters with multiple related genes. All of the genes consist of at least three exons separated by two introns. Although the introns differ dramatically in size, they are in homologous locations. The *CYGB* and *NGB* genes each have an additional exon. The conservation of intron position in vertebrate globin genes has been proposed to facilitate the shuffling of exons during protein evolution (Gilbert 1978). However, intron positions differ considerably in globin genes outside vertebrates, suggesting that the conservation of intron position could simply reflect an ancestral state that has not changed over vertebrate evolution (Hardison 1998).

Species in an early diverging branch of vertebrates, the cyclostomes (represented by hagfish and lampreys), also use a heme-containing globin for oxygen transport, but surprisingly, it is more closely related to CYGB than to the gnathostome hemoglobins (Fig. 1) (Hoffmann et al. 2010). This suggests that the oxygen transport function of heme-containing globins arose by independent, convergent evolution in the two major branches of vertebrates. For cyclostomes, it appears that oxygen transport is derived by cooption of the widely expressed *CYGB* gene.

HEMOGLOBIN GENE CLUSTERS IN JAWED VERTEBRATES

Developmental Regulation of Expression within α-Globin and β-Globin Gene Clusters

In all jawed vertebrates, erythrocytes produced at distinct developmental stages contain different forms of hemoglobin. All species examined make embryonic-specific hemoglobins in primitive erythroid cells derived from the yolk sac, some species make a fetal-specific form in the liver, and all species produce an "adult" hemoglobin in erythroid cells produced in the bone marrow (Maniatis et al. 1980; Karlsson and Nienhuis 1985). Like the major adult hemoglobin A, each of these is a heterotetramer of two α-like globins and two β-like globins, each bound by heme. The α-like globins are paralogous, meaning that they are homologous genes generated by gene duplication. ζ-globin is made in embryonic red cells, and α-globin is produced in fetal and adult red cells (Fig. 2) (Higgs et al. 2005). Likewise, the paralogous β-like globin genes are also expressed at progressive stages of gestation. In humans, ε-globin is made in embryonic erythrocytes, γ-globins are produced in fetal liver erythroid cells, and the δ- and β-globins are made in erythroid cells from adult bone marrow (Grosveld et al. 1993). The hemoglobins

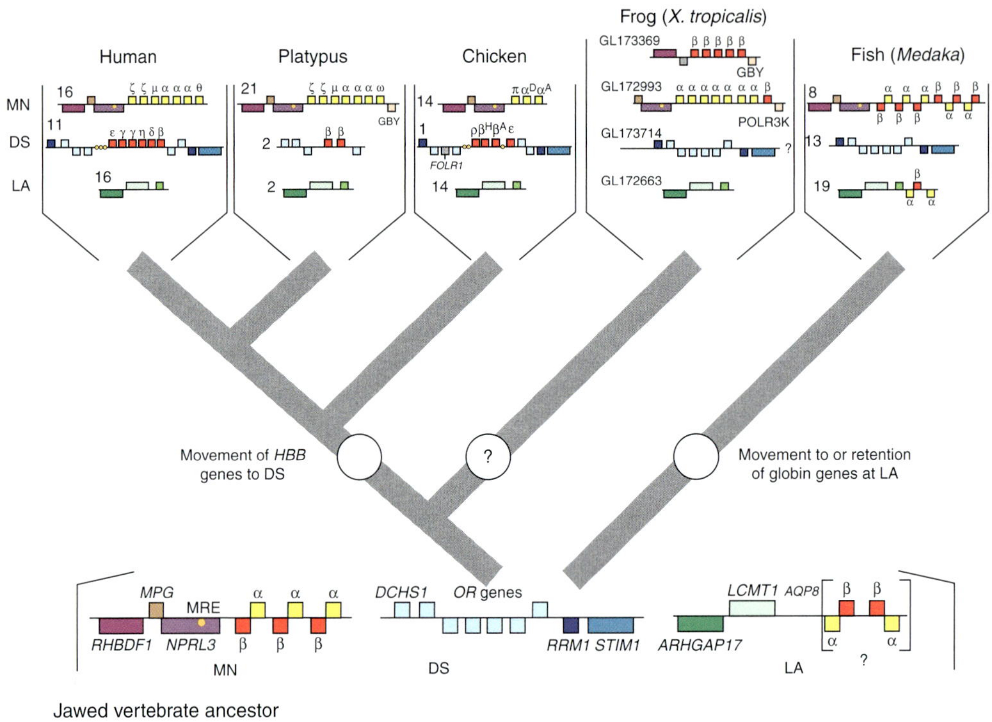

Figure 2. Models for evolution of hemoglobin gene complexes in jawed vertebrates. Gene clusters in contemporary species are diagrammed at the *top* of the figure, and inferred gene arrangements in the last common ancestor to jawed vertebrates are diagrammed at the *bottom*. Genes in brackets with a question mark could either be present in the LCA and lost in one or more descendant lineages, or they could be absent in the LCA but acquired via transposition in some descendant lineages. (Thick gray lines) The major bifurcations during evolution of the lineages exemplified by species at the *top*. Maps of the gene clusters were derived from a combination of viewing annotations in the UCSC Genome Browser (Kent et al. 2002) of assembled genomes of human (Lander et al. 2001), platypus (Warren et al. 2008), chicken (Hillier et al. 2004), frog *Xenopus tropicalis* (Hellsten et al. 2010), and fish *Medaka* (Kasahara et al. 2007), and from recent publications (Fuchs et al. 2006; Opazo et al. 2008b; Patel et al. 2008). The genes are represented by boxes, with those above the line transcribed from left to right and those below the line in the opposite orientation. (Red) β-like globin genes; (yellow) α-like globin genes; (light blue) *OR* genes; others have colors distinctive for each locus (e.g., shades of green for nonglobin genes at the LA locus). (Small orange circles) Major regulatory regions. The gene maps are not complete, nor are they to scale; the genes shown here were chosen to illustrate logic for the proposed ancestral arrangements and transpositions in two clades. The number to the left of each cluster specifies the chromosome on which it is located; for the frog gene clusters, the scaffold identifier is given. The Greek letter name is specified for hemoglobin genes in human, platypus, and chicken, but generic "α-globin" or "β-globin" is used for frog and fish because the genes are less well characterized. (This diagram is adapted from Hardison 2008.)

produced at distinct developmental stages have different affinities for oxygen and are subject to complex regulation by cofactors, favoring an overall movement of oxygen from the maternal bloodstream to that of the fetus or embryo.

The multiple, developmentally regulated genes in the gnathostome α-globin gene clusters are derived from a common ancestral gene cluster (Flint et al. 2001). However, the β-like globin genes in mammals are more similar to each other than they are to the multiple β-like globin genes in birds (Hardison and Miller 1993; Reitman et al. 1993). This implies that the β-like globin gene clusters were generated by independent gene duplications in the bird and mammal lineages. Because differential regulation during development is a consistent property of these independently derived gene clusters, either an ancestral developmental regulatory mechanism was enforced on the newly duplicated genes or the mechanism evolved by convergence. Regulatory mechanisms are complex, and as is discussed in the last section, the current mechanisms are combinations of conserved and acquired features.

Coordinated Regulation between α-Globin and β-Globin Gene Clusters

Expression of α-like and β-like globin genes must be strictly coordinated. A balanced production of α-globin and β-globin in erythroid cells is required for the efficient formation of hemoglobin, and an imbalance leads to the pathological phenotypes of inherited anemias called thalassemias (Weatherall and Clegg 2001). The separation of α-like and β-like globin gene clusters in amniotes requires coordination of expression between different chromosomes.

Fish species show an interesting contrast, in that the gene cluster orthologous (homologous genes generated by a speciation event) to that of the mammalian α-globin gene cluster contains both α-like and β-like globin genes (Fig. 2). Some of the genes in the larger globin gene cluster in fish are expressed in larvae, and others are expressed in adults (Chan et al. 1997). Thus within this fish globin gene cluster, genes are regulated coordinately (balancing α-globin and β-globin synthesis) and differentially during development (larval vs. adult).

Evolution of Multiple Globin Gene Clusters in Vertebrates

As just discussed, differential expression of paralogous globin genes within a cluster and coordinated regulation between gene clusters on different chromosomes are consistent properties in amniotes (birds and mammals). The unfolding story regarding how this arose during vertebrate evolution is dynamic and complex. Analysis of the maps of globin genes and surrounding genes in contemporary vertebrate species suggests a model featuring movement to new locations or differential retention of globin genes but still leading to multiple hemoglobin gene clusters in most if not all vertebrates examined (Fig. 2).

Genes diagnostic for a particular genomic region can be found flanking hemoglobin gene clusters (Bulger et al. 1999; Flint et al. 2001; Gillemans et al. 2003). The diagram in Figure 2 focuses on single-copy, flanking diagnostic genes for clarity (Hardison 2008). One globin gene cluster is found in all gnathostomes examined; it is flanked on one side by the genes *MPG* and *NPRL3* (Flint et al. 2001), and the locus can be called "MN," the acronym for these two genes (Fig. 2). The major DNA region regulating expression of the globin genes (MRE) is located in an intron of *NPRL3* (Higgs et al. 1990). Frequently, the gene *RHBDF1* is adjacent to the *MPG* gene. In contrast to placental mammals and chickens, which have only α-like globin genes at the *MN* locus, the orthologous loci in the monotreme platypus and in marsupials have a set of α-like globin genes plus a globin gene related to β-globin, the ω-globin gene (Wheeler et al. 2004; Patel et al. 2008). In addition, the platypus *MN* locus contains a homolog to the globin Y gene (*GBY*), a globin discovered in amphibians. Direct molecular cloning from the genome of the frog *Xenopus laevis* (Jeffreys et al. 1980) and examination of the genome assembly of *Xenopus tropicalis* (Fuchs et al. 2006; Hellsten et al. 2010) reveal a different β-globin gene linked to several α-globin genes at the *MN* locus. Given the presence of globin genes at this locus in all gnathostomes examined, one can infer with considerable confidence that the *MN* locus contained globin genes in the last common ancestor (LCA) of vertebrates (Fig. 2).

A second locus contains α- and β-globin genes in the pufferfish *Fugu rubripes* (Gillemans et al. 2003), and examination of the genome assemblies of zebrafish and *Medaka* shows a similar arrangement (Fig. 2). The globin genes in this locus are flanked by the genes *LCMT1* and *AQP8*, and the locus can be called "LA." The gene *ARHGAP17* is also part of this locus in many species. These three nonglobin genes are in the same arrangement and order in the tetrapods (human, platypus, chicken, and frog), but the *LA* locus is devoid of globin genes in these species. This suggests two different models for this locus in the LCA of jawed vertebrates. One model posits that the LCA had globin genes at the *LA* locus (Gillemans et al. 2003), and these globin genes were retained in fish but lost in tetrapods. The converse posits that the globin genes were not present at the *LA* locus of the LCA, but moved into it during the lineage to fish.

A third locus contains only β-like globin genes in amniotes. The β-like globin genes in amniotes are flanked by olfactory receptor (OR) genes (Bulger et al. 1999; Patel et al. 2008). In placental mammals, hundreds of OR genes are in this locus, along with additional multigene families such as *TRIM* genes. Thus one has to look several megabases away from the β-like globin genes to find single-copy genes that are distinctive for this locus, which are *DCHS1* on one side and *STIM1* on the other. Hence this locus can be called *DS* (Fig. 2); the *RRM1* gene is adjacent to *STIM1* in many species. The presence of β-like globin genes in the *DS* locus in amniotes but absence in both fish and amphibians is most easily explained by transposition of the β-like globin genes into the *DS* locus in the stem amniote (Fig. 2) (Patel et al. 2008). A proposal that they were present at the *DS* in the LCA of jawed vertebrates also requires independent deletions in the fish and amphibian lineages; thus, parsimony favors the transposition model. One possible source for the β-like globin genes could be the *MN* locus (Patel et al. 2008), but it also could be from the *LA* locus (Hardison 2008).

No globin genes have been mapped to the *LA* or *DS* loci in the current assembly of *X. tropicalis*, but one contig covers a cluster of β-like globin genes linked to *RHBDF1* (Fig. 2). Further work is needed to ascertain whether this cluster is linked to the *MN* locus (Fuchs et al. 2006) or if they are on different chromosomes.

In summary, the history of the gene clusters encoding hemoglobins is dynamic and complex. The *MN* locus now contains only α-globin genes in eutherians; it retained these and non-globin flanking genes since the gnathostome LCA, while losing β-globin genes in many vertebrate lineages. β-like globin genes were acquired at the *DS* locus in the stem amniote, and subsequently they duplicated and acquired differential developmental expression independently in the avian and mammalian lineages. The *LA* locus has undergone dramatic losses or gains of globin genes.

Relative Stability of the α-Like Globin Gene Cluster

The consistent location of α-like globin genes in the *MN* locus in gnathostomes indicates a more stable history than that of the β-like globin genes. This greater stability is also seen in the composition and expression patterns of the α-like globin genes. Extensive phylogenetic comparisons indicate that this gene cluster in the LCA of tetrapods contained orthologs to ζ-globin, μ-globin (also called α^D), and α-globin (or α^A) genes (Hoffmann and Storz 2007; Hoffmann et al. 2008), and this arrangement is still seen in chickens (Fig. 3). Before the divergence of the three major subclasses of mammals (monotremes, marsupials, and placentals), both the ζ-globin and α-globin genes duplicated. Most contemporary mammals retain at least two copies of these genes (in some cases, they are pseudogenes). The θ-globin gene appears to have been generated by a duplication of an α-globin gene after the divergence of monotremes from the other mammals (Hoffmann et al. 2008). The μ-globin and θ-globin genes each are present in only a single copy, and although they are transcribed, no evidence has been found for polypeptide products of either in mammals (Clegg 1987; Hsu et al. 1988; Leung et al. 1989; Goh et al. 2005; Cooper et al. 2006). The ortholog of the μ-globin gene is expressed in adult erythroid cells in birds, producing α^D-globin.

The expression timing of the genes encoding α-like globins is remarkably consistent. In all species examined, including birds, the active orthologs to the ζ-globin gene are expressed in embryonic erythroid cells, and the orthologs of the α-globin gene are expressed in fetal and adult erythroid cells (Fig. 3) (Higgs et al. 1989; Whitelaw et al. 1990).

The α-like globin gene cluster *does* show some dynamic features. Genes are lost and

Cite this article as *Cold Spring Harb Perspect Med* doi: 10.1101/cshperspect.a011627

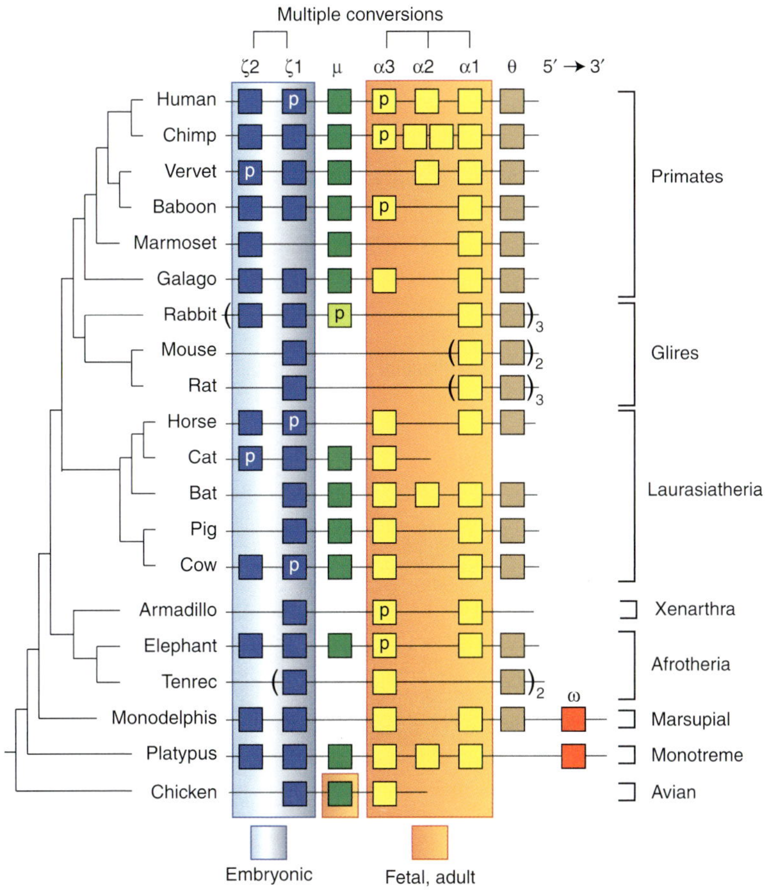

Figure 3. Maps of orthologous α-like globin genes and expression timing in amniotes. Each gene is shown as a rectangle, colored by the orthology relationship to the human genes, labeled by the Greek letters at the *top*. The timing of expression is indicated by distinctive background shading as indicated in the legend. The letter "p" denotes a pseudogene. Sizes of and spacing between genes are not to scale. The gene clusters or parts of them are duplicated or triplicated in rabbits, mouse, rat, and tenrec, indicated by the parentheses and subscripts. The ω-globin gene encodes a β-like globin that has been identified in marsupials and monotremes. The assignments of orthologous relationships are based on grouping within phylogenetic comparisons of coding sequences and flanking regions (Hardison and Gelinas 1986; Cheng et al. 1987; Hoffmann et al. 2008) and automated determination of orthologs using a method that recognizes gene conversions (Song et al. 2012). The "junction" sequence of the rabbit *HBA* cluster, associated with recombination breakpoints, is assigned to the μ position in this diagram, but it contains only a remnant of a globin gene.

gained in specific lineages (Hoffmann et al. 2008), and some of the genes have undergone multiple conversion events during mammalian evolution (Hess et al. 1984; Song et al. 2011, 2012). However, the genomic context, that is, the *MN* locus, has been a constant across gnathostome evolution, and the expression patterns of the ζ- and α-globin genes are strikingly consistent in amniotes.

Lineage-Specific Gains and Losses of β-Like Globin Genes

Within the three major subclasses of mammals, the β-like globin genes at the DS locus have been duplicated and lost in specific lineages (Fig. 4). Both monotremes and marsupials have two β-like globin genes. In marsupials the ε-globin ortholog is expressed in embryonic erythrocytes,

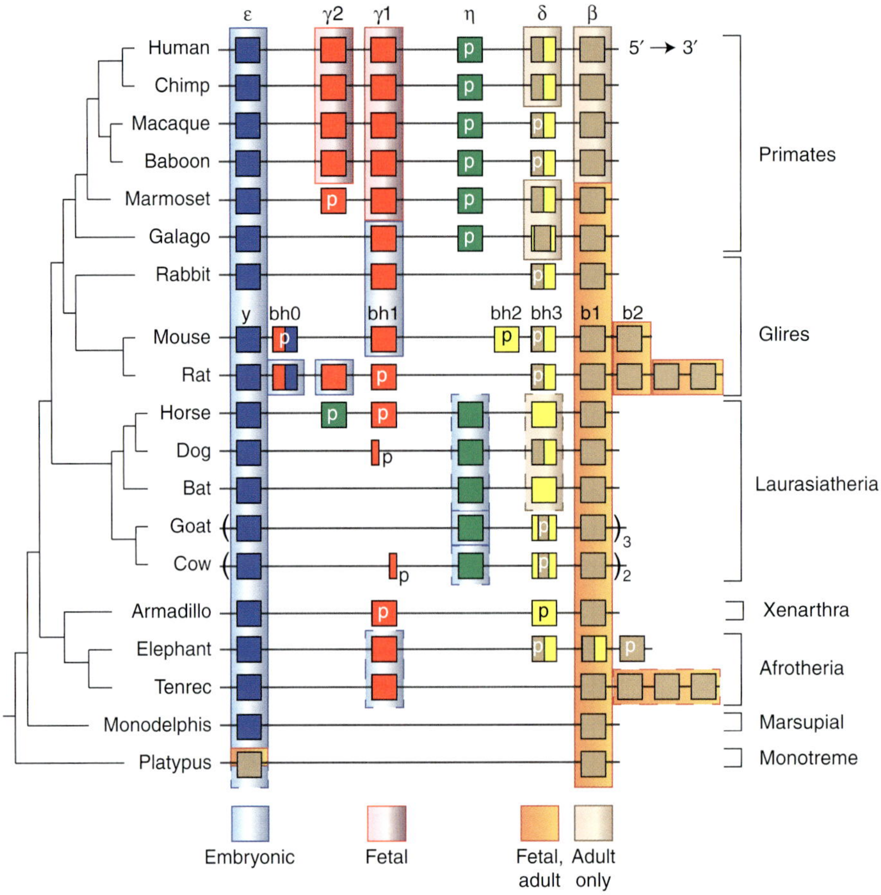

Figure 4. Maps of orthologous β-like globin genes and expression timing in mammals. The symbols and backgrounds are similar to those in Figure 3. When the timing of expression is predicted rather than experimentally determined (or highly likely, as in the case of embryonic expression of ε-globin gene orthologs), the background shading is outlined with a dashed line. The gene clusters are triplicated in goats and duplicated in cows, indicated by the parentheses and subscripts. The orthology and expression assignments for these are based largely on the first copy of the segmental duplication; the three β-globin orthologs are expressed in juvenile, adult, and fetal goats (Townes et al. 1984). The assignments of orthologous relationships are based on grouping within phylogenetic comparisons of coding sequences (Koop and Goodman 1988; Opazo et al. 2008a,b, 2009; Patel et al. 2008) and automated determination of orthologs using a method that recognizes gene conversions (Song et al. 2012). The timing of expression is based on multiple reports in the literature (Stockell et al. 1961; LeCrone 1970; Efstratiadis et al. 1980; Rohrbaugh and Hardison 1983; Shapiro et al. 1983; Townes et al. 1984; Schimenti and Duncan 1985; Koop and Goodman 1988; Tagle et al. 1988; Whitelaw et al. 1990; Johnson et al. 1996, 2000; Satoh et al. 1999; Patel et al. 2008).

whereas the β-globin ortholog is expressed in fetal and adult erythroid cells (Koop and Goodman 1988). The two β-globin genes in platypus are more similar to each other than to other β-like globin genes, consistent with either a gene conversion event (Patel et al. 2008) or a gene duplication independent of the one that established therian ε-globin and β-globin genes (Opazo et al. 2008b). Both the β-globin genes in platypus are expressed in adults (Patel et al. 2008), but no information is available currently on whether the leftmost β-globin gene in platypus (in the orientation in Fig. 4) is also expressed in embryonic erythroid cells, as is expected based on its position.

A proposed cluster of five β-like globin genes, in the orientation 5′-ε-γ-η-δ-β-3′, in the stem eutherian is consistent with the gene arrangements in contemporary species (Goodman et al. 1984; Hardies et al. 1984; Hardison 1984). The relative similarities among orthologous genes indicate that this gene cluster was formed by a series of duplications, first to make the ancestor to β- and δ-globin genes and the ancestor to ε-, γ-, and η-globin genes, followed by duplications to generate the proposed five-gene cluster (Hardison and Miller 1993). The initial duplication established two major lineages of β-like globin genes that differ in their positions in the gene clusters and in their timing of expression (Fig. 4). Genes in the β- and δ-globin lineage are located to the right in the gene clusters and, if active, are expressed in fetal and/or adult erythroid cells. Genes in the ε-, γ-, and η-globin gene lineage are toward the left in the gene clusters and are expressed in embryonic erythroid cells, except for the γ-globin genes in anthropoid primates, which were coopted for fetal-specific expression.

The full set of five β-like globin genes is not used in any extant mammal examined. At least one pseudogene is found in this gene cluster for almost every eutherian species (Fig. 4), and any exceptions to this could reflect a lack of detailed characterization of the genes. Pseudogenes are DNA segments with sequences homologous to those of actively expressed globin genes, but they harbor mutations, such as frameshifts or chain terminators, that preclude expression to form a globin protein.

Deletion can completely inactivate a gene, and gene loss has also occurred widely in the β-like globin gene clusters of eutherians. Some gene losses tend to be consistent across the members of each eutherian order (Fig. 4). No ortholog for the η-globin gene is found in the species sampled from the order Glires (rodents and lagomorphs), but it is present in the sister clades Primates and Laurasiatherians (represented by dog, horse, bat, goat, and cow). This strongly suggests that the η-globin gene was lost in the LCA for Glires (Opazo et al. 2008a). The η-globin gene is also absent from sampled members of the superorders Xenarthra or Afrotheria, which can be explained either by gene loss in the LCA, or perhaps the duplication to form η-globin occurred after these superorders diverged from the other eutherians. No active γ-globin gene has been identified in Laurasiatherians, with the gene either being absent, partially deleted, or harboring inactivating mutations. Note that the loss of the γ-globin gene was not in the stem Laurasiatherian, but rather different losses and inactivations have occurred in the lineages to each species.

All species examined within the therians (marsupial and placental mammals) have an ortholog of the ε-globin gene. This gene has the most consistent features across species of any of the paralogous β-like globin genes. It is always present at the left end of the gene cluster, it is almost always a single gene, and in all species examined, it is expressed only in embryonic erythroid cells (Fig. 4).

The γ-globin genes of both the prosimian primate galago and species in order Glires (rabbit, mouse, and rat) are expressed in embryonic erythroid cells (Rohrbaugh and Hardison 1983; Whitelaw et al. 1990; Satoh et al. 1999), whereas the γ-globin genes of anthropoid primates (monkeys, apes, and humans) are expressed only in fetal erythroid cells (Fig. 4) (Johnson et al. 1996, 2000). One common interpretation is that the embryonic expression pattern was ancestral, and the recruitment to fetal expression was an adaptation in the anthropoids, coinciding with a duplication of the γ-globin gene (Johnson et al. 1996). γ-globin genes are also present in Afrotherians, but the developmental timing of their expression has not been reported.

The η-globin gene homolog in goats is expressed embryonically (Shapiro et al. 1983). Currently, this is the only example of an active η-globin gene, but studies of expression in other Laurasiatherians would reveal whether it is active in other species, and if the timing of expression is embryonic. Fetal and adult hemoglobins were found to be identical for horse (Stockell et al. 1961) and dog (LeCrone 1970), and based on the absence of evidence for a fetal-specific hemoglobin, the η-globin homologs are predicted to be expressed embryonically in Figure 4. The η-globin gene is a pseudogene in all primates.

The δ-globin gene is present in almost all eutherian species examined, but it is frequently a pseudogene (Fig. 4). In every case examined in sufficient detail, the δ-globin gene has been involved in a gene conversion, with sequence from the paralogous β-globin gene copied into the δ-globin gene locus (Spritz et al. 1980; Martin et al. 1983; Hardies et al. 1984; Hardison and Margot 1984; Song et al. 2012). The boundaries of the conversions are different in each species, indicating that these are independent gene conversions. The structural and mechanistic bases for this propensity for conversion are not understood. In galago, the replacement with β-globin gene sequences extends into the promoter region, leading to high-level expression from this gene (Tagle et al. 1991). That, in turn, led to efforts to engineer a form of the δ-globin gene that would express at sufficiently high levels to provide potential therapy (Tang et al. 1997).

In most eutherian species, the β-globin gene is expressed in fetal and adult erythroid cells (Fig. 4). Concomitantly with the recruitment of γ-globin genes to fetal expression in anthropoid primates, the onset of expression of the β-globin gene was delayed to shortly before birth in catarrhine primates (Old World monkeys, apes, and humans) (Johnson et al. 2000). The onset of β-globin gene expression is earlier in fetal life in the New World monkeys (Johnson et al. 1996), perhaps representing a transitional state intermediate between the fetal onset seen in most eutherians and the prenatal onset observed in humans.

This overview of the evolution of β-globin genes illustrates the diversity of events that have been inferred, including duplications, deletions, inactivations, and reactivations. It shows that the ε-globin genes have been stable over eutherian evolution, whereas the γ-, η-, and δ-globin genes have been gained and lost frequently, sometimes in entire orders of mammals. Furthermore, the timing of expression can change dramatically between clades, notably the delay in γ-globin (fetal) and β-globin (adult) gene expression in anthropoid primates. Strategies being pursued to reactivate γ-globin gene expression in adult erythroid cells, either pharmacologically or by gene therapy, in a sense are attempts to modulate expression patterns in humans that recapitulate expression changes that have occurred during eutherian evolution.

INSIGHTS INTO GENE REGULATION

The previous discussion shows that expression of globin genes is tightly regulated. Hemoglobin gene expression is restricted to erythroid cells. The genes are expressed at extremely high levels late in erythroid differentiation, with balanced production of α-globin and β-globin. Paralogous globin genes are expressed at progressive developmental stages. This exquisite regulation is exerted, at least in part, by the binding of specific transcription factors to DNA sequences that serve as *cis*-regulatory modules (CRMs), such as promoters and enhancers (Maniatis et al. 1987).

Detailed studies over the past three decades have led to the discovery of numerous CRMs in both the α-globin gene (*HBA*) and β-globin gene (*HBB*) clusters (Fig. 5). Some are located proximal to and within the genes, such as promoters and internal enhancers (Mellon et al. 1981; Wright et al. 1984; Myers et al. 1986; Antoniou et al. 1988; Wall et al. 1988), and others are located distal to the genes (Grosveld et al. 1987; Talbot et al. 1989; Higgs et al. 1990). For instance, the major regulatory element (MRE) of the *HBA* gene complex is located distal to the adult *HBA* genes (~60 kb upstream in human), residing in an intron of the large *NPRL3* gene (Fig. 5A). Several additional CRMs are present around the MRE (Anguita et al. 2004) both in human and mouse (Fig. 5A,B). A cluster of CRMs called the locus control region (LCR) is found 50–70 kb upstream of the *HBB* gene (Grosveld et al. 1987; Talbot et al. 1989; Moon and Ley 1990) in human and mouse (Fig. 5C,D). These distal regulatory regions are enhancers (Tuan et al. 1989; Ney et al. 1990; Pondel et al. 1992) required for high-level expression of the globin genes (Grosveld et al. 1987; Talbot et al. 1989; Higgs et al. 1990; Bender et al.

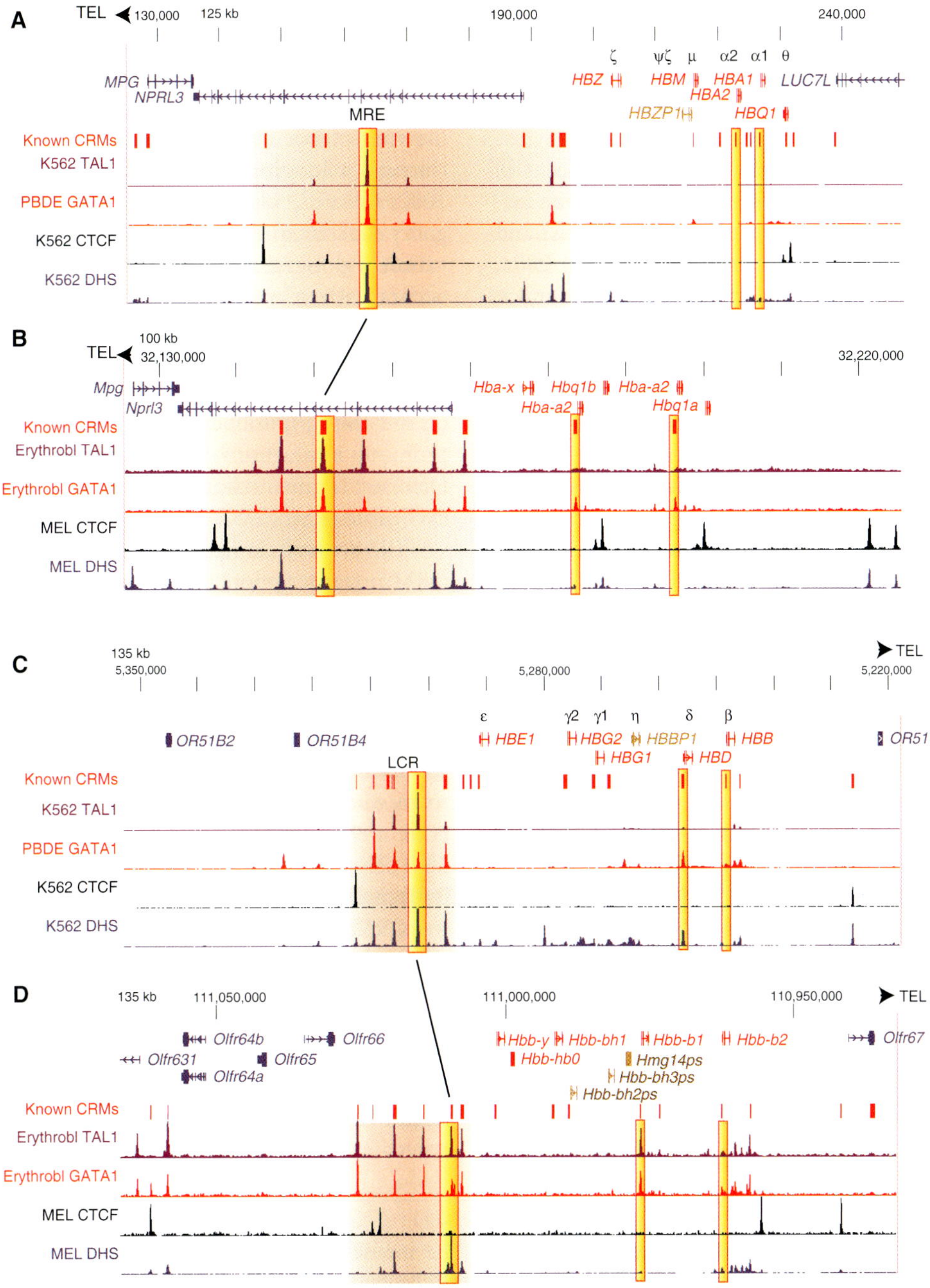

Figure 5. Genes and regulatory modules in the human and mouse hemoglobin gene clusters. Each panel shows chromosomal positions (human genome assembly GRCh37/hg19 and mouse genome assembly NCBI37/mm9), the direction to the telomere (TEL), genes (from UCSC genes unless otherwise indicated) with rectangles for exons and lines with the direction of transcription shown as arrowheads, known *cis*-regulatory modules (CRMs), and signals for occupancy of TAL1, GATA1, and CTCF based on genome-wide ChIP-seq and the signal for DNase hypersensitive sites (DHSs) based on genome-wide DNase-seq. Strong enhancers (MRE in the *HBA* clusters and HS2 in the *HBB* clusters) and adult globin gene promoters are in gold, and groups of distal CRMs (such as the LCR) are in tan. (*See following page for legend.*)

reveals formation of chromatin globules. *Nat Struct Mol Biol* **18:** 107–114.

Bender MA, Bulger M, Close J, Groudine M. 2000a. β-*globin* gene switching and DNase I sensitivity of the endogenous β-*globin* locus in mice do not require the locus control region. *Mol Cell* **5:** 387–393.

Bender MA, Mehaffey MG, Telling A, Hug B, Ley TJ, Groudine M, Fiering S. 2000b. Independent formation of DnaseI hypersensitive sites in the murine β-globin locus control region. *Blood* **95:** 3600–3604.

Bender MA, Roach JN, Halow J, Close J, Alami R, Bouhassira EE, Groudine M, Fiering SN. 2001. Targeted deletion of 5′HS1 and 5′HS4 of the β-globin locus control region reveals additive activity of the DNaseI hypersensitive sites. *Blood* **98:** 2022–2027.

Bourque G, Leong B, Vega VB, Chen X, Lee YL, Srinivasan KG, Chew JL, Ruan Y, Wei CL, Ng HH, et al. 2008. Evolution of the mammalian transcription factor binding repertoire via transposable elements. *Genome Res* **18:** 1752–1762.

Bulger M, von Doorninck JH, Saitoh N, Telling A, Farrell C, Bender MA, Felsenfeld G, Axel R, Groudine M. 1999. Conservation of sequence and structure flanking the mouse and human β-globin loci: The β-globin genes are embedded within an array of odorant receptor genes. *Proc Natl Acad Sci* **96:** 5129–5134.

Bulger M, Schubeler D, Bender MA, Hamilton J, Farrell CM, Hardison RC, Groudine M. 2003. A complex chromatin "landscape" revealed by patterns of nuclease sensitivity and histone modification within the mouse β-globin locus. *Mol Cell Biol* **23:** 5234–5244.

Bungert J, Dave U, Lim K-C, Kieuw KH, Shavit JA, Liu Q, Engel JD. 1995. Synergistic regulation of human β-globin gene switching by locus control region elements HS3 and HS4. *Genes Dev* **9:** 3083–3096.

Burmester T, Weich B, Reinhardt S, Hankeln T. 2000. A vertebrate globin expressed in the brain. *Nature* **407:** 520–523.

Burmester T, Ebner B, Weich B, Hankeln T. 2002. Cytoglobin: A novel globin type ubiquitously expressed in vertebrate tissues. *Mol Biol Evol* **19:** 416–421.

Carter D, Chakalova L, Osborne CS, Dai YF, Fraser P. 2002. Long-range chromatin regulatory interactions in vivo. *Nat Genet* **32:** 623–626.

Caterina JJ, Ryan TM, Pawlik KM, Palmiter RD, Brinster RL, Behringer RR, Townes TM. 1991. Human β-globin locus control region: Analysis of the 5′ DNaseI hypersensitive site HS2 in transgenic mice. *Proc Natl Acad Sci* **88:** 1626–1630.

Chan F, Robinson J, Brownlie A, Shivdasani R, Donovan A, Brugnara C, Kim J, Lau B, Witkowska H, Zon L. 1997. Characterization of adult α- and β-globin genes in the zebrafish. *Blood* **89:** 688–700.

Cheng J-F, Raid L, Hardison RC. 1987. Block duplications of α ζ–ζ–α–τ gene set in the rabbit α-like globin gene cluster. *J Biol Chem* **262:** 5414–5421.

Cheng Y, Wu W, Kumar SA, Yu D, Deng W, Tripic T, King DC, Chen KB, Zhang Y, Drautz D, et al. 2009. Erythroid GATA1 function revealed by genome-wide analysis of transcription factor occupancy, histone modifications, and mRNA expression. *Genome Res* **19:** 2172–2184.

Clegg JB. 1987. Can the product of the θ gene be a real globin? *Nature* **329:** 465–466.

Cooper SJ, Wheeler D, De Leo A, Cheng JF, Holland RA, Marshall Graves JA, Hope RM. 2006. The mammalian α^{D}-globin gene lineage and a new model for the molecular evolution of α-globin gene clusters at the stem of the mammalian radiation. *Mol Phylogenet Evol* **38:** 439–448.

de Laat W, Grosveld F. 2003. Spatial organization of gene expression: The active chromatin hub. *Chromosome Res* **11:** 447–459.

Dietz GP. 2011. Protection by neuroglobin and cell-penetrating peptide-mediated delivery in vivo: A decade of research. Comment on Cai et al.: TAT-mediated delivery of neuroglobin protects against focal cerebral ischemia in mice. Exp Neurol. 2011; 227(1): 224–31. *Exp Neurol* **231:** 1–10.

Dostie J, Richmond TA, Arnaout RA, Selzer RR, Lee WL, Honan TA, Rubio ED, Krumm A, Lamb J, Nusbaum C, et al. 2006. Chromosome Conformation Capture Carbon Copy (5C): A massively parallel solution for mapping interactions between genomic elements. *Genome Res* **16:** 1299–1309.

Efstratiadis A, Posakony JW, Maniatis T, Lawn RM, O'Connell C, Spritz RA, DeRiel JK, Forget BG, Weissman SM, Slightom JL, et al. 1980. The structure and evolution of the human β-globin gene family. *Cell* **21:** 653–668.

Elnitski L, Miller W, Hardison R. 1997. Conserved E boxes function as part of the enhancer in hypersensitive site 2 of the β-globin locus control region: Role of basic helix–loop–helix proteins. *J Biol Chem* **272:** 369–378.

ENCODE Project Consortium, Myers RM, Stamatoyannopoulos J, Snyder M, Dunham I, Hardison RC, Bernstein BE, Gingeras TR, Kent WJ, Birney E, et al. 2011. A user's guide to the Encyclopedia of DNA Elements (ENCODE). *PLoS Biol* **9:** e1001046.

ENCODE Project Consortium. 2012. An integrated encyclopedia of DNA elements in the human genome. *Nature* doi: 10.1038/nature11247.

Fago A, Hundahl C, Dewilde S, Gilany K, Moens L, Weber RE. 2004. Allosteric regulation and temperature dependence of oxygen binding in human neuroglobin and cytoglobin. Molecular mechanisms and physiological significance. *J Biol Chem* **279:** 44417–44426.

Flint J, Tufarelli C, Peden J, Clark K, Daniels RJ, Hardison R, Miller W, Philipsen S, Tan-Un KC, McMorrow T, et al. 2001. Comparative genome analysis delimits a chromosomal domain and identifies key regulatory elements in the α globin cluster. *Hum Mol Genet* **10:** 371–382.

Forrester WC, Thompson C, Elder JT, Groudine M. 1986. A developmentally stable chromatin structure in the human β-globin gene cluster. *Proc Natl Acad Sci* **83:** 1359–1363.

Forrester WC, Epner E, Driscoll MC, Enver T, Brice M, Papayannopoulou T, Groudine M. 1990. A deletion of the human β-globin locus activation region causes a major alteration in chromatin structure and replication across the entire β-globin locus. *Genes Dev* **4:** 1637–1649.

Fuchs C, Burmester T, Hankeln T. 2006. The amphibian globin gene repertoire as revealed by the *Xenopus* genome. *Cytogenet Genome Res* **112:** 296–306.

Fujiwara T, O'Geen H, Keles S, Blahnik K, Linnemann AK, Kang YA, Choi K, Farnham PJ, Bresnick EH. 2009. Discovering hematopoietic mechanisms through genome-wide analysis of GATA factor chromatin occupancy. *Mol Cell* **36:** 667–681.

Gilbert W. 1978. Why genes in pieces? *Nature* **271:** 501.

Gillemans N, McMorrow T, Tewari R, Wai AW, Burgtorf C, Drabek D, Ventress N, Langeveld A, Higgs D, Tan-Un K, et al. 2003. A functional and comparative analysis of globin loci in pufferfish and man. *Blood* **101:** 2842–2849.

Gladwin MT, Kim-Shapiro DB. 2008. The functional nitrite reductase activity of the heme-globins. *Blood* **112:** 2636–2647.

Goh SH, Lee YT, Bhanu NV, Cam MC, Desper R, Martin BM, Moharram R, Gherman RB, Miller JL. 2005. A newly discovered human α-globin gene. *Blood* **106:** 1466–1472.

Goodman M, Koop BF, Czelusniak J, Weiss ML. 1984. The η-globin gene. Its long evolutionary history in the β-globin gene family of mammals. *J Mol Biol* **180:** 803–823.

Gourdon G, Sharpe JA, Higgs DR, Wood WG. 1995. The mouse α-globin locus regulatory element. *Blood* **86:** 766–775.

Grass JA, Jing H, Kim SI, Martowicz ML, Pal S, Blobel GA, Bresnick EH. 2006. Distinct functions of dispersed GATA factor complexes at an endogenous gene locus. *Mol Cell Biol* **26:** 7056–7067.

Grosveld F, van Assendelft GB, Greaves D, Kollias G. 1987. Position-independent, high-level expression of the human β-globin gene in transgenic mice. *Cell* **51:** 975–985.

Grosveld F, Antoniou M, Berry M, de Boer E, Dillon N, Ellis J, Fraser P, Hanscombe O, Hurst J, Imam A, et al. 1993. The regulation of human globin gene switching. *Philos Trans R Soc Lond* **339:** 183–191.

Gumucio DL, Heilstedt-Williamson H, Gray TA, Tarle SA, Shelton DA, Tagle D, Slightom J, Goodman M, Collins FS. 1992. Phylogenetic footprinting reveals a nuclear protein which binds to silencer sequences in the human γ and ε globin genes. *Mol Cell Biol* **12:** 4919–4929.

Gumucio D, Shelton D, Zhu W, Millinoff D, Gray T, Bock J, Slightom J, Goodman M. 1996. Evolutionary strategies for the elucidation of *cis* and *trans* factors that regulate the developmental switching programs of the β-like globin genes. *Mol Phylog Evol* **5:** 18–32.

Hardies SC, Edgell MH, Hutchison CA III. 1984. Evolution of the mammalian β-globin gene cluster. *J Biol Chem* **259:** 3748–3756.

Hardison RC. 1984. Comparison of the β-like globin gene families of rabbits and humans indicates that the gene cluster 5′ ε-γ-δ-β 3′ predates the mammalian radiation. *Mol Biol Evol* **1:** 390–410.

Hardison R. 1998. Hemoglobins from bacteria to man: Evolution of different patterns of gene expression. *J Exp Biol* **201:** 1099–1117.

Hardison R. 1999. The evolution of hemoglobin: Studies of a very ancient protein suggest that changes in gene regulation are an important part of the evolutionary story. *Am Sci* **87:** 126–137.

Hardison RC. 2000. Conserved noncoding sequences are reliable guides to regulatory elements. *Trends Genet* **16:** 369–372.

Hardison RC. 2008. Globin genes on the move. *J Biol* **7:** 35.

Hardison RC, Gelinas R. 1986. Assignment of orthologous relationships among mammalian α-globin genes by examining flanking regions reveals a rapid rate of evolution. *Mol Biol Evol* **3:** 243–261.

Hardison RC, Margot JB. 1984. Rabbit globin pseudogene pseudoβ2 is a hybrid of δ- and β-globin gene sequences. *Mol Biol Evol* **1:** 302–316.

Hardison R, Miller W. 1993. Use of long sequence alignments to study the evolution and regulation of mammalian globin gene clusters. *Mol Biol Evol* **10:** 73–102.

Hardison RC, Taylor J. 2012. Genomic approaches towards finding *cis*-regulatory modules in animals. *Nat Rev Genet* **13:** 469–483.

Hardison R, Xu J, Jackson J, Mansberger J, Selifonova O, Grotch B, Biesecker J, Petrykowska H, Miller W. 1993. Comparative analysis of the locus control region of the rabbit β-like globin gene cluster: HS3 increases transient expression of an embryonic ε-globin gene. *Nucl Acids Res* **21:** 1265–1272.

Hardison R, Oeltjen J, Miller W. 1997a. Long human–mouse sequence alignments reveal novel regulatory elements: A reason to sequence the mouse genome. *Genome Res* **7:** 959–966.

Hardison R, Slightom JL, Gumucio DL, Goodman M, Stojanovic N, Miller W. 1997b. Locus control regions of mammalian β-globin gene clusters: Combining phylogenetic analyses and experimental results to gain functional insights. *Gene* **205:** 73–94.

Hedges SB, Dudley J, Kumar S. 2006. TimeTree: A public knowledge-base of divergence times among organisms. *Bioinformatics* **22:** 2971–2972.

Hellsten U, Harland RM, Gilchrist MJ, Hendrix D, Jurka J, Kapitonov V, Ovcharenko I, Putnam NH, Shu S, Taher L, et al. 2010. The genome of the Western clawed frog *Xenopus tropicalis*. *Science* **328:** 633–636.

Hendgen-Cotta UB, Merx MW, Shiva S, Schmitz J, Becher S, Klare JP, Steinhoff HJ, Goedecke A, Schrader J, Gladwin MT, et al. 2008. Nitrite reductase activity of myoglobin regulates respiration and cellular viability in myocardial ischemia-reperfusion injury. *Proc Natl Acad Sci* **105:** 10256–10261.

Hess J, Schmid C, Shen C. 1984. A gradient of sequence divergence in the human adult α-globin duplication units. *Science* **226:** 67–70.

Higgs DR, Vickers MA, Wilkie AO, Pretorius IM, Jarman AP, Weatherall DJ. 1989. A review of the molecular genetics of the human α-globin gene cluster. *Blood* **73:** 1081–1104.

Higgs D, Wood W, Jarman A, Sharpe J, Lida J, Pretorious IM, Ayyub H. 1990. A major positive regulatory region located far upstream of the human α-globin gene locus. *Genes Dev* **4:** 1588–1601.

Higgs DR, Garrick D, Anguita E, De Gobbi M, Hughes J, Muers M, Vernimmen D, Lower K, Law M, Argentaro A, et al. 2005. Understanding α-globin gene regulation: Aiming to improve the management of thalassemia. *Ann NY Acad Sci* **1054:** 92–102.

Hillier LW, Miller W, Birney E, Warren W, Hardison RC, Ponting CP, Bork P, Burt DW, Groenen MA, Delany ME, et al. 2004. Sequence and comparative analysis of the chicken genome provide unique perspectives on vertebrate evolution. *Nature* **432:** 695–716.

Hoffmann FG, Storz JF. 2007. The α^D-globin gene originated via duplication of an embryonic α-like globin gene in the ancestor of tetrapod vertebrates. *Mol Biol Evol* **24:** 1982–1990.

Hoffmann FG, Opazo JC, Storz JF. 2008. Rapid rates of lineage-specific gene duplication and deletion in the α-globin gene family. *Mol Biol Evol* **25:** 591–602.

Hoffmann FG, Opazo JC, Storz JF. 2010. Gene cooption and convergent evolution of oxygen transport hemoglobins in jawed and jawless vertebrates. *Proc Natl Acad Sci* **107:** 14274–14279.

Hsu S, Marks J, Shaw J, Tam M, Higgs D, Shen C, Shen C. 1988. Structure and expression of the human θ1 globin gene. *Nature* **331:** 94–96.

Hughes JR, Cheng JF, Ventress N, Prabhakar S, Clark K, Anguita E, De Gobbi M, de Jong P, Rubin E, Higgs DR. 2005. Annotation of *cis*-regulatory elements by identification, subclassification, and functional assessment of multispecies conserved sequences. *Proc Natl Acad Sci* **102:** 9830–9835.

Jackson JD, Petrykowska H, Philipsen S, Miller W, Hardison R. 1996. Role of DNA sequences outside the cores of DNase hypersensitive sites (HSs) in functions of the β-globin locus control region: Domain opening and synergism between HS2 and HS3. *J Biol Chem* **271:** 11871–11878.

Jeffreys AJ, Wilson V, Wood D, Simons JP, Kay RM, Williams JG. 1980. Linkage of adult α- and β-globin genes in *X. laevis* and gene duplication by tetraploidization. *Cell* **21:** 555–564.

Johnson RM, Buck S, Chiu C, Schneider H, Sampaio I, Gage DA, Shen TL, Schneider MP, Muniz JA, Gumucio DL, et al. 1996. Fetal globin expression in New World monkeys. *J Biol Chem* **271:** 14684–14691.

Johnson RM, Buck S, Chiu CH, Gage DA, Shen TL, Hendrickx AG, Gumucio DL, Goodman M. 2000. Humans and Old World monkeys have similar patterns of fetal globin expression. *J Exp Zool* **288:** 318–326.

Johnson KD, Grass JA, Boyer ME, Keikhaefer CM, Blobel GA, Weiss MJ, Bresnick EH. 2002. Cooperative activities of hematopoietic regulators recruit RNA polymerase II to a tissue-specific chromatin domain. *Proc Natl Acad Sci* **99:** 11760–11765.

Karlsson S, Nienhuis AW. 1985. Developmental regulation of human globin genes. *Annu Rev Biochem* **54:** 1071–1108.

Kasahara M, Naruse K, Sasaki S, Nakatani Y, Qu W, Ahsan B, Yamada T, Nagayasu Y, Doi K, Kasai Y, et al. 2007. The medaka draft genome and insights into vertebrate genome evolution. *Nature* **447:** 714–719.

Kassouf MT, Hughes JR, Taylor S, McGowan SJ, Soneji S, Green AL, Vyas P, Porcher C. 2010. Genome-wide identification of TAL1's functional targets: Insights into its mechanisms of action in primary erythroid cells. *Genome Res* **20:** 1064–1083.

Kent WJ, Sugnet CW, Furey TS, Roskin KM, Pringle TH, Zahler AM, Haussler D. 2002. The human genome browser at UCSC. *Genome Res* **12:** 996–1006.

King DC, Taylor J, Elnitski L, Chiaromonte F, Miller W, Hardison RC. 2005. Evaluation of regulatory potential and conservation scores for detecting *cis*-regulatory modules in aligned mammalian genome sequences. *Genome Res* **15:** 1051–1060.

King DC, Taylor J, Zhang Y, Cheng Y, Lawson HA, Martin J, ENCODE Groups for Transcriptional Regulation and Multispecies Sequence Analysis, Chiaromonte F, Miller W, Hardison RC. 2007. Finding *cis*-regulatory elements using comparative genomics: Some lessons from ENCODE data. *Genome Res* **17:** 775–786.

Koop B, Goodman M. 1988. Evolutionary and developmental aspects of two hemoglobin β-chain genes (εM and βM) of opossum. *Proc Natl Acad Sci* **85:** 3893–3897.

Lander ES, Linton LM, Birren B, Nusbaum C, Zody MC, Baldwin J, Devon K, Dewar K, Doyle M, FitzHugh W, et al. 2001. Initial sequencing and analysis of the human genome. *Nature* **409:** 860–921.

LeCrone CN. 1970. Absence of special fetal hemoglobin in beagle dogs. *Blood* **35:** 451–452.

Leung S, Whitelaw E, Proudfoot N. 1989. Transcriptional and translational analysis of the human θ globin gene. *Nucleic Acids Res* **17:** 8283–8300.

Maniatis T, Fritsch EF, Lauer J, Lawn RM. 1980. The molecular genetics of human hemoglobins. *Annu Rev Genet* **14:** 145–178.

Maniatis T, Goodbourn S, Fischer JA. 1987. Regulation of inducible and tissue-specific gene expression. *Science* **236:** 1237–1245.

Martin SL, Vincent KA, Wilson AC. 1983. Rise and fall of the δ globin gene. *J Mol Biol* **164:** 513–528.

Mellon P, Parker V, Gluzman Y, Maniatis T. 1981. Identification of DNA sequences required for transcription of the human α1-globin gene in a new SV40 host-vector system. *Cell* **27:** 279–288.

Milot E, Strouboulis J, Trimborn T, Wijgerde M, de Boer E, Langeveld A, Tan-Un K, Vergeer W, Yannoutsos N, Grosveld F, et al. 1996. Heterochromatin effects on the frequency and duration of LCR-mediated gene transcription. *Cell* **87:** 105–114.

Molete JM, Petrykowska H, Bouhassira EE, Feng YQ, Miller W, Hardison RC. 2001. Sequences flanking hypersensitive sites of the β-globin locus control region are required for synergistic enhancement. *Mol Cell Biol* **21:** 2969–2980.

Moon AM, Ley TJ. 1990. Conservation of the primary structure, organization, and function of the human and mouse β-globin locus-activating regions. *Proc Natl Acad Sci* **87:** 7693–7697.

Mouse ENCODE Project Consortium. 2012. An encyclopedia of mouse DNA elements (Mouse ENCODE). *Genome Biol* **13:** 418.

Myers RM, Tilly K, Maniatis T. 1986. Fine structure genetic analysis of a β-globin promoter. *Science* **232:** 613–618.

Ney P, Sorrentino B, McDonagh K, Nienhuis A. 1990. Tandem AP-1-binding sites within the human β-globin dominant control region function as an inducible enhancer in erythroid cells. *Genes Dev* **4:** 993–1006.

Oleksiewicz U, Liloglou T, Field JK, Xinarianos G. 2011. Cytoglobin: Biochemical, functional and clinical perspective of the newest member of the globin family. *Cell Mol Life Sci* **68:** 3869–3883.

Opazo JC, Hoffmann FG, Storz JF. 2008a. Differential loss of embryonic globin genes during the radiation of placental mammals. *Proc Natl Acad Sci* **105:** 12950–12955.

Opazo JC, Hoffmann FG, Storz JF. 2008b. Genomic evidence for independent origins of β-like globin genes in monotremes and therian mammals. *Proc Natl Acad Sci* **105:** 1590–1595.

Opazo JC, Sloan AM, Campbell KL, Storz JF. 2009. Origin and ascendancy of a chimeric fusion gene: The β/δ-globin gene of paenungulate mammals. *Mol Biol Evol* **26:** 1469–1478.

Osborne CS, Chakalova L, Brown KE, Carter D, Horton A, Debrand E, Goyenechea B, Mitchell JA, Lopes S, Reik W, et al. 2004. Active genes dynamically colocalize to shared sites of ongoing transcription. *Nat Genet* **36:** 1065–1071.

Palstra RJ, Tolhuis B, Splinter E, Nijmeijer R, Grosveld F, de Laat W. 2003. The β-globin nuclear compartment in development and erythroid differentiation. *Nat Genet* **35:** 190–194.

Patel VS, Cooper SJ, Deakin JE, Fulton B, Graves T, Warren WC, Wilson RK, Graves JA. 2008. Platypus globin genes and flanking loci suggest a new insertional model for β-globin evolution in birds and mammals. *BMC Biol* **6:** 34.

Pennacchio LA, Ahituv N, Moses AM, Prabhakar S, Nobrega MA, Shoukry M, Minovitsky S, Dubchak I, Holt A, Lewis KD, et al. 2006. In vivo enhancer analysis of human conserved non-coding sequences. *Nature* **444:** 499–502.

Pondel MD, George M, Proudfoot NJ. 1992. The LCR-like α-globin positive regulatory element functions as an enhancer in transiently transfected cells during erythroid differentiation. *Nucleic Acids Res* **20:** 237–243.

Ragoczy T, Bender MA, Telling A, Byron R, Groudine M. 2006. The locus control region is required for association of the murine β-globin locus with engaged transcription factories during erythroid maturation. *Genes Dev* **20:** 1447–1457.

Reitman M, Grasso JA, Blumentahl R, Lewit P. 1993. Primary sequence, evolution and repetitive elements of the *G. gallus* (chicken) β-globin cluster. *Genomics* **18:** 616–626.

Rohrbaugh ML, Hardison RC. 1983. Analysis of rabbit β-like globin gene transcripts during development. *J Mol Biol* **164:** 395–417.

Sankaran VG, Xu J, Ragoczy T, Ippolito GC, Walkley CR, Maika SD, Fujiwara Y, Ito M, Groudine M, Bender MA, et al. 2009. Developmental and species-divergent globin switching are driven by BCL11A. *Nature* **460:** 1093–1097.

Satoh H, Inokuchi N, Nagae Y, Okazaki T. 1999. Organization, structure, and evolution of the nonadult rat β-globin gene cluster. *J Mol Evol* **49:** 122–129.

Schimenti JC, Duncan CH. 1985. Structure and organization of the bovine β-globin genes. *Mol Biol Evol* **2:** 514–525.

Schoenfelder S, Sexton T, Chakalova L, Cope NF, Horton A, Andrews S, Kurukuti S, Mitchell JA, Umlauf D, Dimitrova DS, et al. 2010. Preferential associations between co-regulated genes reveal a transcriptional interactome in erythroid cells. *Nat Genet* **42:** 53–61.

Schubeler D, Francastel C, Cimbora DM, Reik A, Martin DI, Groudine M. 2000. Nuclear localization and histone acetylation: A pathway for chromatin opening and transcriptional activation of the human β-globin locus. *Genes Dev* **14:** 940–950.

Shapiro SG, Schon EA, Townes TM, Lingrel JB. 1983. Sequence and linkage of the goat ε^{I} and ε^{II} β-globin genes. *J Mol Biol* **169:** 31–52.

Soler E, Andrieu-Soler C, de Boer E, Bryne JC, Thongjuea S, Stadhouders R, Palstra RJ, Stevens M, Kockx C, van Ijcken W, et al. 2010. The genome-wide dynamics of the binding of Ldb1 complexes during erythroid differentiation. *Genes Dev* **24:** 277–289.

Song G, Hsu CH, Riemer C, Zhang Y, Kim HL, Hoffmann F, Zhang L, Hardison RC, Green ED, Miller W. 2011. Conversion events in gene clusters. *BMC Evol Biol* **11:** 226.

Song G, Riemer C, Dickins B, Kim HL, Zhang L, Zhang Y, Hsu CH, Hardison RC, NISC Comparative Sequencing Program, Green ED, et al. 2012. Revealing mammalian evolutionary relationships by comparative analysis of gene clusters. *Genome Biol Evol* **4:** 586–601.

Spritz R, DeRiel J, Forget B, Weissman S. 1980. Complete nucleotide sequence of the human δ-globin gene. *Cell* **21:** 639–646.

Stockell A, Perutz MF, Muirhead H, Glauser SC. 1961. A comparison of adult and foetal horse haemoglobins. *J Mol Biol* **3:** 112–116.

Tagle DA, Koop BF, Goodman M, Slightom J, Hess DL, Jones RT. 1988. Embryonic ε and γ globin genes of a prosimian primate (*Galago crassicaudatus*): Nucleotide and amino acid sequences, developmental regulation and phylogenetic footprints. *J Mol Biol* **203:** 7469–7480.

Tagle D, Slightom J, Jones R, Goodman M. 1991. Concerted evolution led to high expression of a prosimian primate δ globin gene locus. *J Biol Chem* **266:** 7469–7480.

Talbot D, Collis P, Antoniou M, Vidal M, Grosveld F, Greaves DR. 1989. A dominant control region from the human β-globin locus conferring integration site-independent gene expression. *Nature* **338:** 352–355.

Tang DC, Ebb D, Hardison RC, Rodgers GP. 1997. Restoration of the CCAAT box and insertion of the CACCC motif activate δ-globin gene expression. *Blood* **90:** 421–427.

Tiso M, Tejero J, Basu S, Azarov I, Wang X, Simplaceanu V, Frizzell S, Jayaraman T, Geary L, Shapiro C, et al. 2011. Human neuroglobin functions as a redox-regulated nitrite reductase. *J Biol Chem* **286:** 18277–18289.

Tolhuis B, Palstra RJ, Splinter E, Grosveld F, de Laat W. 2002. Looping and interaction between hypersensitive sites in the active β-globin locus. *Mol Cell* **10:** 1453–1465.

Townes TM, Fitzgerald MC, Lingrel JB. 1984. Triplication of a four-gene set during evolution of the goat β-globin locus produced three genes now expressed differentially during development. *Proc Natl Acad Sci* **81:** 6589–6593.

Trent JT III, Hargrove MS. 2002. A ubiquitously expressed human hexacoordinate hemoglobin. *J Biol Chem* **277:** 19538–19545.

Tuan D, Solomon W, London I, Lee D. 1989. An erythroid-specific, developmental-stage-independent enhancer far upstream of the human "β-like globin" genes. *Proc Natl Acad Sci* **86:** 2554–2558.

Vakoc CR, Letting DL, Gheldof N, Sawado T, Bender MA, Groudine M, Weiss MJ, Dekker J, Blobel GA. 2005. Proximity among distant regulatory elements at the β-globin locus requires GATA-1 and FOG-1. *Mol Cell* **17:** 453–462.

Vernimmen D, Marques-Kranc F, Sharpe JA, Sloane-Stanley JA, Wood WG, Wallace HA, Smith AJ, Higgs DR. 2009. Chromosome looping at the human α-globin locus is mediated via the major upstream regulatory element (HS-40). *Blood* **114:** 4253–4260.

Vyas P, Vickers MA, Simmons DL, Ayyub H, Craddock CF, Higgs DR. 1992. *Cis*-acting sequences regulating expression of the human α-globin cluster lie within constitutively open chromatin. *Cell* **69:** 781–793.

Wall L, deBoer E, Grosveld F. 1988. The human β-globin gene 3′ enhancer contains multiple binding sites for an erythroid-specific protein. *Genes Dev* **2:** 1089–1100.

Wallace HA, Marques-Kranc F, Richardson M, Luna-Crespo F, Sharpe JA, Hughes J, Wood WG, Higgs DR, Smith AJ. 2007. Manipulating the mouse genome to engineer precise functional syntenic replacements with human sequence. *Cell* **128:** 197–209.

Warren WC, Hillier LW, Marshall Graves JA, Birney E, Ponting CP, Grutzner F, Belov K, Miller W, Clarke L, Chinwalla AT, et al. 2008. Genome analysis of the platypus reveals unique signatures of evolution. *Nature* **453:** 175–183.

Weatherall DJ, Clegg JB. 2001. *Thalassemia syndromes.* Blackwell Science, Oxford.

Wheeler D, Hope RM, Cooper SJ, Gooley AA, Holland RA. 2004. Linkage of the β-like ω-globin gene to α-like globin genes in an Australian marsupial supports the chromosome duplication model for separation of globin gene clusters. *J Mol Evol* **58:** 642–652.

Whitelaw E, Tsai SF, Hogben P, Orkin SH. 1990. Regulated expression of globin chains and the erythroid transcription factor GATA-1 during erythropoiesis in the developing mouse. *Mol Cell Biol* **10:** 6596–6606.

Wijgerde M, Grosveld F, Fraser P. 1995. Transcription complex stability and chromatin dynamics in vivo. *Nature* **377:** 209–213.

Wilson NK, Foster SD, Wang X, Knezevic K, Schutte J, Kaimakis P, Chilarska PM, Kinston S, Ouwehand WH, Dzierzak E, et al. 2010. Combinatorial transcriptional control in blood stem/progenitor cells: Genome-wide analysis of ten major transcriptional regulators. *Cell Stem Cell* **7:** 532–544.

Wittenberg BA, Wittenberg JB. 1987. Myoglobin-mediated oxygen delivery to mitochondria of isolated cardiac myocytes. *Proc Natl Acad Sci* **84:** 7503–7507.

Wright S, Rosenthal A, Flavell R, Grosveld F. 1984. DNA sequences required for regulated expression of β-globin genes in murine erythroleukemia cells. *Cell* **38:** 265–273.

Wu W, Cheng Y, Keller CA, Ernst J, Kumar SA, Mishra T, Morrissey C, Dorman CM, Chen KB, Drautz D, et al. 2011. Dynamics of the epigenetic landscape during erythroid differentiation after GATA1 restoration. *Genome Res* **21:** 1659–1671.

Yu M, Riva L, Xie H, Schindler Y, Moran TB, Cheng Y, Yu D, Hardison R, Weiss MJ, Orkin SH, et al. 2009. Insights into GATA-1-mediated gene activation versus repression via genome-wide chromatin occupancy analysis. *Mol Cell* **36:** 682–695.

Zhang Y, Wu W, Cheng Y, King DC, Harris RS, Taylor J, Chiaromonte F, Hardison RC. 2009. Primary sequence and epigenetic determinants of in vivo occupancy of genomic DNA by GATA1. *Nucleic Acids Res* **37:** 7024–7038.

Zhou GL, Xin L, Song W, Di LJ, Liu G, Wu XS, Liu DP, Liang CC. 2006. Active chromatin hub of the mouse α-globin locus forms in a transcription factory of clustered housekeeping genes. *Mol Cell Biol* **26:** 5096–5105.

Transcriptional Mechanisms Underlying Hemoglobin Synthesis

Koichi R. Katsumura, Andrew W. DeVilbiss, Nathaniel J. Pope, Kirby D. Johnson, and Emery H. Bresnick

Department of Cell and Regenerative Biology, UW-Madison Blood Research Program, Wisconsin Institute for Medical Research, Carbone Cancer Center, University of Wisconsin School of Medicine and Public Health, Madison, Wisconsin 53705

Correspondence: ehbresni@wisc.edu

The physiological switch in expression of the embryonic, fetal, and adult β-like globin genes has garnered enormous attention from investigators interested in transcriptional mechanisms and the molecular basis of hemoglobinopathies. These efforts have led to the discovery of cell type-specific transcription factors, unprecedented mechanisms of transcriptional coregulator function, genome biology principles, unique contributions of nuclear organization to transcription and cell function, and promising therapeutic targets. Given the vast literature accrued on this topic, this article will focus on the master regulator of erythroid cell development and function GATA-1, its associated proteins, and its frontline role in controlling hemoglobin synthesis. GATA-1 is a crucial regulator of genes encoding hemoglobin subunits and heme biosynthetic enzymes. GATA-1-dependent mechanisms constitute an essential regulatory core that nucleates additional mechanisms to achieve the physiological control of hemoglobin synthesis.

The transcriptional switch from embryonic to fetal, and subsequently to adult globin gene transcription constitutes an elegant mechanism to meet the unique developmental requirements for oxygenation of cells and tissues during gestation (Blau and Stamatoyannopoulos 1994; Orkin and Higgs 2010; Sankaran and Orkin 2013). Mechanistic studies on this problem have a dual motivation. Investigators studying transcriptional mechanisms in erythroid cells have accrued an impressive track record with respect to forging principles of gene and chromosome regulation that have broad relevance to diverse systems. Furthermore, prevalent hemoglobin-associated pathologies (sickle cell anemia, additional anemias, thalassemias, and developmental disorders) constitute a major public health problem worldwide (Weatherall 2010). Elevating expression of the developmentally silenced γ-globin gene can supplant mutant or inadequate levels of β-globin in human disease states, thereby suppressing the associated symptoms (Rodgers et al. 1989; Bunn 1999; Atweh and Schechter 2001; Orkin and Higgs 2010). Thus, dissecting mechanisms controlling hemoglobin biosynthesis in exquisite detail is expected to yield novel molecular strategies for the treatment of hemoglobinopathies.

Copyright © 2013 Cold Spring Harbor Laboratory Press; all rights reserved
Cite this article as *Cold Spring Harb Perspect Med* doi: 10.1101/cshperspect.a015412

Other articles in this collection have highlighted the molecular biology of genes encoding hemoglobin subunits (Hardison 2012), hemoglobinopathies (Nienhuis and Nathan 2012; Higgs 2013), and human genetic studies that revealed an extremely important new regulator of hemoglobin switching, the transcriptional coregulator BCL11A (Lettre 2012; Sankaran and Orkin 2013). Undoubtedly, BCL11A is a crucial component of the hemoglobin switching machinery, and major efforts are under way to elucidate how BCL11A functions in physiological and pathological contexts. As transcriptional coregulators almost invariably function in large heteromeric protein complexes, one would expect a host of established and undiscovered BCL11A interactors (either direct or indirect) to have integral roles in hemoglobin switching. Independent of BCL11A, additional factors (e.g., the master regulator of erythrocyte development and function GATA-1) (Evans and Felsenfeld 1989; Tsai et al. 1989; Yamamoto et al. 1990; Zon et al. 1990), regulate hemoglobin synthesis, and initial studies imply potentially important functional links to BCL11A-dependent mechanisms.

Whereas the exciting BCL11A work (Menzel et al. 2007a; Lettre et al. 2008; Sankaran et al. 2008, 2009, 2010b; Uda et al. 2008; Xu et al. 2010) has inarguably energized this extremely important field, more than a decade of studies have highlighted the essential role of GATA-1 as a critical determinant of hemoglobin synthesis. GATA-1 directly regulates expression of genes encoding hemoglobin subunits and heme biosynthetic enzymes (Johnson et al. 2002; Cheng et al. 2009; Fujiwara et al. 2009; Yu et al. 2009). Elucidating mechanisms underlying GATA-1 function/regulation is essential to achieve a holistic perspective of hemoglobin synthesis, which will naturally lead to rational approaches to therapeutically modulate hemoglobin switching. The transcriptional control of hemoglobin synthesis has multiple regulatory layers, all contributing essential functions. GATA-1 establishes the erythroid-specific chromatin structure of the α- and β-like globin gene clusters (Stamatoyannopoulos et al. 1995; Pomerantz et al. 1998; Goodwin et al. 2001; Letting et al. 2003; Anguita et al. 2004; Im et al. 2005; Kim et al. 2007; Fujiwara et al. 2009; Wu et al. 2011) and participates in various mechanistic steps to ensure expression of the globin genes in a developmental stage-appropriate manner. Thus, GATA-1 creates a regulatory core, on which additional mechanisms (e.g., those involving BCL11A), are seamlessly integrated to yield the complete regulatory system. Without GATA-1, the active loci would likely revert into repressive chromatin structures resembling nonerythroid cells, and erythroblasts would lose the capacity to survive.

GATA FACTOR MECHANISMS AND BIOLOGY

Efforts to understand globin gene regulation led to the discovery of a simple DNA sequence, (A/T)GATA(A/G) (WGATAR = GATA motif), that commonly resides at *cis*-regulatory elements of globin promoters and enhancers (Evans et al. 1988). This discovery paved the way for the cloning of the erythroid protein that binds this sequence, the founding member of the GATA factor family, GATA-1 (Evans and Felsenfeld 1989; Tsai et al. 1989; Yamamoto et al. 1990). Impressively, GATA-1 regulates hundreds of genes (Welch et al. 2004; Fujiwara et al. 2009; Yu et al. 2009) that mediate the development and function of erythroid cells (Pevny et al. 1991, 1995; Simon et al. 1992; Fujiwara et al. 1996). The cloning of GATA-1 ushered in the discovery of related GATA factors (GATA-2-6) (Wilson et al. 1990; Yamamoto et al. 1990; Ho et al. 1991; Lee et al. 1991; Dorfman et al. 1992; Arceci et al. 1993; Ito et al. 1993; Kelley et al. 1993; Tsai et al. 1994; Morrisey et al. 1996, 1997; Kuo et al. 1997; Molkentin et al. 1997), which almost invariably differ from GATA-1 in expression patterns and functions in systems including the brain, heart, vasculature, and skin. A hallmark attribute of GATA factors is their diagnostic dual zinc finger motif (Fig. 1) (Liew et al. 2005; Gamsjaeger et al. 2007). The zinc finger closest to the carboxyl terminus of GATA-1 (carboxyl finger) mediates sequence-specific DNA binding (Martin and Orkin 1990; Ko and Engel 1993; Merika and Orkin 1993), whereas the zinc finger closest to the GATA-1 amino terminus (amino finger)

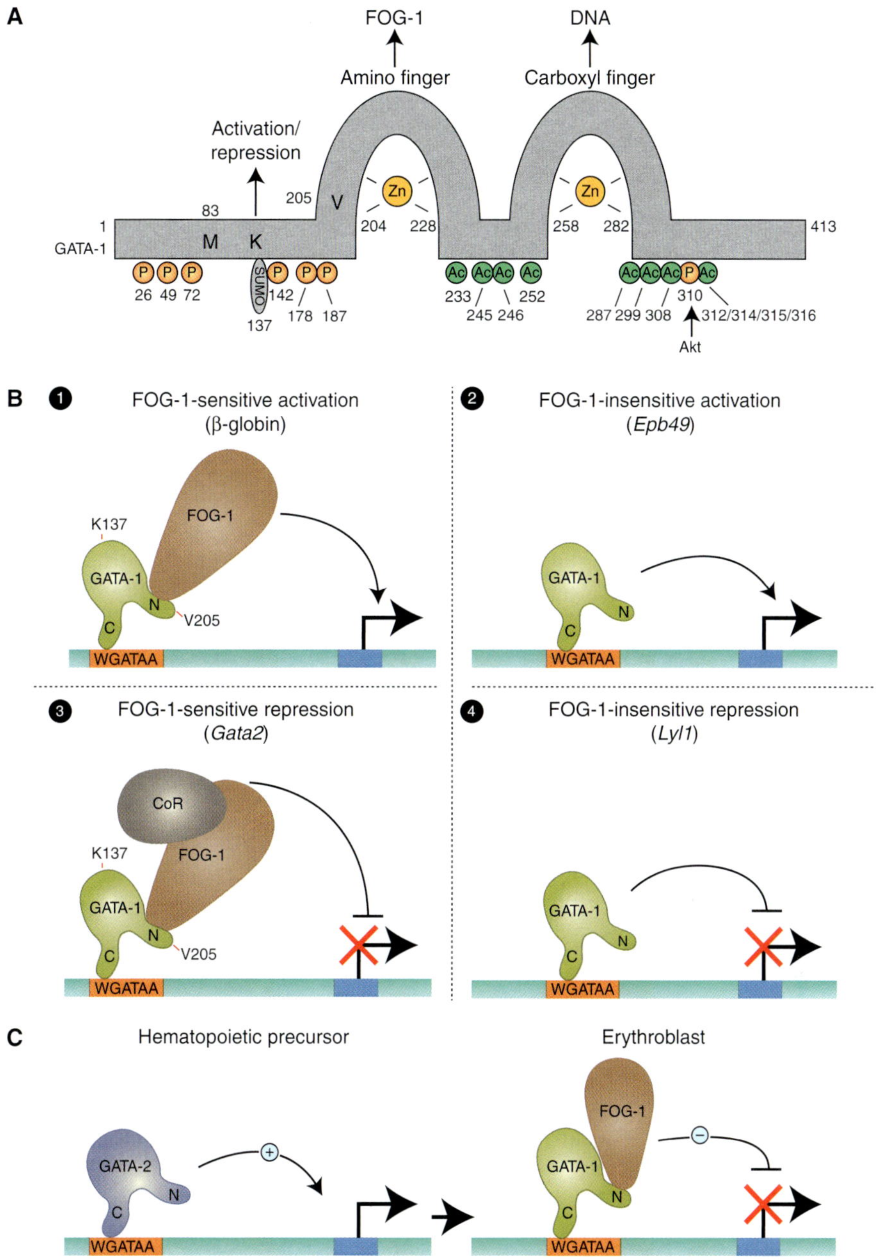

Figure 1. The molecular basis of GATA-1 action. (*A*) Diagram of GATA-1 protein organization illustrating the amino and carboxyl fingers and posttranslational modification sites. The amino finger mediates FOG-1 binding, whereas the carboxyl finger mediates sequence-specific DNA binding to GATA motifs. M83 is an alternative translation start site of a leukemogenic form of GATA-1 in acute megakaryoblastic leukemia (Wechsler et al. 2002; Crispino 2005). V205 is mutated in dyserythropoietic anemia and thrombocytopenia (Nichols et al. 2000), and facilitates FOG-1 binding (Crispino et al. 1999). Phosphorylation (P) (Crossley and Orkin 1994; Towatari et al. 2004; Kadri et al. 2005; Yu et al. 2005; Zhao et al. 2006) and acetylation (Ac) (Hung et al. 1999; Lamonica et al. 2006, 2011) sites are depicted, as well as a sumoylated site (Collavin et al. 2004; Lee et al. 2009). (*B*) Multiple modes of GATA-1 function. GATA-1 activates or represses its target genes with or without FOG-1. (*Legend continues on following page.*)

binds the cell type-restricted transcriptional coregulator Friend of GATA-1 (FOG-1) (Tsang et al. 1997, 1998; Crispino et al. 1999). The amino fingers of other GATA factors (e.g., GATA-4 and GATA-6) bind the FOG-1-related coregulator FOG-2 (Lu et al. 1999; Svensson et al. 1999; Tevosian et al. 1999, 2000). FOG-1 and FOG-2 each contain nine zinc fingers, none of which have been shown to mediate DNA binding (Gamsjaeger et al. 2007). However, four zinc fingers of FOG-1 are implicated in binding the GATA-1 amino finger (Fox et al. 1999; Cantor et al. 2002). FOG-1 interaction partners include the chromatin remodeling/modifying nucleosome remodeling and histone deacetylase (NuRD) complex (Hong et al. 2005; Gregory et al. 2010; Miccio et al. 2010), the corepressor CtBP (Holmes et al. 1999; Katz et al. 2002; Snow et al. 2010a), and the centrosomal protein transforming acidic coiled coiled 3 (TACC3) (Garriga-Canut and Orkin 2004; Simpson et al. 2004).

The intimation that GATA-1 would be a universally important regulator of erythroid gene expression was definitively established through loss-of-function studies in embryonic stem cells and mice (Pevny et al. 1991, 1995; Simon et al. 1992; Weiss et al. 1994; Weiss and Orkin 1995; Fujiwara et al. 1996) and via human disease-associated GATA-1 mutations (Nichols et al. 2000; Crispino 2005). The targeted deletion of GATA-1 yielded embryonic lethality at E10.5–11.5 and a striking disruption of erythroid cell development and maturation (Fujiwara et al. 1996). Consistent with its expression in select additional hematopoietic lineages, GATA-1 is also required for the development and maturation of megakaryocytes, eosinophils, and mast cells (Zon et al. 1991; Vyas et al. 1999; Hirasawa et al. 2002; Yu et al. 2002; Migliaccio et al. 2003; Humbles et al. 2004; Cantor et al. 2008; Sugiyama et al. 2008; Huang et al. 2009). The critical importance of FOG-1 as a GATA-1 coregulator was also established through gene-targeting studies in the mouse. FOG-1 nullizygous mouse embryos die at an embryonic stage similar to that of GATA-1 nullizygous embryos (E10.5–E11.5) and show defective erythropoiesis and megakaryopoiesis (Tsang et al. 1998). Furthermore, a V205M mutation in the GATA-1 amino finger disrupts the GATA-1–FOG-1 interaction, resulting in familial dyserythropoietic anemia and thrombocytopenia (Nichols et al. 2000).

The extraordinarily important biological functions of GATA-1 and FOG-1 provided the impetus for intense efforts to elucidate their mechanisms of action. Analogous to numerous *trans*-acting factors, GATA factors can activate and repress transcription (Kim and Bresnick 2007; Bresnick et al. 2010). Mechanistic studies to drill deeply into GATA-1-dependent transcriptional mechanisms are ongoing. FOG-1 mediates the majority of GATA-1-dependent transcriptional responses (Johnson et al. 2007; Chlon et al. 2012), and FOG-2 is used by other GATA factors. Although many questions remain unanswered regarding FOG-1 and FOG-2 mechanisms, FOG-1 has a chromatin occupancy facilitator activity (Letting et al. 2004; Pal et al. 2004) that increases GATA-1 chromatin occupancy at certain, but not all, chromatin sites and suppresses GATA-1 from accessing another cohort of sites (Johnson et al. 2006; Chlon et al. 2012). GATA-2 frequently occupies chromatin sites before GATA-1 during hematopoiesis (Grass et al. 2003, 2006; Bresnick et al. 2005, 2010, 2012; Martowicz et al. 2005; Kim and Bresnick 2007; Lugus et al. 2007; Fujiwara et al. 2009; Dore et al. 2012). When GATA-1 levels increase during erythropoiesis, it displaces GATA-2 from chromatin, including sites at the *Gata2* gene.

Figure 1. (*Continued*) The precise mode of transcriptional control is locus-specific, and representative target genes that conform to each regulatory mode are depicted (Bresnick et al. 2012). (*C*) GATA switch model. GATA switches are defined as a molecular transition in which one GATA factor replaces another from a chromatin site, which is often associated with an altered transcriptional output. In hematopoietic stem cells, select progenitors, and early-stage erythroblasts, GATA-2 occupies target loci. During erythropoiesis, poorly understood signals induce GATA-1 expression, and GATA-1 acquires the capacity to displace GATA-2 from chromatin, including sites at the *Gata2* locus, which results in transcriptional repression. FOG-1 facilitates GATA-1 occupancy at select chromatin sites and GATA switches.

This "GATA switch" is often associated with altered transcriptional outputs of the respective genes (Bresnick et al. 2010, 2012). For example, GATA switches occur at five sites of the *Gata2* locus and were initially correlated with *Gata2* repression (Grass et al. 2003, 2006; Martowicz et al. 2005). However, targeted deletions of three of these sites individually revealed qualitatively distinct activities. The −1.8 site was essential to maintain *Gata2* repression in late-stage erythroblasts (Snow et al. 2010b). The −2.8 and −1.8 sites conferred maximal *Gata2* expression in hematopoietic precursors (Snow et al. 2010b, 2011). The most interesting of the sites, the intronic +9.5 site, was essential for *Gata2* expression in the fetal liver, for fetal liver hematopoietic stem and progenitor cell activity, and for conferring vascular integrity (Johnson et al. 2012). This *cis*-element targeting study provided the first example of an exceedingly important function for a GATA switch site in vivo. Similar results were obtained using a +9.5 site-driven Cre recombinase to conditionally delete a floxed *Gata2* allele (Lim et al. 2012).

FOG-1 facilitates the GATA switch at the *Gata2* locus, required for GATA-1 to repress *Gata2* (Pal et al. 2004), which is important to ensure normal erythrocyte development and therefore normal hemoglobin synthesis. Constitutive overexpression of GATA-2 in bone marrow represses hematopoiesis (Persons et al. 1999). Analysis of a mutant mouse strain containing a knock-in FOG-1 allele defective in NuRD binding indicated that this interaction is important for activation and repression (Miccio et al. 2010). Furthermore, the FOG-1-NuRD interaction appears to be important for the GATA switch at the *Gata2* locus (Gao et al. 2010). The GATA switch mechanism is applicable to the control of numerous erythroid genes (Pal et al. 2004; Bresnick et al. 2010; Dore et al. 2012), and as discussed later in this review, this mechanism has been implicated in γ-globin transcriptional regulation (Costa et al. 2012; Zhu et al. 2012).

A limited ensemble of direct GATA-1 target genes (occupied by endogenous GATA-1 [Cheng et al. 2009; Fujiwara et al. 2009; Yu et al. 2009; Kang et al. 2012] and regulated by GATA-1 in a genetic complementation assay in GATA-1-null erythroid precursor cells [Weiss et al. 1997]) are not particularly sensitive to a GATA-1 mutation that reduces FOG-1 binding (e.g., V205G or V205M) or to reducing the FOG-1 level (e.g., in a FOG-1 nullizygous hematopoietic precursor cell line [Cantor et al. 2002], or on siRNA-mediated knockdown of FOG-1) (Crispino et al. 1999; Johnson et al. 2006, 2007; Lee et al. 2009). We have referred to these genes as "FOG-1-independent," but examples exist in which there appears to be a partial FOG-1 requirement, rather than absolute FOG-1 independence. This differs from the genes conventionally deemed to be "FOG-1-dependent," such as *Gata2* (repressed) and *β major* (activated), as GATA-1 is largely incapable of regulating these genes without FOG-1. Given the apparent spectrum of FOG-1 sensitivities, we now refer to FOG-1-sensitive and -insensitive, rather than an absolute dependence or independence. The magnitude of the gene expression changes at FOG-1-insensitive genes can be considerable, and these genes include *Lyl1* (repressed) (Johnson et al. 2007), which encodes an important transcriptional regulator of HSCs (Souroullas et al. 2009; Zohren et al. 2012) and *EPB49* (activated) (Kim et al. 2007), which encodes an important component of the red cell cytoskeleton, dematin (Rana et al. 1993; Khanna et al. 2002; Mohseni and Chishti 2008). Whereas it is likely that this mode of GATA-1 transcriptional control has functional importance, why certain target genes lack the crucial FOG-1 requirement remains a black box. Of greatest importance for this review is the FOG-1-sensitive mechanism, as GATA-1 uses FOG-1 to regulate the α- and β-like globin genes (Kim and Bresnick 2007).

Consistent with the conventional mechanism of metazoan transcriptional regulators, GATA factors control chromatin remodeling and modification. GATA-1 binds and recruits the histone/protein acetyltransferase CREB-binding protein (CBP/p300) to chromatin, inducing both GATA-1 acetylation and histone acetylation (Blobel et al. 1998; Hung et al. 1999). GATA-1 recruits the BRG1 subunit of the ATP-dependent chromatin remodeling complex SWI/SNF (Im et al. 2005; Kim et al. 2007,

2009a,b), and *Brg1* hypomorphic mutant mice show significant anemia and impaired erythropoiesis (Bultman et al. 2005). In contrast to CBP/p300 and BRG1, GATA-1 interacts with a component of the mediator complex Med1 (Crawford et al. 2002; Stumpf et al. 2006), which is not known to have chromatin remodeling and/or modifying activity. Med1 is a core component of the broadly expressed 1–2 MDa mediator complex, which consists of more than 30 subunits, the vast majority of which have no ascribed biochemical function (Conaway et al. 2005; Kornberg 2005). Med1 knockout mouse embryos are severely anemic and show reduced progenitor activity in the erythroid, but not myeloid, lineage (Crawford et al. 2002; Stumpf et al. 2006). A conditional Med1 knockout confirmed the cell-autonomous nature of the erythroid requirement for Med1 (Stumpf et al. 2010). Although GATA-1 recruits Med1 and FOG-1 similarly, Med1 and FOG-1 differ quantitatively in their contributions to GATA-1 function. Knockdown studies indicate that Med1 amplifies GATA-1 target gene activation, whereas FOG-1 contributes to a much greater extent to GATA-1 function (Pope and Bresnick 2010).

As an important manifestation of its molecular activities, GATA-1 bound to distal *cis*-regulatory elements can induce chromatin looping, thus bringing a distal site in close proximity to a promoter (Vakoc et al. 2005). GATA-1 binding to the β-globin LCR repositions the LCR such that its proximity to the *β major* promoter increases considerably (Vakoc et al. 2005). GATA-1 can also induce higher-order chromosomal transitions, in which a target locus is expelled from the nuclear periphery, freeing the locus from potential inhibitory constraints imposed by the unique subnuclear environment at the periphery (Lee et al. 2009, 2011). FOG-1 promotes both looping (Vakoc et al. 2005; Kim et al. 2007) and subnuclear relocalization of loci (Lee et al. 2009, 2011). BRG1 (Kim et al. 2009b) and the Lim domain protein Ldb1 (Song et al. 2007) are also required to establish the β-globin locus LCR-promoter loop. Tiling the chromosomal region surrounding the β-globin locus with bacterial artificial chromosome probes and conducting three-dimensional (3D) immunofluorescence in situ hybridization (immunoFISH) revealed that GATA-1 expels the β-globin locus, but not the neighboring chromosomal sequences, from the nuclear periphery (Lee et al. 2011). Both FOG-1 and K137-sumoylated GATA-1 are required for this GATA-1 activity, and it is facilitated by the erythroid Krüppel-like factor KLF1 (Miller and Bieker 1993; Perkins et al. 1995) and the NuRD complex subunit Mi2β (Lee et al. 2011). KLF1 organizes numerous target genes within nuclear neighborhoods (Drissen et al. 2004; Schoenfelder et al. 2010). Looping and subnuclear relocalization of the β-globin locus, as measured by chromosome conformation capture (3C) and 3D-immunoFISH assays, may be related or identical processes. Although it is not our intent to comprehensively review the precise details of GATA-1-dependent molecular and supramolecular transitions, many of which remain unresolved, dissecting these mechanisms in exquisite detail is expected to continue to yield pivotal insights into the control of hemoglobin synthesis and to uncover targets for therapeutic modulation of this crucial process.

EVIDENCE LINKING GATA FACTORS TO THE CONTROL OF HEMOGLOBIN SYNTHESIS

Studies on how GATA-1 regulates genes encoding hemoglobin subunits and heme biosynthetic enzymes have forged principles of GATA factor function and revealed important insights into mechanisms controlling hemoglobin synthesis. The presence of GATA motifs (WGATAR) at globin gene promoters and enhancers, as well as transient transfection studies with globin promoter and enhancer reporter genes, led to the view that GATA-1 controls hemoglobin synthesis (Evans et al. 1990; Orkin 1990). The application of ChIP technology to measure protein occupancy at endogenous loci revealed GATA-1 occupancy at loci encoding globin subunits and heme biosynthetic enzymes (Horak et al. 2002; Johnson et al. 2002; Kiekhaefer et al. 2002; Im et al. 2005; Cheng et al. 2009; Fujiwara et al. 2009; Yu et al. 2009). In a genetic complementation assay in GATA-1-null erythroid precursor cells (Weiss et al. 1997), GATA-1 occupancy of LCR sites precedes occupancy at the adult *β major* promoter (Im et al. 2005; Kim et al. 2007). GATA-1 recruits RNA polymerase II (Pol II) to the LCR and subsequently to the promoter, which occurs concomitantly with transcriptional activation (Johnson et al. 2001, 2002, 2003; Im et al. 2005; Kim et al. 2007). Strikingly, however, numerous genomic sites harbor GATA motifs that bind GATA factors with high affinity in the context of naked DNA, but these sites are not occupied in cells (Bresnick et al. 2006). Despite the ~7 million GATA motifs (WGATAR) in the human genome, genome-wide ChIP-seq analysis showed that GATA-1 discriminates exquisitely among these sites (Fujiwara et al. 2009; Kang et al. 2012). Occupancy was detected at only 0.07% of sites containing WGATAR. Thus, the presence of a GATA motif at a chromosomal site, even when evolutionarily conserved, appears to have little to no predictive value as to whether the associated gene will be bound to or regulated by GATA-1.

Given the low frequency of GATA-1 occupancy at sites containing the conventional GATA motif WGATAR, the sequence requirements for occupancy in vivo might vary from those required for naked DNA binding, and/or additional parameters beyond the presence of a high-affinity GATA motif may dictate GATA motif access in vivo. Genome-wide ChIP-seq studies provided evidence for the existence of a more complex GATA motif at GATA-1-occupied sites in erythroid cells [(C/G)(A/T)GATAA(G/A/C)(G/A/C)], which exists at 297,124 copies in the human genome (Fujiwara et al. 2009). Although the frequency of occupancy at these sites is an order of magnitude greater than at WGATAR, it remains quite low (0.7%). GATA motifs often reside in close proximity to other *cis*-regulatory elements, and it is likely that the neighboring elements constitute an important parameter that dictates GATA motif access in chromatin. However, interrogation of GATA-1-bound DNA sequences isolated by ChIP using de novo motif finding methodology has not revealed individual or combinations of sites with significant utility for predicting GATA-1 occupancy (Fujiwara et al. 2009; Kang et al. 2012). In megakaryocytes (Dore et al. 2012) and endothelial cells (Linnemann et al. 2011), additional *cis*-regulatory elements show highly statistically significant enrichments at GATA-2-occupied sites, respectively, although it remains unclear whether the combinatorial usage of the multiple elements is an important determinant of GATA factor chromatin occupancy.

Analogous to the chromatin site discrimination problem, major work is required to unravel the determinants of whether GATA-1 binding to chromatin activates or represses one or more neighboring target genes. GATA-1 and GATA-2 often colocalize on chromatin with the E-protein Scl/TAL1 (Wozniak et al. 2008), and Scl/TAL1 colocalization correlates with activation in certain contexts (Cheng et al. 2009; Tripic et al. 2009; Yu et al. 2009); however, exceptions to this correlation exist. As noted above, FOG-1 mediates both activation and repression (Crispino et al. 1999). This is a particularly challenging issue, as transient transfection assays do not faithfully recapitulate the activation and repression functions of endogenous GATA-1-associated *cis*-regulatory elements. Sites that confer enhancer activity with reporter genes in transiently transfected cells and in transgenic mice do not invariably exert crucial activities at endogenous loci (Snow et al. 2010b, 2011).

Analyses of the β-globin LCR, a far upstream strong enhancer element consisting of four DNaseI hypersensitive sites (HSs) (Tuan and London 1984; Forrester et al. 1986, 1987; Grosveld. et al. 1987), have been particularly instructive for understanding long-range transcriptional mechanisms in which clusters of *cis*-regulatory elements control distant promoters (Fig. 2). Targeted deletion of the four HSs of the LCR from the endogenous mouse genome severely impairs β-like globin gene expression (Epner et al. 1998; Reik et al. 1998; Bender et al. 2000a). Although GATA motifs confer DNaseI hypersensitivity and contribute to enhancer activities of the HSs (Philipsen et al. 1990, 1993; Pruzina et al. 1991, 1994; Ellis et al. 1993; Stamatoyannopoulos et al. 1995; Pomerantz et al. 1998; Goodwin et al. 2001), targeted deletion of the individual HSs, each containing GATA motifs and additional *cis*-regulatory elements, only incrementally influence β-like glo-

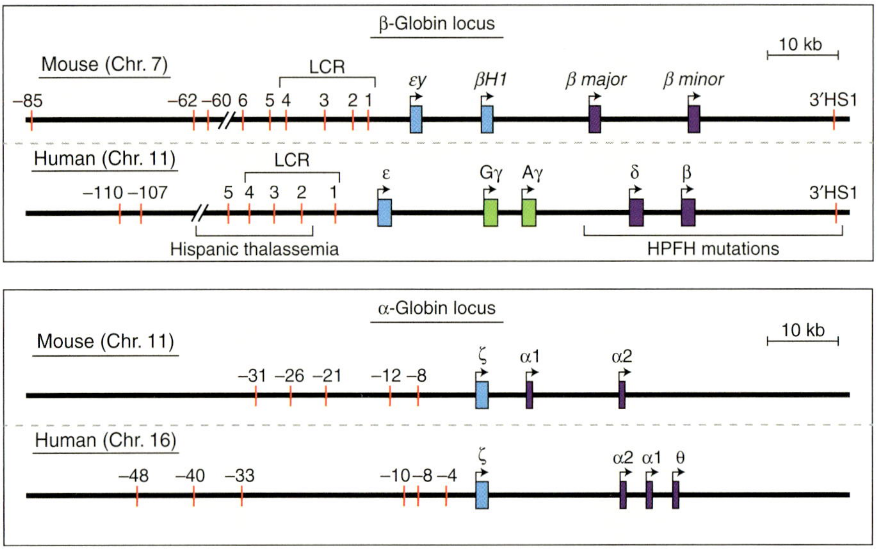

Figure 2. (*Top*) Organization of the human and murine β-globin loci. The human locus contains β-globin genes expressed during embryogenesis (ε), fetal development (Gγ and Aγ), and the adult (δ and β). The mouse locus contains genes expressed during embryogenesis (*εy* and *βH1*) and adult (*β major* and *β minor*). The regions identified in Hispanic thalassemia and hereditary persistence of fetal hemoglobin (HPFH) mutations are described. The genomic structure of the human and murine β-globin and α-globin loci. Multiple DNaseI hypersensitive sites (HSs) are indicated by red bars. The human β-globin locus has five proximal HSs and additional HSs. The mouse locus has six proximal HSs and additional HSs. Among these HSs, HS1–4 are termed the locus control region (LCR). The LCR mediates high-level transcription of the β-globin-like genes. A large deletion of the LCR characterizes Hispanic thalassemia (Driscoll et al. 1989; Forrester et al. 1990). (*Bottom*) Organization of the human and murine α-globin loci. The human locus contains α-globin genes expressed during embryogenesis (ζ) and the adult (α1, α2, and θ). Similarly, the mouse locus contains α-globin genes expressed during embryogenesis (ζ) and the adult (α1 and α2). Analogous to the β-globin locus, there are several DNaseI HSs at the α-globin locus. A large deletion including these HSs and ζ has been identified in several patients (Hatton et al. 1990). Among these HSs, HS-40 (corresponding to HS-26 in the mouse locus) is considered to be functionally important.

bin gene transcription (Fiering et al. 1995; Hug et al. 1996; Bender et al. 2000b). The targeted deletion studies indicated that multiple HSs function additively in vivo (Bender et al. 2001), which was surprising given the qualitatively and quantitatively distinct activities of the individual HSs in reporter gene and transgenic mouse assays (Ryan et al. 1989; Caterina et al. 1994; Bungert et al. 1995, 1999; Pawlik and Townes 1995; Ellis et al. 1996; Peterson et al. 1996; Navas et al. 1998, 2001). It is unclear whether the HS activities are considerably more important in promiscuous reporter assays and at ectopic chromosomal sites versus the endogenous locus, whether human and mouse LCR sequences have distinct activities, or whether compensatory mechanisms mask their important functions when they are removed from the genome, and all are potentially important possibilities to rigorously explore. Regardless, GATA-1 occupies multiple sites at the endogenous β-globin and α-globin loci (Johnson et al. 2002; Anguita et al. 2004; Im et al. 2005; De Gobbi et al. 2007; Fujiwara et al. 2009), and it seems highly likely that some of these sites control endogenous globin gene transcription. Strikingly, a single nucleotide polymorphism upstream of the α-globin locus generates a GATA motif that confers promoter activity, thereby disrupting regulation of the downstream α-globin genes (De Gobbi et al. 2006).

Studies with β-globin locus transgenes have provided further evidence to support a model in which the core GATA-1 regulatory function is a critical determinant of hemoglobin switching. A GATA motif at the ε-globin promoter (−208) suppresses activity of the promoter in a β-globin locus transgene in adult mice (Raich et al. 1995). Similarly, a GATA motif at −566 and −567 relative to the Aγ and Gγ globin genes participates in γ-globin silencing in definitive erythroid cells (Harju-Baker et al. 2008). A hereditary persistence of fetal hemoglobin (HPFH) mutation at the −566 site prevents assembly of the complex and sustains γ-globin expression at developmental stages in which γ-globin should be silenced. Taken together with evidence that GATA-1, FOG-1, and Mi2β occupy this site, and that an erythroid cell-specific Mi2β knockout elevates γ-globin expression from a β-YAC in mice (Costa et al. 2012), these factors assemble a repressive complex that controls hemoglobin switching. In contrast to these results with β-globin locus transgenes, a knock-in mouse strain expressing a mutant FOG-1 defective in NuRD complex binding from the endogenous FOG-1 locus crossed with a human β-globin locus transgenic line shows normal γ-globin silencing (Miccio and Blobel 2010). Thus, the FOG-1-NuRD interaction is apparently not required for silencing the γ-globin genes, although one cannot rule out the possibility that NuRD, and/or specific components of NuRD, are recruited to the locus independent of FOG-1. Studies in the K562 erythroleukemia cell model system and primary erythroid cells indicate that GATA-2 and the CCAAT box binding factor NF-Y participate in γ-globin gene activation, whereas GATA-1, BCL11A, and COUP-TFII confer repression through the same site (Zhu et al. 2012). This finding conforms to the GATA switch mechanism (Bresnick et al. 2005, 2010), in which GATA-1 replacement of GATA-2 from chromatin induces a distinct transcriptional output. In summary, endogenous GATA-1 chromatin occupancy, induction of β-globin gene expression in a genetic complementation assay, and molecular/cellular studies using diverse systems support a role for GATA-1 in establishing a physiological pattern of hemoglobin synthesis.

INTEGRATING GATA FACTOR-DEPENDENT MECHANISMS WITH OTHER MODES OF CONTROLLING HEMOGLOBIN SYNTHESIS

Variations in the level of fetal hemoglobin (HbF) in sickle cell disease (SCD) patients paved the way for the discovery of genetic modulators of hemoglobin switching. Genome-wide association studies identified SNPs that correlate with higher levels of HbF, including QTLs near the *HBS1L-MYB* locus and the β-globin gene cluster, which were described previously, along with several QTLs at the *BCL11A* locus (Menzel et al. 2007b; Lettre et al. 2008; Uda et al. 2008). The *HBS1L-MYB* intergenic region associated with high HbF has HSs that are occupied by GATA-1 in human erythroid precursor cells (Wahlberg et al. 2009). In addition, GATA-1 down-regulates *Myb* expression (Bartunek et al. 2003; Welch et al. 2004). Reduction of *MYB* mRNA by miRNA-15a and -16-1 in primary human erythroid cells results in elevated ε- and γ-globin synthesis (Sankaran et al. 2011). In comparison to the SNPs at the *HBS1L-MYB* and *β-globin* loci, the high-HbF-associated allele at rs4671393 in the *BCL11A* locus was the strongest indicator of fetal hemoglobin levels in a Brazilian cohort with SCD (Lettre et al. 2008). These studies led to the hypothesis that disruption of *BCL11A* modulates HbF levels, and therefore BCL11A represents a target for therapeutic intervention in SCD and β-thalassemia.

Recent studies have made the exceptionally interesting link between BCL11A activity to control hemoglobin switching and GATA-1 function. Proteomic analysis revealed interactions between expressed BCL11A and GATA-1, FOG-1, and the NuRD complex in cultured erythroid cells (Sankaran et al. 2008). BCL11A also interacts with the transcription factor Sox6, a regulator of γ-globin transcription (Yi et al. 2006; Xu et al. 2010). ChIP-chip analysis in primary human erythroid precursors revealed BCL11A occupancy at HSs of the β-globin LCR, which are also occupied by GATA-1 and Pol II. 3C analysis of the human β-globin locus in a β-YAC transgenic mouse crossed to a wild-type or *Bcl11a*$^{-/-}$ mouse indicated that deletion of BCL11A reconfigures the higher-order chromatin structure of the locus (Xu et al. 2010). The LCR-dependent chromatin loop normally positions the LCR in close proximity to the β-globin promoter, but in this case, the LCR engages the Gγ and Aγ promoters. Although this intriguing mechanism has not been described in the context of the endogenous β-globin locus, these studies indicate that BCL11A can interact with GATA-1, and both factors regulate chromatin looping at the β-globin locus.

The conditional knockout of *Bcl11a* using *EpoR-Cre+* mice crossed to β-YAC transgenic mice prevented complete silencing of human γ-globin from the β-YAC (Xu et al. 2011). Genome-wide analysis indicates that this modulation of fetal globin expression is exquisitely specific, as the vast majority of genes were unaffected. Importantly, adult β-globin expression and other regulators of β-like globin genes, including KLF1 (Miller and Bieker 1993; Perkins et al. 1995), are unaffected by loss of BCL11A. *Bcl11a* knockout synergizes with HbF modulators, including the DNA methylation inhibitor 5-aza-2′-deoxycytidine (DAC), and the histone deacetylase (HDAC) inhibitor suberoylanilide hydroxamic acid (SAHA), to further increase γ-globin expression. Crossing the conditional *Bcl11a* knockout to the Berkeley SCD mouse model increases HbF expression, and corrects multiple SCD clinical parameters.

Genome-wide association studies in 27 Maltese family members with HPFH identified a locus with strong association to high HbF levels at 19p13.12–13, which contains the gene encoding the erythroid Krüppel-like factor EKLF (*KLF1*) (Borg et al. 2010). Genomic sequencing of this cohort revealed a deletion that disrupts the zinc finger domain of KLF1, rendering it defective for DNA binding. The resulting KLF1 haploinsufficiency was accompanied by a decrease in BCL11A expression and an increase in the γ-globin to α-globin ratio. A knockout of the HS1 enhancer upstream of the *Klf1* gene greatly reduced KLF1 and BCL11A expression (Zhou et al. 2010). KLF1 occupies the murine *Bcl11a* promoter in bone marrow-derived $Ter119^{+}$ cells, and the human *BCL11A* promoter in erythroid progenitors isolated from peripheral blood (Borg et al. 2010; Zhou et al. 2010). KLF1 does not occupy the human *BCL11A* promoter in fetal liver erythroid progenitors, and KLF1 directly binds and activates the *HBB* gene in adult but not fetal cells. KLF1 expression is elevated in adult erythroid precursors versus fetal erythroid precursors (Zhou et al. 2006). In aggregate, these data suggest that KLF1 directly regulates the switch from fetal to adult hemoglobin by binding to and activating the *HBB* gene and indirectly triggers the switch via activating *BCL11A,* which in turn suppresses fetal globin gene expression. As GATA-1 can directly induce KLF1 expression (Crispino et al. 1999; Kim et al. 2007), this constitutes another critical link between the core GATA-1 regulatory function and hemoglobin switching.

In addition to the obviously crucial KLF1-BCL11A axis for regulating hemoglobin switching, additional factors and mechanisms are implicated in controlling this crucial process (Jane et al. 1995; Filipe et al. 1999; Lopez et al. 2002; Tanabe et al. 2002, 2007; Pace et al. 2003; Basu et al. 2005; Bank 2006; Rupon et al. 2006, 2011; Sangerman et al. 2006; Yi et al. 2006; Bottardi et al. 2009; Sripichai et al. 2009; Yao et al. 2009; Sankaran et al. 2010a; Campbell et al. 2011; Wilber et al. 2011; Funnell et al. 2012; Ross et al. 2012). Many important questions remain, however, regarding the underlying mechanisms and biology, including whether the mechanisms are intricately linked to, or perhaps independent of, GATA-1 and KLF1-BCL11A. Furthermore, it will be important to elucidate the relevance of specific components of the switching machinery to existing and new preclinical and clinical approaches for therapeutically modulating γ-globin expression (Fibach et al. 1993; Rodgers et al. 1993; Fucharoen et al. 1996; Watanapokasin et al. 2005; Weinberg et al. 2005; Sangerman et al. 2006; Lavelle et al. 2008; Mabaera et al. 2008; Macari and Lowrey 2011; Wilber et al. 2011; Kutlar et al. 2012; Macari et al. 2012). Finally, almost undoubtedly, new mechanistic insights will launch leveraging opportunities to develop novel molecularly targeted therapies, although this will require sustained intense experimental efforts, while being particularly cautious not to fast-forward such initiatives in the absence of compelling mechanistic, biological, and/or pharmacological foundations. Clearly, the mechanistic infrastructure is insufficient at the present time to divert major efforts into applied therapeutics development and clinical optimization of existing strategies that are often less efficacious than desired.

As BCL11A has unequivocally emerged as an extremely important constituent in the regulation of hemoglobin switching, it represents an attractive target for therapeutic modulation of this process. However, developing efficacious BCL11A targeting strategies is not without significant challenges. Challenges include those intrinsic to drug design, as BCL11A is a zinc finger transcription factor without obvious binding pockets that can accommodate inhibitory drug docking. Second, as mentioned previously, the whole body knockout of BCL11A results in postnatal death (Liu et al. 2003). Although BCL11A deletion does not affect development of macrophage-granulocyte or erythroid lineages, it severely disrupts lymphopoiesis. Transplantation of *Bcl11a*$^{-/-}$ fetal liver cells into irradiated mice results in T-cell leukemia and almost no donor-derived B-cell production. Conditional *Bcl11a* targeting in the adult also revealed a crucial role in controlling lymphopoiesis (Yu et al. 2012). Thus, an additional challenge relates to establishing efficacy in erythroid cells, while minimizing potential deleterious actions on immune cell function. However, with modern approaches to drug design, potential opportunities for developing BCL11A targeting strategies indirectly through upstream factors, and fundamental pharmacological development, there is no reason to believe that these challenges will be insurmountable. Given the emerging rich flavor of the core GATA-1 regulatory function, in conjunction with additional regulatory layers that will almost certainly implicate previously undiscovered mechanisms, rationale is high for continuing to drill deeply into these mechanisms. This type of work will undoubtedly yield further payoffs vis-a-vis innovative mechanistic insights, attractive therapeutic targets that lead to efficacious therapies for hemoglobinopathies, and serendipitous discoveries that advance science and medicine in disparate fields.

ACKNOWLEDGMENTS

We acknowledge funding from National Institutes of Health (NIH) DK50107. A.W.D. was supported by NIH T32 GM081061. N.J.P. was supported by an American Heart Association Predoctoral Fellowship.

REFERENCES

**Reference is also in this collection.*

Anguita E, Hughes J, Heyworth C, Blobel GA, Wood WG, Higgs DR. 2004. Globin gene activation during haemopoiesis is driven by protein complexes nucleated by GATA-1 and GATA-2. *EMBO J* **23:** 2841–2852.

Arceci RJ, King AA, Simon MC, Orkin SH, Wilson DB. 1993. Mouse GATA-4: A retinoic acid-inducible GATA-binding transcription factor expressed in endodermally derived tissues and heart. *Mol Cell Biol* **13:** 2235–2246.

Atweh GF, Schechter AN. 2001. Pharmacologic induction of fetal hemoglobin: Raising the therapeutic bar in sickle cell disease. *Curr Opin Hematol* **8:** 123–130.

Bank A. 2006. Regulation of human fetal hemoglobin: New players, new complexities. *Blood* **107:** 435–443.

Bartunek P, Kralova J, Blendinger G, Dvorak M, Zenke M. 2003. GATA-1 and c-myb crosstalk during red blood cell differentiation through GATA-1 binding sites in the c-myb promoter. *Oncogene* **22:** 1927–1935.

Basu P, Morris PE, Haar JL, Wani MA, Lingrel JB, Gaensler KM, Lloyd JA. 2005. KLF2 is essential for primitive erythropoiesis and regulates the human and murine embryonic β-like globin genes in vivo. *Blood* **106:** 2566–2571.

Bender MA, Bulger M, Close J, Groudine M. 2000a. β-globin gene switching and DNaseI sensitivity of the endogenous β-globin locus in mice do not require the locus control region. *Mol Cell* **5:** 387–393.

Bender MA, Mehaffey MG, Telling A, Hug B, Ley TJ, Groudine M, Fiering S. 2000b. Independent formation of DnaseI hypersensitive sites in the murine β-globin locus control region. *Blood* **95:** 3600–3604.

Bender MA, Roach JN, Halow J, Close J, Alami R, Bouhassira EE, Groudine M, Fiering SN. 2001. Targeted deletion of 5′HS1 and 5′HS4 of the β-globin locus control region reveals additive activity of the DNaseI hypersensitive sites. *Blood* **98:** 2022–2027.

Blau CA, Stamatoyannopoulos G. 1994. Hemoglobin switching and its clinical implications. *Curr Opin Hematol* **1:** 136–142.

Blobel GA, Nakajima T, Eckner R, Montminy M, Orkin SH. 1998. CREB-binding protein cooperates with transcription factor GATA-1 and is required for erythroid differentiation. *Proc Natl Acad Sci* **95:** 2061–2066.

Borg J, Papadopoulos P, Georgitsi M, Gutierrez L, Grech G, Fanis P, Phylactides M, Verkerk AJ, van der Spek PJ, Scerri CA, et al. 2010. Haploinsufficiency for the erythroid transcription factor KLF1 causes hereditary persistence of fetal hemoglobin. *Nat Genet* **42:** 801–805.

Bottardi S, Ross J, Bourgoin V, Fotouhi-Ardakani N, Affar el B, Trudel M, Milot E. 2009. Ikaros and GATA-1 combinatorial effect is required for silencing of human γ-globin genes. *Mol Cell Biol* **29:** 1526–1537.

Bresnick EH, Martowicz ML, Pal S, Johnson KD. 2005. Developmental control via GATA factor interplay at chromatin domains. *J Cell Physiol* **205:** 1–9.

Bresnick EH, Johnson KD, Kim SI, Im H. 2006. Establishment and regulation of chromatin domains: Mechanistic insights from studies of hemoglobin synthesis. *Prog Nucleic Acid Res Mol Biol* **81:** 435–471.

Bresnick EH, Lee HY, Fujiwara T, Johnson KD, Keles S. 2010. GATA switches as developmental drivers. *J Biol Chem* **285:** 31087–31093.

Bresnick EH, Katsumura KR, Lee HY, Johnson KD, Perkins AS. 2012. Master regulatory GATA transcription factors: Mechanistic principles and emerging links to hematologic malignancies. *Nucleic Acids Res* **40:** 5819–5831.

Bultman SJ, Gebuhr TC, Magnuson T. 2005. A Brg1 mutation that uncouples ATPase activity from chromatin remodeling reveals an essential role for SWI/SNF complexes in β-globin expression and erythroid development. *Genes Dev* **19:** 2849–2861.

Bungert J, Dave U, Lim KC, Lieuw KH, Shavit JA, Liu Q, Engel JD. 1995. Synergistic regulation of human β-globin gene switching by locus control region elements HS3 and HS4. *Genes Dev* **9:** 3083–3096.

Bungert J, Tanimoto K, Patel S, Liu Q, Fear M, Engel JD. 1999. Hypersensitive site 2 specifies a unique function within the human β-globin locus control region to stimulate globin gene transcription. *Mol Cell Biol* **19:** 3062–3072.

Bunn HF. 1999. Induction of fetal hemoglobin in sickle cell disease. *Blood* **93:** 1787–1789.

Campbell AD, Cui S, Shi L, Urbonya R, Mathias A, Bradley K, Bonsu KO, Douglas RR, Halford B, Schmidt L, et al. 2011. Forced TR2/TR4 expression in sickle cell disease mice confers enhanced fetal hemoglobin synthesis and alleviated disease phenotypes. *Proc Natl Acad Sci* **108:** 18808–18813.

Cantor AB, Katz SG, Orkin SH. 2002. Distinct domains of the GATA-1 cofactor FOG-1 differentially influence erythroid versus megakaryocytic maturation. *Mol Cell Biol* **22:** 4268–4279.

Cantor AB, Iwasaki H, Arinobu Y, Moran TB, Shigematsu H, Sullivan MR, Akashi K, Orkin SH. 2008. Antagonism of FOG-1 and GATA factors in fate choice for the mast cell lineage. *J Exp Med* **205:** 611–624.

Caterina JJ, Ciavatta DJ, Donze D, Behringer RR, Townes TM. 1994. Multiple elements in human β-globin locus control region 5′ HS 2 are involved in enhancer activity and position-independent, transgene expression. *Nucleic Acids Res* **22:** 1006–1011.

Cheng Y, Wu W, Kumar SA, Yu D, Deng W, Tripic T, King DC, Chen KB, Zhang Y, Drautz D, et al. 2009. Erythroid GATA1 function revealed by genome-wide analysis of transcription factor occupancy, histone modifications, and mRNA expression. *Genome Res* **19:** 2172–2184.

Chlon TM, Dore LC, Crispino JD. 2012. Cofactor-mediated restriction of GATA-1 chromatin occupancy coordinates lineage-specific gene expression. *Mol Cell* **47:** 608–621.

Collavin L, Gostissa M, Avolio F, Secco P, Ronchi A, Santoro C, Del Sal G. 2004. Modification of the erythroid transcription factor GATA-1 by SUMO-1. *Proc Natl Acad Sci* **101:** 8870–8875.

Conaway RC, Sato S, Tomomori-Sato C, Yao T, Conaway JW. 2005. The mammalian mediator complex and its role in transcriptional regulation. *Trends Biochem Sci* **30:** 250–255.

Costa FC, Fedosyuk H, Chazelle AM, Neades RY, Peterson KR. 2012. Mi2β is required for γ-globin gene silencing: Temporal assembly of a GATA-1-FOG-1-Mi2 repressor complex in β-YAC transgenic mice. *PLoS Genet* **8:** e1003155.

Crawford SE, Qi C, Misra P, Stellmach V, Rao MS, Engel JD, Zhu Y, Reddy JK. 2002. Defects of the heart, eye, and megakaryocytes in peroxisome proliferator activator receptor-binding protein (PBP) null embryos implicate GATA family of transcription factors. *J Biol Chem* **277:** 3585–3592.

Crispino JD. 2005. GATA-1 in normal and malignant hematopoiesis. *Semin Cell Dev Biol* **16:** 137–147.

Crispino JD, Lodish MB, MacKay JP, Orkin SH. 1999. Use of altered specificity mutants to probe a specific protein-protein interaction in differentiation: The GATA-1:FOG complex. *Mol Cell* **3:** 219–228.

Crossley M, Orkin SH. 1994. Phosphorylation of the erythroid transcription factor GATA-1. *J Biol Chem* **269:** 16589–16596.

De Gobbi M, Viprakasit V, Hughes JR, Fisher C, Buckle VJ, Ayyub H, Gibbons RJ, Vernimmen D, Yoshinaga Y, de Jong P, et al. 2006. A regulatory SNP causes a human genetic disease by creating a new transcriptional promoter. *Science* **312:** 1215–1217.

De Gobbi M, Anguita E, Hughes J, Sloane-Stanley JA, Sharpe JA, Koch CM, Dunham I, Gibbons RJ, Wood WG, Higgs DR. 2007. Tissue-specific histone modification and transcription factor binding in α globin gene expression. *Blood* **110:** 4503–4510.

Dore LC, Chlon TM, Brown CD, White KP, Crispino JD. 2012. Chromatin occupancy analysis reveals genome-wide GATA factor switching during hematopoiesis. *Blood* **119:** 3724–3733.

Dorfman DM, Wilson DB, Bruns GA, Orkin SH. 1992. Human transcription factor GATA-2. Evidence for regulation of preproendothelin-1 gene expression in endothelial cells. *J Biol Chem* **267:** 1279–1285.

Driscoll MC, Dobkin CS, Alter BP. 1989. γδβ-Thalassemia due to a de novo mutation deleting the 5′ β-globin gene activation-region hypersensitive sites. *Proc Natl Acad Sci* **86:** 7470–7474.

Drissen R, Palstra RJ, Gillemans N, Splinter E, Grosveld F, Philipsen S, de Laat W. 2004. The active spatial organization of the β-globin locus requires the transcription factor EKLF. *Genes Dev* **18:** 2485–2490.

Ellis J, Talbot D, Dillon N, Grosveld F. 1993. Synthetic human β-globin 5′HS2 constructs function as locus control regions only in multicopy transgene concatamers. *EMBO J* **12:** 127–134.

Ellis J, Tan-Un KC, Harper A, Michalovich D, Yannoutsos N, Philipsen S, Grosveld F. 1996. A dominant chromatin-opening activity in 5′ hypersensitive site 3 of the human β-globin locus control region. *EMBO J* **15:** 562–568.

Epner E, Reik A, Cimbora D, Telling A, Bender MA, Fiering S, Enver T, Martin DI, Kennedy M, Keller G, et al. 1998. The β-globin LCR is not necessary for an open chromatin structure or developmentally regulated transcription of the native mouse β-globin locus. *Mol Cell* **2:** 447–455.

Evans T, Felsenfeld G. 1989. The erythroid-specific transcription factor Eryf1: A new finger protein. *Cell* **58:** 877–885.

Evans T, Reitman M, Felsenfeld G. 1988. An erythrocyte-specific DNA-binding factor recognizes a regulatory sequence common to all chicken globin genes. *Proc Natl Acad Sci* **85:** 5976–5980.

Evans T, Felsenfeld G, Reitman M. 1990. Control of globin gene transcription. *Annu Rev Cell Biol* **6:** 95–124.

Fibach E, Burke KP, Schechter AN, Noguchi CT, Rodgers GP. 1993. Hydroxyurea increases fetal hemoglobin in cultured erythroid cells derived from normal individuals and patients with sickle cell anemia or β-thalassemia. *Blood* **81:** 1630–1635.

Fiering S, Epner E, Robinson K, Zhuang Y, Telling A, Hu M, Martin DI, Enver T, Ley TJ, Groudine M. 1995. Targeted deletion of 5′HS2 of the murine β-globin LCR reveals that it is not essential for proper regulation of the β-globin locus. *Genes Dev* **9:** 2203–2213.

Filipe A, Li Q, Deveaux S, Godin I, Romeo P-H, Stamatopannopoulos G, Mignotte V. 1999. Regulation of embryonic/fetal globin genes by nuclear hormone receptors: A novel perspective on hemoglobin switching. *EMBO J* **18:** 687–697.

Forrester WC, Thompson C, Elder JT, Groudine M. 1986. A developmentally stable chromatin structure in the human β-globin locus. *Proc Natl Acad Sci* **83:** 1359–1363.

Forrester WC, Takegawa S, Papayannopoulou T, Stamatoyannopoulos G, Groudine M. 1987. Evidence for a locus activation region: The formation of developmentally stable hypersensitive sites in globin-expressing hybrids. *Nucleic Acids Res* **15:** 10159–10177.

Forrester WC, Epner E, Driscoll MC, Enver T, Brice M, Papayannopoulou T, Groudine M. 1990. A deletion of the human β-globin locus activation region causes a major alteration in chromatin structure and replication across the entire β-globin locus. *Genes Dev* **4:** 1637–1649.

Fox AH, Liew C, Holmes M, Kowalski K, MacKay JP, Crossley M. 1999. Transcriptional cofactors of the FOG family interact with GATA proteins by means of multiple zinc fingers. *EMBO J* **18:** 2812–2822.

Fucharoen S, Siritanaratkul N, Winichagoon P, Chowthaworn J, Siriboon W, Muangsup W, Chaicharoen S, Poolsup N, Chindavijak B, Pootrakul P, et al. 1996. Hydroxyurea increases hemoglobin F levels and improves the effectiveness of erythropoiesis in β-thalassemia/hemoglobin E disease. *Blood* **87:** 887–892.

Fujiwara Y, Browne CP, Cunniff K, Goff SC, Orkin SH. 1996. Arrested development of embryonic red cell precursors in mouse embryos lacking transcription factor GATA-1. *Proc Natl Acad Sci* **93:** 12355–12358.

Fujiwara T, O'Geen H, Keles S, Blahnik K, Linnemann AK, Kang YA, Choi K, Farnham PJ, Bresnick EH. 2009. Discovering hematopoietic mechanisms through genome-wide analysis of GATA factor chromatin occupancy. *Mol Cell* **36:** 667–681.

Funnell AP, Norton LJ, Mak KS, Burdach J, Artuz CM, Twine NA, Wilkins MR, Power CA, Hung TT, Perdomo J, et al. 2012. The CACCC-binding protein KLF3/BKLF represses a subset of KLF1/EKLF target genes and is required for proper erythroid maturation in vivo. *Mol Cell Biol* **32:** 3281–3292.

Gamsjaeger R, Liew CK, Loughlin FE, Crossley M, Mackay JP. 2007. Sticky fingers: Zinc-fingers as protein-recognition motifs. *Trends Biochem Sci* **32:** 63–70.

Gao Z, Huang Z, Olivey HE, Gurbuxani S, Crispino JD, Svensson EC. 2010. FOG-1-mediated recruitment of NuRD is required for cell lineage re-enforcement during haematopoiesis. *EMBO J* **29:** 457–468.

Garriga-Canut M, Orkin SH. 2004. Transforming acidic coiled-coil protein 3 (TACC3) controls friend of GATA-1 (FOG-1) subcellular localization and regulates the association between GATA-1 and FOG-1 during hematopoiesis. *J Biol Chem* **279:** 23597–23605.

Goodwin AJ, McInerney JM, Glander MA, Pomerantz O, Lowrey CH. 2001. In vivo formation of a human β-globin locus control region core element requires binding sites for multiple factors including GATA-1, NF-E2, erythroid Krüppel-like factor, and Sp1. *J Biol Chem* **276:** 26883–26892.

Grass JA, Boyer ME, Pal S, Wu J, Weiss MJ, Bresnick EH. 2003. GATA-1-dependent transcriptional repression of GATA-2 via disruption of positive autoregulation and domain-wide chromatin remodeling. *Proc Natl Acad Sci* **100:** 8811–8816.

Grass JA, Jing H, Kim SI, Martowicz ML, Pal S, Blobel GA, Bresnick EH. 2006. Distinct functions of dispersed GATA factor complexes at an endogenous gene locus. *Mol Cell Biol* **26:** 7056–7067.

Gregory GD, Miccio A, Bersenev A, Wang Y, Hong W, Zhang Z, Poncz M, Tong W, Blobel GA. 2010. FOG1 requires NuRD to promote hematopoiesis and maintain lineage fidelity within the megakaryocytic-erythroid compartment. *Blood* **115:** 2156–2166.

Grosveld F, van Assendelft GB, Greaves DR, Kollias G. 1987. Position-independent high-level expression of the human β-globin gene in trangsenic mice. *Cell* **51:** 975–985.

* Hardison RC. 2012. Evolution of hemoglobin and its genes. *Cold Spring Harb Perspect Med* **2:** a011627.

Harju-Baker S, Costa FC, Fedosyuk H, Neades R, Peterson KR. 2008. Silencing of $^{A}\gamma$-globin gene expression during adult definitive erythropoiesis mediated by GATA-1-FOG-1-Mi2 complex binding at the −566 GATA site. *Mol Cell Biol* **28:** 3101–3113.

Hatton CS, Wilkie AO, Drysdale HC, Wood WG, Vickers MA, Sharpe J, Ayyub H, Pretorius IM, Buckle VJ, Higgs DR. 1990. α-Thalassemia caused by a large (62 kb) deletion upstream of the human α globin gene cluster. *Blood* **76:** 221–227.

* Higgs DR. 2013. The molecular basis of α-thalassemia. *Cold Spring Harb Perspect Med* **3:** a011718.

Hirasawa R, Shimuzu R, Takahashi S, Osawa M, Takayanagi S, Kato Y, Onodera M, Minegishi N, Yamamoto M, Fukao K, et al. 2002. Essential and instructive roles of GATA factors in eosinophil development. *J Exp Med* **195:** 1379–1386.

Ho IC, Vorhees P, Marin N, Oakley BK, Tsai SF, Orkin SH, Leiden JM. 1991. Human GATA-3: A lineage-restricted transcription factor that regulates the expression of the T cell receptor α gene. *EMBO J* **10:** 1187–1192.

Holmes M, Turner J, Fox A, Chisholm O, Crossley M, Chong B. 1999. hFOG-2, a novel zinc finger protein, binds the co-repressor mCtBP2 and modulates GATA-mediated activation. *J Biol Chem* **274:** 23491–23498.

Hong W, Nakazawa M, Chen YY, Kori R, Vakoc CR, Rakowski C, Blobel GA. 2005. FOG-1 recruits the NuRD repressor complex to mediate transcriptional repression by GATA-1. *EMBO J* **24:** 2367–2378.

Horak CE, Mahajan MC, Luscombe NM, Gerstein M, Weissman SM, Snyder M. 2002. GATA-1 binding sites mapped in the β-globin locus using mammalian chIp-chip analysis. *Proc Natl Acad Sci* **99:** 2924–2929.

Huang Z, Dore LC, Li Z, Orkin SH, Feng G, Lin S, Crispino JD. 2009. GATA-2 reinforces megakaryocyte development in the absence of GATA-1. *Mol Cell Biol* **29:** 5168–5180.

Hug BA, Wesselschmidt RL, Fiering S, Bender MA, Epner E, Groudine M, Ley TJ. 1996. Analysis of mice containing a targeted deletion of β-globin locus control region 5′ hypersensitive site 3. *Mol Cell Biol* **16:** 2906–2912.

Humbles AA, Lloyd CM, McMillan SJ, Friend DS, Xanthou G, McKenna EE, Ghiran S, Gerard NP, Yu C, Orkin SH, et al. 2004. A critical role for eosinophils in allergic airways remodeling. *Science* **305:** 1776–1779.

Hung HL, Lau J, Kim AY, Weiss MJ, Blobel GA. 1999. CREB-binding protein acetylates hematopoietic transcription factor GATA-1 at functionally important sites. *Mol Cell Biol* **19:** 3496–3505.

Im H, Grass JA, Johnson KD, Kim SI, Boyer ME, Imbalzano AN, Bieker JJ, Bresnick EH. 2005. Chromatin domain activation via GATA-1 utilization of a small subset of dispersed GATA motifs within a broad chromosomal region. *Proc Natl Acad Sci* **102:** 17065–17070.

Ito E, Toki T, Ishihara H, Ohtani H, Gu L, Yokoyama M, Engel JD, Yamamoto M. 1993. Erythroid transcription factor GATA-1 is abundantly transcribed in mouse testis. *Nature* **362:** 466–468.

Jane SM, Nienhuis AW, Cunningham JM. 1995. Hemoglobin switching in man and chicken is mediated by a heteromeric complex between the ubiquitous transcription factor CP2 and a developmentally specific protein. *EMBO J* **14:** 97–105.

Johnson KD, Christensen HM, Zhao B, Bresnick EH. 2001. Distinct mechanisms control RNA polymerase II recruitment to a tissue-specific locus control region and a downstream promoter. *Mol Cell* **8:** 465–471.

Johnson KD, Grass JD, Boyer ME, Kiekhaefer CM, Blobel GA, Weiss MJ, Bresnick EH. 2002. Cooperative activities of hematopoietic regulators recruit RNA polymerase II to a tissue-specific chromatin domain. *Proc Natl Acad Sci* **99:** 11760–11765.

Johnson KD, Grass JA, Park C, Im H, Choi K, Bresnick EH. 2003. Highly restricted localization of RNA polymerase II within a locus control region of a tissue-specific chromatin domain. *Mol Cell Biol* **23:** 6484–6493.

Johnson KD, Kim SI, Bresnick EH. 2006. Differential sensitivities of transcription factor target genes underlie cell type-specific gene expression profiles. *Proc Natl Acad Sci* **103:** 15939–15944.

Johnson KD, Boyer ME, Kang JA, Wickrema A, Cantor AB, Bresnick EH. 2007. Friend of GATA-1-independent transcriptional repression: A novel mode of GATA-1 function. *Blood* **109:** 5230–5233.

Johnson KD, Hsu AP, Ryu MJ, Wang J, Gao X, Boyer ME, Liu Y, Lee Y, Calvo KR, Keles S, et al. 2012. Cis-element mutated in GATA2-dependent immunodeficiency governs hematopoiesis and vascular integrity. *J Clin Invest* **122:** 3692–3704.

Kadri Z, Maouche-Chretien L, Rooke HM, Orkin SH, Romeo PH, Mayeux P, Leboulch P, Chretien S. 2005. Phosphatidylinositol 3-kinase/Akt induced by erythropoietin renders the erythroid differentiation factor GATA-1 competent for TIMP-1 gene transactivation. *Mol Cell Biol* **25:** 7412–7422.

Kang YA, Sanalkumar R, O'Geen H, Linnemann AK, Chang CJ, Bouhassira EE, Farnham PJ, Keles S, Bresnick EH. 2012. Autophagy driven by a master regulator of hematopoiesis. *Mol Cell Biol* **32:** 226–239.

Katz SG, Cantor AB, Orkin SH. 2002. Interaction between FOG-1 and the corepressor C-terminal binding protein is dispensable for normal erythropoiesis in vivo. *Mol Cell Biol* **22:** 3121–3128.

Kelley C, Blumberg H, Zon LI, Evans T. 1993. GATA-4 is a novel transcription factor expressed in endocardium of the developing heart. *Development* **118:** 817–827.

Khanna R, Chang SH, Andrabi S, Azam M, Kim A, Rivera A, Brugnara C, Low PS, Liu SC, Chishti AH. 2002. Headpiece domain of dematin is required for the stability of the erythrocyte membrane. *Proc Natl Acad Sci* **99:** 6637–6642.

Kiekhaefer CM, Grass JA, Johnson KD, Boyer ME, Bresnick EH. 2002. Hematopoietic-specific activators establish an overlapping pattern of histone acetylation and methylation within a mammalian chromatin domain. *Proc Natl Acad Sci* **99:** 14309–14314.

Kim SI, Bresnick EH. 2007. Transcriptional control of erythropoiesis: Emerging mechanisms and principles. *Oncogene* **26:** 6777–6794.

Kim SI, Bultman SJ, Jing H, Blobel GA, Bresnick EH. 2007. Dissecting molecular steps in chromatin domain activation during hematopoietic differentiation. *Mol Cell Biol* **27:** 4551–4565.

Kim SI, Bresnick EH, Bultman SJ. 2009a. BRG1 directly regulates nucleosome structure and chromatin looping of the α globin locus to activate transcription. *Nucleic Acids Res* **37:** 6019–6027.

Kim SI, Bultman SJ, Kiefer CM, Dean A, Bresnick EH. 2009b. BRG1 requirement for long-range interaction of a locus control region with a downstream promoter. *Proc Natl Acad Sci* **106:** 2259–2264.

Ko LJ, Engel JD. 1993. DNA-binding specificities of the GATA transcription factor family. *Mol Cell Biol* **13:** 4011–4022.

Kornberg RD. 2005. Mediator and the mechanism of transcriptional activation. *Trends Biochem Sci* **30:** 235–239.

Kuo CT, Morrisey EE, Anandappa R, Sigrist K, Lu MM, Parmacek MS, Soudais C, Leiden JM. 1997. GATA4 transcription factor is required for ventral morphogenesis and heart tube formation. *Genes Dev* **11:** 1048–1060.

Kutlar A, Ataga K, Reid M, Vichinsky EP, Neumayr L, Blair-Britt L, Labotka R, Glass J, Keefer JR, Wargin WA, et al. 2012. A phase 1/2 trial of HQK-1001, an oral fetal globin inducer, in sickle cell disease. *Am J Hematol* **87:** 1017–1021.

Lamonica JM, Vakoc CR, Blobel GA. 2006. Acetylation of GATA-1 is required for chromatin occupancy. *Blood* **108:** 3736–3738.

Lamonica JM, Deng W, Kadauke S, Campbell AE, Gamsjaeger R, Wang H, Cheng Y, Billin AN, Hardison RC, Mackay JP, et al. 2011. Bromodomain protein Brd3 associates with acetylated GATA1 to promote its chromatin occupancy at erythroid target genes. *Proc Natl Acad Sci* **108:** E159–E168.

Lavelle D, Saunthararajah Y, Desimone J. 2008. Comment on DNA methylation and mechanism of action of 5-azacytidine. *Blood* **111:** 2485.

Lee ME, Temizer DH, Clifford JA, Quertermous T. 1991. Cloning of the GATA binding protein that regulates endothelin-1 gene expression in endothelial cells. *J Biol Chem* **266:** 16188–16192.

Lee HY, Johnson KD, Fujiwara T, Boyer ME, Kim SI, Bresnick EH. 2009. Controlling hematopoiesis through sumoylation-dependent regulation of a GATA factor. *Mol Cell* **36:** 984–995.

Lee HY, Johnson KD, Boyer ME, Bresnick EH. 2011. Relocalizing genetic loci into specific subnuclear neighborhoods. *J Biol Chem* **286:** 18834–18844.

Letting DL, Rakowski C, Weiss MJ, Blobel GA. 2003. Formation of a tissue-specific histone acetylation pattern by the hematopoietic transcription factor GATA-1. *Mol Cell Biol* **23:** 1334–1340.

Letting DL, Chen YY, Rakowski C, Reedy S, Blobel GA. 2004. Context-dependent regulation of GATA-1 by friend of GATA-1. *Proc Natl Acad Sci* **101:** 476–481.

* Lettre G. 2012. The search for genetic modifiers of disease severity in the β-hemoglobinopathies. *Cold Spring Harb Perspect Med* **2:** a015032.

Lettre G, Sankaran VG, Bezerra MA, Araujo AS, Uda M, Sanna S, Cao A, Schlessinger D, Costa FF, Hirschhorn JN, et al. 2008. DNA polymorphisms at the BCL11A, HBS1L-MYB, and β-globin loci associate with fetal hemoglobin levels and pain crises in sickle cell disease. *Proc Natl Acad Sci* **105:** 11869–11874.

Liew CK, Simpson RJY, Kwan AHY, Crofts LA, Loughlin FE, Matthews JM, Crossley M, MacKay JP. 2005. Zinc fingers as protein recognition motifs: Structural basis for the GATA-1/Friend of GATA interaction. *Proc Natl Acad Sci* **102:** 583–588.

Lim KC, Hosoya T, Brandt W, Ku CJ, Hosoya-Ohmura S, Camper SA, Yamamoto M, Engel JD. 2012. Conditional Gata2 inactivation results in HSC loss and lymphatic mispatterning. *J Clin Invest* **122:** 3705–3717.

Linnemann AK, O'Geen H, Keles S, Farnham PJ, Bresnick EH. 2011. Genetic framework for GATA factor function in vascular biology. *Proc Natl Acad Sci* **108:** 13641–13646.

Liu P, Keller JR, Ortiz M, Tessarollo L, Rachel RA, Nakamura T, Jenkins NA, Copeland NG. 2003. Bcl11a is essential for normal lymphoid development. *Nat Immunol* **4:** 525–532.

Lopez RA, Schoetz S, DeAngelis K, O'Neill D, Bank A. 2002. Multiple hematopoietic defects and delayed globin switching in Ikaros null mice. *Proc Natl Acad Sci* **99:** 602–607.

Lu JR, McKinsey TA, Xu H, Wang DZ, Richardson JA, Olson EN. 1999. FOG-2, a heart- and brain-enriched cofactor for GATA transcription factors. *Mol Cell Biol* **19:** 4495–4502.

Lugus JJ, Chung YS, Mills JC, Kim SI, Grass JA, Kyba M, Doherty JM, Bresnick EH, Choi K. 2007. GATA2 functions at multiple steps in hemangioblast development and differentiation. *Development* **134:** 393–405.

Mabaera R, Greene MR, Richardson CA, Conine SJ, Kozul CD, Lowrey CH. 2008. Neither DNA hypomethylation nor changes in the kinetics of erythroid differentiation explain 5-azacytidine's ability to induce human fetal hemoglobin. *Blood* **111:** 411–420.

Macari ER, Lowrey CH. 2011. Induction of human fetal hemoglobin via the NRF2 antioxidant response signaling pathway. *Blood* **117:** 5987–5997.

Macari ER, Schaeffer EK, West RJ, Lowrey CH. 2012. Simvastatin and t-butylhydroquinone suppress KLF1 and BCL11A gene expression and additively increase fetal hemoglobin in primary human erythroid cells. *Blood* **121:** 830–839.

Martin DI, Orkin SH. 1990. Transcriptional activation and DNA binding by the erythroid factor GF-1/NF-E1/Eryf 1. *Genes Dev* **4:** 1886–1898.

Martowicz ML, Grass JA, Boyer ME, Guend H, Bresnick EH. 2005. Dynamic GATA factor interplay at a multicomponent regulatory region of the GATA-2 locus. *J Biol Chem* **280:** 1724–1732.

Menzel S, Garner C, Gut I, Matsuda F, Yamaguchi M, Heath S, Foglio M, Zelenika D, Boland A, Rooks H, et al. 2007a. A QTL influencing F cell production maps to a gene encoding a zinc-finger protein on chromosome 2p15. *Nat Genet* **39:** 1197–1199.

Menzel S, Jiang J, Silver N, Gallagher J, Cunningham J, Surdulescu G, Lathrop M, Farrall M, Spector TD, Thein SL. 2007b. The HBS1L-MYB intergenic region on chromosome 6q23.3 influences erythrocyte, platelet, and monocyte counts in humans. *Blood* **110:** 3624–3626.

Merika M, Orkin SH. 1993. DNA-binding specificity of GATA family transcription factors. *Mol Cell Biol* **13:** 3999–4010.

Miccio A, Blobel GA. 2010. Role of the GATA-1/FOG-1/NuRD pathway in the expression of human β-like globin genes. *Mol Cell Biol* **30:** 3460–3470.

Miccio A, Wang Y, Hong W, Gregory GD, Wang H, Yu X, Choi JK, Shelat S, Tong W, Poncz M, et al. 2010. NuRD mediates activating and repressive functions of GATA-1 and FOG-1 during blood development. *EMBO J* **29:** 442–456.

Migliaccio AR, Rana RA, Sanchez M, Lorenzini R, Centurione L, Bianchi L, Vannucchi AM, Migliaccio G,

Orkin SH. 2003. GATA-1 as a regulator of mast cell differentiation revealed by the phenotype of the GATA-1^{low} mouse mutant. *J Exp Med* **197:** 281–296.

Miller IJ, Bieker JJ. 1993. A novel, erythroid cell-specific murine transcription factor that binds to the CACCC element and is related to the Krüppel family of nuclear proteins. *Mol Cell Biol* **13:** 2776–2786.

Mohseni M, Chishti AH. 2008. The headpiece domain of dematin regulates cell shape, motility, and wound healing by modulating RhoA activation. *Mol Cell Biol* **28:** 4712–4718.

Molkentin JD, Lin Q, Duncan SA, Olson EN. 1997. Requirement of the transcription factor GATA4 for heart tube formation and ventral morphogenesis. *Genes Dev* **11:** 1061–1072.

Morrisey EE, Ip HS, Lu MM, Parmacek MS. 1996. GATA-6: A zinc finger transcription factor that is expressed in multiple cell lineages derived from lateral mesoderm. *Dev Biol* **177:** 309–322.

Morrisey EE, Ip HS, Tang Z, Lu MM, Parmacek MS. 1997. GATA-5: A transcriptional activator expressed in a novel temporally and spatially-restricted pattern during embryonic development. *Dev Biol* **183:** 21–36.

Navas PA, Peterson KR, Li Q, Skarpidi E, Rohde A, Shaw SE, Clegg CH, Asano H, Stamatoyannopoulos G. 1998. Developmental specificity of the interaction between the locus control region and embryonic or fetal globin genes in transgenic mice with an HS3 core deletion. *Mol Cell Biol* **18:** 4188–4196.

Navas PA, Peterson KR, Li Q, McArthur M, Stamatoyannopoulos G. 2001. The 5′HS4 core element of the human β-globin locus control region is required for high-level globin gene expression in definitive but not in primitive erythropoiesis. *J Mol Biol* **312:** 17–26.

Nichols KE, Crispino JD, Poncz M, White JG, Orkin SH, Maris JM, Weiss MJ. 2000. Familial dyserythropoietic anaemia and thrombocytopenia due to an inherited mutation in GATA1. *Nat Genet* **24:** 266–270.

* Nienhuis AW, Nathan DG. 2012. Pathophysiology and clinical manifestations of the β-thalassemias. *Cold Spring Harb Perspect Med* **2:** a011726.

Orkin SH. 1990. Globin gene regulation and switching: Circa 1990. *Cell* **63:** 665–672.

Orkin SH, Higgs DR. 2010. Medicine. Sickle cell disease at 100 years. *Science* **329:** 291–292.

Pace BS, Qian XH, Sangerman J, Ofori-Acquah SF, Baliga BS, Han J, Critz SD. 2003. p38 MAP kinase activation mediates γ-globin gene induction in erythroid progenitors. *Exp Hematol* **31:** 1089–1096.

Pal S, Cantor AB, Johnson KD, Moran TB, Boyer ME, Orkin SH, Bresnick EH. 2004. Coregulator-dependent facilitation of chromatin occupancy by GATA-1. *Proc Natl Acad Sci* **101:** 980–985.

Pawlik KM, Townes TM. 1995. Autonomous, erythroid-specific DNase I hypersensitive site formed by human β-globin locus control region (LCR) 5′ HS 2 in transgenic mice. *Dev Biol* **169:** 728–732.

Perkins AC, Sharpe AH, Orkin SH. 1995. Lethal β-thalassaemia in mice lacking the erythroid CACCC-transcription factor EKLF. *Nature* **375:** 318–322.

Persons DA, Allay JA, Allay ER, Ashmun RA, Orlic D, Jane SM, Cunningham JM, Nienhuis AW. 1999. Enforced expression of the GATA-2 transcription factor blocks normal hematopoiesis. *Blood* **93:** 488–499.

Peterson KR, Clegg CH, Navas PA, Norton EJ, Kimbrough TG, Stamatoyannopoulos G. 1996. Effect of deletion of 5′HS3 or 5′HS2 of the human β-globin locus control region on the developmental regulation of globin gene expression in β-globin locus yeast artificial chromosome transgenic mice. *Proc Natl Acad Sci* **93:** 6605–6609.

Pevny L, Simon MC, Robertson E, Klein WH, Tsai SF, D'Agati V, Orkin SH, Costantini F. 1991. Erythroid differentiation in chimaeric mice blocked by a targeted mutation in the gene for transcription factor GATA-1. *Nature* **349:** 257–260.

Pevny L, Lin CS, D'Agati V, Simon MC, Orkin SH, Costantini F. 1995. Development of hematopoietic cells lacking transcription factor GATA-1. *Development* **121:** 163–172.

Philipsen S, Talbot D, Fraser P, Grosveld F. 1990. The β-globin dominant control region: Hypersensitive site 2. *EMBO J* **9:** 2159–2167.

Philipsen S, Pruzina S, Grosveld F. 1993. The minimal requirements for activity in transgenic mice of hypersensitive site 3 of the β globin locus control region. *Embo J* **12:** 1077–1085.

Pomerantz O, Goodwin AJ, Joyce T, Lowrey CH. 1998. Conserved elements containing NF-E2 and tandem GATA binding sites are required for erythroid-specific chromatin structure reorganization within the human β-globin locus control region. *Nucleic Acids Res* **26:** 5684–5691.

Pope NJ, Bresnick EH. 2010. Differential coregulator requirements for function of the hematopoietic transcription factor GATA-1 at endogenous loci. *Nucleic Acids Res* **38:** 2190–2200.

Pruzina S, Hanscombe O, Whyatt D, Grosveld F, Philipsen S. 1991. Hypersensitive site 4 of the human β globin locus control region. *Nucleic Acids Res* **19:** 1413–1419.

Pruzina S, Antoniou M, Hurst J, Grosveld F, Philipsen S. 1994. Transcriptional activation by hypersensitive site three of the human β-globin locus control region in murine erythroleukemia cells. *Biochim Biophys Acta* **1219:** 351–360.

Raich N, Clegg CH, Grofti J, Romeo PH, Stamatoyannopoulos G. 1995. GATA1 and YY1 are developmental repressors of the human ε-globin gene. *EMBO J* **14:** 801–809.

Rana AP, Ruff P, Maalouf GJ, Speicher DW, Chishti AH. 1993. Cloning of human erythroid dematin reveals another member of the villin family. *Proc Natl Acad Sci* **90:** 6651–6655.

Reik A, Telling A, Zitnik G, Cimbora D, Epner E, Groudine M. 1998. The locus control region is necessary for gene expression in the human β-globin locus but not the maintenance of an open chromatin structure in erythroid cells. *Mol Cell Biol* **18:** 5992–6000.

Rodgers GP, Dover GJ, Noguchi CT, Schechter AN, Nienhuis AW. 1989. Induction of fetal hemoglobin in sickle cell patients by hydroxyurea: The N.I.H. experience. *Prog Clin Biol Res* **316B:** 281–293.

Rodgers GP, Dover GJ, Uyesaka N, Noguchi CT, Schechter AN, Nienhuis AW. 1993. Augmentation by erythropoietin of the fetal-hemoglobin response to hydroxyurea in sickle cell disease. *N Engl J Med* **328:** 73–80.

Ross J, Mavoungou L, Bresnick EH, Milot E. 2012. GATA-1 utilizes Ikaros and polycomb repressive complex 2 to suppress Hes1 and to promote erythropoiesis. *Mol Cell Biol* **32:** 3624–3638.

Rupon JW, Wang SZ, Gaensler K, Lloyd J, Ginder GD. 2006. Methyl binding domain protein 2 mediates γ-globin gene silencing in adult human βYAC transgenic mice. *Proc Natl Acad Sci* **103:** 6617–6622.

Rupon JW, Wang SZ, Gnanapragasam M, Labropoulos S, Ginder GD. 2011. MBD2 contributes to developmental silencing of the human ε-globin gene. *Blood Cells Mol Dis* **46:** 212–219.

Ryan TM, Behringer RR, Martin NC, Townes TM, Palmiter RD, Brinster RL. 1989. A single erythroid-specific DNase I super-hypersensitive site activates high levels of human β-globin gene expression in transgenic mice. *Genes Dev* **3:** 314–323.

Sangerman J, Lee MS, Yao X, Oteng E, Hsiao CH, Li W, Zein S, Ofori-Acquah SF, Pace BS. 2006. Mechanism for fetal hemoglobin induction by histone deacetylase inhibitors involves γ-globin activation by CREB1 and ATF-2. *Blood* **108:** 3590–3599.

* Sankaran VG, Orkin SH. 2013. The switch from fetal to adult hemoglobin. *Cold Spring Harb Perspect Med* **3:** a011643.

Sankaran VG, Menne TF, Xu J, Akie TE, Lettre G, Van Handel B, Mikkola HK, Hirschhorn JN, Cantor AB, Orkin SH. 2008. Human fetal hemoglobin expression is regulated by the developmental stage-specific repressor BCL11A. *Science* **322:** 1839–1842.

Sankaran VG, Xu J, Ragoczy T, Ippolito GC, Walkley CR, Maika SD, Fujiwara Y, Ito M, Groudine M, Bender MA, et al. 2009. Developmental and species-divergent globin switching are driven by BCL11A. *Nature* **460:** 1093–1097.

Sankaran VG, Xu J, Orkin SH. 2010a. Advances in the understanding of haemoglobin switching. *Br J Haematol* **149:** 181–194.

Sankaran VG, Xu J, Orkin SH. 2010b. Transcriptional silencing of fetal hemoglobin by BCL11A. *Ann NY Acad Sci* **1202:** 64–68.

Sankaran VG, Menne TF, Scepanovic D, Vergilio JA, Ji P, Kim J, Thiru P, Orkin SH, Lander ES, Lodish HF. 2011. MicroRNA-15a and -16-1 act via MYB to elevate fetal hemoglobin expression in human trisomy 13. *Proc Natl Acad Sci* **108:** 1519–1524.

Schoenfelder S, Sexton T, Chakalova L, Cope NF, Horton A, Andrews S, Kurukuti S, Mitchell JA, Umlauf D, Dimitrova DS, et al. 2010. Preferential associations between co-regulated genes reveal a transcriptional interactome in erythroid cells. *Nat Genet* **42:** 53–61.

Simon MC, Pevny L, Wiles MV, Keller G, Costantini F, Orkin SH. 1992. Rescue of erythroid development in gene targeted GATA-1^{-} mouse embryonic stem cells. *Nat Genet* **1:** 92–98.

Simpson RJ, Yi Lee SH, Bartle N, Sum EY, Visvader JE, Matthews JM, MacKay JP, Crossley M. 2004. A classic zinc finger from friend of GATA mediates an interaction with the coiled-coil of transforming acidic coiled-coil 3. *J Biol Chem* **279:** 39789–39797.

Snow JW, Kim J, Currie CR, Xu J, Orkin SH. 2010a. Sumoylation regulates interaction of FOG1 with C-terminal-binding protein (CTBP). *J Biol Chem* **285:** 28064–28075.

Snow JW, Trowbridge JJ, Fujiwara T, Emambokus NE, Grass JA, Orkin SH, Bresnick EH. 2010b. A single *cis* element maintains repression of the key developmental regulator Gata2. *PLoS Genet* **6:** e1001103.

Snow JW, Trowbridge JJ, Johnson KD, Fujiwara T, Emambokus NE, Grass JA, Orkin SH, Bresnick EH. 2011. Context-dependent function of "GATA" sites in vivo. *Blood* **117:** 4769–4772.

Song SH, Hou C, Dean A. 2007. A positive role for NLI/Ldb1 in long-range β-globin locus control region function. *Mol Cell* **28:** 810–822.

Souroullas GP, Salmon JM, Sablitzky F, Curtis DJ, Goodell MA. 2009. Adult hematopoietic stem and progenitor cells require either Lyl1 or Scl for survival. *Cell Stem Cell* **4:** 180–186.

Sripichai O, Kiefer CM, Bhanu NV, Tanno T, Noh SJ, Goh SH, Russell JE, Rognerud CL, Ou CN, Oneal PA, et al. 2009. Cytokine-mediated increases in fetal hemoglobin are associated with globin gene histone modification and transcription factor reprogramming. *Blood* **114:** 2299–2306.

Stamatoyannopoulos JA, Goodwin A, Joyce T, Lowrey CH. 1995. NF-E2 and GATA binding motifs are required for the formation of DNase I hypersensitive site 4 of the human β-globin locus control region. *EMBO J* **14:** 106–116.

Stumpf M, Waskow C, Krotschel M, van Essen D, Rodriguez P, Zhang X, Guyot B, Roeder RG, Borggrefe T. 2006. The mediator complex functions as a coactivator for GATA-1 in erythropoiesis via subunit Med1/TRAP220. *Proc Natl Acad Sci* **103:** 18504–18509.

Stumpf M, Yue X, Schmitz S, Luche H, Reddy JK, Borggrefe T. 2010. Specific erythroid-lineage defect in mice conditionally deficient for mediator subunit Med1. *Proc Natl Acad Sci* **107:** 21541–21546.

Sugiyama D, Tanaka M, Kitajima K, Zheng J, Yen H, Murotani T, Yamatodani A, Nakano T. 2008. Differential context-dependent effects of friend of GATA-1 (FOG-1) on mast-cell development and differentiation. *Blood* **111:** 1924–1932.

Svensson EC, Tufts RL, Polk CE, Leiden JM. 1999. Molecular cloning of FOG-2: A modulator of transcription factor GATA-4 in cardiomyocytes. *Proc Natl Acad Sci* **96:** 956–961.

Tanabe O, Katsuoka F, Campbell AD, Song W, Yamamoto M, Tanimoto K, Engel JD. 2002. An embryonic/fetal β-type globin gene repressor contains a nuclear receptor TR2/TR4 heterodimer. *EMBO J* **21:** 3434–3442.

Tanabe O, McPhee D, Kobayashi S, Shen Y, Brandt W, Jiang X, Campbell AD, Chen YT, Chang C, Yamamoto M, et al. 2007. Embryonic and fetal β-globin gene repression by the orphan nuclear receptors, TR2 and TR4. *EMBO J* **26:** 2295–2306.

Tevosian SG, Deconinck AE, Cantor AB, Rieff HI, Fujiwara Y, Corfas G, Orkin SH. 1999. FOG-2: A novel GATA-family cofactor related to multitype zinc-finger proteins Friend of GATA-1 and U-shaped. *Proc Natl Acad Sci* **96:** 950–955.

Tevosian SG, Deconinck AE, Tanaka M, Schinke M, Litovsky SH, Izumo S, Fujiwara Y, Orkin SH. 2000. FOG-2, a cofactor for GATA transcription factors, is essential for heart morphogenesis and development of coronary vessels from epicardium. *Cell* **101:** 729–739.

Towatari M, Ciro M, Ottolenghi S, Tsuzuki S, Enver T. 2004. Involvement of mitogen-activated protein kinase in the cytokine-regulated phosphorylation of transcription factor GATA-1. *Hematol J* **5:** 262–272.

Tripic T, Deng W, Cheng Y, Zhang Y, Vakoc CR, Gregory GD, Hardison RC, Blobel GA. 2009. SCL and associated proteins distinguish active from repressive GATA transcription factor complexes. *Blood* **113:** 2191–2201.

Tsai SF, Martin DI, Zon LI, D'Andrea AD, Wong GG, Orkin SH. 1989. Cloning of cDNA for the major DNA-binding protein of the erythroid lineage through expression in mammalian cells. *Nature* **339:** 446–451.

Tsai FY, Keller G, Kuo FC, Weiss M, Chen J, Rosenblatt M, Alt FW, Orkin SH. 1994. An early haematopoietic defect in mice lacking the transcription factor GATA-2. *Nature* **371:** 221–226.

Tsang AP, Visvader JE, Turner CA, Fujiwara Y, Yu C, Weiss MJ, Crossley M, Orkin SH. 1997. FOG, a multitype zinc finger protein, acts as a cofactor for transcription factor GATA-1 in erythroid and megakaryocytic differentiation. *Cell* **90:** 109–119.

Tsang AP, Fujiwara Y, Hom DB, Orkin SH. 1998. Failure of megakaryopoiesis and arrested erythropoiesis in mice lacking the GATA-1 transcriptional cofactor FOG. *Genes Dev* **12:** 1176–1188.

Tuan D, London IM. 1984. Mapping of DNase I-hypersensitive sites in the upstream DNA of human embryonic ε-globin gene in K562 leukemia cells. *Proc Natl Acad Sci* **81:** 2718–2722.

Uda M, Galanello R, Sanna S, Lettre G, Sankaran VG, Chen W, Usala G, Busonero F, Maschio A, Albai G, et al. 2008. Genome-wide association study shows BCL11A associated with persistent fetal hemoglobin and amelioration of the phenotype of β-thalassemia. *Proc Natl Acad Sci* **105:** 1620–1625.

Vakoc CR, Letting DL, Gheldof N, Sawado T, Bender MA, Groudine M, Weiss MJ, Dekker J, Blobel GA. 2005. Proximity among distant regulatory elements at the β-globin locus requires GATA-1 and FOG-1. *Mol Cell* **17:** 453–462.

Vyas P, Ault K, Jackson CW, Orkin SH, Shivdasani RA. 1999. Consequences of GATA-1 deficiency in megakaryocytes and platelets. *Blood* **93:** 2867–2875.

Wahlberg K, Jiang J, Rooks H, Jawaid K, Matsuda F, Yamaguchi M, Lathrop M, Thein SL, Best S. 2009. The HBS1L-MYB intergenic interval associated with elevated HbF levels shows characteristics of a distal regulatory region in erythroid cells. *Blood* **114:** 1254–1262.

Watanapokasin Y, Chuncharunee S, Sanmund D, Kongnium W, Winichagoon P, Rodgers GP, Fucharoen S. 2005. In vivo and in vitro studies of fetal hemoglobin induction by hydroxyurea in β-thalassemia/hemoglobin E patients. *Exp Hematol* **33:** 1486–1492.

Weatherall DJ. 2010. The inherited diseases of hemoglobin are an emerging global health burden. *Blood* **115:** 4331–4336.

Wechsler J, Greene M, McDevitt MA, Anastasi JJEK, Le-Beau MM, Crispino JD. 2002. Acquired mutations in GATA-1 in the megakaryoblastic leukemia of Down syndrome. *Nat Genet* **32:** 148–152.

Weinberg RS, Ji X, Sutton M, Perrine S, Galperin Y, Li Q, Liebhaber SA, Stamatoyannopoulos G, Atweh GF. 2005. Butyrate increases the efficiency of translation of γ-globin mRNA. *Blood* **105:** 1807–1809.

Weiss MJ, Orkin SH. 1995. Transcription factor GATA-1 permits survival and maturation of erythroid precursors by preventing apoptosis. *Proc Natl Acad Sci* **92:** 9623–9627.

Weiss MJ, Keller G, Orkin SH. 1994. Novel insights into erythroid development revealed through in vitro differentiation of GATA-1 embryonic stem cells. *Genes Dev* **8:** 1184–1197.

Weiss MJ, Yu C, Orkin SH. 1997. Erythroid-cell-specific properties of transcription factor GATA-1 revealed by phenotypic rescue of a gene-targeted cell line. *Mol Cell Biol* **17:** 1642–1651.

Welch JJ, Watts JA, Vakoc CR, Yao Y, Wang H, Hardison RC, Blobel GA, Chodosh LA, Weiss MJ. 2004. Global regulation of erythroid gene expression by transcription factor GATA-1. *Blood* **104:** 3136–3147.

Wilber A, Nienhuis AW, Persons DA. 2011. Transcriptional regulation of fetal to adult hemoglobin switching: New therapeutic opportunities. *Blood* **117:** 3945–3953.

Wilson DB, Dorfman DM, Orkin SH. 1990. A nonerythroid GATA-binding protein is required for function of the human preproendothelin-1 promoter in endothelial cells. *Mol Cell Biol* **10:** 4854–4862.

Wozniak RJ, Keles S, Lugus JJ, Young KH, Boyer ME, Tran TM, Choi K, Bresnick EH. 2008. Molecular hallmarks of endogenous chromatin complexes containing master regulators of hematopoiesis. *Mol Cell Biol* **28:** 6681–6694.

Wu W, Cheng Y, Keller CA, Ernst J, Kumar SA, Mishra T, Morrissey C, Dorman CM, Chen KB, Drautz D, et al. 2011. Dynamics of the epigenetic landscape during erythroid differentiation after GATA1 restoration. *Genome Res* **21:** 1659–1671.

Xu J, Sankaran VG, Ni M, Menne TF, Puram RV, Kim W, Orkin SH. 2010. Transcriptional silencing of γ-globin by BCL11A involves long-range interactions and cooperation with SOX6. *Genes Dev* **24:** 783–798.

Xu J, Peng C, Sankaran VG, Shao Z, Esrick EB, Chong BG, Ippolito GC, Fujiwara Y, Ebert BL, Tucker PW, et al. 2011. Correction of sickle cell disease in adult mice by interference with fetal hemoglobin silencing. *Science* **334:** 993–996.

Yamamoto M, Ko LJ, Leonard MW, Beug H, Orkin SH, Engel JD. 1990. Activity and tissue-specific expression of the transcription factor NF-E1 multigene family. *Genes Dev* **4:** 1650–1662.

Yao X, Kodeboyina S, Liu L, Dzandu J, Sangerman J, Ofori-Acquah SF, Pace BS. 2009. Role of STAT3 and GATA-1 interactions in γ-globin gene expression. *Exp Hematol* **37:** 889–900.

Yi Z, Cohen-Barak O, Hagiwara N, Kingsley PD, Fuchs DA, Erickson DT, Epner EM, Palis J, Brilliant MH. 2006. Sox6 directly silences ε globin expression in definitive erythropoiesis. *PLoS Genet* **2:** e14.

Yu C, Cantor AB, Yang H, Browne C, Wells RA, Fukiwara Y, Orkin SH. 2002. Targeted deletion of a high-affinity GATA-binding site in the GATA-1 promoter leads to selective loss of the eosinophil lineage in vivo. *J Exp Med* **195:** 1387–1395.

Yu YL, Chiang YJ, Chen YC, Papetti M, Juo CG, Skoultchi AI, Yen JJ. 2005. MAPK-mediated phosphorylation of GATA-1 promotes Bcl-XL expression and cell survival. *J Biol Chem* **280:** 29533–29542.

Yu M, Riva L, Xie H, Schindler Y, Moran TB, Cheng Y, Yu D, Hardison R, Weiss MJ, Orkin SH, et al. 2009. Insights into GATA-1-mediated gene activation versus repression via genome-wide chromatin occupancy analysis. *Mol Cell* **36:** 682–695.

Yu Y, Wang J, Khaled W, Burke S, Li P, Chen X, Yang W, Jenkins NA, Copeland NG, Zhang S, et al. 2012. Bcl11a is essential for lymphoid development and negatively regulates p53. *J Exp Med* **209:** 2467–2483.

Zhao W, Kitidis C, Fleming MD, Lodish HF, Ghaffari S. 2006. Erythropoietin stimulates phosphorylation and activation of GATA-1 via the PI3-kinase/AKT signaling pathway. *Blood* **107:** 907–915.

Zhou D, Pawlik KM, Ren J, Sun CW, Townes TM. 2006. Differential binding of erythroid Krupple-like factor to embryonic/fetal globin gene promoters during development. *J Biol Chem* **281:** 16052–16057.

Zhou D, Liu K, Sun CW, Pawlik KM, Townes TM. 2010. KLF1 regulates BCL11A expression and γ- to β-globin gene switching. *Nat Genet* **42:** 742–744.

Zhu X, Wang Y, Pi W, Liu H, Wickrema A, Tuan D. 2012. NF-Y recruits both transcription activator and repressor to modulate tissue- and developmental stage-specific expression of human γ-globin gene. *PLoS ONE* **7:** e47175.

Zohren F, Souroullas GP, Luo M, Gerdemann U, Imperato MR, Wilson NK, Gottgens B, Lukov GL, Goodell MA. 2012. The transcription factor Lyl-1 regulates lymphoid specification and the maintenance of early T lineage progenitors. *Nat Immunol* **13:** 761–769.

Zon LI, Tsai SF, Burgess S, Matsudaira P, Bruns GA, Orkin SH. 1990. The major human erythroid DNA-binding protein (GF-1): Primary sequence and localization of the gene to the X chromosome. *Proc Natl Acad Sci* **87:** 668–672.

Zon LI, Gurish MF, Stevens RL, Mather C, Reynolds DS, Austen KF, Orkin SH. 1991. GATA-binding transcription factors in mast cells regulate the promoter of the mast cell carboxypeptidase A gene. *J Biol Chem* **266:** 22948–22953.

The Switch from Fetal to Adult Hemoglobin

Vijay G. Sankaran[1,2,3] **and Stuart H. Orkin**[1,4,5]

[1]Division of Hematology/Oncology, Children's Hospital Boston, Harvard Medical School, Boston, Massachusetts 02115

[2]Whitehead Institute for Biomedical Research, Cambridge, Massachusetts 02142

[3]Broad Institute, Cambridge, Massachusetts 02142

[4]Dana-Farber Cancer Institute, Boston, Massachusetts 02115

[5]Howard Hughes Medical Institute, Boston, Massachusetts 02115

Correspondence: sankaran@broadinstitute.org; stuart_orkin@dfci.harvard.edu

The fetal-to-adult hemoglobin switch and silencing of fetal hemoglobin (HbF) have been areas of long-standing interest among hematologists, given the fact that clinical induction of HbF production holds tremendous promise to ameliorate the clinical symptoms of sickle cell disease (SCD) and β-thalassemia. In this article, we discuss historic attempts to induce HbF that have resulted in some therapeutic approaches to manage SCD and β-thalassemia. We then go on to discuss how more recent molecular studies that have identified regulators, including BCL11A, MYB, and KLF1, hold great promise to develop targeted and more effective approaches for HbF induction. We go on to discuss strategies by which such approaches may be developed. Older studies in this field can provide important lessons for future studies aimed at developing more effective strategies for HbF induction, and we therefore chronologically cover the work accomplished as this field has evolved over the course of the past four decades.

As with many facets of the history of the hemoglobin field, the study of the fetal-to-adult hemoglobin switch and work on fetal hemoglobin silencing has a rich and storied history that spans the course of several decades. In this work, we begin with a historic overview of this field and then move on to discuss more recent molecular findings that are providing important new insight into this process. There is great hope that these new molecular studies may lead to important targeted therapeutic advances for fetal hemoglobin (HbF) induction. However, the historic work in this field suggests that the road to effective therapies may not always be straightforward and reconsidering seemingly simple observations can often yield important new insights. There are many valuable lessons that can be learned from both the historic and more recent work in this field and therefore we attempt to cover the chronology of findings that have resulted in our current understanding of the fetal-to-adult hemoglobin switch.

The disorders of β-hemoglobin, sickle cell disease (SCD) and β-thalassemia, are major sources of morbidity and mortality worldwide (Weatherall et al. 2006). These diseases are the

Copyright © 2013 Cold Spring Harbor Laboratory Press; all rights reserved
Cite this article as *Cold Spring Harb Perspect Med* doi: 10.1101/cshperspect.a011643

most common genetic disorders in the world. It is currently estimated that at least 300,000 children with these diseases are born every year, giving rise to a worldwide population of tens of millions of patients with these diseases (Weatherall et al. 2006). With the epidemiological transition that most developing countries are currently undergoing—with significant decreases in mortality from infectious and nutritional causes—there will likely be an increase in the prevalence of these diseases in the ensuing years. Therefore, developing improved therapies for these disorders will be of paramount importance for this important global health burden.

In SCD, a single base substitution causes a missense mutation of a valine for glutamic acid at amino acid six of the β-globin protein chain, which leads to a propensity of the sickle hemoglobin to polymerize. This in turn results in deformation of erythrocytes containing this hemoglobin, which can block small blood vessels, leading to impaired oxygen delivery to tissues. This can result in significant clinical complications including pain crises, respiratory complications, and organ damage. In β-thalassemia, insufficient production of the β-globin molecule results in an excess of unpaired α-globin chains that can precipitate within erythroid precursors. The precipitation of these free α-globin chains impairs the maturation and leads to death of these precursors, causing ineffective production of erythroid cells. As a result, a significant anemia occurs and the consequent expansion of erythroid precursors can lead to secondary problems in bones and other organs.

The only potentially curative therapies that are available for these diseases are bone marrow transplantation (Michlitsch and Walters 2008) and gene therapy (Persons 2009), but there are numerous challenges that exist to implement these treatments and they have significant limitations for widespread use, particularly in the developing world (Sankaran and Nathan 2010b). These curative therapies still remain largely experimental, although important scientific and clinical advances in these approaches are being made (Hsieh et al. 2009; Cavazzana-Calvo et al. 2010). Currently, the major treatment for these disorders involves symptomatic care and transfusion of red blood cells as clinically necessary. However, the use of regular blood transfusions can lead to significant problems. Most notably, iron overload can lead to major clinical complications, even when the strictest iron chelation regimens are used (Porter and Shah 2010).

THE FETAL-TO-ADULT GLOBIN SWITCH

The hemoglobin molecule is a tetramer composed of two subunits of α-like globin peptide chains and two subunits of the β-like globin peptides, along with heme moieties necessary for this molecule's oxygen-carrying capacity. A variety of β-like globin molecules are produced as a result of the fact that the human β-globin locus on chromosome 11 is developmentally regulated (Fig. 1) (Sankaran and Nathan 2010a; Sankaran et al. 2010). In the early part of the first trimester, there is robust expression of an embryonic form of a β-like globin known as ε-globin within the yolk sac-derived primitive lineage of erythrocytes (Peschle et al. 1985). Soon thereafter, when production of the first enucleated definitive erythrocytes commences from stem and progenitor cells in the fetal liver (McGrath and Palis 2008), the predominant β-like globin molecule produced is γ-globin (Sankaran et al. 2010). This molecule is encoded by two duplicated *γ-globin* genes found within the β-globin gene cluster (Fig. 1). The γ-globin chains combine with adult α-globin chains into a stable tetramer forming HbF. This remains the predominant hemoglobin for much of gestation. Shortly after the time of birth there is a switch from predominant expression of HbF to adult hemoglobin (HbA), which is mediated by a transcriptional switch in definitive erythroid progenitors from *γ-* to *β-globin* (Fig. 1).

A variety of clinical observations have shown that the severity of SCD and β-thalassemia is ameliorated via increased production of HbF. Children with SCD were noted to be asymptomatic until after infancy, which was postulated to be attributable to elevated fetal hemoglobin (HbF) levels (Watson 1948). This notion was substantiated by observations of rare patients with compound heterozygosity for sickle cell

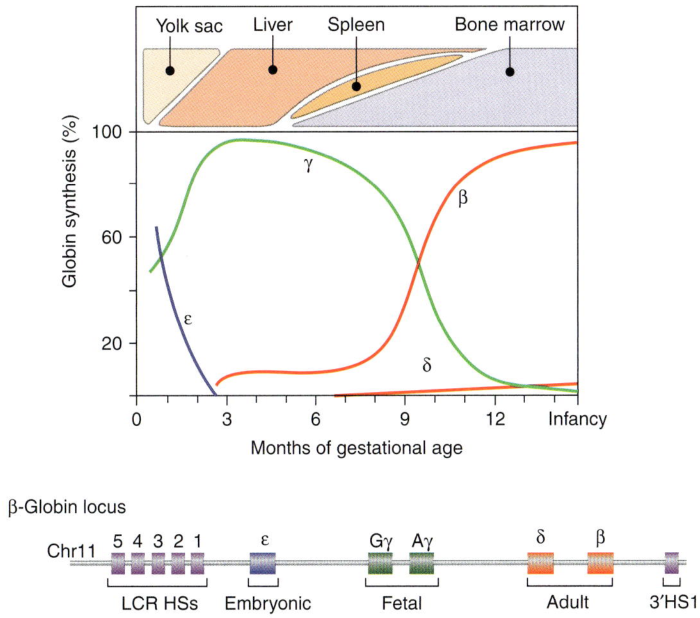

Figure 1. The fetal-to-adult hemoglobin switch. This illustration depicts the normal timing of the developmental hemoglobin switches in humans. In the *top* panel, the sites and levels of various β-like globin molecules are shown with colors corresponding to the various developmental groups of genes shown below it in a model of the human β-globin locus (embryonic in blue, fetal in green, and adult in red). The *bottom* illustration also depicts the upstream enhancer of the β-globin locus, known as the locus control region (LCR), with its corresponding DNAse I hypersensitivity sites (HSs) and a downstream HS known as the 3′HS1.

disease and hereditary persistence of fetal hemoglobin (HPFH) mutations, who were largely asymptomatic (Weatherall and Clegg 2001). Moreover, certain populations of patients with SCD, such as those from parts of Saudi Arabia and India, had unusually high levels of HbF associated with a less symptomatic clinical course (Perrine et al. 1972; Kar et al. 1986; Weatherall and Clegg 2001). These observations were subsequently confirmed with larger epidemiological studies in SCD, which showed that increased HbF levels can significantly and quantitatively ameliorate the clinical severity and decrease mortality in SCD (Platt et al. 1991, 1994; Castro et al. 1994). Similar observations have been made in patients with β-thalassemia. Clinical observations in rare β-thalassemia patients with elevated production of HbF showed that these increased levels resulted in a milder clinical course and infants with β-thalassemia only begin to show symptoms after the expression of HbF declines in the months following birth (Weatherall 2001; Weatherall and Clegg 2001). Larger epidemiological studies of thalassemia populations have confirmed such findings (Premawardhena et al. 2005; Galanello et al. 2009; Nuinoon et al. 2010). While it has been difficult to tease apart the extent to which increased HbF production can ameliorate symptoms in thalassemia because it is difficult to ascertain the extent of γ-globin production directly in patients with the need for frequent blood transfusions and such a great extent of ineffective erythropoiesis, a recent analysis in untransfused patients with β-thalassemia shows the quantitative ameliorating effect of HbF levels on clinical severity in this disease as well (Musallam et al. 2011).

EARLY MOLECULAR STUDIES LEAD TO THERAPEUTIC APPROACHES FOR HbF INDUCTION

Given the variety of robust clinical observations demonstrating the ability of HbF to ameliorate the severity of both SCD and β-thalassemia, there has been a long-standing interest in devising therapeutic approaches to stimulate the production of this beneficial form of hemoglobin. With the advent of molecular biology, the cloning of the globin genes and mapping of the entire β-globin gene cluster became a reality (Fig. 1) (Fritsch et al. 1980; Sankaran and Nathan 2010b). In the course of this work, it was noted that the *γ-globin* genes underwent DNA methylation when silenced in adult erythroid progenitors, while not having DNA methylation present when transcriptionally active in fetal erythroid progenitor cells (van der Ploeg and Flavell 1980). Around this time, the DNA methylation inhibitor 5-azacytidine was first being characterized and was shown to have potent activity to reverse DNA methylation in a variety of mammalian cell types (Jones and Taylor 1980; Groudine et al. 1981). Given the similarity in the regulation of the primate β-globin gene cluster compared with that of humans, 5-azacytidine was tested and showed a potent ability to induce the silenced *γ-globin* genes in adult erythroid cells of primates (DeSimone et al. 1982). This led to small clinical trials of 5-azacytidine in patients with SCD and β-thalassemia in which robust responses were initially seen in the ability of this molecule to stimulate HbF production and reduce clinical markers of disease (Ley et al. 1982, 1983). There was great hope and promise that this would lead to a broadly applicable targeted molecular approach for HbF induction (Benz 1982), but there was significant concern over the mutagenic potential of 5-azacytidine.

Soon after this work was performed, a number of investigators began to question whether the HbF-inductive activity of 5-azacytidine may be directly attributable to its DNA demethylating activity or whether its activity as a cell-cycle inhibitor, specifically an S-phase inhibitor, may play a role in this activity. As a result, a number of S-phase inhibitors that had varying mechanisms of action were tested and nearly all of these compounds showed robust activity at inducing HbF in primate models (Letvin et al. 1984, 1985; Papayannopoulou et al. 1984). This led to small clinical trials with the safest of these compounds, hydroxyurea, in SCD patients in which HbF-inductive effects were observed (Platt et al. 1984; Platt 2008). Subsequent clinical trials showed that treatment with hydroxyurea could ameliorate clinical symptoms in SCD (Charache et al. 1995; Steinberg et al. 2003; Platt 2008) leading to its FDA approval for the management of adult patients with SCD. The exact molecular mechanisms by which hydroxyurea is able to induce HbF are still poorly understood (Platt 2008) and much more work is needed to both investigate these mechanisms and to further investigate whether hydroxyurea or other S-phase inhibitors may be useful as therapeutic agents in β-thalassemia.

Around the same period of time, it was shown that infants of diabetic mothers have a delayed fetal-to-adult hemoglobin switch (Bard and Prosmanne 1985; Perrine et al. 1985). Because it was known that β-hydroxybutyrate is elevated in these diabetic mothers, a series of experiments was performed to test whether butyrate or other similar short-chain fatty acids may be effective as inducers of HbF. Although initial trials showed promise (Perrine et al. 1993, 1994), these therapies did not show consistent effectiveness in larger clinical trials (Sher et al. 1995; Sankaran and Nathan 2010b). It should be noted nonetheless that a subset of patients did show responses in these trials (Sher et al. 1995; Sankaran and Nathan 2010b). However, additional concerns were raised over possible toxic side effects and the difficulty in delivery of these medications, which limited further clinical trials and drug development efforts. It has been suggested that these short-chain fatty acids are likely to induce HbF through the inhibition of histone deacetylases (HDACs) (Fathallah et al. 2007). A variety of HDAC inhibitors are potent inducers of HbF in vitro and further work is needed to study these molecules (Cao et al. 2004; Atweh and Fathallah 2010; Bradner et al. 2010). Although the majority of studies have been focused on nonspecific HDAC inhibitors, recent work suggests that inhibitors of specific HDACs, such as HDAC 1 and 2, may allow improved efficacy of HbF induction while potentially minimizing toxicity (Bradner et al. 2010).

THE PROMISE OF IMPROVED THERAPIES THROUGH MOLECULAR STUDIES OF FETAL HEMOGLOBIN REGULATION

Although the nontargeted therapeutic approaches discussed above have shown promise and some success at inducing HbF in clinical settings, an improved mechanistic understanding of the molecular underpinnings necessary for the normal fetal-to-adult hemoglobin switch holds great promise for allowing more effective and specifically targeted approaches for HbF induction to be developed. In the decades following the molecular cloning of the globin genes, a variety of transcription factors that played roles in globin gene regulation were identified (Cantor and Orkin 2002). This included transcription factors such as GATA1, KLF1, and SCL/TAL1. However, studying the role of these factors in globin gene regulation was confounded by the broad role of these factors in differentiation and their pleiotropic roles as regulators of globin and erythroid gene regulation. None of these factors appeared to be specific regulators of the fetal-to-adult hemoglobin switch. It took nearly three decades from the initial cloning of the globin genes before specific regulators of this process would be identified.

Important clues into the identity of such factors came about from the study of natural human genetic variation. A number of groups pursued the basis for common genetic variation in HbF levels using both targeted and genome-wide association studies (GWAS) (Menzel et al. 2007; Thein et al. 2007; Lettre et al. 2008; Uda et al. 2008). These studies resulted in the identification of three genomic loci harboring common variants that influenced HbF levels. This included a region of chromosome 2 within the *BCL11A* gene, a region intergenic between the genes *HBS1L* and *MYB* on chromosome 6, and variants within the β-globin locus on chromosome 11. Recent studies that have more finely mapped these genetic variants suggest that >50% of the variation in HbF levels may be explained by common variation at these three loci (Galarneau et al. 2010). Although HbF levels are thought to have a heritable genetic component in the range of ~80%, it is important to bear in mind that the additive genetic variation found from common genetic variants ignores potential higher-order genetic interactions that may not be detected and therefore the extent to which HbF levels are genetically determined remains to be explored in future studies (Zuk et al. 2012).

The initial observation of variants associated with HbF levels within the zinc-finger transcription factor *BCL11A* led to an initial study that pursued the hypothesis that BCL11A may be a regulator of HbF expression (Sankaran et al. 2008). Previous work had suggested that BCL11A was a critical transcriptional regulator involved in B lymphopoiesis and neurogenesis (Sankaran et al. 2010). BCL11A protein levels appeared to correlate with the developmental stage of expression, such that primitive and fetal liver definitive erythroid cells that expressed high levels of *γ-globin*, had low or absent expression of the full-length forms of BCL11A. This result suggested that this gene product was acting as a repressor of the *γ-globin* genes. To directly test this, knockdown of BCL11A using short-hairpin RNA (shRNA) approaches was performed in primary adult erythroid progenitors and γ-globin expression could be robustly induced on such a knockdown. The degree of γ-globin induction appeared to be related to the extent of knockdown of BCL11A. Interestingly, major perturbations in erythropoiesis did not appear to occur despite the robust γ-globin induction seen. Follow-up studies have shown similar results when knocking down the expression of BCL11A using shRNAs (Borg et al. 2010; Zhou et al. 2010; Wilber et al. 2011). It was additionally shown that BCL11A directly interacts with chromatin at the human β-globin locus in primary erythroid cells and that it appeared to act as part of a complex with the transcription factor GATA1 and the NuRD chromatin remodeling and repressor complex (Sankaran et al. 2008). Of interest, the NuRD complex contains HDACs 1 and 2, which have been suggested to be the critical HDACs necessary for HbF silencing (Bradner et al. 2010). In addition, it has been suggested that the transcription factor SOX6 may cooperate with BCL11A to help silence the *γ-globin* genes in humans and it may be essential for binding the proximal promoter of these globin genes (Xu et al. 2010).

Although hemoglobin switching in mouse models, even those harboring a human β-globin locus transgene, appears to diverge from the normal ontogeny of globin expression seen in humans, an evolutionarily conserved role of BCL11A in globin gene silencing and switching was shown in mice (Sankaran et al. 2009; McGrath et al. 2011). Mice lacking BCL11A appear to have normal erythropoiesis, but fail to fully silence the embryonic globin genes in definitive erythroid cells and allow some persistent expression of γ-globin when the intact human β-globin locus is present. These findings reinforce the importance of BCL11A as a critical mediator of globin switching in mammals. Although this initial study addressed the role of BCL11A in the developmental process of globin switching in mice (Sankaran et al. 2009), a more recent study used conditional inactivation of BCL11A to show that inducible inactivation can result in robust and similar extents of *γ-globin* gene induction as occurs when inactivation occurs at earlier time points (Xu et al. 2011). Moreover, although there is differential regulation of the globin genes between humans and mice, inactivation of BCL11A was shown to be sufficient to ameliorate the hematological and pathological features seen in mouse models of sickle cell disease, providing an important proof of principal for the potential efficacy of targeting BCL11A to induce HbF (Xu et al. 2011).

The exact mechanisms by which BCL11A silences *γ-globin* gene expression remain unclear. A recent study suggests that this may be mediated through both interactions with transcription factors, such as SOX6 that bind chromatin at the proximal *γ-globin* promoters, as well as through long-range interactions with a variety of regions throughout the β-globin gene cluster (Xu et al. 2010). When BCL11A is absent, the conformation of the β-globin locus changes such that the upstream enhancer known as the locus control region is juxtaposed with the transcriptionally activated *γ-globin* genes. A similar phenomenon occurs when cells are treated with HDAC inhibitors that induce *γ-globin* gene expression (Bradner et al. 2010). However, it is unclear if these conformational alterations are directly mediated by BCL11A or if these alterations occur secondarily to the HbF-inductive effect of BCL11A (or HDAC) inhibition. Nonetheless, these findings strongly support the notion that BCL11A appears to have a direct role in silencing *γ-globin* expression within the β-globin locus. By mapping a variety of deletions within the human β-globin locus that either result in δβ-thalassemia, with modest increases in HbF and some remaining globin chain imbalance present, or HPFH, with robust increases in HbF and balanced globin chain synthesis, it was shown that an N3-kb region upstream of the *δ-globin* gene is necessary for silencing of the *γ-globin* genes (Fig. 2) (Sankaran et al. 2011c). Interestingly, this region harbors binding sites for BCL11A, along with its partners such as GATA1 and HDAC1. Of note, this region has also been independently implicated as being important for γ-globin silencing through studies of the Corfu thalassemia deletion in humans (Chakalova et al. 2005). This study provides important mechanistic insight about how BCL11A functions to silence HbF and reinforces the importance of natural human mutations in understanding this developmental process that is unique to humans (Fig. 2) (Sankaran et al. 2011c).

Given these findings, BCL11A is likely to be an important therapeutic target. The fact that its inactivation induces HbF without resulting in a major perturbation in erythropoiesis is very promising, although it is known to have important effects in other lineages such as B lymphocytes, stressing the importance of in vivo modeling and analysis as part of ongoing efforts to target BCL11A for HbF induction. Further studies exploring the mechanisms of action by which BCL11A functions could lead to better and more specifically targeted approaches for HbF induction (Sankaran et al. 2011c). SOX6 may also be a potentially promising target for HbF induction (Xu et al. 2010), although it is known to be necessary for normal erythropoiesis (Dumitriu et al. 2006). However, a patient has been reported with a heterozygous disruption of *SOX6* who did not have elevated HbF levels, suggesting that the *SOX6* gene may have a

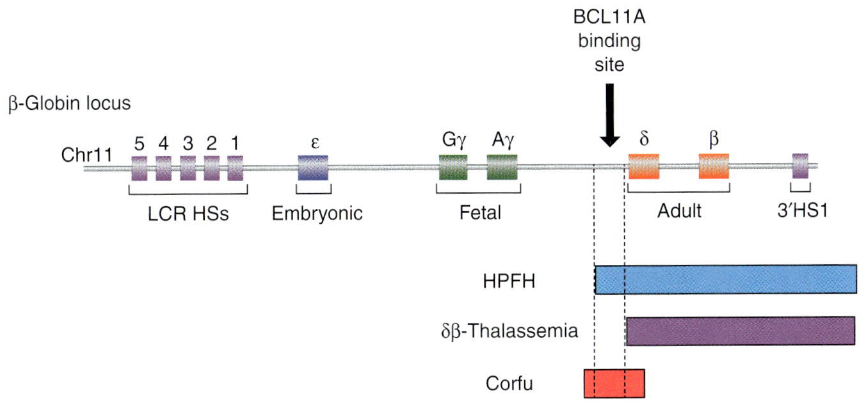

Figure 2. A model for regulation of γ-globin silencing in the human β-globin locus. This illustration depicts the human β-globin locus as shown in Figure 1 with an ~3-kb region upstream of the *δ-globin* gene that was found by comparing the regions removed in various hereditary persistence of fetal hemoglobin (HPFH) deletions with the regions removed by δβ-thalassemia deletions (Sankaran et al. 2011c). Typical deletions are illustrated in the model below the locus. In addition, the Corfu thalassemia deletion is also known to remove this region, as shown by the model below. BCL11A has been shown to bind to chromatin within this 3-kb region, along with its partners GATA1 and HDAC1 (Sankaran et al. 2011c).

mechanism for dosage compensation or that there may be a particular threshold necessary for reducing expression of this gene to have robust HbF induction (Sankaran et al. 2011a). This suggests that there may be limitations in considering SOX6 as a potential HbF induction target.

Following the initial studies on BCL11A, two studies suggested that the expression of BCL11A appears to be controlled by the erythroid specific transcription factor KLF1 using independent and complimentary approaches. In one study, a mouse with a hypomorphic allele of *KLF1* resulted in elevations of embryonic globin expression and transgenic mice with the human β-globin locus showed persistent expression of *γ-globin* (Zhou et al. 2010). As a result, the investigators tested whether the same regulation could occur in primary human erythroid cells and could show a similar connection using shRNA approaches. The investigators then showed that this effect occurs through both direct effects of KLF1 at the β-globin locus, but also via indirect effects mediated through reduced expression of *BCL11A* on KLF1 knockdown. This finding showed that KLF1 was a direct positive transcriptional regulator of *BCL11A* expression. A second group examined the genetic basis of an unlinked form of HPFH that was suggested to result from a *KLF1* missense mutation in this family (Borg et al. 2010). The investigators could show, using primary cells from these patients and from unaffected controls, that the effect seen was in part attributable to a silencing effect of KLF1 on *BCL11A*. One area of ongoing uncertainty relates to the human phenotypes seen with various cases of *KLF1* mutations. Although in the initial report, all recipients of the missense mutation had elevations in HbF expression, it should be noted that this actually varied between 3% and 19% of total hemoglobin levels. In addition, other reports of heterozygous *KLF1* mutations in humans either show concomitant disruption of erythropoiesis or show little effect on HbF expression (Arnaud et al. 2010; Satta et al. 2011). More recent studies suggest that rare variants in *KLF1* are indeed associated with elevations in HbF, but this does not appear to occur consistently or to the same extent even with similar mutations (Gallienne et al. 2012). The basis of this variation remains to be determined and will be important to better understand the mechanisms by which KLF1 acts both directly and indirectly, to affect HbF expression. This will be critical for any future work attempting to target KLF1 for HbF induction, particularly if the untoward effects of such inhibition on erythroid differentiation are to be avoided. However, given the specificity of KLF1 within the erythroid lineage and if the globin gene regulating activity of KLF1 could be specifically targeted, it should still be considered a candidate for inducing HbF.

In addition to the common genetic variants identified in *BCL11A* associated with HbF levels in humans, the GWAS studies showed that there were variants on chromosome 6 intergenic between the genes *HBS1L* and *MYB* that appear to have a dramatic effect on HbF levels in humans (Menzel et al. 2007; Thein et al. 2007; Lettre et al. 2008; Uda et al. 2008). Understanding the mechanism of action by which these variants result in alterations in HbF levels is important, as the genetic variants at this locus appear to have as great or perhaps even a greater effect on clinical morbidity in the β-hemoglobinopathies as those variants at the *BCL11A* locus (Galanello et al. 2009; Nuinoon et al. 2010). Interestingly, this region contains a variety of regulatory elements that have been suggested to have an important role in regulating expression of *MYB* in erythroid progenitors (Mukai et al. 2006; Wahlberg et al. 2009; Stadhouders et al. 2011). Although overexpression of *HBS1L* did not appear to effect γ-globin expression in K562 erythroleukemia cells, overexpression of *MYB* did appear to affect the level of γ-globin produced in these cells (Jiang et al. 2006). Moreover, cultures of primary erythroid progenitors from humans with higher expression of HbF more often had decreased expression of *MYB* (Jiang et al. 2006).

By following up on the classic clinical observation that patients with a trisomy of chromosome 13 have a delayed fetal-to-adult hemoglobin switch and continue to have persistent expression of HbF (Huehns et al. 1964; Sankaran and Sapp 2012), recent studies have provided further evidence for a role of *MYB* in regulating HbF expression. By fine mapping and performing integrative genomic analysis of genes in a region on chromosome 13 implicated as being necessary for the elevation of HbF seen in patients with partial trisomies of chromosome 13, it was found that there were two small ∼22 nucleotide RNA molecules that were top candidates to carry out such an activity, microRNAs 15a and 16-1 (Sankaran et al. 2011b). Overexpression of these microRNAs in primary adult erythroid cells in culture resulted in increases in γ-globin production. By examining targets of these microRNAs, it was noted that one major target in erythroid cells was *MYB*. Direct knockdown of *MYB* in primary adult erythroid progenitors resulted in dramatic increases in γ-globin production (Sankaran et al. 2011b), connecting this historic observation from a rare human aneuploidy syndrome to the more recent work examining molecular mechanisms regulating HbF levels.

The mechanism by which *MYB* may act to regulate HbF levels remains unclear. This may be owing to an effect on the kinetics of erythropoiesis or alternatively may also occur as a result of a direct effect within the β-globin locus (Higgs and Wood 2008). Further work will be necessary to explore such mechanisms and hold great promise in attempts to therapeutically target this molecule for HbF induction. There is concern that targeting of *MYB* could have undesirable side effects, particularly given the pleiotropic role of *MYB* in hematopoiesis (Emambokus et al. 2003; Carpinelli et al. 2004). However, recent work suggests that such strategies may actually hold promise, because partial knockdown of *Myb* in mice had little effect on normal in vivo hematopoiesis while dramatically blocking progression of leukemias (Zuber et al. 2011a). Therefore, partial inhibition of *MYB* may hold promise as a valuable strategy for HbF induction. The additivity of the effects of variants in the *HBS1L-MYB* intergenic region along with variants at the *BCL11A* locus (Lettre et al. 2008; Galanello et al. 2009; Nuinoon et al. 2010) suggests that targeting both of these pathways together could yield even more robust effects than targeting either pathway alone.

Although the regulators of HbF regulation discussed above have been found through human genetic studies, thus confirming their in vivo relevance in humans, a variety of other molecules have also been suggested to play roles in *γ-globin* gene regulation through various studies using cell culture approaches or with mouse models (Table 1). These molecules will be discussed below. It is important to bear in mind that there are limitations for most of the currently available experimental systems used to study *γ-globin* gene regulation. Primary human erythroid cells seem to be permissive to manipulation that will allow increases in γ-globin production, which may not always be relevant in humans in vivo. In addition, many of the stimuli that are known to induce γ-globin production in vivo in humans, including various types of stress erythropoiesis and treatment with hydroxyurea, do not function to induce γ-globin production in humanized mouse models (Pace et al. 1994; Sankaran et al. 2009) stressing an important limitation for interpreting negative findings in this experimental system. Thus, caution needs to be exercised when interpreting experimental findings not supported by evidence from human genetic studies or studies performed in humans or primates in vivo.

By examining proteins that bind the direct repeat elements found within the promoter of the *γ-globin* genes, the orphan nuclear hormone receptors TR2 and TR4 have been suggested to play a role as repressors of the expression of the *γ-globin* genes (Tanabe et al. 2002, 2007; Cui et al. 2011). Paradoxically, overexpression of TR2 and TR4 in transgenic mouse models results in elevated expression of γ-globin and when overexpressed in sickle cell disease mouse models can result in partial amelioration of hematologic and pathologic symptoms of these mice (Campbell et al. 2011). The mechanisms underlying these observations and the relevance to human globin gene regulation remains to be established for TR2 and TR4. In addition, the orphan nuclear hormone nuclear receptor COUP-TFII has also been suggested to bind to the direct repeats and repress the *γ-globin* promoter in humans (Filipe et al. 1999). Using in vitro cultures of primary adult erythroid cells, it was shown that cytokines such as SCF appear to down-regulate expression and chromatin occupancy of COUP-TFII at the β-globin locus, resulting in increased γ-globin expression (Aerbajinai et al. 2009). Moreover, direct down-regulation of COUP-TFII using small interfering RNAs (siRNAs) could result in increases in γ-globin production (Aerbajinai et al. 2009). Further studies on the role of COUP-TFs in the regulation of HbF will be needed to better understand its physiologic role in humans.

Recent studies have also suggested a role for the gene named friend of PRMT1 (FOP1) in repression of *γ-globin* gene expression (van Dijk et al. 2010). Knockdown of FOP1 results in elevated HbF expression in human adult erythroid cells in culture. It was suggested that this was mediated through reduced expression of SOX6, knockdown of which is known to also result in increases in γ-globin production (Xu et al. 2010). Further work will be needed to

Table 1. A partial list of fetal hemoglobin regulators

Regulator	Direction of modulation needed to increase HbF	Human genetic evidence supporting role in HbF regulation	Human or primate studies modulating factor involved in HbF regulation	Cell culture data supporting a role in HbF regulation	Evidence from mouse models suggesting a role in HbF regulation
BCL11A	↓	×		×	×
KLF1	↓	×		×	×
MYB	↓	×		×	
MicroRNAs 15a/16-1	↑	×		×	
SOX6	↓			×	
HDACs 1/ 2	↓		×	×	×
DNMT1	↓		×	×	×
TR2/TR4	↓ or ↑			×	×
COUP-TFII	↓			×	
FOP	↓			×	
NF-E4	↑			×	

confirm these findings, understand whether this gene also has a role in erythropoiesis more broadly, and examine the mechanism underlying these observations.

Some studies have also suggested a role for the stage-selector protein NF-E4 as an activator of γ-globin expression in humans (Jane et al. 1995; Zhao et al. 2006). It has been suggested that a short form of NF-E4 may have a role in repressing the *γ-globin* genes by inhibiting the normal activating function of NF-E4 at the promoter of these genes (Zhou et al. 2004). These studies have all been based on experiments performed in cultured cells and biochemical purification of protein complexes from K562 cells and therefore further work will be needed to confirm the physiological role of this gene in γ-globin expression and better understand the mechanisms through which this complex may be acting to alter human globin gene regulation.

TRANSLATING THE INSIGHT FROM MOLECULAR STUDIES OF HbF REGULATION INTO THERAPIES

The majority of molecules that have been identified as regulators of the fetal-to-adult hemoglobin switch and HbF silencing are transcription factors, including BCL11A, MYB, and KLF1 (Table 1). These proteins are notoriously difficult to target using traditional small molecule approaches (Koehler 2010). With the increasing diversity of chemical libraries that are being synthesized, it is possible that such small molecule modulators of these targets may be found (Dandapani and Marcaurelle 2010). An alternative and perhaps simpler approach, would involve targeting of enzymatic partner proteins of these transcription factors. For example, BCL11A appears to interact with the NuRD chromatin remodeling and repressor complex that contains HDACs 1 and 2 (Sankaran et al. 2008). A number of small molecule inhibitors of these molecules exist (Bradner et al. 2010), suggesting a potential strategy for indirectly targeting such factors. In addition, other epigenetic modifier enzymes, such as the DNA methyltransferase DNMT1, have been suggested to be potentially good targets for HbF induction, which is supported by experiments using direct knockdown of the enzyme and from the known efficacy of 5-azacytidine (Banzon et al. 2011). However, further mechanistic studies of these HbF regulators will be needed before such strategies can be successfully used.

Recent studies have suggested some other creative approaches to targeting transcription factors. One strategy involves the use of stapled peptides to inhibit transcription factor complexes by interfering with protein–protein interaction interfaces, as has been successfully performed for the NOTCH transcriptional complex involved in the pathogenesis of certain forms of leukemia (Moellering et al. 2009). An alternative strategy would be to target "druggable" molecules that would interfere with the expression of the target transcription factor. Recent work has shown that such a strategy can be successfully used to down-regulate expression of the oncogene *MYC* in a number of cancers by targeting a bromodomain protein necessary for its transcription (Delmore et al. 2011; Zuber et al. 2011b).

A more immediate strategy to target some of these transcription factors such as BCL11A, KLF1, or potentially MYB would be to use siRNA or shRNA delivery either through exogenous means or through the use of gene therapy approaches. The targeting of exogenous siRNAs to the desired cell type remains a major challenge, although ongoing advances in siRNA delivery suggest that specific approaches to target erythroid cells may be developed in the future (Leuschner et al. 2011). shRNAs targeting the transcriptional regulators of HbF expression could be introduced via gene therapy vectors (Wilber et al. 2011) and these may have advantages over introducing globin genes into hematopoietic cells because there is the potential for such approaches to both down-regulate the mutated adult β-globin molecule, while also up-regulating endogenous γ-globin production.

Finally, once strategies to target these specific regulators of HbF expression are developed, it is possible that there may be synergistic interactions with existing inducers of HbF. For example, combining hydroxyurea with specific targeting of factors such as BCL11A or MYB could lead to synergistic effects. Such a synergistic effect was observed when 5-azacytidine treatment was used in a BCL11A knockout mouse model (Xu et al. 2011), suggesting that such strategies could potentially also be effective in humans. A great deal of additional work will be needed to explore what strategies may work best. Clearly innovation will be necessary for translating the knowledge gained about HbF regulation into therapies.

CONCLUSIONS

Here we have outlined historic and more recent studies that have provided insight into how the fetal-to-adult hemoglobin switch occurs and how HbF silencing can be manipulated. There is great promise that newer molecular studies will lead to targeted therapeutic approaches for HbF induction. However, as illustrated by historic work aimed at identifying HbF inducers, the path to effective therapies may not always be straightforward. Had it not been for studies questioning the original observation of the mechanism by which 5-azacytidine acted to induce HbF in primates and humans, hydroxyurea may not have been assessed and eventually found to be a clinically effective modulator of HbF levels in humans. Even with the increased understanding provided by more recent molecular studies, leading to promising therapeutic targets such as BCL11A, MYB, and KLF1, the best approach by which such molecules can be therapeutically manipulated remains unclear. It is likely that further studies of these and other molecules that regulate this process will lead to mechanistically based approaches for HbF induction. Trials targeting as many of these candidates as possible will eventually yield strategies that can be clinically effective, as it is difficult to predict a priori what strategies will work best. In addition to the basic science necessary for these translational efforts, a significant investment in clinical investigation will also be necessary, as it is still unclear what clinical end points are best to use in studying both existing and potentially newly developed therapies for HbF induction. However, this significant effort will be extremely worthwhile, because development of such strategies for HbF induction will be of broad importance for the numerous patients affected by sickle cell disease and β-thalassemia worldwide.

ACKNOWLEDGMENTS

We thank T. DiCesare for assistance in preparing illustrations. V.G.S. received support from National Institutes of Health grant T32 HL007574-30 and from funding provided by Boston Children's Hospital. S.H.O. is an investigator of the Howard Hughes Medical Institute.

REFERENCES

Aerbajinai W, Zhu J, Kumkhaek C, Chin K, Rodgers GP. 2009. SCF induces γ-globin gene expression by regulating downstream transcription factor COUP-TFII. *Blood* **114:** 187–194.

Arnaud L, Saison C, Helias V, Lucien N, Steschenko D, Giarratana MC, Prehu C, Foliguet B, Montout L, de Brevern AG, et al. 2010. A dominant mutation in the gene encoding the erythroid transcription factor KLF1 causes a congenital dyserythropoietic anemia. *Am J Hum Genet* **87:** 721–727.

Atweh G, Fathallah H. 2010. Pharmacologic induction of fetal hemoglobin production. *Hematol Oncol Clin North Am* **24:** 1131–1144.

Banzon V, Ibanez V, Vaitkus K, Ruiz MA, Peterson K, DeSimone J, Lavelle D. 2011. siDNMT1 increases γ-globin expression in chemical inducer of dimerization (CID)-dependent mouse βYAC bone marrow cells and in baboon erythroid progenitor cell cultures. *Exp Hematol* **39:** 26–36 e21.

Bard H, Prosmanne J. 1985. Relative rates of fetal hemoglobin and adult hemoglobin synthesis in cord blood of infants of insulin-dependent diabetic mothers. *Pediatrics* **75:** 1143–1147.

Benz Jr. EJJr. 1982. Clinical management of gene expression. *N Engl J Med* **307:** 1515–1516.

Borg J, Papadopoulos P, Georgitsi M, Gutierrez L, Grech G, Fanis P, Phylactides M, Verkerk AJ, van der Spek PJ, Scerri CA, et al. 2010. Haploinsufficiency for the erythroid transcription factor KLF1 causes hereditary persistence of fetal hemoglobin. *Nat Genet* **42:** 801–805.

Bradner JE, Mak R, Tanguturi SK, Mazitschek R, Haggarty SJ, Ross K, Chang CY, Bosco J, West N, Morse E, et al. 2010. Chemical genetic strategy identifies histone deacetylase 1 (HDAC1) and HDAC2 as therapeutic targets in sickle cell disease. *Proc Natl Acad Sci* **107:** 12617–12622.

Campbell AD, Cui S, Shi L, Urbonya R, Mathias A, Bradley K, Bonsu KO, Douglas RR, Halford B, Schmidt L, et al. 2011. Forced TR2/TR4 expression in sickle cell disease mice confers enhanced fetal hemoglobin synthesis and alleviated disease phenotypes. *Proc Natl Acad Sci* **108:** 18808–18813.

Cantor AB, Orkin SH. 2002. Transcriptional regulation of erythropoiesis: An affair involving multiple partners. *Oncogene* **21:** 3368–3376.

Cao H, Stamatoyannopoulos G, Jung M. 2004. Induction of human γ globin gene expression by histone deacetylase inhibitors. *Blood* **103:** 701–709.

Carpinelli MR, Hilton DJ, Metcalf D, Antonchuk JL, Hyland CD, Mifsud SL, Di Rago L, Hilton AA, Willson TA, Roberts AW, et al. 2004. Suppressor screen in Mpl$^{-/-}$ mice: c-Myb mutation causes supraphysiological production of platelets in the absence of thrombopoietin signaling. *Proc Natl Acad Sci* **101:** 6553–6558.

Castro O, Brambilla DJ, Thorington B, Reindorf CA, Scott RB, Gillette P, Vera JC, Levy PS. 1994. The acute chest syndrome in sickle cell disease: Incidence and risk factors. The Cooperative Study of Sickle Cell Disease. *Blood* **84:** 643–649.

Cavazzana-Calvo M, Payen E, Negre O, Wang G, Hehir K, Fusil F, Down J, Denaro M, Brady T, Westerman K, et al. 2010. Transfusion independence and HMGA2 activation after gene therapy of human β-thalassaemia. *Nature* **467:** 318–322.

Chakalova L, Osborne CS, Dai YF, Goyenechea B, Metaxotou-Mavromati A, Kattamis A, Kattamis C, Fraser P. 2005. The Corfu δβ thalassemia deletion disrupts γ-globin gene silencing and reveals post-transcriptional regulation of HbF expression. *Blood* **105:** 2154–2160.

Charache S, Terrin ML, Moore RD, Dover GJ, Barton FB, Eckert SV, McMahon RP, Bonds DR. 1995. Effect of hydroxyurea on the frequency of painful crises in sickle cell anemia. Investigators of the Multicenter Study of Hydroxyurea in Sickle Cell Anemia. *N Engl J Med* **332:** 1317–1322.

Cui S, Kolodziej KE, Obara N, Amaral-Psarris A, Demmers J, Shi L, Engel JD, Grosveld F, Strouboulis J, Tanabe O. 2011. Nuclear receptors TR2 and TR4 recruit multiple epigenetic transcriptional corepressors that associate specifically with the embryonic β-type globin promoters in differentiated adult erythroid cells. *Mol Cell Biol* **31:** 3298–3311.

Dandapani S, Marcaurelle LA. 2010. Current strategies for diversity-oriented synthesis. *Curr Opin Chem Biol* **14:** 362–370.

Delmore JE, Issa GC, Lemieux ME, Rahl PB, Shi J, Jacobs HM, Kastritis E, Gilpatrick T, Paranal RM, Qi J, et al. 2011. BET bromodomain inhibition as a therapeutic strategy to target c-Myc. *Cell* **146:** 904–917.

DeSimone J, Heller P, Hall L, Zwiers D. 1982. 5-Azacytidine stimulates fetal hemoglobin synthesis in anemic baboons. *Proc Natl Acad Sci* **79:** 4428–4431.

Dumitriu B, Patrick MR, Petschek JP, Cherukuri S, Klingmuller U, Fox PL, Lefebvre V. 2006. Sox6 cell-autonomously stimulates erythroid cell survival, proliferation, and terminal maturation and is thereby an important enhancer of definitive erythropoiesis during mouse development. *Blood* **108:** 1198–1207.

Emambokus N, Vegiopoulos A, Harman B, Jenkinson E, Anderson G, Frampton J. 2003. Progression through key stages of haemopoiesis is dependent on distinct threshold levels of c-Myb. *Embo J* **22:** 4478–4488.

Fathallah H, Weinberg RS, Galperin Y, Sutton M, Atweh GF. 2007. Role of epigenetic modifications in normal globin gene regulation and butyrate-mediated induction of fetal hemoglobin. *Blood* **110:** 3391–3397.

Filipe A, Li Q, Deveaux S, Godin I, Romeo PH, Stamatoyannopoulos G, Mignotte V. 1999. Regulation of embryonic/fetal globin genes by nuclear hormone receptors: A novel perspective on hemoglobin switching. *EMBO J* **18:** 687–697.

Fritsch EF, Lawn RM, Maniatis T. 1980. Molecular cloning and characterization of the human β-like globin gene cluster. *Cell* **19:** 959–972.

Galanello R, Sanna S, Perseu L, Sollaino MC, Satta S, Lai ME, Barella S, Uda M, Usala G, Abecasis GR, et al. 2009. Amelioration of Sardinian β0 thalassemia by genetic modifiers. *Blood* **114:** 3935–3937.

Galarneau G, Palmer CD, Sankaran VG, Orkin SH, Hirschhorn JN, Lettre G. 2010. Fine-mapping at three loci known to affect fetal hemoglobin levels explains additional genetic variation. *Nat Genet* **42:** 1049–1051.

Gallienne AE, Dreau HM, Schuh A, Old J, Henderson S. 2012. Ten novel mutations in the erythroid transcription factor KLF1 gene associated with increased fetal hemoglobin levels in adults. *Haematol* **97:** 340–343.

Groudine M, Eisenman R, Weintraub H. 1981. Chromatin structure of endogenous retroviral genes and activation by an inhibitor of DNA methylation. *Nature* **292:** 311–317.

Higgs DR, Wood WG. 2008. Genetic complexity in sickle cell disease. *Proc Natl Acad Sci* **105:** 11595–11596.

Hsieh MM, Kang EM, Fitzhugh CD, Link MB, Bolan CD, Kurlander R, Childs RW, Rodgers GP, Powell JD, Tisdale JF. 2009. Allogeneic hematopoietic stem-cell transplantation for sickle cell disease. *N Engl J Med* **361:** 2309–2317.

Huehns ER, Hecht F, Keil JV, Motulsky AG. 1964. Developmental hemoglobin anomalies in a chromosomal triplication: D1 trisomy syndrome. *Proc Natl Acad Sci* **51:** 89–97.

Jane SM, Nienhuis AW, Cunningham JM. 1995. Hemoglobin switching in man and chicken is mediated by a heteromeric complex between the ubiquitous transcription factor CP2 and a developmentally specific protein. *EMBO J* **14:** 97–105.

Jiang J, Best S, Menzel S, Silver N, Lai MI, Surdulescu GL, Spector TD, Thein SL. 2006. cMYB is involved in the regulation of fetal hemoglobin production in adults. *Blood* **108:** 1077–1083.

Jones PA, Taylor SM. 1980. Cellular differentiation, cytidine analogs and DNA methylation. *Cell* **20:** 85–93.

Kar BC, Satapathy RK, Kulozik AE, Kulozik M, Sirr S, Serjeant BE, Serjeant GR. 1986. Sickle cell disease in Orissa State, India. *Lancet* **2:** 1198–1201.

Koehler AN. 2010. A complex task? Direct modulation of transcription factors with small molecules. *Curr Opin Chem Biol* **14:** 331–340.

Lettre G, Sankaran VG, Bezerra MA, Araujo AS, Uda M, Sanna S, Cao A, Schlessinger D, Costa FF, Hirschhorn JN, et al. 2008. DNA polymorphisms at the BCL11A, HBS1L-MYB, and β-globin loci associate with fetal hemoglobin levels and pain crises in sickle cell disease. *Proc Natl Acad Sci* **105:** 11869–11874.

Letvin NL, Linch DC, Beardsley GP, McIntyre KW, Nathan DG. 1984. Augmentation of fetal-hemoglobin production in anemic monkeys by hydroxyurea. *N Engl J Med* **310:** 869–873.

Letvin NL, Linch DC, Beardsley GP, McIntyre KW, Miller BA, Nathan DG. 1985. Influence of cell cycle phase-specific agents on simian fetal hemoglobin synthesis. *J Clin Invest* **75:** 1999–2005.

Leuschner F, Dutta P, Gorbatov R, Novobrantseva TI, Donahoe JS, Courties G, Lee KM, Kim JI, Markmann JF, Marinelli B, et al. 2011. Therapeutic siRNA silencing in inflammatory monocytes in mice. *Nat Biotechnol* **29:** 1005–1010.

Ley TJ, DeSimone J, Anagnou NP, Keller GH, Humphries RK, Turner PH, Young NS, Keller P, Nienhuis AW. 1982. 5-Azacytidine selectively increases γ-globin synthesis in a patient with β^+ thalassemia. *N Engl J Med* **307:** 1469–1475.

Ley TJ, DeSimone J, Noguchi CT, Turner PH, Schechter AN, Heller P, Nienhuis AW. 1983. 5-Azacytidine increases γ-globin synthesis and reduces the proportion of dense cells in patients with sickle cell anemia. *Blood* **62:** 370–380.

McGrath K, Palis J. 2008. Ontogeny of erythropoiesis in the mammalian embryo. *Curr Top Dev Biol* **82:** 1–22.

McGrath KE, Frame JM, Fromm GJ, Koniski AD, Kingsley PD, Little J, Bulger M, Palis J. 2011. A transient definitive erythroid lineage with unique regulation of the β-globin locus in the mammalian embryo. *Blood* **117:** 4600–4608.

Menzel S, Garner C, Gut I, Matsuda F, Yamaguchi M, Heath S, Foglio M, Zelenika D, Boland A, Rooks H, et al. 2007. A QTL influencing F cell production maps to a gene encoding a zinc-finger protein on chromosome 2p15. *Nat Genet* **39:** 1197–1199.

Michlitsch JG, Walters MC. 2008. Recent advances in bone marrow transplantation in hemoglobinopathies. *Curr Mol Med* **8:** 675–689.

Moellering RE, Cornejo M, Davis TN, Del Bianco C, Aster JC, Blacklow SC, Kung AL, Gilliland DG, Verdine GL, Bradner JE. 2009. Direct inhibition of the NOTCH transcription factor complex. *Nature* **462:** 182–188.

Mukai HY, Motohashi H, Ohneda O, Suzuki N, Nagano M, Yamamoto M. 2006. Transgene insertion in proximity to the c-myb gene disrupts erythroid-megakaryocytic lineage bifurcation. *Mol Cell Biol* **26:** 7953–7965.

Musallam KM, Sankaran VG, Cappellini MD, Duca L, Nathan DG, Taher AT. 2011. Fetal hemoglobin levels and morbidity in untransfused patients with β-thalassemia intermedia. *Blood* **119:** 364–367.

Nuinoon M, Makarasara W, Mushiroda T, Setianingsih I, Wahidiyat PA, Sripichai O, Kumasaka N, Takahashi A, Svasti S, Munkongdee T, et al. 2010. A genome-wide association identified the common genetic variants influence disease severity in β0-thalassemia/hemoglobin E. *Hum Genet* **127:** 303–314.

Pace B, Li Q, Peterson K, Stamatoyannopoulos G. 1994. α-Amino butyric acid cannot reactivate the silenced γ gene of the β locus YAC transgenic mouse. *Blood* **84:** 4344–4353.

Papayannopoulou T, Torrealba de Ron A, Veith R, Knitter G, Stamatoyannopoulos G. 1984. Arabinosylcytosine induces fetal hemoglobin in baboons by perturbing erythroid cell differentiation kinetics. *Science* **224:** 617–619.

Perrine RP, Brown MJ, Clegg JB, Weatherall DJ, May A. 1972. Benign sickle-cell anaemia. *Lancet* **2:** 1163–1167.

Perrine SP, Greene MF, Faller DV. 1985. Delay in the fetal globin switch in infants of diabetic mothers. *N Engl J Med* **312:** 334–338.

Perrine SP, Ginder GD, Faller DV, Dover GH, Ikuta T, Witkowska HE, Cai SP, Vichinsky EP, Olivieri NF. 1993. A short-term trial of butyrate to stimulate fetal-globin-gene expression in the β-globin disorders. *N Engl J Med* **328:** 81–86.

Perrine SP, Olivieri NF, Faller DV, Vichinsky EP, Dover GJ, Ginder GD. 1994. Butyrate derivatives. New agents for stimulating fetal globin production in the β-globin disorders. *Am J Pediatr Hematol Oncol* **16:** 67–71.

Persons DA. 2009. Hematopoietic stem cell gene transfer for the treatment of hemoglobin disorders. *Hematology* **2009:** 690–697.

Peschle C, Mavilio F, Care A, Migliaccio G, Migliaccio AR, Salvo G, Samoggia P, Petti S, Guerriero R, Marinucci M, et al. 1985. Haemoglobin switching in human embryos: Asynchrony of $\zeta \rightarrow \alpha$ and $\varepsilon \rightarrow \gamma$-globin switches in primitive and definite erythropoietic lineage. *Nature* **313:** 235–238.

Platt OS. 2008. Hydroxyurea for the treatment of sickle cell anemia. *N Engl J Med* **358:** 1362–1369.

Platt OS, Orkin SH, Dover G, Beardsley GP, Miller B, Nathan DG. 1984. Hydroxyurea enhances fetal hemoglobin production in sickle cell anemia. *J Clin Invest* **74:** 652–656.

Platt OS, Thorington BD, Brambilla DJ, Milner PF, Rosse WF, Vichinsky E, Kinney TR. 1991. Pain in sickle cell disease. Rates and risk factors. *N Engl J Med* **325:** 11–16.

Platt OS, Brambilla DJ, Rosse WF, Milner PF, Castro O, Steinberg MH, Klug PP. 1994. Mortality in sickle cell disease. Life expectancy and risk factors for early death. *N Engl J Med* **330:** 1639–1644.

Porter JB, Shah FT. 2010. Iron overload in thalassemia and related conditions: Therapeutic goals and assessment of response to chelation therapies. *Hematol Oncol Clin North Am* **24:** 1109–1130.

Premawardhena A, Fisher CA, Olivieri NF, de Silva S, Arambepola M, Perera W, O'Donnell A, Peto TE, Viprakasit V, Merson L, et al. 2005. Haemoglobin E β thalassaemia in Sri Lanka. *Lancet* **366:** 1467–1470.

Sankaran VG, Nathan DG. 2010a. Reversing the hemoglobin switch. *N Engl J Med* **363:** 2258–2260.

Sankaran VG, Nathan DG. 2010b. Thalassemia: An overview of 50 years of clinical research. *Hematol Oncol Clin North Am* **24:** 1005–1020.

Sankaran VG, Sapp MV. 2012. Persistence of fetal hemoglobin expression in an older child with trisomy 13. *J Pediatr* **160:** 352.

Sankaran VG, Menne TF, Xu J, Akie TE, Lettre G, Van Handel B, Mikkola HK, Hirschhorn JN, Cantor AB, Orkin SH. 2008. Human fetal hemoglobin expression is regulated by the developmental stage-specific repressor BCL11A. *Science* **322:** 1839–1842.

Sankaran VG, Xu J, Ragoczy T, Ippolito GC, Walkley CR, Maika SD, Fujiwara Y, Ito M, Groudine M, Bender MA, et al. 2009. Developmental and species-divergent globin switching are driven by BCL11A. *Nature* **460:** 1093–1097.

Sankaran VG, Xu J, Orkin SH. 2010. Advances in the understanding of haemoglobin switching. *Brit J Haematol* **149:** 181–194.

Sankaran VG, Menne J, Heller R. 2011a. Heterozygous disruption of human SOX6 is insufficient to impair erythropoiesis or silencing of fetal hemoglobin. *Blood* **117:** 4396–4397.

Sankaran VG, Menne TF, Scepanovic D, Vergilio JA, Ji P, Kim J, Thiru P, Orkin SH, Lander ES, Lodish HF. 2011b. MicroRNA-15a and -16-1 act via MYB to elevate fetal hemoglobin expression in human trisomy 13. *Proc Natl Acad Sci* **108:** 1519–1524.

Sankaran VG, Xu J, Byron R, Greisman HA, Fisher C, Weatherall DJ, Sabath DE, Groudine M, Orkin SH, Premawardhena A, et al. 2011c. A functional element necessary for fetal hemoglobin silencing. *N Engl J Med* **365:** 807–814.

Satta S, Perseu L, Moi P, Asunis I, Cabriolu A, Maccioni L, Demartis FR, Manunza L, Cao A, Galanello R. 2011. Compound heterozygosity for KLF1 mutations associated with remarkable increase of fetal hemoglobin and red cell protoporphyrin. *Haematologica* **96:** 767–770.

Sher GD, Ginder GD, Little J, Yang S, Dover GJ, Olivieri NF. 1995. Extended therapy with intravenous arginine butyrate in patients with β-hemoglobinopathies. *N Engl J Med* **332:** 1606–1610.

Stadhouders R, Thongjuea S, Andrieu-Soler C, Palstra RJ, Bryne JC, van den Heuvel A, Stevens M, de Boer E, Kockx C, van der Sloot A, et al. 2011. Dynamic long-range chromatin interactions control Myb proto-oncogene transcription during erythroid development. *EMBO J* **31:** 986–999.

Steinberg MH, Barton F, Castro O, Pegelow CH, Ballas SK, Kutlar A, Orringer E, Bellevue R, Olivieri N, Eckman J, et al. 2003. Effect of hydroxyurea on mortality and morbidity in adult sickle cell anemia: Risks and benefits up to 9 years of treatment. *JAMA* **289:** 1645–1651.

Tanabe O, Katsuoka F, Campbell AD, Song W, Yamamoto M, Tanimoto K, Engel JD. 2002. An embryonic/fetal β-type globin gene repressor contains a nuclear receptor TR2/TR4 heterodimer. *EMBO J* **21:** 3434–3442.

Tanabe O, McPhee D, Kobayashi S, Shen Y, Brandt W, Jiang X, Campbell AD, Chen YT, Chang C, Yamamoto M, et al. 2007. Embryonic and fetal β-globin gene repression by the orphan nuclear receptors, TR2 and TR4. *EMBO J* **26:** 2295–2306.

Thein SL, Menzel S, Peng X, Best S, Jiang J, Close J, Silver N, Gerovasilli A, Ping C, Yamaguchi M, et al. 2007. Intergenic variants of HBS1L-MYB are responsible for a major quantitative trait locus on chromosome 6q23 influencing fetal hemoglobin levels in adults. *Proc Natl Acad Sci* **104:** 11346–11351.

Uda M, Galanello R, Sanna S, Lettre G, Sankaran VG, Chen W, Usala G, Busonero F, Maschio A, Albai G, et al. 2008. Genome-wide association study shows BCL11A associated with persistent fetal hemoglobin and amelioration of the phenotype of β-thalassemia. *Proc Natl Acad Sci* **105:** 1620–1625.

van der Ploeg LH, Flavell RA. 1980. DNA methylation in the human γδβ-globin locus in erythroid and nonerythroid tissues. *Cell* **19:** 947–958.

van Dijk TB, Gillemans N, Pourfarzad F, van Lom K, von Lindern M, Grosveld F, Philipsen S. 2010. Fetal globin expression is regulated by Friend of Prmt1. *Blood* **116:** 4349–4352.

Wahlberg K, Jiang J, Rooks H, Jawaid K, Matsuda F, Yamaguchi M, Lathrop M, Thein SL, Best S. 2009. The HBS1L-MYB intergenic interval associated with elevated HbF levels shows characteristics of a distal regulatory region in erythroid cells. *Blood* **114:** 1254–1262.

Watson J. 1948. The significance of the paucity of sickle cells in newborn Negro infants. *Am J Med Sci* **215:** 419–423.

Weatherall DJ. 2001. Phenotype-genotype relationships in monogenic disease: Lessons from the thalassaemias. *Nat Rev* **2:** 245–255.

Weatherall DJ, Clegg JB. 2001. *The thalassaemia syndromes*, 4th ed. Blackwell Science, Malden, MA.

Weatherall D, Akinyanju O, Fucharoen S, Olivieri N, Musgrove P. 2006. Inherited disorders of hemoglobin. In *Disease control priorities in developing countries*, 2nd ed., pp. 663–680. Oxford University Press, New York.

Wilber A, Hargrove PW, Kim YS, Riberdy JM, Sankaran VG, Papanikolaou E, Georgomanoli M, Anagnou NP, Orkin SH, Nienhuis AW, et al. 2011. Therapeutic levels of fetal hemoglobin in erythroid progeny of β-thalassemic $CD34^+$ cells after lentiviral vector-mediated gene transfer. *Blood* **117:** 2817–2826.

Xu J, Sankaran VG, Ni M, Menne TF, Puram RV, Kim W, Orkin SH. 2010. Transcriptional silencing of γ-globin by BCL11A involves long-range interactions and cooperation with SOX6. *Genes Dev* **24:** 783–798.

Xu J, Peng C, Sankaran VG, Shao Z, Esrick EB, Chong BG, Ippolito GC, Fujiwara Y, Ebert BL, Tucker PW, et al. 2011. Correction of sickle cell disease in adult mice by interference with fetal hemoglobin silencing. *Science* **334:** 993–996.

Zhao Q, Zhou W, Rank G, Sutton R, Wang X, Cumming H, Cerruti L, Cunningham JM, Jane SM. 2006. Repression of human γ-globin gene expression by a short isoform of the NF-E4 protein is associated with loss of NF-E2 and RNA polymerase II recruitment to the promoter. *Blood* **107:** 2138–2145.

Zhou W, Zhao Q, Sutton R, Cumming H, Wang X, Cerruti L, Hall M, Wu R, Cunningham JM, Jane SM. 2004. The role of p22 NF-E4 in human globin gene switching. *J Biol Chem* **279:** 26227–26232.

Zhou D, Liu K, Sun CW, Pawlik KM, Townes TM. 2010. KLF1 regulates BCL11A expression and γ- to β-globin gene switching. *Nat Genet* **42:** 742–744.

Zuber J, Rappaport AR, Luo W, Wang E, Chen C, Vaseva AV, Shi J, Weissmueller S, Fellmann C, Taylor MJ, et al. 2011a. An integrated approach to dissecting oncogene addiction implicates a Myb-coordinated self-renewal program as essential for leukemia maintenance. *Genes Dev* **25:** 1628–1640.

Zuber J, Shi J, Wang E, Rappaport AR, Herrmann H, Sison EA, Magoon D, Qi J, Blatt K, Wunderlich M, et al. 2011b. RNAi screen identifies Brd4 as a therapeutic target in acute myeloid leukaemia. *Nature* **478:** 524–528.

Zuk O, Hechter E, Sunyaev SR, Lander ES. 2012. The mystery of missing heritability: Genetic interactions create phantom heritability. *Proc Natl Acad Sci* **109:** 1193–1198.

Iron Metabolism: Interactions with Normal and Disordered Erythropoiesis

Tomas Ganz and Elizabeta Nemeth

Department of Medicine and Department of Pathology and Laboratory Medicine, David Geffen School of Medicine, University of California, Los Angeles, California 90095

Correspondence: tganz@mednet.ucla.edu

Hemoglobinopathies and other disorders of erythroid cells are often associated with abnormal iron homeostasis. We review the molecular physiology of intracellular and systemic iron regulation, and the interactions between erythropoiesis and iron homeostasis. Finally, we discuss iron disorders that affect erythropoiesis as well as erythroid disorders that cause iron dysregulation.

Iron overload is a common complication of hemoglobinopathies treated by erythrocyte transfusions (1 mL of packed erythrocytes contains about 1 mg of iron) and those associated with ineffective erythropoiesis, which stimulates the hyperabsorption of dietary iron. With the increasing use of transfusion therapy, iron overload has become a major cause of morbidity and premature mortality. More recently, the effective treatment of iron overload by iron chelation has dramatically improved survival (Cunningham 2008; Telfer 2009). This work reviews recent advances in our understanding of the molecular basis of iron homeostasis and its disorders.

IRON BIOLOGY AND HOMEOSTASIS

Iron Intake

Iron is the most abundant element on Earth by mass and the fourth most abundant in the Earth's crust but it readily oxidizes into insoluble compounds with poor bioavailability. In this environment, biological organisms evolved to conserve iron. Quantitative analysis of tissue iron distribution and fluxes in humans illustrates how this is accomplished (Finch 1994). The typical adult human male contains about 4 g of iron of which about 2.5 g is in hemoglobin, 1 g is stored predominantly in hepatocytes and hepatic and splenic macrophages, and most of the rest is distributed in myoglobin, cytochromes, and other ferroproteins. Only about 1–2 mg/d, or <0.05%/d, is lost from the body predominantly through desquamation and minor blood loss. In the steady state, this amount is replaced through intestinal iron absorption. Although the loss of iron may increase slightly with increasing iron stores, these changes do not significantly contribute to homeostasis; intestinal iron absorption is by far the predominant determinant of the iron content of the body. A typical Western diet provides about 15 mg of iron per day and only ~10% is absorbed. Recovery from blood loss causes an increase in iron absorption up to 20-fold, indicating that the

Copyright © 2013 Cold Spring Harbor Laboratory Press; all rights reserved
Cite this article as *Cold Spring Harb Perspect Med* doi: 10.1101/cshperspect.a011668

duodenum where iron absorption takes place has a large reserve capacity for iron absorption. Pathological increase of intestinal iron absorption is a common cause of iron overload, accounting for the excess iron in hereditary hemochromatosis and untransfused β-thalassemia. Blood transfusions and parenteral administration of iron compounds bypass the regulatory bottleneck of iron absorption and constitute the other major cause of iron overload.

Iron Recycling

Under normal circumstances, the reutilization of iron recycled from senescent cells accounts for most of the iron flux in humans. With the erythrocyte lifespan of 120 d, 20–25 mg of iron is required to replace the 20–25 mL of erythrocytes that must be produced every day to maintain a steady state. Other cell types also turn over but their much lower iron content contributes relatively little to the iron flux. Macrophages in the liver, spleen, and marrow (formerly called the reticuloendothelial system) phagocytose senescent or damaged erythrocytes, degrade their hemoglobin to release heme, extract iron from heme using heme oxygenase (Poss and Tonegawa 1997), and recycle the iron to the extracellular fluid and plasma. Steady-state iron flux from recycling can increase up to 150 mg/d in conditions with ineffective erythropoiesis in which the number of erythroid precursors is increased and accompanied by the apoptosis of hemoglobinized erythrocyte precursors in the marrow and shortened erythrocyte survival (Beguin et al. 1988).

Iron Distribution and Storage

Free iron is highly reactive and causes cell and tissue injury through its ability to catalyze the production of reactive oxygen species. In living organisms, iron is complexed with proteins or small organic molecules (citrate, acetate), which mitigate its reactivity. Transferrin is the physiological carrier of iron in plasma. Normally, only 20%–40% of the available binding sites on transferrin molecules are occupied by ferric iron. The iron content of plasma is only 2–3 mg so this compartment must turn over every few hours. Erythrocyte precursors take up iron almost exclusively through transferrin receptors (TfR1) so the iron supply to erythrocyte precursors is completely dependent on plasma transferrin. In contrast, hepatocytes and other nonerythroid cells can also take up iron that is not bound to transferrin (nontransferrin-bound iron or NTBI), a process that becomes important during iron overload when plasma transferrin saturation reaches 100% (Breuer et al. 2000). The predominant cellular storage form of iron is the hollow spherical protein ferritin whose cavity contains iron in ferric form complexed with hydroxide and phosphate anions.

Regulation of Plasma Iron Concentrations

Despite varying dietary iron intake and changes in erythropoietic activity owing to occasional or periodic blood loss, iron concentrations in plasma normally remain in the 10–30 μM range. Chronically low concentrations decrease iron supply to erythropoiesis and other processes leading to anemia and dysfunction of other cell types sensitive to iron deprivation. Chronically high iron concentrations lead to intermittent or steady-state saturation of transferrin with iron and the generation of NTBI with consequent deposition of excess iron in the liver, endocrine glands, cardiac myocytes, and other tissues. Excess cellular iron may cause tissue injury by catalyzing the generation of reactive oxygen species, which can cause DNA damage, lipid peroxidation, and oxidation of proteins.

Systemic Iron Homeostasis

Phenomenological description of systemic iron homeostasis was developed starting in the 1930s (Finch 1994). Homeostatic mechanisms regulate dietary iron absorption and iron deposition into or withdrawal from stores depending on the amount of stored iron ("stores regulator") and the requirements of erythropoiesis ("erythropoietic regulator"). The description of the molecular processes that underlie iron homeostasis has progressed rapidly in the last two decades but is still not complete.

 Cite this article as *Cold Spring Harb Perspect Med* doi: 10.1101/cshperspect.a011668

CELLULAR IRON REGULATION

Cellular Iron

Cells require iron predominantly for incorporation into various ferroproteins, where iron exists in iron–sulfur clusters, in heme or hemelike prosthetic moieties, or in other more loosely associated forms. It now appears that most cell types in the body autonomously regulate their iron uptake solely to meet their individual requirements for iron. These cells do not export appreciable amounts of iron and are presumed to give up their iron only when they undergo cell death and are recycled by macrophages. In contrast, several specialized cell types supply or store iron to meet the needs of the entire organism, and are therefore equipped to export iron into extracellular fluid and plasma. Iron-exporting cells include duodenal enterocytes that absorb dietary iron, macrophages that recycle iron from senescent or dead cells, and macrophages and hepatocytes that store iron and release it to meet systemic demand. During pregnancy, the placental syncytiotrophoblast must transport maternal iron into the fetal circulation to meet the iron requirements of fetal growth and development. The endothelial cells that form the blood–brain barrier must also selectively transport iron as it now appears that the iron concentrations in the brain are not appreciably increased in systemic iron overload disorders. Finally, erythroid precursors need much more iron than any other cell type as each cell synthesizes more than a billion heme molecules, therefore facing greater iron-homeostatic challenges.

Cellular Iron Uptake

Transferrin-mediated iron uptake is the best understood mechanism of cellular iron import. Although the transferrin receptor (TfR1) is expressed in many cell types, erythrocyte precursors contain most of the TfR1 molecules and take up the great majority of iron-transferrin in the organism. Iron-transferrin is endocytosed via the cell membrane TfR1 and internalized into endosomal recycling vesicles. As the vesicle acidifies, the low pH releases the transferrin-bound ferric iron and the iron-free (apo)transferrin-TfR1 complex returns to the cell membrane (Fig. 1). The neutral pH at the membrane causes the apotransferrin to dissociate from TfR1, whereupon apotransferrin diffuses away to be loaded with iron again, repeating the cycle. From the vesicle, iron is delivered to mitochondria where it is incorporated into protoporphyrin IX to form heme, or incorporated into nascent iron–sulfur clusters. Alternatively, iron can be exported from the vesicle into the cytoplasm where it is incorporated into cytoplasmic ferroproteins or stored in cytoplasmic ferritin.

Nontransferrin-bound iron (NTBI) (Breuer et al. 2000) usually accumulates when the iron-binding capacity of transferrin is exceeded and then circulates complexed mostly with citrate or acetate. Hemoglobinopathies such as β-thalassemia major and intermedia are associated with particularly high levels of NTBI. Some cells, including hepatocytes, cardiomyocytes, and cells of endocrine glands can take up NTBI, although the precise mechanism is not well understood. Candidate NTBI transporters include L-type voltage-gated calcium channels, DMT-1 and Zip14.

Intracellular Iron Transport

To undergo transport to the cytoplasm or mitochondria, ferric iron must be reduced to its ferrous form through the action of ferrireductases. Recent studies indicate that Steap (six-transmembrane epithelial antigen of the prostate) proteins 1–4 are among the relevant ferrireductases, with Steap3 having a particular importance in erythroid precursors (Fig. 1), assisted perhaps by Steap2 and Steap4 (Ohgami et al. 2006). To reach the cytoplasm, ferrous iron must cross the membrane of the vesicle. In many cells, the proton-dependent ferrous iron transporter divalent metal transporter-1 (DMT1) appears essential for iron transport from the vacuole into the cytoplasm but in macrophages its homolog natural resistance-associated macrophage protein (Nramp1) may also contribute (Soe-Lin et al. 2009). Because of its chemical reactivity, iron is chaperoned in the cytoplasm, at least in part by multifunctional poly(RC)-binding proteins (PCBPs) (Shi et al. 2008). In

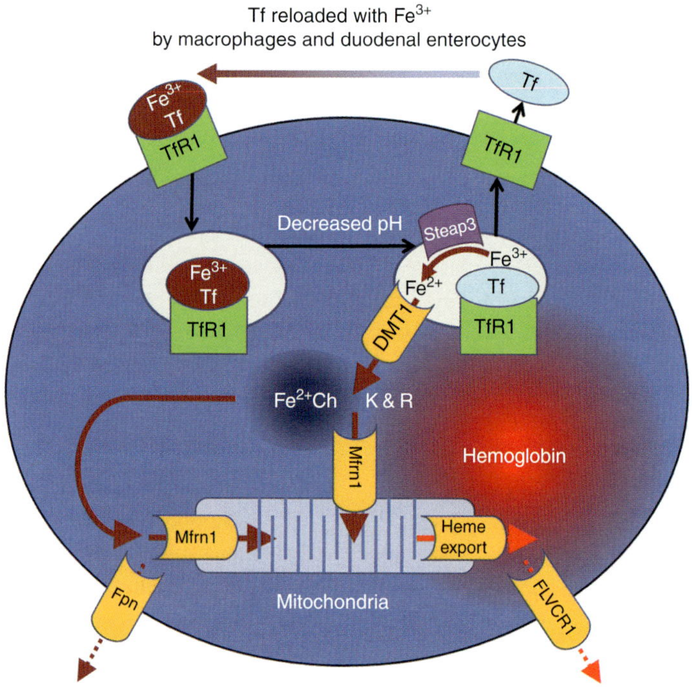

Figure 1. Iron traffic in erythrocyte precursors synthesizing hemoglobin. Iron is taken up as diferric transferrin by the transferrin receptor (TfR1). Acidification of the endocytic vesicle releases ferric iron from transferrin, and the membrane ferrireductase Steap3 reduces it to ferrous iron, which is then exported to the cytoplasm by DMT1. The complex of iron-free apotransferrin (Tf) and TfR1 is returned to the plasma membrane where the neutral pH causes Tf to dissociate from its receptor. The transferrin cycle is completed when Tf is reloaded with ferric iron by duodenal enterocytes or iron-recycling macrophages. Ferrous iron exported by DMT1 may be delivered to mitochondrial mitoferrin-1 (Mfrn1) by direct contact (the kiss-and-run mechanism, K&R) or through intermediate transport by as-yet uncharacterized cytoplasmic chaperones (Fe^{2+}Ch). Mitoferrin-1 imports iron into mitochondria where iron is incorporated into newly synthesized heme. Heme is exported via an unknown exporter (Heme export) and incorporated into globin chains to generate hemoglobin. Under some circumstances, iron is exported as ferrous iron via ferroportin (Fpn) or as heme via feline leukemia virus C receptor (FLVCR1).

particular, PCBP1 mediates the delivery of iron to cytoplasmic ferritin and both PCBP1 and 2 are involved in the delivery of iron to cytoplasmic iron-dependent prolyl and asparaginyl hydroxylases that mediate oxygen sensing (Nandal et al. 2011). It is not known how iron is transported to mitochondria. In erythroid cells, there is evidence for a "kiss-and-run" mechanism (Fig. 1) whereby iron could be transferred from endosomal vesicles directly to mitochondria (Sheftel et al. 2007) but it is not clear how much this mechanism contributes to the iron flux into mitochondria and whether it also functions in nonerythroid cell types.

Mitochondria and Iron

Consistent with their autonomous evolutionary origin, mitochondria are equipped with distinct iron transporters. Iron uptake into mitochondria depends on the inner mitochondrial membrane proteins mitoferrin 1 and 2, with the former predominantly expressed in erythroid cells and the latter ubiquitously (Paradkar et al. 2009; Troadec et al. 2011). In erythroid cells, mitoferrin 1 interacts with the ATP-binding transporter Abcb10 and with ferrochelatase to form a plausible pathway for the delivery of iron for heme formation (Chen et al. 2010). How heme is

Cite this article as *Cold Spring Harb Perspect Med* doi: 10.1101/cshperspect.a011668

exported from mitochondria for incorporation into hemoglobin and other hemoproteins is not known. Mitochondria also contain a distinct ferritin, mitochondrial ferritin, for local iron storage.

Cellular Iron Homeostasis

Cellular, as opposed to systemic, iron homeostasis assures that sufficient but not excessive amounts of iron are taken up by each cell to meet its individual requirement for ferroprotein synthesis. The system that has evolved relies on posttranscriptional regulation through the interaction of iron-regulatory proteins (IRP1 and IRP2) with iron-regulatory elements (IREs) in messenger RNAs (mRNAs) that encode key iron transporters, ferroproteins, and enzymes involved in iron-utilizing pathways. The IRE/IRP system effectively regulates iron uptake, provides for the storage of excess iron in ferritin, and coordinates the synthesis of heme, iron–sulfur clusters, and ferroproteins with the availability of iron. The system in effect acts to decrease wasted expenditure of synthetic energy and substrates, and to prevent accumulation of toxic forms of iron. Target mRNAs contain IREs that form characteristic stem-loop structures either in the 5′ region, where IRP binding represses translation and decreases protein synthesis, or in the 3′ region where IRP binding prevents endonucleases from cleaving sensitive regions of the mRNA, thus stabilizing mRNA and increasing protein synthesis (Casey et al. 1988). IRP1 and IRP2 are structurally related but interact with iron in distinct ways. Both proteins bind to IREs when cellular iron levels are low. In the presence of iron, IRP1 incorporates an iron–sulfur cluster, does not bind IREs, and acts as an aconitase enzyme converting citrate to isocitrate in the Krebs cycle. In contrast, IRP2 is ubiquitinated by a complex iron-dependent process and then degraded in proteasomes (Salahudeen et al. 2009; Vashisht et al. 2009). The dual specificity of IRP1/aconitase may have a functional role in the regulation of erythropoiesis by iron availability, as the provision of the aconitase product isocitrate reverses some of the suppressive effect of iron deprivation on erythropoiesis and the enzymatic inhibition of aconitase has the opposite effect (Bullock et al. 2010; Talbot et al. 2011). Many mRNAs are regulated by the IRP/IRE system (Sanchez et al. 2011) and fall into three classes: (1) iron acquisition, generally with IREs in the 3′ region resulting in increased protein synthesis during cellular iron deprivation; (2) iron utilization and storage, with IREs in the 5′ region, resulting in repressed protein synthesis during iron deprivation; and (3) iron export, with IREs also in the 5′ region and protein synthesis repressed during iron deprivation. Proteins subject to IRE/IRP regulation include TfR1 and DMT1 involved in cellular iron uptake, aminolevulinic acid synthase 2, which catalyzes the first step of the heme synthesis pathway in erythroid cells, the heavy and light subunits of ferritin involved in iron storage, and ferroportin, the iron exporter expressed in tissues and cells involved in iron export to plasma. The net effect of the IRE/IRP response during cellular iron deficiency is to increase cellular iron uptake, mobilize iron from cellular storage, decrease iron utilization, and, when iron becomes sufficient or excessive to reverse these responses and direct more iron into cellular storage and export. Further fine-tuning of iron import and export is achieved by differential splicing of target mRNAs in different tissues to either include or exclude IREs. As an example, systemic adaptation to iron deficiency may be facilitated by a ferroportin mRNA isoform that lacks IRE, which may allow iron-transporting duodenal enterocytes to deliver iron to plasma for systemic needs even if the enterocyte is sensing iron deficiency, and may transfer iron from erythroid cells to other tissues more critically dependent on iron (Zhang et al. 2009).

Generalized Regulation of Protein Synthesis by Iron in Erythroid Cells

In addition to the regulation of the synthesis of individual proteins by iron, erythroid cells also contain a mechanism for a generalized adaptive response to iron deficiency. This response is affected by the heme-regulated inhibitor kinase (HRI) belonging to a class of kinases activated by cellular stress, including nutrient deprivation, viral infection, and endoplasmic reticulum stress (Chen 2007). During iron deficiency as heme concentrations drop, heme dissociates from HRI, causing it to undergo specific autophosphorylation to become a catalytically active kinase targeting the α subunit of eukaryotic translational initiation factor 2 (eIF2α). Activated HRI inhibits translational initiation by phoshorylating eIF2α. Not all protein synthesis is inhibited however, as activated HRI may promote the synthesis of transcription factors that are protective during iron-deficient erythropoiesis (Liu et al. 2008). A priori, it is not obvious how iron deficiency results in the production of smaller, less-hemoglobinized erythrocytes rather than fewer normally sized and hemoglobinized cells. Studies with HRI-deficient mice showed that HRI protects erythroid precursors from apoptosis induced by excessive production of globin chains and contributes to the microcytosis and hypochromia seen in iron deficiency, erythropoietic protoporphyria, and β-thalassemia.

Iron and Hypoxia Sensing

The hypoxia-sensing pathway may also contribute to cellular iron homeostasis. Prolyl and asparaginyl hydroxylases, which inactivate the HIF transcription factors, are not only sensitive to oxygen tension but also to iron concentrations because they use iron as a catalytic cofactor. In support of the potential role of HIF in iron regulation, tissue-specific deletion of HIF2α in mouse enterocytes decreased intestinal iron absorption as well as the expression of DMT1 in enterocytes (Mastrogiannaki et al. 2009). HIF2α bound to the DMT1 promoter and transactivated it. The broader physiologic function of HIF in cellular iron homeostasis still remains to be established and may vary in different tissues depending on oxygen tension and other factors.

SYSTEMIC IRON HOMEOSTASIS

The Central Role of Hepcidin

Systemic iron homeostasis encompasses the regulatory circuitry that controls the absorption of dietary iron, the concentration of iron in extracellular fluid and blood plasma, and the release of iron from macrophages involved in iron recycling and from iron-storing hepatocytes. It now appears that there is a single systemic regulator of iron, the hepatic peptide hormone hepcidin. The hormone inhibits iron delivery to plasma and extracellular fluid thereby controlling the concentration of iron in plasma. Hepcidin inhibits the transfer of dietary iron from duodenal enterocytes to plasma, the release of recycled iron from macrophages to plasma, and the release of stored iron from hepatocytes (Fig. 2). Fetal hepcidin inhibits the transfer of maternal iron across the placenta to the fetal circulation. At the molecular level, hepcidin acts by binding to its receptor, ferroportin, and causing its endocytosis and proteolysis, which results in decreased iron release from cells to plasma and extracellular fluid. Ferroportin is found at very low concentrations in most cell types but much higher amounts in professional iron-transporting tissues, including the duodenal enterocytes and splenic macrophages. Intermediate concentrations of ferroportin are detectable in hepatocytes.

Regulation of Hepcidin Synthesis

Hepatocytes are the main source of hepcidin, with much lower amounts produced by macrophages, adipocytes, and perhaps other cells. Hepcidin synthesis is controlled predominantly at the transcriptional level and is increased by plasma iron-transferrin as well as by iron stored in hepatocytes (the "stores" regulator), is suppressed in response to increased iron requirements of erythroid precursors (the "erythroid" regulator), and is potently stimulated by inflammation (Fig. 2).

Regulation of Hepcidin Synthesis by Iron

Iron regulation of hepcidin is mediated by the canonical bone morphogenetic protein pathway adapted for hepcidin regulation by iron-specific molecular components (Fig. 3). Normal iron regulation in the murine model requires the iron-specific ligand bone morphogenetic protein 6 (BMP6), interacting with the

Cite this article as *Cold Spring Harb Perspect Med* doi: 10.1101/cshperspect.a011668

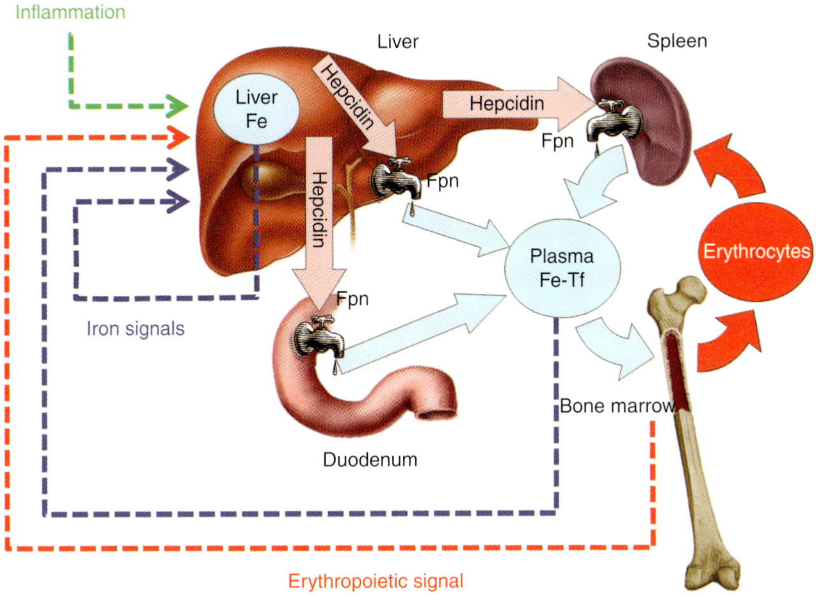

Figure 2. Iron homeostasis. Through membrane ferroportin (Fpn), iron flows into plasma (pale blue arrows) from duodenal enterocytes, iron-storing hepatocytes, and iron-recycling macrophages predominantly in the spleen. Iron-transferrin (Fe-Tf) is mostly delivered to the marrow (pale blue arrow) where iron is incorporated into erythrocyte hemoglobin (red). When the erythrocytes live out their lifespan (normally 120 d in humans), their hemoglobin and heme are degraded in the macrophages, mostly in the spleen, and iron is returned into the plasma iron pool. Hepatocytes secrete hepcidin under the control of stimulatory signals that reflect liver iron stores and plasma iron concentrations (blue), inhibitory signals reflecting erythropoietic activity (red), and inflammatory cytokines (green). Hepcidin causes the degradation of Fpn and thereby inhibits iron delivery to plasma and the erythropoietic bone marrow.

BMP receptor. Another iron-specific component required for normal hepcidin regulation by iron is the membrane protein hemojuvelin, which interacts with BMPs as well as with the BMP receptor. The BMP receptor is a ligand-activated serine/threonine kinase, which phosphorylates the cytoplasmic proteins Smad1, 5, and 8. Together with the common Smad4, the phosphorylated Smads form heterodimers, which reach the nucleus and enhance the transcription of hepcidin. The synthesis of the BMP6 ligand appears to be responsive predominantly to iron stores rather than transferrin saturation, and compared to the large changes in hepcidin expression, BMP6 changes in a relatively narrow range. In contrast, changes in extracellular iron-transferrin concentration affect signal transduction by the BMP receptor even in the absence of changes in BMP6 mRNA concentration, indicating that iron-transferrin modulates the sensitivity of the receptor to its ligands. It is not known with certainty how the concentration of iron-transferrin is sensed but ablation of HFE or TfR2, and especially the combined loss of both molecules, decreases hepcidin expression and interferes with the sensing of transient changes in iron-transferrin while preserving the increase in BMP6 and the chronic hepcidin response to iron loading (Wallace et al. 2009; Ramos et al. 2011; Corradini et al. 2011). A plausible current model postulates that iron-transferrin is sensed by TfR1 and TfR2, with HFE shuttling between the two molecules depending on iron-transferrin concentrations. At higher iron-transferrin concentrations, the association of TfR2 and HFE somehow potentiates BMP receptor complex signaling. Two other proteins, GPI-linked hemojuvelin (Papanikolaou et al. 2004) and the

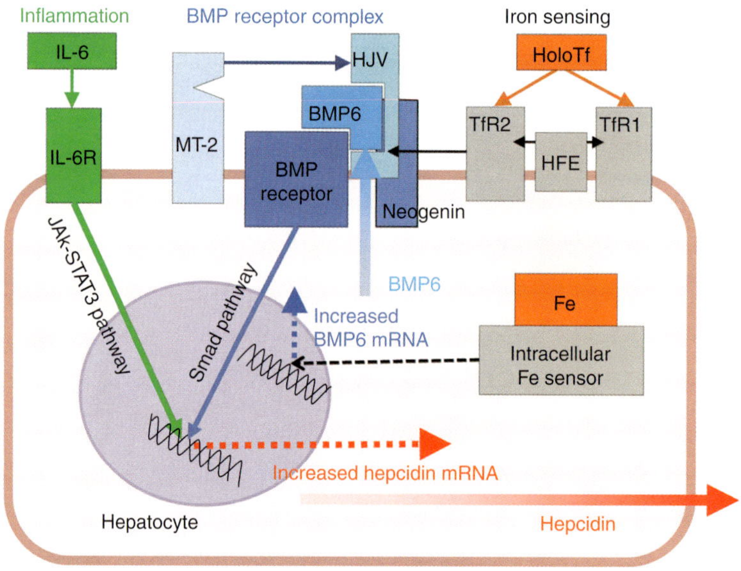

Figure 3. Regulation of hepcidin by iron and inflammation. Hepcidin synthesis is transcriptionally regulated by iron through the BMP receptor complex and its Smad pathway (shades of blue) and by inflammation predominantly via the IL-6 receptor and its JAK-STAT3 pathway (green). Extracellular iron is sensed by transferrin receptors (TfR1 and TfR2) aided by HFE, which can associate with either TfR but is displaced from TfR1 when TfR1 binds diferric transferrin (HoloTf). When HoloTf concentrations are high, HFE is associated mostly with TfR2 and stabilizes it. HFE-TfR2 then potentiates BMP receptor signaling through an unknown mechanism. Stored hepatic intracellular iron increases the concentrations of BMP6 mRNA and presumably BMP6 protein in the liver thereby stimulating the BMP receptor, its Smad pathway, and hepcidin transcription.

membrane protease matriptase 2 (MT2, also called transmembrane serine proteinase 6, TMPRSS6) (Du et al. 2008) respectively enhance and dampen BMP signaling, with hemojuvelin acting as a BMP pathway coreceptor, and MT2 exerting its effect by an inactivating cleavage of hemojuvelin (Silvestri et al. 2008).

Regulation of Hepcidin by Erythropoiesis

Hepcidin mRNA is suppressed during anemia or hypoxia (Nicolas et al. 2002) but it now appears that this is an indirect effect dependent on increased erythropoietin production and the resulting expansion of erythroid precursors in the marrow (Pak et al. 2006; Vokurka et al. 2006; Mastrogiannaki et al. 2011) and not a direct effect of hypoxia-regulated pathways on the hepcidin promoter. In normal volunteers, the administration of erythropoietin was sufficient to lower serum hepcidin profoundly within less than 1 day, in the absence of any significant changes in serum iron (Ashby et al. 2010). Apparently, stimulated erythrocyte precursors produce one or more hepcidin-suppressive factors but the molecular nature of this putative physiological erythroid regulator of hepcidin is not yet known. The suppressive effect on hepcidin is even more prominent under pathological conditions of expanded but ineffective erythropoiesis, seen in β-thalassemia and congenital dyserythropoietic anemias (Adamsky et al. 2004; Papanikolaou et al. 2005) where the large number of apoptosing erythrocyte precursors could generate additional suppressive factors. Two members of the BMP family, growth differentiation factor (GDF) 15 and twisted gastrulation (TWSG) 1, have been proposed to play a role in pathological hepcidin suppression during ineffective erythropoiesis (Tanno et al. 2007, 2009; Casanovas et al. 2011) but their specific regulatory role in iron homeostasis or pathology remains to be established.

Cite this article as *Cold Spring Harb Perspect Med* doi: 10.1101/cshperspect.a011668

Regulation of Hepcidin by Inflammation

During infections and inflammation, the synthesis and serum concentrations of hepcidin are greatly increased (Pigeon et al. 2001; Nicolas et al. 2002; Nemeth et al. 2003, 2004; Ganz et al. 2008). This regulatory circuitry is thought to be related to the possible role of hepcidin in host defense whereby hepcidin-mediated iron restriction may limit microbial growth. Multiple cytokines stimulate hepcidin transcription during inflammation, chief among them IL-6 (Nemeth et al. 2003, 2004) and the members of the BMP family (Maes et al. 2010). Interleukin-6 activates the JAK-STAT3 pathway (Fig. 3), with STAT3 binding to canonical binding sites in the hepcidin promoter, leading to transcriptional stimulation of hepcidin synthesis (Wrighting and Andrews 2006; Pietrangelo et al. 2007; Verga Falzacappa et al. 2007). The BMP and IL-6 pathways are synergistic through a mechanism that is not yet fully defined (Verga Falzacappa et al. 2008; Maes et al. 2010). Inflammation may contribute to elevated serum hepcidin levels seen in many adult patients with sickle cell anemia (Kroot et al. 2009; Porter 2009).

DISORDERS OF IRON HOMEOSTASIS

Iron Deficiency

Worldwide, iron deficiency is the most common iron disorder (see Miller 2013). Although it is thought of as a predominantly acquired problem caused by blood loss and inadequate iron intake, genetic predisposition may modulate the susceptibility to this condition as illustrated by genome-wide association studies (Benyamin et al. 2009a,b; Chambers et al. 2009; Tanaka et al. 2010). Not surprisingly, associations have been reported between serum iron concentrations and polymorphisms and mutations in transferrin, HFE, and MT2 (TMPRSS6).

Anemia of Inflammation (Anemia of Chronic Disease)

Infections and inflammatory disorders are a common cause of iron maldistribution, mediated by increased plasma hepcidin concentrations that restrict the release of recycled iron from macrophages and hepatocyte stores, and interfere with the absorption of dietary iron. Iron restriction limits hemoglobin synthesis and contributes to anemia although other factors (inadequate erythropoietin production, cytokine effects on the marrow, decreased erythrocyte lifespan) may also participate, depending on the underlying disease. Clinically, this disorder is manifested most often as a normocytic normochromic anemia with hypoferremia, but the anemia can be microcytic, especially in children or very chronic inflammatory disorders.

Iron-Refractory Iron Deficiency Anemia

This relatively rare condition is detected in children who present with an unexplained hypoferremia and microcytic anemia resistant to oral iron administration, and partially resistant even to intravenous iron supplementation. The patients have elevated or high normal hepcidin levels (Finberg et al. 2008) in stark contrast to common iron deficiency in which serum hepcidin is very low or undetectable (Ganz et al. 2008).

Iron Overload from Transfusions

Blood transfusions deliver ~1 mg of iron for each mL of packed erythrocytes or more than 200 mg per each unit transfused, effectively bypassing the regulatory mechanisms that control iron intake. Excess iron may eventually cause toxicity and organ damage, and can only be removed by phlebotomy (contraindicated if patient is still anemic) or by treatment with chelating agents. Extrapolating from clinical data for iron-related toxicity in hereditary hemochromatosis, chelation therapy is recommended after 10–20 transfusions for those patients who need chronic erythrocyte transfusions (Brittenham 2011).

Hereditary Hemochromatosis and Related Disorders

Hereditary hemochromatosis is a group of genetic disorders that impede either hepcidin production or its regulation by iron (Nicolas et al.

2001; Papanikolaou et al. 2005) or, very rarely, cause the resistance of ferroportin to internalization by hepcidin (Fernandes et al. 2009; Sham et al. 2009). In the order of increasing severity of hepcidin deficiency and iron overload, autosomal recessive forms can result from mutations in genes encoding HFE, TfR2, hemojuvelin, and hepcidin. Ferroportin resistance to hepcidin is attributable to autosomal-dominant mutations in ferroportin that either interfere with hepcidin binding or with ferroportin internalization. Additional genes that caused iron overload in transgenic mouse models but have not yet been implicated in humans include BMP6 (Andriopoulos Jr. et al. 2009; Meynard et al. 2009) and neogenin (Lee et al. 2010). Hepatic iron overload can also develop as a part of more complex genetic diseases, including deficiencies of transferrin and ceruloplasmin and loss-of-function mutations in DMT1 (Pietrangelo et al. 2011). In these disorders, iron-restrictive anemia develops because of diminished iron release from macrophages (ceruloplasmin deficiency), iron delivery to erythrocytes (transferrin deficiency), or iron utilization by erythrocyte precursors (DMT1 loss of function). In contrast, simple hereditary hemochromatosis in humans has only modest effects on erythropoiesis, limited to a slight increase in mean corpuscular volume (McLaren et al. 2007).

Iron Overload Associated with Ineffective Erythropoiesis

Ineffective erythropoiesis is a condition in which large numbers of marrow erythrocyte precursors undergo apoptosis and so fail to complete maturation into erythrocytes. Genetic lesions that cause ineffective erythropoiesis prominently include thalassemias and congenital dyserythropoietic anemias. In these disorders, hepcidin is suppressed (Adamsky et al. 2004; Papanikolaou et al. 2005; Kearney et al. 2007; Casanovas et al. 2011) causing hyperabsorption of dietary iron and iron overload even in the absence of erythrocyte transfusions (Origa et al. 2007). Transfusions partially relieve the erythropoietin-driven expansion of ineffective erythropoiesis and raise hepcidin concentrations (Kearney et al. 2007) but cause iron overload owing to the iron content of transfused blood.

Modulation of Ineffective Erythropoiesis by Iron Availability

Recent studies in mouse models of thalassemia suggest that increased concentrations of plasma iron in this condition may further unbalance heme and globin synthesis and worsen ineffective erythropoeisis, and conversely, that restricting the iron supply through the administration of apotransferrin or hepcidin may improve erythropoiesis (Gardenghi et al. 2010a,b; Li et al. 2010). It remains to be seen whether similar interventions will be helpful in human disease.

CONCLUDING REMARKS

Iron homeostasis is intimately intertwined with erythropoiesis, the main destination of iron in humans and other vertebrates. Iron overload is a clinically important aspect of various hemoglobinopathies because of transfusion-induced iron overload and because of pathological suppression of hepcidin synthesis with resulting hyperabsorption of dietary iron. Advances in the understanding of iron homeostasis and its interactions with erythropoiesis should translate into improved outcomes for patients with hemoglobinopathies.

REFERENCES

**Reference is also in this collection.*

Adamsky K, Weizer O, Amariglio N, Breda L, Harmelin A, Rivella S, Rachmilewitz E, Rechavi G. 2004. Decreased hepcidin mRNA expression in thalassemic mice. *Br J Haematol* **124:** 123–124.

Andriopoulos B Jr, Corradini E, Xia Y, Faasse SA, Chen S, Grgurevic L, Knutson MD, Pietrangelo A, Vukicevic S, Lin HY, et al. 2009. BMP6 is a key endogenous regulator of hepcidin expression and iron metabolism. *Nat Genet* **41:** 482–487.

Ashby DR, Gale DP, Busbridge M, Murphy KG, Duncan ND, Cairns TD, Taube DH, Bloom SR, Tam FW, Chapman R, et al. 2010. Erythropoietin administration in humans causes a marked and prolonged reduction in circulating hepcidin. *Haematologica* **95:** 505–508.

Beguin Y, Stray SM, Cazzola M, Huebers HA, Finch CA. 1988. Ferrokinetic measurement of erythropoiesis. *Acta Haematol* **79:** 121–126.

 Cite this article as *Cold Spring Harb Perspect Med* doi: 10.1101/cshperspect.a011668

Benyamin B, Ferreira MA, Willemsen G, Gordon S, Middelberg RP, McEvoy BP, Hottenga JJ, Henders AK, Campbell MJ, Wallace L, et al. 2009a. Common variants in TMPRSS6 are associated with iron status and erythrocyte volume. *Nat Genet* **41:** 1173–1175.

Benyamin B, McRae AF, Zhu G, Gordon S, Henders AK, Palotie A, Peltonen L, Martin NG, Montgomery GW, Whitfield JB, et al. 2009b. Variants in TF and HFE explain approximately 40% of genetic variation in serum-transferrin levels. *Am J Hum Genet* **84:** 60–65.

Breuer W, Ronson A, Slotki IN, Abramov A, Hershko C, Cabantchik ZI. 2000. The assessment of serum nontransferrin-bound iron in chelation therapy and iron supplementation. *Blood* **95:** 2975–2982.

Brittenham GM. 2011. Iron-chelating therapy for transfusional iron overload. *N Engl J Med* **364:** 146–156.

Bullock GC, Delehanty LL, Talbot AL, Gonias SL, Tong WH, Rouault TA, Dewar B, Macdonald JM, Chruma JJ, Goldfarb AN. 2010. Iron control of erythroid development by a novel aconitase-associated regulatory pathway. *Blood* **116:** 97–108.

Casanovas G, Swinkels DW, Altamura S, Schwarz K, Laarakkers CM, Gross HJ, Wiesneth M, Heimpel H, Muckenthaler MU. 2011. Growth differentiation factor 15 in patients with congenital dyserythropoietic anaemia (CDA) type II. *J Mol Med* **89:** 811–816.

Casey JL, Hentze MW, Koeller DM, Caughman SW, Rouault TA, Klausner RD, Harford JB. 1988. Iron-responsive elements: Regulatory RNA sequences that control mRNA levels and translation. *Science* **240:** 924–928.

Chambers JC, Zhang W, Li Y, Sehmi J, Wass MN, Zabaneh D, Hoggart C, Bayele H, McCarthy MI, Peltonen L, et al. 2009. Genome-wide association study identifies variants in TMPRSS6 associated with hemoglobin levels. *Nat Genet* **41:** 1170–1172.

Chen JJ. 2007. Regulation of protein synthesis by the heme-regulated eIF2α kinase: Relevance to anemias. *Blood* **109:** 2693–2699.

Chen W, Dailey HA, Paw BH. 2010. Ferrochelatase forms an oligomeric complex with mitoferrin-1 and Abcb10 for erythroid heme biosynthesis. *Blood* **116:** 628–630.

Corradini E, Meynard D, Wu Q, Chen S, Ventura P, Pietrangelo A, Babitt JK. 2011. Serum and liver iron differently regulate the bone morphogenetic protein 6 (BMP6)-SMAD signaling pathway in mice. *Hepatology* **54:** 273–284.

Cunningham MJ. 2008. Update on thalassemia: Clinical care and complications. *Pediatr Clin North Am* **55:** 447–460.

Du X, She E, Gelbart T, Truksa J, Lee P, Xia Y, Khovananth K, Mudd S, Mann N, Moresco EM, et al. 2008. The serine protease TMPRSS6 is required to sense iron deficiency. *Science* **320:** 1088–1092.

Fernandes A, Preza GC, Phung Y, De Domenico I, Kaplan J, Ganz T, Nemeth E. 2009. The molecular basis of hepcidin-resistant hereditary hemochromatosis. *Blood* **114:** 437–443.

Finberg KE, Heeney MM, Campagna DR, Aydinok Y, Pearson HA, Hartman KR, Mayo MM, Samuel SM, Strouse JJ, Markianos K, et al. 2008. Mutations in TMPRSS6 cause iron-refractory iron deficiency anemia (IRIDA). *Nat Genet* **40:** 569–571.

Finch C. 1994. Regulators of iron balance in humans. *Blood* **84:** 1697–1702.

Ganz T, Olbina G, Girelli D, Nemeth E, Westerman M. 2008. Immunoassay for human serum hepcidin. *Blood* **112:** 4292–4297.

Gardenghi S, Grady RW, Rivella S. 2010a. Anemia, ineffective erythropoiesis, and hepcidin: Interacting factors in abnormal iron metabolism leading to iron overload in beta-thalassemia. *Hematol Oncol Clin North Am* **24:** 1089–1107.

Gardenghi S, Ramos P, Marongiu MF, Melchiori L, Breda L, Guy E, Muirhead K, Rao N, Roy CN, Andrews NC, et al. 2010b. Hepcidin as a therapeutic tool to limit iron overload and improve anemia in β-thalassemic mice. *J Clin Invest* **120:** 4466–4477.

Kearney SL, Nemeth E, Neufeld EJ, Thapa D, Ganz T, Weinstein DA, Cunningham MJ. 2007. Urinary hepcidin in congenital chronic anemias. *Pediatr Blood Cancer* **48:** 57–63.

Kroot JJ, Laarakkers CM, Kemna EH, Biemond BJ, Swinkels DW. 2009. Regulation of serum hepcidin levels in sickle cell disease. *Haematologica* **94:** 885–887.

Lee DH, Zhou LJ, Zhou Z, Xie JX, Jung JU, Liu Y, Xi CX, Mei L, Xiong WC. 2010. Neogenin inhibits HJV secretion and regulates BMP-induced hepcidin expression and iron homeostasis. *Blood* **115:** 3136–3145.

Li H, Rybicki AC, Suzuka SM, von BL, Breuer W, Hall CB, Cabantchik ZI, Bouhassira EE, Fabry ME, Ginzburg YZ. 2010. Transferrin therapy ameliorates disease in β-thalassemic mice. *Nat Med* **16:** 177–182.

Liu S, Bhattacharya S, Han A, Suragani RN, Zhao W, Fry RC, Chen JJ. 2008. Haem-regulated eIF2α kinase is necessary for adaptive gene expression in erythroid precursors under the stress of iron deficiency. *Br J Haematol* **143:** 129–137.

Maes K, Nemeth E, Roodman GD, Huston A, Esteve F, Freytes C, Callander N, Katodritou E, Tussing-Humphreys L, Rivera S, et al. 2010. In anemia of multiple myeloma, hepcidin is induced by increased bone morphogenetic protein 2. *Blood* **116:** 3635–3644.

Mastrogiannaki M, Matak P, Keith B, Simon MC, Vaulont S, Peyssonnaux C. 2009. HIF-2α, but not HIF-1α, promotes iron absorption in mice. *J Clin Invest* **119:** 1159–1166.

Mastrogiannaki M, Matak P, Mathieu JR, Delga S, Mayeux P, Vaulont S, Peyssonnaux C. 2011. Hepatic HIF-2 down-regulates hepcidin expression in mice through epo-mediated increase in erythropoiesis. *Haematologica* doi: 10.3324/haematol.2011.056119.

McLaren CE, Barton JC, Gordeuk VR, Wu L, Adams PC, Reboussin DM, Speechley M, Chang H, Acton RT, Harris EL, et al. 2007. Determinants and characteristics of mean corpuscular volume and hemoglobin concentration in white HFE C282Y homozygotes in the hemochromatosis and iron overload screening study. *Am J Hematol* **82:** 898–905.

Meynard D, Kautz L, Darnaud V, Canonne-Hergaux F, Coppin H, Roth MP. 2009. Lack of the bone morphogenetic protein BMP6 induces massive iron overload. *Nat Genet* **41:** 478–481.

* Miller JL. 2013. Iron deficiency anemia: A common and curable disease. *Cold Spring Harb Perspect Med* doi: 10.1101/cshperspect.a011866.

Nandal A, Ruiz JC, Subramanian P, Ghimire-Rijal S, Sinnamon RA, Stemmler TL, Bruick RK, Philpott CC. 2011. Activation of the HIF prolyl hydroxylase by the iron chaperones PCBP1 and PCBP2. *Cell Metab* **14:** 647–657.

Nemeth E, Valore EV, Territo M, Schiller G, Lichtenstein A, Ganz T. 2003. Hepcidin, a putative mediator of anemia of inflammation, is a type II acute-phase protein. *Blood* **101:** 2461–2463.

Nemeth E, Rivera S, Gabayan V, Keller C, Taudorf S, Pedersen BK, Ganz T. 2004. IL-6 mediates hypoferremia of inflammation by inducing the synthesis of the iron regulatory hormone hepcidin. *J Clin Invest* **113:** 1271–1276.

Nicolas G, Bennoun M, Devaux I, Beaumont C, Grandchamp B, Kahn A, Vaulont S. 2001. Lack of hepcidin gene expression and severe tissue iron overload in upstream stimulatory factor 2 (USF2) knockout mice. *Proc Natl Acad Sci* **98:** 8780–8785.

Nicolas G, Chauvet C, Viatte L, Danan JL, Bigard X, Devaux I, Beaumont C, Kahn A, Vaulont S. 2002. The gene encoding the iron regulatory peptide hepcidin is regulated by anemia, hypoxia, and inflammation. *J Clin Invest* **110:** 1037–1044.

Ohgami RS, Campagna DR, McDonald A, Fleming MD. 2006. The Steap proteins are metalloreductases. *Blood* **108:** 1388–1394.

Origa R, Galanello R, Ganz T, Giagu N, Maccioni L, Faa G, Nemeth E. 2007. Liver iron concentrations and urinary hepcidin in β-thalassemia. *Haematologica* **92:** 583–588.

Pak M, Lopez MA, Gabayan V, Ganz T, Rivera S. 2006. Suppression of hepcidin during anemia requires erythropoietic activity. *Blood* **108:** 3730–3735.

Papanikolaou G, Samuels ME, Ludwig EH, MacDonald ML, Franchini PL, Dube MP, Andres L, MacFarlane J, Sakellaropoulos N, Politou M, et al. 2004. Mutations in HFE2 cause iron overload in chromosome 1q-linked juvenile hemochromatosis. *Nat Genet* **36:** 77–82.

Papanikolaou G, Tzilianos M, Christakis JI, Bogdanos D, Tsimirika K, MacFarlane J, Goldberg YP, Sakellaropoulos N, Ganz T, Nemeth E. 2005. Hepcidin in iron overload disorders. *Blood* **105:** 4103–4105.

Paradkar PN, Zumbrennen KB, Paw BH, Ward DM, Kaplan J. 2009. Regulation of mitochondrial iron import through differential turnover of mitoferrin 1 and mitoferrin 2. *Mol Cell Biol* **29:** 1007–1016.

Pietrangelo A, Dierssen U, Valli L, Garuti C, Rump A, Corradini E, Ernst M, Klein C, Trautwein C. 2007. STAT3 is required for IL-6-gp130-dependent activation of hepcidin in vivo. *Gastroenterology* **132:** 294–300.

Pietrangelo A, Caleffi A, Corradini E. 2011. Non-HFE hepatic iron overload. *Semin Liver Dis* **31:** 302–318.

Pigeon C, Ilyin G, Courselaud B, Leroyer P, Turlin B, Brissot P, Loreal O. 2001. A new mouse liver-specific gene, encoding a protein homologous to human antimicrobial peptide hepcidin, is overexpressed during iron overload. *J Biol Chem* **276:** 7811–7819.

Porter JB. 2009. Pathophysiology of transfusional iron overload: Contrasting patterns in thalassemia major and sickle cell disease. *Hemoglobin* **33:** S37–S45.

Poss KD, Tonegawa S. 1997. Heme oxygenase 1 is required for mammalian iron reutilization. *Proc Natl Acad Sci* **94:** 10919–10924.

Ramos E, Kautz L, Rodriguez R, Hansen M, Gabayan V, Ginzburg Y, Roth MP, Nemeth E, Ganz T. 2011. Evidence for distinct pathways of hepcidin regulation by acute and chronic iron loading in mice. *Hepatology* **53:** 1333–1341.

Salahudeen AA, Thompson JW, Ruiz JC, Ma HW, Kinch LN, Li Q, Grishin NV, Bruick NK. 2009. An E3 ligase possessing an iron-responsive hemerythrin domain is a regulator of iron homeostasis. *Science* **326:** 722–726.

Sanchez M, Galy B, Schwanhaeusser B, Blake J, Bahr-Ivacevic T, Benes V, Selbach M, Muckenthaler MU, Hentze MW. 2011. Iron regulatory protein-1 and -2: Transcriptome-wide definition of binding mRNAs and shaping of the cellular proteome by iron regulatory proteins. *Blood* **118:** e168–e179.

Sham RL, Phatak PD, Nemeth E, Ganz T. 2009. Hereditary hemochromatosis due to resistance to hepcidin: High hepcidin concentrations in a family with C326S ferroportin mutation. *Blood* **114:** 493–494.

Sheftel AD, Zhang AS, Brown C, Shirihai OS, Ponka P. 2007. Direct interorganellar transfer of iron from endosome to mitochondrion. *Blood* **110:** 125–132.

Shi H, Bencze KZ, Stemmler TL, Philpott CC. 2008. A cytosolic iron chaperone that delivers iron to ferritin. *Science* **320:** 1207–1210.

Silvestri L, Pagani A, Nai A, De Domenico I, Kaplan J, Camaschella C. 2008. The serine protease matriptase-2 (TMPRSS6) inhibits hepcidin activation by cleaving membrane hemojuvelin. *Cell Metab* **8:** 502511.

Soe-Lin S, Apte SS, Andriopoulos B Jr, Andrews MC, Schranzhofer M, Kahawita T, Garcia-Santos D, Ponka P. 2009. Nramp1 promotes efficient macrophage recycling of iron following erythrophagocytosis in vivo. *Proc Natl Acad Sci* **106:** 5960–5965.

Talbot AL, Bullock GC, Delehanty LL, Sattler M, Zhao ZJ, Goldfarb AN. 2011. Aconitase regulation of erythropoiesis correlates with a novel licensing function in erythropoietin-induced ERK signaling. *PLoS ONE* **6:** e23850.

Tanaka T, Roy CN, Yao W, Matteini A, Semba RD, Arking D, Walston JD, Fried LP, Singleton A, Guralnik J, et al. 2010. A genome-wide association analysis of serum iron concentrations. *Blood* **115:** 94–96.

Tanno T, Bhanu NV, Oneal PA, Goh SH, Staker P, Lee YT, Moroney JW, Reed CH, Luban NL, Wang RH, et al. 2007. High levels of GDF15 in thalassemia suppress expression of the iron regulatory protein hepcidin. *Nat Med* **13:** 1096–1101.

Tanno T, Porayette P, Sripichai O, Noh SJ, Byrnes C, Bhupatiraju A, Lee YT, Goodnough JB, Harandi O, Ganz T, et al. 2009. Identification of TWSG1 as a second novel erythroid regulator of hepcidin expression in murine and human cells. *Blood* **114:** 181–186.

Telfer P. 2009. Update on survival in thalassemia major. *Hemoglobin* **33:** S76–S80.

Troadec MB, Warner D, Wallace J, Thomas K, Spangrude GJ, Phillips J, Khalimonchuk O, Paw BH, Ward DM, Kaplan J. 2011. Targeted deletion of the mouse Mitoferrin1 gene: From anemia to protoporphyria. *Blood* **117:** 5494–5502.

Vashisht AA, Zumbrennen KB, Huang X, Powers DN, Durazo A, Sun D, Bhaskaran N, Persson A, Uhlen M, Sangfelt O, et al. 2009. Control of iron homeostasis by an iron-regulated ubiquitin ligase. *Science* **326:** 718–721.

Verga Falzacappa MV, Vujic SM, Kessler R, Stolte J, Hentze MW, Muckenthaler MU. 2007. STAT3 mediates hepatic hepcidin expression and its inflammatory stimulation. *Blood* **109:** 353–358.

Verga Falzacappa MV, Casanovas G, Hentze MW, Muckenthaler MU. 2008. A bone morphogenetic protein (BMP)-responsive element in the hepcidin promoter controls HFE2-mediated hepatic hepcidin expression and its response to IL-6 in cultured cells. *J Mol Med* **86:** 531–540.

Vokurka M, Krijt J, Sulc K, Necas E. 2006. Hepcidin mRNA levels in mouse liver respond to inhibition of erythropoiesis. *Physiol Res* **55:** 667–674.

Wallace DF, Summerville L, Crampton EM, Frazer DM, Anderson GJ, Subramaniam VN. 2009. Combined deletion of Hfe and transferrin receptor 2 in mice leads to marked dysregulation of hepcidin and iron overload. *Hepatology* **50:** 1992–2000.

Wrighting DM, Andrews NC. 2006. Interleukin-6 induces hepcidin expression through STAT3. *Blood* **108:** 3204–3209.

Zhang DL, Hughes RM, Ollivierre-Wilson H, Ghosh MC, Rouault TA. 2009. A ferroportin transcript that lacks an iron-responsive element enables duodenal and erythroid precursor cells to evade translational repression. *Cell Metab* **9:** 461–473.

Erythroid Heme Biosynthesis and Its Disorders

Harry A. Dailey[1] and Peter N. Meissner[2]

[1]Department of Microbiology, Department of Biochemistry and Molecular Biology, Biomedical and Health Sciences Institute, University of Georgia, Athens, Georgia 30602

[2]Division of Medical Biochemistry, Department of Clinical Laboratory Sciences, Institute of Infectious Disease and Molecular Medicine, University of Cape Town, South Africa

Correspondence: hdailey@uga.edu

Heme, which is composed of iron and the small organic molecule protoporphyrin, is an essential component of hemoglobin as well as a variety of physiologically important hemoproteins. During erythropoiesis, heme synthesis is induced before, and is essential for, globin synthesis. Although all cells possess the ability to synthesize heme, there are distinct differences between regulation of the pathway in developing erythroid cells and all other types of cells. Disorders that compromise the ability of the developing red cell to synthesize heme can have profound medical implications. The biosynthetic pathway for heme and key regulatory features are reviewed herein, along with specific human genetic disorders that arise from defective heme synthesis such as X-linked sideroblastic anemia and erythropoietic protoporphyria.

Hemoglobin is recognized for its brilliant red color and ability to reversibly bind oxygen. The molecule owes both of these characteristics to the presence of the small organic iron tetrapyrrole protoheme IX (heme). Indeed, the very name "hemoglobin" derives from the components of the molecule, heme and globin. During erythropoiesis, cells coordinate the synthesis of both heme and globins, not only for efficiency but also because imbalance in the synthesis of either has profoundly negative physiological consequences. Heme-free globin chains are nonfunctional, and free heme is highly toxic to the cell. As is clearly shown in this collection, erythroid cell differentiation is a widely examined issue, with considerable attention given to regulation of globin synthesis, cytoskeleton construction, and mechanisms for the elimination of intracellular organelles. During differentiation of erythroid cells, the induction of heme precedes that of globin synthesis. This is in part attributable to the requirement for heme to inactivate the heme-regulated eukaryotic initiation factor 2α kinase (HRI) and, thereby, allow for protein translation initiation to proceed. In the absence of sufficient cellular heme, HRI is activated and protein synthesis, including that of globin proteins, is prevented. Given the obligatory need for heme, disorders of heme synthesis can have profound implications for the organism. Below we have briefly reviewed heme synthesis, its regulation, and diseases associated with genetic disorders that have particular relevance for erythroid cells.

Copyright © 2013 Cold Spring Harbor Laboratory Press; all rights reserved
Cite this article as *Cold Spring Harb Perspect Med* doi: 10.1101/cshperspect.a011676

HEME BIOSYNTHETIC PATHWAY

Enzymes of heme biosynthesis are nuclear encoded and cytoplasmically synthesized (see Ajioka et al. 2006; Medlock and Dailey 2009; Layer et al. 2010). The pathway is composed of eight enzymes that are divided between the mitochondria and cytoplasm (Fig. 1). The overall reaction to synthesize heme in metazoan cells is: 8 glycine + 8 succinyl-CoA + 5 O_2 + Fe^{2+} $\rightarrow$ Heme + 8 CoA + 4 NH_4 + 14 CO_2 + 3 H_2O_2 + 2 H^+ + 11 H_2O. Of the products, 8 CoA (coenzyme A), 8 CO_2, 2 H^+, and heme are generated in the mitochondrial matrix and 2 CO_2, 3 H_2O_2, and 2 H_2O are generated in the mitochondrial inner membrane space. All remaining products are cytosolic. The pathway and the individual enzymes have been well researched and frequently reviewed. Regulation of the pathway is tissue- and cell-specific, subject to modulation by a myriad of developmental and environmental pressures through diverse transcriptional factors, and is beyond the scope of the current review (Medlock and Dailey 2009). In general, however, synthesis of the first committed intermediate, 5-aminolevulinate (ALA), is considered rate-limiting.

Aminolevulinate Synthase

The first step in mammalian heme biosynthesis is catalyzed by the enzyme ALA synthase (ALAS) (EC 2.3.1.37) (Hunter and Ferreira 2011). ALAS, which catalyzes the condensation of glycine with succinyl-CoA to form ALA and CO_2, is a homodimeric, pyridoxal phosphate-containing enzyme that is a member of the large and well-characterized α-oxoamine synthase family. The mature protein is located in the mitochondrial matrix. Much is known about the mammalian enzyme from kinetic and site-directed mutagenesis studies, but at present only the crystal structure of ALAS from the bacterium *Rhodobacter capsulatus* has been determined (Astner et al. 2005). However, it is clear that the mature mammalian ALAS is structurally homologous to the bacterial enzyme except that it possesses an additional carboxy-terminal sequence of approximately 25 residues. Interestingly, recent studies have shown that this region

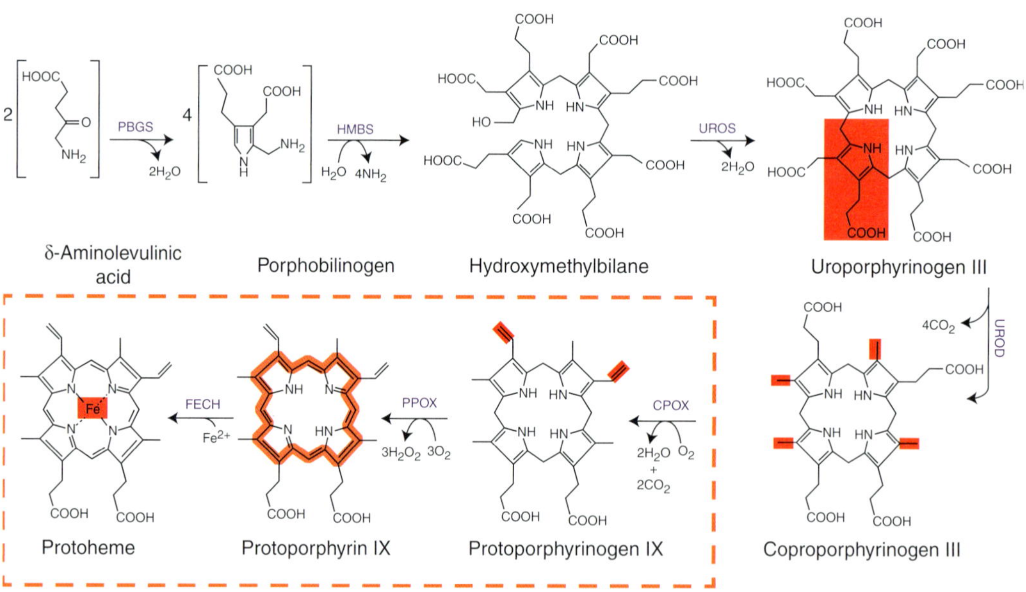

Figure 1. The mammalian heme biosynthetic pathway. The diagram presents the steps and structures of intermediates in the pathway from ALA to heme. Steps that occur in the mitochondrion are enclosed in the dashed box, and those present in the cytosol are outside the box. Synthesis of ALA from glycine and succinyl-CoA, which is the first committed step and occurs in the mitochondrion, is not shown.

of mammalian ALAS somehow modulates the activity of the enzyme (Whatley et al. 2008). Truncation or selected amino acid mutations in this region result in a form of ALAS that has significantly higher activity than the normal ALAS. Such alterations result in dysfunctional cellular heme biosynthesis (see below).

In mammals, one finds two isozymes of ALAS, one specific for differentiating erythroid cells (ALAS-2 or ALAS-E) and one expressed in all other cell types (ALAS-1 or ALAS-N) (May et al. 1995; Medlock and Dailey 2009; Hunter and Ferreira 2011). The genes for ALAS, as for all heme biosynthetic enzymes, are nuclear, although the final destination for ALAS is the mitochondrion. The genes for ALAS-1 and ALAS-2 are located on human chromosomes 3 and X, respectively, and are differentially regulated (see below). ALAS-1 and -2 clearly represent gene duplication events because the proteins are highly similar in structure and catalytic abilities. For human ALAS-1 and ALAS-2, one finds alternative mRNA splice variants. For ALAS-1, two known splice variants occur in the untranslated region of exon 1 (Roberts and Elder 2001). Via an uncharacterized mechanism, these variants possess altered sensitivity to heme-mediated decay of the message. One splice variant of human ALAS-2 has been found in which exon 4 is absent (Cox et al. 2004). This variant represents 35%–45% of total *ALAS2* mRNA in the cell and was shown to encode a protein with slightly reduced activity. This variant ALAS-2 is translocated into the mitochondrial matrix, where it was shown to interact with succinyl CoA synthetase, just as the full-length enzyme does (Furuyama and Sassa 2000), thereby suggesting that it contributes to erythroid heme synthesis.

The precursor proteins of both ALAS-1 and -2 possess mitochondrial targeting sequences that are proteolytically removed after being translocated into the mitochondrial matrix. Both proteins possess three heme regulatory motifs (HRMs) composed of a canonical Cys-Pro sequence (Zhang and Guarente 1995). Two of these HRMs are found in the targeting leader sequence and the third is near the amino terminus of the mature processed ALAS. Evidence from multiple groups has shown that heme binds to the ALAS HRM in vitro and in vivo and inhibits the translocation of the protein into the mitochondrion (Yamauchi et al. 1980; Lathrop and Timko 1993; Dailey et al. 2005). A series of mutagenesis experiments clearly showed the significance of all three ALAS HRMs for heme inhibition of ALAS translocation into mitochondria in vivo (Dailey et al. 2005). Whereas such heme regulation of apoprotein translocation in nonerythroid cells can easily be rationalized, a similar occurrence in differentiating erythroid cells in which massive quantities of heme are being synthesized in a relatively short period of time is less easily justified. Recently, a pair of HRM motifs have been shown to serve as a redox-sensitive switch for heme oxygenase-2 (Yi and Ragsdale 2007), but there is no evidence to suggest such a role with ALAS.

Porphobilinogen Synthase

Once ALA is produced by ALAS, it is exported out of the mitochondrial matrix to reach the second pathway enzyme. The exact mechanism for this transport is not completely defined, but evidence suggests that the mitochondrial inner membrane solute transport protein, SLC25A38, may be responsible for this function (Guernsey et al. 2009). The next enzymatic step is catalyzed by porphobilinogen synthase (PBGS) (EC 4.2.1.24) (previously named ALA dehydratase) (Schubert et al. 2009). PBGS catalyzes the condensation of two molecules of ALA to form one molecule of the monopyrrole PBG. The crystal structure of the soluble, cytoplasmic PBGS reveals a homo-octomer that can best be described as a tetramer of homodimers (Erskine et al. 1997). Each monomer binds one zinc atom for a total of eight zinc atoms. Four metal atoms are essential for catalysis and four are involved in stabilization of tertiary structure. In individuals suffering from chronic lead exposure, one finds that these zinc ions may be replaced by lead, resulting in an inactive enzyme. It has been shown that PBGS exists in alternate quaternary structures named morpheeins (Jaffe and Lawrence 2012). The morpheeins represent a dynamic change in oligomerization of PBGS between the high-activity octomer and a low-activity hexamer. This change in quaternary structure is the basis of allosteric regulation of PBGS. However, given that ALAS is considered rate-limiting to heme synthesis, the role of allosteric regulation at PBGS is something of an enigma.

Although two distinct genes on separate chromosomes exist for ALAS-1 and -2, only single genes exist for the remaining pathway enzymes. However, for all genes one finds distinct erythroid and housekeeping-specific promoter elements. For PBGS one also finds erythroid versus housekeeping differential splice variants (Kaya et al. 1994). Human PBGS possesses two noncoding exons, 1A and 1B. The translational start site for housekeeping PBGS mRNA is exon 1A, while erythroid PBGS mRNA starts in exon 1B. Thus, an additional 5′ untranslated region is present in the mRNA for the housekeeping variant that is lacking in the erythroid splice variant. Given that these variants are in the noncoding region, the housekeeping and erythroid forms of the PBGS enzymes are identical.

Hydroxymethylbilane Synthase

Following formation of PBG, the enzyme hydroxymethylbilane synthase (HMBS, previously called PBG deaminase or PBGD) (EC 2.5.1.61) catalyzes the head-to-tail synthesis of four PBG molecules to form the linear tetrapyrrole HMB and releases four molecules of ammonium (Schubert et al. 2009). HMBS has been purified from a variety of sources, and the crystal structures of the *Escherichia coli* (Louie et al. 1992) and human (Gill et al. 2009) enzymes have been determined. The cytoplasmically located monomer is synthesized as an apoprotein that in its first complete catalytic cycle synthesizes a hexameric linear polypyrrole that is covalently bound to the HMBS protein. The distal linear tetrapyrrole, HMB, is cleaved and released, leaving behind a covalently bound dipyrromethane. The dipyrromethane serves as a cofactor for future turnovers. The HMB produced by HMBS is chemically reactive and will spontaneously cyclize to form uroporphyrinogen I in the absence of the next pathway enzyme. Uroporphyrinogen I cannot be converted into protoporphyrin IX. As with ALAS and PBGS, HMBS has erythroid and housekeeping-specific splice variants (Grandchamp et al. 1987). The mRNA of the erythroid form of HMBS skips exon 1 and starts transcription with the noncoding exon 2. The erythroid-specific translational initiation site is in exon 3. The housekeeping HMBS starts with exon 1, which contains the housekeeping-specific initiation site, and skips the noncoding exon 2. The result is that the housekeeping HMBS has an additional 17 amino acid residues at the amino terminus that are lacking in the erythroid form of the protein.

Uroporphyrinogen Synthase

Conversion of HMB to the physiological uroporphyrinogen III isomer requires the action of uroporphyrinogen synthase (UROS) (EC 4.2.1.75) (Schubert et al. 2009). The reaction, catalyzed without a cofactor, is a spiro inversion of the final, or D, ring of HMB followed by cyclization to yield the III isomer of uroporphyrinogen (Fig. 2). UROS is a monomeric protein; the crystal structure has been determined for both the human protein (Mathews et al. 2001) and the bacterial protein from *Thermus thermophilus* with bound substrate (Schubert et al. 2008). Its structure is a bit unusual because it is composed of two distinct structural domains connected by a flexible linker region. Crystal structures have been obtained for a variety of domain orientations, suggesting that the molecule is highly flexible in solution. Splice variants also exist for UROS (Aizencang et al. 2000). The gene for UROS possesses distinct erythroid and housekeeping promoters. The erythroid promoter elements are located in the intron between exons 1 and 2 and drive transcription from an initiation site in exon 2. The housekeeping promoter region is present upstream of exon 1 and drives transcription from an initiation site in exon 1. However, because exon 1 is noncoding, both housekeeping and erythroid UROS proteins are identical.

Uroporphyrinogen Decarboxylase

The final cytoplasmic enzyme in the pathway is uroporphyrinogen decarboxylase (UROD) (EC

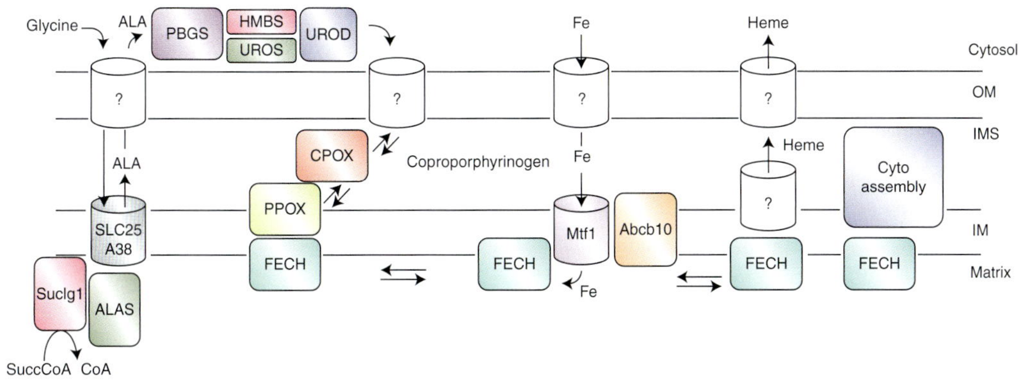

Figure 2. Proposed model for components involved in heme synthesis. Details are discussed in the text, and in cases in which specific data exist for a particular component, its name is shown. Components for which no solid data exist, but for which there is good reason to expect they exist, are denoted with "?". IM, Inner membrane; IMS, intramembrane space; OM, outer membrane.

4.1.1.37) (Shoolingin-Jordan 2003). UROD contains no cofactors or metal ions and catalyzes the decarboxylation of the four pyrrole acetic acid side chains to yield coproporphyrinogen and four molecules of CO_2. This homodimeric enzyme has been crystallized (Phillips et al. 2003) and its catalytic mechanism well studied (Lewis and Wolfenden 2008). UROD will use both uroporphyrinogen I and III, converting them into coproporphyrinogen I and III, respectively. In the presence of high concentrations of uroporphyrinogen, UROD will decarboxylate in a random fashion, but it is believed that in situ the reaction starts with the decarboxylation of the D ring acetate and proceeds sequentially in a clockwise fashion (i.e., D, A, B, C). No evidence exists to suggest that mRNA splice variants exist for UROD.

Mitochondrial Import of Coproporphyrinogen

In mammals, coproporphyrinogen III produced by UROD is next transported into the mitochondrial intermembrane space. A series of studies have proposed that ABCB6, a mitochondrial outer membrane ATP-dependent transporter, is responsible for coproporphyrinogen transport (Krishnamurthy et al. 2006; Krishnamurthy and Schuetz 2011). However, this claim must be viewed with skepticism because only the transport of the fully conjugated, planar macrocycle coproporphyrin, not the physiologically relevant nonplanar coproporphyrinogen, has ever been experimentally shown. ABCB6 has also been shown to exist as two different-size variants that localize in not only the mitochondrial outer membrane but also the Golgi and plasma membranes (Paterson et al. 2007; Tsuchida et al. 2008). Two recent reports regarding the role of ABCB6 as a coproporphyrinogen transporter raise additional questions. One identified ABCB6 as the blood group Langereis (Helias et al. 2012), and a second reported that a defect in ABCB6 is the cause of an inherited developmental defect of the eye known as ocular coloboma (Wang et al. 2012). These defects have no known relationship to heme imbalance, and affected individuals do not have free porphyrin accumulations, as are found in porphyrias. However, given that ABCB6 has been observed to be one of a relatively few genes induced at high levels during erythropoiesis (Nilsson et al. 2009), it seems likely that it plays some role in erythropoiesis. Given the size of coproporphyrinogen, the possibility that it transits the outer membrane via a porin rather than specific transporters cannot be ruled out by current data.

Coproporphyrinogen Oxidase

The next enzyme, coproporphyrinogen oxidase (CPOX) (EC 1.3.3.3), is synthesized in the cytoplasm with an unusually long (~120-amino-acid) mitochondrial leader sequence that targets it to the mitochondrial intermembrane space (Dailey 1990; Akhtar 2003). The enzyme is weakly associated with the outside of the inner membrane, possibly in association with the next pathway enzyme. The mature protein is a homodimer without bound cofactor. The structures of both human (Lee et al. 2005) and yeast (Phillips et al. 2004) (the latter of which is a cytoplasmic protein) CPOX have been solved and were found to possess unique folds. The reaction catalyzed is an unusual oxidative decarboxylation of the A and B ring propionates to yield the vinyl groups of protoporphyrinogen IX. CPOX will use only the coproporphyrinogen III isomer and proceeds in a stepwise fashion that requires two molecules of molecular oxygen and generates two molecules of CO_2. The reaction catalyzed by CPOX has been extensively studied both experimentally and in silico (Lash 2005; Silva and Ramos 2011). Studies on the regulation of the CPOX gene indicate that erythroid and housekeeping promoter elements exist (Takahashi et al. 1998), although our knowledge of how these operate is far from complete. Studies in differentiating murine erythroleukemia cell culture have provided data that are difficult to fit within a simple model. Whereas mRNA levels increase rapidly from the onset of erythropoiesis, measured CPOX enzyme activity first decreases (as coproporphyrin concentrations in the culture medium increase) and then increases (Conder et al. 1991). Additionally, constitutive expression of an externally introduced CPOX results in a more rapid and overall greater increase in heme synthesis by cells undergoing erythropoiesis (Taketani et al. 2001). These observations suggest a potential role for CPOX in pathway regulation during erythroid differentiation.

Protoporphyrinogen Oxidase

The penultimate step in heme synthesis is the oxidation of protoporphyrinogen IX to protoporphyrin IX (Dailey 1990; Akhtar 2003). This is catalyzed by protoporphyrinogen oxidase (PPOX) (EC 1.3.3.4) and requires three molecules of molecular oxygen and generates three molecules of hydrogen peroxide. Crystal structures of the plant (Koch et al. 2004), bacterial (Corradi et al. 2006; Qin et al. 2010), and human (Qin et al. 2011) enzyme have been determined and show that the protein is a homodimer with one noncovalently bound FAD per subunit. PPOX is synthesized in the cytoplasm in its mature size and translocated to the outer surface of the inner mitochondrial membrane via a mechanism that requires both an amino-terminal and internal mitochondrial targeting sequence (von und zu Fraunberg et al. 2003; Morgan et al. 2004; Dailey et al. 2005; Davids et al. 2006). Interestingly, the active site is proposed to be situated in the middle of a tunnel that passes through the protein. The gene for PPOX, like that for CPOX, has been modestly characterized and found to possess both erythroid and housekeeping promoter elements (Taketani et al. 1995; Dailey et al. 2002). There is no evidence to suggest that mRNA splice variants exist for PPOX. Although details are lacking, available data suggest that PPOX, along with the terminal pathway enzyme, are regulated in a fashion distinct from the earlier pathway enzyme genes during erythroid differentiation (Conder et al. 1991). Full erythroid induction of these proteins appears to require the presence of elevated cellular heme (Yin and Dailey 1998).

Ferrochelatase

The terminal step of heme synthesis is the insertion of ferrous iron into the protoporphyrin IX macrocycle to produce protoheme IX (heme). This is catalyzed by the enzyme ferrochelatase (FECH) (EC 4.99.1.1) (Dailey and Dailey 2003). This enzyme is synthesized in the cytoplasm as a preprotein and is translocated to the mitochondrial matrix, where it is associated with the inner mitochondrial membrane. The mature, processed protein is a homodimer with each subunit possessing a [2Fe–2S] cluster (Wu et al. 2001). The presence of the [2Fe–2S] cluster is necessary for enzyme activity, but there is no evidence to suggest that it participates directly in catalysis. Nevertheless, the fact that the cluster must be present for activity makes the cluster a de facto iron regulatory feature for the enzyme (Crooks et al. 2010). Interestingly, recent studies suggest that the cluster may serve a regulatory or sensory role (DI Shah, N Takahashi-Makise, JD Cooney, et al., unpubl.). The crystal structures of human ferrochelatase with and without substrate and product have been determined, and it has been shown that the molecule undergoes considerable active-site remodeling during catalysis (Medlock et al. 2007a,b, 2009). Next to the ALAS and HMBS genes, regulatory features of the FECH gene are probably the best characterized in the pathway. Distinct erythroid and housekeeping elements exist in the promoter region (Tugores et al. 1994; Magness et al. 1998, 2000), and there is evidence for alternate splicing in the noncoding 3′ region of the mouse mRNA (Chan et al. 1993).

HEME BIOSYNTHESIS PATHWAY REGULATION

As noted above, pathway regulation is generally ascribed to modulation of ALAS activity. The ALAS-1 gene is subject to tissue-specific regulation by diverse factors frequently associated with xenobiotic, hormone, and drug metabolism (May et al. 1995; Medlock and Dailey 2009). It has an estimated half-life of only a few hours and may be induced as much as 100-fold in some cells, thereby allowing for rapid and vigorous response to cellular heme needs. ALAS-1 is present in erythroid precursor cells before they begin erythroid differentiation, but the gene is turned off as ALAS-2 is turned on when these cells begin to synthesize heme for hemoglobin (Conder et al. 1991; Yin and Dailey 1998). Thus, ALAS-1 cannot substitute for a deficiency of ALAS-2 and vice versa.

ALAS-2 is transcriptionally regulated by erythroid-specific factors such as GATA-1 and possesses an iron regulatory element (IRE) located in the mRNA 5′ untranslated region that allows for translational regulation by cellular iron concentration (Dierks 1990; Bhasker et al. 1993; Melefors et al. 1993). As with similarly regulated systems, such as for ferritin synthesis, the IRE binds the iron-free form of the iron regulatory protein (IRP), thereby preventing translation of the ALAS-2 message when cellular iron concentrations are insufficient for optimal heme synthesis. Thus, maximal ALAS-2 activity requires gene induction by erythroid-specific transcription factors as well as sufficient cellular iron levels to support heme synthesis. Of note are studies (Schranzhofer et al. 2006) that suggest the IRE–IRP system may not be entirely applicable during the later, accelerated hemoglobinization phase of erythropoiesis because elevated iron would be expected to diminish transferrin receptor expression and increase ferritin synthesis, both of which would diminish iron availability for heme synthesis. The data gathered with differentiating erythroblasts showed that ALAS-2 translational regulation is "uncoupled" from that of transferrin receptors (which are increased, not decreased) and ferritin (whose protein levels do not increase). Additionally, it was postulated that during maximal hemoglobinization, ALAS2 mRNA increases disproportionately to IRP synthesis, thus circumventing the IRE–IRP regulatory mechanism at this time. It is suggested that the "kiss and run" model (Zhang and Guarente 1995) may offer a possible explanation. In this hypothesis, during terminal erythropoiesis, vesicular iron acquired by receptor-mediated endocytosis is targeted directly to the mitochondrion, thereby bypassing the cytosolic iron pool and the IRE–IRP system. Unlike ALAS-2, ALAS-1 is not regulated via the IRE–IRP system.

Some reviews have presented tables and figures based on compilations of published data suggesting that the activity of pathway enzymes from PBGS to FECH are in excess over ALAS activity (see Ponka 1999; Puy et al. 2010). However, these calculations are based on in vitro assays of individual enzymes under what are considered optimal conditions and then extrapolated back to in vivo cell conditions. Interestingly, if these calculations are to be believed, the relative enzyme activities for pathway enzymes appear to vary considerably. Whereas there is no clear evidence for the induction of biosynthetic pathway enzymes other than ALAS-1 in nonerythroid cells, there is up-regulation of all pathway enzymes during erythroid differentiation. Even so, current dogma relegates pathway regulation in all cells to ALAS-1 or -2 with the proposition that all later pathway enzymes are present in excess. This model conveniently ignores the observation that coproporphyrin is routinely excreted by mammals (Kappas et al. 1995), and that in cancer cells one finds an elevation of CPOX and diminishment of FECH that results in the accumulation of protoporphyrin during ALA-based phototherapy (Sopena et al. 2008; Takahashi et al. 2011). Additional recent data may cause a reevaluation of this model for differentiating erythroid cells since it has been discovered that a mutation in the carboxyl terminus of ALAS-2, which results in a hyperactive enzyme, results in vivo in the accumulation of free protoporphyrin and zinc-bound protoporphyrin (Whatley et al. 2008). This is thus similar to what is found in the disease erythropoietic protoporphyria (EPP), although generally protoporphyrin is in the free form. The name currently given this disorder is X-linked protoporphyria (XLPP) to reflect the X chromosome location of ALAS-2. The observation of protoporphyrin accumulation in XLPP is particularly interesting because levels of ferrochelatase have been generally assumed to be present in considerable excess over those of ALAS in the cell. Indeed, EPP occurs only when ferrochelatase levels decrease to approximately one-fourth of normal (Elder et al. 2009). This suggests that either the erythroid pathway is designed so that the amount of ferrochelatase available (or iron transport mechanisms for heme synthesis) is normally just slightly above the maximum possible levels of ALAS-2, or that additional regulatory steps or limiting mechanisms exist at the end of the pathway that we currently do not recognize or understand.

Interestingly, it has been proposed that during the course of normal erythropoiesis, heme synthesis in developing erythroid cells overproduces heme to an extent that it is toxic to the cell unless it is exported by the plasma membrane heme transporter feline leukemia virus subtype C receptor (FLVCR) (Khan and Quigley 2011). Although it is clear that disabling mutations in FLVCR result in cell death, presumably as a result of the toxic effects of excessive heme, it seems counterintuitive that nature would have evolved such intricate regulatory mechanisms for erythroid heme synthesis (i.e., IRE–IRP, erythroid-specific transcription factors, and complex iron supply regulatory schemes) that all allow for excessive synthesis of heme so that it necessitates export and degradation by macrophages and liver. Indeed, given the "cost" of heme synthesis, it seems more likely that the "excessive" heme produced is a planned synthesis of heme by erythroid cells (which must be considered the ultimate heme-synthesizing factories in the body) for orderly export and transit to other organs and cells whose heme-synthesizing capabilities may be physiologically limited. The presence of specific heme transporters, such as hemopexin (Shipulina et al. 2000), and the observation that exogenous administration of heme to acute porphyric patients both down-regulates heme synthesis and is used for hemoproteins by the patient receiving the infusion (Anderson et al. 2005) argue for this hypothesis.

MULTIENZYME COMPLEXES IN HEME BIOSYNTHESIS

Given the reactivity of the pathway intermediates, it is highly unlikely that the intracellular concentrations of any of them attain the micromolar concentrations of the measured enzyme K_ms. Thus it is reasonable to assume that "free intermediates" do not exist in the cell. This topic was first approached experimentally in the 1980s when data were presented in support of the hypothesis that the terminal three pathway enzymes, which are all mitochondrial membrane associated, form a transient complex to facilitate the transfer of intermediates (Ferreira et al. 1988; Dailey 1990; Proulx et al. 1993). Biochemical data clearly showed that while obligate, tight substrate channeling, such as occurs in tryptophan synthesis, does not exist, under normal circumstances equilibration of products/substrates with the bulk medium does not occur. For the potential interaction between the terminal two pathway enzymes, data gleaned from PPOX and FECH crystal structures provide good support for the possible interaction (Koch et al. 2004; Medlock et al. 2007b). The presence and nature of multienzyme complexes involving the terminal three membrane-associated enzymes is considerably advanced over what is known, or not known, about the earlier pathway enzymes, but is still extremely rudimentary. Radiolabeling experiments using isolated mitochondrial fragments clearly support transient interactions in situ, although reconstitution of this process with purified components was not accomplished (Proulx et al. 1993). In silico structural studies show that an interaction across the inner mitochondrial membrane to transport protoporphyrin from PPOX to FECH is feasible and highly likely (Koch et al. 2004). Interestingly, in the docked PPOX–FECH complex, the openings of the active-site tunnels of the PPOX dimer not only coincide with the position of the openings to the active-site pockets of the dimeric FECH, but are spatially juxtaposed to surface-bound porphyrin molecules observed in some FECH crystal structures. The observation that binding of substrate and product induce changes in the surface contour and charge distribution around the active-site pocket opening of FECH provide an explanation for how PPOX and postcatalytic heme-accepting proteins "recognize" the appropriate form of FECH with which to interact (Medlock et al. 2007b). Given the need to acquire coproporphyrinogen from the cytosol and movement of at least some heme out of the mitochondrion, the existence of a complex of an outer membrane coproporphyringen transporter, CPOX, PPOX, the inner membrane iron transporter mitoferrin (Shaw et al. 2006), FECH, and a heme chaperone/outer membrane heme transporter at a mitofilin-mediated junction between outer and inner mitochondrial membranes is an intriguing possibility (Fig. 2).

Other than data for the terminal, membrane-associated enzymes, there exists little experimental evidence to support multienzyme complexes of earlier pathway enzymes. For the first step, an interaction of ALAS-2 with succinyl CoA synthetase on the inner mitochondrial membrane has been demonstrated by two groups (Furuyama and Sassa 2000; Cox et al. 2004). With the recent identification of SLC25A38 as the putative glycine/ALA transporter (Guernsey et al. 2009), it is reasonable to anticipate the existence of transient complexes between ALAS, succinyl CoA synthetase, and SLC25A38 on the mitochondrial inner membrane. The possibility for a stable, rigid complex for these and most other heme synthesis enzymes seems unlikely given that they possess active-site pockets with a single entrance/exit. The necessity for substrate(s) and product(s) to enter and exit via a single route would require movement of one enzyme between donor and acceptor molecules in the complex, much as cytochrome *c* physically cycles between electron donors and acceptors. The only enzyme for which this may not be the case is PPOX, which appears to have a channel through the protein, with the active site being located within this feature.

For the synthesis of uroporphyrinogen III from PBG, individuals have long believed that close proximity of HMBS and UROS would be likely given the chemical reactivity of the linear tetrapyrrolic intermediate, HMB. Unfortunately, at present, no data exist to support the presence of a multienzyme complex involving PBGS and HMBS. Given that ALA must be exported out and coproporphyrinogen imported into the mitochondria, logic would suggest that not only PBGS and HMBS but also UROS and UROD exist in a supramolecular complex that is spatially close to the mitochondrion. To date, high-resolution microscopic or biochemical/biophysical studies that may identify the intracellular distribution of these enzymes are lacking. Approaches that rely on cell disruption and physical isolation of complexes or in vitro reconstitution of complexes from isolated components assume a strength of protein–protein interaction in solution that may not exist and may not be necessary in the highly concentrated cytosolic milieu. One intriguing possibility is that PBGS, which forms a large complex best categorized as a tetramer of homodimers, could, because of its size, serve as a scaffold for a multienzyme complex. The observation that this octomer can also assume a hexameric shape may hint at its plastic role as a supramolecular assembly nucleation site. Resolution of these issues will require masterful research, but the results should be interesting.

Although tetrapyrrole synthesis and iron metabolism must be intimately linked to prevent the inappropriate accumulation of either molecule, there have been limited studies that address this issue at either the genetic or molecular level. The role of the iron-responsive element has been examined by a number of groups (Dierks 1990; Bhasker et al. 1993; Melefors et al. 1993; Wang and Pantopoulos 2011), and the cellular machinery for iron sulfur cluster assembly (Lill and Kispal 2000) has likewise garnered significant interest. Whole-body and intercellular iron trafficking is fairly well understood, but only recently has there been significant advancement with the identification of mitoferrin as the mitochondrial inner membrane iron transporter that supplies iron to ferrochelatase for heme synthesis (Shaw et al. 2006; Troadec et al. 2011). Specifically, mitoferrin 1 is responsible for iron transport for heme synthesis during erythropoiesis, but the presence of another inner membrane transport protein, ABCB10, is required to stabilize mitoferrin 1 (Chen et al. 2009). The exact role that ABCB10 plays has yet to be elucidated. Evidence has also been presented showing that mitoferrin forms a complex with ferrochelatase that might make possible a direct transfer of transported ferrous iron for heme synthesis to ferrochelatase (Chen et al. 2010). If this does occur, structural data from human ferrochelatase and its orientation relative to the inner membrane would indicate that iron would enter the active site of ferrochelatase via a solvent-filled channel whose outside entrance is on the back side of the enzyme (Sellers et al. 2001; Medlock et al. 2007b). It will be of interest to learn the molecular details of this process and how it is regulated.

Transit of heme away from its site of synthesis is unexplored territory. It has been shown that in in vitro assays the release of heme from the enzyme postmetalation is the rate-limiting step (Hoggins et al. 2007). This probably reflects the absence of the native heme acceptor in the in vitro assay. Given that heme is used throughout the cell in a variety of compartments, it is clear that a variety of heme chaperones may exist. For transit from the mitochondrion to other membranous compartments, it appears likely that transfer may occur at direct contact points rather than by diffusion through the cytoplasm (Schultz et al. 2010), but even that will require moving heme from FECH to the site of transfer between membranes. As yet unknown is what molecule first accepts heme from ferrochelatase. Given its location on the inner membrane, it would seem that for respiratory cytochrome maturation direct transfer from FECH to cytochrome assembly machinery could occur. Some experimental systems hint strongly at this possibility (Richard-Fogal and Kranz 2010), but definitive data are still lacking. For cytoplasmic hemoproteins, including hemoglobin, there are no identified chaperones. Putative heme/porphyrin binding/carrier proteins have been suggested and even biochemically characterized (Blackmon et al. 2002; Dias et al. 2006), but none of these is supported by data as in vivo players in heme trafficking.

DISORDERS OF ERYTHROID HEME BIOSYNTHESIS

In discussing disorders of heme metabolism, one may consider both heme synthesis and degradation. However, with particular regard to red cell metabolism, the only concern is heme synthesis because heme degradation is not a significant property of red cells. Thus, below we briefly consider only disorders of synthesis and focus only on those disorders that have significant red cell involvement. The chemical intermediates in tetrapyrrole synthesis (typically porphyrinogens in their oxidized porphyrin state) are relatively reactive and can be cytotoxic, through their inherent photoreactive and chemical properties. The physiological proof of this statement is amply shown by the fact that a deficiency in any one of the pathway enzymes results in a clinically distinct disease in animals (Elder et al. 2009) and inhibition of some later steps can result in photo-induced death of plants or microorganisms. Other than a defect in ALAS-2, which results in X-linked sideroblastic anemia, the problem in these disorders, named the porphyrias, is generally not anemia, but is attributable to the accumulation of heme pathway intermediates proximal to the deficiency.

 Cite this article as *Cold Spring Harb Perspect Med* doi: 10.1101/cshperspect.a011676

Inheritance of these diseases may be recessive or dominant depending on the natural level of enzyme activity. However, most dominantly inherited porphyrias have variable penetrence so that only about 30%–40% of individuals with a gene defect have symptoms. The two X-linked diseases (see below) are dominant in males, but variable in females owing to lyonization of the X chromosome. In the case of early pathway enzyme deficiencies, the porphyrias are neurological in nature and are believed to result from the accumulation of ALA and/or PBG. For the enzymatic steps after the synthesis of the first tetrapyrrole, the most profound symptoms are cutaneous in nature because of the chemical reactivity and photoreactivity of the tetrapyrrole intermediates (Table 1). These diseases and their molecular basis have been well studied and frequently reviewed by others. Interestingly, most heme biosynthetic disorders can be considered nonerythroid, and indeed, early publications categorized the porphyrias as hepatic versus erythroid. Below we have focused only on those diseases associated with erythroid cells.

Erythropoietic Protoporphyria (EPP)

EPP results from a partial deficiency of FECH (Elder et al. 2009). This manifests biochemically as an increased concentration of the heme precursor protoporphyrin (in its free, unbound-to-iron form) in erythrocytes and a secondary protoporphyrin accumulation in the skin and liver. This buildup of protoporphyrin may have clinical sequelae in the form of burning, stinging photocutaneous sensitivity, most often from infancy, and in approximately 3% of patients, liver disease.

Inheritance of EPP is a bit unusual in that about 94% of cases have a primary FECH mutation in *trans* to a hypomorphic low-expression FECH polymorphism (termed IVS3-48C) (Gouya et al. 2002). The IVS3-48C polymorphism increases the use of an aberrant splice site and is present in normal populations with a frequency varying between 1% and 45% (Gouya et al. 2006). This form of inheritance is most appropriately termed recessive because there is a required coinheritance of two molecular defects that result in a decrease in FECH activity to 25%–35% of normal activity and in overt clinical disease. Patients who have inherited only one such FECH defect generally remain asymptomatic throughout their lives. Approximately 4% of EPP cases are due to true homozygous or compound heterozygous FECH mutations (Gouya et al. 2006). Acquired EPP has been reported in the setting of hematological malignancy and cytogenetic FECH deletions associated with clonal expansion of FECH-deficient bone marrow cells (Aplin et al. 2001; Bharati et al. 2006; Goodwin et al. 2006).

Typically, porphyrin-induced phototoxicity is reported seasonally and acutely, that is, within minutes of sun exposure. Mild edema and erythema may arise immediately after sun exposure, and chronic lesions such as thickening of the skin on the hands and waxlike scarring on the face may occur. Manifest liver disease is a rare complication, occurring in approximately 3% of EPP patients (Anstey and Hift 2007). Free protoporphyrin is capable of diffusing out of red cells, with reticulocytes containing proportionally more protoporphyrin than mature red cells (Piomelli et al. 1975). Protoporphyrin is excreted into bile by the liver and undergoes extensive enterohepatic circulation because of its lipophilic nature. Increasing concentrations may lead to protoporphyrin insolubility, cholestasis, and cholelithiasis. Hepatocyte injury is manifest with inflammation, which may eventually lead to cirrhosis (Poh-Fitzpatrick 1986). In the terminal phase of EPP-related liver disease, increasing cholestasis impairs hepatic protoporphyrin excretion. Oxidative stress-induced hemolysis occurs with increasing protoporphyrin concentrations, driving erythropoiesis and increasing protoporphyrin load presented to the liver—propagating a cycle of clinical deterioration (Key et al. 1992; Anstey and Hift 2007). The precise pathogenesis of hepatopathy in EPP, however, remains to be fully elucidated. Although photosensitivity may be painful, distressing, or disruptive to an affected individual's lifestyle (Todd 1994), liver disease may be fatal and only liver transplant may effect control of the disease progression. Avoidance of sunlight is the backbone of the management of skin

Table 1. Summary of the heme biosynthetic enzymes and associated porphyrias, noting their predominant mode of inheritance and clinical features

Enzyme	Disorder	Inheritance	Clinical features
Aminolevulinate synthase	X-linked protoporphyria	X-linked	Immediate photosensitivity, liver disease secondary to protoporphyrin accumulation
	X-linked sideroblastic anemia	X-linked	Anemia with ringed sideroblasts, some pyridoxine responsive
Porphobilinogen synthase	Aminolevulinate dehydratase porphyria	Recessive	Acute neurological attacks
Hydroxymethylbilane synthase	Acute intermittent porphyria	Dominant	Acute neurological attacks
Uroporphyrinogen-III synthase	Congenital erythropoietic porphyria	Recessive	Severe vesiculoerosive skin disease
Uroporphyrinogen decarboxylase	Porphyria cutanea tarda	Acquired, but may be associated with a dominantly inherited mutation in up to approximately 30%	Vesiculoerosive skin disease; usually coexists with iron overload and a number of precipitating factors, particularly alcohol exposure, liver disease, and renal failure
Coproporphyrinogen oxidase	Hereditary coproporphyria	Dominant	Acute neurological attacks and vesiculoerosive skin disease
Protoporphyrinogen oxidase	Variegate porphyria	Dominant	Acute neurological attacks and vesiculoerosive skin disease
Ferrochelatase	Erythropoietic protoporphyria	Recessive, but more common than most recessive disorders because the disease-associated mutation is frequently inherited with a low-expression polymorphism that is common in populations of European extraction	Immediate photosensitivity

symptoms, with some forms of medical treatment, such as oral β-carotene, reportedly effective in about a third of patients.

Apart from the clinical characteristics, which are diagnostic of EPP, the hallmark of the biochemical diagnosis is elevated free protoporphyrin in the red cells and stool (and/or bile). Further, many patients may have a slight microcytic, hypochromic anemia, although this is not invariant. Interestingly, trace amounts of zinc-chelated protoporphyrin are normally formed during heme biosynthesis, but during systemic iron deficiency, zinc protoporphyrin accumulates in erythrocytes (Langer et al. 1972). Similarly, during diverse states of de facto intramitochondrial iron deficiency, as occurs in lead poisoning (Lamola and Yamane 1974), in the anemia of chronic disorders (Hastka et al.

1993), and during accelerated (especially in hemolytic anemia) or ineffective erythropoiesis (Finch 1994), zinc protoporphyrin may increase.

X-Linked Protoporphyria (XLPP)

Recently, a small number of families in Europe and South Africa with an atypical protoporphyria have been investigated. Inexplicably these families presented with a highly penetrant and severe form of protoporphyria, with a relatively high incidence of protoporphyria-associated liver disease and extremely elevated protoporphyrin, with 30%–40% zinc protoporphyrin. Yet no abnormalities in FECH, its gene, or its regulatory components were apparent. There was no anemia or evidence of iron overload. The inheritance appeared X-linked. A collaborative study showed that these families are suffering from a novel form of porphyria, now termed X-linked protoporphyria (XLPP; also known as X-linked dominant protoporphyria, or XLDPP) (Whatley et al. 2008). This arises pathogenetically from a gain-of-function mutation in the erythroid form of ALAS-2. This disorder is, thus, similar, but not identical, to that observed when ALAS-2 is overproduced in *Ireb2*$^{-/-}$ mice (Cooperman et al. 2005). All of the eight families studied displayed one of two *ALAS2* deletions, either c.1706–1709 delAGTG or c.1699–1700 delAT, resulting in frameshifts that lead to replacement or deletion of the 19–20 carboxy-terminal residues of the enzyme. The c.1706–1709 delAGTG mutation is predicted to cause a frameshift in 24 codons distal to the 4-bp mutation (p.E569GfsX24), resulting in the extension of ALAS-2 by four amino acids. In contrast, the c.1699–1700 delAT mutation leads to a frameshift mutation in two codons (p.M567EfsX2) and a truncation of ALAS-2. It is interesting that a single-base-pair substitution in this same region (S568G) reportedly causes a loss of function in ALAS-2, manifesting as X-linked sideroblastic anemia (Harigae et al. 1999) (see below). This mutation is located between two XLPP-causing mutations. It is surprising that mutations in such close proximity can have diametrically opposite effects.

The elevated ALAS-2 activity in XLPP results in a greatly increased flux of porphyrin(ogen)s through the heme pathway, FECH becomes rate-limiting (probably at the level of iron availability), and free and zinc protoporphyrin accumulate. This accounts for the resemblance to classic EPP and presence of photosensitive skin disease and potentially severe liver disease. Management is similar to that of EPP, and investigations are under way to establish a potential role for iron therapy in these patients.

X-Linked Sideroblastic Anemia (XLSA)

Sideroblastic anemia (SA) is characterized by anemia with the presence of "ringed" sideroblasts in the bone marrow. This morphology results from iron-loaded mitochondria in the erythroblasts (Fleming 2002). The resulting mitochondrial iron toxicity results in ineffective erythropoiesis and is a feature of severe forms of SA. In SA, intestinal iron absorption is increased and the affected patients may develop liver and systemic iron overload (Bottomley 2006). For this reason SA is considered an iron-loading anemia.

Inherited SA is a rare and heterogeneous disease caused by mutations of genes involved in heme biosynthesis (*ALAS2*) (Cotter et al. 1992; Bottomley 2006), Fe–S cluster biogenesis (*ABCB7*) (Pondarre et al. 2006), Fe–S cluster transport (Wingert et al. 2005), or mitochondrial transporters (*SLC25A38*) (Kannengiesser et al. 2011). The most common form of SA, X-linked SA (XLSA), results from one of a number of mutations in the *ALAS2* gene (Cotter et al. 1992; Bottomley 2006). A much rarer form, XLSA with ataxia, has also been described in which the X-linked ATP-binding cassette transporter B7 gene, *ABCB7* (Allikmets et al. 1999), is defective. ABCB7 is involved in Fe–S cluster transport, and its functionality is essential for erythroid cells and the central nervous system—hence the presence of anemia and spinocerebellar ataxia and cerebellar hypoplasia in these patients (Pondarre et al. 2007). Interestingly a form of non-X-linked SA resulting from mutations of the *GLRX5* gene, which encodes a mitochondrial protein involved in iron–sulfur cluster biogenesis, has relatively recently been described (Rouault and Tong 2008).

Most XLSA-associated mutations in *ALAS2* are missense substitutions resulting in loss of ALAS-2 functionality, but there are rarer *ALAS2* promoter mutations (Bergmann et al. 2010; Ducamp et al. 2011). XLSA prevalently affects males, but anemia may also be present in females because of either X chromosome inactivation (Cazzola et al. 2000) or excessive skewing that occurs in hematopoietic tissue with aging (Aivado et al. 2006). Thus, XLSA may present early in infancy or not until later in adult life (Camaschella 2009). A number of *ALAS2* mutations decrease the binding of the pyridoxal phosphate cofactor, which neatly accounts for the pyridoxine responsiveness in patients carrying those mutations (Cox et al. 1994; May and Bishop 1998). Patients unresponsive to pyridoxine possess a mutation in an area not connected with pyridoxal phosphate cofactor binding (Furuyama et al. 1997) or impaired interaction with succinyl CoA synthetase (Furuyama and Sassa 2000). Asymptomatic or mildly anemic patients seldom require intervention, although oral pyridoxine supplementation should be considered because in some cases partial to full restoration of hemoglobin levels may be achieved. Severely pyridoxine-unresponsive individuals may require transfusions of red blood cell concentrates in conjunction with iron chelation. Iron status should be monitored even in nontransfused patients and appropriate treatment should be started if iron overload is observed. Iron overload can be treated by phlebotomy (Bottomley 2006; Camaschella 2009). Interestingly, the study of these rare diseases has enlightened our understanding of the relationship between the two main pathways of mitochondrial iron metabolism—those of heme and iron–sulfur biosynthesis (Camaschella 2009).

ACKNOWLEDGMENTS

H.A.D. is supported by grant DK32303 from the National Institutes of Health. P.N.M. is supported by funding from the SA Medical Research Council, the National Research Foundation, the Harry Crossley Foundation, and the University of Cape Town Research Committee. The authors acknowledge T.A. Dailey (University of Georgia) and C.E. Haumann (University of Cape Town) for reviewing and editing the manuscript.

REFERENCES

Aivado M, Gattermann N, Rong A, Giagounidis AA, Prall WC, Czibere A, Hildebrandt B, Haas R, Bottomley SS. 2006. X-linked sideroblastic anemia associated with a novel *ALAS2* mutation and unfortunate skewed X-chromosome inactivation patterns. *Blood Cells Mol Dis* **37:** 40–45.

Aizencang G, Solis C, Bishop DF, Warner C, Desnick RJ. 2000. Human uroporphyrinogen-III synthase: Genomic organization, alternative promoters, and erythroid-specific expression. *Genomics* **70:** 223–231.

Ajioka RS, Phillips JD, Kushner JP. 2006. Biosynthesis of heme in mammals. *Biochim Biophys Acta* **1763:** 723–736.

Akhtar M. 2003. Coproporphyrinogen III and protoporphyrinogen IX oxidase. In *The porphyrin handbook* (ed. Kadish KM, Smith KM, Guilard R), pp. 75–92. Academic, San Diego.

Allikmets R, Raskind WH, Hutchinson A, Schueck ND, Dean M, Koeller DM. 1999. Mutation of a putative mitochondrial iron transporter gene (*ABC7*) in X-linked sideroblastic anemia and ataxia (XLSA/A). *Hum Mol Genet* **8:** 743–749.

Anderson KE, Bloomer JR, Bonkovsky HL, Kushner JP, Pierach CA, Pimstone NR, Desnick RJ. 2005. Recommendations for the diagnosis and treatment of the acute porphyrias. *Ann Intern Med* **142:** 439–450.

Anstey AV, Hift RJ. 2007. Liver disease in erythropoietic protoporphyria: Insights and implications for management. *Gut* **56:** 1009–1018.

Aplin C, Whatley SD, Thompson P, Hoy T, Fisher P, Singer C, Lovell CR, Elder GH. 2001. Late-onset erythropoietic porphyria caused by a chromosome 18q deletion in erythroid cells. *J Invest Dermatol* **117:** 1647–1649.

Astner I, Schulze JO, van den Heuvel J, Jahn D, Schubert WD, Heinz DW. 2005. Crystal structure of 5-aminolevulinate synthase, the first enzyme of heme biosynthesis, and its link to XLSA in humans. *EMBO J* **24:** 3166–3177.

Bergmann AK, Campagna DR, McLoughlin EM, Agarwal S, Fleming MD, Bottomley SS, Neufeld EJ. 2010. Systematic molecular genetic analysis of congenital sideroblastic anemia: Evidence for genetic heterogeneity and identification of novel mutations. *Pediatr Blood Cancer* **54:** 273–278.

Bharati A, Badminton MN, Whatley SD, O'Brien DV, Bell HK. 2006. Late-onset erythropoietic protoporphyria in association with haematological malignancy. *Clin Exp Dermatol* **31:** 668–670.

Bhasker CR, Burgiel G, Neupert B, Emery-Goodman A, Kuhn LC, May BK. 1993. The putative iron-responsive element in the human erythroid 5-aminolevulinate synthase mRNA mediates translational control. *J Biol Chem* **268:** 12699–12705.

Blackmon JB, Dailey TA, Lianchun X, Dailey HA. 2002. Characterization of a human and mouse tetrapyrrole-binding protein. *Arch Biochem Biophys* **407:** 196–201.

Bottomley SS. 2006. Congenital sideroblastic anemias. *Curr Hematol Rep* **5:** 41–49.

Camaschella C. 2009. Hereditary sideroblastic anemias: Pathophysiology, diagnosis, and treatment. *Semin Hematol* **46:** 371–377.

Cazzola M, May A, Bergamaschi G, Cerani P, Rosti V, Bishop DF. 2000. Familial-skewed X-chromosome inactivation as a predisposing factor for late-onset X-linked sideroblastic anemia in carrier females. *Blood* **96:** 4363–4365.

Chan RY, Schulman HM, Ponka P. 1993. Expression of ferrochelatase mRNA in erythroid and non-erythroid cells. *Biochem J* **292:** 343–349.

Chen W, Paradkar PN, Li L, Pierce EL, Langer NB, Takahashi-Makise N, Hyde BB, Shirihai OS, Ward DM, Kaplan J, et al. 2009. Abcb10 physically interacts with mitoferrin-1 (Slc25a37) to enhance its stability and function in the erythroid mitochondria. *Proc Natl Acad Sci* **106:** 16263–16268.

Chen W, Dailey HA, Paw BH. 2010. Ferrochelatase forms an oligomeric complex with mitoferrin-1 and Abcb10 for erythroid heme biosynthesis. *Blood* **116:** 628–630.

Conder LH, Woodard SI, Dailey HA. 1991. Multiple mechanisms for the regulation of haem synthesis during erythroid cell differentiation. Possible role for coproporphyrinogen oxidase. *Biochem J* **275:** 321–326.

Cooperman SS, Meyron-Holtz EG, Olivierre-Wilson H, Ghosh MC, McConnell JP, Rouault TA. 2005. Microcytic anemia, erythropoietic protoporphyria, and neurodegeneration in mice with targeted deletion of iron-regulatory protein 2. *Blood* **106:** 1084–1091.

Corradi HR, Corrigall AV, Boix E, Mohan CG, Sturrock ED, Meissner PN, Acharya KR. 2006. Crystal structure of protoporphyrinogen oxidase from *Myxococcus xanthus* and its complex with the inhibitor acifluorfen. *J Biol Chem* **281:** 38625–38633.

Cotter PD, Baumann M, Bishop DF. 1992. Enzymatic defect in "X-linked" sideroblastic anemia: Molecular evidence for erythroid δ-aminolevulinate synthase deficiency. *Proc Natl Acad Sci* **89:** 4028–4032.

Cox TC, Bottomley SS, Wiley JS, Bawden MJ, Matthews CS, May BK. 1994. X-linked pyridoxine-responsive sideroblastic anemia due to a Thr388-to-Ser substitution in erythroid 5-aminolevulinate synthase. *N Engl J Med* **330:** 675–679.

Cox TC, Sadlon TJ, Schwarz QP, Matthews CS, Wise PD, Cox LL, Bottomley SS, May BK. 2004. The major splice variant of human 5-aminolevulinate synthase-2 contributes significantly to erythroid heme biosynthesis. *Int J Biochem Cell Biol* **36:** 281–295.

Crooks DR, Ghosh MC, Haller RG, Tong WH, Rouault TA. 2010. Posttranslational stability of the heme biosynthetic enzyme ferrochelatase is dependent on iron availability and intact iron-sulfur cluster assembly machinery. *Blood* **115:** 860–869.

Dailey HA. 1990. Conversion of coproporphyrinogen to protoheme in higher eukaryotes and bacteria: Terminal three enzymes. In *Biosynthesis of heme and chlorophylls* (ed. Dailey HA), pp. 123–161. McGraw-Hill, New York.

Dailey HA, Dailey TA. 2003. Ferrochelatase. In *The porphyrin handbook* (ed. Kadish KM, Smith KM, Guilard R), pp. 93–122. Academic, San Diego.

Dailey TA, McManus JF, Dailey HA. 2002. Characterization of the mouse protoporphyrinogen oxidase gene. *Cell Mol Biol (Noisy-le-grand)* **48:** 61–69.

Dailey TA, Woodruff JH, Dailey HA. 2005. Examination of mitochondrial protein targeting of haem synthetic enzymes: In vivo identification of three functional haem-responsive motifs in 5-aminolaevulinate synthase. *Biochem J* **386:** 381–386.

Davids LM, Corrigall AV, Meissner PN. 2006. Mitochondrial targeting of human protoporphyrinogen oxidase. *Cell Biol Int* **30:** 416–426.

Dias JS, Macedo AL, Ferreira GC, Peterson FC, Volkman BF, Goodfellow BJ. 2006. The first structure from the SOUL/HBP family of heme-binding proteins, murine P22HBP. *J Biol Chem* **281:** 31553–31561.

Dierks P. 1990. Molecular biology of eukaryotic 5-aminolevulinate synthase. In *Biosynthesis of heme and chlorophylls* (ed. Dailey HA), pp. 201–233. McGraw-Hill, New York.

Ducamp S, Kannengiesser C, Touati M, Garcon L, Guerci-Bresler A, Guichard JF, Vermylen C, Dochir J, Poirel HA, Fouyssac F, et al. 2011. Sideroblastic anemia: Molecular analysis of the *ALAS2* gene in a series of 29 probands and functional studies of 10 missense mutations. *Hum Mutat* **32:** 590–597.

Elder GH, Gouya L, Whatley SD, Puy H, Badminton MN, Deybach JC. 2009. The molecular genetics of erythropoietic protoporphyria. *Cell Mol Biol (Noisy-le-grand)* **55:** 118–126.

Erskine PT, Senior N, Awan S, Lambert R, Lewis G, Tickle IJ, Sarwar M, Spencer P, Thomas P, Warren MJ, et al. 1997. X-ray structure of 5-aminolaevulinate dehydratase, a hybrid aldolase. *Nat Struct Biol* **4:** 1025–1031.

Ferreira GC, Andrew TL, Karr SW, Dailey HA. 1988. Organization of the terminal two enzymes of the heme biosynthetic pathway. Orientation of protoporphyrinogen oxidase and evidence for a membrane complex. *J Biol Chem* **263:** 3835–3839.

Finch C. 1994. Regulators of iron balance in humans. *Blood* **84:** 1697–1702.

Fleming MD. 2002. The genetics of inherited sideroblastic anemias. *Semin Hematol* **39:** 270–281.

Furuyama K, Sassa S. 2000. Interaction between succinyl CoA synthetase and the heme-biosynthetic enzyme ALAS-E is disrupted in sideroblastic anemia. *J Clin Invest* **105:** 757–764.

Furuyama K, Fujita H, Nagai T, Yomogida K, Munakata H, Kondo M, Kimura A, Kuramoto A, Hayashi N, Yamamoto M. 1997. Pyridoxine refractory X-linked sideroblastic anemia caused by a point mutation in the erythroid 5-aminolevulinate synthase gene. *Blood* **90:** 822–830.

Gill R, Kolstoe SE, Mohammed F, Al d-Bass A, Mosely JE, Sarwar M, Cooper JB, Wood SP, Shoolingin-Jordan PM. 2009. Structure of human porphobilinogen deaminase at 2.8 Å: The molecular basis of acute intermittent porphyria. *Biochem J* **420:** 17–25.

Goodwin RG, Kell WJ, Laidler P, Long CC, Whatley SD, McKinley M, Badminton MN, Burnett AK, Williams GT, Elder GH. 2006. Photosensitivity and acute liver injury in myeloproliferative disorder secondary to late-onset protoporphyria caused by deletion of a ferrochelatase gene in hematopoietic cells. *Blood* **107:** 60–62.

Gouya L, Puy H, Robreau AM, Bourgeois M, Lamoril J, Da Silva V, Grandchamp B, Deybach JC. 2002. The penetrance of dominant erythropoietic protoporphyria is modulated by expression of wildtype *FECH*. *Nat Genet* **30:** 27–28.

Gouya L, Martin-Schmitt C, Robreau AM, Austerlitz F, Da Silva V, Brun P, Simonin S, Lyoumi S, Grandchamp B, Beaumont C, et al. 2006. Contribution of a common single-nucleotide polymorphism to the genetic predisposition for erythropoietic protoporphyria. *Am J Hum Genet* **78:** 2–14.

Grandchamp B, De Verneuil H, Beaumont C, Chretien S, Walter O, Nordmann Y. 1987. Tissue-specific expression of porphobilinogen deaminase. Two isoenzymes from a single gene. *Eur J Biochem* **162:** 105–110.

Guernsey DL, Jiang H, Campagna DR, Evans SC, Ferguson M, Kellogg MD, Lachance M, Matsuoka M, Nightingale M, Rideout A, et al. 2009. Mutations in mitochondrial carrier family gene *SLC25A38* cause nonsyndromic autosomal recessive congenital sideroblastic anemia. *Nat Genet* **41:** 651–653.

Harigae H, Furuyama K, Kimura A, Neriishi K, Tahara N, Kondo M, Hayashi N, Yamamoto M, Sassa S, Sasaki T. 1999. A novel mutation of the erythroid-specific δ-aminolaevulinate synthase gene in a patient with X-linked sideroblastic anaemia. *Br J Haematol* **106:** 175–177.

Hastka J, Lasserre JJ, Schwarzbeck A, Strauch M, Hehlmann R. 1993. Zinc protoporphyrin in anemia of chronic disorders. *Blood* **81:** 1200–1204.

Helias V, Saison C, Ballif BA, Peyrard T, Takahashi J, Takahashi H, Tanaka M, Deybach JC, Puy H, Le Gall M, et al. 2012. ABCB6 is dispensable for erythropoiesis and specifies the new blood group system Langereis. *Nat Genet* **44:** 170–173.

Hoggins M, Dailey HA, Hunter CN, Reid JD. 2007. Direct measurement of metal ion chelation in the active site of human ferrochelatase. *Biochemistry* **46:** 8121–8127.

Hunter GA, Ferreira GC. 2011. Molecular enzymology of 5-aminolevulinate synthase, the gatekeeper of heme biosynthesis. *Biochim Biophys Acta* **1814:** 1467–1473.

Jaffe EK, Lawrence SH. 2012. Allostery and the dynamic oligomerization of porphobilinogen synthase. *Arch Biochem Biophys* **519:** 144–153.

Kannengiesser C, Sanchez M, Sweeney M, Hetet G, Kerr B, Moran E, Fuster Soler JL, Maloum K, Matthes T, Oudot C, et al. 2011. Missense *SLC25A38* variations play an important role in autosomal recessive inherited sideroblastic anemia. *Haematologica* **96:** 808–813.

Kappas A, Sassa S, Galbraith RA, Nordmann Y. 1995. The porphyrias. In *The metabolic and molecular basis of inherited disease* (ed. Scriver CR, Beaudet AL, Sly WS, Valle D), pp. 2103–2159. McGraw-Hill, New York.

Kaya AH, Plewinska M, Wong DM, Desnick RJ, Wetmur JG. 1994. Human δ-aminolevulinate dehydratase (ALAD) gene: Structure and alternative splicing of the erythroid and housekeeping mRNAs. *Genomics* **19:** 242–248.

Key NS, Rank JM, Freese D, Bloomer JR, Hammerschmidt DE. 1992. Hemolytic anemia in protoporphyria: Possible precipitating role of liver failure and photic stress. *Am J Hematol* **39:** 202–207.

Khan AA, Quigley JG. 2011. Control of intracellular heme levels: Heme transporters and heme oxygenases. *Biochim Biophys Acta* **1813:** 668–682.

Koch M, Breithaupt C, Kiefersauer R, Freigang J, Huber R, Messerschmidt A. 2004. Crystal structure of protoporphyrinogen IX oxidase: A key enzyme in haem and chlorophyll biosynthesis. *EMBO J* **23:** 1720–1728.

Krishnamurthy P, Schuetz JD. 2011. The role of ABCG2 and ABCB6 in porphyrin metabolism and cell survival. *Curr Pharm Biotechnol* **12:** 647–655.

Krishnamurthy PC, Du G, Fukuda Y, Sun D, Sampath J, Mercer KE, Wang J, Sosa-Pineda B, Murti KG, Schuetz JD. 2006. Identification of a mammalian mitochondrial porphyrin transporter. *Nature* **443:** 586–589.

Lamola AA, Yamane T. 1974. Zinc protoporphyrin in the erythrocytes of patients with lead intoxication and iron deficiency anemia. *Science* **186:** 936–938.

Langer EE, Haining RG, Labbe RF, Jacobs P, Crosby EF, Finch CA. 1972. Erythrocyte protoporphyrin. *Blood* **40:** 112–128.

Lash TD. 2005. The enigma of coproporphyrinogen oxidase: How does this unusual enzyme carry out oxidative decarboxylations to afford vinyl groups? *Bioorg Med Chem Lett* **15:** 4506–4509.

Lathrop JT, Timko MP. 1993. Regulation by heme of mitochondrial protein transport through a conserved amino acid motif. *Science* **259:** 522–525.

Layer G, Reichelt J, Jahn D, Heinz DW. 2010. Structure and function of enzymes in heme biosynthesis. *Protein Sci* **19:** 1137–1161.

Lee DS, Flachsova E, Bodnarova M, Demeler B, Martasek P, Raman CS. 2005. Structural basis of hereditary coproporphyria. *Proc Natl Acad Sci* **102:** 14232–14237.

Lewis CA Jr, Wolfenden R. 2008. Uroporphyrinogen decarboxylation as a benchmark for the catalytic proficiency of enzymes. *Proc Natl Acad Sci* **105:** 17328–17333.

Lill R, Kispal G. 2000. Maturation of cellular Fe-S proteins: An essential function of mitochondria. *Trends Biochem Sci* **25:** 352–356.

Louie GV, Brownlie PD, Lambert R, Cooper JB, Blundell TL, Wood SP, Warren MJ, Woodcock SC, Jordan PM. 1992. Structure of porphobilinogen deaminase reveals a flexible multidomain polymerase with a single catalytic site. *Nature* **359:** 33–39.

Magness ST, Tugores A, Diala ES, Brenner DA. 1998. Analysis of the human ferrochelatase promoter in transgenic mice. *Blood* **92:** 320–328.

Magness ST, Tugores A, Brenner DA. 2000. Analysis of ferrochelatase expression during hematopoietic development of embryonic stem cells. *Blood* **95:** 3568–3577.

Mathews MA, Schubert HL, Whitby FG, Alexander KJ, Schadick K, Bergonia HA, Phillips JD, Hill CP. 2001. Crystal structure of human uroporphyrinogen III synthase. *EMBO J* **20:** 5832–5839.

May A, Bishop DF. 1998. The molecular biology and pyridoxine responsiveness of X-linked sideroblastic anaemia. *Haematologica* **83:** 56–70.

May BK, Dogra SC, Sadlon TJ, Bhasker CR, Cox TC, Bottomley SS. 1995. Molecular regulation of heme biosynthesis in higher vertebrates. *Prog Nucleic Acid Res Mol Biol* **51:** 1–51.

Medlock AE, Dailey HA. 2009. Regulation of mammalian heme biosynthesis. In *Tetrapyrroles: Birth, life, and death* (ed. Warren MJ, Smith AG), pp. 116–127. Landes Bioscience and Springer Science+Business Media, New York.

Medlock A, Swartz L, Dailey TA, Dailey HA, Lanzilotta WN. 2007a. Substrate interactions with human ferrochelatase. *Proc Natl Acad Sci* **104:** 1789–1793.

Medlock AE, Dailey TA, Ross TA, Dailey HA, Lanzilotta WN. 2007b. A π-helix switch selective for porphyrin deprotonation and product release in human ferrochelatase. *J Mol Biol* **373:** 1006–1016.

Medlock AE, Carter M, Dailey TA, Dailey HA, Lanzilotta WN. 2009. Product release rather than chelation determines metal specificity for ferrochelatase. *J Mol Biol* **393:** 308–319.

Melefors O, Goossen B, Johansson HE, Stripecke R, Gray NK, Hentze MW. 1993. Translational control of 5-aminolevulinate synthase mRNA by iron-responsive elements in erythroid cells. *J Biol Chem* **268:** 5974–5978.

Morgan RR, Errington R, Elder GH. 2004. Identification of sequences required for the import of human protoporphyrinogen oxidase to mitochondria. *Biochem J* **377:** 281–287.

Nilsson R, Schultz IJ, Pierce EL, Soltis KA, Naranuntarat A, Ward DM, Baughman JM, Paradkar PN, Kingsley PD, Culotta VC, et al. 2009. Discovery of genes essential for heme biosynthesis through large-scale gene expression analysis. *Cell Metab* **10:** 119–130.

Paterson JK, Shukla S, Black CM, Tachiwada T, Garfield S, Wincovitch S, Ernst DN, Agadir A, Li X, Ambudkar SV, et al. 2007. Human ABCB6 localizes to both the outer mitochondrial membrane and the plasma membrane. *Biochemistry* **46:** 9443–9452.

Phillips JD, Whitby FG, Kushner JP, Hill CP. 2003. Structural basis for tetrapyrrole coordination by uroporphyrinogen decarboxylase. *EMBO J* **22:** 6225–6233.

Phillips JD, Whitby FG, Warby CA, Labbe P, Yang C, Pflugrath JW, Ferrara JD, Robinson H, Kushner JP, Hill CP. 2004. Crystal structure of the oxygen-dependant coproporphyrinogen oxidase (Hem13p) of *Saccharomyces cerevisiae*. *J Biol Chem* **279:** 38960–38968.

Piomelli S, Lamola AA, Poh-Fitzpatrick MF, Seaman C, Harber LC. 1975. Erythropoietic protoporphyria and lead intoxication: The molecular basis for difference in cutaneous photosensitivity. I. Different rates of disappearance of protoporphyrin from the erythrocytes, both in vivo and in vitro. *J Clin Invest* **56:** 1519–1527.

Poh-Fitzpatrick MB. 1986. The erythropoietic porphyrias. *Dermatol Clin* **4:** 291–296.

Pondarre C, Antiochos BB, Campagna DR, Clarke SL, Greer EL, Deck KM, McDonald A, Han AP, Medlock A, Kutok JL, et al. 2006. The mitochondrial ATP-binding cassette transporter Abcb7 is essential in mice and participates in cytosolic iron–sulfur cluster biogenesis. *Hum Mol Genet* **15:** 953–964.

Pondarre C, Campagna DR, Antiochos B, Sikorski L, Mulhern H, Fleming MD. 2007. *Abcb7*, the gene responsible for X-linked sideroblastic anemia with ataxia, is essential for hematopoiesis. *Blood* **109:** 3567–3569.

Ponka P. 1999. Cell biology of heme. *Am J Med Sci* **318:** 241–256.

Proulx KL, Woodard SI, Dailey HA. 1993. In situ conversion of coproporphyrinogen to heme by murine mitochondria: Terminal steps of the heme biosynthetic pathway. *Protein Sci* **2:** 1092–1098.

Puy H, Gouya L, Deybach JC. 2010. Porphyrias. *Lancet* **375:** 924–937.

Qin X, Sun L, Wen X, Yang X, Tan Y, Jin H, Cao Q, Zhou W, Xi Z, Shen Y. 2010. Structural insight into unique properties of protoporphyrinogen oxidase from *Bacillus subtilis*. *J Struct Biol* **170:** 76–82.

Qin X, Tan Y, Wang L, Wang Z, Wang B, Wen X, Yang G, Xi Z, Shen Y. 2011. Structural insight into human variegate porphyria disease. *FASEB J* **25:** 653–664.

Richard-Fogal C, Kranz RG. 2010. The CcmC:heme:CcmE complex in heme trafficking and cytochrome *c* biosynthesis. *J Mol Biol* **401:** 350–362.

Roberts AG, Elder GH. 2001. Alternative splicing and tissue-specific transcription of human and rodent ubiquitous 5-aminolevulinate synthase (*ALAS1*) genes. *Biochim Biophys Acta* **1518:** 95–105.

Rouault TA, Tong WH. 2008. Iron–sulfur cluster biogenesis and human disease. *Trends Genet* **24:** 398–407.

Schranzhofer M, Schifrer M, Cabrera JA, Kopp S, Chiba P, Beug H, Mullner EW. 2006. Remodeling the regulation of iron metabolism during erythroid differentiation to ensure efficient heme biosynthesis. *Blood* **107:** 4159–4167.

Schubert HL, Phillips JD, Heroux A, Hill CP. 2008. Structure and mechanistic implications of a uroporphyrinogen III synthase–product complex. *Biochemistry* **47:** 8648–8655.

Schubert HL, Erskine PT, Cooper JB. 2009. 5-Aminolaevulinic acid dehydratase, porphobilinogen deaminase and uroporphyrinogen III synthase. In *Tetrapyrroles: Birth, life and death* (ed. Warren MJ, Smith AG), pp. 43–73. Landes Bioscience and Springer Science+Business Media, New York.

Schultz IJ, Chen C, Paw BH, Hamza I. 2010. Iron and porphyrin trafficking in heme biogenesis. *J Biol Chem* **285:** 26753–26759.

Sellers VM, Wu CK, Dailey TA, Dailey HA. 2001. Human ferrochelatase: Characterization of substrate–iron binding and proton-abstracting residues. *Biochemistry* **40:** 9821–9827.

Shaw GC, Cope JJ, Li L, Corson K, Hersey C, Ackermann GE, Gwynn B, Lambert AJ, Wingert RA, Traver D, et al. 2006. Mitoferrin is essential for erythroid iron assimilation. *Nature* **440:** 96–100.

Shipulina N, Smith A, Morgan WT. 2000. Heme binding by hemopexin: Evidence for multiple modes of binding and functional implications. *J Protein Chem* **19:** 239–248.

Shoolingin-Jordan PM. 2003. The biosynthesis of coproporphyrinogen III. In *The porphyrin handbook* (ed. Kadish KM, Smith KM, Guilard R), pp. 33–92. Academic, San Diego.

Silva PJ, Ramos MJ. 2011. Computational characterization of the substrate-binding mode in coproporphyrinogen III oxidase. *J Phys Chem B* **115:** 1903–1910.

Sopena YE, Ferramola de Sancovich AM, Sancovich HA. 2008. Hexachlorobenzene treatment on hepatic mitochondrial function parameters and intracellular coproporphyrinogen oxidase location. *Int J Toxicol* **27:** 455–465.

Takahashi S, Taketani S, Akasaka JE, Kobayashi A, Hayashi N, Yamamoto M, Nagai T. 1998. Differential regulation of coproporphyrinogen oxidase gene between erythroid and nonerythroid cells. *Blood* **92:** 3436–3444.

Takahashi K, Ikeda N, Nonoguchi N, Kajimoto Y, Miyatake S, Hagiya Y, Ogura S, Nakagawa H, Ishikawa T, Kuroiwa T. 2011. Enhanced expression of coproporphyrinogen oxidase in malignant brain tumors: CPOX expression and 5-ALA-induced fluorescence. *Neuro Oncol* **13:** 1234–1243.

Taketani S, Inazawa J, Abe T, Furukawa T, Kohno H, Tokunaga R, Nishimura K, Inokuchi H. 1995. The human protoporphyrinogen oxidase gene (PPOX): Organization and location to chromosome 1. *Genomics* **29:** 698–703.

Taketani S, Furukawa T, Furuyama K. 2001. Expression of coproporphyrinogen oxidase and synthesis of hemoglobin in human erythroleukemia K562 cells. *Eur J Biochem* **268:** 1705–1711.

Todd DJ. 1994. Erythropoietic protoporphyria. *Br J Dermatol* **131:** 751–766.

Troadec MB, Warner D, Wallace J, Thomas K, Spangrude GJ, Phillips J, Khalimonchuk O, Paw BH, Ward DM, Kaplan J. 2011. Targeted deletion of the mouse *Mitoferrin1* gene: From anemia to protoporphyria. *Blood* **117:** 5494–5502.

Tsuchida M, Emi Y, Kida Y, Sakaguchi M. 2008. Human ABC transporter isoform B6 (ABCB6) localizes primarily in the Golgi apparatus. *Biochem Biophys Res Commun* **369:** 369–375.

Tugores A, Magness ST, Brenner DA. 1994. A single promoter directs both housekeeping and erythroid preferential expression of the human ferrochelatase gene. *J Biol Chem* **269:** 30789–30797.

von und zu Fraunberg M, Nyroen T, Kauppinen R. 2003. Mitochondrial targeting of normal and mutant protoporphyrinogen oxidase. *J Biol Chem* **278:** 13376–13381.

Wang J, Pantopoulos K. 2011. Regulation of cellular iron metabolism. *Biochem J* **434:** 365–381.

Wang L, He F, Bu J, Liu X, Du W, Dong J, Cooney JD, Dubey SK, Shi Y, Gong B, et al. 2012. *ABCB6* mutations cause ocular coloboma. *Am J Hum Genet* **90:** 40–48.

Whatley SD, Ducamp S, Gouya L, Grandchamp B, Beaumont C, Badminton MN, Elder GH, Holme SA, Anstey AV, Parker M, et al. 2008. C-terminal deletions in the *ALAS2* gene lead to gain of function and cause X-linked dominant protoporphyria without anemia or iron overload. *Am J Hum Genet* **83:** 408–414.

Wingert RA, Galloway JL, Barut B, Foott H, Fraenkel P, Axe JL, Weber GJ, Dooley K, Davidson AJ, Schmid B, et al. 2005. Deficiency of *glutaredoxin 5* reveals Fe–S clusters are required for vertebrate haem synthesis. *Nature* **436:** 1035–1039.

Wu CK, Dailey HA, Rose JP, Burden A, Sellers VM, Wang BC. 2001. The 2.0 Å structure of human ferrochelatase, the terminal enzyme of heme biosynthesis. *Nat Struct Biol* **8:** 156–160.

Yamauchi K, Hayashi N, Kikuchi G. 1980. Translocation of δ-aminolevulinate synthase from the cytosol to the mitochondria and its regulation by hemin in the rat liver. *J Biol Chem* **255:** 1746–1751.

Yi L, Ragsdale SW. 2007. Evidence that the heme regulatory motifs in heme oxygenase-2 serve as a thiol/disulfide redox switch regulating heme binding. *J Biol Chem* **282:** 21056–21067.

Yin X, Dailey HA. 1998. Erythroid 5-aminolevulinate synthase is required for erythroid differentiation in mouse embryonic stem cells. *Blood Cells Mol Dis* **24:** 41–53.

Zhang L, Guarente L. 1995. Heme binds to a short sequence that serves a regulatory function in diverse proteins. *EMBO J* **14:** 313–320.

Classification of the Disorders of Hemoglobin

Bernard G. Forget[1] and H. Franklin Bunn[2]

[1]Section of Hematology, Department of Medicine, Yale School of Medicine, New Haven, Connecticut 06520-8028

[2]Hematology Division, Department of Medicine, Brigham and Women's Hospital, Harvard Medical School, Boston, Massachusetts 02115

Correspondence: hfbunn@rics.bwh.harvard.edu

Over the years, study of the disorders of hemoglobin has served as a paradigm for gaining insights into the cellular and molecular biology, as well as the pathophysiology, of inherited genetic disorders. To date, more than 1000 disorders of hemoglobin synthesis and/or structure have been identified and characterized. Study of these disorders has established the principle of how a mutant genotype can alter the function of the encoded protein, which in turn can lead to a distinct clinical phenotype. Genotype/phenotype correlations have provided important understanding of pathophysiological mechanisms of disease. Before presenting a brief overview of these disorders, we provide a summary of the structure and function of hemoglobin, along with the mechanism of assembly of its subunits, as background for the rationale and basis of the different categories of disorders in the classification.

An impressive degree of molecular "engineering" was necessary for the evolution of a multisubunit protein that higher organisms require for optimal oxygen homeostasis and buffering of acidic metabolic waste products. Each globin subunit must form a stable linkage with heme (ferroprotoporphyrin IX) situated on the external surface of the protein so that oxygen in the RBC cytosol can bind reversibly to the hemes' iron atoms. Moreover, the hydrophobic cleft into which the heme is inserted must be able to protect the Fe^{2+} heme iron from oxidation to Fe^{3+}, which is incapable of binding oxygen.[3] Delicate noncovalent interactions between unlike globin subunits are required for the hemoglobin tetramer $\alpha_2\beta_2$ to bind and unload oxygen in a cooperative manner, thereby assuring maximal transport to actively metabolizing cells. This phenomenon is reflected by a sigmoid oxygen-binding curve that depends on hemoglobin tetramer having two quaternary structures: the T or deoxy conformer that has low oxygen affinity and the R or oxy conformer that has high oxygen affinity. Further fine-

[3]For more detailed information, see Dailey and Meissner (2013).

Copyright © 2013 Cold Spring Harbor Laboratory Press; all rights reserved
Cite this article as *Cold Spring Harb Perspect Med* doi: 10.1101/cshperspect.a011684

tuning of hemoglobin function comes from its allosteric behavior triggered by the binding of two small effector molecules 2,3-BPG, and protons to specific sites on the T structure distant from the heme groups.[4] To endow the blood with high oxygen carrying capacity hemoglobin must be stuffed into flexible circulating RBCs. A remarkably high degree of solubility is required for hemoglobin to achieve an intracellular concentration of ~34 g/dL or 5 mM (tetramer).

To reach such a high corpuscular hemoglobin concentration it is essential that α-globin as well as β-globin (or γ-globin) mRNA be expressed at very high levels during erythroid differentiation. Moreover, α- and non-α-globin synthesis must be closely matched. As explained in Nienhuis and Nathan (2012), subunit imbalance is central to the pathophysiology of the thalassemias. Free α-globin subunits are particularly toxic to erythroid cells. This threat is alleviated by the presence of α-hemoglobin stabilizing protein (AHSP), a chaperone that is expressed at high levels in erythroid cells and binds specifically and tightly to heme-intact α-globin subunits (Kihm et al. 2002; Feng et al. 2004; Mollan et al. 2010, 2012). The AHSP protects the cell from potentially toxic oxidized (Fe^{3+}) heme until it is reduced to functional Fe^{2+} heme by cytochrome b5 reductase. On encountering a free heme-intact β-globin subunit, the α-globin dissociates from AHSP to form the extremely stable αβ dimer. As discussed later in this work, this process is facilitated by electrostatic attraction between positively charged α-globin subunits and negatively charged β-globin subunits.

In view of the multiple molecular constraints that are required for high-level production of fully functional and highly soluble hemoglobin, it is no surprise that Murphy's law is in full force: whatever can go wrong will. As discussed briefly here, and in much more detail in Thein (2013), Higgs (2013), Nienhuis and Nathan (2012), and Musallam et al. (2012), mutations of globin genes that impair synthesis give rise to thalassemia and anemia of varying degree. In addition, well-defined clinical and hematologic phenotypes are associated with mutations that alter the structure of globin subunits, discussed in more detail in Thom et al. (2013). Impairment of hemoglobin solubility can be caused either by the formation of intracellular polymers (sickle cell disease) or by the development of amorphous precipitates (congenital Heinz body hemolytic anemia). Abnormalities of oxygen binding can lead either to erythrocytosis (high O_2 affinity mutants) or to cyanosis (low O_2 affinity mutants). Some globin mutants have structural alterations within the heme pocket that result in oxidation of the heme iron and pseudocyanosis because of methemoglobinemia.

[4]For more detailed information on hemoglobin function, see Thom et al. (2013).

GENERAL CLASSIFICATION OF HEMOGLOBIN DISORDERS

Hemoglobin disorders can be broadly classified into two general categories (as listed in Table 1):

1. Those in which there is a quantitative defect in the production of one of the globin subunits, either total absence or marked reduction. These are called the thalassemia syndromes.
2. Those in which there is a structural defect in one of the globin subunits.

The majority of human hemoglobin mutants were discovered as an incidental finding, unassociated with any hematologic or clinical phenotype. However, a number of α- and β-globin mutants are associated with distinct clinical phenotypes. These fall into five broad categories: the sickle syndromes (SS, SC, $S\beta^0$-thalassemia, and $S\beta^+$-thalassemia); unstable mutants causing congenital Heinz body hemolytic anemia; mutants with high oxygen affinity resulting in erythrocytosis; low oxygen affinity mutants and the M hemoglobins causing cyanosis; and mutants associated with a thalassemia phenotype. Most of these disorders are inherited genetic defects, but there are some defects that are acquired or occur de novo.

Table 1. Classification of hemoglobin disorders

I. QUANTITATIVE DISORDERS OF GLOBIN CHAIN SYNTHESIS/ACCUMULATION

The thalassemia syndromes

A. β-Thalassemia

Clinical classification:

β-Thalassemia minor or trait
β-Thalassemia major
β-Thalassemia intermedia

Biochemical/genetic classification:

β^0-Thalassemia
β^+-Thalassemia
δ-Thalassemia
γ-Thalassemia
Lepore fusion gene
δβ-Thalassemia
εγδβ-Thalassemia
HPFH
"Dominant" β-thalassemia (structural variants with β-thalassemia phenotype)

β-Thalassemia with other variants:

HbS/β-thalassemia
HbE/β-thalassemia
Other

B. α-Thalassemia

Deletions of α-globin genes:

One gene: α^+-thalassemia
Two genes in *cis*: α^0-thalassemia
Two genes in *trans*: homozygous α^+-thalassemia (phenotype of α^0-thalassemia)
Three genes: HbH disease
Four genes: Hydrops fetalis with Hb Bart's

Nondeletion mutants:

Hb Constant Spring
Other

C. De novo and acquired α-thalassemia

α-Thalassemia with mental retardation syndrome (ATR):

Due to large deletions on chromosome 16 involving the α-globin genes
Due to mutations of the ATRX transcription factor gene on chromosome X

α-Thalassemia associated with myelodysplastic syndromes (ATMDS):

Due to mutations of the ATRX gene

II. QUALITATIVE DISORDERS OF GLOBIN STRUCTURE: STRUCTURAL VARIANTS OF HEMOGLOBIN

A. Sickle cell disorders

SA, sickle cell trait
SS, sickle cell anemia/disease
SC, HbSC disease
S/β thal, sickle β-thalassemia disease
S with other Hb variants: D, O-Arab, other
SF, Hb S/HPFH

B. Hemoglobins with decreased stability (unstable hemoglobin variants)

Mutants causing congenital Heinz body hemolytic anemia
Acquired instability—oxidant hemolysis: Drug-induced, G6PD deficiency

C. Hemoglobins with altered oxygen affinity

High/increased oxygen affinity states:

Fetal red cells
Decreased RBC 2,3-BPG
Carboxyhemoglobinemia, HbCO
Structural variants

Low/decreased oxygen affinity states:

Increased RBC 2,3-BPG
Structural variants

D. Methemoglobinemia

Congenital methemoglobinemia:

Structural variants
Cytochrome b5 reductase deficiency

Acquired (toxic) methemoglobinemia

E. Posttranslational modifications

Nonenzymatic glycosylation
Amino-terminal acetylation
Amino-terminal carbamylation
Deamidation

THE THALASSEMIA SYNDROMES

The thalassemia syndromes are inherited disorders characterized by absence or markedly decreased accumulation of one of the globin subunits of hemoglobin. In the alpha (α)-thalassemias, there is absent or decreased production of α-globin subunits, whereas in the beta (β)-thalassemias, there is absent or reduced production of β-globin subunits. Rare thalassemias affecting the production of delta (δ)- or gamma (γ)-globin subunits have also been described but are not clinically significant disorders. Combined deficiency of δ + β-globin subunits, or of all of the β-like globin subunits also occurs. These disorders are described in detail in Thein (2013) and Higgs (2013).

THE β-THALASSEMIAS

The β-thalassemias can be subclassified into those in which there is total absence of normal β-globin subunit synthesis or accumulation, the β^0-thalassemias, and those in which some structurally normal β-globin subunits are synthesized, but in markedly decreased amounts, the β^+-thalassemias. The molecular basis of the β-thalassemias is very heterogeneous (see Thein 2013), with over 200 different mutations having been described. In general, the mutations causing β-thalassemia are point mutations affecting a single nucleotide, or a small number of nucleotides, in the β-globin gene. Rare deletion forms of β-thalassemia have also been described. One of these deletions is caused by "unequal" crossing over between the linked and partially homologous δ- and β-globin genes, resulting in the formation of a fusion δβ-globin gene, the Lepore gene, that has a low level of expression. Large deletions involving part or all of the β-globin gene cluster are responsible for the δβ-thalassemias, the εγδβ-thalassemias, and the hereditary persistence of fetal hemoglobin (HPFH) syndromes.

Despite the marked heterogeneity in the molecular basis of the β-thalassemias, the clinical phenotype of these disorders is relatively homogeneous because of their common pathophysiology: deficiency of HbA tetramers and excess accumulation of free α-subunits incapable of forming hemoglobin tetramers because of deficiency of β-like globin subunits (see Nienhuis and Nathan 2012). In heterozygotes (β-thalassemia trait or β-thalassemia minor), there is mild to moderate hypochromic microcytic anemia, without evidence of hemolysis, whereas in homozygotes or compound heterozygotes (β-thalassemia major), there is usually a severe transfusion-dependent hemolytic anemia associated with marked ineffective erythropoiesis resulting in destruction of erythroid precursor cells in the bone marrow.

A less common clinical phenotype is referred to as β-thalassemia intermedia. In this clinical disorder, there is a moderate to severe, partially compensated, hemolytic anemia that does not require chronic transfusion therapy to maintain a satisfactory circulating hemoglobin level in the affected patient, although occasional transfusions may be required if the anemia worsens because of associated complications. Such patients have a milder disease because there is less severe α- to non-α-globin subunit imbalance than in typical β-thalassemia major patients, resulting in less accumulation of free α-subunits that cause the ineffective erythropoiesis. There are different possible causes for such a lessened α- to non-α-globin subunit imbalance, including: inheritance of milder β^+-thalassemia mutations with less severe than usual deficiency of β-globin subunit production; coinheritance of a form of α-thalassemia; or coinheritance of another genetic trait associated with increased production of the γ-subunit of fetal hemoglobin. Most patients with β-thalassemia intermedia carry two mutant β-globin genes: they have a genotype typical of β-thalassemia major, but the phenotype is modified by one of the factors listed above. However, rare cases of β-thalassemia intermedia are caused by heterozygosity for a single mutant β-globin gene associated with the production of a highly unstable β-globin subunit that causes RBC damage in a fashion similar to excess free α-subunits; this is the so-called "dominant β-thalassemia" (Thom et al. 2013).

The δβ-thalassemias are associated with total deficiency of β-globin subunit production, but are clinically milder than typical cases of β^0-thalassemia, because there is an associated persistent high level of expression of the γ-subunit of fetal hemoglobin that decreases the degree of α-subunit excess. The εγδβ-thalassemias are associated with neonatal hemolytic anemia, but this resolves during the first few months

 Cite this article as *Cold Spring Harb Perspect Med* doi: 10.1101/cshperspect.a011684

of life and the associated phenotype in adults is that of β-thalassemia trait or β-thalassemia minor. The syndrome of hereditary persistence of fetal hemoglobin (HPFH) is not, strictly speaking, a form of β-thalassemia because it is not associated with significant α- to non-α-globin subunit imbalance, but is characterized by high levels of persistent γ-globin production and is frequently considered within the spectrum of δβ-thalassemia.

THE α-THALASSEMIAS

In contrast to the β-thalassemias, which are usually caused by point mutations of the β-globin gene, the α-thalassemia syndromes are usually caused by the deletion of one or more α-globin genes and are subclassified according to the number of α-globin genes that are deleted (or mutated): one gene deleted (α^+-thalassemia); two genes deleted on the same chromosome or in *cis* (α^0-thalassemia); three genes deleted (HbH disease); or four genes deleted (hydrops fetalis with Hb Bart's). Nondeletion forms of α-thalassemia have also been characterized but are relatively uncommon. These disorders are discussed in detail in Higgs (2013).

Clinically, the deletion of only one of the four α-globin genes is not associated with significant hematologic abnormalities and is sometimes called the "silent carrier" state for α-thalassemia. The deletion of two α-globin genes can occur in two forms: (1) on the same chromosomes, or in *cis*; or (2) on opposite chromosomes, or in *trans*, which is the homozygous state for the single gene deletion, or homozygous α^+-thalassemia. The *cis* genotype is particularly common in Asian populations, whereas the *trans* genotype is highly prevalent in persons of black/African ancestry. The clinical phenotype is similar with both genotypes and consists of mild hypochromic, microcytic anemia, without hemolysis, somewhat analogous to that of β-thalasssemia trait, but somewhat less severe. The deletion (or markedly decreased expression) of three α-globin genes is associated with the syndrome of HbH disease, a compensated hemolytic anemia that usually does not require treatment by RBC transfusion. The basis of the hemolysis is the excess accumulation of β-globin subunits that self-associate to form β-chain tetramers or HbH. In contrast to the situation in the β-thalssemias in which excess α-globin subunits rapidly form insoluble aggregates, excess β-globin chains can form soluble tetramers (HbH). However, HbH is relatively unstable and does precipitate as RBCs age, forming inclusion bodies that damage RBCs and shorten their lifespan (see Higgs 2013). The deletion of all four α-globin genes is usually fatal during late pregnancy or shortly after birth. This condition is called hydrops fetalis with Hb Bart's. Hb Bart's is a tetramer of four γ-globin subunits and is ineffective as an oxygen transporter: it has a very high oxygen affinity, similar to that of myoglobin, and does not release oxygen to the tissues under physiologic conditions. Therefore, the infant, whose RBCs lack HbF or HbA and contain mostly Hb Bart's, suffers severe hypoxia resulting in hydrops fetalis. Rare cases have been rescued by intrauterine transfusions, but these children subsequently require lifelong transfusion support similar to that required by children with β-thalassemia major.

One form of nondeletion α-thalassemia, called Hb Constant Spring (HbCS), is particularly prevalent in southeast Asia. It is caused by a chain termination mutation, which results in the synthesis of an elongated α-globin subunit that accumulates at very low levels in RBCs of affected individuals. When coinherited with the α^0-*cis* deletion on the other chromosome, there results a form of HbH disease that is more severe than the typical HbH disease associated with full deletion of three α-globin genes (see Higgs 2013).

In addition to these forms of α-thalassemia inherited in a pattern consistent with Mendelian genetics, there are two α-thalassemia syndromes that are caused by de novo or acquired mutations affecting expression of the α-globin genes: (1) the α-thalassemia with mental retardation syndrome (ATR); and (2) acquired α-thalassemia (HbH disease) associated with myelodysplastic syndromes (ATMDS). These syndromes are described in detail in Gibbons (2012).

THE α-THALASSEMIA WITH MENTAL RETARDATION SYNDROME

There are two subtypes of this syndrome: (1) one is associated with very large deletions involving the α-globin genes and adjacent genes on chromosome 16, that are not inherited from a parent and appear to have arisen de novo during embryogenesis (the ATR-16 syndrome); and (2) the other is associated with structurally normal α-globin genes and is caused by germline mutations in a transcription factor gene located on the X chromosome (the ATR-X syndrome). The encoded transcription factor has been called ATRX and is an important regulator of α-globin gene expression.

ACQUIRED α-THALASSEMIA (HbH DISEASE) ASSOCIATED WITH MYELODYSPLASTIC SYNDROMES (MDS)

A small number of patients with MDS, a clonal bone marrow disorder characterized by disturbed hematopoiesis with cytopenias and abnormal myeloid differentiation, develop RBC abnormalities consistent with acquired HbH disease. The α-globin genes in such patients are structurally normal, but their expression is severely impaired. The molecular basis for the decreased α-globin gene expression in this condition has been determined to be acquired somatic mutations of the ATRX gene, the same gene that is mutated in the syndrome of α-thalassemia with mental retardation. The impairment of α-globin gene expression is more severe in the ATMDS syndrome than in the ATRX syndrome.

QUALITATIVE DISORDERS OF GLOBIN STRUCTURE

Sickle Cell Disorders

Sickle hemoglobin (HbS) results from an amino acid substitution at the sixth residue of the β-globin subunit: β^6-Glu $\rightarrow$ Val. Approximately 8% of African Americans are heterozygous for this hemoglobin variant, a condition called sickle cell trait or HbAS. In equatorial Africa, where malaria is endemic, the prevalence of HbAS is much higher and can reach over 30% in some populations because of survival advantage of HbAS heterozygotes from complications of falciparum malaria. RBCs of persons with HbAS typically have 40% HbS and 56%–58% HbA. Individuals with HbAS are typically asymptomatic; severe hypoxia is required for them to experience manifestations of sickle cell disease, called sickling.

The basis of sickling in patients homozygous for the disorder, called sickle cell anemia or HbSS, is polymerization of deoxy-HbS resulting in the formation of multistranded fibers that create a gel and change the shape of RBCs from biconcave discs to elongated crescents. The polymerization/sickling reaction is reversible following reoxygenetion of the hemoglobin. Thus, an RBC can undergo repeated cycles of sickling and unsickling. There are two major pathophysiological consequences of sickling:

1. Repeated cycles of sickling damage the red blood cell membrane leading to abnormalities of permeability and cellular dehydration, eventually causing premature destruction of RBCs and a chronic hemolytic anemia.
2. Sickled RBCs are rigid, increase blood viscosity and obstruct capillary flow, causing tissue hypoxia and, if prolonged, cell death, tissue necrosis/infarction, and progressive organ damage. Acute episodes are often called vaso-occlusive pain crises.

The pathophysiology and clinical manifestations of sickle cell disease are described in detail in Serjeant (2013). There can be a great deal of variability in the severity of the clinical manifestations of the various sickle cell syndromes in patients with the same β-globin gene genotype and even between siblings in the same family. This variability can be caused by the coinheritance of a number of different genetic modifiers of the disease, including coinheritance of α-thalassemia, quantitative trait loci affecting the level of production of fetal hemoglobin and other traits, as discussed in Lettre (2012). In general, the clinical severity of the different sickle cell syndromes correlates directly with the amount and concentration of HbS in the red cell. The kinetics of sickling are markedly concentration dependent, the speed of the reaction being inversely proportional to about the 30th power of the starting HbS concentration. Thus, a small increase in intracellular HbS concentration, such as that resulting from loss of intracellular water and cellular dehydration, can have a major effect on the speed of the polymerization reaction. The amount and type of other non-S hemoglobins in the red cell can also influence the extent or rate of sickling, and thus clinical severity. HbSC disease is associated with significant clinical manifestations. Why do SC patients often have vaso-occlusive manifestations as dramatic as those with SS disease, whereas AS individuals have virtually no morbidity? HbC and HbA copolymerize to an identical degree with HbS (Bunn et al. 1982). Two independent factors conspire to make HbSC a disease. The presence of HbC in the red cell greatly enhances potassium efflux and red cell dehydration, which increases corpuscular hemoglobin concentration (Bunn et al. 1982) and promotes polymerization. Sickling in SC patients is further enhanced by the fact that they have ~50% HbS, whereas HbAS individuals have ~40% HbS. This difference is because α-globin subunits bind more readily to negatively charged β^A-subunits than to positively charged β^C-subunits, as explained further at the end of this work. Other clinically significant sickle cell syndromes include: sickle/β^0-thalassemia, sickle/β^+-thalassemia, HbSD disease, and HbSO-Arab disease. HbD and HbO-Arab participate more readily than HbA in copolymer formation with HbS.

A striking example of how another hemoglobin can dramatically influence sickling is the condition that results from coinheritance of HbS with hereditary persistence of fetal hemoglobin: HbSF or S/HFPH. This condition, in which there is ~70% HbS and ~30% HbF, is clinically asymptomatic and is a notable exception to the general rule that clinical severity is directly related to the amount (or %) of HbS in the RBC. In this condition, the HbF is distributed in a pancellular fashion so that every RBC contains ~30% HbF and, because HbF does not participate at all in polymer formation, this is sufficient to prevent sickling of all RBCs under physiologic conditions. This situation is in contrast to what occurs when variable amounts of HbF are present in HbSS disease or the sickle/β-thalassemia syndromes. In the latter disorders, the HbF is distributed in a heterocellular fashion, in which the HbF is present in only a subset of cells called F cells, leaving the other cells that lack HbF unprotected from the antisickling effects of HbF. Nevertheless, a marked increase in the proportion of F cells in such cases can have a beneficial effect on the clinical manifestation of these sickling disorders. Therefore, there is a great deal of interest and ongoing research to develop strategies for enhancing HbF synthesis and F-cell production as a therapeutic modality for sickling disorders (see Sankaran and Orkin 2013).

UNSTABLE HEMOGLOBIN VARIANTS

A substantial minority of Hb mutants have substitutions that alter the solubility of the molecule in the red cell. The intraerythrocytic precipitated material derived from the unstable abnormal Hb is detectable by a supravital stain as dark globular aggregates called Heinz bodies (HBs). These intracellular inclusions reduce the life span of the red cell and generate a hemolytic process of varied severity called congenital Heinz body hemolytic anemia. When a red cell hemolysate of an affected individual is heated to 50°C or treated with 17% isopropanol, a precipitate usually develops. The hemoglobin electrophoresis often reveals an abnormal banding pattern. Definitive diagnosis requires analysis of either globin structure or DNA sequence. For more information on congenital Heinz body hemolytic anemia and other hemoglobin variants summarized in this section, see Thom et al. (2013).

HEMOGLOBIN VARIANTS WITH ALTERED OXYGEN AFFINITY

Over 25 hemoglobin variants have been encountered in individuals with erythrocytosis. Amino acid substitutions cause an increase in oxygen affinity usually because of impairment in the stability of the T or "deoxy" quaternary structure. Erythrocytosis is usually familial and is the only well-documented clinical finding. Increased oxygen affinity results in a decrease in hemoglobin's ability to release oxygen during blood flow in the microcirculation. Unlike individuals with secondary erythrocytosis caused by high altitude exposure or to cardiac or pulmonary hypoxemia, those having hemoglobin with high oxygen affinity have normal arterial oxygen tension. Therefore, they are not hypoxemic. However, the reduction in unloading of oxygen to tissues results in cellular hypoxia. Thus, the cells in the kidney that produce erythropoietin sense hypoxia and up-regulate *Epo* gene expression leading to expansion of the red cell mass. About half of the high-affinity variants can be detected by hemoglobin electrophoresis. Definitive diagnosis is established by demonstration of a "shift to the left" in the oxyhemoglobin binding curve.

Less often, individuals and their relatives will present with low O_2 saturation and cyanosis owing to a hemoglobin variant with a marked reduction in oxygen affinity. The variants that have been most thoroughly characterized entail amino acid replacements that destabilize the R or "oxy" quaternary structure. Affected individuals have normal hemoglobin levels and no clinical manifestations other than the dusky skin color.

THE M HEMOGLOBINS

Rare families have attracted medical attention because affected individuals have cyanosis owing to the inheritance of a hemoglobin variant in which an amino acid substitution within the heme pocket results in stabilization of the heme iron in the oxidized Fe^{3+} form. Among the most highly conserved amino acids within hemoglobin subunits are two histidines, one at helix position F8, which binds to the heme iron and one at E7 in the pocket where oxygen and other ligands bind to heme. Replacement of either of these histidines by tyrosine in either the α-, β-, or γ-globin subunit results in an alteration in the electronic environment of the heme iron so that it is stabilized as Fe^{3+}. Thus, the M hemoglobins include α58 His → Tyr, α87 His → Tyr, β63 His → Tyr, β92 His → Tyr, γ63 His → Tyr, and γ92 His → Tyr. In addition, β67 Val → Glu results in very similar biochemical features and clinical presentation. Affected individuals have cyanosis similar in hue to those with methemoglobinemia in which the heme iron in normal hemoglobin has been oxidized to Fe^{3+}. They have either normal hemoglobin levels or mild anemia and no clinical manifestations other than their skin color.

IMPACT OF SUBUNIT ASSEMBLY ON HEMOGLOBIN PHENOTYPE

The composition of the normal and mutant hemoglobins in red cells of heterozygotes provides insight into the assembly of human hemoglobins (Bunn and McDonald 1983). The great majority of β-subunit mutants are synthesized at the same rate as β^A and have normal stability. Therefore, heterozygotes would be expected to have equal amounts of normal and mutant hemoglobin. However, measurements of the proportion of normal and abnormal hemoglobins in heterezygotes have revealed unexpected variability. Figure 1 shows a comparison of 105 stable β-subunit mutants. Positively charged mutants, such as hemoglobins S, C, D-Los Angeles, and E[5], constitute significantly less than half of the total hemoglobin in heterozygotes and are reduced further in the presence of α-thalassemia, because of increased competition between normal and mutant β-subunits for limiting amounts of α-subunits. In contrast, many of the negatively charged mutants are present in amounts exceeding that of HbA. In two heterozygotes who had a negatively charged mutant (HbJ-Baltimore or HbJ-Iran) in conjunction with α-thalassemia, the proportion of the mutant hemoglobin was further increased because these β^J-subunits out-compete normal β^A-subunits for limiting amounts of α-subunits.

[5]Levels of HbE in AE heterozygotes are lower than the others because the synthesis of β^E-globin is impaired owing to a defect in mRNA splicing (see Fucharoen and Weatherall 2012 and Thein 2013).

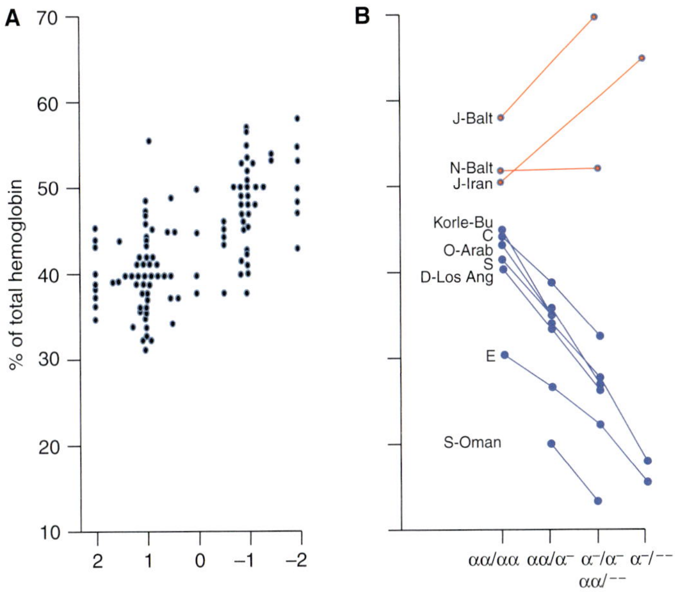

Figure 1. Impact of surface charge on hemoglobin assembly. (*A*) Effect of charge on the proportion of abnormal hemoglobin in individuals heterozygous for 105 stable β-globin variants. Each data point represents a mean value for a given variant. Substitutions involving a histidine residue were scored as a change of one half charge. The "−1" group differs significantly from the "+1" group ($P < 0.001$). (*B*) Effect of α-thalassemia on the proportion of seven positively charged β-subunit variants (blue) and three negatively charged variants (red). (Updated from Bunn and McDonald 1983 and Bunn 1987.)

Analyses of the proportion of β-subunit mutant in heterozygotes suggest that alterations in surface charge contribute to different rates of assembly of the hemoglobin tetramer. This hypothesis is supported by in vitro mixing experiments with normal and mutant subunits showing that when α-subunits are present in limiting amounts (mimicking α-thalassemia), negatively charged mutants are formed much more readily than are positively charged mutants (Mrabet et al. 1986; Adachi et al. 1998).

This electrostatic model of hemoglobin assembly has clinical implications. Differences in rates of assembly explain the higher proportion in HbS in sickle C (SC) disease than in sickle trait (AS), a difference that contributes significantly to HbS polymerization and consequent vaso-occlusion in SC patients (Bunn et al. 1982). HbS-Oman is a rare stable mutant with both the sickle (β6Glu → Val) substitution and β121Glu → Lys, a substitution that stabilizes the sickle polymer. Thus, β-subunits of HbS-Oman have a gain of three positive charges. Accordingly, in red cells of heterozygotes, it is present in much less abundance than HbA. However, because of its markedly enhanced polymerization, heterozygotes with only 20% HbS-Oman and one-gene deletion α-thalassemia (α-/αα) have hemolytic anemia and signs and symptoms of sickle vaso-occlusion. In contrast, HbS-Oman heterozygotes with more severe α-thalassemia (α-/α-) have only 13% HbS-Oman and a normal clinical and hematologic phenotype (Nagel et al. 1998).

The impact of surface charge on hemoglobin assembly also explains the differences in the levels of HbA_2 (Bunn 1987) and HbF (Adams et al. 1985; Chui et al. 1990) that accompany certain hematological disorders. For example, α-hemoglobin subunits bind more readily to β than to the positively charged δ. Accordingly, HbA_2 levels are decreased in the presence of limiting levels of α-hemoglobin, e.g., α-thalassemia as well as iron deficiency in which there is an acquired impairment of α-globin synthesis.

POSTTRANSLATIONAL ALTERATIONS IN HEMOGLOBIN STRUCTURE

A wide range of posttranslational modifications of hemoglobin structure have been identified, some of which result in significant perturbations of function.

Nonenzymatic Glycation

The concentration of glucose in the RBC cytosol is nearly that of the plasma. Glucose is in equilibrium between the cyclic hemiacetal structure and the open aldehyde structure. The latter is able to react covalently with amino groups of a wide range of proteins. Because this reaction favors amino groups with relatively low pKas, the most abundant glycated hemoglobin is at the amino termini of α- and β-globin. In normal red cells, ~5% of hemoglobin has a glucose adduct at the β-amino terminus designated HbAIc:

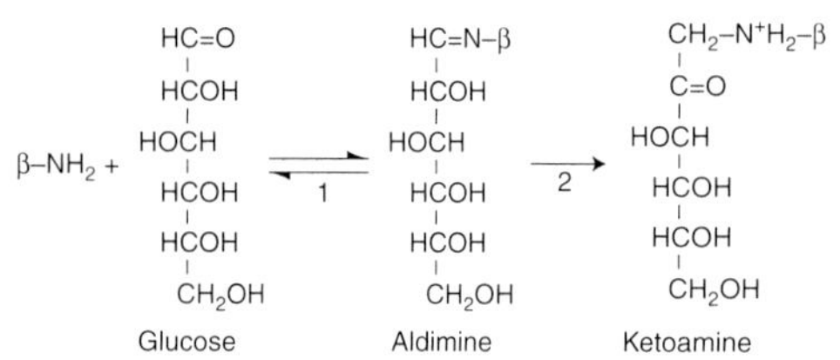

Reaction 1 shows the formation of the Schiff base or aldimine adduct. This product then rearranges to the more stable (virtually irreversible) ketoamine linkage (reaction 2). HbAIc and other less abundant glucose adducts are formed in a linear manner over the 120-day life span of the red cell. Therefore, HbAIc levels are low in individuals with hemolysis. Because the level of HbAIc is an accurate reflection of the average concentration of plasma glucose over the preceding 2 months, it has proven to be very useful in monitoring therapeutic control in patients with diabetes.

Amino-Terminal Acetylation

Shortly following translation of most proteins, the amino-terminal (initiator) methionine is cleaved and often the adjacent amino acid is acetylated at the amino terminus. These posttranslational modifications are highly dependent on the sequence of neighboring amino acids. All human globins undergo cleavage of the amino-terminal methionine but none undergo amino-terminal acetylation except for ~20% of γ-globin subunits (HbF_I). However, rare human hemoglobin variants have been identified that have amino acid replacements in the amino-terminal region that cause both retention of the initiator methionine and *N*-acetylation.

Amino-Terminal Carbamylation

Urea in the plasma is in equilibrium with ammonium and cyanate ions. Cyanate is able to bind covalently to the amino termini of α- and β-globin subunits to form adducts. Carbamylated hemoglobin is not detectable in normal red cells but can accumulate to measurable levels in patients with uremia.

Deamidation

In many proteins, the amino group on certain asparagine residues, particularly when adjacent to a histidine, dissociates leaving aspartic acid. Deamidation does not occur on asparagines in normal globin subunits. However, seven human variants have been identified in which partial or complete deamidation occurs. Four of these entail replacements by asparagine. The deamidation does not appear to impact function in any of these variants.

ACQUIRED ALTERATIONS IN HEMOGLOBIN FUNCTION

Methemoglobinemia

Certain drugs and toxins can cause enhanced oxidation of heme iron to form methemoglobin. Above levels of 10%, methemoglobin individuals appear cyanotic. The extent of hemoglobin oxidation is determined by spectrophotometric analysis of the blood. Methemoglobinemia impairs oxygenation of tissues because the Fe^{3+} heme is incapable of binding to oxygen, and, even more importantly, the presence of one or two oxidized hemes per tetramer stabilizes the R quaternary structure and thereby increases the affinity of the remaining hemes for oxygen. Central nervous and cardiovascular symptoms and signs begin to appear when methemoglobin exceeds 30%. Much lower levels are seen in individuals with congenital methemoglobinemia owing to deficiency in cytochrome b5 reductase. Both toxic and congenital methemoglobinemia can be effectively treated by administration of the redox dye methylene blue.

Carboxyhemoglobinemia

Carbon monoxide (CO) binds to heme iron of hemoglobin with an affinity 210 greater than that of oxygen. Normal red cells contain a very small amount of carboxyhemoglobin owing to the release of CO from macrophages following the uptake of senescent red cells and degradation of heme by heme oxygenase. Higher, albeit nontoxic, levels of HbCO are present in patients with hemolytic anemias. Inhalation of CO from incomplete combustion of hydrocarbons can result in toxic carboxyhemoglobinemia. As in methemoglobinemia, CO poisoning causes morbidity and mortality by ischemia owing to a left shift in the oxyhemoglobin binding curve, thereby impairing the unloading of oxygen to cells and tissues. Patients with acute CO poisoning generally develop CNS and cardiovascular symptoms when levels exceed 20%. Levels of 40%–60% HbCO result in stupor, coma, and death. All symptomatic patients should be promptly removed from further exposure to CO and treated with oxygen. If possible, those with severe toxicity should be treated with hyperbaric oxygen administration or exchange transfusion.

Alterations in Oxygen Affinity

As mentioned above, protons (H^+) and 2,3-BPG are potent allosteric modulators of oxygen affinity in normal RBC. Because of their preferential binding to the T or "deoxy" quaternary structure of hemoglobin, they lower the affinity for oxygen and therefore shift the oxyhemoglobin binding curve "to the right." Except under clinical circumstances resulting in either acidosis or alkalosis, plasma pH and red cell pH are tightly regulated at 7.4 and 7.2, respectively. In contrast, levels of red cell 2,3-BPG can vary considerably in a wide range of clinical settings. Patients with low levels, such as those who have received large amounts of stored RBCs will have increased oxygen affinity, which could compromise tissue oxygenation. Increased levels of RBC 2,3-BPG are much more commonly observed, primarily in patients with some form of hypoxia. As a result, the "right shift" in the oxyhemoglobin dissociation curve significantly enhances the unloading of oxygen to tissues and is therefore a major mode of compensation, particularly in patients with severe anemia.

REFERENCES

*Reference is also in this collection.

Adachi K, Yamaguchi T, Pang J, Surrey S. 1998. Effects of increased anionic charge in the β-globin chain on assembly of hemoglobin in vitro. *Blood* **91:** 1438–1445.

Adams JD, Coleman MB, Hayes J, Morrison WT, Steinberg MH. 1985. Modulation of fetal hemoglobin synthesis by iron deficiency. *N Engl J Med* **313:** 1402–1405.

Bunn HF. 1987. Subunit assembly of hemoglobin: An important determinant of hematologic phenotype. *Blood* **69:** 1–6.

Bunn HF, McDonald MJ. 1983. Electrostatic interactions in the assembly of haemoglobin. *Nature* **306:** 498–500.

Bunn HF, Noguchi CT, Hofrichter J, Schechter GP, Schechter AN, Eaton WA. 1982. Molecular and cellular pathogenesis of hemoglobin SC disease. *Proc Natl Acad Sci* **79:** 7527–7531.

Chui DH, Patterson M, Dowling CE, Kazazian HJ, Kendall AG. 1990. Hemoglobin Bart's disease in an Italian boy. Interaction between α-thalassemia and hereditary persistence of fetal hemoglobin. *N Engl J Med* **323:** 179–182.

* Dailey HA, Meissner PN. 2013. Erythroid heme biosynthesis and its disorders. *Cold Spring Harb Perspect Med* **3:** a011676.

Feng L, Gell DA, Zhou S, Gu L, Kong Y, Li J, Hu M, Yan N, Lee C, Rich AM, et al. 2004. Molecular mechanism of AHSP-mediated stabilization of α-hemoglobin. *Cell* **119:** 629–640.

* Fucharoen S, Weatherall DJ. 2012. The hemoglobin E thalassemias. *Cold Spring Harb Perspect Med* **2:** a011734.

* Higgs DR. 2013. The molecular basis of α-thalassemia. *Cold Spring Harb Perspect Med* **3:** a011718.

Kihm AJ, Kong Y, Hong W, Russell JE, Rouda S, Adachi K, Simon MC, Blobel GA, Weiss MJ. 2002. An abundant erythroid protein that stabilizes free α-haemoglobin. *Nature* **417:** 758–763.

* Lettre G. 2012. The search for genetic modifiers of disease severity in the β-hemoglobinopathies. *Cold Spring Harb Perspect Med* **2:** a015032.

Mollan TL, Yu X, Weiss MJ, Olson JS. 2010. The role of α-hemoglobin stabilizing protein in redox chemistry, denaturation, and hemoglobin assembly. *Antioxid Redox Signal* **12:** 219–231.

Mollan TL, Khandros E, Weiss MJ, Olson JS. 2012. The kinetics of α-globin binding to α hemoglobin stabilizing protein (AHSP) indicate preferential stabilization of a hemichrome folding intermediate. *J Biol Chem* **287:** 11338–11350.

Mrabet NT, McDonald MJ, Turci S, Sarkar R, Szabo A, Bunn HF. 1986. Electrostatic attraction governs the dimer assembly of human hemoglobin. *J Biol Chem* **261:** 5222–5228.

* Musallam KM, Tahar AT, Rachmilewitz EA. 2012. β-Thalassemia intermedia: A clinical perspective. *Cold Spring Harb Perspect Med* **2:** a013482.

Nagel RL, Daar S, Romero JR, Suzuka SM, Gravell D, Bouhassira E, Schwartz RS, Fabry M, Krishnamoorthy R. 1998. HbS-Oman heterozygote: A new dominant sickle syndrome. *Blood* **92:** 4375–4382.

* Nienhuis AW, Nathan DG. 2012. Pathophysiology and clinical manifestations of the β-thalassemias. *Cold Spring Harb Perspect Med* **2:** a011726.

* Sankaran VG, Orkin SH. 2013. The switch from fetal to adult hemoglobin. *Cold Spring Harb Perspect Med* **3:** a011643.

* Serjeant GR. 2013. The natural history of sickle cell disease. *Cold Spring Harb Perspect Med* doi: 10.1101/cshperspect.a011783.

* Thein SL. 2013. The molecular basis of β-thalassemia. *Cold Spring Harb Perspect Med* doi: 10.1101/cshperspect.a011700.

* Thom CS, Dickson CF, Gell DA, Weiss MJ. 2013. Hemoglobin variants: Biochemical properties and clinical correlates. *Cold Spring Harb Perspect Med* **3:** a011858.

World Distribution, Population Genetics, and Health Burden of the Hemoglobinopathies

Thomas N. Williams[1] and David J. Weatherall[2]

[1]Kenya Medical Research Institute/Wellcome Trust Programme, Centre for Geographical Research, Kilifi District Hospital, PO Box 230, Kilifi, Kenya

[2]Weatherall Institute of Molecular Medicine, University of Oxford, John Radcliffe Hospital, Oxford OX39DU, United Kingdom

Correspondence: twilliams@kilifi.kemri-wellcome.org

Although information about the precise world distribution and frequency of the inherited hemoglobin disorders is still limited, there is no doubt that they are going to pose an increasing burden on global health resources in the future. Their high frequency is a reflection of natural selection combined with a high frequency of consanguineous marriages in many countries, together with an epidemiological transition; whereby, as public health measures improve in the poorer countries of the world, more babies with these disorders are surviving to present for treatment.

The inherited disorders of hemoglobin are by far the most common monogenic diseases. Recent surveys suggest that between 300,000 and 400,000 babies are born with a serious hemoglobin disorder each year and that up to 90% of these births occur in low- or middle-income countries (Christianson et al. 2006). Here we will discuss the world distribution of these conditions and what is known about the mechanisms that have led to their extremely high frequency in many tropical countries.

DISTRIBUTION AND FREQUENCY OF THE HEMOGLOBIN DISORDERS

An estimate of the annual births of the important hemoglobin disorders is shown in Table 1. These figures are only a very approximate assessment and are based, in many cases, on only relatively small samples from localized regions of different countries. It has been found recently that the distribution of all the important hemoglobin disorders is extremely heterogeneous within different countries, even within small geographical distances (Weatherall 2010). Hence, a great deal more work needs to be done to provide anything like an accurate estimate of their true frequency.

Maps showing the world distribution of the sickle cell disorders, hemoglobin (Hb)E and the different forms of thalassemia are shown in Figures 1 and 2. Sickle cell anemia occurs throughout sub-Saharan Africa and in small pockets in the Mediterranean region, the Middle East, and the Indian subcontinent. By the use of haplotype analysis, that is the pattern of single nucleotide polymorphisms in the β-globin gene cluster, it has been determined that the sickle cell

Copyright © 2013 Cold Spring Harbor Laboratory Press; all rights reserved
Cite this article as *Cold Spring Harb Perspect Med* doi: 10.1101/cshperspect.a011692

Table 1. A breakdown of the annual number of births with the different hemoglobin disorders

Annual births with major hemoglobin disorders	
β-thalassemia major	22,989
HbE β-thalassemia	19,128
HbH disease	9,568
Hb Bart's hydrops (α^0/α^0)	5,183
SS disease	217,331
S β-thalassemia	11,074
SC disease	54,736

From available data (Modell and Darlison 2008; Weatherall 2010).

gene arose at least twice, once in Africa and once in India or the Middle East (Kulozik et al. 1986). Similar evidence suggests that the gene may have also arisen on several different occasions in sub-Saharan Africa although these data have to be reviewed with caution because it is possible that the different haplotype patterns could have resulted from gene conversion events (Flint et al. 1998). Hemoglobin SC disease is more restricted to parts of west and north Africa whereas HbS β-thalassemia occurs in localized parts of sub-Saharan Africa and sporadically throughout the Middle East and Indian subcontinent. Unlike some forms of α-thalassemia and HbE, which are discussed below, the gene frequency for HbS rarely rises much above 20%–25% of a particular population although there are occasional exceptions.

Hemoglobin E, the other particularly common structural hemoglobin variant, occurs widely throughout the eastern half of the Indian subcontinent, Bangladesh, Myanmar, and east and southeast Asia. It occurs at varying frequencies but in some parts of Asia, notably the northern parts of Thailand and Cambodia, called the "hemoglobin E triangle," up to 70% of the population are carriers (see Fucharoen and Weatherall 2012).

The mild forms of α-thalassemia (see Higgs 2013), which result from a single α gene deletion ($-\alpha/\alpha\alpha$), occur in a broad tropical belt stretching from sub-Saharan Africa through the Mediterranean regions and Middle East to the Indian subcontinent and the whole of east and southeast Asia. In this band, they occur at a frequency of 10%–25%, although in a few localized populations such as those of north India and Papua New Guinea, they are found in up to 80% of the population and the gene appears to be going to fixation. The more severe forms of α-thalassemia due to loss of both α chain genes—α^0-thalassemia ($-/\alpha\alpha$)—have a much more restricted occurrence, reaching high frequencies

Figure 1. The world distribution of the origins of hemoglobin S and hemoglobin E. (From Weatherall and Clegg 2001; reprinted, with permission, from the author.)

 Cite this article as *Cold Spring Harb Perspect Med* doi: 10.1101/cshperspect.a011692

Figure 2. The world distribution of the origins of the α- and β-thalassemias. (From Weatherall and Clegg 2001; reprinted, with permission, from the author.)

only in southeast Asia and in some of the Mediterranean islands. They occur sporadically throughout other populations, however, although their true frequency has not been determined.

Although the hemoglobinopathies all occur at particularly high frequencies in these tropical regions, they have been transported to most countries of the world by population migrations over many years. For example, the sickle cell gene is extremely common in many of the Caribbean Islands and in parts of North America, and occurs in most countries to a varying degree. The same phenomenon has occurred in the case of all the common forms of thalassemia. Interestingly, there are no common hemoglobin disorders among the American Indian population, probably because these conditions were not established in Asia at the time of the early population movements across the Bering Strait.

WHY ARE THE INHERITED DISORDERS OF HEMOGLOBIN SO COMMON?

There are several reasons why the hemoglobin disorders have reached such extraordinarily high gene frequencies in many tropical countries. First and foremost, it appears that their frequency reflects natural selection through protection of heterozygotes against severe malaria, a mechanism first suggested by Haldane (1949) and later confirmed independently in the case of sickle cell anemia by Allison (1954a,b,c). We will consider the possible mechanisms later in this work.

Natural selection is not the only mechanism involved in maintaining these high gene frequencies. Another important factor is the relatively high frequency of consanguineous marriages in many of the high frequency countries; this mechanism has an important effect on increasing the gene frequency of any recessively inherited disorder. Although accurate data on the frequency of consanguineous marriage are lacking, there is no doubt from such that are available that this is an important factor in helping to maintain the global health problem posed by these conditions. Another important factor is the epidemiological transition, whereby as public health and nutritional standards improve in the poorer countries, babies with these conditions who would otherwise have died in early life are now living long enough to present for diagnosis and management. As evidenced by what happened in Cyprus after it went through this transition after the Second World War, there is no doubt that this mechanism will cause a major increase in the number of patients with the more severe hemoglobin disorders in the future (Weatherall and Clegg 2001). Finally, the varying distribution of some of the hemoglobin disorders in different populations reflects strong founder effects by their original inhabitants. This mechanism is clearly illustrated in some of the Pacific Island populations (O'Shaughnessy et al. 1990).

It should be emphasized that even if some of the selective factors are reduced—the eradication of malaria, for example—it will take many generations before the frequency of these diseases starts to decline; a period during which there will be a tendency for their increase because of the epidemiological transition outlined above.

NATURAL SELECTION AND THE HEMOGLOBIN DISORDERS

During the six decades since Haldane first proposed the malaria hypothesis (Haldane 1949), evidence has steadily grown to confirm that malaria is indeed the primary force behind the high frequency of inherited hemoglobin disorders that is seen in many tropical and subtropical populations. The evidence supporting this assertion comes from four main sources: the similarity between the distributions of malaria and specific hemoglobin disorders at local, regional, and global scales; from population genetic predictions of their historic age; from clinical studies conducted in malaria-endemic areas; and from mechanistic studies conducted both in-vitro and ex-vivo. Although such evidence is strongest for HbS and for the α-thalassemias, which have been the focus of the most research, there can now be little doubt that malaria is responsible for the current distributions of all the major hemoglobin disorders. In the following sections we will review the evidence as it stands for each hemoglobin disorder individually. The proposed mechanisms for malaria protection by these conditions are summarized in Table 2. However, later in this work we will also discuss emerging data that suggest that selection for hemoglobin disorders may be competitive, and might be influenced by negative effects when inherited in combination with one another.

Hemoglobin S

The heterozygous state for hemoglobin S (HbAS) is the best described of all malaria-protective traits and is used as the classic example of balanced polymorphism in schools and colleges throughout the world. The close resemblance between the geographic distribution of HbAS and that of malaria, documented by a number of different investigators more than 60 years ago (Beet 1946, 1947; Brain 1952; Foy et al. 1952, 1954), gave the first clues. But it is Allison who first articulated the malaria hypothesis with regard to HbS most definitively (Allison 1954a,b,c). Not only, as others before him (Beet 1946, 1947; Brain 1952), did he note that "sicklers" were significantly less likely than "non-sicklers" to carry malaria parasites under conditions of natural exposure but he also showed that, when inoculated intravenously with large volumes of parasite-infected blood, sicklers were less likely to develop a clinical malaria infection. Many studies have since been conducted that put the malaria hypothesis with regard to HbS beyond reasonable doubt. First, the close correspondence between the frequencies of the β^S allele and the historic prevalence of malaria

Table 2. Summary of some proposed mechanisms of action of the malaria-resistance polymorphisms involving the inherited hemoglobin disorders

Hemoglobin disorder	Potential mechanisms for malaria protection	Key references
Sickle cell trait (HbAS)	Impairment of *P. falciparum* red cell invasion and growth under conditions of low oxygen tension	Friedman 1978; Pasvol et al. 1978; Friedman et al. 1979
	Enhanced removal of parasite-infected HbAS red blood cells	Luzzatto et al. 1970; Friedman 1978; Roth et al. 1978; Shear et al. 1993; Ayi et al. 2004
	Reduced pathogenicity of *P. falciparum*-infected red blood cells because of reduced expression of PfEMP1	Cholera et al. 2008; Cyrklaff et al. 2011
	Improved acquisition of malaria-specific immunity	Edozien et al. 1960; Cornille-Brogger et al. 1979; Guggenmoos-Holzmann et al. 1981; Marsh et al. 1989; Bayoumi et al. 1990; Abu-Zeid et al. 1992; Odegbemi and Williams 1995; Le Hesran et al. 1999; Cabrera et al. 2005; Williams et al. 2005a; Verra et al. 2007b
Hemoglobin C (HbAC and HbCC)	Impairment of *P. falciparum* red cell invasion and growth under conditions of low oxygen tension	Friedman et al. 1979; Pasvol and Wilson 1982; Olson and Nagel 1986; Fairhurst et al. 2003
	Improved acquisition of malaria-specific immunity	Ringelhann et al. 1976; Verra et al. 2007b
	Reduced pathogenicity of *P. falciparum*-infected red blood cells because of reduced expression of PfEMP1	Fairhurst et al. 2005; Brittain et al. 2007; Cyrklaff et al. 2011
Hemoglobin E (HbAE and HbEE)	Impairment of *P. falciparum* red cell invasion and growth	Chotivanich et al. 2002
α-Thalassemia	Specific protection against malaria-induced anaemia	Allen et al. 1997; Williams et al. 2005d; Wambua et al. 2006; May et al. 2007; Fowkes et al. 2008; Veenemans et al. 2008
	Reduced pathogenicity through reduced cytoadherence or rosetting	Udomsangpetch et al. 1993; Carlson et al. 1994; Cockburn et al. 2004
	Immuological priming through cross-species immunity between *P. vivax* and *P. falciparum*	Williams et al. 1996; Veenemans et al. 2011
β-Thalassemia	Enhanced removal of parasite-infected red blood cells	Ayi et al. 2004
	Reduced invasion and growth of *P. falciparum* parasites	Kaminsky et al. 1986; Senok et al. 1997
	Reduced pathogenicity through reduced cytoadherence or rosetting	Udomsangpetch et al. 1993; Carlson et al. 1994

has been confirmed through a growing number of surveys at both the local (Enevold et al. 2007) and global scales (Livingstone 1973, 1985; Cavalli-Sforza et al. 1994). Recently, this relationship has been quantified statistically, an analysis that found strong geographical support for the malaria hypothesis in Africa but not in either the Americas or in Asia (Piel et al. 2010). Hemoglobin S is absent from indigenous populations in the Americas, probably because malaria did not reach the continent until relatively recently, whereas in Asia it follows a rather restricted distribution, being confined to a small number of tribal populations in India (Piel et al. 2010).

Direct clinical support for a protective effect of HbAS has been provided by case control, cohort, and family-based association studies conducted in multiple populations over the last 30 years (for summary, see Williams et al. 2005b; Jallow et al. 2009). In general, such studies have shown that HbAS is ~50% protective against uncomplicated and >80% protective against severe *P. falciparum* malaria. Although few individual studies have been powered to detect an effect on malaria-specific mortality, HbAS was associated with 55% protection against all-cause mortality in children 6–16 months old in a single cohort study conducted in Western Kenya (Aidoo et al. 2002). Intriguingly, HbAS has also been associated with protection against invasive bacterial infections in two studies conducted in different parts of Africa (Eeckels et al. 1967; Scott et al. 2011). This observation appears to be explained by the protective effect of HbAS against malaria and provides evidence for a causal association between malaria and bacterial infections, indicating that in malaria-endemic environments, genetic factors like HbAS may confer an additional benefit of protection against death from bacterial diseases (Scott et al. 2011). Recently, Gouagna and colleagues working in Burkina Faso, made an additional intriguing observation regarding the malaria-protective effect of HbS (Gouagna et al. 2010). Although HbAS protects against clinical malaria infections, parasite-infected blood from HbAS subjects is several times more infectious to the *Anopheles* vector than that of normal subjects (Gouagna et al. 2010). This observation, which is echoed by a similar observation from Senegal (Lawaly et al. 2010), suggests that the personal advantage of HbAS is not only balanced by the cost of potential homozygosity in offspring but also by that of increased malaria transmission to the general population.

Whereas the malaria-protective effect of HbAS is clear, the effect of the homozygous state, HbSS, on malaria risk is more controversial. Biologically, there are hypothetical reasons why subjects with HbSS might either enjoy a greater degree of protection than those with HbAS or, conversely, might be at more increased risk than normal subjects (Williams and Obaro 2011). Few controlled studies of this question have been reported but the balance of evidence suggests that both scenarios might be true. Subjects with HbSS appear to be less susceptible than normals to developing malaria infections (Makani et al. 2010; McAuley et al. 2010) but are highly susceptible to the catastrophic consequences of malaria, particularly severe anemia, if they do become infected with the disease (Makani et al. 2010; McAuley et al. 2010; Williams and Obaro 2011). The net result is that malaria is almost certainly a major cause of premature mortality in children born with HbSS in malaria-endemic areas and that the early detection of HbSS in association with active malaria prevention could have a major impact on survival (Serjeant 2005).

The Mechanism of Malaria Protection Afforded by HbAS

Although the fact that HbAS confers malaria protection is now well established, the mechanism by which it does so remains a matter of some speculation. Through early studies, conducted soon after it became possible to grow malaria parasites in culture, investigators converged on the general hypothesis that protection resulted from impairment in the invasion and growth of *P. falciparum* parasites into HbAS red cells under conditions of low oxygen tension that were physiologically representative of in vivo conditions (Friedman 1978; Pasvol et al. 1978; Friedman et al. 1979). However, a number of alternative hypotheses have subsequently developed. Several investigators have suggested that the mechanism may relate to the enhanced removal of parasite-infected HbAS red blood cells. Some have proposed that this may relate to the fact that such cells sickle under low oxygen tension (Luzzatto et al. 1970; Friedman 1978; Roth et al. 1978) and that this leads to their premature destruction in the spleen (Friedman 1978; Shear et al. 1993), whereas Ayi and colleagues have proposed a mechanism that is generic to a number of red cell disorders and involves the enhanced opsonization of red cells infected with ring-stage parasites through a process that involves increased oxidative stress (Ayi et al. 2004).

An alternative hypothesis has emerged more recently. Cholera and colleagues have shown that in vitro, relative to parasitized normal HbAA red blood cells, the binding of parasitized HbAS red blood cells to both microvascular endothelial cells and to blood monocytes was significantly reduced and that this correlated with altered surface display of the parasite-encoded protein *P. falciparum* Erythrocyte Protein-1 (PfEMP1) (Cholera et al. 2008). Because the PfEMP1-mediated sequestration of mature parasites in the post-capillary venules of critical tissues such as the brain have been implicated in both the pathogenesis of severe malaria and the evasion of parasite-infected red cells from immune clearance by the spleen, this provides an attractive explanation for the protective effect of HbAS. Very recently, Cyrklaff and colleagues have elaborated on this hypothesis by showing, again in vitro, that HbAS affects the trafficking system that directs PfEMP1 to the surface of infected erythrocytes. Using cryo-electron tomography they showed that within the cytoplasm of normal red blood cells the parasite proteins are transported to the surface via a parasite-generated host-derived actin cytoskeleton but that hemoglobin oxidation products disrupted this process in HbAS red cells (Cyrklaff et al. 2011).

Whereas all the hypotheses discussed so far imply that HbAS protects against malaria entirely through innate mechanisms, data from both epidemiological and clinical studies suggest that this may not be the entire story. The protective effect of HbAS in naturally exposed populations increases with age (Cornille-Brogger et al. 1979; Guggenmoos-Holzmann et al. 1981; Le Hesran et al. 1999; Williams et al. 2005a), indicating that the mechanism might not be entirely innate but might also include an acquired, immunological component. This hypothesis is supported by a number of studies that have reported enhanced malaria-specific immune responses in HbAS individuals (Edozien et al. 1960; Marsh et al. 1989; Bayoumi et al. 1990; Abu-Zeid et al. 1992; Odegbemi and Williams 1995; Cabrera et al. 2005; Verra et al. 2007a), and potentially by recent studies using a mouse model that suggest an immuno-modulatory mechanism mediated through hemoxygenase-1 (Ferreira et al. 2011). However, the mouse model of the sickling disorders is metabolically very different to the human sickle-cell traits. Furthermore, whereas such studies appear to support an immune component to the malaria protective effects of HbAS they are balanced by others that show no apparent differences between HbAS and normal subjects with regard to a range of immunological responses (Cornille-Brogger et al. 1979; Storey et al. 1979; Le Hesran et al. 1999; Tan et al. 2011), leaving the balance between innate and acquired mechanisms unresolved.

From the above discussion, it is clear that HbAS is associated with strong protection against all forms of clinical *P. falciparum* malaria but that the mechanism for this protection remains somewhat speculative. Although many plausible mechanisms have been proposed during the last 40 years, which (if any) provides the true explanation remains unresolved. One possibility is that there is no single mechanism for the protective effect of HbAS but that protection results from the composite of many of the mechanisms described or by different mechanisms in different populations; more work will be required, both in vitro and ex vivo, to resolve this question definitively.

Hemoglobin C

Fewer studies have focused on the malaria-protective effects of HbC than for HbS. The geographic range of HbC is considerably more limited than that of HbS, being centered on West and North-West Africa with the exception of a low-frequency corridor between West Africa and Egypt that appears to reflect patterns of human migration (FB Piel, AP Patil, RE Howes, et al., unpubl.). Although historically HbC has been restricted to malaria-endemic communities, to the best of our knowledge the relationship between the population frequency of HbC and the endemicity of malaria has not been formally tested. Early studies investigating the association between HbC and clinical protection were either inconclusive (Thompson 1962, 1963; Gilles et al. 1967; Guinet et al. 1997) or suggested only a marginal effect of HbC in comparison to the marked protection afforded by HbS (Ringelhann et al. 1976). This was confirmed most recently in both a large cross-sectional survey (Danquah et al. 2010) and in a birth cohort study (Kreuels et al. 2010), both conducted among children living in areas of high malaria transmission in Ghana. This may well be explained by more recent studies suggesting that protection might be specific to particular categories of strictly defined severe malaria (May et al. 2007) and that it is greater in homozygotes (with HbCC) than in heterozygotes (with HbAC) (Agarwal et al. 2000; Modiano et al. 2001; Mockenhaupt et al. 2004a). The strongest evidence for a homozygous advantage of HbC comes from a large case-control study conducted by Modiano and colleagues in Burkino Faso, in which they found a 29% reduction in the risk of clinical malaria among children with HbAC compared to a 93% reduction among children with HbCC (Modiano et al. 2001). Finally, like HbS, it appears that HbC has a marked effect on the transmissibility of malaria by the *Anopheles* vector (Gouagna et al. 2010).

Mechanistically, similar hypotheses have been pursued for HbC as for HbS. As for HbAS, early studies suggested that the protective effect of HbC might result from a reduced ability of *P. falciparum* parasites to grow and multiply in red blood cells containing HbC (Friedman et al. 1979; Pasvol and Wilson 1982; Olson and Nagel 1986; Fairhurst et al. 2003). Subsequently, however, it has been suggested that the mechanism might involve an immunological component (Ringelhann et al. 1976). This is supported by the finding of higher immune responses to PfEMP1 and various malaria antigens among children expressing the HbC allele in a low transmission urban area of Burkina Faso (Verra et al. 2007b), although no differences were found in a higher transmission rural population in the same study or in a more recent study conducted in Mali (Tan et al. 2011). As for HbAS, an alternative hypothesis has recently been developed: that HbC exerts its protection through a specific effect on cytoadherence, mediated by the altered display of surface-expressed parasite proteins (Fairhurst et al. 2005; Brittain et al. 2007; Cyrklaff et al. 2011).

Hemoglobin E

Much less is known about the effects of HbE on malaria than for the other common structural variants of hemoglobin. Like most of the other disorders, HbE is restricted to populations that have traditionally been endemic for malaria and there is some evidence for a correlation between the prevalence of the two conditions (Flatz et al. 1965; Kar et al. 1992; Win et al. 2005). Moreover, in one Thai population, it has recently been estimated on the basis of genetic linkage analysis that the most frequent variant, HbE $\beta 26$ Glu $\rightarrow$ lys, has reached its current frequency in <5000 years, a period compatible with the malaria hypothesis (Ohashi et al. 2004). Nevertheless, few clinical studies have reported the relative risk of clinical malaria in subjects with HbE. In one such study, conducted in adults with malaria who were admitted to the hospital in Thailand, the manifestations of malaria were less severe in patients with HbE trait (HbAE) (Chotivanich et al. 2002). However, in a similar study, Oo and colleagues found no significant relationship between either HbAE or HbEE and the severity of malaria in adults admitted to the hospital in Burma (Oo et al. 1995). Similarly, a small recent study found no relationship between HbE and the risk of cerebral malaria in Thai adults (Naka et al. 2008). To the best of our knowledge, the relative risk of malaria in such subjects has not been reported through case-control or cohort studies, although a single report from Sri Lanka found subjects with the clinically significant condition HbE/β-thalassemia to be at greater risk than normals from *P. vivax* malaria. This observation, which may be explained by the rapid turnover of red cells seen in patients with this condition—as *P. vivax* favors the youngest, most metabolically active red blood cells—suggests that patients with HbE/β-thalassemia should be actively protected from malaria (O'Donnell et al. 2009).

Very few studies have examined the potential mechanisms by which HbE might confer malaria protection. In one such study, Chotivanich et al. (2002) found that compared to normal subjects, a smaller proportion of red cells from subjects with HbAE were susceptible to invasion by *P. falciparum* parasites in culture, raising the possibility that HbAE might protect against severe malaria by limiting the ability of infections to achieve high parasite densities. The degree to which HbE protects against different forms of malaria and any potential mechanisms remain, therefore, a subject for further investigation.

THE THALASSEMIAS

There is strong evidence from population data that malaria selection explains the current distribution of the thalassemias. First, at a global scale, both α- and β-thalassemia follow remarkably similar distribution to that of malaria (Livingstone 1973, 1985; Cavalli-Sforza et al. 1994), an observation that also holds true at a microepidemiological scale. For example, in early studies, Siniscalco and colleagues noted a strong correlation between the population prevalence of β-thalassemia and the historic incidence of malaria among villages in Sardinia in southern Italy (Siniscalco et al. 1961), whereas Hill and colleagues drew similar conclusions in Melanesia (Hill et al. 1988). Similar observations have subsequently been made regarding the distribution of α-thalassemia in South Asia (Modiano et al. 1991), Tanzania (Enevold et al. 2007), and the Pacific (Flint et al. 1986), where in some populations these conditions have virtually reached fixation (Flint et al. 1998). Second, unlike many of the other disorders of hemoglobin, the molecular defects that result in the thalassemias are extremely diverse (see Higgs 2013), having arisen separately and been locally amplified, in multiple populations throughout the malaria-endemic regions of the world. The one notable exception is South America, where in evolutionary terms malaria was introduced only relatively recently.

Despite the overwhelming epidemiological evidence for malaria selection, the protective effect of β-thalassemia has been the subject of few clinical studies. Nevertheless, in one case-control study conducted in northern Liberia, Willcox and colleagues estimated a protective effect of ~50% (Willcox et al. 1983). Whereas many more studies have recorded, the clinical relationship between α-thalassemia and malaria, the results have not been entirely consistent. The incidence of uncomplicated clinical malaria has been lower in α-thalassemic than normal subjects in some cohort studies (Wambua et al. 2006; Enevold et al. 2008), whereas in others the incidence has either been equal (Lin et al. 2010; Veenemans et al. 2011) or even higher (Williams et al. 1996; Veenemans et al. 2011). Conversely, studies focused on severe malaria have shown consistent evidence for a strongly protective effect that is generally more marked in homozygous than heterozygous subjects (Allen et al. 1997; Mockenhaupt et al. 2004b; Williams et al. 2005d; Wambua et al. 2006; May et al. 2007). Available data, therefore, provide good evidence for a protective effect of α-thalassemia against severe and fatal malaria, but the data on uncomplicated malaria are somewhat perplexing. It seems likely that these observations are pointing to important clues about the mechanism by which α-thalassemia protects against severe malaria.

Several studies suggest that, as opposed to protecting against all forms of clinical malaria, α-thalassemia might specifically protect against the anemia that commonly complicates infections caused by *P. falciparum* malaria (Allen et al. 1997; Williams et al. 2005d; Wambua et al. 2006; May et al. 2007). Furthermore, unlike HbAS, there is little evidence that α-thalassemia has any effect at the level of malaria parasite density (Allen et al. 1997; Mockenhaupt et al. 2004b; Migot-Nabias et al. 2006; Wambua et al. 2006; Enevold et al. 2008; Veenemans et al. 2008, 2011). Such clinical observations suggest that the mechanisms by which α-thalassemia protects against malaria might well be different from those involved in HbAS. Two possible explanations for a specific effect against malaria-induced anemia have been put forward: first, that the higher red cell counts that are associated with α-thalassemia may buffer against the development of anemia during *P. falciparum* infections (Fowkes et al. 2008), and second, that this protection might be immunologically mediated (Veenemans et al. 2008). With regard to the latter, Veenemans and colleagues found that in two cohorts of children in East Africa, the greatest falls in hemoglobin occurred when malaria parasitemia was associated with inflammation in the form of a raise C-reactive protein concentration in blood and that α-thalassemia was strongly protective against this "inflammatory" form of the disease (Veenemans et al. 2008). A further characteristic of α-thalassemic red blood cells may provide an alternative explanation. In studies conducted in Papua New Guinea, Cockburn and colleagues noted that the expression of complement receptor-1 (CR1) is reduced on the surface of α-thalassemic red blood cells (Cockburn et al. 2004). CR1 is a ligand for rosetting in which uninfected red blood cells adhere to parasite-infected red cells in vitro (Figs. 3 and 4). Because rosetting has been associated with severe forms of malaria, including severe anemia, the reduced expression of CR1 might be relevant in the protection afforded by α-thalassemia. Finally, it is also possible that immune factors might be involved. In two clinical studies, the incidence of uncomplicated clinical malaria has been paradoxically higher in the youngest α-thalassemic children (Williams et al. 1996; Veenemans et al. 2011), raising the possibility that α-thalassemia might be acting like a natural vaccine, boosting exposure and immunity to malaria during early childhood. As for all the hemoglobin disorders, the story is more complicated than one might have thought at one point.

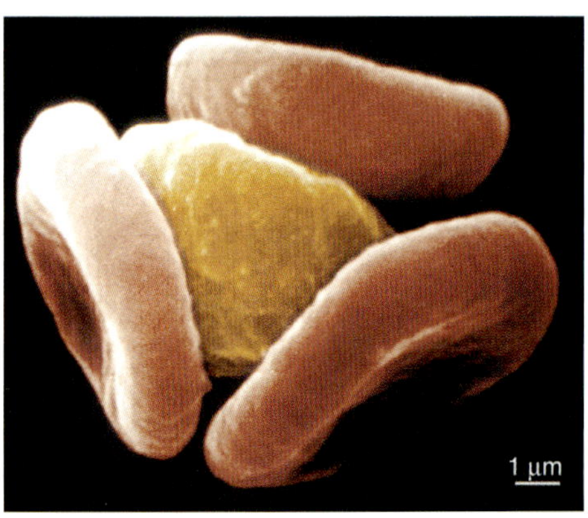

Figure 3. Three-dimensional photograph of normal red cells aggregating around a malaria-infected red cell to form a rosette. (Image supplied by D.J.P. Ferguson, Oxford University.)

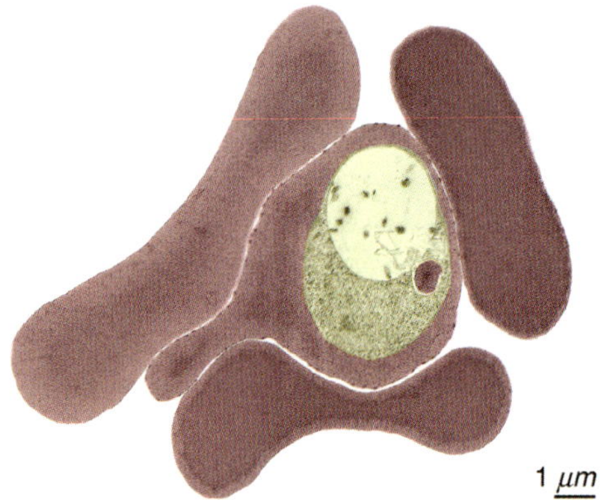

Figure 4. A cross-sectional photograph of a rosette as shown in Figure 3. (Image supplied by D.J.P. Ferguson, Oxford University.)

EPISTATIC INTERACTIONS BETWEEN THE HEMOGLOBIN DISORDERS

If, as discussed above, malaria is responsible for the Mendelian selection of many of the inherited disorders of hemoglobin, a question remains about why they have not all become common throughout the whole of the malaria-endemic world. This is a question that has been asked particularly from the perspective of pairs of hemoglobin disorders by a number of different investigators. Studies conducted both in Kenya and in Ghana have recently shown that when inherited alone, both HbAS and α-thalassemia are strongly protective against *P. falciparum* malaria but that the protective effects of each are lost when both conditions are inherited together (Williams et al. 2005c; May et al. 2007). From the perspective of HbAS, this may be explained by the changes in the intracellular concentration of HbAS that accompany the coinheritance of α-thalassemia (Brittenham et al. 1980). This negative interaction between the two malaria-protective phenotypes could well explain why both are held at intermediate frequencies in populations where both occur together (Williams et al. 2005c; Hedrick 2011; Penman et al. 2011). A similar hypothesis has more recently been advanced to explain the relative frequencies of HbAS and the α- and β-thalassaemias within the Mediterranean region (Penman et al. 2009).

CONCLUSION

There is no doubt that Haldane's "malaria hypothesis" has been fully vindicated over the years, although his suggested mechanisms for malaria protection in thalassemia carriers, that is, the small size of their red cells, has not stood the test of time. Further work is still required toward a better understanding of the protective mechanisms, information which is of potential value not just in the field of evolutionary biology but also for its potential in directing approaches for the better prevention and management of the different forms of malaria. Similarly, further work on the remarkable epistatic interactions between various malaria-protective polymorphisms should provide invaluable information about the mechanisms for the distribution of the different forms of inherited hemoglobin disorders, in particular high-frequency populations.

ACKNOWLEDGMENTS

We are extremely grateful to Liz Rose for her help in the preparation of this manuscript. We also thank D.J.P. Ferguson of Oxford University for supplying the photographs shown in Figures 3 and 4.

REFERENCES

**Reference is also in this collection.*

Abu-Zeid YA, Theander TG, Abdulhadi NH, Hviid L, Saeed BO, Jepsen S, Jensen JB, Bayoumi RA. 1992. Modulation of the cellular immune response during *Plasmodium falciparum* infections in sickle cell trait individuals. *Clin Exp Immunol* **88:** 112–118.

Agarwal A, Guindo A, Cissoko Y, Taylor JG, Coulibaly D, Kone A, Kayentao K, Djimde A, Plowe CV, Doumbo O, et al. 2000. Hemoglobin C associated with protection from severe malaria in the Dogon of Mali, a West African population with a low prevalence of hemoglobin S. *Blood* **96:** 2358–2363.

Aidoo M, Terlouw DJ, Kolczak MS, McElroy PD, ter Kuile FO, Kariuki S, Nahlen BL, Lal AA, Udhayakumar V. 2002. Protective effects of the sickle cell gene against malaria morbidity and mortality. *Lancet* **359:** 1311–1312.

Allen SJ, O'Donnell A, Alexander ND, Alpers MP, Peto TE, Clegg JB, Weatherall DJ. 1997. α^+-Thalassemia protects children against disease caused by other infections as well as malaria. *Proc Natl Acad Sci* **94:** 14736–14741.

Allison AC. 1954a. The distribution of the sickle-cell trait in East Africa and elsewhere, and its apparent relationship to the incidence of subtertian malaria. *Trans R Soc Trop Med Hyg* **48:** 312–318.

Allison AC. 1954b. Notes on sickle-cell polymorphism. *Ann Hum Genet* **19:** 39–51.

Allison AC. 1954c. Protection afforded by sickle cell trait against subtertian malarial infection. *Br Med J* **1:** 290–295.

Ayi K, Turrini F, Piga A, Arese P. 2004. Enhanced phagocytosis of ring-parasitized mutant erythrocytes. A common mechanism that may explain protection against falciparum-malaria in sickle-trait and β-thalassemia-trait. *Blood* **104:** 3364–3371.

Bayoumi RA, Abu-Zeid YA, Abdulhadi NH, Saeed BO, Theander TG, Hviid L, Ghalib HW, Nugud AH, Jepsen S, Jensen JB. 1990. Cell-mediated immune responses to *Plasmodium falciparum* purified soluble antigens in sickle-cell trait subjects. *Immunol Lett* **25:** 243–249.

Beet EA. 1946. Sickle cell disease in the Balovale District of Northern Rhodesia. *East Afr Med J* **23:** 75–86.

Beet EA. 1947. Sickle cell disease in northern Rhodesia. *East Afr Med J* **24:** 212–222.

Brain P. 1952. Sickle-cell anaemia in Africa. *Br Med J* **2:** 880.

Brittain NJ, Erexson C, Faucette L, Ward J, Fujioka H, Wellems TE, Fairhurst RM. 2007. Non-opsonising aggregates of IgG and complement in haemoglobin C erythrocytes. *Br J Haematol* **136:** 491–500.

Brittenham G, Lozoff B, Harris JW, Kan YW, Dozy AM, Nayudu NV. 1980. α globin gene number: Population and restriction endonuclease studies. *Blood* **55:** 706–709.

Cabrera G, Cot M, Migot-Nabias F, Kremsner PG, Deloron P, Luty AJ. 2005. The sickle cell trait is associated with enhanced immunoglobulin G antibody responses to *Plasmodium falciparum* variant surface antigens. *J Infect Dis* **191:** 1631–1638.

Carlson J, Nash GB, Gabutti V, al-Yaman F, Wahlgren M. 1994. Natural protection against severe *Plasmodium falciparum* malaria due to impaired rosette formation. *Blood* **84:** 3909–3914.

Cavalli-Sforza LL, Menozzi P, Piazza A. 1994. *The history and geography of human genes.* Princeton University Press, Princeton, NJ.

Cholera R, Brittain NJ, Gillrie MR, Lopera-Mesa TM, Diakite SA, Arie T, Krause MA, Guindo A, Tubman A, Fujioka H, et al. 2008. Impaired cytoadherence of *Plasmodium falciparum*-infected erythrocytes containing sickle hemoglobin. *Proc Natl Acad Sci* **105:** 991–996.

Chotivanich K, Udomsangpetch R, Pattanapanyasat K, Chierakul W, Simpson J, Looareesuwan S, White N. 2002. Hemoglobin E: A balanced polymorphism protective against high parasitemias and thus severe *P falciparum* malaria. *Blood* **100:** 1172–1176.

Christianson A, Howson CP, Modell B. 2006. *March of Dimes global report on birth defects*. March of Dimes Birth Defects Foundation, New York.

Cockburn IA, Mackinnon MJ, O'Donnell A, Allen SJ, Moulds JM, Baisor M, Bockarie M, Reeder JC, Rowe JA. 2004. A human complement receptor 1 polymorphism that reduces *Plasmodium falciparum* rosetting confers protection against severe malaria. *Proc Natl Acad Sci* **101:** 272–277.

Cornille-Brogger R, Fleming AF, Kagan I, Matsushima T, Molineaux L. 1979. Abnormal haemoglobins in the Sudan savanna of Nigeria. II: Immunological response to malaria in normals and subjects with sickle cell trait. *Ann Trop Med Parasitol* **73:** 173–183.

Cyrklaff M, Sanchez CP, Kilian N, Bisseye C, Simpore J, Frischknecht F, Lanzer M. 2011. Hemoglobins S and C interfere with actin remodeling in Plasmodium falciparum-infected erythrocytes. *Science* **334:** 1283–1286.

Danquah I, Ziniel P, Eggelte TA, Ehrhardt S, Mockenhaupt FP. 2010. Influence of haemoglobins S and C on predominantly asymptomatic Plasmodium infections in northern Ghana. *Trans R Soc Trop Med Hyg* **104:** 713–719.

Edozien JC, Boyo AE, Morley DC. 1960. The relationship of serum γ-globulin concentration to malaria and sickling. *J Clin Pathol J* **13:** 118–123.

Eeckels R, Gatti F, Renoirte AM. 1967. Abnormal distribution of haemoglobin genotypes in Negro children with severe bacterial infections. *Nature* **216:** 382.

Enevold A, Alifrangis M, Sanchez JJ, Carneiro I, Roper C, Borsting C, Lusingu J, Vestergaard LS, Lemnge MM, Morling N, et al. 2007. Associations between α^+-thalassemia and *Plasmodium falciparum* malarial infection in northeastern Tanzania. *J Infect Dis* **196:** 451–459.

Enevold A, Lusingu JP, Mmbando B, Alifrangis M, Lemnge MM, Bygbjerg IC, Theander TG, Vestergaard LS. 2008. Reduced risk of uncomplicated malaria episodes in children with α^+-thalassemia in northeastern Tanzania. *Am J Trop Med Hyg* **78:** 714–720.

Fairhurst RM, Fujioka H, Hayton K, Collins KF, Wellems TE. 2003. Aberrant development of *Plasmodium falciparum* in hemoglobin CC red cells: Implications for the malaria protective effect of the homozygous state. *Blood* **101:** 3309–3315.

Fairhurst RM, Baruch DI, Brittain NJ, Ostera GR, Wallach JS, Hoang HL, Hayton K, Guindo A, Makobongo MO, Schwartz OM, et al. 2005. Abnormal display of PfEMP-1 on erythrocytes carrying haemoglobin C may protect against malaria. *Nature* **435:** 1117–1121.

Ferreira A, Marguti I, Bechmann I, Jeney V, Chora A, Palha NR, Rebelo S, Henri A, Beuzard Y, Soares MP. 2011. Sickle hemoglobin confers tolerance to Plasmodium infection. *Cell* **145:** 398–409.

Flatz G, Pik C, Sringam S. 1965. Haemoglobin E and β-thalassaemia: Their distribution in Thailand. *Ann Hum Genet* **29:** 151–170.

Flint J, Hill AV, Bowden DK, Oppenheimer SJ, Sill PR, Serjeantson SW, Bana-Koiri J, Bhatia K, Alpers MP, Boyce AJ, et al. 1986. High frequencies of α-thalassaemia are the result of natural selection by malaria. *Nature* **321:** 744–750.

Flint J, Harding RM, Boyce AJ, Clegg JB. 1998. The population genetics of the haemoglobinopathies. *Baillieres Clin Haematol* **11:** 1–51.

Fowkes FJ, Allen SJ, Allen A, Alpers MP, Weatherall DJ, Day KP. 2008. Increased microerythrocyte count in homozygous α^+-thalassaemia contributes to protection against severe malarial anaemia. *PLoS Med* **5:** e56.

Foy H, Kondi A, Rebello A, Martins F. 1952. The distribution of sickle cell trait and the incidence of sickle cell anaemia in the Negro tribes of Portuguese East Africa. *East Afr Med J* **29:** 247–251.

Foy H, Kondi A, Timms GL, Brass W, Bushra F. 1954. The variability of sickle-cell rates in the tribes of Kenya and the Southern Sudan. *Br Med J* **1:** 294–297.

Friedman MJ. 1978. Erythrocytic mechanism of sickle cell resistance to malaria. *Proc Natl Acad Sci* **75:** 1994–1997.

Friedman MJ, Roth EF, Nagel RL, Trager W. 1979. The role of hemoglobins C, S, and Nbalt in the inhibition of malaria parasite development in vitro. *Am J Trop Med Hyg* **28:** 777–780.

* Fucharoen S, Weatherall DJ. 2012. The hemoglobin E thalassemias. *Cold Spring Harb Perspect Med* **2:** a011734.

Gilles HM, Fletcher KA, Hendrickse RG, Lindner R, Reddy S, Allan N. 1967. Glucose-6-phosphate-dehydrogenase deficiency, sickling, and malaria in African children in South Western Nigeria. *Lancet* **1:** 138–140.

Gouagna LC, Bancone G, Yao F, Yameogo B, Dabire KR, Costantini C, Simpore J, Ouedraogo JB, Modiano D. 2010. Genetic variation in human HBB is associated with *Plasmodium falciparum* transmission. *Nat Genet* **42:** 328–331.

Guggenmoos-Holzmann I, Bienzle U, Luzzatto L. 1981. *Plasmodium falciparum* malaria and human red cells. II: Red cell genetic traits and resistance against malaria. *Int J Epidemiol* **10:** 16–22.

Guinet F, Diallo DA, Minta D, Dicko A, Sissoko MS, Keita MM, Wellems TE, Doumbo O. 1997. A comparison of the incidence of severe malaria in Malian children with normal and C-trait hemoglobin profiles. *Acta Trop* **68:** 175–182.

Haldane JBS. 1949. The rate of mutation of human genes. *Hereditas* **35:** 267–273.

Hedrick PW. 2011. Selection and mutation for α thalassemia in nonmalarial and malarial environments. *Ann Hum Genet* **75:** 468–474.

* Higgs DR. 2013. The molecular basis of α- thalassemia. *Cold Spring Harb Perspect Med* **3:** a011718.

Hill AV, Bowden DK, O'Shaughnessy DF, Weatherall DJ, Clegg JB. 1988. β Thalassemia in Melanesia: Association with malaria and characterization of a common variant (IVS-1 nt 5 G---C). *Blood* **72:** 9–14.

Jallow M, Teo YY, Small KS, Rockett KA, Deloukas P, Clark TG, Kivinen K, Bojang KA, Conway DJ, Pinder M, et al. 2009. Genome-wide and fine-resolution association analysis of malaria in West Africa. *Nat Genet* **41:** 657–665.

Kaminsky R, Kruger N, Hempelmann E, Bommer W. 1986. Reduced development of *Plasmodium falciparum* in β-thalassaemic erythrocytes. *Z Parasitenkd* **72:** 553–556.

Kar S, Seth S, Seth PK. 1992. Prevalence of malaria in Ao Nagas and its association with G6PD and HbE. *Hum Biol* **64:** 187–197.

Kreuels B, Kreuzberg C, Kobbe R, Ayim-Akonor M, Apiah-Thompson P, Thompson B, Ehmen C, Adjei S, Langefeld I, Adjei O, et al. 2010. Differing effects of HbS and HbC traits on uncomplicated falciparum malaria, anemia, and child growth. *Blood* **115:** 4551–4558.

Kulozik AE, Wainscoat JS, Serjeant GR, Kar BC, Al-Awamy B, Essan GJF, Falusi AG, Haque SK, Hilali AM, Kate S, et al. 1986. Geographical survey of β^S-globin gene haplotypes: Evidence for an independent Asian origin of the sickle-cell mutation. *Am J Hum Genet* **39:** 239–244.

Lawaly YR, Sakuntabhai A, Marrama L, Konate L, Phimpraphi W, Sokhna C, Tall A, Sarr FD, Peerapittayamongkol C, Louicharoen C, et al. 2010. Heritability of the human infectious reservoir of malaria parasites. *PLoS ONE* **5:** e11358.

Le Hesran JY, Personne I, Personne P, Fievet N, Dubois B, Beyeme M, Boudin C, Cot M, Deloron P. 1999. Longitudinal study of *Plasmodium falciparum* infection and immune responses in infants with or without the sickle cell trait. *Int J Epidemiol* **28:** 793–798.

Lin E, Tavul L, Michon P, Richards JS, Dabod E, Beeson JG, King CL, Zimmerman PA, Mueller I. 2010. Minimal association of common red blood cell polymorphisms with *Plasmodium falciparum* infection and uncomplicated malaria in Papua New Guinean school children. *Am J Trop Med Hyg* **83:** 828–833.

Livingstone FB. 1973. *Data on the abnormal hemoglobin and glucose-6-phosphate dehydrogenase deficiency in human populations.* Museum of Anthropology, University of Michigan, Ann Arbor, MI.

Livingstone FB. 1985. *Frequencies of hemoglobin variants.* Oxford University Press, New York.

Luzzatto L, Nwachuku-Jarrett ES, Reddy S. 1970. Increased sickling of parasitised erythrocytes as mechanism of resistance against malaria in the sickle-cell trait. *Lancet* **1:** 319–321.

Makani J, Komba AN, Cox SE, Oruo J, Mwamtemi K, Kitundu J, Magesa P, Rwezaula S, Meda E, Mgaya J, et al. 2010. Malaria in patients with sickle cell anemia: Burden, risk factors, and outcome at the outpatient clinic and during hospitalization. *Blood* **115:** 215–220.

Marsh K, Otoo L, Hayes RJ, Carson DC, Greenwood BM. 1989. Antibodies to blood stage antigens of *Plasmodium falciparum* in rural Gambians and their relation to protection against infection. *Trans R Soc Trop Med Hyg* **83:** 293–303.

May J, Evans JA, Timmann C, Ehmen C, Busch W, Thye T, Agbenyega T, Horstmann RD. 2007. Hemoglobin variants and disease manifestations in severe falciparum malaria. *JAMA* **297:** 2220–2226.

McAuley CF, Webb C, Makani J, Macharia A, Uyoga S, Opi DH, Ndila C, Ngatia A, Scott JA, Marsh K, et al. 2010. High mortality from *Plasmodium falciparum* malaria in children living with sickle cell anemia on the coast of Kenya. *Blood* **116:** 1663–1668.

Migot-Nabias F, Pelleau S, Watier L, Guitard J, Toly C, De Araujo C, Ngom MI, Chevillard C, Gaye O, Garcia A. 2006. Red blood cell polymorphisms in relation to *Plasmodium falciparum* asymptomatic parasite densities and morbidity in Senegal. *Microbes Infect* **8:** 2352–2358.

Mockenhaupt FP, Ehrhardt S, Cramer JP, Otchwemah RN, Anemana SD, Goltz K, Mylius F, Dietz E, Eggelte TA, Bienzle U. 2004a. Hemoglobin C and resistance to severe malaria in Ghanaian children. *J Infect Dis* **190:** 1006–1009.

Mockenhaupt FP, Ehrhardt S, Gellert S, Otchwemah RN, Dietz E, Anemana SD, Bienzle U. 2004b. α^+ Thalassemia protects African children from severe malaria. *Blood* **104:** 2003–2006.

Modell B, Darlison M. 2008. Global epidemiology of haemoglobin disorders and derived service indicators. *Bull World Health Organ* **86:** 480–487.

Modiano G, Morpurgo G, Terrenato L, Novelletto A, Di Rienzo A, Colombo B, Purpura M, Mariani M, Santachiara-Benerecetti S, Brega A, et al. 1991. Protection against malaria morbidity: Near-fixation of the α-thalassemia gene in a Nepalese population. *Am J Hum Genet* **48:** 390–397.

Modiano D, Luoni G, Sirima BS, Simpore J, Verra F, Konate A, Rastrelli E, Olivieri A, Calissano C, Paganotti GM, et al. 2001. Haemoglobin C protects against clinical *Plasmodium falciparum* malaria. *Nature* **414:** 305–308.

Naka I, Ohashi J, Nuchnoi P, Hananantachai H, Looareesuwan S, Tokunaga K, Patarapotikul J. 2008. Lack of association of the HbE variant with protection from cerebral malaria in Thailand. *Biochem Genet* **46:** 708–711.

Odegbemi JO, Williams AI. 1995. Immunoglobulin levels in malaria infected Nigerians with and without abnormal haemoglobin. *Afr J Med Med Sci* **24:** 21–25.

O'Donnell A, Premawardhena A, Arambepola M, Samaranayake R, Allen SJ, Peto TE, Fisher CA, Cook J, Corran PH, Olivieri NF, et al. 2009. Interaction of malaria with a common form of severe thalassemia in an Asian population. *Proc Natl Acad Sci* **106:** 18716–18721.

Ohashi J, Naka I, Patarapotikul J, Hananantachai H, Brittenham G, Looareesuwan S, Clark AG, Tokunaga K. 2004. Extended linkage disequilibrium surrounding the hemoglobin E variant due to malarial selection. *Am J Hum Genet* **74:** 1198–1208.

Olson JA, Nagel RL. 1986. Synchronized cultures of *P. falciparum* in abnormal red cells: The mechanism of the inhibition of growth in HbCC cells. *Blood* **67:** 997–1001.

Oo M, Tin S, Marlar T, O'Sullivan WJ. 1995. Genetic red cell disorders and severity of falciparum malaria in Myanmar. *Bull World Health Organ* **73:** 659–665.

O'Shaughnessy DF, Hill AVS, Bowden DK, Weatherall DJ, Clegg JB. 1990. Globin genes in Micronesia: Origins and affinities of Pacific Island peoples. *Am J Hum Genet* **46:** 144–155.

Pasvol G, Wilson RJ. 1982. The interaction of malaria parasites with red blood cells. *Br Med Bull* **38:** 133–140.

Pasvol G, Weatherall DJ, Wilson RJ. 1978. Cellular mechanism for the protective effect of haemoglobin S against *P. falciparum* malaria. *Nature* **274:** 701–703.

Penman BS, Pybus OG, Weatherall DJ, Gupta S. 2009. Epistatic interactions between genetic disorders of hemoglobin can explain why the sickle-cell gene is uncommon in the Mediterranean. *Proc Natl Acad Sci* **106:** 21242–21246.

Penman BS, Habib S, Kanchan K, Gupta S. 2011. Negative epistasis between α^+ thalassaemia and sickle cell trait can explain interpopulation variation in South Asia. *Evolution* **65:** 3625–3632.

Piel FB, Patil AP, Howes RE, Nyangiri OA, Gething PW, Williams TN, Weatherall DJ, Hay SI. 2010. Global distribution of the sickle cell gene and geographical confirmation of the malaria hypothesis. *Nat Commun* **1:** 104.

Ringelhann B, Hathorn MK, Jilly P, Grant F, Parniczky G. 1976. A new look at the protection of hemoglobin AS and AC genotypes against *Plasmodium falciparum* infection: A census tract approach. *Am J Hum Genet* **28:** 270–279.

Roth EF Jr, Friedman M, Ueda Y, Tellez I, Trager W, Nagel RL. 1978. Sickling rates of human AS red cells infected *in vitro* with *Plasmodium falciparum* malaria. *Science* **202:** 650–652.

Scott JA, Berkley JA, Mwangi I, Ochola L, Uyoga S, Macharia A, Ndila C, Lowe BS, Mwarumba S, Bauni E, et al. 2011. Relation between falciparum malaria and bacteraemia in Kenyan children: A population-based, case-control study and a longitudinal study. *Lancet* **378:** 1316–1323.

Senok AC, Li K, Nelson EA, Yu LM, Tian LP, Oppenheimer SJ. 1997. Invasion and growth of *Plasmodium falciparum* is inhibited in fractionated thalassaemic erythrocytes. *Trans R Soc Trop Med Hyg* **91:** 138–143.

Serjeant GR. 2005. Mortality from sickle cell disease in Africa. *BMJ* **330:** 432–433.

Shear HL, Roth EF Jr, Fabry ME, Costantini FD, Pachnis A, Hood A, Nagel RL. 1993. Transgenic mice expressing human sickle hemoglobin are partially resistant to rodent malaria. *Blood* **81:** 222–226.

Siniscalco M, Bernini L, Latte B. 1961. Favism and thalassaemia in Sardinia and their relationship to malaria. *Nature* **190:** 1179–1180.

Storey J, Fleming AF, Cornille-Brogger R, Molineaux L, Matsushima T, Kagan I. 1979. Abnormal haemoglobins in the Sudan savanna of Nigeria. IV: Malaria, immunoglobulins and antimalarial antibodies in haemoglobin AC individuals. *Ann Trop Med Parasitol* **73:** 311–315.

Tan X, Traore B, Kayentao K, Ongoiba A, Doumbo S, Waisberg M, Doumbo OK, Felgner PL, Fairhurst RM, Crompton PD. 2011. Hemoglobin S and C heterozygosity enhances neither the magnitude nor breadth of antibody responses to a diverse array of *Plasmodium falciparum* antigens. *J Infect Dis* **204:** 1750–1761.

Thompson GR. 1962. Significance of haemoglobins S and C in Ghana. *Br Med J* **5279:** 682–685.

Thompson GR. 1963. Malaria and stress in relation to haemoglobins S and C. *Br Med J* **5363:** 976–978.

Udomsangpetch R, Sueblinvong T, Pattanapanyasat K, Dharmkrong-at A, Kittikalayawong A, Webster HK. 1993. Alteration in cytoadherence and rosetting of *Plasmodium falciparum*-infected thalassemic red blood cells. *Blood* **82:** 3752–3759.

Veenemans J, Andang'o PE, Mbugi EV, Kraaijenhagen RJ, Mwaniki DL, Mockenhaupt FP, Roewer S, Olomi RM, Shao JF, van der Meer JW, et al. 2008. α^+-Thalassemia protects against anemia associated with asymptomatic malaria: Evidence from community-based surveys in Tanzania and Kenya. *J Infect Dis* **198:** 401–408.

Veenemans J, Jansen EJ, Baidjoe AY, Mbugi EV, Demir AY, Kraaijenhagen RJ, Savelkoul HF, Verhoef H. 2011. Effect of α^+-thalassaemia on episodes of fever due to malaria and other causes: A community-based cohort study in Tanzania. *Malar J* **10:** 280.

Verra F, Bancone G, Avellino P, Blot I, Simpore J, Modiano D. 2007a. Haemoglobin C and S in natural selection against Plasmodium falciparum malaria: A plethora or a single shared adaptive mechanism? *Parassitologia* **49:** 209–213.

Verra F, Simpore J, Warimwe GM, Tetteh KK, Howard T, Osier FH, Bancone G, Avellino P, Blot I, Fegan G, et al. 2007b. Haemoglobin C and S role in acquired immunity against *Plasmodium falciparum* malaria. *PLoS ONE* **2:** e978.

Wambua S, Mwangi TW, Kortok M, Uyoga SM, Macharia AW, Mwacharo JK, Weatherall DJ, Snow RW, Marsh K, Williams TN. 2006. The effect of α^+-thalassaemia on the incidence of malaria and other diseases in children living on the coast of Kenya. *PLoS Med* **3:** e158.

Weatherall DJ. 2010. Thalassemia as a global health problem: Recent progress towards its control in the developing countries. *Ann NY Acad Sci* **1202:** 17–23.

Weatherall DJ, Clegg JB. 2001. *The thalassaemia syndromes.* Blackwell Science, Oxford.

Willcox M, Bjorkman A, Brohult J. 1983. Falciparum malaria and β-thalassaemia trait in northern Liberia. *Ann Trop Med Parasitol* **77:** 335–347.

Williams TN, Obaro SK. 2011. Sickle cell disease and malaria morbidity: A tale with two tails. *Trends Parasitol* **27:** 315–320.

Williams TN, Maitland K, Bennett S, Ganczakowski M, Peto TE, Newbold CI, Bowden DK, Weatherall DJ, Clegg JB. 1996. High incidence of malaria in α-thalassaemic children. *Nature* **383:** 522–525.

Williams TN, Mwangi TW, Roberts DJ, Alexander ND, Weatherall DJ, Wambua S, Kortok M, Snow RW, Marsh K. 2005a. An immune basis for malaria protection by the sickle cell trait. *PLoS Med* **2:** e128.

Williams TN, Mwangi TW, Wambua S, Alexander ND, Kortok M, Snow RW, Marsh K. 2005b. Sickle cell trait and the risk of *Plasmodium falciparum* malaria and other childhood diseases. *J Infect Dis* **192:** 178–186.

Williams TN, Mwangi TW, Wambua S, Peto TEA, Weatherall DJ, Gupta S, Recker M, Penman BS, Uyoga S, Macharia A, et al. 2005c. Negative epistasis between the malaria-protective effects of α^+-thalassemia and the sickle cell trait. *Nat Genet* **37:** 1253–1257.

Williams TN, Wambua S, Uyoga S, Macharia A, Mwacharo JK, Newton CR, Maitland K. 2005d. Both heterozygous and homozygous α^+ thalassemias protect against severe and fatal *Plasmodium falciparum* malaria on the coast of Kenya. *Blood* **106:** 368–371.

Win N, Lwin AA, Oo MM, Aye KS, Soe S, Okada S. 2005. Hemoglobin E prevalence in malaria-endemic villages in Myanmar. *Acta Med Okayama* **59:** 63–66.

The Molecular Basis of β-Thalassemia

Swee Lay Thein

Department of Haematological Medicine, King's College London School of Medicine/King's College Hospital NHS Foundation Trust, London SE5 9NU, United Kingdom

Correspondence: sl.thein@kcl.ac.uk

The β-thalassemias are characterized by a quantitative deficiency of β-globin chains underlaid by a striking heterogeneity of molecular defects. Although most of the molecular lesions involve the structural β gene directly, some down-regulate the gene through distal *cis* effects, and rare *trans*-acting mutations have also been identified. Most β-thalassemias are inherited in a Mendelian recessive fashion but there is a subgroup of β-thalassemia alleles that behave as dominant negatives. Unraveling the molecular basis of β-thalassemia has provided a paradigm for understanding of much of human genetics.

This article outlines the molecular mechanisms underlying the quantitative reduction in β-globin production. Mutations that completely inactivate the β gene resulting in no β-globin production cause β^0-thalassemia. Other mutations allow the production of some β globin, and depending on the degree of quantitative reduction in the output of the β chains, are classified as β^+- or β^{++}- ("silent") thalassemia. A quantitative reduction in β globin results in the accumulation of excess α-globin chains that are responsible for the pathophysiology of the disorder (Thein 2005). Thus, the severity of the phenotype is usually related to the degree of imbalance between α- and non-α-globin chain synthesis, and the size of the free α-chain pool. The primary determinant of β-thalassemia severity is the type of β allele (β^0, β^+, β^{++}), ameliorated by coinheritance of interacting α-thalassemia (which reduces the pool of free α chains) and coinheritance of an innate ability to increase production of γ chains. The additional γ-globin chains will partner the excess α globin to form fetal hemoglobin ($\alpha_2\gamma_2$).

Some structurally abnormal β-chain variants are also quantitatively reduced, with a phenotype of β-thalassemia, in which case they are referred to as "thalassemic hemoglobinopathies" (e.g., HbE [β26 Glu → Lys]) (Weatherall and Clegg 2001). Other β-globin chain variants, although synthesized in normal amounts, are extremely unstable and are not capable of forming stable hemoglobin tetramers, causing a functional deficiency of β globin, and a thalassemia phenotype (Thein 1999). These β alleles are rare and result in moderately severe anemia even when present in a single copy. They are thus inherited as dominant negatives unlike the common forms of β-thalassemias that are inherited as haploinsufficient Mendelian recessives where inheritance of two copies of β-thalassemia alleles are required to produce disease.

Almost all the mutations down-regulating the β-globin gene are physically linked to the

Copyright © 2013 Cold Spring Harbor Laboratory Press; all rights reserved
Cite this article as *Cold Spring Harb Perspect Med* doi: 10.1101/cshperspect.a011700

gene and behave as alleles of the β-globin locus (i.e., they are *cis*-acting), but mutations that alter β-globin gene expression and segregate independently of the β-globin cluster (i.e., *trans*-acting) have also been identified in recent years (Viprakasit et al. 2001; Yu et al. 2002; Cantor 2005).

Almost 300 β-thalassemia alleles have now been characterized (http://globin.cse.psu.edu). Unlike α-thalassemia, in which deletions in the α-globin gene cluster account for most of the mutations, the vast majority of β-thalassemias are caused by mutations involving one (or a limited number of nucleotides) within the β gene or its immediate flanking regions (Giardine et al. 2011). Figure 1 summarizes the mechanisms underlying β-thalassemia.

NONDELETION FORMS OF β-THALASSEMIA

These defects account for the vast majority of the β-thalassemia alleles (Thein and Wood 2009). They include single base substitutions, small insertions, or deletions within the gene or its immediate flanking sequences and affect almost every known stage of gene expression (Fig. 2). They are listed in Table 1 according to the mechanism by which they affect gene function: transcription, RNA processing or translation of β-globin mRNA.

Transcriptional Mutants

Point mutations involving the conserved DNA sequences that form the β-globin promoter (from 100 bp upstream to the site of the initiation of transcription, including the functionally important CACCC, CCAAT, and ATAA boxes) and the stretch of 50 nucleotides in the 5′ UTR have been identified in patients of different ethnic groups. In general, mutations of the β-globin gene promoter within the CCAAT box cause mild β-thalassemia phenotypes, consistent with the relatively minor deficit in β-globin production and minor globin-chain imbalance. This finding is also consistent with transient expression studies of the mutant genes in tissue culture cells, which show that the defects allow an output of β-globin mRNA ranging from 10% to 25% of normal (Treisman et al. 1983).

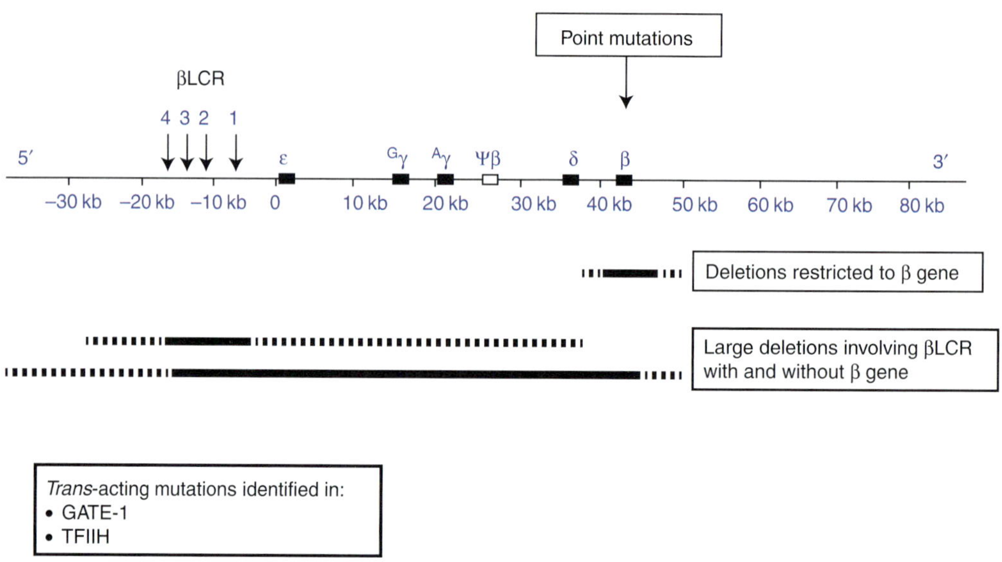

Figure 1. Mutations causing β-thalassemia. The *upper* panel depicts the β-globin gene cluster with the upstream locus control region (βLCR). The mutations can be *cis*-acting and include point mutations affecting the structural β gene, deletions restricted to the β gene, and large deletions involving the βLCR with or without the β gene. The dashed lines represent variations in the amount of flanking DNA removed by the deletions—detailed in Figures 3 and 4, respectively.

 Cite this article as *Cold Spring Harb Perspect Med* doi: 10.1101/cshperspect.a011700

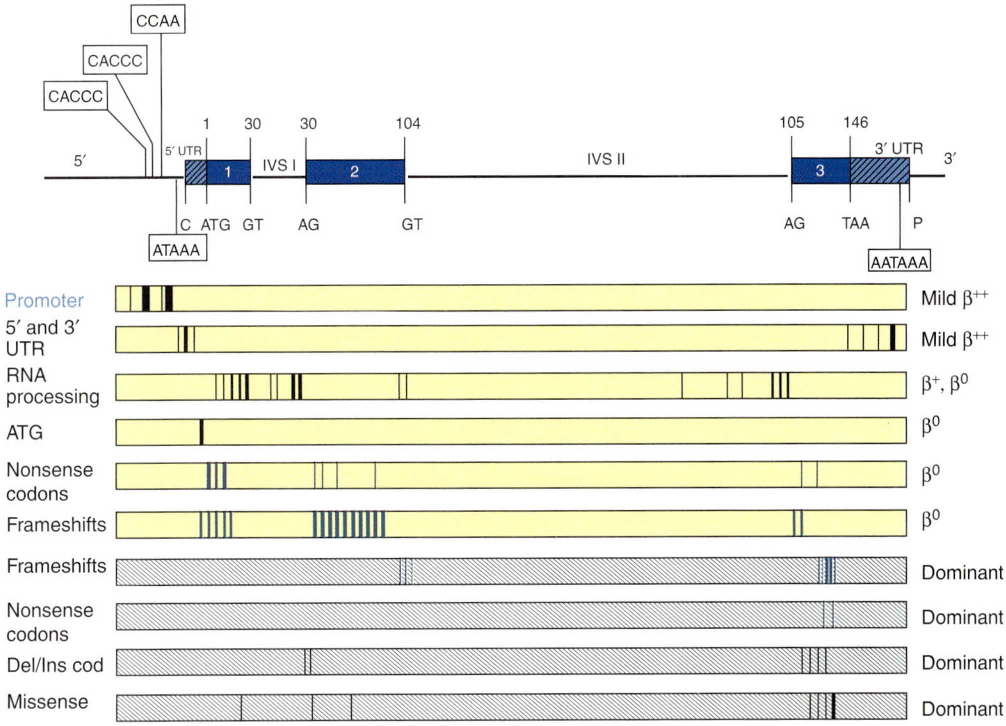

Figure 2. Point mutations causing β-thalassemia. The β-globin gene is depicted in the *upper* panel with conserved sequences in the 5′ and 3′ UTRs, and the invariant dinucleotides in exon–intron junctions of the gene, important in the control of gene expression. The boxes represent the different categories of mutants. The vertical lines within the boxes represent the sites of the different mutations. Dominantly inherited mutants are found within shaded boxes.

Some of these β-thalassemia alleles are so mild that heterozygotes (carriers) are "silent" with near normal red cell indices and HbA_2 levels, the only abnormality being an imbalanced globin-chain synthesis (Gonzalez-Redondo et al. 1989). Overall, the "silent" β-thalassemia alleles are uncommon except for the −101 C → T mutation, which has been observed fairly frequently in the Mediterranean region in which it interacts with a variety of more severe β-thalassemia mutations to produce milder forms of β-thalassemia (Maragoudaki et al. 1999).

Other "silent" mutations include those in the 5′ UTR. Because the description of the CAP +1 A-C allele (Wong et al. 1987), other molecular defects including single base substitutions and minor deletions distributed along the stretch of 50 nucleotides have been identified. As in the −101 C → T mutation, heterozygotes for this class of mutations are "silent"; the extremely mild phenotype is exemplified in a homozygote for the +1 A → C mutation who has the hematologic values of a thalassemia carrier (Wong et al. 1987). It is not known whether the CAP mutation causes β-thalassemia by decreasing β-globin gene transcription or by decreasing the efficiency of capping (posttranscriptional addition of m^7G) and mRNA translation. In vivo and in vitro studies show that the +33 C → G mutation leads to a reduction of β mRNA that is 33% of the output from a normal β gene, milder than the mutations involving the promoter elements (Ho et al. 1996).

Within this group of transcriptional mutants, ethnic variation in phenotype has been observed. Black individuals homozygous for the −29 A → G mutation have an extremely mild disease (Safaya et al. 1989), whereas a Chinese individual homozygous for the same

Table 1. Point mutations that cause β-thalassemia

Mutation	Type	Distribution
I. Transcriptional mutations		
Promoter regulatory elements		
1) −101 (C → T)	β^{++} (silent)	Mediterranean
2) −101 (C → G)	β^{++} (silent)	Ashkenazi Jew
3) −92 (C → T)	β^{++} (silent)	Mediterranean
4) −90 (C → T)	β^{+}	Portuguese
5) −88 (C → T)	β^{++}	U.S. Blacks, Asian Indians
6) −88 (C → A)	β^{+}	Kurds
7) −87 (C → G)	β^{++}	Mediterranean
8) −87 (C → T)	β^{++}	German, Italian
9) −87 (C → A)	β^{++}	U.S. Blacks
10) −86 (C → G)	β^{+}	Thai, Lebanese
11) −86 (C → A)	β^{++}	Italian
12) −73 (A → T)	β^{++}	Chinese
13) −32 (C → A)	β^{+}	Taiwanese
14) −32 (C → T)	β^{+}	Hispanic
15) −31 (A → G)	β^{+}	Japanese
16) −31 (A → C)	β^{+}	Italian
17) −30 (T → A)	β^{+}	Mediterranean, Bulgarian
18) −30 (T → C)	β^{+}	Chinese
19) −29 (A → G)	β^{+}	U.S. Blacks, Chinese
20) −29 (A → C)	β^{+}	Jordanian
21) −29 (G → A)	β^{+}	Turkish
22) −28 (A → C)	β^{+}	Kurds
23) −28 (A → G)	β^{+}	Blacks, SE Asians
24) −27 (A → T)	β^{+}	Corsican
25) −27 to −26 (−AA)	β^{+}	African American
26) −25 (G → C)	β^{+}	African American
5′ UTR		
27) CAP +1 (A → C)	β^{++} (silent)	Asian Indian
28) CAP +8 (C → T)	β^{++} (silent)	Chinese
29) CAP +10 (−T)	β^{++} (silent)	Greeks
30) CAP +20 (C → T)[a]	?	Bulgarian
31) CAP +22 (G → A)	β^{++}	Mediterranean, Bulgarian
32) CAP +33 (C → G)	β^{++} (silent)	Greek Cypriot
33) CAP +40 to +43 (−AAAC)	β^{+}	Chinese
34) CAP +45 (G → C)	β^{+}	Italian
II. RNA processing		
Splice junction		
1) IVS1-(−2) CD30 (AGG → **G**GG)	β^{0}	Sephardic Jews
2) IVS1-(−2) CD30 (AGG → **C**GG)	β^{0}	Italian Canadian
3) IVS1-(−1) CD30 (A**G**G → A**C**G) (Arg → Thr)	β^{0}	Mediterranean, U.S. Blacks, N. African, Kurds, UAE
4) IVS1-(−1) CD30 (A**G**G → A**A**G)	β^{0}	Bulgaria, UAE
5) IVS1-1 (G → A)	β^{0}	Mediterranean
6) IVS1-1 (G → T)	β^{0}	Asian Indian, SE Asian, Chinese
7) IVS1-1 (G → C)	β^{0}	Italian Canadian, Japanese
8) IVS1-2 (T → G)	β^{0}	Tunisian
9) IVS1-2 (T → C)	β^{0}	U.S. Blacks

Continued

Table 1. *Continued*

Mutation	Type	Distribution
10) IVS1-2 (T → A)	β^0	Algerian, Italian
11) IVS2-1 (G → A)	β^0	Mediterranean, U.S. Blacks
12) IVS2-1 (G → C)	β^0	Iranian
13) IVS2-2 (T → A)	? β^0	Turkish
14) IVS2-2 (−T)	β^0	Chinese
15) IVS1-3′ del 17 bp	β^0	Kuwaiti
16) IVS1-3′ end del 25 bp	β^0	Asian Indian, UAE
17) IVS1-3′ end del 44 bp	β^0	Mediterranean
18) IVS1-3′ end duplication 22 bp	β^0	Thai
19) IVS1-130 (G → C)	β^0	Italian, Japanese, UAE
20) IVS1-130 G → A	β^0	Egyptian
21) IVS1-130 (+1) CD30 (AG**G** → AG**C**) (Arg → Ser)	β^0	Middle East
22) IVS2-849 (A → G)	β^0	U.S. Blacks
23) IVS2-849 (A → C)	β^0	U.S. Blacks
24) IVS2-850 (G → C)	β^0	Yugoslavian
25) IVS2-850 (G → A)	β^0	N. European
26) IVS2-850 (G → T)	β^0	Japanese
27) IVS2-850 (−G)	β^0	Italian
Consensus splice sites		
28) IVS1-5 (G → C)	β^0	Asian Indian, SE Asian, Melanesian
29) IVS1-5 (G → T)	β^+	Mediterranean, N. European
30) IVS1-5 (G → A)[b]	β^+	Mediterranean, Algerian
31) IVS1-6 (T → C)	β^{++}	Mediterranean
32) IVS1- (−3) CD29 (GG**C** → GG**T**)	β^+	Lebanese
33) IVS1-128 (T → G)	β^+	Saudi Arabian
34) IVS1-129 (A → G)		German
35) IVS2-5 (G → C)	β^+	Chinese
36) IVS2-843 (T → G)	β^+	Algerian
37) IVS2-844 (C → G)	β^{++} (silent)	Italian
38) IVS2-844 (C → A)	β^{++} (silent)	Ghanaian
39) IVS2-848 (C → A)	β^+	UB Blacks, Egyptian, Iranian
40) IVS2-848 (C → G)	β^+	Japanese
Cryptic splice sites		
41) IVS1-110 (G → A)	β^+	Mediterranean
42) IVS1-116 (T → G)	β^0	Mediterranean
43) IVS2-654 (C → T)	β^0/β^+	Chinese, SE Asians, Japanese
44) IVS2-705 (T → G)	β^+	Mediterranean
45) IVS2-745 (C → G)	β^+	Mediterranean
46) IVS2-837 (T → G)	?	Asian Indian
47) CD10 (GCC → GCA)		Asian Indian
48) CD19 (AAC → AGC) Hb Malay (Asn → Ser)	β^{++}	SE Asian
49) CD24 (GGT → GGA)	β^{++}	U.S. Black, Japanese
50) CD26 (GAG → AAG) (Glu → Lys, Hb E)	β^+	SE Asian, European
51) CD26 (GAG → GCG) (Glu → Ala, Hb Tripoli)	β^+	Libyan

Continued

Table 1. *Continued*

Mutation	Type	Distribution
52) CD27 (GCC → TCC) (Ala → Ser, Knossos)[c]	β^{+}	Mediterranean
RNA cleavage—Poly A signal		
53) AATAAA → AACAAA	β^{++}	U.S. Blacks
54) AATAAA → AATGAA	β^{++}	Mediterranean
55) AATAAA → AATAGA	β^{++}	Malay
56) AATAAA → AATAAG	β^{++}	Kurd
57) AATAAA → AA–AA	β^{+}	French, U.S. Blacks
58) AATAAA → A——	β^{+}	Kurd, UAE
59) AATAAA → AAAAAA	β^{+}	Tunisian
60) AATAAA → CATAAA	β^{++} (silent)	Chinese
61) AATAAA → GATAAA	β^{+}	Czechoslovakian, Mediterranean Yugoslavian, Canadian
62) AATAAA → ——	β^{+}	Nigerian
Others in 3′ UTR		
63) Term CD +6, C → G	β^{++} (silent)	Greek
64) Term CD +90, del 13 bp	β^{++} (silent)	Turkish, Persian (Hamid and Akbari 2011)
65) Term CD +47 (C → G	β^{++}	Armenian
III. RNA translation		
Initiation codon		
1) ATG → GTG	β^{0}	Japanese
2) ATG → CTG	β^{0}	Northern Irish
3) ATG → ACG	β^{0}	Yugoslavian
4) ATG → AGG	β^{0}	Chinese
5) ATG → AAG	β^{0}	N. European
6) ATG → ATC	β^{0}	Japanese
7) ATG → ATA	β^{0}	Italian, Swedish
8) ATG → ATT	β^{0}	Iranian
9) 45 bp insertion (−22 to +23)	?	Maori, Polynesian
Nonsense codons		
1) CD6 GAG → TAG	β^{0}	Brazilian
2) CD7 GAG → TAG	β^{0}	English
3) CD15 TGG → TAG	β^{0}	Asian Indian, Japanese
4) CD15 TGG → TGA	β^{0}	Portuguese, Japanese
5) CD17 AAG → TAG	β^{0}	Chinese, Japanese
6) CD22 GAA → TAA	β^{0}	Reunion Island
7) CD26 GAG → TAG	β^{0}	Thai
8) CD35 TAC → TAA	β^{0}	Thai
9) CD37 TGG → TGA	β^{0}	Saudi Arabian
10) CD39 CAG → TAG	β^{0}	Mediterranean
11) CD43 GAG → TAG	β^{0}	Chinese, Thai
12) CD59 AAG → TAG	β^{0}	Italian-American
13) CD61 AAG → TAG	β^{0}	Black
14) CD90 GAG → TAG	β^{0}	Japanese
15) CD112 TGT → TGA	β^{0}	Slovenian
16) CD121 GAA → TAA	β^{0}	Czechoslovakian

Continued

Table 1. *Continued*

Mutation	Type	Distribution
Frameshift		
1) CD1 −G	β^0	Mediterranean
2) CD2/3/4 (−9 bp, +31 bp)	β^0	Algerian
3) CD2−4, 5−9, 7, 10	β^0	Algerian
4) CD5-CT	β^0	Mediterranean
5) CD6 −A	β^0	Mediterranean, U.S. Blacks
6) CD8 −AA	β^0	Mediterranean
7) CD8/9 +G	β^0	Asian Indian, Japanese
8) CD9 +TA	? β^0	Tunisian
9) CD9/10 +T	β^0	Greek, Arab
10) CD11 −T	β^0	Mexican
11) CD14/15 +G	β^0	Chinese
12) CD15 −T	β^0	Malay
13) CD15/16 −G	β^0	German
14) CD15/16 +G	β^0	Chinese
15) CD16 −C	β^0	Asian Indian
16) CD22/23/24 −7 bp (−AAGTTGG)	β^0	Turkish
17) CD24 −G; +CAC	β^0	Egyptian
18) CD24/25 −GGT	?	No additional information
19) CD25/26 +T	β^0	Tunisian
20) CD26 +T	β^0	Japanese
21) CD27/28 +C	β^0	Chinese, Thai
22) CD28 −C	β^0	Egyptian
23) CD28/29 −G	β^0	Japanese, Egyptian
24) CD31 −C	β^0	Chinese
25) CD35 −C	β^0	Malay
26) CD36/37 −T	β^0	Kurd, Iranian
27) CD37/38/39 del 7 bp (−GACCCAG)	β^0	Turkish
28) CD38/39 −C	β^0	Czechoslovkian
29) CD38/39 −CC	β^0	Belgian
30) CD40 −G	β^0	Japanese
31) CD40/41 +T	β^0	Chinese
32) CD41 −C	β^0	Thai
33) CD41/42 −TTCT	β^0	Chinese, SE Asian, Indian
34) CD42/43 +T	β^0	Japanese
35) CD42/43 +G	β^0	Japanese
36) CD44 −C	β^0	Kurdish
37) CD45 −T	β^0	Pakistani
38) CD45 +T	β^0	Turkish
39) CD47 +A	β^0	Surinamese
40) CD47/48 +ATCT	β^0	Asian Indian
41) CD49 −C	β^0	Jordanian
42) CD51 −C	β^0	Hungarian
43) CD53/54 +G	β^0	Japanese
44) CD54 −T	β^0	Swedish
45) CD54/58 (−T ATG GGC AAC CCT)	β^0	Chinese (Li et al. 2009)
46) CD55 −A	β^0	Asian Indian

Continued

Table 1. *Continued*

Mutation	Type	Distribution
47) CD54/55 +A	β^0	Asian Indian
48) CD56-60 +14 bp	β^0	Iranian
49) CD57/58 +C	β^0	Asian Indian
50) CD59 −A	β^0	Italian
51) CD62/63/64 del 7 bp (−TCATGGC)	β^0	Asian Indian
52) CD64 −G	β^0	Swiss
53) CD67 −TG	β^0	Filipino
54) CD71/72 +T	β^0	Chinese
55) CD71/72 +A	β^0	Chinese
56) CD72/73 −AGTGA, +T	β^0	British
57) CD74/75 −C	β^0	Turkish
58) CD76 GCT → −T	β^0	North African
59) CD76 −C	β^0	Italian
60) CD82/83 −G	β^0	Czech, Azerbaijani
61) CD81-87 (−22 bp)	β^0	Asian Indian
62) CD83-86 del 8 bp (−CACCTTTG)	β^0	Japanese
63) CD84/85 +C	β^0	Japanese
64) CD84/85/86 +T	β^0	Japanese
65) CD88 +T	β^0	Asian Indian
66) CD88 −TG	β^0	Japanese
67) CD89/90 −GT	β^0	Japanese
68) CD95 +A	β^0	SE Asian
69) CD106/107 +G	β^0	U.S. Black, Egyptian
70) CD109 (GTG → GT−)	?	Irish
71) CD120/121 +A[d]	β^0	Philippino
72) CD130/131 +GCCT	? β^0	German
73) CD142/143 (−CC)	?	French Caucasian

References to these mutations can be found in Forget (2001), Weatherall and Clegg (2001), and Thein and Wood (2009).
[a]Occurs in *cis* to the IVS2-745 C → G mutation.
[b]Also occurs in *cis* to 7201 bp deletion involving δ gene.
[c]Occurs in *cis* to δ59 −A.
[d]This frameshift leads to predicted truncated β variant of 138aa with an abnormal carboxyl terminal. Heterozygotes do not appear to have an unusually severe phenotype.

mutation had thalassemia major (Huang et al. 1986). The cause of this striking difference in phenotype is not known but may be related to the different chromosomal backgrounds on which the apparently identical mutations have arisen or to the C-T polymorphism at position −158 upstream of the $^{G}\gamma$-globin gene (*Xmn*1-$^{G}\gamma$ site), which is associated with increased fetal hemoglobin production under conditions of erythropoietic stress. The *Xmn*1-$^{G}\gamma$ site is present in the β chromosome carrying the −29 A → G mutation in Blacks but absent in that of the Chinese.

Mutants Affecting RNA Processing

A wide variety of mutations interfere with processing of the primary mRNA transcript. Those that affect the invariant dinucleotides GT or AG at the exon–intron splice junction completely abolish normal splicing and produce the phenotype of β^0-thalassemia. These mutations can be base substitutions that change one or the other of invariant dinucleotides or short deletions that remove them. Flanking the invariant dinucleotides are sequences that are fairly well conserved and a consensus sequence can be recognized at the exon–intron boundaries. They encompass the last three nucleotides of the exon and the first six nucleotides of the intron for the 5′ donor site; and the last 10 nucleotides of the intron and the first nucleotide of the exon for the 3′ acceptor site.

Mutations within the consensus sequences at the splice junctions reduce the efficiency of normal splicing to varying degrees and produce a β-thalassemia phenotype that ranges from mild to severe. For example, mutations at position 5 IVS1 G → C, T or A, considerably reduce splicing at the mutated donor site compared with normal. The mutations appear to activate the use of three "cryptic" donor sites, two in exon 1 and one in IVS1, which are used preferentially to the mutated donor site (Treisman et al. 1983). On the other hand, the substitution of C for T in the adjacent nucleotide, intron 1 position 6, only mildly affects normal RNA splicing even though it activates the same three cryptic donor sites as seen in the IVS1-5 mutants. Although the IVS1-6 T → C mutation is generally associated with milder β-thalassemia, studies have shown that in some cases, apparently identical mutations can be severe; and again this is presumably related to the chromosomal background on which the mutations have arisen (Rund et al. 1997).

Both exons and introns also contain "cryptic" splice sites, which are sequences very similar to the consensus sequence for a splice site but are not normally used. Mutations can occur in these sites creating a sequence that resembles more closely the normal splice site. During RNA processing the newly created site is used preferentially, leading to aberrant splicing. Such mutations have been identified in both introns 1 and 2, and exon 1 (Tables 1 and 2). The associated phenotype may be either β^+- or β^0- thalassemia, depending on the site and nature of the mutation.

The IVS1-110 G to A mutation was the first base substitution identified in a β-thalassemia gene (Spritz et al. 1981; Westaway and Williamson 1981). It is one of the most common forms of β-thalassemia in the Mediterranean population. The G to A substitution creates an alternative acceptor AG, 19 bp 5′ to the normal acceptor AG of IVS1. In vitro expression studies have shown that this newly created alternative splice site is preferentially used in 80% to 90% of the transcripts (Busslinger et al. 1981). The mutant mRNA is hardly detected in affected erythroid cells presumably because the 19 bp segment of retained intronic sequence contains an in-phase premature termination codon. Only 10%–20% of the transcripts are normally spliced, hence the severe β^+-thalassemia. In the case of the T → G substitution in position 116 of IVS1, however, all transcripts are aberrantly spliced utilizing the newly created 3′ acceptor site, resulting in little or no normal β mRNA and a β^0-thalassemia phenotype (Metherall et al. 1986).

Other β-thalassemia genes have substitutions within intron 2 that generate new donor sites. They include the IVS2 position 654 C → T, 705 T → G, and 745 C → G (Orkin et al. 1982a; Dobkin et al. 1983; Cheng et al. 1984). In each case, an identical upstream acceptor site at position 579 is activated such that the normal 5′ donor site at exon 2/IVS2 is spliced to the activated site at position 579 while the newly created donor site is spliced to the normal 3′ acceptor site at IVS1/exon 3. This two-stage splicing results in the retention of 73 bp, 121 bp and 151 bp of IVS2 in the misspliced β mRNA for the IVS2-654, IVS2-705, and IVS2-745 mutations, respectively. Variable amounts of splicing from the normal donor to the normal acceptor also occurs, resulting in phenotypes that range from β^+- to β^0- thalassemia. The β^0 versus β^+ phenotype associated with these different mutations must be related to different affinities of the enzymatic splicing complex for a given mutant splice site versus the normal splice sites.

Four mutations have been identified in exon 1 that are associated with activation of alternative splice sites. Three of these mutations modify the cryptic splice site spanning codons 24 to 27 in exon 1 so that it more closely resembles the consensus splice sequence AAGGTGAGT. The codon 24 GGT-GGA mutation is translationally silent (Goldsmith et al. 1983), whereas codon 26 GAG-AAG and codon 27 GCC-TCC result in the β^E and $\beta^{Knossos}$ variants, respectively (Orkin et al. 1982b, 1984). The mutation in codons 26 and 27 lead to a minor use of the alternative

Table 2. Dominantly inherited β-thalassemia alleles

Mutations	Exon	Hb variant	Ethnic group
I. Missense mutations			
1) CD 28 (CTG → CGG) Leu to Arg	1	Hb Chesterfield[a]	English
2) CD 32 (CTG → CAG) Leu to Glu in *cis* with CD 98 (GTG → ATG) Val to Met, Hb Köln	2	Hb Medicine Lake[a]	U.S. Caucasian
3) CD 60 (GTG → GAG) Val to Glu	2	Hb Cagliari[a]	Italian
4) CD 106 (CTG → CGG) Leu to Arg	3	Hb Terre Haute	N. European, French
5) CD 110 (CTG → CCG) Leu to Pro	3	Hb Showa-Yakushiji	Japanese
6) CD 114 (CTG → CCG) Leu to Pro	3	Hb Durham NC/Hb Brescia[a]	U.S. Irish, Italian
7) CD 115 (GCC → GAC) Ala to Asp	3	Hb Hradec Kralove	Czech
8) CD 127 (CAG → CCG) Gln to Pro	3	Hb Houston	U.S. English
9) CD 127 (CAG → CGG) Gln to Arg	3	Hb Dieppe	French
10) CD 128 Ala to Pro	3	Hb Mont Saint Aignan	French Caucasian
II. Deletion or insertion of intact codons → destabilization			
1) CD 3 (+T), CD5 (−C) Leu-Thr-Pro to Ser-Asp-Ser	1	Hb Antalya	Turkish
2) CD 30-31 (+CGG) +Arg	1/2		Spanish
3) CD 32-34 (−GGT) Val-Val to Val	2	Hb Korea[a]	Korean
4) CD 33-35 (−TGGTCT) Val-Val-Tyr to 0-0-Asp	2	Hb Dresden	German
5) CD 108-112 (−12 bp) Asn-Val-Leu-Val-Cys to Ser	3		Swedish
6) CD 124-126 (+CCA) +Pro	3		Armenian
7) CD 127-128 (−AGG) Gln-Ala to Pro	3	Hb Gunma	Japanese
8) CD 134-137 (−12, +6) Val-Ala-Gly-Val to Gly-Arg	3		Portuguese
9) CD 137-139 (−TGGCTA) Val-Ala-Asn to Asp	3	Hb Stara Zagnora[a]	Bulgarian
III. Premature termination			
1) CD 121 (GAA → TAA) Glu to Term (120aa)[b,c]	3		N. European Japanese
2) CD 127 (CAG → TAG) Gln to Term (127aa)	3		English, French
IV. Frameshift mutations			
1) IVSII: 2,3 (+11, −2)	IVSII		Iranian
2) IVSII: 4, 5 (−AG) → aberrant splicing	IVSII		Portuguese
3) IVSII: 535 to CD108 (+23, −310, +28) to CD105-108 Leu-Leu-Glu-Asn to Val-Pro-Ser-Val-Thr-Leu-Phe-Phe-Asp	IVSII (Exon 3)	Hb Jambol[a]	Bulgarian
4) CD 91 (CTG → CG) → 156aa	2	Hb Morgantown	Irish
5) CD 94 (+TG) → 156aa	2	Hb Agnana[a]	S. Italian
6) CD 100 (−CTT, +TCTGAGAACTT) → 158aa			S. African
7) CD 104 (AGG → AG) → 156aa[c]	2		German Caucasian
8) CD 109 (GTG → TG) → 156aa	3	Hb Manhatten	Lithuanian
9) CD 113 (GTG → TG) → 156aa	3		Canadian, N. European
10) CD 114 (−CT, +G) → 156aa	3	Hb Geneva	Swiss-French
11) CD 118 (−T)	3	Hb Sainte Seve	French Caucasian
12) CD 120-121 (+a) → 138aa	3		Philippino

Continued

Table 2. *Continued*

Mutations	Exon	Hb variant	Ethnic group
13) CD 123 (−A) → 156aa	3	Hb Makabe	Japanese
14) CD 123-125 (−ACCCCACC) → 135aa	3	Hb Khon Kaen[d]	Thai
15) CD 124 (−A) → 156aa	3		Russian
16) CD 124-126 (+CCA) → Pro-Pro-Val to Pro-Pro-Val	3		Armenian
17) CD 125 (−A) → 156aa	3		Japanese
18) CD 126 (−T) → 156aa	3	Hb Vercelli[a]	N. Italian
19) CD 126-131 (−17 bp) → 132aa	3	Hb Westdale[d]	Asian Indian, Pakistani
20) CD 128-129 (−4, +5, −11) → 153aa	3		Irish
21) CD 131-132 (−GA) → 138aa	3		Swiss
22) CD 131-134 (−11 bp) → 134aa	3		Spanish
23) CD 140-141 (−C) → 156aa	3	Hb Florida	Argentinian Spanish

Some of these variants are not associated with elevated A_2 in heterozygous state (e.g., Hb Dresden, Hb Jambol, Hb Morgantown, Hb Stara Zagnora). References to these mutations can be found in Thein (2001) and Thein and Wood (2009).

[a]Spontaneous mutations.

[b]Several families reported including one spontaneous mutation.

[c]Coinheritance of extra α-globin gene (genotype ααα/αα) contributed to unusually severe thal intermedia in proband.

[d]Difficult to evaluate phenotypes of heterozygotes as only homozygote and compound heterozygotes reported.

pathway so that there is a reasonable level of normally spliced products that result in the mild β^+-thalassemic phenotype of the β^E and $\beta^{Knossos}$ alleles, respectively.

Similarly, the A → G mutation in codon 19 activates another cryptic donor site spanning codons 17 to 19 resulting in a reduced level of normally spliced β mRNA with the codon 19 mutation encoding Hb Malay (Yang et al. 1989).

Mutations Causing Abnormal Posttranscriptional Modification

The nascent precursor globin mRNA molecule has to be modified at both of its ends to be functional; a methylated (m^7G) cap structure is added at the 5′ end, and a string of adenylic acid residues (poly-A) added at the 3′ end. Proper cleavage of the primary RNA transcript and polyadenylation of the 3′ end of the mRNA is guided by a consensus hexanucleotide sequence (AATAAA) about 20 nucleotides upstream of the poly-A tail. Mutations affecting the AATAAA sequence include seven base substitutions at different locations; two short deletions of 2 and 5 bp each, and one deletion of the total AATAAA sequence (Table 1). These mutations markedly decrease the efficiency of the cleavage-polyadenylation process, only about 10% of the mRNA is properly modified (Orkin et al. 1985). Therefore, the associated phenotype is that of β^+-thalassemia of moderate severity. The remainder of the transcripts extend far beyond the normal polyadenylation site and are probably cleaved and polyadenylated after the next AATAAA consensus sequences, which occur about 0.9–3 kb downstream. Mutations affecting other sites in the 3′ UTR, a C → G substitution at nucleotide 6, and a 13 bp deletion at nucleotides 90 downstream from the termination codon, also result in β^+-thalassemia (Rund et al. 1992; Hamid and Akbari 2011).

Mutants Affecting Translation of β-Globin mRNA

Mutations Affecting the Initiation Codon

Mutations affecting the initiation codon (ATG) all produce β^0-thalassemia (Table 1). One mutation involves an insertion of 45 bp between positions −22 to +23, thus affecting the initiation codon. The rest are single base substitutions, two affecting the first (A), three the second (T) and, three the third (G) nucleotide of ATG (Jankovic et al. 1990; Lam et al. 1990; Hattori et al. 1991; Saba et al. 1992; Ohba et al. 1997; Forget 2001; Blacklock et al. 2005; Thein and Wood 2009). It is theoretically possible for mutant β mRNAs to be initiated at the next downstream initiation codons, which are located at codons 21 and 22, or codon 55. However, it is predicted that these alternative initiation codons would result in premature termination, and that mutant mRNAs would be nonfunctional and subjected to nonsense mediated decay surveillance.

Premature Termination Codons

Approximately half the β-thalassemia alleles result from the introduction of premature termination codons, either because of direct mutations creating a stop codon or a change in the reading frame by insertion or deletion of a single to a few nucleotides. These frameshifts lead to premature termination further downstream when the next nonsense codon is reached.

One of the first nonsense mutations to be characterized and extensively studied was the mutation at codon 39 (CAG to TAG) (Humphries et al. 1984; Takeshita et al. 1984; Huang and Benz 2001). This mutation is the second most common cause of β-thalassemia in the Mediterranean population and accounts for most of the cases of β-thalassemia in Sardinia. An interesting feature of this and other nonsense mutations is the finding of very low levels of the mutant β-globin mRNA in affected erythroid cells. Initial studies of this phenomenon revealed that the gene was transcribed normally, but there appeared to be defective β-mRNA stability in the nucleus or defective processing and/or transport of the mRNA from nucleus to cytoplasm; mRNA stability in the cytoplasm appeared to be normal. It is now clear that reduced levels of the mutant β mRNA is a result of the nonsense-mediated decay quality-control mechanism (Isken and Maquat 2007; Schoenberg and Maquat 2012).

The frameshift and nonsense mutations that cause recessively inherited β^0-thalassemia typically result in premature termination within exon 1 and 2 with minimal steady-state levels of abnormal β mRNA (Fig. 2). In heterozygotes for such cases, no β chain is produced from the mutant allele, resulting in a typical asymptomatic phenotype (Hall and Thein 1994). In contrast, the mutant mRNA associated with mutations that produce in-phase termination later in the β sequence, in exon 3, does not undergo nonsense-mediated decay and presumably, gets translated into variant β chains. The abnormal β chains together with the concomitant excess α chains overcome the proteolytic machinery of the cell, increasing the ineffective erythropoiesis resulting in a severe phenotype. These mutants are usually dominantly inherited, in contrast to the typical recessive inheritance of β-thalassemia (see below).

GENE DELETIONS

In contrast to α-thalassemia, the β-thalassemias are rarely caused by major gene deletions with two exceptions: a group of deletions that are restricted to the β-globin gene (Fig. 3) and a second group of larger deletions affecting the upstream βLCR with or without the β-globin gene (Fig. 4).

Deletions Restricted to the β-Globin Gene

Deletions affecting only the β-globin gene ranged from 105 bp to ~67 kb in size (Fig. 3). The phenotype associated with these deletions is that of β^0-thalassemia (Thein and Wood 2009).

Two deletions remove the 3′ end but preserve the integrity of the 5′ end of the β-globin gene. The 0.6 kb deletion involving the 3′ end of the β-globin gene is a relatively common cause of β^0-thalassemia in Asian Indians and accounts for about one-third of the cases of β-thalassemia in this population (Thein et al. 1984; Varawalla et al. 1991). The second deletion was recently described in compound heterozygosity with HbS in a woman from Cape Verde Islands who presented with sickle cell disease (Andersson et al. 2007). The deletion removes 7.7 kb, starting in IVS2 of the β-globin gene and extending 7.1 kb downstream, in the midst of the *Kpn I* family of L1 repeat elements. The other dele-

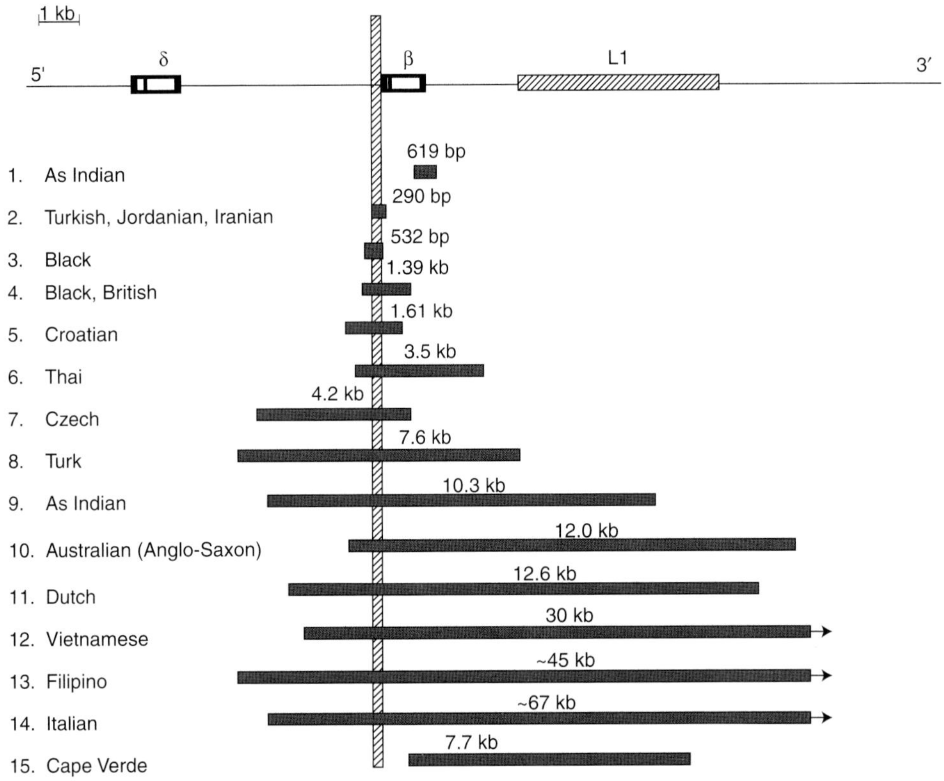

Figure 3. β-thalassemia deletions that are restricted to the gene. The vertical bar indicates the promoter region that is removed in common by these deletions except for the 619 bp Asian Indian and the 7.7 kb Cape Verde deletions.

tions differ widely in size, but remove in common a region in the β promoter (from positions −125 to +78 relative to the mRNA cap site), which includes the CACCC, CCAAT, and TATA elements. They are associated with unusually high levels of HbA_2 and variable increases of HbF in heterozygotes. It has been proposed that deletion of the β promoter removes competition for the upstream βLCR and limiting transcription factors, allowing greater interaction of the LCR with the *cis* δ and γ genes, thus enhancing their expression. Indeed, studies of an individual heterozygous for the 1.39 kb β-thalassemia deletion and a δ-chain variant showed that there is a disproportionate increase of HbA_2 derived from the δ gene in *cis* to the β-gene deletion (Codrington et al. 1990). The β promoter can also be inactivated by point mutations and again, carriers have unusually high HbA_2 and HbF levels (Huisman 1997). Although the increases in Hb F are variable and modest in heterozygotes for such mutations, they can be sufficiently increased to partially compensate for the complete absence of β globin in homozygotes; two individuals homozygous for different deletions in this group are not transfusion dependent with a mild disease (Schokker et al. 1966; Gilman 1987; Craig et al. 1992).

Upstream Deletions and $(\varepsilon\gamma\delta\beta)^0$-Thalassemia

This group of deletions affects the upstream regulatory locus control region (βLCR) and expression of the β-globin gene is down-regulated together with all the linked globin genes in the cluster on chromosome 11p, as part of $(\varepsilon\gamma\delta\beta)^0$-thalassemia (Fig. 4). The deletions are

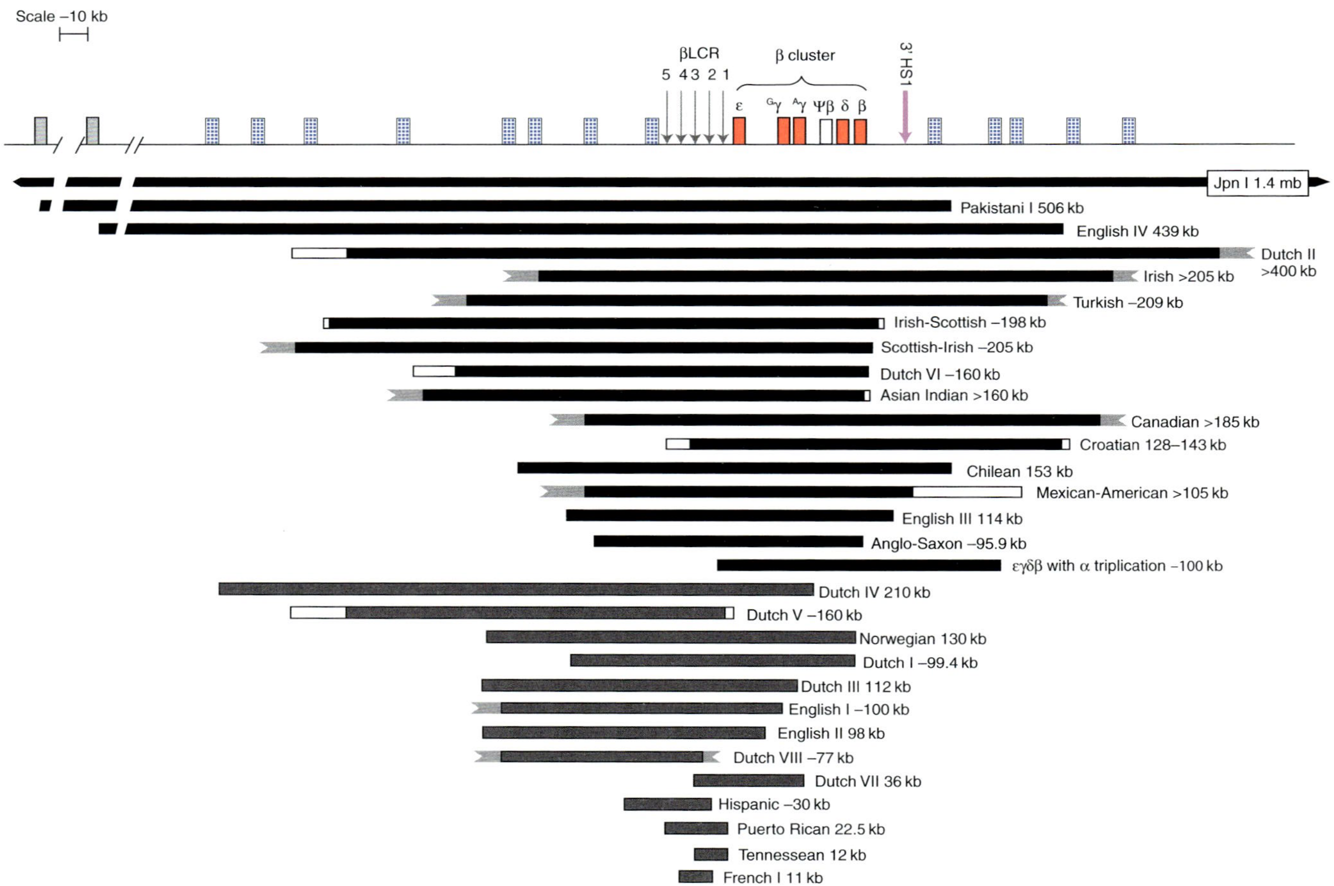

Figure 4. Deletions causing β-thalassemia as part of $(\varepsilon\gamma\delta\beta)^0$-thalassemia. The deletions can be classified into two groups: group I deletions (in black) remove all or most of the cluster including the βLCR and the β-globin gene, whereas group II deletions remove all or part of βLCR leaving the β gene intact. Only about one-third of the breakpoints of these deletions have been defined; the white boxes and jagged ends indicate undefined breakpoints.

classified molecularly into two categories: a larger group (I) that removes all or most of the complex, including the *HBB* gene and the βLCR, and group II that removes the upstream LCR but leaves the *HBB* gene itself intact (Thein and Wood 2009; Rooks et al. 2012). It was the characterization of the upstream deletions that indicated the importance of long-range regulatory elements in the control of the β-globin locus (Kioussis et al. 1983; Grosveld et al. 1987). Adult heterozygotes for these deletions have a hematological phenotype similar to that of β-thalassemia trait, but the HbA_2 and HbF levels are within normal limits and the red blood cells tend to be relatively more hypochromic microcytic. Newborns have anemia and hemolysis; some requiring intrauterine and perinatal blood transfusions to tide them over the perinatal period. The severity of anemia and hemolysis is variable (even within a family with identical mutations) (Verhovsek et al. 2012), and appears to show no correlation with type (group I or II) or size of deletions. Only heterozygotes have been identified; homozygotes, presumably would not survive early gestation. These deletions are rare and unique to the families in which they have been described.

DOMINANTLY INHERITED β-THALASSEMIA

The syndrome comprises a distinct set of mutations affecting the *HBB* gene that are associated with typical hematological features of β-thalassemia (i.e., increased HbA_2 levels and imbalanced α-/β-globin-chain biosynthesis in heterozygotes), yet cause a disease phenotype when present in single copy (Thein 1999, 2001). Unlike the common recessive forms of β-thalassemia, which are prevalent in malaria-endemic regions, "dominantly inherited" β-thalassemia has been described in dispersed geographical regions, a large number of the mutations are spontaneous. Severe dyserythropoiesis associated with inclusion bodies in bone marrow erythroblasts and peripheral red blood cells after splenectomy were frequently observed in affected individuals prompting the term "inclusion body β-thalassemia" (Weatherall et al. 1973; Fei et al. 1989).

Globin-chain biosynthesis studies of both bone marrow erythroblasts and peripheral blood reticulocytes show imbalanced synthesis of α- and β-globin chains typical of those found in heterozygous β-thalassemia (Weatherall et al. 1973; Ho et al. 1997). As the molecular lesions of an increasing number of such cases were characterized, it became apparent that the defects were extremely heterogeneous; some mutations were highly complex involving deletions interrupted by insertions similar to the complex arrangement originally described in the Irish family (Weatherall et al. 1973). All mutations were associated with the synthesis or predicted synthesis of extremely unstable β-globin-chain variants (Fig. 2). In the majority of cases, the abnormal β variant is not detectable but there is a significant pool of free α chains, which presumably form the characteristic cytoplasmic inclusions (Weatherall et al. 1973).

The predicted synthesis is supported by the presence of substantial amounts of mutant β-globin mRNA in the peripheral blood reticulocytes, comparable in amounts to that of the other normal β-globin allele (Hall and Thein 1994). In dominantly inherited β-thalassemia, not only is there a functional deficiency of β globin, but precipitation of the β-chain variants with the concomitant excess α chains overload the proteolytic intracellular mechanisms increasing ineffective erythropoiesis. Indeed, the large intraerythroblastic inclusions, that are so characteristic of this form of β-thalassemia, have subsequently been shown to be composed of both α- and β-globin chains (Ho et al. 1997). In contrast, the inclusion bodies in homozygous β-thalassemia consisted only of precipitated α globin. The molecular mechanisms underlying the instability include: substitution of the critical amino acids in the hydrophobic heme pocket displacing heme leading to aggregation of the globin variant; disruption of secondary structure because of replacement of critical amino acids; substitution or deletion of amino acids involved in αβ dimer formation; and elongation of subunits by a hydrophobic tail.

The molecular defects include missense mutations, deletions or insertion of intact codons, nonsense mutations causing premature termination codons in exon 3, which leads to a failure of the nonsense mediated decay of the RNA (Fig. 2; Table 2). Frameshifts may also result in aberrant splicing producing elongated or truncated β-globin-chain variants with abnormal carboxy-terminal ends similar to the variant predicted in the Irish family (Weatherall et al. 1973).

Missense Mutations

An example of a missense mutation causing β-thalassemia intermedia is Hb Terre Haute (β106 Leu → Arg) (Coleman et al. 1991). This mutant was initially described as Hb Indianapolis (β112 Cys → Arg) (Adams et al. 1979), which was subsequently reported in two families (Spanish and Italian) (Baiget et al. 1986; De Biasi et al. 1988) in whom affected members had evidence of mild hemolytic anemia with 2%–4% reticulocytosis. In two patients heterozygous for Hb Terre Haute, globin-chain biosynthesis studies showed an α:non-α ratio of ~1.0 in bone marrow erythroblasts compared with a ratio of ~2.0 in peripheral blood reticulocytes. Although the variant β-globin chain was synthesized at a level almost equal to that of the normal β-globin chain, most of it was rapidly precipitated on the red cell membrane. The half-life of this globin variant was less than 10 min, and the abnormal hemoglobin was not detectable by standard techniques.

Other examples of missense variants include Hb Chesterfield (Thein et al. 1991), Hb Cagliari (Podda et al. 1991), Hb Showa-Yakushiji (Kobayashi et al. 1987), Hb Durham NC/Brescia (de Castro et al. 1994), Hb Houston (Kazazian et al. 1992), and more recently, Hb Mont Saint Aignan (Wajcman et al. 2001). In the example of Hb Chesterfield, an abnormal peak in the position expected for the β-globin-chain variant but without detectable corresponding protein was demonstrated by globin-chain biosynthesis studies. Hb Mont Saint Aignan (β128 [Hb] Ala → Pro), in comparison, appeared mildly unstable; the abnormal β chain could be isolated by selective isopropanol precipitation and the structure determined by protein chemistry methods. Mass spectrometry electrospray analysis estimated Hb Mont Saint Aignan to be ~20% of total hemoglobin.

Most of the other abnormal hemoglobins were not detected by routine hemoglobin electrophoresis.

Deletion or Insertion of Intact Codons

Deletions or insertions of entire codons allow the reading frame to remain in phase, and the remaining amino acids are normal. Both Hb Korea (Park et al. 1991) and Hb Gunma (Fucharoen et al. 1990a) have 145 amino acid residues each; in Hb Gunma, the β127–128 Gln-Ala dipeptide is replaced by a proline residue because of the deletion of three bases (AGG), whereas in Hb Korea, the deletion of three bases (GGT) removes one of the Val residues from codons 32–34. Other β-globin-chain variants have extra residues, and include the insertion of Arg in codons 30–31 in a Spanish family (Arjona et al. 1996) and insertion of a single proline in codons 124–126 in an Armenian patient (Çürük et al. 1994).

In all cases, no trace of abnormal hemoglobin could be detected by the standard techniques of IEF, HPLC, or heat stability tests. One mechanism that could explain the lack of detection of these structural β-globin-chain variants is that the affected amino acids are involved in α1β1 contacts. In the normal β-globin chain, β30 Arg (B 12), β33 Val (B 15), β34 Val (B 16), β108 Asn (G 10), β112 Cys (G 14), β124 Pro (H2), β125 Pro (H3), β127 Gln (H5), and β128 Ala (H6) are essential for α1β1 dimer formation (Bunn and Forget 1986). Deletion or substitution of these critical amino acids would be likely to prevent the formation of αβ subunits and, effectively, lead to a functional loss of half of the β-globin chains.

Another example in this category is a deletion of 12 in combination with an insertion of six nucleotides, leading to the substitution of the normal Val-Ala-Gly-Val by Gly-Arg in codons 134–137 and a β-globin subunit that was two amino acids shorter than normal. Affected individuals of this Portuguese family had moderately severe anemia, splenomegaly, and leg ulcers (Öner et al. 1991).

Hb Stara Zagnora, found in a 2-year-old Bulgarian boy, results from the deletion of 6 bp spanning β codons 137–139 such that Val-Ala-Asn is replaced by Asp (Efremov 2007). Hb Stara Zagnora was hyperunstable; the affected 2-year-old had hemolysis and dyserythropoietic anemia with mild globin-chain imbalance (α/β synthesis ratio 1.40) although HbA_2 levels were within normal limits.

Premature Termination (Nonsense Mutation)

All the dominantly inherited β alleles are rare and unique to the families in which they have been described with one exception. The GAA → TAA termination codon at codon 121, which leads to the synthesis of a truncated β-globin chain, has been described in several families of different ethnic backgrounds (Stamatoyannopoulos et al. 1974; Kazazian et al. 1986; Fei et al. 1989; Thein et al. 1990; Ohba et al. 1997), and in some families, heterozygotes do not have an unusually severe phenotype (i.e., the mutation is not dominantly inherited). Substantial amounts of mutant β-globin mRNA could be demonstrated in individuals in whom the mutation is dominantly inherited, but demonstration of the presence of the truncated β-globin variant has been difficult. Presence of the predicted truncated variant, however, was implicated from a large difference between the total radioactivity incorporated into newly synthesized chains and the total amount of protein in globin biosynthesis studies. In one study, the truncated β-globin chain was estimated to comprise only 0.05% to 0.1% of the total non-α globin (Ho et al. 1997).

Heterozygosity for a premature stop codon in β127 (CAG TO TAG) has also been to cause thalassemia intermedia in an English woman (Hall et al. 1991), and a 29-year-old French Caucasian woman (Prehu et al. 2005).

Elongated or Truncated Variants with Abnormal Carboxy-Terminal Ends

In general, the elongated or truncated β-globin gene variants in this group have arisen from frameshift mutations that generated distal premature termination codons. Again, the abnormal β-globin variants were not detected in any of the cases by hemoglobin electrophoresis or globin biosynthesis studies, but in all cases there was imbalanced synthesis of α- and β-globin chains.

Elongated β-globin subunits could also arise from aberrant splicing of precursor mRNA, as described in a deletion of two nucleotides affecting the IVS2 consensus donor splice site in a Portuguese family (Faustino et al. 1998). The mutation was associated with unusually severe anemia and intraerythroblastic inclusions, transmitted as a single allele in five generations of the family.

In Hb Jambol, an insertion of 23 nucleotides and a deletion of 310 nucleotides in βIVS2 extending to exon 3 of the *HBB* gene results in the replacement of Leu-Leu-Glu-Asn at codon 105 to 108 with nine residues and was associated with severe hemolytic anemia and mildly imbalanced globin synthesis ratio (Efremov 2007).

Recessive versus Dominant Inheritance

Frameshift mutations and premature termination codons (PTCs) that are recessively inherited terminate in exon 1 or 2, whereas those that are dominantly inherited, terminate much later in the sequence of the β-globin gene, in the 3′ part of exon 2 and exon 3 (Fig. 2). These in-phase termination mutants appear to have differential effects on triggering the surveillance mechanism of nonsense mediated decay (NMD). NMD is an in-house mRNA quality-control mechanism that degrades abnormal mRNAs that arise from mistakes in gene expression such as those caused by premature termination codons. In mammalian cells, termination codons are recognized as "premature" if it is located upstream of a boundary of 50–55 nucleotides 5′ to the final exon-exon junction (Nagy and Maquat 1998; Schoenberg and Maquat 2012). In the β-globin gene this boundary corresponds to a position 54 bp upstream of exon 2–exon 3 junction in intron 2; premature termination mutants that reside upstream are associated with minimal mutant β mRNA. However, exceptions to the "50–55 nt boundary rule" have been reported. β mRNA with nonsense mutations in codons 5, 15, and 17, all within exon 1, were detected at high levels, similar to those of wild type β-globin mRNA. It is possible that early PTCs within β-globin mRNA and proximity to the translation initiation codon (ATG) can override the "50–55 nt boundary rule" (Romao et al. 2000).

Similarly, frameshift mutations that occur later in the sequences, terminate later and tend to lead to accumulation of the mutant message and the synthesis of elongated β-globin variants. These elongated variants have abnormal carboxy-terminal ends made up of hydrophobic sequences causing their instability. Furthermore, these β-globin variants would not be able to form αβ dimers as most of the α1β1 contact residues would have been removed. Because the heme contact site codons—mostly located in exon 2—are retained, these elongated variants should have some tertiary structure, be less susceptible to proteolytic degradation, and presumably, form the characteristic inclusion bodies. Indeed, prominent inclusions were noted in individuals heterozygous for Hb Geneva (Beris et al. 1988), Hb Makabe (Fucharoen et al. 1990b), Hb Agnana (Ristaldi et al. 1990), Hb Vercelli (Murru et al. 1991), and the frameshift mutation at codon 128 in the Irish family (Weatherall et al. 1973; Thein et al. 1990).

VARIANTS OF β-THALASSEMIA

Normal HbA$_2$ β-Thalassemias

Normal HbA$_2$ β-thalassemias (previously referred to as type 2) refers to the form in which the blood picture is typical of heterozygous β-thalassemia (i.e., hypochromic microcytic red blood cells) except for the normal levels of Hb A$_2$. Most cases of normal HbA$_2$ β-thalassemia result from coinheritance of δ-thalassemia (δ^0 or δ^+) in *cis* or *trans* to a β^0- or β^+-thalassemia gene (Thein 1998).

One relatively common form of normal Hb A$_2$-thalassemia in the Middle East and Mediterranean is that associated with Hb Knossos (β27 Ala → Ser). Like Hb E, the mutation β27(GCC → TCC) activates an alternative splice site reducing the amount of normal transcript that contains the variant. Unlike Hb E, the Hb A$_2$ level is not elevated in heterozygotes as there is a δ^0-thalassemia (Cd59-A) mutation in *cis* to the β27 Ala → Ser mutation (Olds et al. 1991).

Another relatively common cause of normal Hb A$_2$ β-thalassemia phenotype in the Greek population is the Corfu form of δβ-thalassemia, a 7.2 kb deletion that includes the 5′ part of the δ gene (Wainscoat et al. 1985; Traeger-Synodinos et al. 1991). The β-globin gene in *cis* is down-regulated by a G → A mutation in position 5 of the IVS1 (Kulozik et al. 1988). Heterozygotes for the Corfu β-thalassemia allele have a slight increase in Hb F (1.1%–2.8%) and low to normal Hb A$_2$ levels but homozygotes have a milder than expected phenotype of thalassemia intermedia. They have almost 100% Hb F with no Hb A$_2$ and trace levels of Hb A, suggesting that the effect of the deletion is to allow increased γ-chain production under the stress of anemia. Studies in primary erythroid cell cultures from heterozygotes, homozygotes and compound heterozygotes for the Corfu deletion suggest that γ-mRNA accumulation and HbF expression is indirectly dependent on the total amount of viable β mRNA (Chakalova et al. 2005). Reduction of β mRNA below a critical threshold, as in compound heterozygotes and homozygotes, allows full expression of HbF, and hence the unusually high HbF in such individuals. The Corfu mutation has been described as separate lesions in two different populations. The β gene in *cis* to the 7.2 kb deletion in an Italian individual is normal and expressed at normal levels (Galanello et al. 1990) whereas Algerian homozygotes for the βIVS1-5 G → A mutation have a severe transfusion-dependent anemia.

The phenotype of normal Hb A$_2$ β-thalassemia is also seen in heterozygotes for εγδβ-thalassemia (see above).

"Silent" β-Thalassemia (See also Transcriptional Mutants)

Heterozygotes for "silent" β-thalassemias do not have any evident hematological phenotype;

 Cite this article as *Cold Spring Harb Perspect Med* doi: 10.1101/cshperspect.a011700

the only abnormality being a mild imbalance of globin-chain synthesis. "Silent" β-thalassemia alleles are not common except for the C → T mutation at position −101 of the β-globin gene, which accounts for most of the milder forms of β-thalassemia in the Mediterranean (Maragoudaki et al. 1999). It has been noted that carriers for this mutation have highly variable HbA_2 levels despite similar hematological parameters and globin-chain synthesis ratios. Coinheritance of δ-thalassemia mutations were implicated but sequence analysis of the δ-globin genes in one family excluded this possibility (Ristaldi et al. 2001). A C → G transversion has also been reported in the same −101 position and heterozygotes have a "silent" phenotype (Moi et al. 2004). Several other mutations in the 5′ and 3′ UTRs are also "silent" (Table 1).

It has been suggested that the $[TA]_x[T]_y$ sequence variation at position −530 of the β-globin gene may be responsible for some "silent" β-thalassemia carriers and that the reduced β-globin expression may be related to increased binding of the BP1 repressor protein (Berg et al. 1991). However, population surveys and clinical studies do not show a consistent correlation between the $[TA]_x[T]_y$ variants and a β-thalassemia phenotype, suggesting that it is a neutral polymorphism (Wong et al. 1989).

β-Thalassemia Trait with Unusually High HbA_2

Despite the vast heterogeneity of mutations, the increased levels of HbA_2 observed in heterozygotes for the different β-thalassemia alleles in different ethnic groups are remarkably uniform, usually 3.5%–5.5% and rarely exceeding 6%. Unusually high levels of HbA_2 (over 6.5%) seem to characterize the subgroup of β-thalassemias caused by lesions (point mutations or deletions) that inactivate the regulatory elements in the β promoter (Huisman 1997; Thein and Wood 2009). The unusually high HbA_2, often accompanied by modest increases in HbF, may be related to the removal of competition for the upstream LCR and transcription factors, allowing an increased interaction with the *cis* δ and γ genes.

UNUSUAL CAUSES OF β-THALASSEMIA

Insertion of a Transposable Element Causing β-Thalassemia

Transposable elements may occasionally disrupt human genes and result in their inactivation. The insertion of such an element, a retrotransposon of the L1 family has been reported with the phenotype of β^+-thalassemia (Kimberland et al. 1999). Despite the insertion of 6–7 kb DNA into its IVS2, the affected gene expresses full length β-globin transcripts at a level corresponding to about 15% of normal β-globin mRNA.

Trans-Acting Mutations

Population studies have shown that ~1% of β-thalassemias remain uncharacterized despite extensive sequence analysis, including the flanking regions of the β-globin genes. In the last 10 years several rare *trans*-acting mutations that affect *HBB* and its linked *HBD* gene have been identified. Mutations in *XPD* that cause trichothiodystrophy (TTD) are frequently associated with a phenotype of β-thalassemia trait, supported by reduced levels of β-globin synthesis and reduced β-globin mRNA (Viprakasit et al. 2001). The *XPD* protein is a subunit of the general transcription factor TF11H, which is involved in basal transcription and DNA repair. Some mutations in the erythroid-specific transcription factor *GATA-1* on the X-chromosome have also been reported to cause β-thalassemia in association with thrombocytopenia (Yu et al. 2002). In other cases of β-thalassemia, no mutations have been detected in the *HBB* gene, and intrafamilial segregation implicates *trans*-acting regulatory factors (Murru et al. 1992; Thein et al. 1993; Pacheco et al. 1995).

More recently, rare variants in *KLF1*, an erythroid-specific transcription factor, have been described in association with isolated borderline increases in HbA_2 ($\alpha_2\delta_2$) levels in the absence of mutations in *HBB* gene (Perseu et al. 2011). It has been suggested that the increased *HBD* expression is indirect, via impaired interaction of the *KLF1* variant with *HBB* gene in favor of the competing *HBD* gene.

Somatic Deletion of β-Globin Gene/Uniparental Isodidomy

Somatic deletions involving *HBB* gene have been described in three unrelated families of French (Badens et al. 2002) and Italian (Galanello et al. 2004) origins. Affected individuals had moderately severe anemia with hepatosplenomegaly despite being constitutionally heterozygous for a typical β-thalassemia mutation (β^0 39 C → T) with a normal α genotype. Subsequent investigations revealed a somatic deletion of chromosome 11p15, including the β-globin complex, in *trans* to the mutation, in a subpopulation of erythroid cells. This results in a somatic mosaic—10% to 20% of erythroid cells were heterozygous with one normal copy of β-globin gene, and the rest hemizygous (i.e., without any normal β-globin gene). The implication for gene therapy is that expression in ~20% of erythroid cells may be sufficient to convert a transfusion-dependent state to transfusion independence.

Uniparental isodisomy of a segment of chromosome 11p containing the *HBB* cluster has also been described in two cases. In one family of Chinese origin (Chang et al. 2008), uniparental isodisomy in a carrier for β-thalassemia has resulted in homozygosity for the β-thalassemia allele and thalassemia major. In another, paternal isodisomy of a segment of chromosome 11p in a carrier for β^S has resulted in a subpopulation of erythroid cells homozygous for HbS and sickle cell disease (Swensen et al. 2010).

CONCLUDING REMARKS

A complete spectrum of genetic lesions affecting the β-globin gene giving rise to a spectrum of phenotypic severity has been described. Characterization of the β-thalassemia alleles has involved a multitude of methodologies, mirroring the changing DNA technology.

ACKNOWLEDGMENTS

I thank Claire Steward for help with preparation of the manuscript and Helen Rooks for help in preparation of some of the figures.

REFERENCES

Adams JGI, Boxer LA, Baehner RL, Forget BG, Tsistrakis GA, Steinberg MH. 1979. Hemoglobin Indianapolis (β112[G14] Arginine). An unstable β-chain variant producing the phenotype of severe β-thalassemia. *J Clin Invest* **63:** 931–938.

Andersson BA, Wering ME, Luo HY, Basran RK, Steinberg MH, Smith HP, Chui DH. 2007. Sickle cell disease due to compound heterozygosity for Hb S and a novel 7.7-kb β-globin gene deletion. *Eur J Haematol* **78:** 82–85.

Arjona SN, Maldonado Eloy-Garcia J, Gu L-H, Smetanina NS, Huisman THJ. 1996. The dominant β-thalassaemia in a Spanish family is due to a frameshift that introduces an extra CGG codon (=arginine) at the 5′ end of the second exon. *Br J Haematol* **93:** 841–844.

Badens C, Mattei MG, Imbert AM, Lapoumérouliee C, Martini N, Michel G, Lena-Russo D. 2002. A novel mechanism for thalassaemia intermedia. *Lancet* **359:** 132–133.

Baiget M, Gomez Pereira C, Jue DL, Johnson MH, McGuffey JE, Moo-Penn WF. 1986. A case of hemoglobin Indianapolis [β112(G14) Cys → Arg] in an individual from Cordoba, Spain. *Hemoglobin* **10:** 483–494.

Berg P, Mittleman M, Elion J, Labie D, Schechter AN. 1991. Increased protein binding to a −530 mutation of the human β-globin gene associated with decreased β-globin synthesis. *Am J Hematol* **36:** 42–47.

Beris P, Miescher PA, Diaz-Chico JC, Han IS, Kutlar A, Hu H, Wilson JB, Huisman THJ. 1988. Inclusion-body β-thalassemia trait in a Swiss family is caused by an abnormal hemoglobin (Geneva) with an altered and extended β-chain carboxy-terminus due to a modification in codon β114. *Blood* **72:** 801–805.

Blacklock HA, Case J, Chan T, Raizis T, Doocey R, Fellowes A, Royle G, Jackson S, Brennan S, George P. 2005. Novel sequence insertion in a Maori patient with transfusion-dependent β-thalassaemia. *Br J Haematol* **131:** 400–402.

Bunn HF, Forget BG. 1986. *Hemoglobin: Molecular, genetic and clinical aspects.* W.B. Saunders Company, Philadelphia, PA.

Busslinger M, Moschonas N, Flavell RA. 1981. β^+ thalassemia: Aberrant splicing results from a single point mutation in an intron. *Cell* **27:** 289–298.

Cantor AB. 2005. GATA transcription factors in hematologic disease. *Int J Hematol* **81:** 378–384.

Chakalova L, Osborne CS, Dai YF, Goyenechea B, Metaxotou-Mavromati A, Kattamis A, Kattamis C, Fraser P. 2005. The Corfu δβ thalassemia deletion disrupts γ-globin gene silencing and reveals post-transcriptional regulation of HbF expression. *Blood* **105:** 2154–2160.

Chang JG, Tsai WC, Chong IW, Chang CS, Lin CC, Liu TC. 2008. β-thalassemia major evolution from β-thalassemia minor is associated with paternal uniparental isodisomy of chromosome 11p15. *Haematologica* **93:** 913–916.

Cheng TC, Orkin SH, Antonarakis SE, Potter MJ, Sexton JP, Giardina PJV, Li A, Kazazian HHJ. 1984. β-thalassemia in Chinese: Use of in-vivo RNA analysis and oligonucleotide hybridization in systematic characterization of molecular defects. *Proc Natl Acad Sci* **81:** 2821.

Codrington JF, Li H-W, Kutlar F, Gu L-H, Ramachandran M, Huisman THJ. 1990. Observations on the levels of Hb A_2 in patients with different β-thalassemia mutations and a δ chain variant. *Blood* **76:** 1246–1249.

Coleman MB, Steinberg MH, Adams JG III. 1991. Hemoglobin Terre Haute arginine β106. *J Biol Chem* **266:** 5798–5800.

Craig JE, Kelly SJ, Barnetson R, Thein SL. 1992. Molecular characterization of a novel 10.3 kb deletion causing β-thalassaemia with unusually high Hb A_2. *Br J Haematol* **82:** 735–744.

Çürük MA, Molchanova TP, Postnikov YV, Pobedimskaya DD, Liang R, Baysal E, Kolodey S, Smetanina NS, Tokarev YN, Rumyantsev AG, et al. 1994. β-Thalassemia alleles and unstable hemoglobin types among Russian pediatric patients. *Am J Hematol* **46:** 329–332.

De Biasi R, Spiteri D, Caldora M, Iodice R, Pucci P, Malorni A, Ferranti P, Marino G. 1988. Identification by fast atom bombardment mass spectrometry on Hb Indianapolis [b112(G14)Cys → Arg] in a family from Naples, Italy. *Hemoglobin* **12:** 323–336.

de Castro CM, Devlin B, Fleenor DE, Lee ME, Kaufman RE. 1994. A novel β-globin mutation, β Durham-NC [β114 Leu → Pro], produces a dominant thalassemia-like phenotype. *Blood* **83:** 1109–1116.

Dobkin C, Pergolizzi RG, Bahre P, Bank A. 1983. Abnormal splice in a mutant β-globin gene not at the site of a mutation. *Proc Natl Acad Sci* **80:** 1184.

Efremov GD. 2007. Dominantly inherited β-thalassemia. *Hemoglobin* **31:** 193–207.

Faustino P, Osório-Almeida L, Romão L, Barbot J, Fernandes B, Justiça B, Lavinha J. 1998. Dominantly transmitted β-thalassaemia arising from the production of several aberrant mRNA species and one abnormal peptide. *Blood* **91:** 685–690.

Fei YJ, Stoming TA, Kutlar A, Huisman TH, Stamatoyannopoulos G. 1989. One form of inclusion body β-thalassemia is due to a GAA → TAA mutation at codon 121 of the β chain. *Blood* **73:** 1075–1077.

Forget BG. 2001. Molecular mechanisms of β thalassemia. In *Disorders of hemoglobin: Genetics, pathophysiology, and clinical management* (ed. Steinberg MH, Forget BG, Higgs DR, Nagel RL), pp. 252–276. Cambridge University Press, Cambridge, UK.

Fucharoen S, Fucharoen G, Fukumaki Y, Nakayama Y, Hattori Y, Yamamoto K, Ohba Y. 1990a. Three-base deletion in exon 3 of the β-globin gene produced a novel variant (β Gunma) with a thalassemia-like phenotype. *Blood* **76:** 1894–1896.

Fucharoen S, Kobayashi Y, Fucharoen G, Ohba Y, Miyazono, Fukumaki Y, Takaku F. 1990b. A single nucleotide deletion in codon 123 of the β-globin gene causes an inclusion body β-thalassaemia trait: A novel elongated globin chain β^{Makabe}. *Br J Haematol* **75:** 393–399.

Galanello R, Melis MA, Podda A, Monne M, Persey L, Loudianos G, Cao A, Pirastu M, Piga A. 1990. Deletion δ-thalassemia: The 7.2 kb deletion of Corfu δβ-thalassemia in a non-β-thalassemia chromosome. *Blood* **75:** 1747–1749.

Galanello R, Perseu L, Perra C, Maccioni L, Barella S, Longinotti M, Cao A, Cazzola M. 2004. Somatic deletion of the normal β-globin gene leading to thalassaemia intermedia in heterozygous β-thalassaemic patients. *Br J Haematol* **127:** 604–606.

Giardine B, Borg J, Higgs DR, Peterson KR, Philipsen S, Maglott D, Singleton BK, Anstee DJ, Basak AN, Clark B, et al. 2011. Systematic documentation and analysis of human genetic variation in hemoglobinopathies using the microattribution approach. *Nat Genet* **43:** 295–301.

Gilman JG. 1987. The 12.6 kilobase DNA deletion in Dutch β^0-thalassaemia. *Br J Haematol* **67:** 369–372.

Goldsmith ME, Humphries RK, Ley T, Cline A, Kantor JA, Nienhuis AW. 1983. Silent substitution in β^+-thalassemia gene activating a cryptic splice site in β-globin RNA coding sequence. *Proc Natl Acad Sci* **80:** 2318.

Gonzalez-Redondo JM, Stoming TA, Kutlar A, Kutlar F, Lanclos KD, Howard EF, Fei YJ, Aksoy M, Altay C, Gurgey A, et al. 1989. A C → T substitution at nt −101 in a conserved DNA sequence of the promoter region of the β-globin gene is associated with "silent" β-thalassemia. *Blood* **73:** 1705–1711.

Grosveld F, van Assendelft GB, Breaves DR, Kollias G. 1987. Position-independent, high-level expression of the human γ-globin gene in transgenic mice. *Cell* **51:** 975–985.

Hall GW, Thein SL. 1994. Nonsense codon mutations in the terminal exon of the β-globin gene are not associated with a reduction in β-mRNA accumulation: A mechanism for the phenotype of dominant β-thalassemia. *Blood* **83:** 2031–2037.

Hall GW, Franklin IM, Sura T, Thein SL. 1991. A novel mutation (nonsense β127) in exon 3 of the β-globin gene produces a variable thalassaemia phenotype. *Br J Haematol* **79:** 342–344.

Hamid M, Akbari MT. 2011. A 13-bp deletion in the 3′ untranslated region of the β-globin gene causes β-thalassemia major in compound heterozygosity with IVSII-1 mutation. *Med Princ Pract* **20:** 488–490.

Hattori Y, Yamashiro Y, Ohba Y, Miyaji T, Morishita M, Yamamoto K, Yamamoto K, Narai S, Kimura A. 1991. A new β-thalassemia mutation (initiation codon ATG → GTG) found in the Japanese population. *Hemoglobin* **15:** 317.

Ho PJ, Rochette J, Fisher CA, Wonke B, Jarvis MK, Yardumian A, Thein SL. 1996. Moderate reduction of β globin gene transcript by a novel mutation in the 5′ untranslated region: A study of its interaction with other genotypes in two families. *Blood* **87:** 1170–1178.

Ho PJ, Wickramasinghe SN, Rees DC, Lee MJ, Eden A, Thein SL. 1997. Erythroblastic inclusions in dominantly inherited β thalassaemias. *Blood* **89:** 322–328.

Huang S-C, Benz EJJ. 2001. Posttranscriptional factors influencing the hemoglobin content of the red cells. In *Disorders of hemoglobin: Genetics, pathophysiology, and clinical management* (ed. Steinberg MH, Forget BG, Higgs DR, Nagel RL), pp. 252–276. Cambridge University Press, Cambridge, UK.

Huang S-Z, Wong C, Antonarakis SE, Ro-Lein T, Lo WHY, Kazazian HHJ. 1986. The same TATA box β-thalassemia mutation in Chinese and US Blacks: Another example of independent origins of mutation. *Hum Genet* **74:** 162–164.

Huisman THJ. 1997. Levels of Hb A_2 in heterozygotes and homozygotes for β-thalassemia mutations: Influence of mutations in the CACCC and ATAAA motifs of the β-globin gene promoter. *Acta Haematol* **98:** 187–194.

Humphries RK, Ley TJ, Anagnou NP, Baur AW, Nienhuis AW. 1984. β^0-39 thalassemia gene: A premature termination codon causes β-mRNA deficiency without affecting cytoplasmic β-mRNA stability. *Blood* **64:** 23–32.

Isken O, Maquat LE. 2007. Quality control of eukaryotic mRNA: Safeguarding cells from abnormal mRNA function. *Genes Dev* **21:** 1833–1856.

Jankovic L, Efremov GD, Josifovska O, Juricic D, Stoming TA, Kutlar A, Huisman THJ. 1990. An initiation codon mutation as a cause of a β^0-thalassemia. *Hemoglobin* **14:** 169.

Kazazian HHJ, Orkin SH, Boehm CD, Goff SC, Wong C, Dowling CE, Newburger PE, Knowlton RG, Brown V, Donis-Keller H. 1986. Characterisation of a spontaneous mutation to a β-thalassaemia allele. *Am J Human Genet* **38:** 860–867.

Kazazian HHJ, Dowling CE, Hurwitz RL, Coleman M, Stopeck A, Adams JGI. 1992. Dominant thalassemia-like phenotypes associated with mutations in exon 3 of the β-globin gene. *Blood* **79:** 3014–3018.

Kimberland ML, Divoky V, Prchal J, Schwahn U, Berger W, Kazazian HH Jr. 1999. Full-length human L1 insertions retain the capacity for high frequency retrotransposition in cultured cells. *Hum Mol Genet* **8:** 1557–1560.

Kioussis D, Vanin E, de Lange T, Flavell RA, Grosveld FG. 1983. β-globin gene inactivation by DNA translocation in γβ-thalassaemia. *Nature* **306:** 662–666.

Kobayashi Y, Fukumaki Y, Komatsu N, Ohba Y, Miyaji T, Miura Y. 1987. A novel globin structural mutant, Showa-Yakushiji (β110 Leu-Pro) causing a β-thalassemia phenotype. *Blood* **70:** 1688–1691.

Kulozik A, Yarwood N, Jones RW. 1988. The Corfu $\delta\beta^0$ thalassemia: A small deletion acts at a distance to selectively abolish β globin gene expression. *Blood* **71:** 457–462.

Lam VMS, Xie SS, Tam JWO, Woo YK, Gu YL, Li AMC. 1990. A new single nucleotide change at the initiation codon (ATG → AGG) identified in amplified genomic DNA of a Chinese β-thalassemic patient. *Blood* **75:** 1207–1208.

Li DZ, Liao C, Li J. 2009. A novel β-thalassemic allele due to a thirteen nucleotide deletion: Codons 54–58 (-T ATG GGC AAC CCT). *Ann Hematol* **88:** 799–801.

Maragoudaki E, Kanavakis E, Trager-Synodinos J, Vrettou C, Tzetis M, Metxotou-Mavrommati A, Kattamis C. 1999. Molecular, haematological and clinical studies of the −101 C → T substitution in the β-globin gene promoter in 25 β-thalassaemia intermedia patients and 45 heterozygotes. *Br J Haematol* **107:** 699–706.

Moi P, Faa V, Marini MG, Asunis I, Ibba G, Cao A, Rosatelli MC. 2004. A novel silent β-thalassemia mutation in the distal CACCC box affects the binding and responsiveness to EKLF. *Br J Haematol* **126:** 881–884.

Metherall JE, Collins FS, Pan J, Weissman SM, Forget BG. 1986. β^0 thalassemia caused by a base substitution that creates an alternative splice acceptor site in an intron. *EMBO J* **5:** 2551–2557.

Murru S, Loudianos G, Deiana M, Camaschella C, Sciarratta GV, Agosti S, Parodi MI, Cerruti P, Cao A, Pirastu M. 1991. Molecular characterization of β-thalassemia intermedia in patients of Italian descent and identification of three novel β-thalassemia mutations. *Blood* **77:** 1342–1347.

Murru S, Loudianos G, Porcu S, Sciarratta GV, Agosti S, Parodi MI, Cao A, Pirastu M. 1992. A β-thalassaemia phenotype not linked to the β-globin cluster in an Italian family. *Br J Haematol* **81:** 283–287.

Nagy E, Maquat LE. 1998. A rule for termination-codon position within intron-containing genes: When nonsense affects RNA abundance. *Trends Biochem Sci* **23:** 198–199.

Ohba Y, Hattori Y, Harano T, Harano K, Fukumaki Y, Ideguchi H. 1997. β-Thalassemia mutations in Japanese and Koreans. *Hemoglobin* **21:** 191–200.

Olds RJ, Sura T, Jackson B, Wonke B, Hoffbrand AV, Thein SL. 1991. A novel δ^0 mutation in *cis* with Hb Knossos: A study of different genetic interactions in three Egyptian families. *Br J Haematol* **78:** 430–436.

Öner R, Öner C, Wilson JB, Tamagnini GP, Ribeiro LML, Huisman THJ. 1991. Dominant β-thalassaemia trait in a Portuguese family is caused by a deletion of (G) TGGCTGGTGT(G) and an insertion of (G)GCAG(G) in codons 134, 135, 136 and 137 of the β-globin gene. *Br J Haematol* **79:** 306.

Orkin SH, Kazazian HHJ, Antonarakis SE, Goff SC, Boehm CD, Sexton JP, Waber PG, Giardina PJV. 1982a. Linkage of β-thalassaemia mutations and β-globin gene polymorphisms with DNA polymorphisms in human β-globin gene cluster. *Nature* **296:** 627–631.

Orkin SH, Kazazian HHJ, Antonarakis SE, Ostrer H, Goff SC, Sexton JP. 1982b. Abnormal RNA processing due to the exon mutation of β^E-globin gene. *Nature* **300:** 768–769.

Orkin S, Antonarakis S, Loukopoulos D. 1984. Abnormal processing of $\beta^{Knossos}$ RNA. *Blood* **64:** 311–313.

Orkin SH, Cheng T-C, Antonarakis SE, Kazazian HHJ. 1985. Thalassemia due to a mutation in the cleavage-polyadenylation signal of the human β-globin gene. *EMBO J* **4:** 453–456.

Pacheco P, Peres MJ, Faustino P, Pischedda C, Gonçalves J, Carvajales-Ramos M, Seixas T, Martins MC, Moi P, Lavinha J. 1995. β-thalassaemia unlinked to the β-globin gene interacts with sickle-cell trait in a Portuguese family. *Br J Haematol* **91:** 85–89.

Park SS, Barnetson R, Kim SW, Weatherall DJ, Thein SL. 1991. A spontaneous deletion of b33/34 Val in exon 2 of the b globin gene (Hb Korea) produces the phenotype of dominant β thalassaemia. *Br J Haematol* **78:** 581.

Perseu L, Satta S, Moi P, Demartis FR, Manunza L, Sollaino MC, Barella S, Cao A, Galanello R. 2011. KLF1 gene mutations cause borderline HbA2. *Blood* **118:** 4454–4458.

Podda A, Galanello R, Maccioni L, Melis MA, Rosatelli C, Perseu L, Cao A. 1991. Hemoglobin Cagliari (β60[E4] VAL → Glu): A novel unstable thalassemic hemoglobinopathy. *Blood* **77:** 371–375.

Prehu C, Pissard S, Al-Sheikh M, Le Niger C, Bachir D, Galacteros F, Wajcman H. 2005. Two French Caucasian families with dominant thalassemia-like phenotypes due to hyper unstable hemoglobin variants: Hb Sainte Seve [codon 118 (-T)] and codon 127 [CAG → TAG (Gln → stop]). *Hemoglobin* **29:** 229–233.

Ristaldi MS, Pirastu M, Murru S, Casula L, Loudianos G, Cao A, Sciarratta GV, Agosti S, Parodi MI, Leone D, et al. 1990. A spontaneous mutation produced a novel elongated β-globin chain structural variant (Hb Agnana) with a thalassemia-like phenotype. *Blood* **75:** 1378–1380.

Ristaldi MS, Casula S, Porcu S, Cao A. 2001. Normal δ globin gene sequence in carrier of the silent-101 (C-T) β-thalassemia mutation with normal HbA2 level. *Am J Hematol* **67:** 58.

Romao L, Inacio A, Santos S, Avila M, Faustino P, Pacheco P, Lavinha J. 2000. Nonsense mutations in the human β-globin gene lead to unexpected levels of cytoplasmic mRNA accumulation. *Blood* **96:** 2895–2901.

Rooks H, Clark B, Best S, Rushton P, Oakley M, Thein OS, Cuthbert AC, Britland A, Ruf A, Thein SL. 2012. A novel 506 kb deletion causing εγδβ thalassemia. *Blood Cells Mol Dis* **49:** 121–127.

Rund D, Dowling C, Najjar K, Rachmilewitz EA, Kazazian HH Jr, Oppenheim A. 1992. Two mutations in the β-globin polyadenylylation signal reveal extended transcripts and new RNA polyadenylylation sites. *Proc Natl Acad Sci* **89:** 4324–4328.

Rund D, Oron-Karni V, Filon D, Goldfarb A, Rachmilewitz E, Oppenheim A. 1997. Genetic analysis of β-thalassaemia intermedia in Israel: Diversity of mechanisms and unpredictability of phenotype. *Am J Hematol* **54:** 16–22.

Saba L, Meloni A, Sardu R, Travi M, Primignani P, Rosatelli MC, Cao A. 1992. A novel β-thalassemia mutation (G → A) at the initiation codon of the β-globin gene. *Human Mutation* **1:** 420–422.

Safaya S, Rieder RF, Dowling CE, Kazazian HHJ, Adams JGI. 1989. Homozygous β-thalassemia without anemia. *Blood* **73:** 324–328.

Schoenberg DR, Maquat LE. 2012. Regulation of cytoplasmic mRNA decay. *Nat Rev Genet* **13:** 246–259.

Schokker RC, Went LN, Bok J. 1966. A new genetic variant of β-thalassaemia. *Nature* **209:** 44–46.

Spritz RA, Jagadeeswaran P, Choudary PV, Biro PA, Elder JT, deRiel JK, Manley JL, Forget BG, Weissman SM. 1981. Base substitution in an intervening sequence of a β^+-thalassemic human globin gene. *Proc Natl Acad Sci* **78:** 2455.

Stamatoyannopoulos G, Woodson R, Papayannopoulou T, Heywood D, Kurachi MS. 1974. Inclusion-body β-thalassemia trait. A form of β thalassemia producing clinical manifestation in simple heterozygotes. *N Engl J Med* **290:** 939–943.

Swensen JJ, Agarwal AM, Esquilin JM, Swierczek S, Perumbeti A, Hussey D, Joiner CH, Pont-Kingdon G, Lyon E, Prchal JT. 2010. Sickle cell disease due to uniparental disomy in a child who inherited sickle cell trait. *Blood* **116:** 2822–2825.

Takeshita K, Forget BG, Scarpa A, Benz EJ Jr. 1984. Intranuclear defect in β-globin mRNA accumulation due to a premature translation termination codon. *Blood* **64:** 13–22.

Thein SL. 1998. β thalassaemia. In *Baillière's clinical haematology, sickle cell disease and thalassaemia* (ed. Rodgers GP), pp. 91–126. Baillière Tindall, London.

Thein SL. 1999. Is it dominantly inherited β thalassaemia or just a β-chain variant that is highly unstable? *Br J Haematol* **107:** 12–21.

Thein SL. 2001. Structural Variants with a β-Thalassemia Phenotype. In *Disorders of hemoglobin: Genetics, pathophysiology, and clinical management* (ed. Steinberg MH, Forget BG, Higgs DR, Nagel RL), pp. 342–355. Cambridge University Press, Cambridge, UK.

Thein SL. 2005. Pathophysiology of β thalassemia—A guide to molecular therapies. *Hematology Am Soc Hematol Educ Program* 31–37.

Thein SL, Wood WG. 2009. The molecular basis of β thalassemia, δβ thalassemia, and hereditary persistence of fetal hemoglobin. In *Disorders of hemoglobin: Genetics, pathophysiology, and clinical management* (ed. Steinberg MH, Forget BG, Higgs DR, Nagel RL), pp. 323–356. Cambridge University Press, Cambridge, UK.

Thein SL, Old JM, Wainscoat JS, Petrou M, Modell B, Weatherall DJ. 1984. Population and genetic studies suggest a single origin for the Indian deletion β^0 thalassaemia. *Br J Haematol* **57:** 271–278.

Thein SL, Hesketh C, Taylor P, Temperley IJ, Hutchinson RM, Old JM, Wood WG, Clegg JB, Weatherall DJ. 1990. Molecular basis for dominantly inherited inclusion body β-thalassemia. *Proc Natl Acad Sci* **87:** 3924–3928.

Thein SL, Best S, Sharpe J, Paul B, Clark DJ, Brown MJ. 1991. Hemoglobin Chesterfield (β28 Leu → Arg) produces the phenotype of inclusion body β thalassemia. *Blood* **77:** 2791–2793.

Thein SL, Wood WG, Wickramasinghe SN, Galvin MC. 1993. β-Thalassemia unlinked to the β-globin gene in an English family. *Blood* **82:** 961–967.

Traeger-Synodinos J, Tzetis M, Kanavakis E, Metaxotou-Mavromati A, Kattamis C. 1991. The Corfu δβ thalassaemia mutation in Greece: Haematological phenotype and prevalence. *Br J Haematol* **79:** 302–305.

Treisman R, Orkin SH, Maniatis T. 1983. Specific transcription and RNA splicing defects in five cloned β-thalassaemia genes. *Nature* **302:** 591–596.

Varawalla NY, Old JM, Weatherall DJ. 1991. Rare β-thalassaemia mutations in Asian Indians. *Br J Haematol* **79:** 640–644.

Verhovsek M, Shah NR, Wilcox I, Koenig SC, Barros T, Thornburg CD, Steinberg MH, Luo HY, Chui DH. 2012. Severe fetal and neonatal hemolytic anemia due to a 198 kb deletion removing the complete β-globin gene cluster. *Pediatr Blood Cancer* **59:** 941–944.

Viprakasit V, Gibbons RJ, Broughton BC, Tolmie JL, Brown D, Lunt P, Winter RM, Marinoni S, Stefanini M, Brueton L, et al. 2001. Mutations in the general transcription factor TFIIH result in β-thalassaemia in individuals with trichothiodystrophy. *Hum Mol Genet* **10:** 2797–2802.

Wainscoat JS, Thein SL, Wood WG, Weatherall DJ, Tzotos S, Kanavakis E, Metaxatou-Mavromati A, Kattamis C. 1985. A novel deletion in the b globin gene complex. *Ann NY Acad Sci* **445:** 20–27.

Wajcman H, Lahary A, Prome D, Kister J, Riou J, Godart C, Prehu C, Traeger-Synodinos J, Papassotiriou I, Galacteros F. 2001. Hb Mont Saint Aignan [β128(H6)Ala →

Pro]: A new unstable variant leading to chronic microcytic anemia. *Hemoglobin* **25:** 57–65.

Weatherall DJ, Clegg JB. 2001. *The thalassaemia syndromes.* Blackwell Science, Oxford.

Weatherall DJ, Clegg JB, Knox-Macaulay HHM, Bunch C, Hopkins CR, Temperley IJ. 1973. A genetically determined disorder with features both of thalassaemia and congenital dyserythropoietic anaemia. *Br J Haematol* **24:** 681–702.

Westaway D, Williamson R. 1981. An intron nucleotide sequence variant in a cloned β^+-thalassemia globin gene. *Nucl Acid Res* **9:** 1777.

Wong C, Dowling CE, Saiki RK, Higuchi RG, Erlich HA, Kazazian HHJ. 1987. Characterization of β-thalassaemia mutations using direct genomic sequencing of amplified single copy DNA. *Nature* **330:** 384–386.

Wong SC, Stoning TA, Efremov GD, Huisman THJ. 1989. High frequencies of a rearrangement (+ATA; −T) at −530 to the β-globin gene in different populations indicate the absence of a correlation with a silent β-thalassemia determinant. *Hemoglobin* **13:** 1–5.

Yang KG, Kutlar F, George E, Wilson JB, Kutlar A, Stoming TA, Gonzalez-Redondo JM, Huisman THJ. 1989. Molecular characterization of β-globin gene mutations in Malay patients with Hb E-β-thalassaemia and thalassaemia major. *Br J Haematol* **72:** 73–80.

Yu C, Niakan KK, Matsushita M, Stamatoyannopoulos G, Orkin SH, Raskind WH. 2002. X-linked thrombocytopenia with thalassemia from a mutation in the amino finger of GATA-1 affecting DNA binding rather than FOG-1 interaction. *Blood* **100:** 2040–2045.

The Molecular Basis of α-Thalassemia

Douglas R. Higgs

MRC Molecular Haematology Unit, Weatherall Institute of Molecular Medicine, University of Oxford, John Radcliffe Hospital, Headington, Oxford OX3 9DS, United Kingdom

Correspondence: doug.higgs@imm.ox.ac.uk

The globin gene disorders including the thalassemias are among the most common human genetic diseases with more than 300,000 severely affected individuals born throughout the world every year. Because of the easy accessibility of purified, highly specialized, mature erythroid cells from peripheral blood, the hemoglobinopathies were among the first tractable human molecular diseases. From the 1970s onward, the analysis of the large repertoire of mutations underlying these conditions has elucidated many of the principles by which mutations occur and cause human genetic diseases. This work will summarize our current knowledge of the α-thalassemias, illustrating how detailed analysis of this group of diseases has contributed to our understanding of the general molecular mechanisms underlying many orphan and common diseases.

The molecular basis of α-thalassemia has been extensively reviewed many times in the past (Higgs et al. 1989; Harteveld and Higgs 2010), so why should we take the opportunity to revisit this topic; what can this contribute to our further understanding of how globin synthesis may be perturbed and how this relates more generally to human genetic disease? In my view, despite the extensive new knowledge of human genetic disease that is being obtained from the application of new generation sequencing of material from patients with a wide range of inherited and acquired human genetic diseases, many of the *principles* by which mutations cause human disease have been established for many years since the application of modern molecular and cellular biology to the globin gene disorders. It is underrecognized that what we have learned from the hemoglobinopathies applies to most human genetic diseases and further analysis of these uniquely well-characterized gene clusters will continue to elucidate important new mechanisms by which mammalian genes are normally switched on and off during cell fate decisions, and how this is perturbed in both common and orphan genetic diseases.

An argument that is commonly put forward is that the globin genes are very unusual, atypical genes and what we learn from them may be exceptions to the rule. On the contrary, it is hard to think of a single example in which what has been established from analyzing the globin genes (involving enhancers, locus control regions, boundary elements, promoters, RNA processing, and numerous aspects of protein translation, structure, and function) has been limited to these genes rather than establishing general principles of mammalian genome organization and how

Copyright © 2013 Cold Spring Harbor Laboratory Press; all rights reserved
Cite this article as *Cold Spring Harb Perspect Med* doi: 10.1101/cshperspect.a011718

this relates to gene expression. Furthermore, the molecular mechanisms by which globin gene mutations arise (chromosomal rearrangements, telomere truncations, homologous and illegitimate recombination, gene conversion, copy number variation, nucleotide polymorphism, changes in tandem repeats, abnormal methylation, the involvement of antisense RNAs, and many more) are now known to be common to many human diseases and yet were first recognized in the globin gene disorders. Their analysis has been at the forefront of identifying the mechanisms by which mutations at every level of gene expression may cause human disease. Rather than being unusual outliers, the globin genes are in most ways, typical mammalian genes, which when mutated cause typical human genetic diseases by common mechanisms and have pioneered our understanding of how natural mutations cause human genetic disease.

In the future, it will be important to continue using the knowledge gained from the well-characterized globin gene disorders to further develop our understanding of the mechanisms by which both common and orphan human genetic diseases may arise. In this article, I summarize some of what we have learned from studying mutations of the α-globin cluster and how this may provide insight into how we may understand the mechanisms underlying human genetic disease in general.

IDENTIFYING INDIVIDUALS WITH α-THALASSEMIA

Down-regulation of α-globin synthesis causes α-thalassemia, which is characterized by underproduction of fetal (HbF, $\alpha_2\gamma_2$) and adult (HbA, $\alpha_2\beta_2$) hemoglobin (Higgs et al. 1989). The considerable diversity of mutations that have been identified is explained by the fact that carriers for α-thalassemia are (to some extent) protected from falciparum malaria in a manner that we still do not fully understand (reviewed in Higgs 2009a; Williams and Weatherall 2012). Consequently, in tropical and subtropical regions of the world (where malaria is endemic), α-thalassemia reaches very high frequencies, making it one of the most common of all human genetic disease traits (Higgs 2009a; Williams and Weatherall 2012). Among people originating from these areas, compound heterozygotes and homozygotes for some mutations may have severe hematologic phenotypes. When α-globin synthesis is reduced to ~25% or less, patients may suffer from a moderately severe hemolytic anemia associated with readily detectable excess β-globin chains in the form of β4 tetramers (referred to as HbH); hence, the condition is called HbH disease (Chui et al. 2003; Vichinsky 2012). When α-globin synthesis is even further reduced (or even abolished), this gives rise to a severe intrauterine anemia associated with excess γ-globin chains in the form of γ4 tetramers (referred to as Hb Bart's); hence, this condition is referred to as the Hb Bart's hydrops fetalis syndrome (BHFS) (Lorey et al. 2001; Vichinsky 2012). The high frequency of α-thalassemia trait, HbH disease, and BHFS means that tens of thousands of patient samples have been analyzed and that probably most of the natural mutations affecting these genes are now known. Consequently, in most instances, accurate genetic counseling and (when appropriate) prenatal testing can be made available in countries where the health care infrastructure is able to support this. However, even when all known mutations have been excluded, new, often rare but informative genetic variants are still being found.

THE NORMAL STRUCTURE AND EXPRESSION OF THE α-GLOBIN CLUSTER

Expression of the α-like and β-like globin chains is regulated by clusters of genes on chromosomes 16 and 11, respectively (Higgs and Wood 2008; Noordermeer and de Laat 2008). The α-like gene cluster is located close to the telomere of chromosome 16 (16p13.3) including an embryonic gene (ζ) and two fetal/adult genes arranged along the chromosome in the order telomere-ζ-α2-α1-centromere, surrounded by widely expressed genes (Fig. 1). The normal cluster is denoted αα. Approximately 25–65 kb upstream of the α-globin genes there are four highly conserved noncoding sequences, or multispecies conserved sequences (MCS), called MCS-R1 to -R4, which

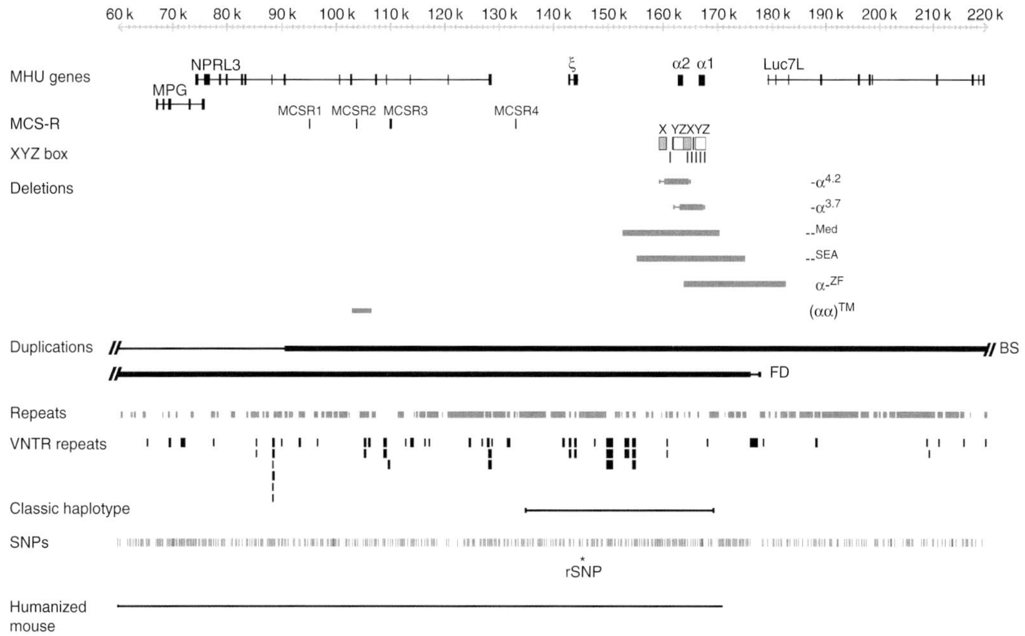

Figure 1. The human α-globin cluster (5′-ζ-α2-α1-3′) surrounded by widely expressed genes (MPG, NPRL3, and Luc7L) on chromosome 16 (16p13.3). Below this, the multispecies conserved elements (MCS-Rs) are shown. The X, Y, and Z boxes are the regions of duplication that play a part in the generation of the common α-thalassemias, as discussed in the text. The most common deletions removing one (-$\alpha^{3.7}$ and -$\alpha^{4.2}$) or both (--^{Med} and --^{SEA}) α genes and causing thalassemia are shown as horizontal bars, as are the unusual deletions α-^{ZF} and $(\alpha\alpha)^{TM}$. Duplications of the cluster BS and FD are also indicated. For a full catalog of all deletions that cause α-thalassemia, see Higgs (2009a). At the bottom of the figure the positions of common repeats, variable number random repeats (VNTRs), and single-nucleotide polymorphisms (SNPs) are shown. The region containing all SNPs and VNTRs corresponding to the classic haplotype used in population studies (as discussed in the text) is illustrated as a thin horizontal line. The regulatory SNP (rSNP) that creates new functional GATA1-binding sites seen in the Melanesian nondeletional α-thalassemia is shown with an asterisk. The extent of the α-globin locus present in the humanized mouse is shown as a thin line.

are thought to be involved in the regulation of the α-like globin genes (see Fig. 1). It is of interest that three of these elements (MCSR1–3) lie within the introns of the adjacent widely expressed gene NPRL3 (Kowalczyk et al. 2012b). Of these elements to date, only MCS-R2 (also known as HS-40) has been shown to be essential for α-globin expression (Higgs and Wood 2008).

As multipotent hemopoietic progenitors commit to the erythroid lineage and begin to differentiate to form mature red blood cells, key erythroid transcription factors (e.g., GATA-2, GATA-1, SCL, Sp/XKLF, NF-E2) and various cofactors (e.g., FOG, pCAF, p300) progressively bind to the upstream MCS-R elements and the promoters of the α-like globin genes (Fig. 2). Binding of these factors is associated with widespread modifications of the associated chromatin reflecting activation (e.g., histone acetylation). Finally, RNA polymerase II (PolII) is recruited to both the upstream regions and the globin promoters as transcription starts in early erythroid progenitors (Anguita et al. 2004; Vernimmen et al. 2007). At the same time, the upstream elements and promoters of the globin genes interact with each other via the formation of chromatin loops. It has recently been shown that HS-40 is critically required for looping to occur and for the stable recruitment of PolII to the α-globin promoters (Vernimmen et al. 2007, 2009).

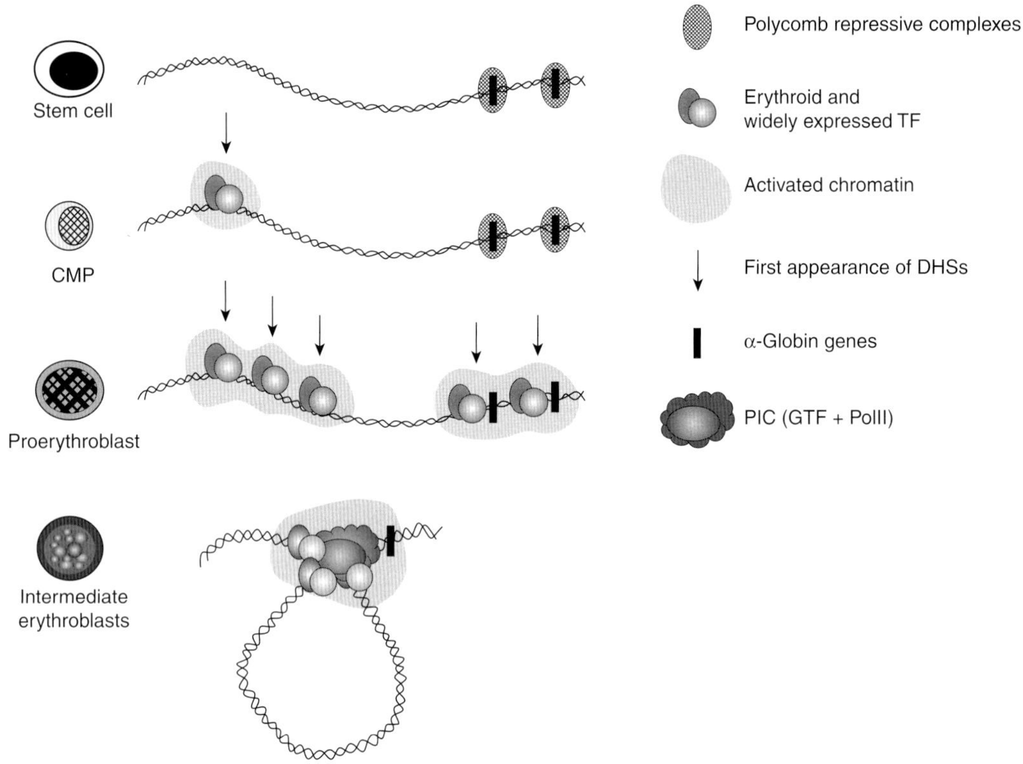

Figure 2. In stem cells and early progenitors, the cluster is silenced by the Polycomb repressive complex. In multipotent cells (CMP), the cluster is primed in the upstream region (MCSR-2) by multiprotein complexes containing SCL and NF-E2 nucleated by GATA-2. In committed erythroid progenitors (U-MEL, proerythroblast stage), additional remote regulatory sequences are bound by multiprotein complexes containing various combinations of SCL, NF-E2, and GATA1 replacing GATA2. At this stage, the α-globin promoter is also occupied by a combination of factors including NF-Y and is poised for expression. In differentiating erythroid cells, the preinitiation complex (PIC), including PolII, is recruited to the enhancers in a cooperative manner but independently of the promoter. Krüppel-like transcription factors are also recruited, independently of the upstream elements and to the promoter. At this final stage, the α-globin promoter is now occupied by a multiprotein complex that represents a docking site for the recruitment of the PIC, which is entirely dependent on the presence of the upstream elements that interact with the promoter, forming a loop.

This series of up-regulatory events taking place in erythroid cells is only a part of the complete story of α-globin regulation. Although the α-globin genes are expressed in a strictly tissue-specific manner, they are contained in a large chromosomal region, which is broadly transcriptionally active and bears the hallmarks of constitutively active chromatin (Flint et al. 1997; Daniels et al. 2001). Therefore, it was predicted that within such a region, mechanisms should exist to maintain the α-globin genes in a silent state in nonerythroid cells. An extensive survey of chromatin modifications in erythroid and nonerythroid cells showed that the α-globin cluster is marked by a specific signature (H3K27me3) in nonerythroid cells but less so in erythroid cells (Garrick et al. 2008). The H3K27me3 mark is imposed by a histone methyltransferase (EZH2), which forms part of a repressive Polycomb complex called PRC2 (Morey and Helin 2010). Consistent with this, components of this complex (EZH2 and SUZ12) have also been found at the α-globin cluster in nonerythroid cells. The PRC2 complex recruits histone deacetylases (HDACs) and another repressive Polycomb complex (called PRC1) (Lynch et al. 2012). Experimental data suggest that together, these complexes (PRC1, PRC2, and HDACs) maintain the silence of the α genes in multipotent cells and differentiated nonerythroid cells (Garrick et al. 2008; Lynch et al. 2012) but that they are cleared from the α-globin promoter before the activation steps described (see Fig. 2).

Activation of the α-globin genes can thus be thought of as a multistep process in which transcription of α globin is at first somewhat leaky, albeit expressed at very low levels, in hematopoietic stem cells when the α genes are partially repressed by the Polycomb system. As cells differentiate down the nonerythroid pathway, the genes become increasingly silenced by Polycomb so that in mature nonerythroid cells no transcription can be detected. In contrast, as cells differentiate along the erythroid pathway, Polycomb silencing is removed as the activating events are played out on the cluster (see Fig. 2) (De Gobbi et al. 2011). All results to date have been obtained by analyzing populations of cells so that we only see the average effects. Seeing the true order of events in single cells is a challenge for the future.

NORMAL VARIATION IN THE α-GLOBIN CLUSTER

Over the past few years, whole genome analysis has revealed the scale of variation across the human genome in apparently normal individuals (Lupski and Stankiewicz 2005; Day 2010). Such variation results from single-nucleotide polymorphisms (SNPs), variations in the numbers of tandem repeats (VNTRs), variations in microsatellites, copy number variants (CNVs, caused by homologous or illegitimate recombination), segmental duplications and deletions, and chromosomal translocations (Lupski and Stankiewicz 2005; Day 2010). Although heralded as a surprise, in fact the extent of variation and mechanisms underlying such polymorphic changes were anticipated from detailed characterization of the regions containing the globin loci in large numbers (thousands) of nonthalassemic individuals. In addition to SNPs forming ancestral haplotypes (Higgs et al. 1986), it had been shown that extensive polymorphic variations caused by genomic rearrangements were found in hematologically normal individuals. These variants are summarized in detail by Higgs (2009a) but the most notable small-scale variations in the α cluster occur in the many G-rich VNTRs in this region (Flint et al. 1997) together with the small- and large-scale variations commonly observed in the subtelomeric region of chromosome 16 (Wilkie et al. 1991). Studies of the different arrangements of these polymorphisms and VNTRs in various population samples have made it possible to define a series of α-globin gene haplotypes that have been of considerable value both in the analysis of evolutionary aspects of the gene clusters and in defining the origins of many of the α-thalassemia mutations (Higgs et al. 1986; Higgs 2009b). In addition to their value as genetic markers, these variants have been useful in distinguishing functionally important areas of the α-globin cluster from regions that appear to be of little functional significance (Rugless et al. 2008).

α-THALASSEMIA CAUSED BY SEQUENCE VARIATIONS IN THE STRUCTURAL GENES

Although most sequence variants in the α-globin cluster (including variation within HS-40) are caused by neutral SNPs, we currently know of 69 point mutations or oligonucleotide variants that alter gene expression, referred to as nondeletional forms of α-thalassemia (denoted $\alpha^T\alpha$ or $\alpha\alpha^T$ depending on whether the α2 or α1 gene is affected). As for many other human genetic diseases, these mutations may affect the canonical sequences that control gene expression, including the CCAAT and TATA box sequences associated with the promoter, the initiation codon (ATG), splicing signals (GT/AG), the termination codon (TAA), and the polyadenylation (polyA) signal (AATAAA). In addition to these mutations, α-thalassemia may also be caused by in-frame deletions, frame-shift mutations, and nonsense mutations (often leading to nonsense-mediated decay of the RNA) and/

or to the production of abnormal protein. Some variants alter the structure of the hemoglobin molecule making the dimer ($\alpha\beta$) or tetramer ($\alpha_2\beta_2$) unstable. Such molecules may precipitate in the red cell, forming insoluble inclusions that damage the red cell membrane. Over the past few years, it has become apparent that some α-globin structural variants are so unstable that they undergo very rapid postsynthetic degradation and thereby cause the phenotype of α-thalassemia. A full list of sequence variations in the structural genes that cause α-thalassemia is available (Higgs 2009a). An even greater number of similar point mutations have been described for the β-like globin genes (summarized in Giardine et al. 2011). Of all such variants, a major class of mutations that is not found in the globin genes is trinucleotide expansions (Lopez Castel et al. 2010). Furthermore, to date, no mutations affecting mRNAs or long noncoding RNAs have been found.

CHROMOSOMAL TRANSLOCATIONS AND DUPLICATIONS AFFECTING THE α-GLOBIN CLUSTER

Rarely, chromosomal translocations involving 16p13.3 place the α-globin locus at the tip of another chromosome, as seen, for example, in some relatives of patients with the ATR-16 syndrome (Gibbons 2012) (see below). To date, we know of 16 individuals with such balanced translocations and none of them has α-thalassemia (Steinberg et al. 2009). Because the closest centromeric breakpoint of these chromosomal translocations lies only 1.2 Mb from the α-globin genes, these findings show that the *cis*-acting sequences required for full α-globin regulation are contained within this region and that expression is not perturbed by rearrangements on this scale. In two individuals with unbalanced translocations and three copies of 16p13.3, the α/β-globin chain synthesis ratios were 1.5 and 1.6 (Wainscoat et al. 1981; Buckle et al. 1988), again indicating that the additional, mislocalized copy of the α complex is expressed even though its genomic position has changed.

Two large, internal duplications of the terminal region of chromosome 16 have also been described in patients with β-thalassemia intermedia. Because these patients are simply heterozygotes for the β-thalassemia mutations, the implication is that their relatively severe phenotype results from the production of excess α-globin chains from the α genes in the duplicated regions. In one pedigree, three α clusters (αα:αα:αα) are present on one copy of chromosome 16 (Fichera et al. 1994). Provisional data suggest that at least two and possibly all three clusters in the duplicated region are fully active. A carrier for this abnormal chromosome (αα:αα:αα/αα), with a total of eight α genes, has an α/β-globin chain synthesis ratio of 2.7 (Fichera et al. 1994). A recent study has more fully characterized what appears to be a very similar rearrangement in another Italian family with β-thalassemia intermedia (Harteveld et al. 2008) revealing a duplication of ~260 kb (see BS in Fig. 1). These investigators also characterized another duplication of ~175 kb of chromosome 16 lying between the end of the α cluster and the telomere (see FD in Fig. 1). Again the phenotype of a compound heterozygote for this rearrangement (αα:αα/αα) and β-thalassemia trait suggested that the additional α clusters in the duplicated segment are fully active, indicating that all sequences required for fully regulated α-globin expression lie in this duplicated segment of chromosome 16. These findings, delimiting the region required to direct fully regulated expression of the human α-globin cluster, are consistent with experimental data from a mouse model in which the mouse α-globin cluster was replaced with ~135 kb of the human α-globin cluster (Wallace et al. 2007). This region contains all sequences within a region of conserved synteny between the human and mouse, including the globin genes and their regulatory elements. The pattern and levels of expression of the human transgenes in this segment of DNA suggest that this region of ~135 kb contains all of the sequences required to express the α-globin genes correctly from an appropriate chromosomal environment, although, as discussed previously (Wallace et al. 2007), the level of expression in a mouse environment is less than in the human.

α-THALASSEMIA CAUSED BY DELETIONS REMOVING ONE OR MORE OF THE DUPLICATED STRUCTURAL GENES

Heteroduplex and DNA sequence analysis has shown that the duplicated α-globin genes (αα) are embedded within two highly homologous, 4-kb duplication units whose sequence identity appears to have been maintained throughout evolution by gene conversion and unequal crossover events (Lauer et al. 1980; Zimmer et al. 1980; Michelson and Orkin 1983; Hess et al. 1984). These regions are divided into homologous subsegments (X, Y, and Z) by nonhomologous elements (I, II, and III) (see Figs. 1 and 3). Reciprocal homologous recombination between Z segments, which are 3.7 kb apart, produces chromosomes with only one α gene ($-\alpha^{3.7}$, rightward deletion) (see Figs. 1 and 3) (Embury et al. 1980) that cause α-thalassemia and others with three α genes ($\alpha\alpha\alpha^{anti3.7}$) (Goossens et al. 1980). Recombination between homologous X boxes, which are 4.2 kb apart, also gives rise to an α-thalassemia determinant ($-\alpha^{4.2}$, leftward deletion) (see Figs. 1 and 3) (Embury et al. 1980) and an $\alpha\alpha\alpha^{anti4.2}$ chromosome (Trent et al. 1981). Further recombination events between the resulting chromosomes (α, αα, and ααα) may give rise to quadruplicated α genes (αααα) (De Angioletti et al. 1992), or quintuplicated (ααααα) (Cook et al. 2006) or other unusual patchwork rearrangements. Although these long-standing observations have pointed to the mechanism by which -α and ααα chromosomes arise, it has now been shown (using single DNA molecule polymerase chain reaction) how they may occur in vivo (Lam and Jeffreys 2007). From this work, the overall picture is one of reciprocal recombination (and unequal exchange) occurring in mitosis (premeiotic) in the germ line. The estimated frequencies of -α and ααα arrangements in sperm are in the order of $1-5 \times 10^5$. Deletions and duplications may also occur in somatic tissues by related mechanisms, although deletions detected in blood occur by intrachromosomal rather than interchromosomal recombination.

In addition to the common -α chromosomes, several rare deletions that remove either the α1 or α2 gene (leaving one gene intact, -α) have been described. In general, these remove

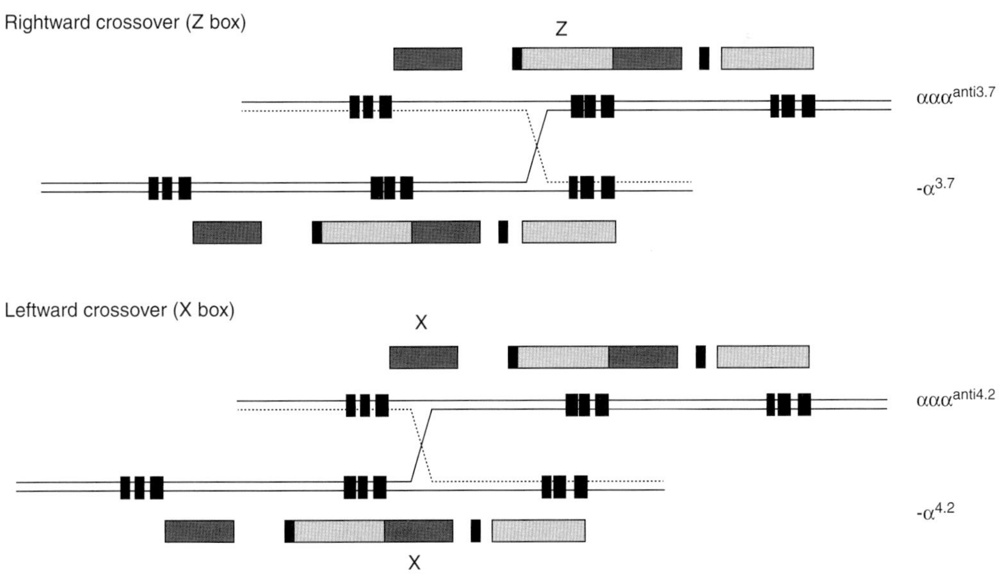

Figure 3. The mechanism by which the common deletions underlying α-thalassemia occur. Crossovers between misaligned Z boxes give rise to the $-\alpha^{3.7}$ and $\alpha\alpha\alpha^{anti3.7}$ chromosomes. Crossovers between misaligned X boxes give rise to $-\alpha^{4.2}$ and $\alpha\alpha\alpha^{anti4.2}$ chromosomes.

either the α1 or α2 gene by nonhomologous recombination. All of these deletions leave one α gene intact, and it is therefore possible to assess the influence of these deletions on expression of the remaining gene. When all of the data from these deletions that cause α-thalassemia are considered alongside the observations from nonthalassemic variants (see above), it appears that large segments of the α-globin cluster are not essential for α-globin expression (Rugless et al. 2008). Figures showing the full list of currently known deletions removing a single α gene are presented by Rugless and colleagues (2008).

It is now known that many human diseases (referred to as copy number variants) ranging from color blindness to inherited neuropathies (Carvalho et al. 2010) result from unequal exchange of low copy number duplicated sequences in exactly the same way as described in the globin clusters. In these diseases, too, the phenotypes associated with gain or loss of gene product may be quite different.

α-THALASSEMIA CAUSED BY DELETIONS REMOVING BOTH OF THE DUPLICATED STRUCTURAL GENES

There are currently approximately 50 deletions from the α-globin cluster that either completely or partially delete both α-globin genes, and consequently no α-chain synthesis is directed by these chromosomes in vivo. Examples of two common deletions --^{MED} and --^{SEA} are shown in Fig. 1. Homozygotes for such chromosomes (--/--) have the Hb BHFS. Compound heterozygotes for these deletions and deletions removing a single α gene (see above) have HbH disease (--/-α). With completion of the DNA sequence of 16p13.3 and beyond (International Human Genome Sequencing Consortium 2004), it has been possible to define the full extent of many of the deletions that remove both α genes (--). These deletions can be grouped into those (like --^{MED} and --^{SEA}) (see Fig. 1) that lie entirely within the α-globin cluster and deletions that extend up to ~800 kb beyond the α cluster to include the flanking genes. Although these deletions remove other genes, affected heterozygotes appear phenotypically normal apart from their α-thalassemia: in some patients, α-thalassemia trait (--/αα) and, in others, HbH disease (--/-α). In patients with more extensive deletions, with monosomy for a large segment of 16p13.3, α-thalassemia is associated with developmental abnormalities and mental retardation (so-called ATR-16 syndrome; see below) (Gibbons 2012).

It is interesting to note that all of the α-thalassemia deletions that occur at polymorphic frequencies in human populations are limited to the α cluster and do not extend into the surrounding genes, suggesting that deletion of these genes (even in heterozygotes) may result in a selective disadvantage. Detailed analysis of several of these determinants of α-thalassemia indicates that they often result from illegitimate or nonhomologous recombination events (e.g., Nicholls et al. 1987; Rugless et al. 2008). Such events may involve short regions of partial sequence homology at the breakpoints of the molecules that are rejoined, but they do not involve the extensive sequence matching required for homologous recombination as described in the previous section. Sequence analysis has shown that members of the dispersed family of Alu repeats are frequently found at or near the breakpoints of these deletions. Alu-family repeats occur frequently in the genome (3×10^5 copies) and seem to be particularly common in and around the α-globin cluster, where they make up ~25% of the entire sequence. These repeats may simply provide partially homologous sequences that promote DNA-strand exchanges during replication, or possibly a subset of Alu sequences may be more actively involved in the process. Detailed sequence analysis of the junctions of the α-globin deletions has revealed several interesting features including palindromes, direct repeats, and regions of weak homology. Some deletions involve more complex rearrangements that introduce new pieces of DNA bridging the two breakpoints of the deletion. In two deletions, this inserted DNA originates from upstream of the α cluster, and appears to have been incorporated into the junction in a manner suggesting that the upstream segment lies close to the breakpoint regions during replication (Nicholls et al. 1987;

Rugless et al. 2008). Orphan sequences from unknown regions of the genome are frequently found bridging the sequence breakpoints of other α-thalassemia deletions. Such complex deletion events are now being recognized with increasing frequency by new generation sequencing (Zhang et al. 2009). At least two of the deletions result from chromosomal breaks in the 16p telomeric region that have been "healed" by the direct addition of telomeric repeats (TTAGGG)n (Flint et al. 1994). This mechanism is described below in further detail. All currently known deletions that remove both α genes are summarized in Higgs (2009a).

LARGE DELETIONS EXTENDING BEYOND THE α-GLOBIN CLUSTER

Nearly all patients with large deletions (up to 900 kb) from the end of chromosome 16p (removing one copy of up to 52 genes) appear phenotypically normal apart from the presence of α-thalassemia. However, in 40 patients analyzed to date, deletions of >900 kb have been associated with a variety of developmental abnormalities; such patients are said to have the ATR-16 syndrome (Wilkie et al. 1990; Gibbons 2012). Relating phenotype to genotype is difficult in these cases because of the rather heterogeneous nature of the underlying abnormalities. Using a combination of conventional cytogenetics, fluorescent in situ hybridization, and molecular analysis, at least three types of chromosomal rearrangements (translocation, inversion/deletion, and truncation) have now been found in ATR-16 patients. To date, few breakpoints have been fully characterized. However, in some cases, telomeric truncations have been documented; in these cases, it appears that the affected chromosomes have been broken, truncated, and "healed" by the direct addition of telomeric repeats (TTAGGG)n, as described above for some less extensive 16p deletions in patients with α-thalassemia.

Altogether, 11 cases of ATR-16 have been shown to have pure monosomy for 16p13.3 and deletions of 900–1700 kb with various developmental abnormalities (e.g., facial dysmorphism, speech delay, and skeletal abnormalities). How might monosomy for 16p13.3 cause such developmental abnormalities? One possibility is that deletion of a large number of genes from one copy of chromosome 16 may unmask mutations in its homolog; the more genes that are deleted the greater the probability of this occurring. A further possibility is that some genes in 16p are imprinted (Reik and Walter 2001) so that deletions could remove the only active copy of the gene. At present there is no evidence for imprinting of the 16p region (reviewed in Schneider et al. 1996), and in the relatively few ATR-16 cases analyzed there appear to be no major clinical differences between patients with deletions of the maternally or paternally derived chromosomes (Gibbons 2012). It therefore seems more likely that there are some genes in the 16p region that encode proteins whose effect is critically determined by the amount produced—so-called dosage-sensitive genes. Examples of such genes include those encoding proteins that form heterodimers, those required at a critical level for a rate-determining step of a regulatory pathway, and tumor suppressor genes (e.g., TSC2). Removal of genes from one copy of 16p13.3 consistently reduces their levels of expression to ~50% of normal (V Buckle and colleagues, unpubl.). If the deletion includes one or more dosage-sensitive genes, this could account for the clinical effects seen in ATR-16 patients. The region lying between 900 and 1700 kb from the 16p telomere, deleted in all patients with the characteristic features of ATR-16 syndrome, contains genes and gene families of known function that have been implicated in a wide range of disorders with few or no features in common with ATR-16. One of these (SOX8) was considered as a strong candidate because it is involved in the regulation of embryonic development and is strongly expressed in the brain (Pfeifer et al. 2000). However, a recently described Brazilian patient with a deletion that removes both the α-globin locus and SOX8 was not associated with mental retardation (MR) or any dysmorphism (Bezerra et al. 2008). It is clear that further examples of ATR-16 owing to monosomy for 16p13.3 would have to be characterized to identify the gene(s) responsible for the MR and other developmental abnormalities associated with this condition. However, we cannot rule out the fact that similar deletions may cause different phenotypes owing to variations in the genetic background. The ATR-16 syndrome has served as an important model for improving our general understanding of the molecular basis for mental retardation (Gibbons 2012). The ATR-16 syndrome provided the first examples of mental retardation caused by a cryptic chromosomal translocation and truncation. Further work showed that such telomeric rearrangements may underlie a significant proportion of unexplained mental retardation (Flint et al. 1995). The current challenge is to understand in detail the mechanisms by which monosomy causes developmental abnormalities; the ATR-16 syndrome provides an excellent model for addressing this issue. All deletions currently associated with ATR-16 syndrome are shown in Higgs (2009a).

A RARE MUTATION CAUSING α-THALASSEMIA VIA AN ANTISENSE RNA

During a study to identify thalassemia in families from the Czech Republic, Indrak and colleagues (1993) reported a novel deletion (>18 kb) involving the α1 and θ gene (denoted α-^{ZF}) (see Figs. 1 and 4). Heterozygotes for this deletion have a mild hypochromic microcytic anemia with a reduced α/β-globin chain biosynthesis ratio and Hb H inclusions. These findings suggested that although the α2 gene appeared to be intact, it had been inactivated by the deletion. Barbour and colleagues (2000) and Tufarelli and colleagues (2003) further characterized this mutation and showed that the deletion juxtaposes a downstream gene (Luc7L) next to the structurally normal α2-globin gene. Although this α2 gene retains all of its local (e.g., promoter) and remote (e.g., MCS-R2/HS-40) *cis*-acting elements, its expression is silenced and its associated CpG island (see Fig. 4) becomes completely methylated during early development, and the chromatin associated with the promoter remains inactive and inaccessible, even in erythroid cells. From the analysis of experimental models recapitulating this deletion and from further characterization of the affected individual (Tufarelli et al. 2003), it was shown that transcription of antisense mRNA from Luc7L through the α2-globin gene was responsible for methylation of the associated CpG island and silencing of α-globin expression. Since this original report, the authors have identified two additional individuals (also from Poland) with the same mutation. The mutation is not only important for understanding the molecular basis for this rare form of thalassemia, but also illustrates a new mechanism underlying human genetic disease. Since this description, it has become increasingly clear that antisense RNAs may play important roles in regulating mammalian gene expression (Kowalczyk et al. 2012a). In terms of causing human genetic disease, there has been a report of a similar mechanism silencing the mismatch repair gene MSH2 in patients with susceptibility

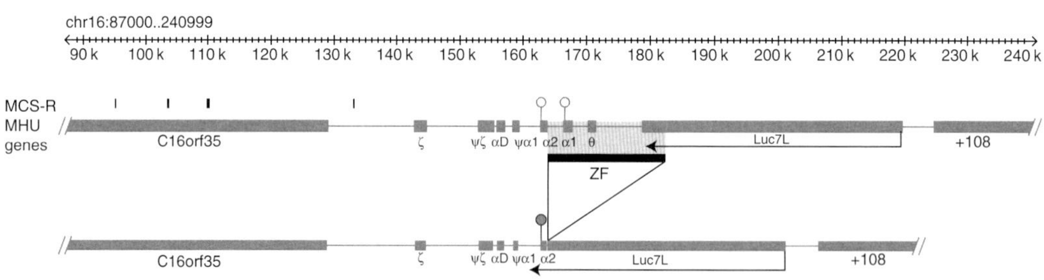

Figure 4. The key features of the α-^{ZF} mutation. In the normal cluster, the promoters of the α-globin genes lie in unmethylated CpG-rich islands. Partial deletion of Luc7L juxtaposes this truncated gene next to the remaining α2-globin gene and RNA transcripts from Luc7L extend through the α2-globin gene. This process is thought to attract de novo DNA methylases early in development, methylating the α2 CpG island and silencing it.

Cite this article as *Cold Spring Harb Perspect Med* doi: 10.1101/cshperspect.a011718

to colorectal cancer (Lynch syndrome) (Ligtenberg et al. 2009).

DELETIONS REMOVING THE UPSTREAM REGULATORY ELEMENTS OF THE α-GLOBIN CLUSTER

As discussed earlier, expression of the α genes is critically dependent on a multispecies conserved, noncoding regulatory sequence (MCS-R) that lies 40 kb upstream of the ζ2-globin gene. This region (called MCS-R2) is associated with an erythroid-specific DNase I hypersensitive site, referred to as HS-40. Detailed analysis of MCS-R2 has shown that it contains multiple binding sites for the erythroid-restricted *trans*-acting factors GATA-1 and NF-E2, and binding sites for the ubiquitously expressed Sp/XKLF family of transcription factors (Higgs et al. 2008).

The first indication that remote regulatory sequences controlling α-globin expression might exist came from observations on a patient with α-thalassemia (Hatton et al. 1990). Analysis of the abnormal chromosome $(\alpha\alpha)^{RA}$ from this patient showed a 62-kb deletion from upstream of the α complex that includes HS-40. Although both α genes on this chromosome are intact and entirely normal, they appear to be nonfunctional. Since this original observation, many more patients with α-thalassemia caused by deletions of HS-40 and a variable amount of the flanking DNA have been described, summarized in Higgs and Wood (2008). Until recently, the smallest of these deletions, which completely abolishes α-globin expression, removed both MCS-R1 and MCS-R2. Recently, an even smaller deletion, which removes ~3.3 kb of DNA including MCS-R2 but no other MCS-R element, was described (see Fig. 1) (Phylipsen et al. 2010). The phenotype (HbH disease) of the proband, which carries the same small deletion on both copies of chromosome 16, is consistent with expression of both *cis*-linked α genes being down-regulated but not completely abolished (Coelho et al. 2010). This suggests that other *cis*-acting elements on this chromosome (possibly MCS-R1) can activate α-globin expression. Clearly, these new observations demand further analysis of the other upstream MCS elements (MCS-R1, -R3, and -R4), which contain similar combinations of binding sites to MCS-R2 but whose functions are not yet clear. Together with similar deletions removing the regulatory elements controlling the β-globin cluster (Kioussis et al. 1983; Grosveld et al. 1987), these observations were among the first to illustrate in detail how mutation of distal regulatory elements can cause human genetic disease.

The mechanisms by which these natural mutations have arisen are quite diverse. In one, the deletion resulted from a recombination event between partially homologous Alu repeats that are normally 62 kb apart (Hatton et al. 1990). In another, the deletion arose via a subtelomeric rearrangement (Flint et al. 1996). The chromosomal breakpoint was found in an Alu element located ~105 kb from the 16p subtelomeric region. The broken chromosome was stabilized with a new telomere acquired by recombination between this Alu element and a subtelomeric Alu repeat associated with the newly acquired chromosome end. In at least five cases, the chromosomes appear to have been broken and then stabilized by the direct addition of telomeric repeats to nontelomeric DNA (Flint et al. 1994). Sequence analysis suggests that these chromosomes are "healed" via the action of telomerase, an enzyme that is normally involved in maintaining the integrity of telomeres. In the remaining cases, the mechanism has not yet been established. However, it is interesting that some (e.g., $[\alpha\alpha]^{IJ}$, $[\alpha\alpha]^{Sco}$) (Liebhaber et al. 1990; Viprakasit et al. 2003), but not all, of these mutations appear to have arisen de novo, because neither parent has the abnormal chromosome.

α-THALASSEMIA RESULTING FROM COMPETITION FOR THE UPSTREAM REGULATORY ELEMENTS

α-Thalassemia is common throughout Melanesia and is frequently caused by the known $-\alpha^{3.7}$ and $-\alpha^{4.2}$ mutations. However, it is also documented that in some Melanesian patients (from Papua New Guinea and Vanuatu) with α-thalassemia (α-thalassemia trait and HbH disease), the α-globin genes are intact, suggesting a nondeletional form of α-thalassemia ($\alpha^{T}\alpha/\alpha\alpha$ or

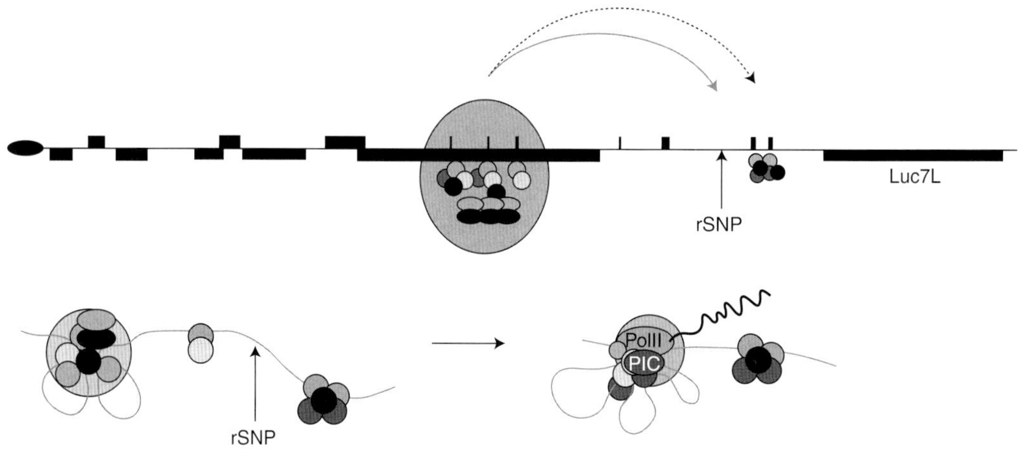

Figure 5. The key features of the Melanesian form of α-thalassemia. A regulatory SNP (rSNP) located between the ζ- and α2-globin genes creates a new GATA1-binding site, which in turn also creates a new promoter. This promoter preferentially interacts with the upstream enhancer elements (gray curved arrow) and steals activity from the α genes (dashed curved arrow). In the hypothetical looping model (*below*), the upstream element interacts with the new promoter but not the α genes.

$\alpha^T\alpha/\alpha^T\alpha$) (Fig. 5). In these patients, detailed mapping and DNA sequence analysis of the α genes and all of the upstream MCS elements was normal, and yet further studies showed that this form of α-thalassemia is linked to the α-globin cluster at 16p13.3. To identify the mutation responsible for this unusual form of α-thalassemia, De Gobbi and colleagues (2006) cloned this region from an affected homozygote and resequenced ∼213 kb of DNA containing and flanking the α-globin cluster, identifying 283 SNPs. The SNP responsible for the mutation was identified when the SNPs were aligned with a tiled microarray analyzed using labeled RNA from the patient's erythroid cells. This configuration revealed a new peak of mRNA expression (located between the ζ and ψζ genes), which coincides with a SNP that creates a GATA-1 binding site. In association studies, this SNP is always linked to the phenotype of α-thalassemia. Like the MCS-R elements, this new GATA site binds erythroid transcription factors in vivo and becomes activated in erythroid cells. Why should creating a new promoter-like sequence between the α genes and their regulatory elements cause α-thalassemia? Perhaps the most likely explanation is that, because it lies closer to the MCS elements, this new promoter competes with the α-globin promoters and "steals" the activity of the upstream regulatory elements, thus resulting in α-thalassemia. Several lines of evidence suggest that the α-globin promoters may indeed compete for the activity of the upstream regulatory elements, in particular, MCS-R2(HS-40). First, it has been noted that although the sequences of the duplicated α genes and their promoters are identical, the gene (α2) nearest MCS-R2 is expressed at a higher level than the more distal gene (α1). Furthermore, additional duplications of the α genes (ααα, αααα, and ααααα) do not lead to a linear increase in α-globin expression1; genes located further from MCS-R2 are expressed at progressively lower levels. When one α gene is deleted from the chromosome (-α), the remaining α gene appears to recruit more PolII and is expressed at increased levels (Rugless et al. 2008). It was recently shown that MCS-R2 also regulates expression of another gene (NME4) located 300 kb away from this element on chromosome 16 (Lower et al. 2009). The α-globin genes lie between MCS-R2 and NME4. When both α genes are deleted, expression of NME4 increases by a factor of eightfold. Together these findings suggest that promoters may compete for the activity associated with enhancers, and this may provide the explanation for this form of α-thalassemia seen in patients from the south Pacific. It seems likely that other similar "decoy promoters" will be found to cause changes in gene expression during the course of genome-wide studies.

SUMMARY

The α-globin cluster provides one of the best characterized models to develop our understanding of how the integration of a transcriptional program and an epigenetic program via a key set of *cis*-acting elements dispersed across a large chromosomal region, switch genes on and off at the correct time and place during development and differentiation. In particular, by analyzing the normal cluster, the natural mutants, and experimental models we should gain significant insight into how long-range regulatory elements control gene expression. There is now sufficient knowledge of the mutations underlying α-thalassemia to develop comprehensive genetic counseling and prenatal diagnosis wherever there is sufficient expertise and resources to support such a program. Further identification of natural mutations of the globin genes should be pursued because these observations continue to elucidate the general principles underlying human molecular genetics.

REFERENCES

Anguita E, Hughes J, Heyworth C, Blobel GA, Wood WG, Higgs DR. 2004. Globin gene activation during haemopoiesis is driven by protein complexes nucleated by GATA-1 and GATA-2. *EMBO J* **23:** 2841–2852.

Barbour VM, Tufarelli C, Sharpe JA, Smith ZE, Ayyub H, Heinlein CA, Sloane-Stanley J, Indrak K, Wood WG, Higgs DR. 2000. α-Thalassemia resulting from a negative chromosomal position effect. *Blood* **96:** 800–807.

Bezerra MA, Araujo AS, Phylipsen M, Balak D, Kimura EM, Oliveira DM, Costa FF, Sonati MF, Harteveld CL. 2008. The deletion of SOX8 is not associated with ATR-16 in an HbH family from Brazil. *Br J Haematol* **142:** 324–326.

Buckle VJ, Higgs DR, Wilkie AO, Super M, Weatherall DJ. 1988. Localisation of human α globin to 16p13.3—pter. *J Med Genet* **25:** 847–849.

Carvalho CM, Zhang F, Lupski JR. 2010. Evolution in health and medicine Sackler colloquium: Genomic disorders: A window into human gene and genome evolution. *Proc Natl Acad Sci* **107:** 1765–1771.

Chui DH, Fucharoen S, Chan V. 2003. Hemoglobin H disease: Not necessarily a benign disorder. *Blood* **101:** 791–800.

Coelho A, Picanco I, Seuanes F, Seixas MT, Faustino P. 2010. Novel large deletions in the human α-globin gene cluster: Clarifying the HS-40 long-range regulatory role in the native chromosome environment. *Blood Cells Mol Dis* **45:** 147–153.

Cook RJ, Hoyer JD, Highsmith WE. 2006. Quintuple α-globin gene: A novel allele in a Sudanese man. *Hemoglobin* **30:** 51–55.

Daniels RJ, Peden JF, Lloyd C, Horsley SW, Clark K, Tufarelli C, Kearney L, Buckle VJ, Doggett NA, Flint J, et al. 2001. Sequence, structure and pathology of the fully annotated terminal 2 Mb of the short arm of human chromosome 16. *Hum Mol Genet* **10:** 339–352.

Day IN. 2010. dbSNP in the detail and copy number complexities. *Hum Mutat* **31:** 2–4.

De Angioletti M, Lacerra G, Castaldo C, Cutolo R, de Bonis C, Buonanno G, Carestia C. 1992. $\alpha\alpha\alpha^{anti\text{-}3.7\ type\ II}$: A new α-globin gene rearrangement suggesting that the α-globin gene duplication could be caused by intrachromosomal recombination. *Hum Genet* **89:** 37–41.

De Gobbi M, Viprakasit V, Hughes JR, Fisher C, Buckle VJ, Ayyub H, Gibbons RJ, Vernimmen D, Yoshinaga Y, de Jong P, et al. 2006. A regulatory SNP causes a human genetic disease by creating a new transcriptional promoter. *Science* **312:** 1215–1217.

De Gobbi M, Garrick D, Lynch M, Vernimmen D, Hughes JR, Goardon N, Luc S, Lower KM, Sloane-Stanley JA, Pina C, et al. 2011. Generation of bivalent chromatin domains during cell fate decisions. *Epigenetics Chromatin* **4:** 9.

Embury SH, Miller JA, Dozy AM, Kan YW, Chan V, Todd D. 1980. Two different molecular organizations account for the single α-globin gene of the α-thalassemia-2 genotype. *J Clin Invest* **66:** 1319–1325.

Fichera M, Rappazzo G, Spalletta A, et al. 1994. Triplicated α-globin gene locus with translocation of the whole telomeric end in association with β-thalassemia trait, results in a severe syndrome. *Blood* **84:** 260a.

Flint J, Craddock CF, Villegas A, Bentley DP, Williams HJ, Galanello R, Cao A, Wood WG, Ayyub H, Higgs DR. 1994. Healing of broken human chromosomes by the addition of telomeric repeats. *Am J Hum Genet* **55:** 505–512.

Flint J, Wilkie AO, Buckle VJ, Winter RM, Holland AJ, McDermid HE. 1995. The detection of subtelomeric chromosomal rearrangements in idiopathic mental retardation. *Nat Genet* **9:** 132–140.

Flint J, Rochette J, Craddock CF, Dode C, Vignes B, Horsley SW, Kearney L, Buckle VJ, Ayyub H, Higgs DR. 1996. Chromosomal stabilisation by a subtelomeric rearrangement involving two closely related Alu elements. *Hum Mol Genet* **5:** 1163–1169.

Flint J, Thomas K, Micklem G, Raynham H, Clark K, Doggett NA, King A, Higgs DR. 1997. The relationship between chromosome structure and function at a human telomeric region. *Nat Genet* **15:** 252–257.

Garrick D, De Gobbi M, Samara V, Rugless M, Holland M, Ayyub H, Lower K, Sloane-Stanley J, Gray N, Koch C, et al. 2008. The role of the polycomb complex in silencing α-globin gene expression in nonerythroid cells. *Blood* **112:** 3889–3899.

Giardine B, Borg J, Higgs DR, Peterson KR, Philipsen S, Maglott D, Singleton BK, Anstee DJ, Basak AN, Clark B, et al. 2011. Systemic documentation and analysis of human genetic variation in hemoglobinopathies using the microattribution approach. *Nat Genet* **43:** 295–301.

Goossens M, Dozy AM, Embury SH, Zachariades Z, Hadjiminas MG, Stamatoyannopoulos G, Kan YW. 1980. Triplicated α-globin loci in humans. *Proc Natl Acad Sci* **77:** 518–521.

Grosveld F, van Assendelft GB, Greaves DR, Kollias G. 1987. Position-independent, high-level expression of the human β-globin gene in transgenic mice. *Cell* **51:** 975–985.

Harteveld CL, Higgs DR. 2010. α-Thalassaemia. *Orphanet J Rare Dis* **5:** 13.

Harteveld CL, Refaldi C, Cassinerio E, Cappellini MD, Giordano PC. 2008. Segmental duplications involving the α-globin gene cluster are causing β-thalassemia intermedia phenotypes in β-thalassemia heterozygous patients. *Blood Cells Mol Dis* **40:** 312–316.

Hatton CS, Wilkie AO, Drysdale HC, Wood WG, Vickers MA, Sharpe J, Ayyub H, Pretorius IM, Buckle VJ, Higgs DR. 1990. α-Thalassemia caused by a large (62 kb) deletion upstream of the human α globin gene cluster. *Blood* **76:** 221–227.

Hess JF, Schmid CW, Shen CK. 1984. A gradient of sequence divergence in the human adult α-globin duplication units. *Science* **226:** 67–70.

Higgs DR. 2009a. The molecular basis of α thalassemia. In *Disorders of hemoglobin* (ed. Steinberg MH, et al.), pp. 241–265. Cambridge University Press, Cambridge.

Higgs DR. 2009b. The pathophysiology and clinical features of α thalassemia. In *Disorders of hemoglobin* (ed. Steinberg MH, et al.), pp. 266–295. Cambridge University Press, Cambridge.

Higgs DR, Wood WG. 2008. Long-range regulation of α globin gene expression during erythropoiesis. *Curr Opin Hematol* **15:** 176–183.

Higgs DR, Wainscoat JS, Flint J, Hill AV, Thein SL, Nicholls RD, Teal H, Ayyub H, Peto TE, Falusi AG, et al. 1986. Analysis of the human α-globin gene cluster reveals a highly informative genetic locus. *Proc Natl Acad Sci* **83:** 5165–5169.

Higgs DR, Vickers MA, Wilkie AO, Pretorius IM, Jarman AP, Weatherall DJ. 1989. A review of the molecular genetics of the human α-globin gene cluster. *Blood* **73:** 1081–1104.

Higgs DR, Vernimmen D, Wood B. 2008. Long-range regulation of α-globin gene expression. *Adv Genet* **61:** 143–173.

Indrak K, Gu YC, Novotny J, Huisman TH. 1993. A new α-thalassemia-2 deletion resulting in microcytosis and hypochromia and in vitro chain imbalance in the heterozygote. *Am J Hematol* **43:** 144–145.

International Human Genome Sequencing Consortium. 2004. Finishing the euchromatic sequence of the human genome. *Nature* **431:** 931–945.

Kioussis D, Vanin E, deLange T, Flavell RA, Grosveld FG. 1983. β-Globin gene inactivation by DNA translocation in γ β-thalassaemia. *Nature* **306:** 662–666.

Kowalczyk MS, Higgs DR, Gingeras TR. 2012a. Molecular biology: RNA discrimination. *Nature* **482:** 310–311.

Kowalczyk MS, Hughes JR, Babbs C, Sanchez-Pulido L, Szumska D, Sharpe JA, Sloane-Stanley JA, Morriss-Kay GM, Smoot LB, Roberts AE, et al. 2012b. Nprl3 is required for normal development of the cardiovascular system. *Mamm Genome* **23:** 404–415.

Lam KW, Jeffreys AJ. 2007. Processes of de novo duplication of human α-globin genes. *Proc Natl Acad Sci* **104:** 10950–10955.

Lauer J, Shen CK, Maniatis T. 1980. The chromosomal arrangement of human α-like globin genes: Sequence homology and α-globin gene deletions. *Cell* **20:** 119–130.

Liebhaber SA, Griese EU, Weiss I, Cash FE, Ayyub H, Higgs DR, Horst J. 1990. Inactivation of human α-globin gene expression by a de novo deletion located upstream of the α-globin gene cluster. *Proc Natl Acad Sci* **87:** 9431–9435.

Ligtenberg MJ, Kuiper RP, Chan TL, Goossens M, Hebeda KM, Voorendt M, Lee TY, Bodmer D, Hoenselaar E, Hendriks-Cornelissen SJ, et al. 2009. Heritable somatic methylation and inactivation of MSH2 in families with Lynch syndrome due to deletion of the 3′ exons of TACSTD1. *Nat Genet* **41:** 112–117.

Lopez Castel A, Cleary JD, Pearson CE. 2010. Repeat instability as the basis for human diseases and as a potential target for therapy. *Nat Rev Mol Cell Biol* **11:** 165–170.

Lorey F, Charoenkwan P, Witkowska HE, Lafferty J, Patterson M, Eng B, Waye JS, Finklestein JZ, Chui DH. 2001. Hb H hydrops foetalis syndrome: A case report and review of literature. *Br J Haematol* **115:** 72–78.

Lower KM, Hughes JR, De Gobbi M, Henderson S, Viprakasit V, Fisher C, Goriely A, Ayyub H, Sloane-Stanley J, Vernimmen D, et al. 2009. Adventitious changes in long-range gene expression caused by polymorphic structural variation and promoter competition. *Proc Natl Acad Sci* **106:** 21771–21776.

Lupski JR, Stankiewicz P. 2005. Genomic disorders: Molecular mechanisms for rearrangements and conveyed phenotypes. *PLoS Genet* **1:** e49.

Lynch MD, Smith AJ, De Gobbi M, Flenley M, Hughes JR, Vernimmen D, Ayyub H, Sharpe JA, Sloane-Stanley JA, Sutherland L, et al. 2012. An interspecies analysis reveals a key role for unmethylated CpG dinucleotides in vertebrate Polycomb complex recruitment. *EMBO J* **31:** 317–329.

Michelson AM, Orkin SH. 1983. Boundaries of gene conversion within the duplicated human α-globin genes. Concerted evolution by segmental recombination. *J Biol Chem* **258:** 15245–15254.

Morey L, Helin K. 2010. Polycomb group protein-mediated repression of transcription. *Trends Biochem Sci* **35:** 323–332.

Nicholls RD, Fischel-Ghodsian N, Higgs DR. 1987. Recombination at the human α-globin gene cluster: Sequence features and topological constraints. *Cell* **49:** 369–378.

Noordermeer D, de Laat W. 2008. Joining the loops: β-Globin gene regulation. *IUBMB Life* **60:** 824–833.

Pfeifer D, Poulat F, Holinski-Feder E, Kooy F, Scherer G. 2000. The SOX8 gene is located within 700 kb of the tip of chromosome 16p and is deleted in a patient with ATR-16 syndrome. *Genomics* **63:** 108–116.

Phylipsen M, Prior JF, Lim E, Lingam N, Vogelaar IP, Giordano PC, Finlayson J, Harteveld CL. 2010. Thalassemia in Western Australia: 11 novel deletions characterized by multiplex ligation-dependent probe amplification. *Blood Cells Mol Dis* **44:** 146–151.

Reik W, Walter J. 2001. Genomic imprinting: Parental influence on the genome. *Nat Rev Genet* **2:** 21–32.

Rugless MJ, Fisher CA, Old JM, Sloane-Stanley J, Ayyub H, Higgs DR, Garrick D. 2008. A large deletion in the human α-globin cluster caused by a replication error is associated with an unexpectedly mild phenotype. *Hum Mol Genet* **17:** 3084–3093.

Schneider AS, Bischoff FZ, McCaskill C, Coady ML, Stopfer JE, Shaffer LG. 1996. Comprehensive 4-year follow-up on a case of maternal heterodisomy for chromosome 16. *Am J Med Genet* **66:** 204–208.

Steinberg MH, Forget BG, Higgs DR, Weatherall DJ, ed. 2009. *Disorders of hemoglobin*. Cambridge University Press, New York.

Trent RJ, Higgs DR, Clegg JB, Weatherall DJ. 1981. A new triplicated α-globin gene arrangement in man. *Br J Haematol* **49:** 149–152.

Tufarelli C, Stanley JA, Garrick D, Sharpe JA, Ayyub H, Wood WG, Higgs DR. 2003. Transcription of antisense RNA leading to gene silencing and methylation as a novel cause of human genetic disease. *Nat Genet* **34:** 157–165.

Vernimmen D, De Gobbi M, Sloane-Stanley JA, Wood WG, Higgs DR. 2007. Long-range chromosomal interactions regulate the timing of the transition between poised and active gene expression. *EMBO J* **26:** 2041–2051.

Vernimmen D, Marques-Kranc F, Sharpe JA, Sloane-Stanley JA, Wood WG, Wallace HA, Smith AJ, Higgs DR. 2009. Chromosome looping at the human α-globin locus is mediated via the major upstream regulatory element (HS -40). *Blood* **114:** 4253–4260.

Viprakasit V, Kidd AM, Ayyub H, Horsley S, Hughes J, Higgs DR. 2003. De novo deletion within the telomeric region flanking the human α globin locus as a cause of α thalassaemia. *Br J Haematol* **120:** 867–875.

Wainscoat JS, Kanavakis E, Weatherall DJ, Walker J, Holmes-Seidle M, Bobrow M, Donnison AB. 1981. Regional localisation of the human α-globin genes. *Lancet* **2:** 301–302.

Wallace HA, Marques-Kranc F, Richardson M, Luna-Crespo F, Sharpe JA, Hughes J, Wood WG, Higgs DR, Smith AJ. 2007. Manipulating the mouse genome to engineer precise functional syntenic replacements with human sequence. *Cell* **128:** 197–209.

Wilkie AO, Buckle VJ, Harris PC, Lamb J, Barton NJ, Reeders ST, Lindenbaum RH, Nicholls RD, Barrow M, Bethlenfalvay NC, et al. 1990. Clinical features and molecular analysis of the α thalassemia/mental retardation syndromes. I: Cases due to deletions involving chromosome band 16p13.3. *Am J Hum Genet* **46:** 1112–1126.

Wilkie AO, Higgs DR, Rack KA, Buckle VJ, Spurr NK, Fischel-Ghodsian N, Ceccherini I, Brown WR, Harris PC. 1991. Stable length polymorphism of up to 260 kb at the tip of the short arm of human chromosome 16. *Cell* **64:** 595–606.

Zhang F, Carvalho CM, Lupski JR. 2009. Complex human chromosomal and genomic rearrangements. *Trends Genet* **25:** 298–307.

Zimmer EA, Martin SL, Beverley SM, Kan YW, Wilson AC. 1980. Rapid duplication and loss of genes coding for the α chains of hemoglobin. *Proc Natl Acad Sci* **77:** 2158–2162.

Pathophysiology and Clinical Manifestations of the β-Thalassemias

Arthur W. Nienhuis[1] and David G. Nathan[2]

[1]Department of Hematology, St. Jude Children's Research Hospital, Memphis, Tennessee 38105

[2]Department of Pediatric Oncology, Dana-Farber Cancer Institute, and the Division of Hematology and Oncology, Childrens Hospital, Boston, Massachusetts 02215

Correspondence: arthur.nienhuis@stjude.org

The β-thalassemia syndromes reflect deficient or absent β-globin synthesis usually owing to a mutation in the β-globin locus. The relative excess of α-globin results in the formation of insoluble aggregates leading to ineffective erythropoiesis and shortened red cell survival. A relatively high capacity for fetal hemoglobin synthesis is a major genetic modifier of disease severity, with polymorphisms in other genes also having a significant role. Iron overload secondary to enhanced absorption and red cell transfusions causes an increase in liver iron and in various other tissues, leading to endocrine and cardiac dysfunction. Modern chelation regimens are effective in removing iron and preserving or restoring organ function.

The β-thalassemias are genetic disorders of hemoglobin synthesis characterized by deficient (β^+) or absent (β^0) synthesis of the β-globin subunit of hemoglobin molecule (Weatherall and Clegg 2001). The vast majority of individuals with thalassemia inherit their disorder as a Mendelian recessive. Heterozygous individuals have mild anemia and microcytosis and are categorized as having thalassemia minor or trait, and homozygous individuals have severe anemia of varying degrees and are characterized as having homozygous β-thalassemia or thalassemia major or intermedia, as discussed in detail below. Much rarer is a dominantly inherited β-thalassemia in which disease occurs in heterozygous individuals because of synthesis of a highly unstable β-globin variant (Thein 1999). Usually the disturbance is only in β-globin synthesis, but rare deletional mutations may remove one or more of the other genes on chromosome 11, resulting in forms of the disease characterized as δβ-, γδβ-, or εγδβ-thalassemia.

HETEROZYGOUS β-THALASSEMIA

The hematological features of thalassemia trait are microcytosis, hypochromia, and usually an increase in the percentage of HbA_2. The hemoglobin composition is 92%–95% HbA, >3.8% HbA_2, and variable amounts of HbF amounting to 0.5%–4%. In addition to microcytosis and hypochromia, there is marked variation in size and shape of red blood cells with the red cells of β^0-thalassemia trait having a lower mean

Copyright © 2013 Cold Spring Harbor Laboratory Press; all rights reserved
Cite this article as *Cold Spring Harb Perspect Med* doi: 10.1101/cshperspect.a011726

corpuscular volume than those of β^+-thalassemia trait (Cao and Galanello 2010). Historically, the mild anemia with microcytic, hypochromic red cells characteristic of thalassemia trait has been thought to lack clinical consequences other than its association with the anemia of pregnancy (White et al. 1985). However, a recently controlled trial performed in Sri Lanka suggested that individuals with thalassemia trait may experience symptoms of anemia including headache, lethargy, fatigue, dizziness, and exercise intolerance despite having hemoglobin levels that overlap the normal range (Premawardhena et al. 2008). There was no difference in the frequency of these symptoms between the two groups that had either mild anemia or hemoglobin levels in the normal range. There was also a significant increase in frequency of infectious episodes in individuals with β-thalassemia trait. Men but not women with thalassemia trait have a reduced frequency of advanced coronary artery disease, and myocardial infarction occurs at an older age in men with thalassemia trait (Tassiopoulos et al. 2005).

HOMOZYGOUS β-THALASSEMIA

The clinical spectrum of patients with homozygous β-thalassemia is highly variable (Weatherall and Clegg 2001; Cao and Galanello 2010). Many individuals present with severe anemia early in life and remain transfusion dependent for their entire lives. Such individuals carry the diagnosis of thalassemia major. Others with homozygous β-thalassemia present with milder anemia and never require transfusion, and some have variable degrees of anemia and may require transfusion intermittently. Such individuals are designated as having thalassemia intermedia. The degree of anemia in those with thalassemia intermedia is from nearly normal levels to sufficiently severe anemia to require occasional blood transfusions. Erythroid hyperplasia leads to medullary expansion with facial deformity and osteoporosis, which may be quite severe. Extramedullary hematopoiesis results in enlargement of the liver and spleen and paraspinal and pulmonary masses of erythroid cells. Diagnostic criteria for the major versus the intermedia syndrome are rather ill defined and are largely based on the hemoglobin level without transfusion. Generally, a cutoff of 7 g/dL is used to distinguish between the two forms, but this criterion is confounded by the fact that the severity of the anemia and associated splenomegaly and defective development may vary in individual patients at different times, and the use of transfusion is partly based on socioeconomic issues as well as access to an adequate blood supply. As noted in a definitive review of this topic, the remarkable phenotypical diversity of β-thalassemias reflects the heterogeneity of mutations of the β-globin locus, the action of many secondary and tertiary modifiers, and a wide range of environmental factors (Weatherall 2001).

HEMOGLOBIN E β-THALASSEMIA

Hemoglobin E has a substitution of lysine for glutamic acid at position 26 of the β-globin chain. This hemoglobin variant is particularly common in Southeast Asia (Weatherall and Clegg 2001). Both heterozygotes and homozygotes for HbE have hypochromic and microcytic cells, and molecular studies have documented deficient accumulation of the β^E mRNA. Molecular analysis indicates that the GAG change to AAG at codon 26, which is responsible for the amino acid substitution, also creates a cryptic splice donor site that slows splicing of intron 1 and results in abnormally spliced mRNA species (Orkin et al. 1982). Double heterozygosity for a β-thalassemia mutation and the β^E mutation account for almost 50% of the patients with severe β-thalassemia worldwide. This genotype is particularly common in certain countries in Southeast Asia (Olivieri et al. 2010a). Because of the Asian migration to the United States, this genotype is common where Asians have settled (e.g., California). The phenotype of HbE β-thalassemia is highly variable, with many patients remaining largely free of transfusion throughout their lifetime, whereas others are started on transfusion at an early age (Olivieri et al. 2010a). For those for whom transfusions were initiated early in life for unclear reasons, discontinuation of transfusion support may be possible later in life (Olivieri et al. 2010b). Recent reviews of the clinical features of HbE β-thalassemia summarize the clinical heterogeneity and the approach to management of these patients (Olivieri et al. 2008, 2010b). This condition is discussed in more detail by Fucharoen and Weatherall (2012).

PATHOPHYSIOLOGY

Erythropoiesis in individuals with β-thalassemia reflects the consequences of excess, unpaired α-globin (Nathan and Gunn 1966; Nathan et al. 1969; Cao and Galanello 2010; Sankaran and Nathan 2010). Indeed, the degree of imbalance in the α-globin versus β + γ-globin biosynthetic ratio is the major determinate of disease severity rather than the underproduction of hemoglobin (Weatherall et al. 1965; Bank and Marks 1966; Nathan and Gunn 1966; Nathan et al. 1969; Weatherall 2001; Rund and Rachmilewitz 2005). In β-thalassemia trait there is a twofold excess in the synthesis of α-globin, which is consistent with fairly normal hematopoiesis with only mild microcytosis and hypochromia of the red cells. The α to non-α biosynthetic ratio in individuals with thalassemia intermedia is typically 3–4/1 because residual capacity for β-globin synthesis along with usually modest but variable γ-globin synthesis mitigates the consequences of excess α-globin production. Individuals with β^0-thalassemia mutations have marked chain biosynthetic imbalance as the underlying basis for their severe phenotype.

Following synthesis, α-globin interacts with its molecular chaperone, α-hemoglobin stabilizing protein (AHSP), to form a protein complex before it is released to interact with β-globin in forming the hemoglobin tetramer (Yu et al. 2007; Weiss and dos Santos 2009). AHSP facilitates folding of α-globin and prevents the formation of misfolded aggregates. α-Globin mutations that impair interaction with AHSP are associated with microcytosis and anemia in humans (Yu et al. 2009). Loss of AHSP has also been shown to impair erythropoiesis in a mouse model of β-thalassemia (Kong et al 2004). Evidence suggests that AHSP levels may influence the phenotype of β-thalassemia (Lai et al. 2006).

Once the capacity of AHSP is exceeded, α-globin forms molecular aggregates, which precipitate, forming inclusions that damage the cell membrane and the membranes of intracellular organelles (Fig. 1). Aggregated α-chains also trigger the formation of reactive oxygen species, which further damage the protein and lipid constituents of cell membranes. As well as heme and iron, one of the most toxic products of unpaired α-chains is hemichromes, which bind to the membrane and promote clustering of band 3, one of its major constituents (Rund and Rachmilewitz 2005). The formation of α-chain inclusions occurs early during erythropoiesis and peaks in the polychromatophilic erythroblasts, leading to cellular apoptosis (Mathias et al. 2000). Thus, anemia in the severe β-thalassemias reflects both ineffective erythropoiesis as well as shortened red cell survival as a consequence of α-globin inclusions (Nathan and Gunn 1966; Nathan et al. 1969). The pathophysiology of β-thalassemia has been compared with other disorders such as Parkinson's disease and Huntington's disease, which are caused by accumulations of unstable, aggregation-prone proteins (Khandros and Weiss 2010; Khandros et al. 2011). Almost all cells have some capacity to detoxify and remove damaging proteins via multiple biochemical pathways termed protein quality control (PQC). The ubiquitin–proteasome system (UPS) and the lysosome-autophagy pathways that function in PQC are thought to participate in the degradation of α-globin, but the capacities of these pathways are exceeded in the erythroid cells of individuals with the severe forms of β-thalassemia.

GENETIC MODIFIERS

The thalassemias are heterogeneous at the molecular level, with more than 200 disease-causing mutations having been identified (Weatherall 2001; Rund and Rachmilewitz 2005; Cao and Galanello 2010; Thein 2013). Alleles characterized by a mild phenotype typically include promoter mutations, although splicing and frameshift mutations have also been discovered in individuals homozygous for β-thalassemia but having a mild phenotype. Coinheritance of an α-

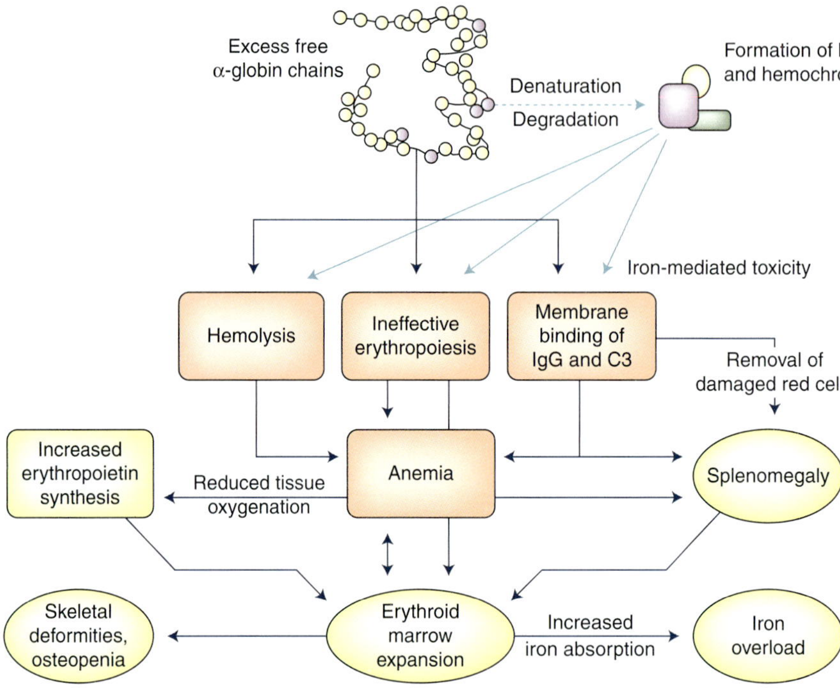

Figure 1. Pathophysiology of β-thalassemia. Because of the imbalance in chain synthesis, an excess of freed α-globin chains accumulates within erythroid cells. Aggregation, denaturation, and degradation of these chains leads to the formation of insoluble precipitates as well as hemichromes, which damage cell membranes. Membrane damage leads to ineffective erythropoiesis within the bone marrow, hemolysis of red cells within the circulation, and binding of immunoglobulin and complement components to red cell membranes, triggering loss of red cells in the spleen. The resulting anemia leads to diminished tissue oxygenation, an increase in erythropoietin levels, and further stimulation of the bone marrow. Bone marrow expansion causes skeletal deformities and osteopenia. Substances released from degenerating red cells increase iron absorption, which contributes to iron overload.

thalassemia gene is another mechanism that mitigates the phenotype of homozygous β-thalassemia (Kan and Nathan 1970; Wainscoat et al. 1983; Cao et al. 1994). Individuals doubly heterozygous for α- and β-thalassemia have microcytosis but essentially normal circulating hemoglobin concentrations. Individuals homozygous for β-thalassemia who inherit a chromosome having a single α-globin gene deletion may have a milder phenotype, whereas deletion of both α-globin genes on one chromosome is typically associated with thalassemia intermedia. Another mechanism for thalassemia intermedia is heterozygous β-thalassemia inherited along with either triplicated α-globin gene arrangement or with homozygosity for a triplicated globin gene arrangement (Premawardhena et al. 2005).

Fetal hemoglobin is a much more common and major modifier of disease severity in individuals with β-thalassemia. In normal individuals, HbF synthesis occurs in a minority of erythroid cells. Only 5%–8% of red cells, called F cells, contain HbF, amounting to 5%–20% of the total hemoglobin in those cells (Boyer et al. 1975). In the context of a severe deficiency of β-globin synthesis, even the low levels of γ-globin in F cells reduce the relative excess of α-globin and provide a potent selective survival advantage for cells making HbF in the context of the ineffective erythropoiesis characteristic of the most severe forms of β-thalassemia. This selective survival is thought to account for the elevated levels of fetal hemoglobin percentage in a significant majority of patients who are homozygous for β-thalassemia (Gabuzda et al. 1962; Weatherall 2001). Certain β-globin promoter mutations are associated with increased γ-chain synthesis from the same chromosome (Cao and Galanella 2010). The number of F cells in normal individuals, and presumably those with β-thalassemia, is under genetic control and varies among individuals (Galarneau et al. 2010). Presumably having F-cell numbers at the higher range of normal is associated with a milder clinical phenotype. Study of a large pedigree has shown that coinheritance of an HPFH gene is associated with higher mean corpuscular hemoglobin, mean corpuscular volume, and HbA_2 levels (Garner et al. 2003). A recent study in untransfused patients with thalassemia intermedia has documented an inverse correlation between HbF levels and the frequencies of morbidities reflecting disease severity (Musallam et al. 2012). Hydroxyurea has been reported to increase HbF in some untransfused patients with severe β-thalassemia and to reduce the transfusion requirement in a minority of patients on regular transfusions (Karimi et al. 2011). A favorable response to hydroxyurea may be more frequent in thalassemic individuals with certain haplotypes (Italia et al. 2009).

Hereditary persistence of HbF (HPFH) was initially observed more than 50 years ago (Weatherall and Clegg 2001). Recent molecular studies have shown that HPFH may result from deletion mutations or point mutations within the β-globin gene cluster. Deletion mutations involving the δ- and β-globin genes may cause the phenotype of δβ-thalassemia, characterized by microcytosis and heterocellular distribution of HbF, or be associated with increased production of HbF in a pancellular distribution with morphologically normal red cells, the phenotype of HPFH. Recent characterization of several deletional mutations in individuals with either the HPFH phenotype or the phenotype of δβ-thalassemia has identified a functional element within a 3.5-kb region between the $^{A}\gamma$- and δ-globin genes that, when present, results in relative silencing of the γ-globin genes and the γβ-thalassemia phenotype and, when absent, allows higher levels of HbF expression and the HPFH phenotype (Sankaran et al. 2011). This region includes sites for binding of the transcription factor BCL11A. Alternatively, augmentation of HbF production may reflect a single nucleotide polymorphism (SNP) in one of the γ-globin gene promoters or elsewhere in the globin locus, which leads to overexpression of the cognate gene. Among the most common is an SNP at −158 bp of the $^{G}\gamma$ gene promoter that creates an *XMNI* site polymorphism. Detailed linkage analysis has shown that this SNP is not the actual causal mechanism of HPFH, but, rather, augmentation of $^{G}\gamma$ gene expression occurs as a consequence of an SNP in linkage disequilibrium with the aforementioned *Xmn-1* site located midway between $^{A}\gamma$- and δ-globin genes, which may affect BCL11A binding (Galarneau et al. 2010). Fetal hemoglobin is normal in normal subjects and in β-thalassemia heterozygotes with this polymorphism, but the variant leads to increased production during hematopoietic stress such as occurs in homozygous β-thalassemia (Labie et al. 1985). Another HPFH SNP, a substitution at −96 bp of the $^{A}\gamma$ promoter, has been found linked to a nonsense mutation in the β-globin gene in individuals with Sardinian β-thalassemia (Pirastu and Kan 1984). Homozygotes for this combination of mutations are clinically normal other than having predominantly HbF in their red cells.

In addition to SNPs within the β-globin locus, genome-wide association studies (GWAS) have identified additional loci that can contribute significantly to the variation in HbF levels (Garner et al. 2000; Uda et al. 2008; Galarneau et al. 2010). An SNP in the BCL11A gene has been shown to contribute to the level of HbF expression. As discussed by Sankaran and Orkin (2013), functional studies have shown that BCL11A is a significant component of the silencing mechanism that turns off the γ-globin genes (Sankaran et al. 2009). Suppression of BCL11A levels in cultured thalassemic erythroid cells markedly increases HbF production (Wilber et al. 2011a). Experimental evidence shows correction of sickle cell disease in a mouse model by suppression of Bclla (Xu et al. 2011). BCL11A binds several points in the region between the $^{A}\gamma$- and δ-globin genes including within the critical region that differentiates HPFH from δβ-thalassemia deletions (Sankaran et al. 2011). The other locus that has been implicated in modulating HbF levels is MYB (Galarneau et al. 2010). An SNP in what is thought to be a regulatory region near the gene is associated with differences in the level of MYB expression. Individuals with lower levels of MYB have higher HbF levels than individuals with higher MYB levels. In another study, a 3-bp deletion in an intragenic region near the MYB gene with enhancer-like activity was associated with higher HbF levels (Farrell et al. 2011). Variants in both the BCL11A and MYB loci have been shown to modify disease severity in individuals homozygous for Sardinian β^0-thalassemia (Galanello et al. 2009). Other transcription factors, including KLF1, have been implicated in regulation of HbF levels (for review, see Wilber et al. 2011b). KLF1 may mediate its effects on HbF through a dual mechanism acting both on the β-globin and BCL11A promoters.

Individuals with thalassemia have hyperbilirubinemia secondary to ongoing hemolysis and ineffective erythropoiesis. Iron loading occurs because of increased iron absorption as well as the administration of blood transfusions. Osteoporosis is common, potentially secondary to hypogonadism and other endocrine abnormalities as well as expansion of marrow cavities. Increased susceptibility to infection is also characteristic of the severe forms of β-thalassemia. Genetic polymorphisms in various genes involved with these clinical features have been described as acting as tertiary modifiers of the phenotype of thalassemia (Weatherall 2001).

IRON OVERLOAD

Patients with the more severe forms of β-thalassemia, both intermedia and major, have increased tissue deposition of iron. Senescence of transfused red cells in patients with transfusion-dependent β-thalassemia results in iron deposition within the reticuloendothelial system. As iron overload progresses, deposited iron also appears within the hepatic parenchyma, various endocrine tissues, and, more slowly, in the myocardium.

In normal humans, iron homeostasis is achieved by controlling absorption (see Ganz and Nemeth 2012). Only ~1 mg is lost from the body each day, largely through shedding of the epithelial cells from the intestine, urinary tract, skin, and other mucosal organs. Each milliliter of transfused blood contains ~1 mg of iron, thus receipt of a unit of packed red cells typically results in the deposition of 200 mg of iron ultimately in the tissues following red cell senescence. Lacking excretory mechanisms, individuals with thalassemia who receive blood transfusion inevitably experience significant iron overload.

In individuals with thalassemia intermedia, increased tissue iron occurs as a consequence of occasional transfusions but mainly reflects increased iron absorption (Ginzburg et al. 2011). This paradoxical increase in iron absorption despite systemic iron overload probably reflects release of erythroid factors during the cellular apoptosis associated with ineffective erythropoiesis that inhibit hepcidin production by the liver. Hepcidin is a 25-amino-acid peptide hormone that negatively regulates iron flow into the plasma by inhibiting the absorption of dietary iron, the release of iron from macrophages, and the release of stored iron from hepatocytes (Ganz and Nemeth 2011). Ferric iron exits from cells via the iron exporter ferroportin, a multipass transmembrane protein. Ferroportin is abundant in duodenal enterocytes, in splenic and hepatic macrophages, and in hepatocytes, all cells known to export iron (Ganz and Nemeth 2011). Hepcidin inhibits iron efflux by binding ferroportin and triggering endocytosis of both molecules with subsequent lysosomal degradation (Nemeth et al. 2004). Factors released from apoptotic erythroblasts include growth differentiation factor 15 (GDF15), a member of the transforming growth factor-β superfamily (Tanno et al. 2007). Serum from thalassemia patients suppressed the production of hepcidin by primary human hepatocytes, and depletion of GDF15 prevented the hepcidin suppression. A second molecule, named twisted gastrulation (TWSGI), was also shown to be highly expressed in erythroblasts and also able to inhibit hepcidin expression in mouse and human erythroid cells (Tanno et al. 2009). The roles of GDF15 and TWSGI remain to be fully defined. The end result of suppression of hepcidin expression is to enhance iron absorption from the intestine and to allow iron release from macrophages. The normal and pathological regulation of hepcidin synthesis has recently been reviewed (Nemeth 2010). Synthetic mini-hepcidins with a longer half-life have been shown to reduce iron absorption in mice after oral administration and have been proposed as potential therapeutic agents in individuals with severe β-thalassemia (Preza et al. 2011).

Iron released from cells is bound to transferrin and transported to the bone marrow and other tissues, where iron is taken up by the transferrin receptors. Less than 1% of the total body iron is found in blood at any one time, although up to 25 mg may circulate as transferrin-bound iron throughout the 24-h cycle. The expanded erythropoiesis in individuals with β-thalassemia results in a dramatic increase in plasma iron turnover of 10-fold to 15-fold over normal (Hershko et al. 2005). Normally, the transferrin saturation is maintained in a relatively low range of ~10%–50%. Iron is stored in tissue in the form of ferritin, a multicomponent protein shell that internalizes iron, thereby protecting cellular constituents from potential oxidative damage. Intracellular ferritin iron is in equilibrium with the cytosolic soluble iron pool. Hemosiderin is an insoluble aggregate of iron that forms in lysosomes when ferritin is degraded (Wu et al. 2010; Kim et al. 2011). Both ferritin and hemosiderin accumulate in the cells of individuals with severe β-thalassemia. Accumulation of iron in reticuloendothelial cells is relatively harmless, whereas accumulation of iron in parenchymal tissues may damage critical cells in the heart, endocrine glands, and liver (Hershko et al. 2005).

As a consequence of iron overload in patients with thalassemia either from blood transfusion or excessive absorption or a combination of the two, transferrin saturation increases to 75%–100% and non-transferrin-bound iron (NTBI) is found in the blood (Hershko et al. 1978). NTBI is very heterogeneous and exists in various forms including in complex with citrate and proteins (Evans et al. 2008). Strategies have been devised to detect labile plasma iron (LPI), a fraction of the NTBI pool that is metabolically active in interacting with membrane constituents, leading to membrane damage via the formation of reactive oxygen species (ROS) (Pootrakul et al. 2004; Hershko 2010). Rather than using the transferrin receptor, NTBI enters cells by various cellular channels in forms that have the potential to damage cells (Oudit et al. 2003; Liuzzi et al. 2006). The liver and myocardium clear NTBI at a rate 200-fold that of transferrin-bound iron (Oudit et al. 2003, 2006). A portion of the NTBI is accessible to iron chelators directly (Breuer et al. 2001), whereas additional iron may be chelated in the presence of iron-mobilizing agents (Esposito et al. 2003). Chelators administered to patients with thalassemia have the potential to reduce NTBI (Porter et al. 1996). Intravenous administration is more effective at reducing NTBI than subcutaneous administration. The reduction and reemergence of NTBI have rather complex kinetics (Porter et al. 1996), although these studies support the use of continuous rather than intermittent chelation in high-risk patients. As discussed in detail below, cardiac myocytes are a critical target for iron-induced ROS. Using an in vitro model of neonatal rat cardiomyocytes, the oral chelators deferasirox and deferiprone were able to effectively access and reduce the intracellular labile iron compartments, whereas desferrioxamine removed iron mainly by eliminating the LPI in plasma (Glickstein et al. 2006). These in vitro studies suggested that the rapid accessibility of oral chelators to intracellular labile iron compartments may render them potentially efficacious for protection and reversal of cardiac damage induced by iron overload (Glickstein et al. 2006).

CLINICAL MANIFESTATIONS

Anemia

Classically, individuals with severe β-thalassemia have presented with variable but often very severe degrees of anemia, expansion of the bone marrow spaces secondary to erythroid hyperplasia, hepatosplenomegaly, and extramedullary hematopoiesis in the chest and abdomen. The external appearance is characterized by pallor and slight jaundice, frontal bossing and other abnormalities of the facies secondary to marrow expansion, and abdominal enlargement due to hepatosplenomegaly. Usually, these manifestations are absent or minimally present in patients with thalassemia major if transfusion therapy is initiated early during the first year of life provided that the hemoglobin levels are maintained at 9–10 g/dL. Where transfusion is readily available, the classical physical findings of thalassemia are now more commonly found in patients with thalassemia intermedia, particularly those who are at the severe end of the phenotypical spectrum. In such patients, the physical manifestations can be suppressed in part with subsequent transfusion. Transfusion practices for thalassemia intermedia vary considerably. Individuals with thalassemia intermedia in developed countries are likely to be transfused, whereas transfusion is used more sparingly in countries where resources are limited (Rachmilewitz and Giardina 2011; Taher et al. 2011). The natural history of thalassemia intermedia is highly variable, although the complications of severe anemia are common (Borgna-Pignatti et al. 2010). Splenectomy may be required because of very large spleen size, with increasing anemia or increasing transfusion requirement with or without a decrease in neutrophil or platelet count.

Iron Overload

The clinical manifestations of iron overload have come to dominate the clinical phenotype of individuals with severe β-thalassemia. Cardiac dysfunction is the main clinical problem that may lead to early death. Endocrine abnormalities, particularly hypogonadism, low growth hormone, hypothyroidism, and diabetes mellitus, are also significant problems. Iron deposition of the liver may be substantial, although functional abnormalities are usually mild unless iron overload is very severe. Fortunately chelation therapy prevents and, when given in intensive combination therapy, may reverse the complications of iron overload (Farmaki et al. 2010).

Noninvasive Measurement of Tissue Iron

Hepatic biopsy and measurement of liver iron concentration long remained the gold standard for estimating the degree of iron overload in patients with severe β-thalassemia. Magnetic susceptibility imaging was the first noninvasive quantitative technique, which established a correlation between liver iron concentration by biopsy and this noninvasive measurement (Brittenham et al. 1982). The instrumentation required for this technique is available in only a few institutions, and therefore this method has not been widely applied. Only when liver iron concentration could be reliably estimated by magnetic resonance imaging (MRI) with clinically available devices that could be standardized from institution to institution, did noninvasive measurement of liver iron supplant hepatic biopsy as the gold standard (Gandon et al. 2004; St Pierre et al. 2005).

MRI has also been applied to estimate cardiac iron. The $T2^*$ parameter and its reciprocal $1/R2^*$ have been shown to correlate with the degree of iron overload in the liver and, by extrapolation, to the concentration of iron in the myocardium (Wood 2011). Limited autopsy studies support a direct correlation between chemically determined myocardial iron and MRI measurements (Carpenter et al. 2011). MRI has also been applied to the evaluation of iron deposition in the pituitary gland and the pancreas. Availability of these noninvasive techniques has been useful in determining the relative rate of mobilization of iron from various tissues during chelation therapy. Iron is mobilized more readily from the liver and much more slowly from the heart and endocrine tissues (Wood 2011).

Cardiac Manifestations

The institution of regular transfusions for patients with thalassemia major prevents the overt consequences of anemia. However, iron accumulation progresses with iron deposition in the myocardium and toxic NTBI in the plasma,

 Cite this article as *Cold Spring Harb Perspect Med* doi: 10.1101/cshperspect.a011726

becoming problematic during the late teenage years or early twenties. Cardiac abnormalities include arrhythmias, both atrial and ventricular, and/or congestive heart failure (Engle 1964). Sudden death often occurred, whereas other patients died of progressive congestive heart failure. Death by the mid twenties was the norm.

This dismal clinical course was modified with the introduction of regular chelation with desferrioxamine, particularly when it was given by continuous subcutaneous infusion over a significant portion of each 24-h period (Propper et al. 1976, 1977; Pippard et al. 1978; Wolfe et al. 1985; Olivieri et al. 1994; Brittenham et al. 1993). Not unexpectedly, compliance with the cumbersome subcutaneous administration via a small infusion pump varied widely among participants, providing an unintended opportunity to compare well-chelated versus poorly chelated individuals (Brittenham et al. 1994; Olivieri et al. 1994). The dramatic reduction in cardiac mortality in well-chelated groups was among the first unequivocal evidence that chelation therapy could modify the clinical phenotype with respect to the cardiac manifestations of iron overload. Subsequent large studies have strongly supported the use of chronic chelation to prevent cardiac death (Borgna-Pignatti et al. 2006; Wood 2011).

Improvement in established cardiac disease was first achieved with continuous interfusion of desferrioxamine (Anderson et al. 2004). Iron was mobilized more readily from the liver than the heart as established by MRI T2* measurements. Some evidence suggests that mobilization of iron from the myocardium is more efficient with the low-molecular-weight chelator deferiprone (Brittenham et al. 2003), and that desferrioxamine and deferiprone administration together are more effective than desferrioxamine alone in reversing cardiac dysfunction as shown in a randomized trial (Tanner et al. 2007). Deferasirox has also been shown to reduce and prevent cardiac iron overload (Pennell et al. 2010, 2011).

Pulmonary hypertension is an increasingly recognized complication in patients with severe β-thalassemia (Morris and Vichinsky 2010). Advancing age and history of splenectomy are the major risk factors that have been identified. Among the pathophysiological factors are red cell membrane pathology, coagulation abnormalities, platelet activation, oxidative stress, and chronic hemolysis with release of thrombogenic vesicles (Morris and Vichinsky 2010). Although the incidence of pulmonary hypertension is more common in patients with thalassemia intermedia, a recent cross-sectional study showed that pulmonary hypertension may occur in both children and adults with age, splenectomy, hepatitis C infection, and smoking as significant univariate risk factors (Morris et al. 2011). When present, pulmonary hypertension may contribute to the severity of cardiovascular symptomatology and complicate heart failure due to left ventricular dysfunction.

Endocrine Abnormalities

Growth retardation secondary in part to growth hormone deficiency and hypogonadism are typically the initial manifestations of iron overload in β-thalassemic patients (Chatterjee and Bajoria 2010). In unchelated patients, failure to develop secondary sex characteristics during the teenage years was very common. Regular chelation therapy has reduced the incidence of hypogonadism, although hormone replacement is often required (Wood 2011). Well-chelated young men with thalassemia are fertile, although potentially requiring hormonal administration, and women with severe β-thalassemia may achieve motherhood either with or without obstetrical intervention. Other endocrine complications include impaired glucose tolerance and diabetes (Noetzli et al. 2011), hypothyroidism, hypoparathyroidism, and growth hormone deficiency (Borgna-Pignatti et al. 2004a).

Hepatic Manifestations

Progressive iron deposition is characteristic in patients with severe β-thalassemia. Invasive liver biopsies and now non-invasive methodologies (see below) have established the correlation between liver iron concentration and histological and functional abnormalities. Iron accumulation to the point of 7 mg/g liver (dry weight) seems well tolerated, but as the liver iron concentration increases with regular transfusion and no or inadequate chelation, fibrosis in the periportal areas and ultimately frank cirrhosis occur. Liver functional abnormalities remain mild until iron overload is severe. Hepatitis C infection was common before the development of effective screening and potentially accelerated the development of cirrhosis and increased the risk for development of hepatocellular carcinoma (Borgna-Pignotti et al. 2004a; Mancuso 2010). Chelation therapy prevents iron accumulation, although the increased risk of hepatocellular carcinoma as a consequence of hepatic iron overload is not completely eliminated.

Other Complications

Osteoporosis often occurs in patients with thalassemia, reflecting marrow expansion, endocrine deficiencies, iron toxicity, and the potential toxicity of chelators (Terpos and Voskaridou 2010). Cortical thinning and subclinical fractures as well as problematic clinical fractures may occur, the latter with minimal trauma. A hypercoagulable state, which increases the risk for thromboembolism, has also been described in patients with thalassemia secondary to platelet activation, red cell membrane damage, and endothelial cell activation (Cappellini et al. 2010). Another clinical manifestation is the occurrence of chronic skin ulceration about the ankles secondary to chronic anemia. All of these complications are more common in untransfused patients with thalassemia intermedia than in frequently transfused patients with thalassemia intermedia or patients with thalassemia major who receive regular transfusions (Borgna-Pignatti et al. 2010). Nutritional deficiencies are common in individuals with severe β-thalassemia and may contribute to other complications such as osteoporosis and diabetes mellitus (Fung 2010). The role of inhibitors of osteoclast function in thalassemia is not firmly established, but small trials suggest that such treatment may be useful (Leung et al. 2009).

REFERENCES

**Reference is also in this collection.*

Anderson LJ, Westwood MA, Holden S, Davis B, Prescott E, Wonke B, Porter JB, Walker JM, Pennell DJ. 2004. Myocardial iron clearance during reversal of siderotic cardiomyopathy with intravenous desferrioxamine: A prospective study using T2* cardiovascular magnetic resonance. *Br J Haematol* **127:** 348–355.

Bank A, Marks PA. 1966. Excess α chain synthesis relative to β chain synthesis in thalassaemia major and minor. *Nature* **212:** 1198–1200.

Borgna-Pignatti C, Rugolotto S, De Stefano P, Zhao H, Cappellini MD, Del Vecchio GC, Romeo MA, Forni GL, Gamberini MR, Ghilardi R, et al. 2004. Survival and complications in patients with thalassemia major treated with transfusion and deferoxamine. *Haematologica* **89:** 1187–1193.

Borgna-Pignatti C, Cappellini MD, De Stefano P, Del Vecchio GC, Fomi GL, Gamberini MR, Ghilard R, Piga A, Romeo MA, Zhao H, et al. 2006. Cardiac morbidity and mortality in deferoxamine- or deferiprone-treated patients with thalassemia major. *Blood* **107:** 3733–3737.

Borgna-Pignatti C, Marsella M, Zanforrin N. 2010. The natural history of thalassemia intermedia. *Ann NY Acad Sci* **1202:** 214–220.

Boyer Sh, Belding TK, Margolet L, Noyes AN. 1975. Fetal hemoglobin restriction to a few erythrocytes (F cells) in normal human adults. *Science* **188:** 361–363.

Breuer W, Ermers MJ, Pootrakul P, Abramov A, Hershko C, Cabantichik ZI. 2001. Desferrioxamine-chelatable iron, a component of serum non-transferrin-bound iron, used for assessing chelation therapy. *Blood* **197:** 792–798.

Brittenham GM, Farrell DE, Harris JW, Feldman ES, Danish EH, Muir WA, Tripp JH, Bellon EM. 1982. Magnetic-susceptibility measurement of human iron stores. *N Engl J Med* **307:** 1671–1675.

Brittenham GM, Cohen AR, McLaren Ce, Martin MB, Griffith PM, Nienhuis AW, Young NS, Allen CJ, Farrell DE, Harris JW. 1993. Hepatic iron stores and plasma ferritin concentration in patients with sickle cell anemia and thalassemia major. *Am J Hematol* **42:** 81–85.

Brittenham GM, Griffith PM, Nienhuis AW, McLaren CE, Young NS, Tucker EE, Allen CJ, Farrell DE, Harris JW. 1994. Efficacy of deferoxamine in preventing complications of iron overload in patients with thalassemia major. *N Engl J Med* **1331:** 567–573.

Brittenham GM, Nathan DG, Olivieri NF, Pippard MJ, Weatherall DJ. 2003. Deferiprone versus desferrioxamine in thalassaemia and T2* validation and utility. *Lancet* **361:** 183.

Cao A, Galanello R. 2010. β-Thalassemia. *Genet Med* **12:** 61–76.

Cao A, Galanello R, Rosatelli MC. 1994. Genotype–phenotype correlations in β-thalassemias. *Blood Rev* **8:** 1–12.

Cappellini MD, Motta I, Musallam KM, Taher AT. 2010. Redefining thalassemia as a hypercoagulable state. *Ann NY Acad Sci* **1202:** 231–236.

Carpenter JP, He T, Kirk P, Roughton M, Anderson LJ, de Noronha SV, Sheppard MN, Porter JB, Walker JM,

Wood JC, et al. 2011. On T2* magnetic resonance and cardiac iron. *Circulation* **123:** 1519–1528.

Chatterjee R, Bajoria R. 2010. Critical appraisal of growth retardation and pubertal disturbances in thalassemia. *Ann NY Acad Sci* **1202:** 100–114.

Engle MA. 1964. Cardiac involvement in Cooley's anemia. *Ann NY Acad Sci* **119:** 694–702.

Esposito BP, Breuer W, Sirankapracha P, Pootrakul P, Hershko C, Cabantchik ZI. 2003. Labile plasma iron in iron overload: Redox activity and susceptibility to chelation. *Blood* **102:** 2670–2677.

Evans RW, Rafique R, Zarea A, Rapisarda C, Cammack R, Evans PJ, Porter JB, Hider RC. 2008. Nature of non-transferrin-bound iron: Studies on iron citrate complexes and thalassemic sera. *J Biol Inorg Chem* **13:** 57–74.

Farmaki T, Tzoumari I, Pappa C, Chouliaras G, Berdoukas V. 2010. Normalisation of total body iron load with very intensive combined chelation reverses cardiac and endocrine complications of thalassaemia major. *Br J Haematol* **148:** 466–475.

Farrell JJ, Sherva RM, Chen Z, Luo H, Chu BF, Ha SY, Li CK, Lee ACW, Li RCH, Li CK, et al. 2011. A 3-bp deletion in the *HBSIl-MYB* intergenic region of chromosome 6q23 is associated with HbF expression. *Blood* **117:** 4935–4945.

* Fucharoen S, Weatherall DJ. 2012. The hemoglobin E thalassemias. *Cold Spring Harb Perspect Med* **2:** a011734.

Fung EB. 2010. Nutritional deficiencies in patients with thalassemia. *Ann NY Acad Sci* **1202:** 188–196.

Gabuzda TG, Nathan DG, Gardner FH. 1962. Comparative metabolism of haemoglobins A and F in thalassaemia. *Nature* **196:** 781–782.

Galanello R, Sanna S, Perseu L, Sollaino MC, Satta S, Lai ME, Barella S, Uda M, Usala G, Abecasis GR, et al. 2009. Amelioration of Sardianian β^0 thalassemia by genetic modifiers. *Blood* **114:** 3935–3937.

Galarneau G, Palmer CD, Sankaran VG, Orkin SH, Hirschhorn JN, Lettre G. 2010. Fine-mapping at three loci known to affect fetal hemoglobin levels explains additional genetic variation. *Nat Genet* **42:** 1049–1051.

Gandon Y, Olivie D, Guyader D, Aube C, Oberti F, Sebille V, Deugnier Y. 2004. Non-invasive assessment of hepatic iron stores by MRI. *Lancet* **363:** 357–362.

Ganz T, Nemeth E. 2011. Hepcidin and disorders of iron metabolism. *Annu Rev Med* **62:** 347–360.

* Ganz T, Nemeth E. 2012. Iron metabolism: Interactions with normal and disordered erythropoiesis. *Cold Spring Harb Perspect Med* **2:** a011668.

Garner C, Tatu T, Reittie JE, Littlewood T, Darley J, Cervino S, Farrall M, Kelly P, Spector TD, Thein SL. 2000. Genetic influences on F cells and other hematologic variables: A twin heritability study. *Blood* **95:** 342–346.

Garner C, Dew TK, Sherwood R, Rees D, Thein SL. 2003. Heterocellular hereditary persistence of fetal haemoglobin affects the haematological parameters of β-thalassemia trail. *Br J Haematol* **123:** 353–358.

Ginzburg Y, Rivella S. 2011. β-Thalassemia: A model for elucidating the dynamic regulation of ineffective erythropoiesis and iron metabolism. *Blood* **118:** 4321–4330.

Glickstein H, Ben El R, Link G, Breuer W, Konijn AM, Hershko C, Nick H, Cabantchik ZI. 2006. Action of chelators in iron-loaded cardiac cells: Accessibility to intracellular labile iron and functional consequences. *Blood* **108:** 3195–3203.

Hershko C. 2010. Pathogenesis and management of iron toxicity in thalassemia. *Ann NY Acad Sci* **1202:** 1–9.

Hershko C, Graham G, Bates GW, Rachmilewitz EA. 1978. Non-specific serum iron in thalassaemia: An abnormal serum iron fraction of potential toxicity. *Br J Haematol* **40:** 255–263.

Hershko C, Link G, Konijn AM, Cabantchik ZI. 2005. Objectives and mechanism of iron chelation therapy. *Ann NY Acad Sci* **1054:** 124–135.

Italia KY, Jijina FJ, Merchant R, Panjwani S, Nadkarni AH, Sawant PM, Nair SB, Ghosh K, Colah RB. 2009. Response to hydroxyurea in β thalassemia major and intermedia: Experience in western India. *Clin Chim Acta* **407:** 10–15.

Kan YW, Nathan DG. 1970. Mild thalassemia: The result of interactions of α and β thalassemia genes. *J Clin Invest* **49:** 635–642.

Karimi M, Haghpanah S, Farhadi A, Yavarian M. 2011. Genotype–phenotype relationship of patients with β-thalassemia taking hydroxyurea: A 13-year experience in Iran. *Int J Hematol* **95:** 51–56.

Khandros E, Weiss MJ. 2010. Protein quality control during erythropoiesis and hemoglobin synthesis. *Hematol Oncol Clin N Am* **24:** 1071–1088.

Khandros E, Thom CS, D'Souza J, Weiss MJ. 2011. Integrated protein quality control pathways degrade free α globin in β-thalassemia. *Blood* **119:** 5265–5275.

Kim D, Jensen JH, Wu EX, Feng L, Au WY, Cheung JS, Ha SY, Sheth SS, Brittenham GM. 2011. Rapid monitoring of iron-chelating therapy in thalassemia major by a new carciovascular MR measure: The reduced transverse relaxation rate. *NRM Biomed* **24:** 771–777.

Kong Y, Zhou S, Kihm A, Katein AM, Yu X, Gell DA, Mackay JP, Adachi K, Foster-Brown L, Louden CS, et al. 2004. Loss of α-hemoglobin-stabilizing protein impairs erythropoiesis and exacerbates β-thalassemia. *J Clin Invest* **114:** 1457–1466.

Labie D, Pagnier J, Lapoumeroulie C, Rauabhi F, Dunda-Belkhodja O, Chardin P, Beldjord C, Wajcman H, Fabry ME, Nagel RL. 1985. Common haplotype dependency of high G γ-globin gene expression and high HbF levels in β-thalassemia and sickle cell anemia patients. *Proc Natl Acad Sci* **82:** 2111–2114.

Lai MI, Jiang J, Silver N, Best S, Menzel S, Mijovic A, Colella S, Ragoussi J, Garncer C, Weiss MJ, et al. 2006. α-haemoglobin stabilizing protein is a quantitative trait gene that modifies the phenotype of β-thalassaemia. *Br J Haematol* **133:** 675–682.

Leung TF, Chu Y, Lee Y, Cheng FW, Leung WK, Shing MM, Li CK. 2009. Long-term effects of pamidronate in thalassemic patients with severe bone mineral density deficits. *Hemoglobin* **33:** 361–369.

Liuzzi JP, Avdemir F, Nam H, Knutson MD, Cousins RJ. 2006. Zip 14 (Slc39a14) mediates non-transferrin-bound iron uptake into cells. *Proc Natl Acad Sci* **103:** 13612–13617.

Mancuso A. 2010. Hepatocellular carcinoma in thalassemia: A critical review. *World J Hepatol* **2:** 171–174.

Mathias LA, Fisher TC, Zeng L, Meiselman HJ, Weinberg KI, Hiti AL, Malik P. 2000. Ineffective erythropoiesis in β-thalassemia major is due to apoptosis at the polychromatophilic normoblast stage. *Exp Hematol* **28:** 1343–1353.

Morris CR, Vichinsky EP. 2010. Pulmonary hypertension in thalassemia. *Ann NY Acad Sci* **1202:** 205–213.

Morris CR, Kim HY, Trachtenberg F, Wood J, Quinn CT, Sweeters N, Kwiatkowski JL, Thompson AA, Giardina PJ, Boudreaux J, et al. 2011. Risk factors and mortality associated with an elevated tricuspid regurgitant jet velocity measured by Doppler-echocardiography in thalassemia: A thalassemia clinical research network report. *Blood* **18:** 3794–3802.

Musallam KM, Sankaran VG, Cappellini MD, Duca L, Nathan DG, Taher AT. 2012. Fetal hemoglobin levels and morbidity in untransfused patients with β-thalassemia intermedia. *Blood* **119:** 364–367.

Nathan DG, Gunn RB. 1966. Thalassemia: The consequences of unbalanced hemoglobin synthesis. *Am J Med* **4:** 815–830.

Nathan DG, Stossel TB, Gunn RB, Zarkowsky HS, Laforet MT. 1969. Influence of hemoglobin precipitation on erythrocyte metabolism in α and β thalassemia. *J Clin Invest* **48:** 33–41.

Nemeth E. 2010. Hepcidin in β-thalassemia. *Ann NY Acad Sci* **1202:** 31–35.

Nemeth E, Tuttle MS, Powelson J, Vaughn MB, Donovan A, Ward DM, Ganz T, Kaplan J. 2004. Hepcidin regulates cellular iron efflux by binding to ferroportin and inducing its internalization. *Science* **306:** 2051–2053.

Noetzli LJ, Mittelman SD, Watanabe RM, Coates TD, Wood JC. 2011. Pancreatic iron and glucose dysregulation in thalassemia major. *Am J Hematol* doi: 10.1002/ajh.22223.

Olivieri NF, Nathan DG, MacMillan JH, Wayne AS, Liu PP, McGee A, Martin M, Koren G, Cohen AR. 1994. Survival in medically treated patients with homozygous β-thalassemia. *N Engl J Med* **331:** 574–578.

Olivieri NF, Muraca GM, O'Donnell A, Premawardhena A, Fisher C, Weatherall DJ. 2008. Studies in haemoglobin E β-thalassaemia. *Br J Haematol* **141:** 388–397.

Olivieri NF, Pakbaz Z, Vichinsky E. 2010a. HbE/β-thalassemia: Basis of marked clinical diversity. *Hematol Oncol Clin North Am* **24:** 1055–1070.

Olivieri NF, Thayalsuthan V, O'Donnell A, Premawardhena A, Rigobon C, Muraca G, Fisher C, Weatherall DJ. 2010b. Emerging insights in the management of hemoglobin E β thalassemia. *Ann NY Acad Sci* **1202:** 155–157.

Orkin SH, Kazzazian HH Jr, Antonarakis SE, Oster H, Goff SC, Sexton JP. 1982. Abnormal RNA processing due to the exon mutation of β E-globin gene. *Nature* **300:** 768–769.

Oudit GY, Sun H, Trivieri MG, Koch SE, Dawood F, Ackerley C, Yazdanpanah M, Wilson GJ, Schwartz A, Liu PP, et al. 2003. L-type Ca^{2+} channels provide a major pathway for iron entry into cardiomyocytes in iron-overload cardiomyopathy. *Nat Med* **9:** 1187–1194.

Oudit GY, Trivieri MG, Khaper N, Liu PP, Backx PH. 2006. Role of L-type Ca^{2+} channels in iron transport and iron-overload cardiomyopathy. *J Mol Med (Berl)* **84:** 349–364.

Pennell DJ, Porter JB, Cappellini MD, El-Beshlawy A, Chan LL, Aydinok Y, Elalfy MS, Sutcharitchan P, Li CK, Ibrahim H, et al. 2010. Efficacy of deferasirox in reducing and preventing cardiac iron overload in β-thalassemia. *Blood* **115:** 2364–2371.

Pennell DJ, Porter JB, Cappellini MD, Chan LL, El-Beshlawy A, Aydinok Y, Ibrahim H, Li CK, Viprakasit V, Elalfy MS, et al. 2011. Continued improvement in myocardial T2* over two years of deferasirox therapy in thalassemia major patients with cardiac iron overload. *Haematologica* **96:** 48–54.

Pippard MJ, Callender ST, Weatherall DJ. 1978. Intensive iron-chelation therapy with desferrioxamine in iron loading patients. *Clin Sci Mol Med* **54:** 99–106.

Pirastu M, Kan YW. 1984. Multiple mutations produce $\delta\beta^0$ thalassemia in Sardinia. *Science* **223:** 929–930.

Pootrakul P, Breuer W, Sametband M, Sirankapracha P, Hershko C, Cabantchik ZI. 2004. Labile plasma iron (LPI) as an indicator of chelatable plasma redox activity in iron-overloaded β-thalassemia/HbE patients treated with an oral chelator. *Blood* **104:** 1504–1510.

Porter JB, Abeysinghe RD, Marshall L, Hider RC, Singh S. 1996. Kinetics of removal and reappearance of non-transferrin-bound plasma iron with deferoxamine therapy. *Blood* **88:** 705–713.

Premawardhena A, Fisher CA, Olivieri NF, de Silva S, Sloane-Stanley J, Wood WG, Weatherall DJ. 2005. A novel molecular basis for β thalassemia intermedia poses new questions about its pathophysiology. *Blood* **106:** 3251–3255.

Premawardhena A, Arambepola M, Katugaha N, Weatherall DJ. 2008. Is the β thalassaemia trait of clinical importance? *Br J Haematol* **141:** 407–410.

Preza GC, Ruchala P, Pinon R, Ramos E, Qiao B, Perallta MA, Sharma S, Waring G, Ganz T, Nameth E. 2011. Minihepcidins are rationally designed small peptides that mimic hepcidin activity in mice and may be useful for treatment of iron overload. *J Clin Invest* **121:** 4880–4888.

Propper RD, Shurin SB, Nathan DG. 1976. Reassessment of the use of desferrioxamine B in iron overload. *N Engl J Med* **294:** 1421–1423.

Propper RD, Cooper B, Rufo RR, Nienhuis AW, Anderson WF, Burn HF, Rosenthal A, Nathan DG. 1977. Continuous subcutaneous administration of deferoxamine in patients with iron overload. *N Engl J Med* **25:** 418–423.

Rachmilewitz EA, Giardina PJ. 2011. How I treat thalassemia. *Blood* **118:** 3479–3488.

Rund D, Rachmilewitz E. 2005. β-Thalassemia. *N Engl J Med* **353:** 1135–1146.

Sankaran VG, Nathan DG. 2010. Thalassemia: An overview of 50 years of clinical research. *Hematol Oncol N Am* **24:** 1005–1020.

* Sankaran VG, Orkin SH. 2013. The switch from fetal to adult hemoglobin. *Cold Spring Harb Perspect Med* **3:** a011643.

Sankaran VG, Xu J, Ragoczy T, Ippolito GC, Walkley CR, Maika SD, Fujiwara Y, Ito M, Groudine M, Bender MA, et al. 2009. Developmental and species-divergent globin switching are driven by BCL11A. *Nature* **460:** 1093–1097.

Sankaran VG, Xu J, Byron R, Greisman HA, Fisher C, Weatherall DJ, Sabath DE, Groudine M, Orkin SH,

Premawardhena A, et al. 2011. A functional element necessary for fetal hemoglobin silencing. *N Engl J Med* **365:** 807–814.

St Pierre TG, Clark PR, Chua-Anusom W, Fleming AJ, Jeffrey GP, Olynyk JK, Pootrakul P, Robins E, Lindeman R. 2005. Noninvasive measurement and imaging of liver iron concentrations using proton magnetic resonance. *Blood* **105:** 855–861.

Taher AT, Musallam KM, Cappellini MD, Weatherall DJ. 2011. Optimal management of β thalassaemia intermedia. *Br J Haematol* **152:** 512–523.

Tanner MA, Galanello R, Dessi C, Smith GC, Westwood MA, Agus A, Roughton M, Assomull R, Nair SV, Walker JM, et al. 2007. A randomized, placebo-controlled, double-blind trial of the effect of combined therapy with deferoxamine and deferiprone on myocardial iron in thalassemia major using cardiovascular magnetic resonance. *Circulation* **115:** 1876–1884.

Tanno T, Bhanu NV, Oneal PA, Goh SH, Staker P, Lee YT, Moroney JW, Reed CH, Luban NL, Wang RH, et al. 2007. High levels of GDF15 in thalassemia suppress expression of the iron regulatory protein hepcidin. *Nat Med* **13:** 1096–1101.

Tanno T, Porayette P, Sripichai O, Noh SJ, Byrnes C, Bhupatiraju A, Lee YT, Goodnough JB, Harandi O, Ganz T, et al. 2009. Identification of TWSG1 as a second novel erythroid regulator of hepcidin expression in murine and human cells. *Blood* **114:** 181–186.

Tassiopoulos S, Deftereos S, Konstantopoulos K, Farmakis D, Tsironi M, Kyriakdis M, Aessopos A. 2005. Does heterozygous β-thalassemia confer a protection against coronary artery disease? *Ann NY Acad Sci* **1054:** 467–470.

Terpos E, Voskaridou E. 2010. Treatment options for thalassemia patients with osteoporosis. *Ann NY Acad Sci* **1202:** 237–243.

Thein SL. 1999. Is it dominantly inherited β thalassemia or just a β-chain variant that is highly unstable? *Br J Haematol* **107:** 12–21.

* Thein SL. 2013. The molecular basis of β-thalassemia. *Cold Spring Harb Perspect Med* **3:** a011700.

Uda M, Galanello R, Sanna S, Lettre G, Sankaran VG, Chen W, Usala G, Busonero F, Maschio A, Albai G, et al. 2008. Genome-wide association study shows BCL11A associated with persistent fetal hemoglobin and amelioration of phenotype of β-thalassemia. *Proc Natl Acad Sci* **105:** 1620–1625.

Wainscoat JS, Kanavakis E, Wood WG, Letsky EA, Huehns ER, Marsh GW, Higgs DR, Clegg JB, Weatherall DJ. 1983. Thalassaemia intermedia in Cyprus: The interaction of α and β thalassaemia. *Br J Haematol* **53:** 411–416.

Weatherall DJ. 2001. Phenotype–genotype relationships in monogenic disease: Lessons from the thalassemias. *Nat Rev Genet* **2:** 245–255.

Weatherall DJ, Clegg JB. 2001. *The thalassemia syndromes*, 4th ed. Blackwell Science, Oxford.

Weatherall DJ, Clegg JB, Naughton MA. 1965. Globin synthesis in thalassemia: An in vitro study. *Nature* **208:** 1061–1065.

Weiss MJ, dos Santos CO. 2009. Chaperoning erythropoiesis. *Blood* **113:** 2136–2144.

White JM, Richards R, Byrne M, Buchanan T, White YS, Jelenski G. 1985. Thalassemia trait and pregnancy. *J Clin Pathol* **38:** 810–817.

Wilber A, Hargrove PW, Kim YS, Riberdy JM, Sankaran VG, Papanikolaou E, Georgomanoli M, Anagnou NP, Orkin SH, Nienhuis AW, et al. 2011a. Therapeutic levels of fetal hemoglobin in erythroid progeny of β-thalassemic $CD34^+$ cells after lentiviral vector–mediated gene transfer. *Blood* **117:** 2817–2826.

Wilber A, Nienhuis AW, Persons DA. 2011b. Transcriptional regulation of fetal to adult hemoglobin switching: New therapeutic opportunities. *Blood* **117:** 3945–3953.

Wolfe L, Olivieri N, Sallan D, Colan S, Rose V, Propper R, Freedman MH, Nathan DG. 1985. Prevention of cardiac disease by subcutaneous deferoxamine in patients with thalassemia major. *N Engl J Med* **312:** 1600–1603.

Wood JC. 2011. Impact of iron assessment by MRI. *Hematology Am Soc Hematol Educ Program* **2011:** 443–450.

Wu EX, Kim D, Tosti CL, Tang H, Jensen JH, Cheung JS, Feng L, Au WY, Ha SY, Sheth SS, et al. 2010. Magnetic resonance assessment of iron overload by separate measurement of tissue ferritin and hemosiderin iron. *Ann NY Acad Sci* **1202:** 115–122.

Xu J, Peng C, Sankaran VG, Shao Z, Esrick EB, Chong BG, Ippolito GC, Fujiwara Y, Ebert BL, Tucker PW, et al. 2011. Correction of sickle cell disease in adult mice by interference with fetal hemoglobin silencing. *Science* **334:** 993–996.

Yu X, Kong Y, Dore LC, Abdulmalik O, Katein AM, Zhou S, Choi JK, Gell D, Mackay JP, Gow AJ, et al. 2007. An erythroid chaperone that facilitates folding of α-globin subunits for hemoglobin synthesis. *J Clin Invest* **117:** 1856–1865.

Yu X, Mollan TL, Butler A, Gow AJ, Olson JS, Weiss MJ. 2009. Analysis of human α globin gene mutations that impair binding to the α hemoglobin stabilizing protein. *Blood* **113:** 5961–5969.

β-Thalassemia Intermedia: A Clinical Perspective

Khaled M. Musallam[1], Ali T. Taher[2], and Eliezer A. Rachmilewitz[3]

[1]Department of Medicine and Medical Specialties, IRCCS Cà Granda Foundation Maggiore Policlinico Hospital, 20122 Milan, Italy

[2]Department of Internal Medicine, American University of Beirut Medical Center, 1107 2020 Beirut, Lebanon

[3]Department of Hematology, Wolfson Medical Center, 58100 Holon, Israel

Correspondence: rachmilewitz@wolfson.health.gov.il

Our understanding of the molecular and pathophysiological mechanisms underlying the disease process in patients with β-thalassemia intermedia has substantially increased over the past decade. Earlier studies observed that patients with β-thalassemia intermedia experience a clinical-complications profile that is different from that in patients with β-thalassemia major. In this article, a variety of clinical morbidities are explored, and their associations with the underlying disease pathophysiology and risk factors are examined. These involve several organs and organ systems including the vasculature, heart, liver, endocrine glands, bone, and the extramedullary hematopoietic system. The effects of some therapeutic interventions on the development of clinical complications are also discussed.

Distinction between the various phenotypes of β-thalassemia relies primarily on the clinical severity of the disease, which should be assessed both at initial presentation and over a period of close follow-up (Rund and Rachmilewitz 2005). The term "β-thalassemia intermedia" (TI) was first suggested to describe patients who had clinical manifestations that are too severe to be termed "β-thalassemia minor" yet too mild to be termed "β-thalassemia major" (TM) (Sturgeon et al. 1955). Patients with TI usually present to medical attention in later childhood or even adulthood. They show mild to moderate anemia and a hemoglobin level ranging between 7 and 10 g/dL, which is sustainable without the need for regular transfusion therapy (Camaschella and Cappellini 1995). Unfortunately, many patients with TI are set on a life of unnecessary regular transfusions, that is, similar to patients with TM, particularly if they present during a period of intercurrent infection requiring a few transfusions. It is essential to evaluate the patient carefully over the first few months after the genetic diagnosis of β-thalassemia is established and not to embark on any treatment modality, especially regular transfusion therapy, too hastily. The patient's well-being, particularly with respect to activity, growth, development, and the early appearance of skeletal changes or other morbidities, is the factor to be taken into consideration before the phenotype is clearly established and the treatment modality is selected (Steinberg et al. 2009; Taher et al. 2011).

In this article, we provide an overview of novel insights regarding the variety of clinical

Copyright © 2013 Cold Spring Harbor Laboratory Press; all rights reserved
Cite this article as *Cold Spring Harb Perspect Med* doi: 10.1101/cshperspect.a013482

morbidities experienced by TI patients and examine their associations with the underlying disease pathophysiology and risk factors. We primarily focus on complications due to primary iron overload and the hypercoagulable state of TI, while also highlighting pertinent aspects of other clinical complications.

MOLECULAR UNDERSTANDING AT A GLANCE

Although the term TI lacks specific molecular correlates and the diagnosis remains largely clinical, a genotype/phenotype association has been described (Galanello and Cao 1998). Most TI patients are homozygotes or compound heterozygotes for β-thalassemia, meaning that both β-globin loci are affected and the disease has a recessive genetic pattern (Galanello and Cao 1998). Less commonly, only a single β-*globin* locus is affected, the other being completely normal; thus, in these instances, TI is dominantly inherited (Weatherall and Clegg 2001). The phenotype of TI may also result from the increased production of α-globin chains by a triplicated or quadruplicated α-genotype associated with β-heterozygosity (Steinberg et al. 2009). In TI, the genetic basis for phenotypic diversity is best explained in terms of primary, secondary, and tertiary genetic modifiers (Weatherall 2001). The primary modifiers represent the broad diversity of mutations that affect the β-globin genes, ranging from mild promoter mutations that cause a slight reduction in β-globin chain production to the many different mutations that result in the β^0-thalassemias, that is, a complete absence of β-globin product (see Thein 2013). Compound heterozygosity for these different mutations can provide a broad spectrum of clinical phenotypes. The secondary genetic modifiers are those that are involved directly in modifying the degree of globin-chain imbalance in β-thalassemia. The coinheritance of α-thalassemia has this effect, and, because there are numerous different molecular forms of α-thalassemia of different severity, this interaction provides further scope for a wide range of different β-thalassemia phenotypes. Similarly, the degree of globin chain imbalance can be reduced by the more effective synthesis of the γ-chains of fetal hemoglobin after birth. There are several genes involved in modifying the γ-chain response, some that are encoded in the β-globin gene cluster, others that are on different chromosomes (Fig. 1) (Sankaran et al. 2010). The tertiary modifiers are those that are not

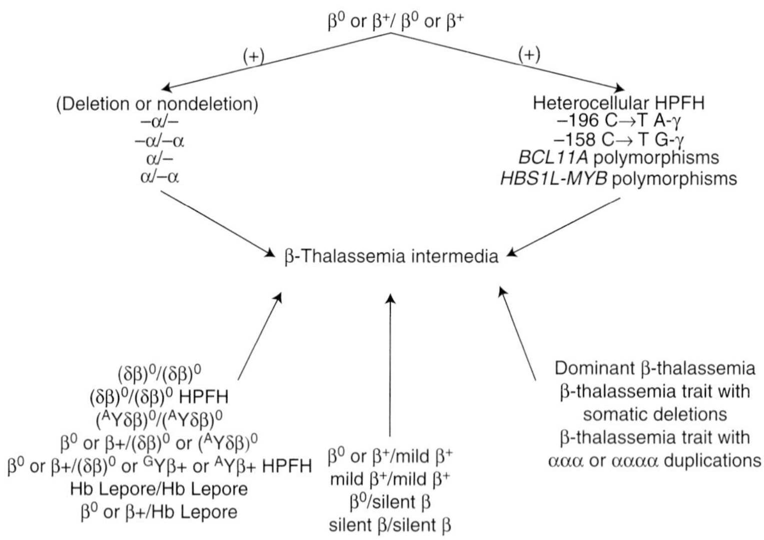

Figure 1. Main genetic profiles that could lead to the β-thalassemia intermedia phenotype. Hb, hemoglobin; HPFH, hereditary persistence of fetal hemoglobin.

 Cite this article as *Cold Spring Harb Perspect Med* doi: 10.1101/cshperspect.a013482

related to globin chain production but that may have an important effect on the complications of the disease (Weatherall 2004).

PATHOPHYSIOLOGY AND CLINICAL MORBIDITY

Current evidence highlights that transfusion independence in TI does not come without its own side effects. Although TI patients sustain levels of anemia that are generally adequate for growth and development without blood transfusions, several other pathogenic mechanisms remain in play. Ineffective erythropoiesis and a secondary drive for increased intestinal iron absorption, extramedullary hematopoiesis (EMH), intra- and extravascular hemolysis, and hypercoagulability have all been described (Taher et al. 2006a). Knowledge of the various clinical morbidities that could emanate from these underlying mechanisms continues to expand, and it is now apparent that TI patients experience a spectrum of morbidities that remain different from those commonly observed in TM (Taher et al. 2006a).

Iron Overload and Target Organ Toxicity

Current models for iron metabolism in TI suggest that the combination of ineffective erythropoiesis, anemia, and hypoxia leads to a compensatory increase in serum levels of erythropoietin, as well as a decrease in serum levels of hepcidin, which control the concentration of ferroportin on the intestinal epithelium (Weizer-Stern et al. 2006; Gardenghi et al. 2010; Melchiori et al. 2010; Tanno and Miller 2010; Taher et al. 2011). Three proposed regulators of hepcidin production are growth differentiation factor 15, secreted by erythroid precursors, twisted gastrulation factor 1, and hypoxia-inducible transcription factors (Tanno et al. 2007; Taher et al. 2009a). Regardless of the signaling mechanism, the end result is suppression of hepcidin levels, increased intestinal iron absorption, and increased release of recycled iron from macrophages within the reticuloendothelial system. This, in turn, leads to relatively low levels of serum ferritin and preferential portal and hepatocyte iron loading (Origa et al. 2007; Taher et al. 2008a, 2009a). Although iron accumulation in patients with TI occurs more slowly than in patients with TM who are regularly transfused, it can reach levels much higher than normal thresholds, especially as patients advance in age (Taher et al. 2008a, 2010a). Considerably elevated liver iron concentration (LIC) and high levels of circulating toxic iron species like non-transferrin-bound iron (NTBI) have been documented in transfusion-independent patients with TI (Pakbaz et al. 2007; Taher et al. 2008a, 2009b).

Iron overload has important clinical consequences in patients with TI. Because iron accumulation primarily occurs in hepatocytes, it can predispose patients to liver fibrosis and cirrhosis, and potentially, hepatocellular carcinoma (Borgna-Pignatti et al. 2004; Taher et al. 2009a; Mancuso 2010; Restivo Pantalone et al. 2010). An association between elevated LIC and the occurrence of endocrinopathy, bone disease, and vascular morbidity has also been reported in TI (Musallam et al. 2011b). In a study of 168 patients with TI and a mean LIC of 8.4 mg Fe/g dry weight (dw), a 1-mg increase in LIC was independently associated with a significantly increased risk of developing thrombosis, pulmonary hypertension (PHT), hypothyroidism, hypogonadism, and osteoporosis. The level associated with a significantly increased risk of developing vascular morbidity was $\geq$7 mg Fe/g dw, and the level for endocrine and bone morbidity was $\geq$6 mg Fe/g dw. These observations were made in both transfusion-naive patients and in patients who received previous transfusions. Moreover, it was apparent that elevated LIC was associated with a steeper increase in the rate of age-related vascular morbidity and earlier onset of endocrine and bone disease compared with patients with low LIC (Musallam et al. 2011b). Recent studies, to be further illustrated in the subsequent section, have also documented a high incidence of silent brain infarction, large cerebral vessel disease, and decreased neuronal function, primarily in the temporal and parietal lobes, among splenectomized patients with TI (Karimi et al. 2010a; Taher et al. 2010d; Musallam et al. 2011a, 2012a). Large-vessel cerebrovascular disease significantly correlated with higher NTBI levels (Musallam et al. 2011a), and decreased neuronal function was observed more frequently in patients with elevated LIC (>15 mg Fe/g dw) (Musallam et al. 2012a).

How iron overload contributes to vascular morbidity in TI remains unclear. Iron-derived reactive oxygen species are implicated in the pathogenesis of several vascular disorders including atherosclerosis, microangiopathic hemolytic anemia, vasculitis, and reperfusion injury (Balla et al. 2003; Malyutin et al. 2012). It has been shown that the presence of NTBI in serum can cause oxidative vessel injury (Auer et al. 2002). Free radicals act directly on the endothelial cells and have a close interaction with lipid peroxidation, causing a modification of low-density lipoprotein and facilitating its deposition, with the consequent formation of atherosclerotic plaques (Aessopos et al. 2009). Thus, iron-mediated endothelial dysfunction and a secondary atherosclerotic process may explain arterial disease in TI patients and echo recent studies supporting the idea that TI patients show a pro-atherogenic biochemical phenotype (Hahalis et al. 2011; Lai et al. 2011). Moreover, iron overload may aggravate ineffective erythropoiesis and the secondary release into the circulation of damaged red blood cells (RBC) with thrombogenic potential, thus putting patients at risk of venous thrombosis (Borenstain-Ben Yashar et al. 1993; Gardenghi et al. 2010).

Several studies have reported absence of cardiac iron overload on T2* magnetic resonance imaging (MRI), even in TI patients who show significant hepatic iron loading (Mavrogeni et al. 2008; Origa et al. 2008; Roghi et al. 2010; Taher et al. 2010e); however, myocardial siderosis has been documented in small subgroups of older TI patients at autopsy (Lombardo et al. 1995; Voskaridou et al. 2004; Au et al. 2009). Moreover, damage to cardiac tissue may result from exposure to NTBI without accumulation of toxic iron species within myocytes (Glickstein et al. 2006; Taher et al. 2010e). This suggests that even without evidence of cardiac siderosis, TI patients may still be at risk for iron-related cardiac dysfunction.

Hypercoagulability and Thromboembolic Disease

The hypercoagulable state in patients with TI has been attributed to several factors (Fig. 2) (Eldor and Rachmilewitz 2002; Cappellini et al. 2010) and can be present from childhood (Eldor et al. 1999). It is often a combination of these factors that leads to thromboembolic events (Ataga et al. 2007; Musallam and Taher 2011). Patients with TI have chronically activated platelets and enhanced platelet aggregation (Winichagoon et al. 1981), as confirmed by the increased expression of CD62P (P-selectin) and CD63, markers of in vivo platelet activation (Del Principe et al. 1993; Ruf et al. 1997). It has been shown that TI patients have four to 10 times higher metabolites of prostacyclin (PG I2) and thromboxane A2, both markers of hemostatic activity, than healthy individuals (Eldor et al. 1991). Splenectomized TI patients also have high platelet counts (Atichartakarn et al. 2003a; Cappellini et al. 2005), but with a shorter life span due to enhanced consumption (Eldor et al. 1989). A recent study showed that increased platelet adhesion is a common finding in splenectomized β-thalassemia patients, which is induced by mechanisms involving both platelets and RBC and could potentially predict clinical thrombosis (Goldschmidt et al. 2008).

The role of the RBCs in the hypercoagulability of TI has received great attention. The oxidation of globin subunits in thalassemia erythroid cells leads to the formation of hemichromes (Rund and Rachmilewitz 2005), which precipitate instigating heme disintegration and the eventual release of toxic NTBI species (Hershko et al. 1978). The free iron, in turn, catalyzes the formation of reactive oxygen species, leading to oxidation of membrane proteins and formation of red cell senescence antigens like phosphatidylserine (Kuypers and de Jong 2004), which cause the thalassemic RBCs to become rigid, deformed, and to aggregate, resulting in premature cell removal (Tavazzi et al. 2001). Thalassemic RBCs with negatively charged phospholipids increase thrombin generation (Borenstain-Ben Yashar et al. 1993; Helley

 Cite this article as *Cold Spring Harb Perspect Med* doi: 10.1101/cshperspect.a013482

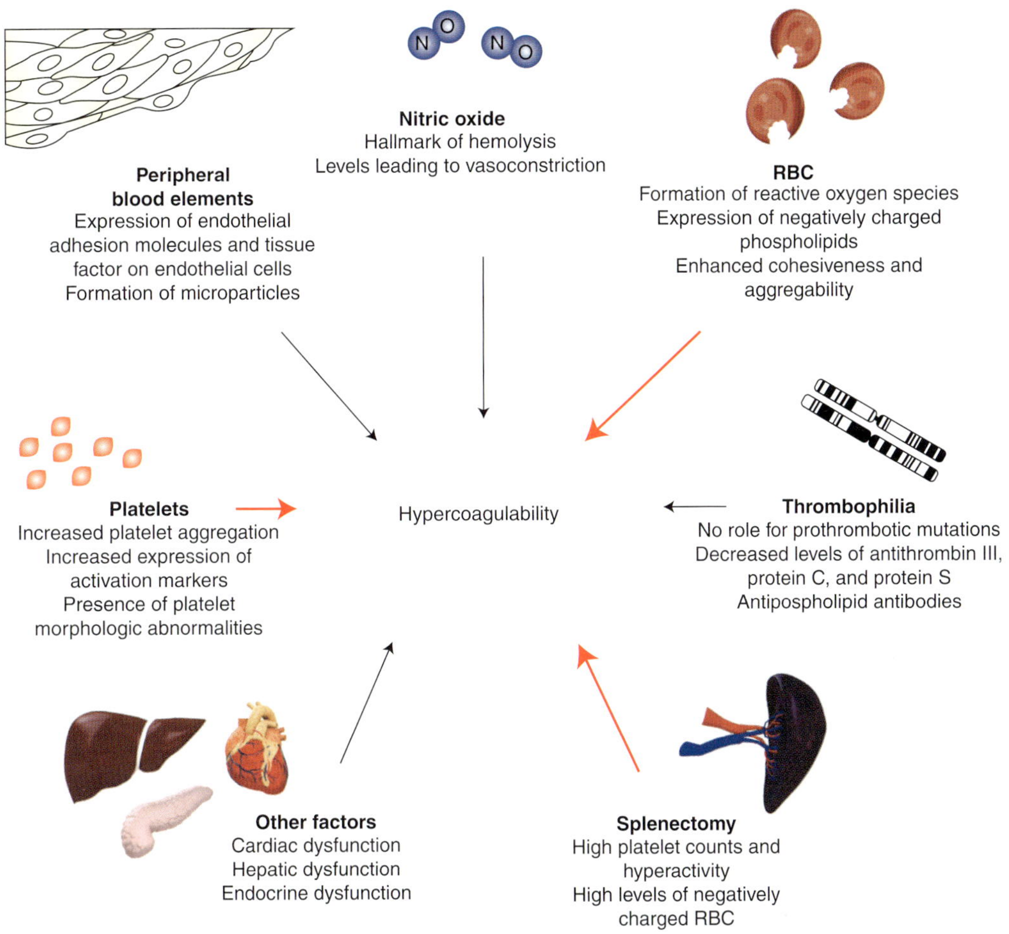

Figure 2. Factors contributing to the hypercoagulable state in β-thalassemia intermedia. RBC, red blood cells. (From Cappellini et al. 2010; reprinted, with permission.)

et al. 1996), as evidenced by studies using annexin V, a protein with high affinity and specificity for anionic phospholipids (Helley et al. 1996). Splenectomized patients have a substantially higher number of these negatively charged RBCs and, in turn, show higher thrombin generation (Cappellini et al. 2000; Atichartakarn et al. 2002). TI patients were also found to have higher levels of pro-coagulant microparticles of RBC, leukocytic, and endothelial origins compared with controls (Habib et al. 2008); the contribution of these fragments to thromboembolic events in TI is under investigation.

The presence of other peripheral blood elements in thalassemics such as E-selectin (ELAM-1), intercellular adhesion molecule-1 (ICAM-1), von Willebrand factor (VWF), and vascular cell adhesion molecule-1 (VCAM-1) indicates that endothelial injury or activation may be an aspect of the disease, aiding in the recruitment of white blood cells and RBCs, and promoting thrombosis (Butthep et al. 1995, 1997). In fact, studies have shown that RBCs from TI patients show increased adhesion to cultured endothelial cells (Hovav et al. 1999). Inherited thrombophilia does not have a role in the hypercoagulable state of TI (Iolascon et al. 2001; Zalloua et al. 2003), but high levels of anti-phospholipid antibodies and low proteins C and S levels have been documented (Taher et al. 2008b). The presence of cardiac, hepatic, or endocrine dysfunction in older patients with severe iron overload may also contribute to hypercoagulability (Taher et al. 2008b).

Data on the incidence of thromboembolic events in TI are limited. In one study including nine Italian pediatric thalassemia centers, 4% of 683 patients with TM and 9.6% of 52 patients with TI had experienced a thromboembolic event (Borgna-Pignatti et al. 1998). In a cohort study including 83 splenectomized patients with TI followed for >10 yr, 29% of patients experienced a venous thromboembolic event (Cappellini et al. 2000). Conventional risk factors (described in the non-thalassemic population) for venous thrombosis were usually absent in such patients (Gillis et al. 1999), further highlighting the unique pathophysiology of hypercoagulability in TI. The largest study to date examined data from 8860 thalassemia patients in the Mediterranean area and Iran and observed that thromboembolic event occurred 4.38 times more frequently in TI than TM, with more venous events occurring in TI and more arterial events occurring in TM (Taher et al. 2006b). It was found that 14% of mortalities in the whole group were attributed to thromboembolic events. Age above 20 yr, splenectomy, and personal or family history of thromboembolic events were identified as the main risk factors for thrombosis in TI. Furthermore, the study showed that 68% of TI patients who had a thrombotic event had an average hemoglobin level of <9 g/dL, and only 33% were receiving regular blood transfusions, whereas 94% were splenectomized. Moreover, patients receiving aspirin therapy had a significantly lower rate of recurrent thromboembolic events (Taher et al. 2006b).

The OPTIMAL CARE (Overview on Practices in Thalassemia Intermedia Management Aiming for Lowering Complication Rates Across a Region of Endemicity) study evaluated 584 patients with TI at six comprehensive care centers (Lebanon, Italy, Iran, Egypt, United Arab Emirates, and Oman) for the associations between patient and disease characteristics, treatment received, and the rate of clinical complications (Taher et al. 2010c). Thromboembolic disease, mostly venous, was the fifth most common complication, affecting 14% of the patient population. Splenectomy was associated with a fivefold increased risk of thrombosis. Conversely, a positive history of transfusion and a hemoglobin level $\geq$9 g/dL were found to be protective against thrombosis (Taher et al. 2010c). A higher occurrence of thrombosis with advancing age was also observed (Taher et al. 2010a). A substudy of the OPTIMAL CARE determined that splenectomized TI patients who experience thrombosis are characterized by high nucleated RBC ($\geq 300 \times 10^6$/L) and platelet counts ($\geq 500 \times 10^9$/L) (Taher et al. 2010b), further confirming the dual role of platelets and RBCs in this setting (Goldschmidt et al. 2008). Moreover, they were more likely to have evidence of PHT and be transfusion-naive (Taher et al. 2010b). As such, it was suggested that splenectomized TI patients at risk of developing thrombosis may be identified early on by these laboratory markers, presence of PHT, and transfusion status (Taher et al. 2010b). The study further examined how long it took for a thrombosis to develop following splenectomy and found the median time to thrombosis to be 8 yr (Taher et al. 2010b). This delay indicates that thrombosis in splenectomized patients with TI is not an acute complication, but a manifestation of a chronic underlying process, further emphasizing the need for long-term treatment preventive strategies (Taher et al. 2010b).

The prevalence of overt strokes in TI patients with a history of thrombosis ranges between 5% and 9% (Taher et al. 2006b, 2010b; Karimi et al. 2008). Few case reports also describe a frequent occurrence of overt strokes in TI patients with moyamoya syndrome (Sanefuji et al. 2006; Marden et al. 2008; Goksel et al. 2010; Oberoi et al. 2010). However, a higher prevalence of silent strokes has been consistently documented. The earliest study was conducted in 1999 and showed a 37.5% rate of ischemic lesions on brain MRI in 16 patients with TI who were neurologically intact and had no conventional stroke-related risk factors (Manfre et al. 1999). More recently, a cross-sectional brain MRI study was conducted in Lebanon on 30 splenectomized adults with TI that were selected from a larger cohort of patients based on absence of neurological or gross cognitive signs or symptoms and any stroke-related risk factors. None of the patients were receiving anti-platelet or anticoagulant therapy. Eighteen patients (60%) had evidence of one or more ischemic lesions on brain MRI, all involving the subcortical white matter. Most patients had evidence of multiple lesions. The frontal subcortical white matter was nearly always involved, followed by the parietal and occipital subcortical white matter. The vast majority of patients (94%) had evidence of small to medium (<1.5 cm) lesions with only one patient showing evidence of a large lesion (>1.5 cm) (Taher et al. 2010d). It was noted that increasing age and transfusion naivety were both independently associated with a higher occurrence and multiplicity of lesions (Taher et al. 2010d). Around the same time, another cross-sectional study was conducted in Iran on 30 randomly selected TI adults who were splenectomized, had a hemoglobin level >7 g/dL, and a platelet count $\geq 500 \times 10^9$/L. The investigators noted eight patients (26.7%) with silent ischemic lesions (Karimi et al. 2010a). The variability in the observed frequency and multiplicity of silent stroke in the three studies could be primarily attributed to the strength of the magnetic field used (Tesla units). Although none of these studies included a control group, the incidence of silent strokes discovered incidentally on brain scans of healthy individuals of a similar age group (<50 yr) ranges from zero to a maximum of 11%, suggesting that the described changes are pathological rather than normal variations (Taher et al. 2010d).

Brain magnetic resonance angiography (MRA) and positron emission tomography-computed tomography (PET-CT) studies have also been recently conducted in TI. In one study including 29 asymptomatic, splenectomized adults, 27.6% had evidence of arterial stenosis on MRA. Two patients had more than one artery involved, and the internal carotid artery was the most commonly involved artery. Among the 12 identified stenotic lesions, two were severe (>75% stenosis), one was moderate (51%–75% stenosis), and the remaining nine were mild ($\leq$50% stenosis). However, there was no association between large-vessel stenosis and silent strokes identified on MRI, leaving small arteriolar pathology as the most likely explanation for the latter, especially that cardiac emboli do not seem to play a role (Musallam et al. 2011a). Although PET-CT scanning was also not helpful in identifying silent strokes in patients with TI, it revealed that decreased neuronal function is a common finding (63.3%) in this patient population, that is, primarily left sided, multiple, and most commonly in the temporal and parietal lobes (Musallam et al. 2012a).

An important question is whether the observed silent brain abnormalities in TI are truly silent or do they require careful consideration and intervention. In the general population and in patients with sickle cell disease, silent strokes, arterial stenosis on MRA, and decreased neuronal function on PET-CT have all been associated with subsequent risk of overt stroke and neurocognitive decline (Taher et al. 2010d; Musallam et al. 2011a, 2012a).

Other Complications

Pulmonary Hypertension

Another complication of TI that was found to occur at a relatively high frequency, especially compared with patients with TM, is PHT (Aessopos et al. 2001, 2005, 2007; Taher et al. 2006a, 2010c; Amoozgar et al. 2011a). One limitation in most available studies, however, is the use of echocardiography instead of cardiac catheterization for the diagnosis of PHT, which may increase the rate of false-positive findings (Parent et al. 2011). Chronic anemia and hypoxia (Aessopos et al. 2001), iron overload (Isma'eel et al. 2008; Mokhtar et al. 2010; Karimi et al. 2011), splenectomy (Phrommintikul et al. 2006; Amoozgar et al. 2010; Karimi et al. 2011), hypercoagulability and microthrombotic disease of the pulmonary circulation (Singer et al. 2006; Karimi et al. 2011), and chronic hemolysis (Aessopos et al. 2007; Karimi et al. 2011) have all been implicated in the pathophysiology of PHT in TI. PHT is neither associated with myocardial siderosis (Roghi et al. 2010; Taher et al. 2010e) nor left ventricular dysfunction in TI, but is a leading cause of right-sided heart failure and thus warrants attention (Aessopos et al. 1995, 2005; Isma'eel et al. 2008). Sildenafil citrate, a potent inhibitor of cyclic guanosine monophosphate-specific phosphodiesterase-5 and a selective smooth muscle relaxant, showed promising results for the management of PHT in small studies in TI patients (Littera et al. 2002; Derchi et al. 2005). The drug is currently being evaluated in a large multicenter trial on patients with thalassemia including TI.

Extramedullary Hematopoietic Pseudotumors

Ineffective red cell production by the bone marrow forces expansion of the hematopoietic tissue outside the marrow medulla and leads to hematopoietic compensatory involvement, mostly in the form of masses, in other regions in the body. This is called extramedullary hematopoiesis (EMH) (Taher et al. 2006a). Almost all body sites may be involved including the spleen, liver, lymph nodes, thymus, heart, breasts, prostate, broad ligaments, kidneys, adrenal glands, pleura, retroperitoneal tissue, skin, peripheral and cranial nerves, and the spinal canal. Among the various body regions reported, paraspinal involvement (11%–15% of cases) receives most attention because of the debilitating clinical consequences secondary to spinal cord compression (Fig. 3) (Haidar et al. 2010). Management options include blood transfusion therapy, which helps decrease the demand for EMH, radiotherapy of the pseudotumors, or fetal hemoglobin induction by hydroxcarbamide. Combinations of these modalities have also been used. There is no evidence as to the best treatment, and this remains individualized depending on the severity of symptoms, the size

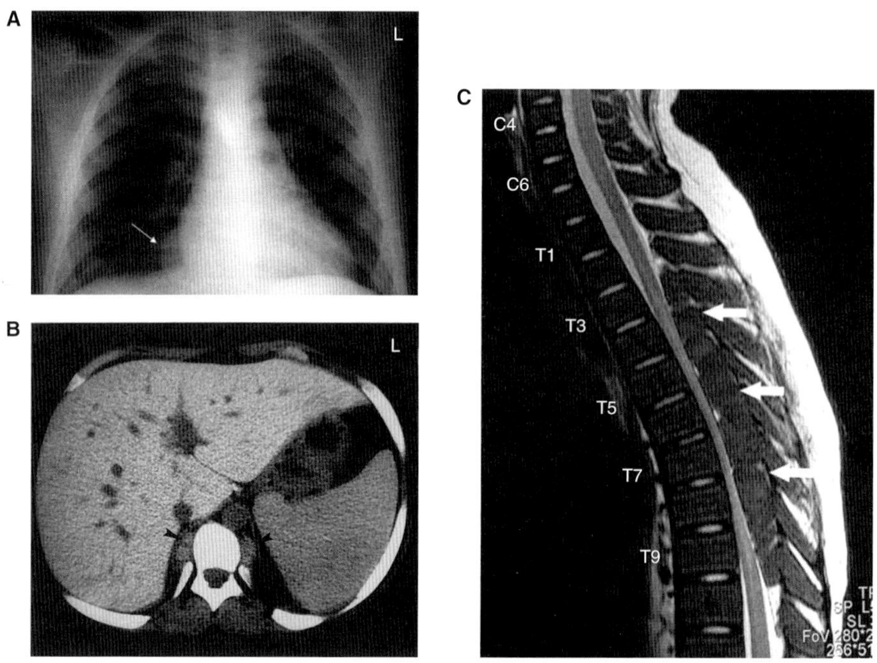

Figure 3. Paraspinal extramedullary hematopoietic pseudotumors. (*A*) Chest X-ray showing expanded anterior rib ends consistent with medullary hyperplasia. A paraspinal mass is seen in the *right lower* zone (white arrow). (*B*) Computed tomography scan showing inactive paraspinal extramedullary hematopoietic lesion with increased density compared with soft tissue, caused by iron deposition (black arrowheads). (*C*) Magnetic resonance image of cervical and thoracic spine. T2-weighted sagittal image showing thoracic cord compression by extramedullary intraspinal epidural hematopoietic mass from T2 to T10 (white arrows). (From Haidar et al. 2010; reprinted, with permission.)

 Cite this article as *Cold Spring Harb Perspect Med* doi: 10.1101/cshperspect.a013482

of the mass, clinical condition, and previous treatment (Chehal et al. 2003; Tai et al. 2006; Haidar et al. 2010). Surgery is not always possible because of the diffuse nature of the mass and the likelihood of recurrence. Furthermore, immediate total resection of extramedullary hematopoietic masses can lead to clinical decompensation and deterioration because these masses play a crucial role in maintaining an adequate hemoglobin level (Jackson et al. 1988).

Leg Ulcers

Leg ulcers are more common in older than in younger patients with TI. It is unclear why ulcers develop in some patients who are maintained at relatively low hemoglobin levels and have the same amount of fetal hemoglobin as others in whom ulcers do not develop. The skin at the extremities of elderly TI patients can be thin because of reduced tissue oxygenation making the subcutaneous tissue fragile and increasing the risk of ulceration after minimal trauma. These ulcers are often very painful and indolent, although blood transfusions may provide some relief (Taher et al. 2010c). Simple measures may be beneficial, such as keeping the patient's legs and feet raised above the level of the heart for 1–2 h during the day or sleeping with the end of the bed raised. Pentoxifylline (Trental, Sanofi-Aventis), which alters the rheological properties of RBCs (Dettelbach and Aviado 1985), can accelerate the healing of ulcers. Hydroxycarbamide also has some benefit, either alone or in combination with erythropoietin (Eprex, Janssen-Cilag) (al-Momen 1991; Taher et al. 2010c). The use of an oxygen chamber can also provide moderate relief where tissue hypoxia may be an underlying cause of the ulceration (Gimmon et al. 1982).

Gallstones

Gallstones are more common in TI than in TM primarily because of increased hemolysis. Unrelated genetic factors such as uridine 5′-diphospho-α-D-glucose (UDPG) deficiency (Gilbert's syndrome) have also been reported to increase gallstone formation in patients with thalassemia (Borgna-Pignatti et al. 2003). The gallbladder is always inspected during splenectomy, and cholecystectomy is performed, particularly if stones are considered symptomatic. This is undertaken to prevent cholecystitis, which can have serious consequences in the splenectomized patient.

Endocrine Disease and Pregnancy

Osteoporosis secondary to bone marrow expansion and 25-hydroxy vitamin D deficiency is highly prevalent in TI patients (Napoli et al. 2006; Taher et al. 2010c). Fractures and bone pain can be devastating consequences. Different regimens of vitamin D and calcium are frequently prescribed to patients with TI, but with careful monitoring of renal function (Borgna-Pignatti 2007; Taher et al. 2011). Although the efficacy and safety of bisphosphonates have been proven in patients with TM, data on patients with TI are limited (Voskaridou et al. 2008). Other endocrine complications can also occur in patients with TI secondary to anemia and iron overload (Taher et al. 2010c; Musallam et al. 2011b). Delayed puberty is common, but fertility is usually normal. In pregnant women with TI, experience reveals an increased risk of abortion, pre-term delivery, intrauterine growth restriction, Caesarean section delivery, and thromboembolic events (Nassar et al. 2008). Although the use of blood transfusions may be required to address these complications, the risk of alloimmunization in never-transfused women should always be taken into consideration. Splenomegaly can interfere with the enlargement of the uterus and can be complicated by hypersplenism. Splenectomy can therefore become necessary during gestation or after delivery. Anticoagulation should be considered especially in women with additional pro-thrombotic risk factors (Nassar et al. 2006).

GENERAL CONSIDERATIONS FOR MANAGEMENT

It is evident that without treatment, TI patients experience more frequent morbidity and poorer health-related quality of life (Taher et al. 2010a; Musallam et al. 2011c, 2011d). Currently, management of patients with TI is almost entirely based on clinical expertise and evidence derived from observational studies.

Transfusion Therapy

In patients with TI, the most challenging therapeutic decisions are whether and when to initiate transfusion therapy (Borgna-Pignatti 2007; Taher et al. 2011). Many patients require intermittent RBC transfusions because of intercurrent infection or pregnancy. Even if few transfusions have been administered in the acute situation, immediate commitment to a transfusion program should not be undertaken before the patient is observed in the non-emergency setting. Regular transfusion therapy is often indicated for growth failure, skeletal deformity, exercise intolerance, or when hemoglobin levels decline because of progressive splenomegaly (Borgna-Pignatti 2007; Borgna-Pignatti et al. 2010; Taher et al. 2011). Moreover, observational studies continue to confirm that transfused patients with TI experience fewer thromboembolic events, PHT, and silent brain infarcts compared with transfusion-naive patients (Taher et al. 2006b, 2010c,d, 2011; Aessopos et al. 2007; Karimi et al. 2011), which may be attributed to correction of the underlying ineffective erythropoiesis and resulting damaged RBCs with thrombogenic potential (Chen et al. 1996).

Iron Chelation

Iron overload in patients with TI can occur as a result of both increased intestinal absorption and transfusion therapy, but regardless of the source, iron overload can be monitored and chelation therapy initiated. Although iron chelator trials in TI patients are limited, the benefits of iron chelation therapy have been reported (Taher et al. 2010c). The initiation of chelation therapy in patients with TI depends primarily on the extent of iron overload and rate of accumulation, but, as with other aspects of the management of TI, clear disease-specific guidelines are not available. Patients with TI may have a positive iron balance by 5 yr of age (Cossu et al. 1981). The main challenge in TI patients is monitoring iron levels because serum ferritin measurement may underestimate the extent of iron overload (Origa et al. 2007). Studies have shown that for the same LIC, patients with TI show considerably lower serum ferritin levels than patients with TM, which is primarily attributed to shunting of iron from the reticuloendothelial system to the hepatocytes during primary iron overload (Taher et al. 2008a). Therefore, direct assessment of LIC either by biopsy or imaging every 1–2 yr is recommended, and chelation therapy should be initiated in patients with elevated indices of iron overload (Taher et al. 2009a, 2011).

Based on evidence that LIC levels of 6–7 mg Fe/g dw are associated with an increased occurrence of morbidity, decreasing LIC levels to ≤5 mg Fe/g dw is suggested (Musallam et al. 2011b). When LIC is reduced to desirable levels, low-dose chelation therapy may be of value to prevent further iron loading. If LIC measurement is not possible, a serum ferritin level of 400–500 µg/L is a reasonable alternative threshold for chelation therapy in this patient population. In most cases, intermittent iron chelation with careful periodic assessment is sufficient in patients with TI. The available chelators have been evaluated in small studies in TI patients (Cossu et al. 1981; Pippard and Weatherall 1988; Olivieri et al. 1992; Ladis et al. 2010; Voskaridou et al. 2010) and showed both efficacy and safety, although results from larger clinical trials are awaited (see Olivieri and Brittenham 2013).

Splenectomy

The role of splenectomy in the management of TI is complex. Clinical observations suggest that splenectomy in patients with TI can contribute to an increased susceptibility to venous thrombosis (Cappellini et al. 2000; Taher et al. 2006b), PHT (Atichartakarn et al. 2003b; Aessopos and Farmakis 2005; Karimi et al. 2011), silent brain infarcts (Karimi et al. 2010a; Taher et al. 2010d), and other complications (Taher et al. 2010c). Based on these considerations, a guarded approach to splenectomy is advised, and the procedure should be delayed unless considered vitally necessary, as in cases with poor growth and development or increased transfusion demand when chelation therapy is not possible, hypersplenism, or symptomatic splenomegaly (Taher et al. 2011).

Fetal Hemoglobin Induction

The clinical picture of TI could be greatly improved by an even partial reduction in the degree of the non-α to α-globin chain imbalance through reactivation of the γ-chain synthesis. Clinical experience suggests that induction of fetal hemoglobin production using hydroxycarbamide is associated with increases in total hemoglobin levels of ~1 g/dL, a decrease in transfusion demand, and amelioration of certain disease complications like PHT, although a decrease in efficacy on long-term therapy is noted (Panigrahi et al. 2005; Karimi et al. 2010b,c, 2011; Rigano et al. 2010; Taher et al. 2010c; Amoozgar et al. 2011b). Treatment with hydroxycarbamide may also reduce the incidence of thrombotic events including silent strokes (M Karimi, S Haghpanah, MH Bagheri, et al., unpubl.). Hydroxycarbamide decreases plasma markers of thrombin generation and coagulation activation by reducing phospholipid expression on the surface of both RBCs and platelets and decreases RBC adhesion to thrombospondin. In addition to being a nitric oxide donor, hydroxycarbamide may also decrease hemostatic activation by its effect in decreasing the white blood cell count and particularly monocytes that express tissue factor (Ataga et al. 2007). Experience with other fetal hemoglobin-inducing agents for the management of TI is limited (Borgna-Pignatti 2007); however, the concept deserves further study, in light of the observation that patients with elevated fetal hemoglobin levels experience fewer morbidities including thrombosis (Musallam et al. 2012b).

Anticoagulation

The roles of anticoagulant or anti-platelet therapy for the prevention of vascular disease in TI patients have not been formally evaluated (Musallam and Taher 2011), although already on the basis of the available data (Karimi et al. 2010a, 2011; Taher et al. 2010b), there is a rationale to give aspirin to splenectomized patients with platelets $>500 \times 10^9$/L and Coumadin to patients who experienced a thromboembolic event.

CONCLUDING REMARKS

The TI phenotype carries greater morbidity than previously recognized. Our knowledge of the various clinical morbidities that TI patients endure has substantially increased over the past decade. Findings confirm that TI should no longer be regarded as a mild form of thalassemia because patients experience serious manifestations involving almost every organ system. Studies that expand our understanding of the mechanisms and risk factors of disease, as well as clinical trials evaluating the roles of available treatments, will help establish management guidelines that transform patient care into optimal standards.

REFERENCES

**Reference is also in this collection.*

Aessopos A, Farmakis D. 2005. Pulmonary hypertension in β-thalassemia. *Ann NY Acad Sci* **1054:** 342–349.

Aessopos A, Stamatelos G, Skoumas V, Vassilopoulos G, Mantzourani M, Loukopoulos D. 1995. Pulmonary hypertension and right heart failure in patients with β-thalassemia intermedia. *Chest* **107:** 50–53.

Aessopos A, Farmakis D, Karagiorga M, Voskaridou E, Loutradi A, Hatziliami A, Joussef J, Rombos J, Loukopoulos D. 2001. Cardiac involvement in thalassemia intermedia: A multicenter study. *Blood* **97:** 3411–3416.

Aessopos A, Farmakis D, Deftereos S, Tsironi M, Tassiopoulos S, Moyssakis I, Karagiorga M. 2005. Thalassemia heart disease: A comparative evaluation of thalassemia major and thalassemia intermedia. *Chest* **127:** 1523–1530.

Aessopos A, Kati M, Farmakis D. 2007. Heart disease in thalassemia intermedia: A review of the underlying pathophysiology. *Haematologica* **92:** 658–665.

Aessopos A, Tsironi M, Andreopoulos A, Farmakis D. 2009. Heart disease in thalassemia intermedia. *Hemoglobin* **33:** S170–S176.

al-Momen AK. 1991. Recombinant human erythropoietin induced rapid healing of a chronic leg ulcer in a patient with sickle cell disease. *Acta Haematol* **86:** 46–48.

Amoozgar H, Farhani N, Karimi M. 2010. Risk factors for pulmonary hypertension in patients with thalassemia intermedia. *Eur J Haematol* **85:** 549–551.

Amoozgar H, Farhani N, Karimi M. 2011a. Early echocardiographic findings in β-thalassemia intermedia patients using standard and tissue Doppler methods. *Pediatr Cardiol* **32:** 154–159.

Amoozgar H, Farhani N, Khodadadi N, Karimi M, Cheriki S. 2011b. Comparative study of pulmonary circulation and myocardial function in patients with β-thalassemia intermedia with and without hydroxyurea, a case-control study. *Eur J Haematol* **87:** 61–67.

Ataga KI, Cappellini MD, Rachmilewitz EA. 2007. β-Thalassaemia and sickle cell anaemia as paradigms of hypercoagulability. *Br J Haematol* **139:** 3–13.

Atichartakarn V, Angchaisuksiri P, Aryurachai K, Onpun S, Chuncharunee S, Thakkinstian A, Atamasirikul K. 2002. Relationship between hypercoagulable state and erythrocyte phosphatidylserine exposure in splenectomized haemoglobin E/β-thalassaemic patients. *Br J Haematol* **118:** 893–898.

Atichartakarn V, Angchaisuksiri P, Aryurachai K, Chuncharunee S, Thakkinstian A. 2003a. In vivo platelet activation and hyperaggregation in hemoglobin E/β-thalassemia: A consequence of splenectomy. *Int J Hematol* **77:** 299–303.

Atichartakarn V, Likittanasombat K, Chuncharunee S, Chandanamattha P, Worapongpaiboon S, Angchaisuksiri P, Aryurachai K. 2003b. Pulmonary arterial hypertension in previously splenectomized patients with β-thalassemic disorders. *Int J Hematol* **78:** 139–145.

Au WY, Lam WW, Chu WW, Tam S, Wong WK, Lau J, Yeung YM, Liu HS, Liang R. 2009. Organ-specific hemosiderosis and functional correlation in Chinese patients with thalassemia intermedia and hemoglobin H disease. *Ann Hematol* **88:** 947–950.

Auer JW, Berent R, Weber T, Eber B. 2002. Iron metabolism and development of atherosclerosis. *Circulation* **106:** e7.

Balla J, Vercellotti GM, Nath K, Yachie A, Nagy E, Eaton JW, Balla G. 2003. Haem, haem oxygenase and ferritin in vascular endothelial cell injury. *Nephrol Dial Transplant* **18:** v8–v12.

Borenstain-Ben Yashar V, Barenholz Y, Hy-Am E, Rachmilewitz EA, Eldor A. 1993. Phosphatidylserine in the outer leaflet of red blood cells from β-thalassemia patients may explain the chronic hypercoagulable state and thrombotic episodes. *Am J Hematol* **44:** 63–65.

Borgna-Pignatti C. 2007. Modern treatment of thalassaemia intermedia. *Br J Haematol* **138:** 291–304.

Borgna-Pignatti C, Carnelli V, Caruso V, Dore F, De Mattia D, Di Palma A, Di Gregorio F, Romeo MA, Longhi R, Mangiagli A, et al. 1998. Thromboembolic events in β thalassemia major: An Italian multicenter study. *Acta Haematol* **99:** 76–79.

Borgna-Pignatti C, Rigon F, Merlo L, Chakrok R, Micciolo R, Perseu L, Galanello R. 2003. Thalassemia minor, the Gilbert mutation, and the risk of gallstones. *Haematologica* **88:** 1106–1109.

Borgna-Pignatti C, Vergine G, Lombardo T, Cappellini MD, Cianciulli P, Maggio A, Renda D, Lai ME, Mandas A, Forni G, et al. 2004. Hepatocellular carcinoma in the thalassaemia syndromes. *Br J Haematol* **124:** 114–117.

Borgna-Pignatti C, Marsella M, Zanforlin N. 2010. The natural history of thalassemia intermedia. *Ann NY Acad Sci* **1202:** 214–220.

Butthep P, Bunyaratvej A, Funahara Y, Kitaguchi H, Fucharoen S, Sato S, Bhamarapravati N. 1995. Alterations in vascular endothelial cell-related plasma proteins in thalassaemic patients and their correlation with clinical symptoms. *Thromb Haemost* **74:** 1045–1049.

Butthep P, Bunyaratvej A, Funahara Y, Kitaguchi H, Fucharoen S, Sato S, Bhamarapravati N. 1997. Possible evidence of endothelial cell activation and disturbance in thalassemia: An in vitro study. *Southeast Asian J Trop Med Public Health* **28**(Suppl): 141A–148A.

Camaschella C, Cappellini MD. 1995. Thalassemia intermedia. *Haematologica* **80:** 58–68.

Cappellini MD, Robbiolo L, Bottasso BM, Coppola R, Fiorelli G, Mannucci AP. 2000. Venous thromboembolism and hypercoagulability in splenectomized patients with thalassaemia intermedia. *Br J Haematol* **111:** 467–473.

Cappellini MD, Grespi E, Cassinerio E, Bignamini D, Fiorelli G. 2005. Coagulation and splenectomy: An overview. *Ann NY Acad Sci* **1054:** 317–324.

Cappellini MD, Motta I, Musallam KM, Taher AT. 2010. Redefining thalassemia as a hypercoagulable state. *Ann NY Acad Sci* **1202:** 231–236.

Chehal A, Aoun E, Koussa S, Skoury H, Taher A. 2003. Hypertransfusion: A successful method of treatment in thalassemia intermedia patients with spinal cord compression secondary to extramedullary hematopoiesis. *Spine (Phila Pa 1976)* **28:** E245–E249.

Chen S, Eldor A, Barshtein G, Zhang S, Goldfarb A, Rachmilewitz E, Yedgar S. 1996. Enhanced aggregability of red blood cells of β-thalassemia major patients. *Am J Physiol* **270:** H1951–H1956.

Cossu P, Toccafondi C, Vardeu F, Sanna G, Frau F, Lobrano R, Cornacchia G, Nucaro A, Bertolino F, Loi A, et al. 1981. Iron overload and desferrioxamine chelation therapy in β-thalassemia intermedia. *Eur J Pediatr* **137:** 267–271.

Del Principe D, Menichelli A, Di Giulio S, De Matteis W, Cianciulli P, Papa G. 1993. PADGEM/GMP-140 expression on platelet membranes from homozygous β thalassaemic patients. *Br J Haematol* **84:** 111–117.

Derchi G, Forni GL, Formisano F, Cappellini MD, Galanello R, D'Ascola G, Bina P, Magnano C, Lamagna M. 2005. Efficacy and safety of sildenafil in the treatment of severe pulmonary hypertension in patients with hemoglobinopathies. *Haematologica* **90:** 452–458.

Dettelbach HR, Aviado DM. 1985. Clinical pharmacology of pentoxifylline with special reference to its hemorrheologic effect for the treatment of intermittent claudication. *J Clin Pharmacol* **25:** 8–26.

Eldor A, Rachmilewitz EA. 2002. The hypercoagulable state in thalassemia. *Blood* **99:** 36–43.

Eldor A, Krausz Y, Atlan H, Snyder D, Goldfarb A, Hy-Am E, Rachmilewitz EA, Kotze HF, Heyns AD. 1989. Platelet survival in patients with β-thalassemia. *Am J Hematol* **32:** 94–99.

Eldor A, Lellouche F, Goldfarb A, Rachmilewitz EA, Maclouf J. 1991. In vivo platelet activation in β-thalassemia major reflected by increased platelet-thromboxane urinary metabolites. *Blood* **77:** 1749–1753.

Eldor A, Durst R, Hy-Am E, Goldfarb A, Gillis S, Rachmilewitz EA, Abramov A, MacLouf J, Godefray YC, De Raucourt E, et al. 1999. A chronic hypercoagulable state in patients with β-thalassaemia major is already present in childhood. *Br J Haematol* **107:** 739–746.

Galanello R, Cao A. 1998. Relationship between genotype and phenotype. Thalassemia intermedia. *Ann NY Acad Sci* **850:** 325–333.

Gardenghi S, Grady RW, Rivella S. 2010. Anemia, ineffective erythropoiesis, and hepcidin: Interacting factors in abnormal iron metabolism leading to iron overload in β-thalassemia. *Hematol Oncol Clin North Am* **24:** 1089–1107.

Gillis S, Cappellini MD, Goldfarb A, Ciceri L, Fiorelli G, Rachmilewitz EA. 1999. Pulmonary thromboembolism in thalassemia intermedia patients. *Haematologica* **84:** 959–960.

Gimmon Z, Wexler MR, Rachmilewitz EA. 1982. Juvenile leg ulceration in β-thalassemia major and intermedia. *Plast Reconstr Surg* **69:** 320–325.

Glickstein H, El RB, Link G, Breuer W, Konijn AM, Hershko C, Nick H, Cabantchik ZI. 2006. Action of chelators in iron-loaded cardiac cells: Accessibility to intracellular labile iron and functional consequences. *Blood* **108:** 3195–3203.

Goksel BK, Ozdogu H, Yildirim T, Oguzkurt L, Asma S. 2010. β-Thalassemia intermedia associated with moyamoya syndrome. *J Clin Neurosci* **17:** 919–920.

Goldschmidt N, Spectre G, Brill A, Zelig O, Goldfarb A, Rachmilewitz E, Varon D. 2008. Increased platelet adhesion under flow conditions is induced by both thalassemic platelets and red blood cells. *Thromb Haemost* **100:** 864–870.

Habib A, Kunzelmann C, Shamseddeen W, Zobairi F, Freyssinet JM, Taher A. 2008. Elevated levels of circulating procoagulant microparticles in patients with β-thalassemia intermedia. *Haematologica* **93:** 941–942.

Hahalis G, Kalogeropoulos A, Terzis G, Tselepis AD, Kourakli A, Mylona P, Grapsas N, Alexopoulos D. 2011. Premature atherosclerosis in non-transfusion-dependent β-thalassemia intermedia. *Cardiology* **118:** 159–163.

Haidar R, Mhaidli H, Taher AT. 2010. Paraspinal extramedullary hematopoiesis in patients with thalassemia intermedia. *Eur Spine J* **19:** 871–878.

Helley D, Eldor A, Girot R, Ducrocq R, Guillin MC, Bezeaud A. 1996. Increased procoagulant activity of red blood cells from patients with homozygous sickle cell disease and β-thalassemia. *Thromb Haemost* **76:** 322–327.

Hershko C, Graham G, Bates GW, Rachmilewitz EA. 1978. Non-specific serum iron in thalassaemia: An abnormal serum iron fraction of potential toxicity. *Br J Haematol* **40:** 255–263.

Hovav T, Goldfarb A, Artmann G, Yedgar S, Barshtein G. 1999. Enhanced adherence of β-thalassaemic erythrocytes to endothelial cells. *Br J Haematol* **106:** 178–181.

Iolascon A, Giordano P, Storelli S, Li HH, Coppola B, Piga A, Fantola E, Forni G, Cianciulli P, Perrotta S, et al. 2001. Thrombophilia in thalassemia major patients: Analysis of genetic predisposing factors. *Haematologica* **86:** 1112–1113.

Isma'eel H, Chafic AH, Rassi FE, Inati A, Koussa S, Daher R, Gharzuddin W, Alam S, Taher A. 2008. Relation between iron-overload indices, cardiac echo-Doppler, and biochemical markers in thalassemia intermedia. *Am J Cardiol* **102:** 363–367.

Jackson DV Jr, Randall ME, Richards FII. 1988. Spinal cord compression due to extramedullary hematopoiesis in thalassemia: Long-term follow-up after radiotherapy. *Surg Neurol* **29:** 389–392.

Karimi M, Khanlari M, Rachmilewitz EA. 2008. Cerebrovascular accident in β-thalassemia major (β-TM) and β-thalassemia intermedia (β-TI). *Am J Hematol* **83:** 77–79.

Karimi M, Bagheri H, Rastgu F, Rachmilewitz EA. 2010a. Magnetic resonance imaging to determine the incidence of brain ischaemia in patients with β-thalassaemia intermedia. *Thromb Haemost* **103:** 989–993.

Karimi M, Cohan N, Moosavizadeh K, Falahi MJ, Haghpanah S. 2010b. Adverse effects of hydroxyurea in β-thalassemia intermedia patients: 10 years' experience. *Pediatr Hematol Oncol* **27:** 205–211.

Karimi M, Mohammadi F, Behmanesh F, Samani SM, Borzouee M, Amoozgar H, Haghpanah S. 2010c. Effect of combination therapy of hydroxyurea with l-carnitine and magnesium chloride on hematologic parameters and cardiac function of patients with β-thalassemia intermedia. *Eur J Haematol* **84:** 52–58.

Karimi M, Musallam KM, Cappellini MD, Daar S, El-Beshlawy A, Belhoul K, Saned MS, Temraz S, Koussa S, Taher AT. 2011. Risk factors for pulmonary hypertension in patients with β thalassemia intermedia. *Eur J Intern Med* **22:** 607–610.

Kuypers FA, de Jong K. 2004. The role of phosphatidylserine in recognition and removal of erythrocytes. *Cell Mol Biol (Noisy-le-grand)* **50:** 147–158.

Ladis V, Berdousi H, Gotsis E, Kattamis A. 2010. Deferasirox administration for the treatment of non-transfusional iron overload in patients with thalassaemia intermedia. *Br J Haematol* **151:** 504–508.

Lai ME, Vacquer S, Carta MP, Spiga A, Cocco P, Angius F, Mandas A, Dessi S. 2011. Thalassemia intermedia is associated with a proatherogenic biochemical phenotype. *Blood Cells Mol Dis* **46:** 294–299.

Littera R, La Nasa G, Derchi G, Cappellini MD, Chang CY, Contu L. 2002. Long-term treatment with sildenafil in a thalassemic patient with pulmonary hypertension. *Blood* **100:** 1516–1517.

Lombardo T, Tamburino C, Bartoloni G, Morrone ML, Frontini V, Italia F, Cordaro S, Privitera A, Calvi V. 1995. Cardiac iron overload in thalassemic patients: An endomyocardial biopsy study. *Ann Hematol* **71:** 135–141.

Malyutin Z, Shai E, Dana M, Rachmilewitz E, Fibach E, Varon D. 2012. Shorter carotid artery occlusion in a thalassemic mouse model: A potential role for oxidative stress affecting both RBCs and platelets. *17th European Haematology Association Congress*, Amsterdam, The Netherlands, June 2012.

Mancuso A. 2010. Hepatocellular carcinoma in thalassemia: A critical review. *World J Hepatol* **2:** 171–174.

Manfre L, Giarratano E, Maggio A, Banco A, Vaccaro G, Lagalla R. 1999. MR imaging of the brain: Findings in asymptomatic patients with thalassemia intermedia and sickle cell–thalassemia disease. *AJR Am J Roentgenol* **173:** 1477–1480.

Marden FA, Putman CM, Grant JM, Greenberg J. 2008. Moyamoya disease associated with hemoglobin Fairfax and β-thalassemia. *Pediatr Neurol* **38:** 130–132.

Mavrogeni S, Gotsis E, Ladis V, Berdousis E, Verganelakis D, Toulas P, Cokkinos DV. 2008. Magnetic resonance evaluation of liver and myocardial iron deposition in thalassemia intermedia and b-thalassemia major. *Int J Cardiovasc Imaging* **24:** 849–854.

Melchiori L, Gardenghi S, Rivella S. 2010. β-Thalassemia: HiJAKing ineffective erythropoiesis and iron overload. *Adv Hematol* **2010:** 938640.

Mokhtar GM, Adly AA, El Alfy MS, Tawfik LM, Khairy AT. 2010. N-terminal natriuretic peptide and ventilation-perfusion lung scan in sickle cell disease and thalassemia patients with pulmonary hypertension. *Hemoglobin* **34:** 78–94.

Musallam KM, Taher AT. 2011. Thrombosis in thalassemia: Why are we so concerned? *Hemoglobin* **35:** 503–510.

Musallam KM, Beydoun A, Hourani R, Nasreddine W, Raad R, Koussa S, Taher AT. 2011a. Brain magnetic resonance angiography in splenectomized adults with β-thalassemia intermedia. *Eur J Haematol* **87:** 539–546.

Musallam KM, Cappellini MD, Wood JC, Motta I, Graziadei G, Tamim H, Taher AT. 2011b. Elevated liver iron concentration is a marker of increased morbidity in patients with β thalassemia intermedia. *Haematologica* **96:** 1605–1612.

Musallam KM, Khoury B, Abi-Habib R, Bazzi L, Succar J, Halawi R, Hankir A, Koussa S, Taher AT. 2011c. Health-related quality of life in adults with transfusion-independent thalassaemia intermedia compared to regularly transfused thalassaemia major: New insights. *Eur J Haematol* **87:** 73–79.

Musallam KM, Taher AT, Duca L, Cesaretti C, Halawi R, Cappellini MD. 2011d. Levels of growth differentiation factor-15 are high and correlate with clinical severity in transfusion-independent patients with β thalassemia intermedia. *Blood Cells Mol Dis* **47:** 232–234.

Musallam KM, Nasreddine W, Beydoun A, Hourani R, Hankir A, Koussa S, Haidar M, Taher AT. 2012a. Brain positron emission tomography in splenectomized adults with β-thalassemia intermedia: Uncovering yet another covert abnormality. *Ann Hematol* **91:** 235–241.

Musallam KM, Sankaran VG, Cappellini MD, Duca L, Nathan DG, Taher AT. 2012b. Fetal hemoglobin levels and morbidity in untransfused patients with β-thalassemia intermedia. *Blood* **119:** 364–367.

Napoli N, Carmina E, Bucchieri S, Sferrazza C, Rini GB, Di Fede G. 2006. Low serum levels of 25-hydroxy vitamin D in adults affected by thalassemia major or intermedia. *Bone* **38:** 888–892.

Nassar AH, Usta IM, Taher AM. 2006. β-Thalassemia intermedia and pregnancy: Should we anticoagulate? *J Thromb Haemost* **4:** 1413–1414.

Nassar AH, Naja M, Cesaretti C, Eprassi B, Cappellini MD, Taher A. 2008. Pregnancy outcome in patients with β-thalassemia intermedia at two tertiary care centers, in Beirut and Milan. *Haematologica* **93:** 1586–1587.

Oberoi S, Bansal D, Singh P, Marwaha RK. 2010. Stroke in a young boy with β-thalassemia intermedia secondary to moyamoya syndrome. *J Pediatr Hematol Oncol* **32:** 568–570.

* Olivieri NF, Brittenham GM. 2013. Management of the thalassemias. *Cold Spring Harb Perspect Med* doi: 10.1101/cshperspect.a011767.

Olivieri NF, Koren G, Matsui D, Liu PP, Blendis L, Cameron R, McClelland RA, Templeton DM. 1992. Reduction of tissue iron stores and normalization of serum ferritin during treatment with the oral iron chelator L1 in thalassemia intermedia. *Blood* **79:** 2741–2748.

Origa R, Galanello R, Ganz T, Giagu N, Maccioni L, Faa G, Nemeth E. 2007. Liver iron concentrations and urinary hepcidin in β-thalassemia. *Haematologica* **92:** 583–588.

Origa R, Barella S, Argiolas GM, Bina P, Agus A, Galanello R. 2008. No evidence of cardiac iron in 20 never- or minimally-transfused patients with thalassemia intermedia. *Haematologica* **93:** 1095–1096.

Pakbaz Z, Fischer R, Fung E, Nielsen P, Harmatz P, Vichinsky E. 2007. Serum ferritin underestimates liver iron concentration in transfusion independent thalassemia patients as compared to regularly transfused thalassemia and sickle cell patients. *Pediatr Blood Cancer* **49:** 329–332.

Panigrahi I, Dixit A, Arora S, Kabra M, Mahapatra M, Choudhry VP, Saxena R. 2005. Do α deletions influence hydroxyurea response in thalassemia intermedia? *Hematology* **10:** 61–63.

Parent F, Bachir D, Inamo J, Lionnet F, Driss F, Loko G, Habibi A, Bennani S, Savale L, Adnot S, et al. 2011. A hemodynamic study of pulmonary hypertension in sickle cell disease. *N Engl J Med* **365:** 44–53.

Phrommintikul A, Sukonthasarn A, Kanjanavanit R, Nawarawong W. 2006. Splenectomy: A strong risk factor for pulmonary hypertension in patients with thalassaemia. *Heart* **92:** 1467–1472.

Pippard MJ, Weatherall DJ. 1988. Iron balance and the management of iron overload in β-thalassemia intermedia. *Birth Defects Orig Artic Ser* **23:** 29–33.

Restivo Pantalone G, Renda D, Valenza F, D'Amato F, Vitrano A, Cassara F, Rigano P, Di Salvo V, Giangreco A, Bevacqua E, et al. 2010. Hepatocellular carcinoma in patients with thalassaemia syndromes: Clinical characteristics and outcome in a long term single centre experience. *Br J Haematol* **150:** 245–247.

Rigano P, Pecoraro A, Calzolari R, Troia A, Acuto S, Renda D, Pantalone GR, Maggio A, Di Marzo R. 2010. Desensitization to hydroxycarbamide following long-term treatment of thalassaemia intermedia as observed in vivo and in primary erythroid cultures from treated patients. *Br J Haematol* **151:** 509–515.

Roghi A, Cappellini MD, Wood JC, Musallam KM, Patrizia P, Fasulo MR, Cesaretti C, Taher AT. 2010. Absence of cardiac siderosis despite hepatic iron overload in Italian patients with thalassemia intermedia: An MRI T2* study. *Ann Hematol* **89:** 585–589.

Ruf A, Pick M, Deutsch V, Patscheke H, Goldfarb A, Rachmilewitz EA, Guillin MC, Eldor A. 1997. In-vivo platelet activation correlates with red cell anionic phospholipid exposure in patients with β-thalassaemia major. *Br J Haematol* **98:** 51–56.

Rund D, Rachmilewitz E. 2005. β-Thalassemia. *N Engl J Med* **353:** 1135–1146.

Sanefuji M, Ohga S, Kira R, Yoshiura T, Torisu H, Hara T. 2006. Moyamoya syndrome in a splenectomized patient with β-thalassemia intermedia. *J Child Neurol* **21:** 75–77.

Sankaran VG, Lettre G, Orkin SH, Hirschhorn JN. 2010. Modifier genes in Mendelian disorders: The example of hemoglobin disorders. *Ann NY Acad Sci* **1214:** 47–56.

Singer ST, Kuypers FA, Styles L, Vichinsky EP, Foote D, Rosenfeld H. 2006. Pulmonary hypertension in thalassemia: Association with platelet activation and hypercoagulable state. *Am J Hematol* **81:** 670–675.

Steinberg MH, Forget BG, Higgs DR, Weatherall DJ. 2009. *Disorders of hemoglobin: Genetics, pathophysiology, and clinical management.* Cambridge University Press, New York.

Sturgeon P, Itano HA, Bergren WR. 1955. Genetic and biochemical studies of intermediate types of Cooley's anaemia. *Br J Haematol* **1:** 264–277.

Taher A, Isma'eel H, Cappellini MD. 2006a. Thalassemia intermedia: Revisited. *Blood Cells Mol Dis* **37:** 12–20.

Taher A, Isma'eel H, Mehio G, Bignamini D, Kattamis A, Rachmilewitz EA, Cappellini MD. 2006b. Prevalence of thromboembolic events among 8,860 patients with thalassaemia major and intermedia in the Mediterranean area and Iran. *Thromb Haemost* **96:** 488–491.

Taher A, El Rassi F, Isma'eel H, Koussa S, Inati A, Cappellini MD. 2008a. Correlation of liver iron concentration determined by R2 magnetic resonance imaging with serum ferritin in patients with thalassemia intermedia. *Haematologica* **93:** 1584–1586.

Taher AT, Otrock ZK, Uthman I, Cappellini MD. 2008b. Thalassemia and hypercoagulability. *Blood Rev* **22:** 283–292.

Taher A, Hershko C, Cappellini MD. 2009a. Iron overload in thalassaemia intermedia: Reassessment of iron chelation strategies. *Br J Haematol* **147:** 634–640.

Taher A, Musallam KM, El Rassi F, Duca L, Inati A, Koussa S, Cappellini MD. 2009b. Levels of non-transferrin-bound iron as an index of iron overload in patients with thalassaemia intermedia. *Br J Haematol* **146:** 569–572.

Taher AT, Musallam KM, El-Beshlawy A, Karimi M, Daar S, Belhoul K, Saned MS, Graziadei G, Cappellini MD. 2010a. Age-related complications in treatment-naive patients with thalassaemia intermedia. *Br J Haematol* **150:** 486–489.

Taher AT, Musallam KM, Karimi M, El-Beshlawy A, Belhoul K, Daar S, Saned M, Cesaretti C, Cappellini MD. 2010b. Splenectomy and thrombosis: The case of thalassemia intermedia. *J Thromb Haemost* **8:** 2152–2158.

Taher AT, Musallam KM, Karimi M, El-Beshlawy A, Belhoul K, Daar S, Saned MS, El-Chafic AH, Fasulo MR, Cappellini MD. 2010c. Overview on practices in thalassemia intermedia management aiming for lowering complication rates across a region of endemicity: The OPTIMAL CARE study. *Blood* **115:** 1886–1892.

Taher AT, Musallam KM, Nasreddine W, Hourani R, Inati A, Beydoun A. 2010d. Asymptomatic brain magnetic resonance imaging abnormalities in splenectomized adults with thalassemia intermedia. *J Thromb Haemost* **8:** 54–59.

Taher AT, Musallam KM, Wood JC, Cappellini MD. 2010e. Magnetic resonance evaluation of hepatic and myocardial iron deposition in transfusion-independent thalassemia intermedia compared to regularly transfused thalassemia major patients. *Am J Hematol* **85:** 288–290.

Taher AT, Musallam KM, Cappellini MD, Weatherall DJ. 2011. Optimal management of β thalassaemia intermedia. *Br J Haematol* **152:** 512–523.

Tai SM, Chan JS, Ha SY, Young BW, Chan MS. 2006. Successful treatment of spinal cord compression secondary to extramedullary hematopoietic mass by hypertransfusion in a patient with thalassemia major. *Pediatr Hematol Oncol* **23:** 317–321.

Tanno T, Miller JL. 2010. Iron loading and overloading due to ineffective erythropoiesis. *Adv Hematol* **2010:** 358283.

Tanno T, Bhanu NV, Oneal PA, Goh SH, Staker P, Lee YT, Moroney JW, Reed CH, Luban NL, Wang RH, et al. 2007. High levels of GDF15 in thalassemia suppress expression of the iron regulatory protein hepcidin. *Nat Med* **13:** 1096–1101.

Tavazzi D, Duca L, Graziadei G, Comino A, Fiorelli G, Cappellini MD. 2001. Membrane-bound iron contributes to oxidative damage of β-thalassaemia intermedia erythrocytes. *Br J Haematol* **112:** 48–50.

* Thein SL. 2013. The molecular basis of β-thalassemia. *Cold Spring Harb Perspect Med* doi: 10.1101/cshperspect.a011700.

Voskaridou E, Douskou M, Terpos E, Papassotiriou I, Stamoulakatou A, Ourailidis A, Loutradi A, Loukopoulos D. 2004. Magnetic resonance imaging in the evaluation of iron overload in patients with β thalassaemia and sickle cell disease. *Br J Haematol* **126:** 736–742.

Voskaridou E, Christoulas D, Antoniadou L, Terpos E. 2008. Continuous increase in erythropoietic activity despite the improvement in bone mineral density by zoledronic acid in patients with thalassemia intermedia-induced osteoporosis. *Acta Haematol* **119:** 40–44.

Voskaridou E, Plata E, Douskou M, Papadakis M, Delaki EE, Christoulas D, Terpos E. 2010. Treatment with deferasirox (Exjade) effectively decreases iron burden in patients with thalassaemia intermedia: Results of a pilot study. *Br J Haematol* **148:** 332–334.

Weatherall DJ. 2001. Phenotype–genotype relationships in monogenic disease: Lessons from the thalassaemias. *Nat Rev Genet* **2:** 245–255.

Weatherall D. 2004. 2003 William Allan Award address. The thalassemias: The role of molecular genetics in an evolving global health problem. *Am J Hum Genet* **74:** 385–392.

Weatherall DJ, Clegg JB. 2001. *The thalassaemia syndromes.* Blackwell Science, Oxford.

Weizer-Stern O, Adamsky K, Amariglio N, Rachmilewitz E, Breda L, Rivella S, Rechavi G. 2006. mRNA expression of iron regulatory genes in β-thalassemia intermedia and β-thalassemia major mouse models. *Am J Hematol* **81:** 479–483.

Winichagoon P, Fucharoen S, Wasi P. 1981. Increased circulating platelet aggregates in thalassaemia. *Southeast Asian J Trop Med Public Health* **12:** 556–560.

Zalloua PA, Shbaklo H, Mourad YA, Koussa S, Taher A. 2003. Incidence of thromboembolic events in Lebanese thalassemia intermedia patients. *Thromb Haemost* **89:** 767–768.

The Hemoglobin E Thalassemias

Suthat Fucharoen[1] and David J. Weatherall[2]

[1]Thalassemia Research Centre, Institute of Science and Technology for Research and Development, Mahidol University, Bangkok, Thailand

[2]Weatherall Institute of Molecular Medicine, University of Oxford, John Radcliffe Hospital, Oxford OX3 9DS, United Kingdom

Correspondence: grsfc@mahidol.ac.th

Hemoglobin E (HbE) is an extremely common structural hemoglobin variant that occurs at high frequencies throughout many Asian countries. It is a β-hemoglobin variant, which is produced at a slightly reduced rate and hence has the phenotype of a mild form of β-thalassemia. Its interactions with different forms of α-thalassemia result in a wide variety of clinical disorders, whereas its coinheritance with β-thalassemia, a condition called hemoglobin E β-thalassemia, is by far the most common severe form of β-thalassemia in Asia and, globally, comprises approximately 50% of the clinically severe β-thalassemia disorders.

As discussed by Williams and Weatherall (2012), HbE occurs at an extremely high frequency in many countries in Asia. Because there is also a high frequency of different β-thalassemia alleles in these populations, the coinheritance of HbE and β-thalassemia, HbE β-thalassemia, occurs very frequently. Similarly, because different forms of α-thalassemia are also very common in these countries, HbE also occurs together with them, producing a complex series of phenotypes.

The first description of HbE β-thalassemia appeared in a paper by Minnich and her colleagues in 1954 under what, at the time, was the rather surprising title "Mediterranean Anaemia: A study of 32 cases in Thailand" (Minnich et al. 1954). In the same year, the first electrophoretic identification of HbE was reported independently (Itano et al. 1954). The first detailed clinical description of HbE β-thalassemia was reported in 1956 by Chernoff and colleagues (1956). Much later, groups in Thailand began a detailed analysis of the interaction of the various forms of α-thalassemia with HbE, which result in a complex series of phenotypes, most of which are much milder than HbE β-thalassemia (Wasi et al. 1969).

MOLECULAR PATHOLOGY AND PROPERTIES OF HEMOGLOBIN E

At least in vitro, HbE appears to be mildly unstable and shows increased sensitivity to oxidants (Frischer and Bowman 1975). However, in vitro studies of hemoglobin synthesis do not show evidence of instability similar to that found in other unstable hemoglobin variants, although HbE is unstable at increased temperatures, similar to those that would occur in a wide range of infective diseases (Rees et al. 1998). The whole-blood oxygen dissociation curves of

Copyright © 2013 Cold Spring Harbor Laboratory Press; all rights reserved
Cite this article as *Cold Spring Harb Perspect Med* doi: 10.1101/cshperspect.a011734

homozygotes for HbE appear to be normal or very slightly right-shifted. HbE is synthesized at a slightly reduced rate and homozygotes show mild globin-chain imbalance, similar to that observed in β-thalassemia heterozygotes. It is caused by a base substitution at codon 26 of the β-globin gene, GAG-AAG, which results in the substitution of lysine for glutamic acid. This mutation also activates a cryptic splice site that causes abnormal messenger RNA processing (Orkin et al. 1982). Because the normal donor site has to compete with this new site, the level of normally spliced β^E messenger RNA is reduced (Traeger et al. 1980), resulting in the clinical phenotype of a mild form of β-thalassemia.

The heterozygous state for HbE is characterized by minimal morphological abnormalities of the red cells and normal red cell indices; HbE constitutes 25%–30% of the hemoglobin (Fig. 1). Homozygotes for HbE have hypochromic microcytic red cells with significant morphological abnormalities including increased numbers of target cells (Fig. 2). They are mildly anemic and the overall hematological findings are very similar to those of heterozygous β-thalassemia.

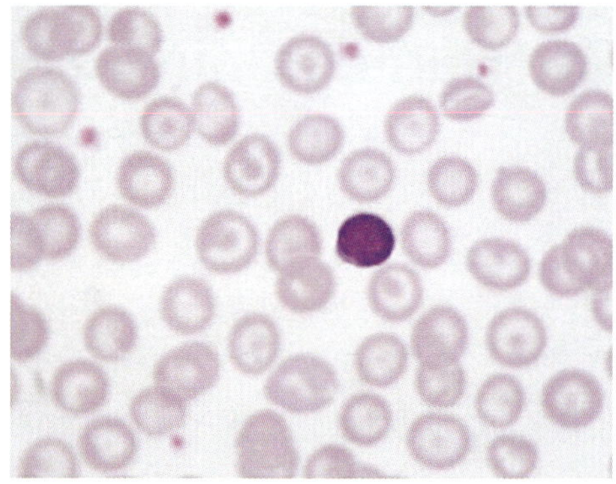

Figure 2. The peripheral blood film in the homozygous state for hemoglobin E showing large numbers of target cells.

THE INTERACTIONS OF HEMOGLOBIN E WITH DIFFERENT FORMS OF THALASSEMIA

Although HbE alone does not cause any significant clinical problems, its interactions with various forms of α- and β-thalassemia produce a very wide range of clinical syndromes of varying severity. The molecular basis for the different forms of α- and β-thalassemia, which are coinherited with HbE, are described by Thein (2013) and Higgs (2013).

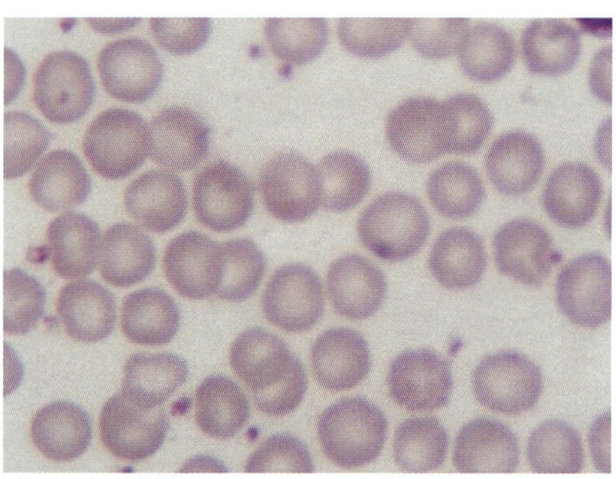

Figure 1. The peripheral blood film in hemoglobin E trait showing normal red cell morphology.

The various interactions of HbE and α thalassemia, which have been defined in the Thai population, and occur elsewhere in SE Asia, are described by their particular hemoglobin constitutions (Table 1). Heterozygotes for HbE, which are also heterozygous for α^+-thalassemia ($-\alpha/\alpha\alpha$), have similar levels of HbE to HbE heterozygotes, whereas those that are heterozygous for α^0-thalassemia ($-/\alpha\alpha$) have mild thalassemic red cell changes and the level of HbE ranges between 19% and 21%. HbE heterozygotes who also inherit the genotype of HbH disease ($-/-\alpha$) have a marked decrease of HbE in the 13%–15% range and a thalassemic disorder of intermediate severity, which is called HbAE Bart's disease.

Hemoglobin E homozygotes who coinherit the heterozygous state for α^+- or α^0-thalassemia have a mild hypochromic microcytic anemia with slightly elevated levels of HbF. Those who coinherit the genotype of HbH disease have a thalassemic disorder of intermediate severity with moderate anemia and elevated levels of HbF and Bart's, a condition called EF Bart's disease.

The compound heterozygous state for HbE and β-thalassemia, HbE β-thalassemia, is a remarkably heterogenous disease with a phenotype ranging from mild anemia to the most severe forms of β-thalassemia major (Weatherall

Table 1. Hematological data and clinical picture of subjects with HbE with different kinds of α-globin gene interactions

HbE	α-Globin gene	Hb (g/dL)	MCV (fl)	Hb typing	HbE (%)	HbBart's (%)	HbF (%)	Clinical
HbE heterozygote	αα/αα	12.8 ± 1.5	84 ± 5	EA	29 ± 2.3	-	0.9 ± 0.7	Normal
	−α/αα	13.1 ± 1.4	88 ± 4	EA	28 ± 1.5	-	0.7 ± 0.6	Normal
	−/αα	12.5 ± 1.4	77 ± 5	EA	21 ± 1.2	-	0.9 ± 0.4	Normal
	−/−α	9.1 ± 1.1	60 ± 3	EFA Bart's	13 ± 2.1	4.5 ± 1.9	2.2 ± 1.9	Thal intermedia (AEBart's disease)
HbE homozygote	αα/αα	10.6 ± 1.2	65 ± 3	EF	88 ± 2.6	-	3.6 ± 1.6	Normal
	−α/αα	11.0 ± 1.6	65 ± 4	EF	87 ± 3.3	-	4.8 ± 3.7	Normal
	−/αα	10.5 ± 2.4	64 ± 7	EF	88 ± 5.7	-	3.8 ± 2.1	Normal
	−/−α	7.5 ± 0.8	60 ± 2	EF Bart's	81 ± 1.5	4.2 ± 1.1	6.4 ± 1.2	Thal intermedia (EFBart's disease)
HbE β-thalassemia	αα/αα	7.1 ± 1.4	59 ± 3	EF	58 ± 9.5	-	38 ± 11.7	Mild to severe disease
	−α/αα	8.5 ± 1.1	55 ± 3	EF	71 ± 7.5	-	24 ± 8.7	Mild disease
	−/αα	9.3 ± 0.5	52 ± 1	EF	84 ± 3.8	-	12 ± 2.5	Mild disease
	−/−α	7.6 ± 1.2	61 ± 2	EF Bart's	82 ± 2.5	1.5 ± 0.3	5.5 ± 0.7	Thal intermedia (EFBart's disease)

and Clegg 2001). This condition may also be coinherited with a variety of different forms of α-thalassemia. The coinheritance of the heterozygous states for α^+- and α^0-thalassemia have an ameliorating effect on the disease, whereas those who also inherit the genotype of HbH disease have a form of HbEF Bart's disease, as described above.

These complex interactions, which are summarized in Table 1, produce three significant clinical disorders; HbAE Bart's disease, HbEF Bart's disease, and HbE β-thalassemia.

HbAE BART'S DISEASE

This thalassemia syndrome is characterized by the presence of HbA, HbE, and Hb Bart's, and results from the interaction of the genotype of HbH disease (see Higgs 2013; Vichinsky 2013) with heterozygous HbE (Wasi et al. 1967; Thonglairuam et al. 1989). Two common subtypes of HbAE Bart's disease have been observed: α^+-thalassemia/α^0-thalassemia—A/E and α^0-thalassemia/Hb Constant Spring—A/E. Clinically, HbAE Bart's disease is similar to HbH disease, and like the latter, the α^0-thalassemia/Hb Constant Spring interaction is more severe (see Vichinsky 2013). However, in this syndrome there are no hemolytic crises during stress similar to those seen in HbH disease. The amount of HbE is decreased to 13%–15%. This is because α-globin chains have a lower affinity for β^E than β^A-globin chains. Small amounts of Hb Bart's are always present in this genotype and intraerythrocytic inclusion bodies (HbH inclusions) can be demonstrated in ~5% of the erythrocytes, indicating the presence of small amounts of HbH that is insufficient to be detected by electrophoresis.

The diagnosis of this disorder requires detailed family studies together with DNA analysis to define the type of α-thalassemia. Management is similar to HbH disease (see Vichinsky 2013).

HbEF BART'S DISEASE

HbEF Bart's disease is characterized by the presence of HbE, HbF, and Hb Bart's (Fucharoen et al. 1988b). HbE constitutes ~80% and HbF 10% of the hemoglobin; the remainder is Hb Bart's. The presence of Hb Bart's indicates that there are excess γ-globin chains. However, no inclusion bodies or HbH are present, probably because the abnormal β^E-globin chains do not form tetramers. Four genotypes of HbEF Bart's disease have been identified. They result from interactions between the genotype for HbH disease, either α^0-thalassemia/α^+-thalassemia or α^0-thalassemia/Hb Constant Spring, with either homozygous HbE or HbE β-thalassemia. Hb Constant Spring and small amounts of HbA may be observed in patients with the α^0-thalassemia/Hb Constant Spring and HbE β^+-thalassemia genotype. To differentiate among these genotypes, family studies and further investigation by DNA analysis are required.

Overall, the interactions of the different genotypes for HbH disease with the homozygous state for HbE produce relatively mild forms of thalassemia intermedia, not dissimilar to HbH disease (see Vichinsky 2013). Their interactions with HbE β-thalassemia are more complex and clinically variable, and are discussed in more detail in the next section.

HEMOGLOBIN E β-THALASSEMIA

In general, HbE β-thalassemia is a thalassemia syndrome of intermediate severity with a very heterogeneous clinical spectrum. Two types have been described, depending on the presence or absence of HbA. In HbE, β^0-thalassemia, β^A-globin chains are not synthesized and the condition is characterized by the production of HbE and HbF without detectable HbA; HbE constitutes between 30% and 70% of the hemoglobin with the remainder HbF. Variable amounts of HbA are detected, in addition to HbE and HbF, in HbE β^+-thalassemia. Different β^+-thalassemia mutations result in variable severity of the disease, reflecting different levels of HbA.

Pathophysiology

The pathophysiology of HbE β-thalassemia reflects both the reduced output of HbE together with the added globin-chain imbalance consequent on the coinheritance of β-thalassemia.

Although early studies suggested that HbE is slightly unstable and may precipitate under conditions of oxidative stress, later biosynthetic analyses showed little evidence of its instability in the red cells of patients with HbE β-thalassemia. The one exception was if the cells were exposed to increased temperatures at the level that might be encountered in severe forms of infection at which there was evidence of instability (Rees et al. 1998). However, the remarkable ameliorating effect on the phenotype that results from the coinheritance of α-thalassemia or other modifiers of the degree of globin chain imbalance (see later section) suggests that defective β chain synthesis is the major factor in the pathophysiology of this condition (reviewed by Weatherall and Clegg 2001). Like other forms of severe β-thalassemia, there is marked expansion of the erythroid bone marrow with ineffective erythropoiesis. Interestingly, using a combination of electron-microscopic and immunocytochemical studies, Wickramasinghe and Lee (1997) demonstrated that the erythroblast inclusions in the bone marrow of patients with this condition consist entirely of precipitated α chains; there is no evidence of coprecipitation of β^E chains. Therefore the mechanisms of damage to red cells and their precursors are similar to those described in other forms of β-thalassemia (see Nienhuis and Nathan 2012).

Recent studies have suggested that there are important differences in the compensatory mechanisms between HbE β-thalassemia and other forms of severe thalassemia intermedia. In particular, patients with HbE β-thalassemia appear to be able to compensate for anemia by a right shift in their oxygen dissociation curves, unlike those with many other forms of β-thalassemia intermedia (Allen et al. 2010). This probably reflects both the properties of HbE and the lower mean levels of HbF that occur in HbE β-thalassemia compared with other forms of β-thalassemia intermedia. Studies of the erythropoietin response to anemia in this condition have shown that hemoglobin level and age are independent variables with respect to the erythropoietin level and that there is a relative reduction in response to a particular hemoglobin level with aging (O'Donnell et al. 2007). This may be at least partly responsible for some of the remarkable phenotypic heterogeneity observed during the early years of development in babies with HbE β-thalassemia.

Clinical Features

One of the most striking features of HbE β-thalassemia is its remarkable clinical heterogeneity. At one end of the spectrum, there are patients whose clinical course is almost indistinguishable from that of severe β-thalassemia major; whereas at the other end, there are patients who grow and develop normally without the need for blood transfusion, albeit often at a relatively low hemoglobin level.

At birth, infants with severe HbE β-thalassemia are asymptomatic because HbF levels are high. As HbF production decreases and is replaced by HbE at 6–12 months of age, anemia with splenomegaly develops. Signs of impaired growth appear during the first decade of life. The initial complaints vary from patient to patient, and several symptoms usually appear simultaneously (Table 2). Most common are the development of a mass in the left upper quadrant and pallor. With time and without transfusions, anemia, jaundice, hepatosplenomegaly, growth

Table 2. Presenting symptoms in 378 HbE β-thalassemia patients in Thailand

Symptoms	Number	%
1. Pallor	150	39.7
2. Fever	72	19.1
3. Abdominal mass	36	9.5
4. Abdominal pain	23	6.1
5. Combined: Abdominal mass and pain	21	5.6
6. Yellow eyes (jaundice)	16	4.2
7. Edema	9	2.4
8. Pregnancy with anemia	3	0.8
9. Bone pain	3	0.8
10. Paraplegia	2	0.5
11. Headache and dizziness	2	0.5
12. Miscellaneous[a]	41	10.8
Total	378	100

[a]Miscellaneous includes lymphadenopathy, skin rashes, chest pain, request for plastic surgery, joint pain, and cases with multiple symptoms.

Table 3. Clinical signs in 378 adult HbE β-thalassemia patients in Thailand

Sign[a]	Number	%
1. Splenomegaly	369	97.6
2. Jaundice	350	92.6
3. Hepatomegaly	337	89.1
4. Thalassemic facies	313	82.8
5. Growth retardation	284	75.1
6. Anemia	152	40.2
7. Abnormal cardiovascular system[b]	70	18.5
8. Respiratory tract infection	44	11.6
9. Arthritis and bone pain	40	10.6
10. Abnormal neurological system[c]	32	8.5
11. Chronic leg ulcer	31	8.2
12. Soft tissue infection	7	1.8

[a]Most patients have more than one clinical sign.

[b]Abnormal cardiovascular manifestations are mainly related to congestive heart failure (50 cases), deep vein thrombosis (five cases), pericarditis (four cases), and others.

[c]Abnormal neurological features are mainly weakness of both hands (14 cases), headache (11 cases), and paraplegia (two cases).

retardation, and thalassemic facies evolve (Table 3). Absence of secondary sexual development is common and chronic leg ulcers are sometimes observed. These manifestations, which are secondary to decreased oxygen delivery to tissue, ineffective erythropoiesis, and iron overload, resemble those of β-thalassemia major.

Patients with the milder forms of HbE β-thalassemia tend to grow and develop reasonably well during early childhood and are fully active. There may be some delay in the pubertal growth spurt and in the appearance of secondary sexual characteristics but they usually attain a reasonable height and sexual maturation. Further work is required to determine whether they develop the later complications that have been described in older patients with other forms of thalassemia intermedia, such as renal disease, pulmonary hypertension, cerebral infarcts, and others (see Musallam et al. 2012). Certainly some of them accumulate iron through increased absorption and may develop associated endocrine complications including diabetes. Furthermore, there is undoubtedly phenotypic instability in the early years of life; in a series of children observed over 15 years with HbE β-thalassemia in Sri Lanka, at least 20 cases changed from a mild to a more severe phenotype during the first 15 years of life.

Laboratory Findings

The steady-state hemoglobin levels in HbE β-thalassemia range widely between the different phenotypes, from 3 g/dL or less to as high as 11 g/dL. The red-cell indices and morphological changes are similar to those described in other forms of severe β-thalassemia (Fig. 3). Hemoglobin analysis reveals a preponderance of HbE with low levels of HbA in those who have inherited a β^+-thalassemia mutation. The mean level of HbF in 200 patients from Sri Lanka was approximately 28%, ranging from less than 10% to 50%; even higher levels of HbF have been reported from Thailand. Family studies reveal the carrier state for HbE in one parent and for β-thalassemia trait in the other. The laboratory findings associated with different complications are discussed in the following sections.

Complications

Hyperspleninism

Splenomegaly, together with pooling of red cells and their increased rate of destruction, is extremely common. In the more severe phenotypes, it often progresses rapidly from the first few years

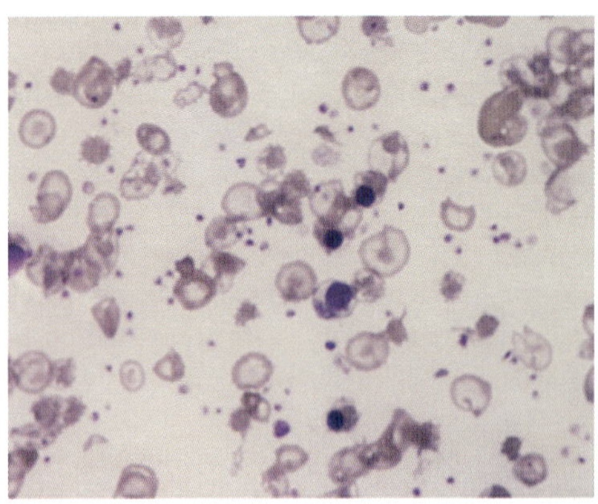

Figure 3. The peripheral blood film in hemoglobin E β-thalassemia after splenectomy showing numerous nucleated red cells and a high platelet count.

of life, whereas in the milder phenotypes, although the spleen is palpable, it usually does not attain a size greater than 5–6 cm below the costal margin. Much less common, and usually in the milder phenotypes, splenomegaly may slowly increase over 10–20 years and only become a problem later in life. As well as resulting in increasing anemia, progressive enlargement of the spleen is quite often associated with pain and discomfort in the left upper quadrant.

Infection

Patients with HbE β-thalassemia are susceptible to viral, bacterial, and fungal infection, which are common causes of mortality (Aswapokee et al. 1988a,b). In splenectomized patients, septicemia can be very acute and overwhelming, leading to death in a short period. Gram-negative and gram-positive bacteria are frequent causes of septicemia. Fungal infection with *Pythium* can lead to arterial occlusion and gangrene of the legs (Sathapatayavongs et al. 1989; Wanachiwanawin et al. 1993). The mechanism that causes increased susceptibility to infections is not known. Recent studies have suggested that patients with HbE β-thalassemia may be more prone to infection by both *P. falciparum* and *P. vivax* malaria, particularly the latter, and that those who have undergone splenectomy may be even more susceptible (O'Donnell et al. 2009).

Cardiac Disease

Approximately half of the patients with HbE β-thalassemia in Thailand die of heart failure. This is associated with failure of other organs, delayed growth and sexual maturation, hepatomegaly, and endocrinopathies. Organ failure results from iron deposition in the heart and other tissues (Vannasaeng et al. 1981; Sonakul et al. 1988; Thakerngpol et al. 1988) resulting from increased absorption and blood transfusion. Myocardial iron deposition is usually limited, occurring primarily as small granules in perinuclear areas, with later accumulation throughout the muscle fibers, predominantly subepicardial and occasionally subendocardial (Sonakul et al. 1984). The small amount of iron deposited in the heart is in marked contrast to enormous iron deposition in the liver and pancreas. Cardiomegaly is proportional to the severity of anemia and systolic murmurs are frequently present (Jootar and Fucharoen 1990). Chronic pericarditis following upper respiratory tract infection is frequently encountered, more commonly in splenectomized patients. A pericardial rub may be detected, often transiently. Intractable pericardial effusion may follow, causing cardiac tamponade and failure, and requires aspiration. In a very few cases, chronic constrictive pericarditis develops, requiring surgical intervention.

Heart rate variability (HRV) has been developed to determine cardiac autonomic function and is applied to investigate patients with thalassemia major (De Chiara et al. 2005). Studies of the HRV in HbE β-thalassemia have obtained similar results to those in thalassemia major (Rutjanaprom et al. 2009). The depressed HRV compared to normal suggests that HRV may be a marker of cardiac sympatho-vagal imbalance as well as an early indicator of cardiac involvement in both thalassemia major and HbE β-thalassemia.

Pulmonary hypertension and right heart failure are discussed in the following sections.

Hypoxemia

The majority of splenectomized HbE β-thalassemia patients in Thailand develop hypoxemia with low arterial pO_2 (Wasi et al. 1982). Platelet counts in these patients are double those of nonsplenectomized patients; young and larger platelets are also observed in the absence of the spleen. Platelet microaggregates have been detected in the circulation of these splenectomized patients. One hypothesis for the pathogenesis of hypoxemia in HbE β-thalassemia is that platelets increase in number, are younger and more active after splenectomy, and aggregate in the circulation and in the pulmonary vasculature. Substances released during platelet aggregation may cause constriction of the terminal bronchioles leading to decreased oxygenation and hypoxemia. Administration of aspirin to inhibit platelet aggregation reduces the degree of hypoxemia in the majority of cases

(Fucharoen et al. 1981), suggesting that aspirin should be routinely given to splenectomized patients with HbE β-thalassemia. Interestingly, this combination of pulmonary hypertension and hypoxemia, together with right ventricular failure, has not been observed so frequently in other populations, suggesting that other factors may be involved in the Thai population.

Thromboembolism

Autopsy findings in a large number of Thai patients with HbE β-thalassemia revealed striking pulmonary artery occlusion (Fig. 4) (Sonakul et al. 1980). Thromboembolism in HbE β-thalassemia seems to involve platelets, a reactive thalassemic red cell surface, coagulation factors, and abnormal endothelium (Butthep et al. 2002; Pattanapanyasat et al. 2004).

Hypertension, Convulsions, and Cerebral Hemorrhage

Some patients in Thailand develop hypertension, convulsions, and cerebral hemorrhage after transfusion of 2 units or more of blood (Wasi et al. 1978). This complication may develop as late as 2 weeks after multiple transfusions, suggesting that blood volume overload is not the cause of hypertension. Monitoring blood pressure during and after blood transfusions with prompt antihypertensive intervention has reduced deaths from this complication. This syndrome has not been reported in other populations.

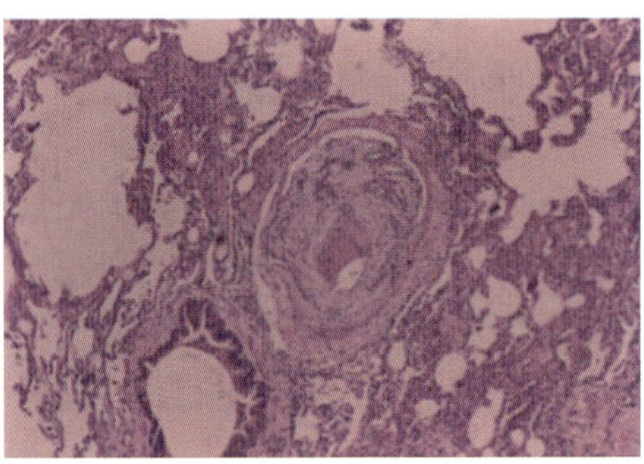

Figure 4. Pulmonary vascular occlusion in hemoglobin E β-thalassemia after splenectomy.

Extramedullary Hemopoiesis

Erythropoiesis is massively increased to 10–15 times normal because anemia stimulates erythropoietin production. Extensive erythropoiesis can be found in the liver, spleen, bone marrow, and in extramedullary sites. Erythropoietic masses in the spinal canal can cause spinal cord compression and paraplegia, and when they occur intracranially, convulsions may result (Issaragrisil et al. 1981). Massive erythropoiesis leads to fragility and distortion of the bones and decreases bone density because of osteoporosis and osteomalacia, as observed in irregularly transfused patients (Pootrakul et al. 1980).

Jaundice, Gallstones, and Cholecystitis

Some HbE β-thalassemia patients have severe and persistent jaundice in the absence of definable liver disease. It turns out that this is a result of the homozygous inheritance of the TA(7) allele of the promoter of the glucuronyltransferase 1 gene, a polymorphism that is particularly common in Sri Lanka (Premawardhena et al. 2001). These patients have a highly significant increase in the incidence of gallstones. Homozygosity for the TA(7) allele occurs in 10%–25% of some populations of Africa and the Indian subcontinent but at a much lower frequency in Southeast Asia (Premawardhena et al. 2003). Stones are found in approximately 50% of HbE β-thalassemia patients in Thailand (Chandcharoensin-Wilde et al. 1988). For the detection of biliary calculi, ultrasonography is more sensitive than oral cholecystography and plain abdominal films. Cholecystitis and ascending cholangitis may occur with abdominal pain, fever, and increasing jaundice (Vathanopas et al. 1988).

Iron Overload

Iron overload occurs commonly (Pootrakul et al. 1981). Excessive iron accumulates because of blood transfusions and enhanced gastrointestinal absorption. The skin is darkened and iron deposition occurs in the bone marrow, liver, spleen, heart, pancreas, and elsewhere (Sonakul et al. 1988; Thakerngpol et al. 1988). The related cardiac complications were discussed earlier.

 Cite this article as *Cold Spring Harb Perspect Med* doi: 10.1101/cshperspect.a011734

Although liver fibrosis from iron overload is common, ascites and other signs of cirrhosis are very rare. Diabetes mellitus, secondary to iron deposition in the pancreas, frequently develops in untreated adult patients (Vannasaeng et al. 1981). A terminal wasting stage occurs in some patients who survive to their fourth and fifth decades. They develop severe skin pigmentation, poor appetite, weight loss, and increasing anemia, and eventually die. This is believed to result from organ failure caused by uncontrolled tissue oxidation from chronic, severe iron overload.

Recently, it has been possible to obtain more direct data about the liver iron concentration in HbE β-thalassemia using spin density projection assisted R2-MRI technology (Olivieri et al. 2011). These studies showed a marked variation in hepatic iron levels even in patients who had received only minimal transfusion. They also underlined the rather poor agreement between hepatic iron levels as measured in this way and serum ferritin values. Clearly, a great deal more work is required to try to determine the reasons for the variable rate of iron accumulation by increased intestinal absorption and, in particular, whether this reflects polymorphisms in one or more of the genes that are involved in the regulation of iron absorption.

Other Endocrine Disorders

As well as diabetes there are other important endocrine disorders that occur as a result of iron loading. In particular hypothyroidism and hypoparathyroidism are quite common and the former is frequently associated with growth retardation. It is vital, therefore, to carry out regular assessments of thyroid and parathyroid function in patients with HbE β-thalassemia, regardless of the severity of their phenotype.

Genotype–Phenotype Interaction

Definition of Severity

Despite seemingly identical genotypes, compound heterozygotes for β-thalassemia and HbE have remarkably variable phenotypes. Notable are variations in the degree of anemia, growth, development, hepatosplenomegaly, and transfusion requirements. A novel scoring system based on six independent parameters—hemoglobin level, age at disease presentation, age at receiving first blood transfusion, requirement for transfusion, spleen size, growth and development—was able to separate patients into three distinctive severity categories: mild, moderate, and severe. The scoring system consisting of six clinical criteria scored as 0, 0.5, 1, or 2, according to clinical presentation. HbE β-thalassemia patients with total scores ranging from 0 to 3.5, 4 to 7, and 7.5 to 10 are grouped as mild, moderate, and severe cases, respectively. The severe patients are very anemic and are usually transfusion dependent; some may have marked growth retardation, whereas the mild cases have mild anemia and usually have normal growth and development (Sripichai et al. 2008).

As indicated by recent studies in Sri Lanka, the application of a clearly defined scoring system for severity combined with a long period of observation and genetic and environmental analysis (Premawardhena et al. 2005; Olivieri et al. 2010) should help us to understand the factors that determine the broad range of severity of HbE β-thalassemia.

β-Thalassemia Mutations

Although β^0-thalassemia is caused by many mutations, all result in absence of β-globin chain production by the abnormal gene. β^0-thalassemia is more severe than β^+-thalassemia, in which a wide range of β-globin chain production is observed. In most countries with a high frequency of HbE β-thalassemia, the common β-thalassemia mutations are either β^0-thalassemia or β^+-thalassemia associated with very small amounts of β-globin chain synthesis. The fact that there is still considerable clinical heterogeneity in these patients is clearly not a result of variation in β-globin chain synthesis directed by the chromosome containing the β-thalassemia mutation. There are, however, certain milder forms of β^+-thalassemia associated with much higher levels of β-chain production and, when inherited together with HbE, produce a much milder form of HbE β-thalassemia. The phenotypes and hemoglobin findings in patients of this type are summarized by Weatherall and Clegg (2001).

Coinheritance of α-Thalassemia

The concomitant inheritance of α-thalassemia or Hb Constant Spring may be responsible for less severe anemia and a milder phenotype in HbE β^0-thalassemia (Winichagoon et al. 1985, 1993). Coinheritance of α^0-thalassemia with HbE β^0-thalassemia may lead to so mild a condition that the individuals do not have a clinical abnormality that requires medical attention. Similar findings have been observed in Sri Lankan populations, in which the coinheritance of the heterozygous state for α^+-thalassemia has been found to result in a remarkable degree of amelioration of the clinical severity of HbE β-thalassemia (Premawardhena et al. 2005).

Association with Increased HbF

Coinheritance of determinants that increase HbF expression can ameliorate the severity of HbE β-thalassemia. Inheritance of a chromosome with the C → T polymorphism that results in an *Xmn-1* cleavage site at position −158 to the $^G\gamma$-globin gene is associated with increased HbF and milder anemia (Winichagoon et al. 1987). Two copies of this allele (*Xmn-1* +/+) are necessary to produce a significant clinical effect. Increased expression of the $^G\gamma$-globin gene was also detected in the *Xmn-1* +/+ patients. This increase of γ-globin gene activity reduces the overall globin chain imbalance and thus ameliorates the anemia. The association between the *Xmn-1* +/+ genotype and a highly significant increase in the absolute level of HbF and a milder phenotype has also been observed in patients with HbE β-thalassemia in Sri Lanka (Premawardhena et al. 2005). It is likely that several other polymorphisms will have this effect (see later section and Sankaran and Orkin 2013).

Amount of Alternatively Spliced β^E-Globin mRNA

An underproduction of β-globin chains from the β^E-globin gene strongly suggests that alternative RNA splicing is of functional significance. The percentage of alternative spliced β^E-globin mRNA was determined by the reverse transcriptase polymerase chain reaction technique in 14 patients with the same thalassemia mutation (Winichagoon et al. 1995). Preliminary results showed abnormally spliced β^E-globin mRNA in patients with severe symptoms and low hemoglobin levels between 2.9% and 6.1%, whereas those with higher hemoglobin levels had values from 1.6% to 2.6%. The majority of patients with the *XmnI*-negative genotype had more severe anemia and a higher percentage of abnormally spliced β^E-globin mRNA. This indicated that the amount of alternatively spliced E-globin mRNA was a more important factor in determining severity of anemia than the pattern of *XmnI* polymorphism or the level of HbF. Recently, Tubsuwan et al. used the allele-specific RT-qPCR to study β^E-globin gene expression and found that the correctly to aberrantly spliced β^E-globin mRNA ratio in 30% of mild HbE β-thalassemia patients was higher than that of the severe patients. It appears therefore that the splicing process of β^E-globin pre-mRNA differs among HbE β-thalassemia patients and serves as one of the modifying factors for disease severity (Tubsuwan et al. 2011). It will be important to determine whether this phenomenon occurs in other ethnic groups.

Pyrimidine 5′ Nucleotidase Deficiency

In a Bangladeshi family, an individual homozygous for both HbE and pyrimidine 5′ nucleotidase deficiency was found. The patient had a severe hemolytic anemia in contrast to HbE homozygotes. Globin chain synthesis experiments showed that the mechanism underlying the interaction between these two genotypes was a marked decrease in the stability of HbE in pyrimidine 5′ nucleotidase-deficient red blood cells. In these cells, free α-globin chains but not β^E-globin chains accumulated on the membrane. It was hypothesized that the marked instability of HbE in the enzyme-deficient cells resulted from oxidant damage to mildly unstable HbE (Rees et al. 1996). Clearly this interaction also has the potential to modify the phenotype of HbE β-thalassemia.

Genome-Wide Association Study

Recently, a genome-wide association study (GWAS) was performed in 618 Thai HbE β^0-thalassemia patients using the Illumina Human 610-Quad BeadChips array (Nuinoon et al. 2010). DNAs were extracted from 383 severe and 235 mild phenotypes, by a validated scoring system, after the exclusion of α-thalassemia. Twenty-three single nucleotide polymorphisms (SNPs) in three independent genes/regions were identified as being significantly associated with the disease severity. The highest association was observed with SNPs in the β-globin gene cluster (chr.11p15). The second was identified in the intergenic region between the *HBS1L* and *MYB* genes (chr.6q23), and the third region was located in the *BCL11A* gene (chr.2p16.1). An association to HbF levels with SNPs in these three regions was observed. This result suggests that several genetic loci act in concert to influence HbF levels of HbE β^0-thalassemia patients (see Sankaran and Orkin 2013).

Variation in Adaptation to Anemia and Environmental Factors

As already mentioned, there appears to be a reduction in erythropoietin production in relationship to a similar hemoglobin level with aging, a finding which may explain some of the phenotypic instability during the early years of life. Whether there is a genetic component to the magnitude of erythropoietin response remains to be determined. Undoubtedly, patients with HbE β-thalassemia adapt more readily to low hemoglobin levels than those with other forms of β-thalassemia intermedia (Allen et al. 2010). Although as is the case for most inherited hemoglobin disorders, the role of the environment in modifying the phenotype has been neglected; recent studies suggest that patients with HbE β-thalassemia are more susceptible to malaria infection, particularly that caused by *P. vivax*, than age-matched controls in the population (O'Donnell et al. 2009). There is also reasonable evidence that those who have been exposed to malaria tend to have larger spleens and fall into the more severe phenotypic categories. Much more work is required to further elucidate environmental factors that may modify the phenotype.

Conclusion

The genotypic factors that can be used to predict a mild phenotype in HbE β-thalassemia are mild β^+-thalassemia mutations, the coinheritance of α-thalassemia, the polymorphisms associated with HbF production such as homozygosity for *Xmn-l* restriction site 5′ to the $^G\gamma$-globin gene and the *BCL11A* gene. Some complications of the disease such as severe jaundice are also affected by genetic modifiers. And it is also clear that at least some factors that modify the response to anemia or the environment of patients with this disease are also responsible for phenotypic diversity. But although progress has been made, it is still only possible to explain part of its wide phenotypic diversity.

Treatment

Because HbE β-thalassemia has such a variable phenotype and patients with this disorder—probably because they have relatively lower levels of HbF and reflecting the oxygen affinity of HbE—are able to adapt to anemia better than patients with other forms of thalassemia intermedia, it is vital to observe babies and young children with this condition after presentation for a reasonable period before deciding on the best approach to management. It is important to remember that they may present with a particularly low hemoglobin level consequent to a recent infection, and it is particularly important therefore not to establish them on a regular transfusion until their steady-state hemoglobin level and level of growth and degree of splenomegaly has been assessed. Particularly in areas where malaria is endemic it is also important to exclude chronic *P. vivax* infection as a possible cause of rapidly progressive splenomegaly.

The hemoglobin level alone should not be the major factor in initiating transfusion. Rather, the broader picture should be considered with particular attention to growth failure, lack of activity, and the earlier appearance of skeletal change. If it is clear that the patient requires regular transfusion the regimen to be followed, including chelation, is similar to that for the management of thalassemia major (see Olivieri and Brittenham 2013). Those who do not require transfusion should be maintained on folic acid supplements and advised about the early treatment of infective episodes. Although some patients with increasing splenomegaly and evidence of hypersplenism may benefit from splenectomy, this should be avoided where possible because of the particularly high risk of infection.

Patients who do not require regular transfusion should have serum ferritin or hepatic MRI estimations at least twice per year. Increased iron levels should be controlled by intermittent courses of chelating agents.

Hydroxyurea therapy may increase HbF levels (Fucharoen et al. 1996), although recent studies in other populations have shown that this effect is not great, even when combined with erythropoietin. For those who present early with severe disease, bone marrow transplantation remains an important option (Issaragrisil 1994; Leelahavarong et al. 2010).

Rapidly expanding extramedullary hemopoietic masses, particularly involving the brain or spinal cord, require urgent treatment by blood transfusion, hydroxyurea, or possibly, radiotherapy. Limited experience in those with profound jaundice as a result of genetic inability to conjugate bilirubin suggest that, at least in some cases, very low doses of phenobarbitone may be helpful.

The first successful report of gene therapy in thalassemia involved a patient with HbE β-thalassemia (Cavazzana-Calvo et al. 2010). Although the patient showed clinical improvement and did not need further blood transfusion, it is not yet clear whether this is mainly because of the action of the inserted "normal" β-globin gene. The improvement also seems to be at least partly a result of an increase in HbF levels by an unknown mechanism.

Prevention of HbE β-Thalassemia

Effective prevention programs for thalassemia have been demonstrated in many countries in which carrier rates for different types of thalassemia are very high (see Olivieri and Brittenham 2013). In Thailand, screening has improved by using an automatic high-performance liquid chromatography (HPLC) system (Fucharoen et al. 1998). In recent years, a nationwide program in Thailand has been developed to prevent homozygous β-thalassemia, HbE β-thalassemia, and Hb Bart's hydrops fetalis, with encouraging results (Tongsong et al. 2000). The prospective screening consisted of osmotic fragility (OF) and HbE screening tests in pregnant women, followed by testing the husbands of the women with a positive result. Subsequently, the OF test was replaced by mean corpuscular volume (MCV) when automated cell counters became available nationwide. If both partners of the couple have a positive result, further diagnostic tests by HPLC and genotyping of the carrier are carried out. A pregnancy in which both partners of the couple are carriers is considered as a couple at risk, and further detailed counseling and prenatal diagnosis is offered for the severe thalassemia syndromes (Fucharoen and Winichagoon 2007).

Since the program began, the number of new cases of thalassemia in Thailand has gradually decreased. In the first few years of the program, its cost-effectiveness was evaluated and it was found that among a total pregnant population of 21,000 individuals that were screened, 80 affected fetuses had been identified and the pregnancy terminated. The total cost of the prevention program was about U.S. $257,140, and the cost of management of these affected cases, if they had been born, would have been U.S. $7,200,000. The cost–benefit ratio was 1:28, which indicates a highly cost-effective project.

CONCLUSIONS

HbE β-thalassemia is a major public health problem in Southeast Asia and in other Asian countries. Although some progress has been made toward a better understanding of its pathophysiology and clinical management a great deal remains to be learned. Recent work has made it absolutely clear that there must be other genetic modifiers to be discovered that are responsible for the variable phenotype. A better approach to predicting the phenotype is urgently required, particularly if prenatal diagnosis is to be widely used for the control of this condition and, even more so, if experimental forms of gene therapy become available in the future. Because it may be some time before there are more definitive forms of treatment, it is important to utilize the information that we already have more effectively. For example, in malarious areas it will be very important to conduct trials of malaria prophylaxis. With particular respect to the phenotype of patients with this condition early in life, and because recent evidence suggests that the erythropoietin response to anemia tends to decline with age, the possibility of transient periods of transfusion during maximum erythroid expansion should be seriously considered. And because genetic evidence indicates that the phenotype in this condition may be improved quite dramatically with only a modest increase in steady-state hemoglobin level, more efforts should be directed at trying to raise the HbF level in these patients.

Other Interactions of HbE

This article has considered the most common and important clinical interactions of HbE. Because it is so common it is not surprising that many rare interactions between this variant and a wide variety of structural hemoglobin variants and related conditions have been reported. The resulting disorders are described briefly in a recent review (Fucharoen and Weatherall 2009).

ACKNOWLEDGMENTS

This work is supported by the Office of the Higher Education Commission and Mahidol University under the National Research University Initiative and the National Center for Genetic Engineering, Thailand and The Wellcome Trust, UK. We thank Liz Rose for her help in preparing this work.

REFERENCES

**Reference is also in this collection.*

Allen A, Fisher C, Premawardena A, Peto T, Allen S, Arambepola K, Thayalsuthan V, Olivieri N, Weatherall D. 2010. Adaptation to anemia in hemoglobin E-β thalassemia. *Blood* **116:** 5368–5370.

Aswapokee N, Aswapokee P, Fucharoen S, Wasi P. 1988a. A study of infective episodes in patients with β-thalassemia/Hb E disease in Thailand. *Birth Defects* **23:** 513–520.

Aswapokee P, Aswapokee N, Fucharoen S, Sukroongreung S, Wasi P. 1988b. Severe infection in thalassemia: A prospective study. *Birth Defects* **23:** 521–526.

Butthep P, Rummavas S, Wisedpanichkij R, Jindadamrongwech S, Fucharoen S, Bunyaratvej A. 2002. Increased circulating activated endothelial cells, vascular endothelial growth factor, and tumor necrosis factor in thalassemia. *Am J Hematol* **70:** 100–106.

Cavazzana-Calvo M, Payen E, Negre O, Wang G, Hehir K, Fusil F, Down J, Denaro M, Brady T, Westerman K, et al. 2010. Transfusion independence and HMGA2 activation after gene therapy of human β-thalassaemia. *Nature* **467:** 318–322.

Chandcharoensin-Wilde C, Chairoongruang S, Jitnuson P, Fucharoen S, Vathanopas V. 1988. Gallstones in thalassemia. *Birth Defects* **23:** 263–267.

Chernoff AI, Minnich V, Na-Nakorn S, Tuchinda S, Kashamsant C, Chernoff RR. 1956. Studies on hemoglobin E: I. The clinical, hematologic, and genetic characteristics of the hemoglobin E syndromes. *J Lab Clin Med* **47:** 490–498.

De Chiara B, Crivellaro W, Sara R, Ruffini L, Parolini M, Fesslova V, Carnelli V, Fiorentini C, Parodi O. 2005. Early detection of cardiac dysfunction in thalassemic patients by radionuclide angiography and heart rate variability analysis. *Eur J Haematol* **74:** 517–522.

Frischer H, Bowman J. 1975. Hemoglobin E, an oxidatively unstable mutation. *J Lab Clin Med* **85:** 531–539.

Fucharoen S, Weatherall DJ. 2009. Hemoglobin E disorders. In *Disorders of hemoglobin* (ed. Steinberg MH, Forget BG, Higgs DR, Weatherall DJ), pp. 417–433. Cambridge University Press, Cambridge.

Fucharoen S, Winichagoon P. 2007. Prevention and control of thalassemia in Asia. *Asian Biomed* **1:** 1–6.

Fucharoen S, Youngchaiyud P, Wasi P. 1981. Hypoxaemia and the effect of aspirin in thalassaemia. *Southeast Asian J Trop Med Public Health* **12:** 90–93.

Fucharoen S, Winichagoon P, Thonglairuam V. 1988a. Beta-thalassemia associated with α-thalassemia in Thailand. *Hemoglobin* **12:** 581–592.

Fucharoen S, Winichagoon P, Thonglairuam V, Wasi P. 1988b. EF Bart's disease: Interaction of the abnormal α- and β-globin genes. *Eur J Haematol* **40:** 75–78.

Fucharoen S, Siritanaratkul N, Winichagoon P, Chowthaworn J, Siriboon W, Muangsup W, Chaicharoen S, Poolsup N, Chindavijak B, Pootrakul P, et al. 1996. Hydroxyurea increases hemoglobin F levels and improves the effectiveness of erythropoiesis in β-thalassemia/hemoglobin E disease. *Blood* **87:** 887–892.

Fucharoen S, Winichagoon P, Wisedpanichkij R, Sae-Ngow B, Sriphanich R, Oncoung W, Muangsapaya W, Chowthaworn J, Kanokpongsakdi S, Bunyaratvej A, et al. 1998. Prenatal and postnatal diagnoses of thalassemias and hemoglobinopathies by HPLC. *Clin Chem* **44:** 740–748.

* Higgs DR. 2013. The molecular basis of α-thalassemia. *Cold Spring Harb Perspect Med* **3:** a011718.

Issaragrisil S. 1994. Bone marrow transplantation in Thailand. *Bone Marrow Transplant* **13:** 721–723.

Issaragrisil S, Piankigagum A, Wasi P. 1981. Spinal cord compression in thalassemia. Report of 12 cases and recommendations for treatment. *Arch Intern Med* **141:** 1033–1036.

Itano HA, Bergren WR, Sturgeon P. 1954. Identification of a fourth abnormal human hemoglobin. *J Am Chem Soc* **76:** 2278.

Jootar P, Fucharoen S. 1990. Cardiac involvement in β-thalassemia/hemoglobin E disease: Clinical and hemodynamic findings. *Southeast Asian J Trop Med Public Health* **21:** 269–273.

Leelahavarong P, Chaikledkaew U, Hongeng S, Kasemsup V, Lubell Y, Teerawattananon Y. 2010. A cost-utility and budget impact analysis of allogeneic hematopoietic stem cell transplantation for severe thalassemic patients in Thailand. *BMC Health Serv Res* **10:** 209.

Minnich V, Na-Nakorn S, Chongchareonsuk S, Kochaseni S. 1954. Mediterranean anemia: A study of 32 cases in Thailand. *Blood* **9:** 1–23.

* Musallam KM, Taher AT, Rachmilewitz EA. 2012. β-Thalassemia intermedia: A clinical perspective. *Cold Spring Harb Perspect Med* **2:** a013482.

* Nienhuis AW, Nathan DG. 2012. Pathophysiology and clinical manifestations of the β-thalassemias. *Cold Spring Harb Perspect Med* **2:** a011726.

Nuinoon M, Makarasara W, Mushiroda T, Setianingsih I, Wahidiyat PA, Sripichai O, Kumasaka N, Takahashi A, Svasti S, Munkongdee T, et al. 2010. A genome-wide association identified the common genetic variants influence disease severity in β0-thalassemia/hemoglobin E. *Hum Genet* **127:** 303–314.

O'Donnell A, Premawardhena A, Arambepola M, Allen SJ, Peto TE, Fisher CA, Rees DC, Olivieri NF, Weatherall DJ. 2007. Age-related changes in adaptation to severe anemia in childhood in developing countries. *Proc Natl Acad Sci* **104:** 9440–9444.

O'Donnell A, Premawardhena A, Arambepola M, Samaranayake R, Allen SJ, Peto TE, Fisher CA, Cook J, Corran PH, Olivieri NF, et al. 2009. Interaction of malaria with a common form of severe thalassemia in an Asian population. *Proc Natl Acad Sci* **106:** 18716–18721.

* Olivieri NF, Brittenham GM. 2013. Management of the thalassemias. *Cold Spring Harb Perspect Med* doi: 10.1101/cshperspect.a011767.

Olivieri NF, Thayalsuthan V, O'Donnell A, Premawardena A, Rigobon C, Muraca G, Fisher C, Weatherall DJ. 2010. Emerging insights in the management of hemoglobin E β thalassemia. *Ann NY Acad Sci* **1202:** 155–157.

Olivieri NF, Thayalasuthan V, Muraca G, Weatherall D, Kim C, Premawardhena A, Perera A, O'Donnell A, St Pierre T. 2011. Relationship between serum ferritin and liver iron concentration in hemoglobin E thalassemia. *16th Congress of European Hematology Association*, London, June 9–12.

Orkin SH, Kazazian HH, Antonarakis SE, Ostrer H, Goff SC, Sexton JP. 1982. Abnormal RNA processing due to the exon mutation of β^E-globin gene. *Nature* **300:** 768–769.

Pattanapanyasat K, Noulsri E, Fucharoen S, Lerdwana S, Lamchiagdhase P, Siritanaratkul N, Webster HK. 2004. Flow cytometric quantitation of red blood cell vesicles in thalassemia. *Cytometry B Clin Cytom* **57:** 23–31.

Pootrakul P, Rugkiatsakul R, Wasi P. 1980. Increased transferrin iron saturation in splenectomized thalassaemia patients. *Brit J Haematol* **46:** 143–145.

Pootrakul P, Vougsmasa V, Laongpanich P, Wasi P. 1981. Serum ferritin levels in thalassemias and the effect of splenectomy. *Acta Haematol* **66:** 244–250.

Premawardhena A, Fisher CA, Fathiu F, de Silva S, Perera W, Peto TE, Olivieri NF, Weatherall DJ. 2001. Genetic determinants of jaundice and gallstones in haemoglobin E β thalassaemia. *Lancet* **357:** 1945–1946.

Premawardhena A, Fisher CA, Liu YT, Verma IC, de Silva S, Arambepola M, Clegg JB, Weatherall DJ. 2003. The global distribution of length polymorphisms of the promoters of the glucuronosyltransferase 1 gene (UGT1A1): Hematologic and evolutionary implications. *Blood Cells Mol Dis* **31:** 98–101.

Premawardhena A, Fisher CA, Olivieri NF, de Silva S, Arambepola M, Perera W, O'Donnell A, Peto TEA, Viprakasit V, Merson L, et al. 2005. Haemoglobin E β thalassaemia in Sri Lanka. *Lancet* **366:** 1467–1470.

Rees DC, Duley J, Simmonds HA, Wonke B, Thein SL, Clegg JB, Weatherall DJ. 1996. Interaction of hemoglobin E and pyrimidine 5′ nucleotidase deficiency. *Blood* **88:** 2761–2767.

Rees DC, Clegg JB, Weatherall DJ. 1998. Is hemoglobin instability important in the interaction between hemoglobin E and β thalassemia? *Blood* **92:** 2141–2146.

Rutjanaprom W, Kanlop N, Charoenkwan P, Sittiwangkul R, Srichairatanakool S, Tantiworawit A, Phrommintikul A, Chattipakorn S, Fucharoen S, Chattipakorn N. 2009. Heart rate variability in β thalassemia patients. *Eur J Haematol* **83:** 483–489.

* Sankaran VG, Orkin SH. 2013. The switch from fetal to adult hemoglobin. *Cold Spring Harb Perspect Med* **3:** a011643.

Sathapatayavongs B, Leelachaikul P, Prachaktam R, Atichartakarn V, Sriphojanart S, Trairatvorakul P, Jirasiritham S, Nontasut S, Eurvilaichit C, Flegel T. 1989. Human pythiosis associated with thalassemia hemoglobinopathy syndrome. *J Infect Dis* **159:** 274–280.

Sonakul D, Pacharee P, Laohapand T, Fucharoen S, Wasi P. 1980. Pulmonary artery obstruction in thalassaemia. *Southeast Asian J Trop Med Public Health* **11:** 516–523.

Sonakul D, Pacharee P, Wasi P, Fucharoen S. 1984. Cardiac pathology in 47 patients with β thalassaemia/haemoglobin E. *Southeast Asian J Trop Med Public Health* **15:** 554–563.

Sonakul D, Pacharee P, Thakerngpol K. 1988. Pathologic findings in 76 autopsy cases of thalassemia. *Birth Defects* **23:** 157–176.

Sripichai O, Makarasara W, Munkongdee T, Kumkhaek C, Nuchprayoon I, Chuansumrit A, Chuncharunee S, Chantrakoon N, Boonmongkol P, Winichagoon P, et al. 2008. A scoring system for the classification of β-thalassemia/Hb E disease severity. *Am J Hematol* **83:** 482–484.

Thakerngpol K, Sonakul D, Fucharoen S, Vathanopas V, Stitnimankarn T. 1988. Histochemical study of liver tissue from thalassemic patients. *Birth Defects* **23:** 193–198.

* Thein SL. 2013. The molecular basis of β-thalassemia. *Cold Spring Harb Perspect Med* doi: 10.1101/cshperspect.a011700.

Thonglairuam V, Winichagoon P, Fucharoen S, Wasi P. 1989. The molecular basis of AE-Bart's disease. *Hemoglobin* **13:** 117–124.

Tongsong T, Wanapirak C, Sirivatanapa P, Sanguansermsri T, Sirichotiyakul S, Piyamongkol W, Chanprapaph P. 2000. Prenatal control of severe thalassaemia: Chiang Mai strategy. *Prenat Diag* **20:** 229–234.

Traeger J, Wood WG, Clegg JB, Weatherall DJ, Wasi P. 1980. Defective synthesis of Hb E is due to reduced levels of βE mRNA. *Nature* **288:** 497–499.

Tubsuwan A, Munkongdee T, Jearawiriyapaisarn N, Boonchoy C, Winichagoon P, Fucharoen S, Svasti S. 2011. Molecular analysis of globin gene expression in different thalassaemia disorders: Individual variation of β(E) pre-mRNA splicing determine disease severity. *Brit J Haematol* **154:** 635–643.

Vannasaeng S, Ploybutr S, Visutkul P, Tandhanand S, Suwanik R, Wasi P. 1981. Endocrine function in thalassaemia. *Clin Endocrinol (Oxf)* **14:** 165–173.

Vathanopas V, Fucharoen S, Chandrcharoensin-Wilde C, Sukroongreung S, Nilakul C. 1988. Cholecystectomy in thalassemia. *Birth Defects* **23:** 269–273.

* Vichinsky EP. 2013. Clinical manifestations of α-thalassemia. *Cold Spring Harb Perspect Med* doi: 10.1101/cshperspect.a011742.

Wanachiwanawin W, Thianprasit M, Fucharoen S, Chaiprasert A, Sudasna N, Ayudhya N, Sirithanaratkul N, Piankijagum A. 1993. Fatal arteritis due to Pythium insidiosum infection in patients with thalassaemia. *Trans R Soc Trop Med Hyg* **87:** 296–298.

Wasi P, Sookanek M, Pootrakul S, Na-Nakorn S, Suingdumrong A. 1967. Haemoglobin E and α-thalassaemia. *Br Med J* **4:** 29–32.

Wasi P, Na-Nakorn S, Pootrakul S, Sookanek M, Disthasongcham P, Pornpatkul M, Manich V. 1969. Alpha- and β-thalassemia in Thailand. *Ann NY Acad Sci* **165:** 60–82.

Wasi P, Na-Nakorn S, Pootrakul P, Sonakul D, Piankijagum A, Pacharee P. 1978. A syndrome of hypertension, convulsions, and cerebral haemorrhage in thalassaemic patients after multiple blood transfusions. *Lancet* **ii:** 602–604.

Wasi P, Fucharoen S, Younghchaiyud P, Sonakul D. 1982. Hypoxemia in thalassemia. *Birth Defects* **18:** 213–217.

Weatherall DJ, Clegg JB. 2001. *The thalassaemia syndromes.* Blackwell Science, Oxford.

Wickramasinghe SN, Lee MJ. 1997. Observations on the relationship between g-globin chain content and globin chain precipitation in thalassaemic erythroblasts and on the composition of erythroblastic inclusions in Hb E/β-thalassaemia. *Eur J Haematol* **59:** 305–309.

* Williams TN, Weatherall DJ. 2012. World distribution, population genetics, and health burden of the hemoglobinopathies. *Cold Spring Harb Perspect Med* **2:** a011692.

Winichagoon P, Fucharoen S, Weatherall DJ, Wasi P. 1985. Concomitant inheritance of α-thalassemia in β^0-thalassemia/Hb E disease. *Am J Hematol* **20:** 217–222.

Winichagoon P, Fucharoen S, Thonglairoam V, Wasi P. 1987. Different severity of homozygous β-thalassemia among siblings. *Hum Genet* **76:** 296–297.

Winichagoon P, Thonglairoam V, Fucharoen S, Wilairat P, Fukimaki Y, Wasi P. 1993. Severity differences in β-thalassaemia haemoglobin E syndromes: Implication of genetic factors. *Brit J Haematol* **83:** 633–639.

Winichagoon P, Fucharown S, Wilairat P, Chihara K, Fukumaki Y. 1995. The role of alternatively spliced β^E-globin mRNA on clinical severity of β-thalassemia/hemoglobin E disease. *Southeast Asian J Trop Med Public Health* **26:** 241–245.

Clinical Manifestations of α-Thalassemia

Elliott P. Vichinsky

Department of Hematology/Oncology, Children's Hospital & Research Center Oakland, Oakland, California 94609

Correspondence: evichinsky@mail.cho.org

α-Thalassemia mutations affect up to 5% of the world's population. The clinical spectrum ranges from an asymptomatic condition to a fatal in utero disease. Hemoglobin H disease results from mutations of three α-globin genes. Deletional forms result in a relatively mild anemia, whereas nondeletional mutations result in a moderate to severe disease characterized by ineffective erythropoiesis, recurrent transfusions, and growth delay. Hemosiderosis develops secondary to increased iron absorption, as well as transfusion burden. Hemoglobin Bart's hydrops fetalis is usually a fatal in utero disease caused by the absence of α genes. Population screening to identify at-risk couples is essential. Affected pregnancies result in severe fetal and maternal complications. Doppler ultrasonography with intrauterine transfusion therapy may improve the fetal prognosis but creates ethical challenges for the family and health providers.

α-Thalassemia syndrome refers to a group of hereditary anemias caused by absent or decreased production of the α-chains of hemoglobin. Up to 5% of the world's population are carriers for these common gene mutations (Weatherall and Clegg 2001; Weatherall 2008). There is marked phenotypic heterogeneity in α-thalassemia resulting from the genotypic diversity. At one end of the spectrum is a completely asymptomatic phenotype resulting from one of the four α genes being dysfunctional. At the other end is a fatal in utero disease called hemoglobin Bart's hydrops fetalis, which results from absence of the four α genes.

The phenotype of the α-thalassemias is directly related to the number of α-globin genes affected. They are divided into the α^+-thalassemias, in which one of the linked pair of α genes is deleted ($-\alpha/\alpha\alpha$); and the α^0-thalassemias, in which both are deleted ($-/\alpha\alpha$). The heterozygous state for α^+-thalassemia is clinically silent, while the heterozygous state for α^0-thalassemia produces the α-thalassemia trait, a condition that is hematologically similar to the β-thalassemia trait. The homozygous state for α^+-thalassemia results in a similar phenotype. The inheritance of both α^+-thalassemia and α^0-thalassemia ($-\alpha/-$) results in hemoglobin H disease. There are also nondeletional forms of α-thalassemia ($-\alpha^{ND}/\alpha\alpha$) caused by various structural abnormalities of an α-globin gene. When these are inherited together with α^0-thalassemia ($-\alpha^{ND}/-$), a more severe form of hemoglobin H disease occurs. Hemoglobin H is a tetramer of β-chains (β_4), which is unstable and has an extremely high oxygen affinity and is thus unable to effectively deliver oxygen. It is also relatively unstable and causes ineffective

Copyright © 2013 Cold Spring Harbor Laboratory Press; all rights reserved
Cite this article as *Cold Spring Harb Perspect Med* doi: 10.1101/cshperspect.a011742

erythropoiesis and a hemolytic anemia. Hemoglobin Bart's, named after the London hospital in which it was discovered, is the fetal counterpart of hemoglobin H and consists of a tetramer of γ-chains of fetal hemoglobin (γ_4). It too has a very high oxygen affinity, although it is less unstable than hemoglobin H.

EPIDEMIOLOGY

Worldwide, α-thalassemia mutations occur in almost all populations (Williams and Weatherall 2012). They are very common in people of Asian descent; in several regions of China and Southeast Asia, up to 40% of the population are carriers (Lau et al. 1997; Li et al. 2006; Michlitsch et al. 2009; Vichinsky 2009, 2012; Lal et al. 2011). The Mediterranean and Middle Eastern regions also have a high frequency (Weatherall and Clegg 2001). Immigration patterns have dramatically changed the geographic distribution of α-thalassemia (Michlitsch et al. 2009). Once rare in North America, it is becoming a major public health problem. There has been a 2000% increase in Asian and other at-risk populations immigrating to the United States in the last three decades (Hoppe 2009). This has resulted in California establishing universal newborn screening for α-thalassemia, and consideration of a national α-thalassemia newborn screening policy (Benz 2011). In the United States, recent studies indicate that in addition to these known at-risk ethnic groups, people of Hispanic and African descent are also at risk (Lam and Tang 2002). In the California Newborn Screening Program, newborns classified as mixed-race are a rapidly growing group born with α-thalassemia syndromes (Lam and Tang 2002; Michlitsch et al. 2009). The marked ethnic diversity in California indicates that universal rather than targeted screening is necessary. α-Thalassemia syndromes have become the most common nonsickle hemoglobinopathy in California (Table 1). α-Thalassemia syndromes are increasingly being diagnosed outside of California. Because most states do not screen for these disorders, the diagnosis is not made until clinical complications occur.

Table 1. California Newborn Screening Program, January 1998–June 2006

Diagnosis	Newborns identified	Incidence (per 100,000 screened)
α-Thalassemia syndromes	502	11.1
Hb H disease	406	9.0
Hb H with Hb E trait	35	0.8
Hb H with Hb EE	16	0.4
Hb H Constant Spring	25	0.6
Hb H with other variants	8	0.2
Hb H with SS	7	0.2
Hb Bart's hydrops fetalis[a]	5	0.1
β-Thalassemia syndromes	79	1.8
β^0-Thalassemia	19	0.4
Hb E/β^0-thalassemia	31	0.7
Hb C/β^0-thalassemia	7	0.2
Hb D/β^0-thalassemia	5	0.1
Hb C/β^+-thalassemia	14	0.3
Hb E/β^+-thalassemia	3	0.1
Other mutations	862	19.1
Hb EE	514	11.4
Hb CC	41	0.9
Hb C-HPFH	3	0.1
Heterozygous variants	304	6.7

Data from Michlitsch et al. 2009.

Hb, hemoglobin; HPFH, hereditary persistence of fetal hemoglobin.

[a]All five cases of Hb Bart's hydrops fetalis were homozygous for the Southeast Asian (SEA) deletion mutation.

The numerous molecular forms of α-thalassemia are reviewed by Higgs (2013). There are two common deletional forms of α^+-thalassemia, described by the size of the particular deletion, $-\alpha^{3.7}$ and $-\alpha^{4.2}$. There are many different forms of α^0-thalassemia that result from different-sized deletions of the α-globin genes. Most of them are named by the place of origin of their discovery. The most common form is the Southeast Asian deletion ($-^{SEA}$) (Chui and Waye 1998; Hoppe 2009). This deletion involves both α genes but not the embryonic globin genes (Chui and Waye 1998). Other α-thalassemias such as the Thai form ($-^{THAI}$) also remove the

Cite this article as *Cold Spring Harb Perspect Med* doi: 10.1101/cshperspect.a011742

embryonic globin genes. There are also many different nondeletional forms of α-thalassemia, many of which are also named after their place of origin. The clinical phenotypes of hemoglobin H disease that result from the interaction of nondeletional mutations ($-/\alpha^{ND}\alpha$) are more severe and less common than the deletional forms ($-\alpha/-$) (Lau et al. 1997; Lorey et al. 2001a; Ne et al. 2008; Michlitsch et al. 2009; Singer et al. 2009; Kidd et al. 2010; Lal et al. 2011; Vichinsky 2012). The most common interaction of this type is with the α-globin chain-termination mutant, Hb Constant Spring ($-/\alpha^{CS}\alpha$), which accounts for approximately 27% of hemoglobin H diseases in California (Table 2) and which occurs frequently in many countries in Southeast Asia.

THE MILDER FORMS OF α-THALASSEMIA

The milder forms of α-thalassemia are reviewed in detail by Weatherall and Clegg (2001). The heterozygous states for deletional ($-\alpha/\alpha\alpha$) and nondeletional ($\alpha^{ND}\alpha/\alpha\alpha$) α-thalassemia are associated with extremely mild reductions in the red cell indices that overlap with normal values. At birth some, but not all, cases have slightly elevated levels of hemoglobin Bart's, which disappears over the first year of life and is not replaced by hemoglobin H. The homozygous state for these conditions is characterized by a mild hypochromic microcytic anemia with significantly reduced levels of mean corpuscular volume (MCV) and mean corpuscular hemoglobin (MCH) and elevated levels of hemoglobin Bart's at birth.

Hence these conditions themselves are of no clinical significance and their major importance is the modifying effect that they have on various severe forms of β-thalassemia and sickle cell anemia. These interactions are summarized by Musallam et al. (2012), Fucharoen and Weatherall (2012), and Lettre (2012). As discussed in a later section, their identification is vital in population screening for couples at risk for the more severe forms of α-thalassemia.

Table 2. Hemoglobin H disease genotypes identified by the California Newborn Screening Program, January 2006–December 2008

Genotype	Newborns identified	Estimated birth prevalence (per 100,000)
$\alpha^{3.7}$/SEA	85	5
$\alpha^{3.7}$/FIL	25	1.5
$\alpha^{4.2}$/SEA	5	0.3
$\alpha^{3.7}$/THAI	1	0.05
$\alpha^{4.2}$/FIL	1	0.05
$\alpha^{CS}\alpha$/SEA	13	0.77
$\alpha^{PolyA}\alpha$/SEA	1	0.05
SEA/SEA	3	0.17
Total	134	8

Data from Hoppe 2009.

HEMOGLOBIN H DISEASE

Clinical Severity

Hemoglobin H–Constant Spring and deletional hemoglobin H disease are two distinct clinical entities (Lal et al. 2011). Patients with hemoglobin H–Constant Spring commonly have complications develop within the first decade of childhood, including significant growth delay, recurrent transfusions, and iron overload (Table 3) (Chui et al. 2003; Fucharoen and Viprakasit 2009; Singer et al. 2009; Benz 2011; Lal et al. 2011; Vichinsky 2012). They are particularly at risk for sudden, severe anemia following acute febrile illness. Often these infections may be mild. Most patients by 10 years of age have received recurrent transfusions. The acute anemic events are usually hemolytic and have associated increases in bilirubin and reticulocyte counts. Hypoplastic acute anemic events secondary to parvovirus and other viral infections account for 20% of acute anemic episodes. In addition to febrile events, pregnancy often is associated with unexpected decreases in hemoglobin levels.

The anemia of hemoglobin H disease results from the deficit of α-globin expression, which is more severe in the nondeletional form (Table 3). Membrane injury from oxidative damage results in shortened red cell survival (Pootrakul et al. 2000; Fucharoen and Viprakasit 2009). The hemolytic anemia may result in the

Table 3. Clinical and hematologic manifestation of hemoglobin H disease

Clinical manifestation	Deletional	Nondeletional
Hemoglobin (g/dL)	8.5 (range, 6.9–10.7)	7.2 (range, 3.8–8.7)
MCV (fL)	54.0 (range, 46.0–76.0)	65.2 (range, 48.7–80.7)
MCH (pg)	16.6 (range, 14.3–24.7)	18.6 (range, 14.8–24.8)
Anemia	Microcytic	Normocytic
Reticulocytosis	+	++
Hypochromia	++	+
Age at first transfusion (yr)	11 ± 5.5	1.5 ± 2.1
History of blood transfusion (%)	3–29	24–80
Splenomegaly	+	+++
Gallstones	+	++
Growth retardation	Rare	Common
Decreased bone density	Rare	Common

Data from Vichinsky 2009.
MCH, mean corpuscular hemoglobin; MCV, mean corpuscular volume.

development of cholelithiasis. There is intramedullary death of erythroblasts resulting in ineffective erythropoiesis. This is supported by the finding of markedly elevated serum transferrin receptor levels in hemoglobin H–Constant Spring. Phosphatidylserine exposure on the red cell surface further increases the morbidity of the disease, including an increased risk of thrombosis (Yuan et al. 1995).

The chronic anemia is more severe in hemoglobin H–Constant Spring than in deletional hemoglobin H disease and often results in transfusion therapy in childhood (Fig. 1) (Lal et al. 2011).

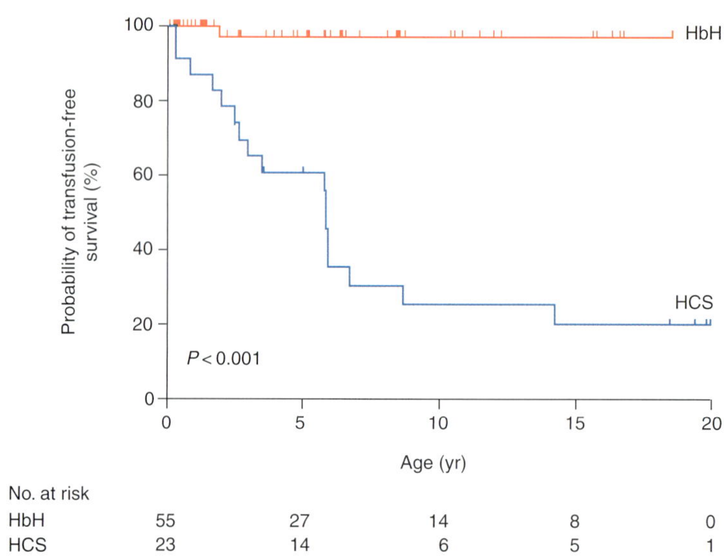

Figure 1. Age at first transfusion for patients with deletional hemoglobin H disease (HbH) and hemoglobin H–Constant Spring disease (HCS). Excluded from the analysis in the HbH group are five patients whose first clinic visit occurred after the age of 20 years and three patients with nondeletional HbH disease. (From Lal et al. 2011; reprinted, with permission, © The Massachusetts Medical Society.)

The mean hemoglobin level in infants with deletional hemoglobin H disease is 8.5 g/dL (range, 6.9–10.7) and increases to 9.4 g/dL with age. In infants with hemoglobin H–Constant Spring, the mean hemoglobin level is 7.0 g/dL (range, 5.8–9.5) and changes little with age (Chui et al. 2003; Fucharoen and Viprakasit 2009; Lal et al. 2011; Vichinsky 2012). Although hemoglobin H–Constant Spring is more severe, red cell indices are less affected. The red cell MCV is markedly reduced in hemoglobin H deletion patients (54.0 fL; range, 46.0–76.0). In contrast, the MCV in the hemoglobin H–Constant Spring group is 65.2 fL (range, 48.7–80.7). Hemoglobin E is often coinherited with hemoglobin H and its effects are variable. It may lower the MCV but decrease the hemolytic anemia (Fucharoen and Weatherall 2012).

Splenomegaly is more common in nondeletional hemoglobin H disease and may correlate with the clinical severity of the disease (Fig. 2) (Styles et al. 1997; Vichinsky 2010). Splenectomy is helpful in decreasing extravascular hemolysis and increasing the hemoglobin. Patients with hemoglobin H–Constant Spring often have a significant increase in hemoglobin following splenectomy. Transfusion-dependent patients may become transfusion-independent following surgery. However, splenectomy increases the rate of thrombosis (Tso et al. 1982). Chronic anticoagulation therapy may be necessary to prevent serious thrombotic complications. Portal vein thrombosis may be a perioperative postsplenectomy complication. Early screening of perioperative patients with abdominal symptoms is warranted. Splenectomy may be necessary in infants, and therefore prophylactic antibiotics are required.

Iron overload is a common, underdiagnosed problem in hemoglobin H disorders (Fucharoen and Viprakasit 2009; Singer et al. 2009; Benz 2011; Lal et al. 2011; Vichinsky 2012). The ferritin increases with age irrespective of the transfusion history and may underestimate body iron stores (Fig. 3) (Lal et al. 2011). Increased dietary iron absorption continues throughout life and results in hemosiderosis-related complications in some adults. Elevated pituitary, pancreatic, and cardiac iron has been observed, particularly in hemoglobin H–Constant Spring. Older patients are at risk for cardiac and endocrine dysfunction (Chui et al. 2003; Chan et al. 2006; Au et al. 2009; Fucharoen and Viprakasit 2009; Singer et al. 2009). Oral iron chelation has been beneficial in patients with elevated body iron stores (Au et al. 2009).

Laboratory Screening and Diagnosis

The hemoglobin and red cell indices are helpful in identifying patients. However, they are nonspecific findings. In addition, in hemoglobin

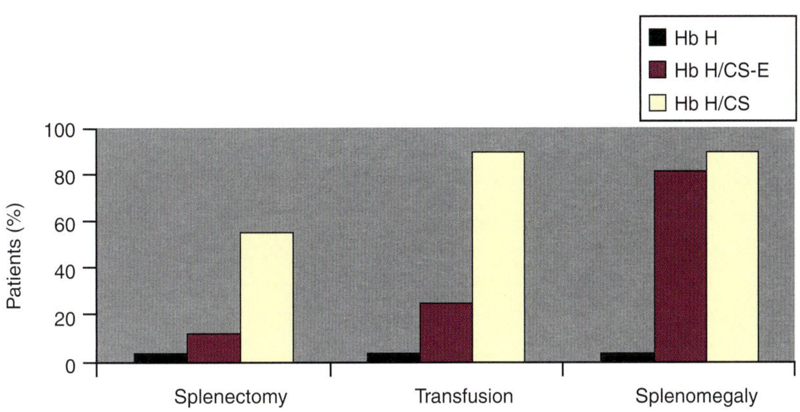

Figure 2. Comparison of the rate of splenectomy, transfusion, and splenomegaly history in three types of hemoglobin (Hb) H disease: Hb H (three-gene deletion), Hb H/CS (Hb H with Constant Spring), and Hb H/CS-E (Hb H/CS + heterozygote for Hb E). (From Styles et al. 1997; reprinted, with permission.)

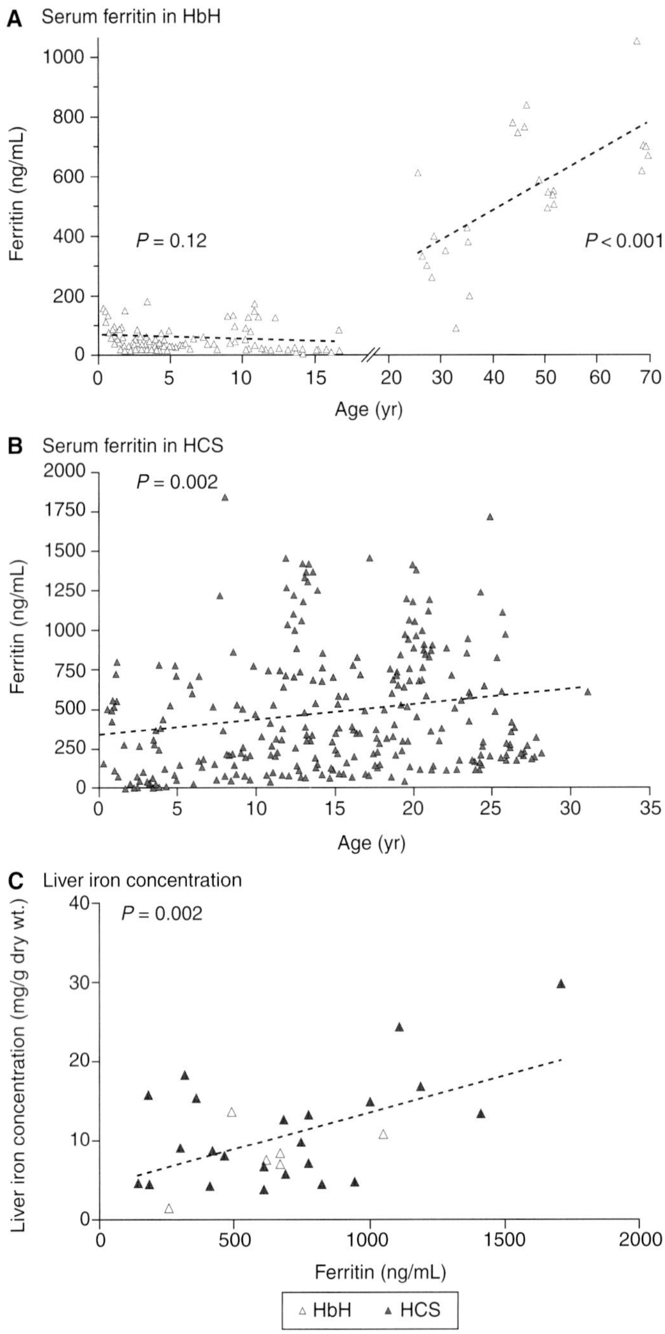

Figure 3. Serum ferritin levels in patients with deletional hemoglobin H disease (HbH) and hemoglobin H–Constant Spring disease (HCS) and correlation with liver iron concentration. The trend for ferritin levels according to age on the basis of linear regression (dashed line) increases after the age of 18 years in patients with HbH (*A*), and increases from birth to the age of 30 years in those with HCS (*B*). A significant correlation between the serum ferritin level and liver iron concentration is shown in patients with HbH and HCS (*C*), with the dashed line indicating the linear regression of liver iron concentration as compared with ferritin. (From Lal et al. 2011; reprinted, with permission, © The Massachusetts Medical Society.)

H–Constant Spring, microcytosis may not be observed. Red cell inclusion bodies may be observed after supravital stain analysis with brilliant cresyl blue (Fig. 4) (Fucharoen and Viprakasit 2009). These inclusions are markedly increased after splenectomy. Hemoglobin electrophoresis may demonstrate a hemoglobin H band. However, hemoglobin H is unstable and is often missed, particularly if the blood sample has been frozen or if organic solvents are used to prepare the hemoglobin for analysis. Hemoglobin A_2 levels are lower in α-thalassemia disorders. Automated high-performance liquid chromatography (HPLC) is an excellent screening method for high-risk populations and in newborn screening programs (Lorey et al. 2001a; Hoppe 2009; Michlitsch et al. 2009; Kidd et al. 2010). In the neonatal period, hemoglobin Bart's is commonly seen on HPLC or hemoglobin electrophoresis. Hemoglobin Bart's is an aggregate of γ-chains that reflects the high fetal hemoglobin observed in neonates. It rapidly disappears following birth. The quantitation of hemoglobin Bart's correlates with the severity of the α-thalassemia disorder. Bart's hemoglobin levels of $>25\%$ indicate hemoglobin H disease. Definitive diagnosis requires molecular genetic testing. Multiplex genotyping assays using gap polymerase chain reaction can rapidly diagnose most deletional and nondeletional mutations (Kidd et al. 2010). Confirmatory parental studies should always be performed in newborns and children identified with hemoglobin H. Patients with α-thalassemia disorders often coinherit other mutations including hemoglobin E, β-thalassemia, and hemoglobin S (Li et al. 2006; Liao et al. 2007; Hoppe 2009). In addition, the same population may be at risk for glucose phosphate dehydrogenase deficiency.

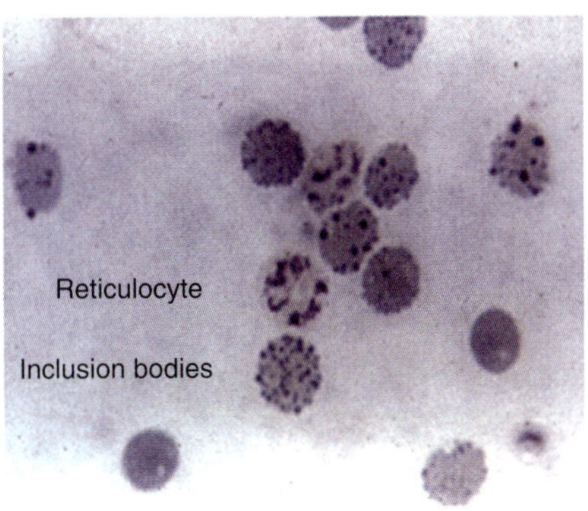

Figure 4. Red cell inclusion body induced after incubation of red cell with brilliant cresyl blue. (From Fucharoen and Viprakasit 2009; reprinted, with permission, © American Society of Hematology.)

Management

Hemoglobin H patients require ongoing monitoring of growth, quality of life, and laboratory markers of organ dysfunction. Dietary supplement with folate and multivitamins without iron is indicated (Lal et al. 2011). Patients may develop calcium and vitamin D deficiency. Monitoring bone density in older patients is indicated. Iron supplementation should be avoided. The decision to chronically transfuse patients is difficult. Brief periods of transfusion during childhood development and pregnancy may be indicated (Fucharoen and Viprakasit 2009; Singer et al. 2009; Lal et al. 2011). Oral iron chelation therapy may be beneficial in patients with hemosiderosis. The dosage indicated in non-transfusion-dependent patients is lower than in transfused patients and requires ongoing reassessment. Splenectomy is a serious undertaking, and its risks and benefits should be reviewed with each case. Aspirin or other anticoagulant therapy should always be considered following splenectomy. Chronic fatigue may develop in older patients, and quality-of-life assessments are helpful in management decisions. Family screening and genetic counseling are important. Preventative education concerning the risk of severe anemia with acute infections is important.

HEMOGLOBIN BART'S HYDROPS FETALIS

Hemoglobin Bart's hydrops fetalis is a devastating and fatal disease. It is an important public health problem in many regions of the world. Although advances in intrauterine therapy have resulted in long-term survival, these cases are rare. Prevention through family screening and education is the most effective treatment.

Homozygosity for $-^{SEA}$ is a common cause of hydrops fetalis. Because the mutation does not affect functional embryonic hemoglobin production, pregnancy continues to late gestation, allowing for the clinical manifestation of hydrops fetalis. Larger mutations that prevent embryonic hemoglobin production are often undiagnosed terminations. The severity of the disease is clearly influenced by the type of α-globin mutations. Nondeletional mutations and hyperunstable mutations increase the severity (Chan et al. 1997; Lorey et al. 2001b; Suwanrath-Kengpol et al. 2005; Siriratmanawong et al. 2007; Fucharoen and Viprakasit 2009; Hoppe 2009; Vichinsky 2009). When severe mutations are involved, hydrops fetalis can result from only three genes being affected.

Early in gestation, functional embryonic hemoglobin is responsible for oxygen delivery. After two months of gestation, a switch to hemoglobin F production occurs. Normal fetal hemoglobin cannot be produced without α-chains. Hemoglobin Bart's, which cannot deliver oxygen to tissues, becomes the major hemoglobin. Therefore, the severity of the disease is underestimated by the intrauterine hemoglobin level. The midtrimester intrauterine hemoglobin analysis demonstrates an average hemoglobin of 6.4 g/dL in hemoglobin Bart's hydrops fetalis (Table 4) (Srisupundit et al. 2008). Here is severe, ineffective erythropoiesis with marked extramedullary hematopoiesis. This leads to massive organomegaly, heart failure, gross body edema, and eventual intrauterine demise. Fetal hypoxia also causes developmental anomalies in approximately 20% of cases (Liang et al. 1985; Singer et al. 2000; Adam et al. 2005; Vichinsky 2009). Most of these anomalies are mild, involving the genitourinary and skeletal systems. However, severe malformations do occur. These are very high-risk pregnancies, and those not receiving close monitoring have a high mortality rate. The majority of pregnancies are complicated by hypertension, preeclampsia, and several other serious complications (Liang et al. 1985; Suwanrath-Kengpol et al. 2005; Vichinsky 2009; Yang and Li 2009).

Prevention and Prenatal Management

α-Thalassemia major is a public health problem requiring community education and screening of at-risk populations. Microcytosis, hemoglobin levels, and electrophoresis are useful initial screening tests. Once an index case is identified, his or her partner should be tested to identify at-risk couples. All identified at-risk couples require molecular diagnosis to determine family planning decisions. In high-risk couples, molecular diagnosis on fetal tissue is obtained early in pregnancy by chorionic villus sampling. Later in pregnancy, hydrops fetalis and anemia can be detected by Doppler ultrasonography of the middle cerebral arteries in 90% of cases (Mari et al. 2000; Lam and Tang 2002; Leung et al. 2002; Zimmerman et al. 2002; Hofstaetter et al. 2006; Oepkes et al. 2006; Moise 2008; Vichinsky 2009). Middle cerebral artery measurements can be monitored after 16 weeks of gestation and reliably determine intrauterine anemia. Fetal

Table 4. Hematology of red blood cell indices in midpregnancy fetuses

Parameter (mean ± SD)	Normal ($n = 22$)	α1 trait[a] ($n = 40$)	Hb Bart's hydrops ($n = 26$)	P
Gestational age (wk)	18.9 + 9.5	18.6 + 7.9	18.8 + 8.8	.422
Hb level (g/dL)	11.34 ± 0.83	10.42 ± 0.91	6.4 ± 1.64	<.001
MCV (fL)	123.40 ± 7.76	105.95 ± 7.35	96.26 ± 8.19	<.001
MCH (pg)	41.39 ± 3.35	33.71 ± 2.74	26.23 ± 2.6	<.001
MCHC (g/dL)	33.44 ± 1.3	31.81 ± 1.09	27.29 ± 1.78	.001

ANOVA test before post hoc multiple comparisons (fetuses, $n = 86$). (Data from Srisupundit et al. 2008.)

Hb, hemoglobin; MCV, mean corpuscular volume; MCH, mean corpuscular hemoglobin; MCHC, mean corpuscular hemoglobin concentration.

[a]α1 trait is the heterozygous state for α^0-thalassemia (–/αα).

survival without intrauterine transfusion is increasingly being identified. These newborns have severe neonatal complications and are at high risk for major neurocognitive dysfunction (Singer et al. 2000; Lucke et al. 2005; Suwanrath-Kengpol et al. 2005; Lee et al. 2007; Vichinsky 2009, 2012). Intrauterine transfusion appears to minimize these complications (Carr et al. 1995; Vichinsky 2009; Weisz et al. 2009; Yi et al. 2009). However, these newborns will require lifetime transfusion therapy or stem cell transplantation (Zhou et al. 2001; Yi et al. 2009; Vichinsky et al. 2010).

SUMMARY

α-Thalassemia syndromes are common and have a wide range of clinical phenotypes. Hemoglobin H disease morbidity is often underappreciated. These patients require early diagnosis and ongoing monitoring. Hemoglobin Bart's hydrops is a tragic, usually fatal complication that can be prevented by adequate screening and counseling. Improved outcome with intrauterine transfusions creates ethical issues for the family and health care providers.

REFERENCES

Reference is also in this collection.

Adam MP, Chueh J, El-Sayed YY, Stenzel A, Vogel H, Weaver DD, Hoyme HE. 2005. Vascular-type disruptive defects in fetuses with homozygous α-thalassemia: Report of two cases and review of the literature. *Prenat Diagn* **25:** 1088–1096.

Au WY, Lam WW, Chu WW, Tam S, Wong WK, Lau J, Yeung YM, Liu HS, Liang R. 2009. Organ-specific hemosiderosis and functional correlation in Chinese patients with thalassemia intermedia and hemoglobin H disease. *Ann Hematol* **88:** 947–950.

Benz EJ Jr. 2011. Newborn screening for α-thalassemia—keeping up with globalization. *N Engl J Med* **364:** 770–771.

Carr S, Rubin L, Dixon D, Star J, Dailey J. 1995. Intrauterine therapy for homozygous α-thalassemia. *Obstet Gynecol* **85:** 876–879.

Chan V, Chan VW, Tang M, Lau K, Todd D, Chan TK. 1997. Molecular defects in Hb H hydrops fetalis. *Br J Haematol* **96:** 224–228.

Chan JC, Chim CS, Ooi CG, Cheung B, Liang R, Chan TK, Chan V. 2006. Use of the oral chelator deferiprone in the treatment of iron overload in patients with Hb H disease. *Br J Haematol* **133:** 198–205.

Chui DH, Waye JS. 1998. Hydrops fetalis caused by α-thalassemia: An emerging health care problem. *Blood* **91:** 2213–2222.

Chui DH, Fucharoen S, Chan V. 2003. Hemoglobin H disease: Not necessarily a benign disorder. *Blood* **101:** 791–800.

Fucharoen S, Viprakasit V. 2009. Hb H disease: Clinical course and disease modifiers. *Hematology Am Soc Hematol Educ Program* 26–34.

* Fucharoen S, Weatherall DJ. 2012. The hemoglobin E thalassemias. *Cold Spring Harb Perspect Med* **2:** a011734.

* Higgs DR. 2013. The molecular basis of α-thalassemia. *Cold Spring Harb Perspect Med* **3:** a011718.

Hofstaetter C, Hansmann M, Eik-Nes SH, Huhta JC, Luther SL. 2006. A cardiovascular profile score in the surveillance of fetal hydrops. *J Matern Fetal Neonatal Med* **19:** 407–413.

Hoppe CC. 2009. Newborn screening for non-sickling hemoglobinopathies. *Hematology Am Soc Hematol Educ Program* 19–25.

Kidd JL, Azimi M, Lubin B, Vichinsky E, Hoppe C. 2010. Application of an expanded multiplex genotyping assay for the simultaneous detection of Hemoglobin Constant Spring and common deletional α-thalassemia mutations. *Int J Lab Hematol* **32:** 373–380.

Lal A, Goldrich ML, Haines DA, Azimi M, Singer ST, Vichinsky EP. 2011. Heterogeneity of hemoglobin H disease in childhood. *N Engl J Med* **364:** 710–718.

Lam YH, Tang MH. 2002. Middle cerebral artery Doppler study in fetuses with homozygous α-thalassaemia-1 at 12–13 weeks of gestation. *Prenat Diagn* **22:** 56–58.

Lau YL, Chan LC, Chan YY, Ha SY, Yeung CY, Waye JS, Chui DH. 1997. Prevalence and genotypes of α- and β-thalassemia carriers in Hong—implications for population screening. *N Engl J Med* **336:** 1298–1301.

Lee SY, Chow CB, Li CK, Chiu MC. 2007. Outcome of intensive care of homozygous α-thalassaemia without prior intra-uterine therapy. *J Paediatr Child Health* **43:** 546–550.

* Lettre G. 2012. The search for genetic modifiers of disease severity in the β-hemoglobinopathies. *Cold Spring Harb Perspect Med* **2:** a015032.

Leung WC, Oepkes D, Seaward G, Ryan G. 2002. Serial sonographic findings of four fetuses with homozygous α-thalassemia-1 from 21 weeks onwards. *Ultrasound Obstet Gynecol* **19:** 56–59.

Li D, Liao C, Li J, Xie X, Huang Y, Zhong H. 2006. Detection of α-thalassemia in β-thalassemia carriers and prevention of Hb Bart's hydrops fetalis through prenatal screening. *Haematologica* **91:** 649–651.

Liang ST, Wong VC, So WW, Ma HK, Chan V, Todd D. 1985. Homozygous α-thalassaemia: Clinical presentation, diagnosis and management. A review of 46 cases. *Br J Obstet Gynaecol* **92:** 680–684.

Liao C, Li Q, Wei J, Feng Q, Li J, Huang Y, Li D. 2007. Prenatal control of Hb Bart's disease in southern China. *Hemoglobin* **31:** 471–475.

Lorey F, Cunningham G, Vichinsky EP, Lubin BH, Witkowska HE, Matsunaga A, Azimi M, Sherwin J, Eastman J, Farina F, et al. 2001a. Universal newborn screening for Hb H disease in California. *Genet Test* **5:** 93–100.

Lorey F, Charoenkwan P, Witkowska HE, Lafferty J, Patterson M, Eng B, Waye JS, Finklestein JZ, Chui DH. 2001b. Hb H hydrops foetalis syndrome: A case report and review of literature. *Br J Haematol* **115:** 72–78.

Lucke T, Pfister S, Durken M. 2005. Neurodevelopmental outcome and haematological course of a long-time survivor with homozygous α-thalassaemia: Case report and review of the literature. *Acta Paediatr* **94:** 1330–1333.

Mari G, Deter RL, Carpenter RL, Rahman F, Zimmerman R, Moise KJ Jr, Dorman KF, Ludomirsky A, Gonzalez R, Gomez R, et al. 2000. Noninvasive diagnosis by Doppler ultrasonography of fetal anemia due to maternal red-cell alloimmunization. Collaborative Group for Doppler Assessment of the Blood Velocity in Anemic Fetuses. *N Engl J Med* **342:** 9–14.

Michlitsch J, Azimi M, Hoppe C, Walters MC, Lubin B, Lorey F, Vichinsky E. 2009. Newborn screening for hemoglobinopathies in California. *Pediatr Blood Cancer* **52:** 486–490.

Moise KJ Jr. 2008. The usefulness of middle cerebral artery Doppler assessment in the treatment of the fetus at risk for anemia. *Am J Obstet Gynecol* **198:** e161–e164.

* Musallam KM, Taher AT, Rachmilewitz EA. 2012. β-Thalassemia intermedia: A clinical perspective. *Cold Spring Harb Perspect Med* **2:** a013482.

Ne W, Harano K, Harano T, Kyaw-S, Aye-Aye M, Khin T-A, Okada S. 2008. Hb Constant Spring [α142, Term → Gln (TAA > CAA in α2)] in the α-thalassemia of anemic patients in Myanmar. *Hemoglobin* **32:** 454–461.

Oepkes D, Seaward PG, Vandenbussche FP, Windrim R, Kingdom J, Beyene J, Kanhai HH, Ohlsson A, Ryan G, DIAMOND Study Group. 2006. Doppler ultrasonography versus amniocentesis to predict fetal anemia. *N Engl J Med* **355:** 156–164.

Pootrakul P, Sirankapracha P, Hemsorach S, Moungsub W, Jumbunlue R, Plangitjagum A, Wasl P, Ma L, Schrier SL. 2000. A correlation of erythrokinetics, ineffective erythropoiesis, and erythroid precursor apoptosis in Thai patients with thalassemia. *Blood* **96:** 2606–2612.

Singer ST, Styles L, Bojanowski J, Quirolo K, Foote D, Vichinsky EP. 2000. Changing outcome of homozygous α-thalassemia: Cautious optimism. *J Pediatr Hematol Oncol* **22:** 539–542.

Singer ST, Kim HY, Olivieri NF, Kwiatkowski JL, Coates TD, Carson S, Neufeld E, Cunningham MJ, Giardina PJ, Mueller BU, et al. 2009. Hemoglobin H-constant spring in North America: An α thalassemia with frequent complications. *Am J Hematol* **84:** 759–761.

Siriratmanawong N, Pinmuang-Ngam C, Fucharoen G, Fucharoen S. 2007. Prenatal diagnosis of Hb Bart's hydrops fetalis caused by a genetic compound heterozygosity for two different α^0-thalassemia determinants. *Fetal Diagn Ther* **22:** 264–268.

Srisupundit K, Piyamongkol W, Tongsong T. 2008. Comparison of red blood cell hematology among normal, α-thalassemia-1 trait, and hemoglobin Bart's fetuses at midpregnancy. *Am J Hematol* **83:** 908–910.

Styles L, Foote D, Kleman K, Vichinsky E. 1997. Hemoglobin H-Constant Spring disease: An under recognized, severe form of α thalassemia. *Int J Pediatr Hematol Oncol* **6:** 69–74.

Suwanrath-Kengpol C, Kor-anantakul O, Suntharasaj T, Leetanaporn R. 2005. Etiology and outcome of non-immune hydrops fetalis in southern Thailand. *Gynecol Obstet Invest* **59:** 134–137.

Tso SC, Chan TK, Todd D. 1982. Venous thrombosis in haemoglobin H disease after splenectomy. *Aust NZ J Med* **12:** 635–638.

Vichinsky EP. 2009. α Thalassemia major—new mutations, intrauterine management, and outcomes. *Hematology Am Soc Hematol Educ Program* 35–41.

Vichinsky E. 2010. Complexity of α thalassemia: Growing health problem with new approaches to screening, diagnosis, and therapy. *Ann NY Acad Sci* **1202:** 180–187.

Vichinsky E. 2012. Advances in the treatment of α-thalassemia. *Blood Rev* **26:** S31–S34.

Vichinsky E, Chu J, Walters M, et al. 2010. Successful hematopoietic cell transplantation for α thalassemia major. *Pediatr Blood Cancer* **54:** 840.

Weatherall DJ. 2008. Hemoglobinopathies worldwide: Present and future. *Curr Mol Med* **8:** 592–599.

Weatherall DJ, Clegg JB, ed. 2001. *The thalassaemia syndromes*, 4th ed. Blackwell Science, Malden, MA.

Weisz B, Rosenbaum O, Chayen B, Peltz R, Feldman B, Lipitz S. 2009. Outcome of severely anaemic fetuses treated by intrauterine transfusions. *Arch Dis Child Fetal Neonatal Ed* **94:** F201–F204.

* Williams TN, Weatherall DJ. 2012. World distribution, population genetics, and health burden of the hemoglobinopathies. *Cold Spring Harbor Perspect Med* **2:** a11692.

Yang Y, Li DZ. 2009. A survey of pregnancies with Hb Bart's disease in Mainland China. *Hemoglobin* **33:** 132–136.

Yi JS, Moertel CL, Baker KS. 2009. Homozygous α-thalassemia treated with intrauterine transfusions and unrelated donor hematopoietic cell transplantation. *J Pediatr* **154:** 766–768.

Yuan J, Bunyaratvej A, Fucharoen S, Fung C, Shinar E, Schrier SL. 1995. The instability of the membrane skeleton in thalassemic red blood cells. *Blood* **86:** 3945–3950.

Zhou X, Ha SY, Chan GC, Luk CW, Chan V, Hawkins B, Lam YH, Liang RH, Lau YL. 2001. Successful mismatched sibling cord blood transplant in Hb Bart's disease. *Bone Marrow Transplant* **28:** 105–107.

Zimmerman R, Carpenter RJ Jr, Durig P, Mari G. 2002. Longitudinal measurement of peak systolic velocity in the fetal middle cerebral artery for monitoring pregnancies complicated by red cell alloimmunisation: A prospective multicentre trial with intention-to-treat. *BJOG* **109:** 746–752.

α-Thalassemia, Mental Retardation, and Myelodysplastic Syndrome

Richard J. Gibbons

MRC Molecular Haematology Unit, Weatherall Institute of Molecular Medicine, University of Oxford, John Radcliffe Hospital, Oxford OX3 9DS, United Kingdom

Correspondence: richard.gibbons@ndcls.ox.ac.uk

This article describes three rare syndromes in which the presence of α-thalassemia provided an important clue to the molecular basis of the underlying condition. It exemplifies how rare diseases allied with careful clinical observation can lead to important biological principles. Two of the syndromes, ATR-16 and ATR-X, are characterized by α-thalassemia in association with multiple developmental abnormalities including mental retardation. The third condition, ATMDS, is an acquired disorder in which α-thalassemia arises in the context of myelodysplasia. Intriguingly, mutations in the chromatin remodeling factor, ATRX, are common to both ATR-X syndrome and ATMDS.

The rare association of α-thalassemia and mental retardation (MR) was recognized over 25 years ago by Weatherall and colleagues (Weatherall et al. 1981). It was known that α-thalassemia arises when there is a defect in the synthesis of the α globin chains of adult hemoglobin (HbA, $\alpha_2\beta_2$). When these investigators encountered three mentally retarded children with α-thalassemia and a variety of developmental abnormalities, their interest was stimulated by the unusual nature of the α-thalassemia. The children were of northern European origin, where α-thalassemia is uncommon, and although one would have expected to find clear signs of this inherited anemia in their parents, it appeared to have arisen de novo in the affected offspring. It was thought that the combination of α-thalassemia with MR (ATR), and the associated developmental abnormalities represented a new syndrome and that a common genetic defect might be responsible for the diverse clinical manifestations. What emerged was the identification of two quite distinct syndromes in which α-thalassemia is associated with mental retardation (Wilkie et al. 1990a, b). In the first condition (ATR-16, OMIM: 141750) there are large (1–2 Mb) chromosomal rearrangements that delete many genes, including the α globin genes from the tip of the short arm of chromosome 16 and this is an example of a contiguous gene syndrome (Schmickel 1986). In the second syndrome (ATR-X, OMIM 301040), a complex phenotype, including α-thalassemia, results from mutations in an X-encoded factor (now called the ATRX protein), which is a putative regulator of gene expression. Mutations in this

Copyright © 2013 Cold Spring Harbor Laboratory Press; all rights reserved
Cite this article as *Cold Spring Harb Perspect Med* doi: 10.1101/cshperspect.a011759

gene down-regulate α gene expression and also perturb the expression of other, as yet unidentified, genes.

THE ATR-16 SYNDROME

ATR-16 is rare and sporadic contiguous gene syndrome. Often one is alerted to this condition by observing the unusual association of α-thalassemia and mental retardation in individuals originating from outside of the areas where thalassemia commonly occurs. However, the widespread use of array-based screening for genomic deletions is identifying cases with little regard for the phenotype. In addition to α-thalassemia, these patients have variable degrees of facial dysmorphism, most commonly featuring a high forehead, mild hypertelorism, and a broad or prominent nasal bridge. A wide spectrum of associated developmental abnormalities has also been described of which clubfoot is the most frequent. In all such cases, initial molecular genetic analyses have shown that affected individuals fail to inherit the entire ζ-α globin cluster from one or other of the parents.

Defining the Genetic Abnormalities in Patients with ATR-16 Syndrome

In some cases, conventional cytogenetic analysis shows the underlying genetic abnormality. Because the α globin complex lies very close to the 16p telomere any chromosomal abnormality affecting this region may give rise to α-thalassemia. In some patients with ATR-16 syndrome, gross chromosomal abnormalities such as ring chromosomes and translocations are visible on the karyotype. However, in the case of deletions and subtelomeric translocations, initial high-resolution cytogenetic analysis appears entirely normal. Although such abnormalities may arise as de novo genetic events, often one parent carries a pre-existing balanced translocation, which the child inherits in an unbalanced fashion resulting in monosomy for 16p and loss of the α cluster.

Fluorescence in situ hybridization (FISH) studies have also been used to analyze ATR-16 families. In this type of study, large segments (~40 kb) of chromosome 16 in cosmid vectors are used as probes to show the presence or absence of the corresponding sequences in the 16p telomeric region using fluorescence microscopy (Buckle and Kearney 1994). By analyzing the chromosomes of both parents and the affected child it has been possible to define the extent of 16p monosomy and the mechanism by which it has arisen. The additional use of panels of subtelomeric probes from each human chromosome can identify trisomic segments in unbalanced translocations.

As a first line investigation, FISH is currently being superceded by technologies enabling high-resolution analysis of the entire human genome. In the first of these types of analyses, the genome is interrogated using microarrays consisting of oligonucleotides spanning the genome. Using a competitive genome hybridization (CGH) approach, comparing DNA from one genome with another, it has been possible to detect large regions of monosomy in patients with ATR-16 syndrome (Price et al. 2005). Such oligo-based microarrays or bead technologies (originally made to identify single nucleotide polymorphisms) with a modified analysis have a resolution in the range 5–50 kb (Redon et al. 2006). One limitation of the technology is that it does not provide information of the structural origin of any detected chromosomal imbalance and FISH is commonly used as a follow-up investigation to supplement this. In addition to these approaches, Harteveld et al. (2007) have developed a rapid and simple high-resolution approach to identify and characterize deletions of the terminal 2 Mb of 16p13.3 using the multiple ligation-dependent probe amplification (MLPA) technique. An established panel of specific synthetic oligonucleotides can now be used to detect deletions in 16p13.3 (Harteveld et al. 2005, 2007). Clearly the resolution of this technique is only limited by the spacing of the appropriate oligonucleotides.

To date, few breakpoints have been fully characterized. However, where this has been fully analyzed it appears that the chromosome has been broken, truncated, and "healed" by the direct addition of telomeric repeats $(TTAGGG)_n$ (Lamb et al. 1993).

 Cite this article as *Cold Spring Harb Perspect Med* doi: 10.1101/cshperspect.a011759

How Do Chromosomal Abnormalities Give Rise to the ATR-16 Syndrome?

Because individuals with ATR-16 syndrome may have quite different degrees of chromosomal imbalance, there is considerable variation in the associated phenotypes. The degree of monosomy in 16p13.3 varies from 0.3–2 Mb but many patients have additional chromosomal aneuploidy and in some cases imbalance of the non-16 material may dominate the clinical picture. For example, in one (DA) loss of material from chromosome 15 while forming the abnormal derivative, t(15:16) chromosome produced the striking phenotype associated with the Prader-Willi syndrome.

Because many patients with ATR-16 have complex genomic rearrangements, to determine the role played by any gene(s) within 16p13.3 in the developmental abnormalities associated with ATR-16 syndrome, future analysis should focus on patients who have pure monosomy for 16p13.3 (Fig. 1). Surprisingly, these studies have shown that patients with deletions of up to 1000 kb of 16p13.3 (including 55 genes) have α-thalassemia, but may be otherwise unaffected. Those with deletions of between 1000 kb and 1200 kb may or may not have mild MR. At the other extreme, patients with 16p monosomy for the entire terminal region of 16p (e.g., BO, HN, LIN, and IM) have a α-thalassemia, mild MR, and dysmorphic features. In a patient whose deletion (associated with unbalanced translocation) extends beyond 2000 kb the clinical picture is dominated by more severe MR, tuberous sclerosis, and polycystic kidney disease (European Polycystic Kidney Disease Consortium 1994).

How might monosomy for 16p13.3 cause such developmental abnormalities? One possibility is that deletion of a large number of genes from one copy of chromosome 16 may unmask mutations in its homolog; the more genes that are deleted the greater the probability of this occurring. However, this is unlikely to be the explanation for most ATR-16 cases because whole genome sequencing indicates that normal individuals carry a relatively small number of harmful mutations of this type. A further possibility is that some genes in 16p are imprinted (Hall 1990) so that deletions could remove the only active copy of the gene. At present, there is no evidence for imprinting of the 16p region (reviewed in Schneider et al. 1996), and in the relatively few ATR-16 cases analyzed there appears to be no major clinical differences between patients with deletions of the maternally or paternally derived chromosomes. It therefore seems more likely that there are some genes in the 16p region that encode proteins whose effect is critically determined by the amount produced, so-called dosage-sensitive genes (see Fisher and Scambler 1994). Examples of such genes include those encoding proteins that form heterodimers, those required at a critical level for a rate-determining step of a regulatory pathway, and tumour suppressor genes (e.g., TSC2). If the deletion includes one or more dosage-sensitive genes this could account for the clinical effects seen in ATR-16 patients.

The region lying between 900–1700 kb from the 16p telomere, deleted in all patients with the characteristic features of ATR-16 syndrome, contains 14 genes and gene families of known function (Daniels et al. 2001) that have been implicated in a wide range of disorders with few or no features in common with ATR-16. One of these (SOX8) was considered a strong candidate because it is involved in the regulation of embryonic development and is strongly expressed in the brain (Pfeifer et al. 2000). However, although deleted in case WG (Fig. 1) (Gibson et al. 2008), the deletion in TN does not extend to SOX8 (Daniels et al. 2001). Furthermore, a recently described Brazilian patient with a deletion that removes both the α globin locus and SOX8 was not associated with MR or any dysmorphism (Bezerra et al. 2008). This suggests that loss of SOX8 is neither necessary nor sufficient to cause MR and dysmorphism. It is conceivable that the genetic background plays a role and this might explain why the deletion in family SH is associated with MR in one individual and a normal phenotype in another. Clearly, further examples of ATR-16 because of monosomy for 16p13.3 must be characterized to identify the gene(s) responsible for the MR and other developmental abnormalities associated with this condition.

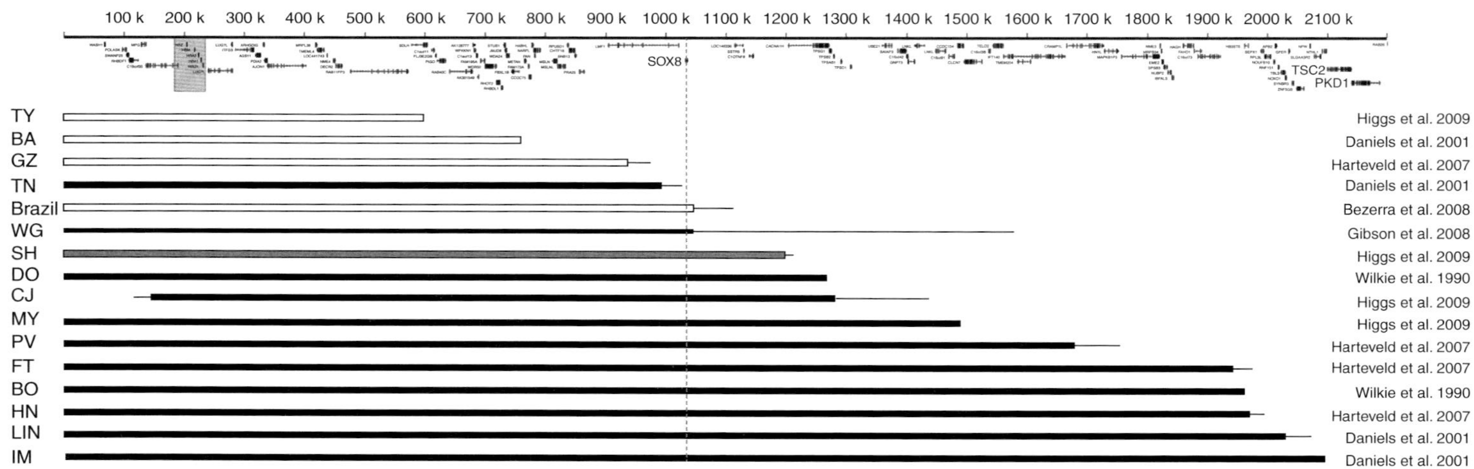

Figure 1. Summary of known ATR-16 deletions that are >500 kb and associated with pure monosomy. Initials represent the patient code. Open boxes represent cases without learning difficulties; gray boxes, families in which some cases have learning difficulties whereas other do not; black boxes represent cases with learning difficulties. The length of the box shows the known extent of the deletion and the fine line indicates the region of uncertainty of the breakpoints. The positions of the genes in this region are shown at the *top*; the α-globin cluster is marked with a pale gray box; SOX8, TSC2, and PKD1 are shown in a larger font.

Cite this article as *Cold Spring Harb Perspect Med* doi: 10.1101/cshperspect.a011759

THE ATR-X SYNDROME

As additional patients with α-thalassemia and mental retardation were identified throughout the 1980s, it became clear that a second group of affected individuals existed in whom no structural abnormalities in the α globin cluster or 16p could be found. In contrast to ATR-16 syndrome, this group of patients was all male, presented with a much more uniform phenotype, and had a remarkably similar facial appearance (Wilkie et al. 1990b). That this group had a distinct and recognizable dysmorphism was underscored when additional cases were identified on the basis of their facial features alone (Gibbons et al. 1991; Wilkie et al. 1991). Ultimately, it was shown that this unusual syndrome of α-thalassemia with severe mental retardation results from an X-linked abnormality (see later) and the condition is now referred to as the ATR-X syndrome (OMIM: 301040).

The Clinical Findings of the ATR-X Syndrome

Cases of ATR-X syndrome from over 180 families have now been characterized, and in contrast to ATR-16 a definite phenotype is emerging. The cardinal features of this condition are severe mental retardation and developmental abnormalities associated with a characteristic facial appearance (Fig. 2) and α-thalassemia. The frequency and nature of the most commonly encountered clinical features are summarized in Tables 1 and 2.

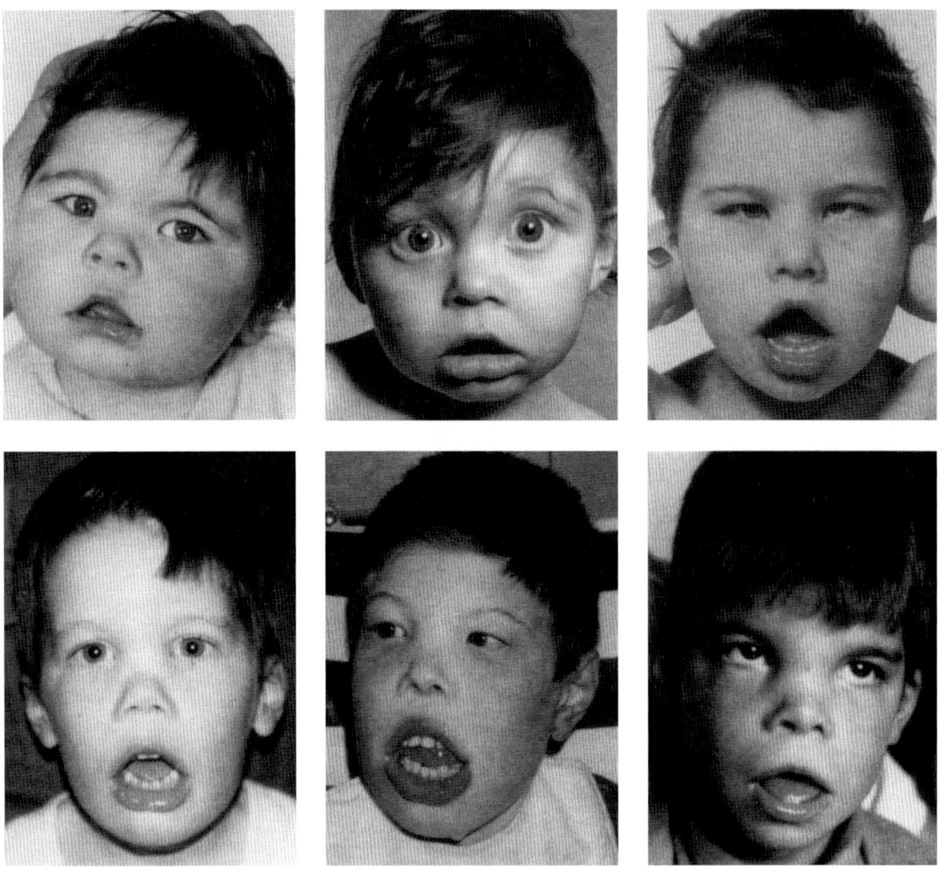

Figure 2. The facial appearance of patients with ATR-X syndrome. (From Gibbons et al. 1995a; adapted, with permission, from the investigator.)

Table 1. Summary of the major clinical manifestations of the ATR-X syndrome

Clinical feature	%
Profound mental retardation	96
Characteristic face	94
Skeletal abnormalities	90
HbH inclusions	88
Neonatal hypotonia	84
Genital abnormalities	79
Microcephaly	77
Gut dysmotility	76
Short stature	65
Seizures	34
Cardiac defects	21
Renal/urinary abnormalities	15

The Hematologic Findings of the ATR-X Syndrome

The presence of α-thalassemia (in the form of thalassemia trait or mild HbH disease) with HbH inclusions was one of the original diagnostic criteria for ATR-X syndrome. However, with the identification of further cases, it became clear that the hematologic findings (e.g., levels of Hb, MCV, and MCH) were different from those seen in the common types of α-thalassemia. Now that the disease can be identified via the clinical phenotype and/or the *ATRX* genotype, it is clear that there is considerable variation in the hematologic manifestations associated with mutations of the *ATRX* gene. In fact, a number of families have been identified in which some or all of the affected members with mutations of ATRX, and the characteristic phenotype described previously, have no signs of α-thalassemia (Villard et al. 1996a,b). Nevertheless, when the family history and phenotype are strongly suspect, a careful search for HbH inclusions should be made in all affected individuals and repeated if necessary as they may be very infrequent.

The hematology is often surprisingly normal considering the presence of α-thalassemia. In many cases, neither the level of hemoglobin nor the mean cell hemoglobin is as severely affected as in the classical forms of α-thalassemia associated with *cis*-acting mutations in the α-globin complex, and this probably reflects the different pathophysiology of the conditions. In most cases of ATR-X, there is insufficient HbH to be detected by electrophoresis and the

Table 2. Clinical manifestations of the ATR-X syndrome

Genital abnormalities	Small/soft testes, cryptorchidism, gonadal dysgenesis, inguinal hernia, micropenis, hypospadias, deficient prepuce, shawl scrotum, hypoplastic scrotum, ambiguous genitalia, female external genitalia (male pseudo-hermaphroditism)
Skeletal abnormalities	Delayed bone age, tapering fingers, drumstick distal phalanges, brachydactyly, clinodactyly, bifid thumb, fixed flexion deformities of joints, overriding toes, vaus or valgus deformities of feet, scoliosis, kyphosis, hemivertebra, spina bifida, coxa valga, chest wall deformity
Renal/urinary abnormalities	Renal agenesis, hydonephrosis, small kidneys, vesico-ureteric reflux, pelvo-ureteric junction obstruction, exstrophy of bladder, urethral diverticulum, urethral stricture
Cardiac defects	Atrial septal defect, ventricular septal defect, patent ductus arteriosus, tetralogy of Fallot, transposition of the great arteries, dextracardia with situs solitus, aortic stenosis, pulmonary stenosis
Gut dysmotility	Discoordinated swallowing, eructation, gastro-esophageal reflux, vomiting, gastric pseudo-volvulus, hiatus hernia, haematemesis, recurrent ileus/small bowel obstruction, volvulus, intermittent diarrhoea, severe constipation, ultra short segment Hirschsprung disease
Miscellaneous	Apneoic episodes, cold/blue extremities, blepharitis, conjunctivitis, entropion, cleft palate, pneumonia, umbilical hernia, encephalitis, iris coloboma, optic atrophy, blindness, sensorineural deafness, prolonged periods of screaming/laughing, self-injury, asplenia

Cite this article as *Cold Spring Harb Perspect Med* doi: 10.1101/cshperspect.a011759

number of HbH inclusions is quite variable between different patients, although relatively constant over time in any affected individual.

The ATR-X Syndrome Is an X-Linked Condition

Evidence that the defect in globin synthesis lay in *trans* to the globin cluster was provided by a study of a family with four affected sibs in whom the condition segregated independent of the α globin cluster (Donnai et al. 1991).

Because affected individuals were always related via the female line, this suggested that the syndrome mapped to the X chromosome and hence the condition was named the ATR-X syndrome. Subsequently linkage analysis in 16 families mapped the disease interval to Xq13.1–q21.1 confirming that the associated α-thalassemia results from a *trans*-acting mutation (Gibbons et al. 1992).

Early genetic studies showed that the ATR-X syndrome behaves as an X-linked recessive disorder; boys are affected almost exclusively. Furthermore, almost all female carriers have a normal appearance and intellect, although approximately one in four carriers has subtle signs of α-thalassemia with very rare cells containing HbH inclusions (Gibbons et al. 1992). The majority of carriers have a highly skewed pattern of X inactivation in leukocytes (derived from mesoderm), hair roots (ectoderm), and buccal cells (endoderm). In each case, the disease-bearing X chromosome is preferentially inactivated. There is evidence from a recently reported mouse model of ATR-X syndrome (Garrick et al. 2006) that skewed X inactivation results from selection, at key steps during development, against cells that are deficient for ATRX (Muers et al. 2007).

Together, these findings showed that ATR-X is an X-linked disease and when the gene is mutated, among many other effects, this leads to down-regulation of expression of the α globin genes on chromosome 16.

Identification of the ATR-X Disease Gene

The isolation of cDNA fragments mapping to Xq13.1–q21.1 provided an opportunity to study candidate genes for ATR-X (Gecz et al. 1994). A number of these cDNA fragments were hybridized to DNA from a group of affected individuals. An absent hybridization signal was noted in one patient when an 84 bp cDNA fragment (E4) was used. E4 was shown to be part of a gene known as XH2/XNP (Gibbons et al. 1995b). Subsequently, a 2-kb genomic deletion was shown in this individual with ATR-X syndrome. Subsequently, analysis of a segment of cDNA corresponding to XH2/XNP identified disease-causing mutations in several individuals with the clinical and hematologic features of ATR-X syndrome. The X-linked gene was thus renamed as the *ATRX* gene.

Characterization of the ATRX Gene and Its Protein Product

We now know that the *ATRX* gene spans about 300 kb of genomic DNA and contains 36 exons (Picketts et al. 1996) although exon 7 may be nonfunctional. It encodes at least two alternatively spliced, 10.5-kb mRNA transcripts that differ at their 5′ ends and are predicted to give rise to slightly different proteins of 265 and 280 kDa (Fig. 3). A further transcript of ~7 kb represents an isoform, which retains intron 11 and truncates at this point. This gives rise to a truncated protein isoform (ATRXt) which is conserved between mouse and man (Garrick et al. 2004).

Within the amino-terminal region lies a complex cysteine-rich segment (called the ADD domain; Fig. 3). This comprises a PHD-like zinc finger and an additional C_2C_2 motif just upstream, which is structurally similar to the GATA1 zinc fingers (Gibbons et al. 1997; Argentaro et al. 2007) and is highly related to the zinc finger domains of DNA methyltransferases (Argentaro et al. 2007). The domain mediates binding to the amino-terminal tail of histone H3 trimethylated at lysine 9 (Dhayalan et al. 2011; Eustermann et al. 2011; Iwase et al. 2011). The functional importance of the ADD segment in *ATRX* is clear. It is highly conserved throughout evolution and it contains over 50% of all mutations found in patients with ATR-X syndrome (Fig. 3 and see below).

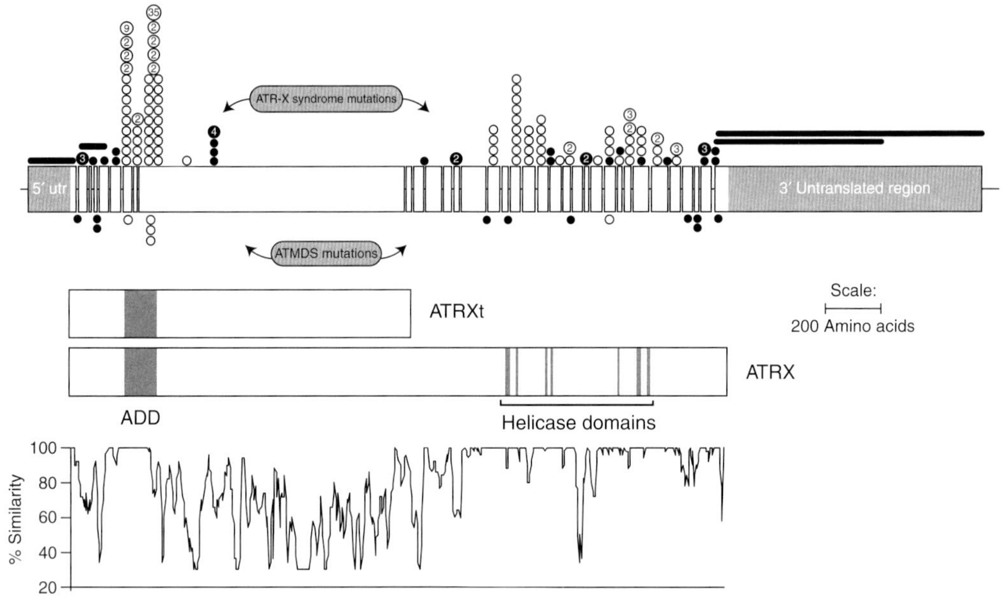

Figure 3. Schematic diagram of the *ATRX* gene: boxes represent the 35 exons (excluding the alternatively spliced exon 7); thin horizontal lines represent introns (not to scale). The 3′ and 5′ untranslated regions (utr) are shown flanking the open reading frame. The two protein products ATRX and ATRXt are shown as rectangles. The principal domains, the zinc finger motif (ADD), and the highly conserved helicase motif are indicated. In the *lower* part of the figure is a graphical representation of the amino acid similarity between human and mouse ATRX proteins. In the *upper* part of the figure is illustrated the spectrum of *ATRX* mutations described in boys with ATR-X syndrome (above the gene) and in ATMDS (below the gene). The positions of the mutations are shown by circles: filled circles represent mutations (nonsense or leading to a frameshift) that would cause protein truncation; open circles represent missense mutations and small deletions that maintain the reading frame; deletions are indicated by horizontal lines. Recurrent mutations are illustrated by larger circles, and the number of independent families is indicated. (From Gibbons et al. 2008; adapted, with permission, from the author.)

The central and carboxy-terminal regions of *ATRX* show the greatest conservation between murine and human sequences (94%) (Fig. 3) (Picketts et al. 1998). The central portion of the molecule contains motifs that identify ATRX as a member of the SNF2 group of proteins. A central property of this group of proteins is to remodel chromatin during processes such as DNA replication, transcription, mitosis, and DNA repair.

Mutations of the ATRX Gene and Their Associated Phenotype

In addition to ATR-X syndrome, mutations in the *ATRX* gene have now been found in many other forms of syndromal X-linked MR (reviewed in Gibbons et al. 2008) and it is also the disease gene associated with the occurrence of α-thalassemia in myelodysplasia (see later section).

To date, 113 different constitutional mutations have been documented in 182 independent families with ATR-X syndrome (reviewed in Gibbons et al. 2008). Missense mutations are clustered in two regions: the ADD domain, and the helicase domain. Analysis of the mutations and their resulting phenotypes allows important conclusions to be drawn. It appears that none of the mutations lead to a true null but rather are hypomorphs.

ATR-X syndrome is frequently caused by missense mutations (Fig. 3) and the levels of ATRX protein were shown to be substantially reduced in patients with such mutations involving the ADD domain (McDowell et al. 1999; Cardoso et al. 2000). The structure of the

ADD domain casts light on how mutations might affect protein folding and stability (Argentaro et al. 2007). Most mutations affect zinc-binding cysteine residues or residues in the tightly packed hydrophobic core thus reducing its stability. Of greater interest, there is a small cluster of surface mutations (e.g., R246, G249D), which are associated with higher levels of stable protein. These mutations may interfere with protein function, possibly by disrupting an important protein/protein interaction.

Most missense mutations affecting the helicase domain lead to protein instability although one interesting mutation, L1746S, appears to decouple the ATPase activity from productive work (Mitson et al. 2011).

Variability in the Severity of α-Thalassemia

There is considerable variability in the degree to which α-globin synthesis is affected in ATR-X syndrome, as judged by the frequency of cells with HbH inclusions. Up to 15% of patients do not have HbH inclusions (Fig. 4) (Villard et al. 1996a,b,c). It is interesting that patients with identical mutations may have very different, albeit stable, degrees of α-thalassemia, suggesting that the effect of the ATRX protein on α-globin expression may be modified by other genetic factors (Fig. 4). This is most clearly illustrated by comparing the hematology of cases with identical mutations. Comparison of the 41 cases from 34 pedigrees with the common 736C > T; R246C mutation shows a wide range in the frequency of cells with HbH inclusions (0%–14%). Recent studies into the basis of this variability have led to an important insight into the function of ATRX. Chromatin immunoprecipitation (ChIP) has indicated the ATRX binds to a G-rich tandem repeat (ψζ VNTR) adjacent to the α-globin cluster (Fig. 5A,B) (Law et al. 2010). The length of the repeat is variable and in individuals with ATR-X syndrome, the severity of the thalassemia (as measured by the frequency of cells with HbH inclusions) was related to the size of the repeat: the longer the repeat, the more down-regulated were the genes of the α-globin cluster (Fig. 5C). Furthermore, the closer the gene to the ψζ VNTR, the more down-regulated was the expression so that HBM was the most affected gene in the cluster (Fig. 5D). It appeared that normally ATRX in some way was preventing the G-rich tandem repeat from perturbing local gene expression.

What Is the Normal Functional Role of ATRX?

In the adult, ATRX mRNA is widely expressed early in development, and continues to be widely expressed throughout development with particularly high expression in the brain, heart, and skeletal muscle (Stayton et al. 1994). Both isoforms, ATRX (280 kDa) and ATRXt (200 kDa) are readily detected on Western blots (McDowell et al. 1999; Garrick et al. 2004). Immunocytochemical analysis and indirect immunolocalization show that both isoforms are nuclear proteins that predominantly associate with heterochromatin (in interphase and metaphase) (McDowell et al. 1999; Garrick et al. 2004). A significant proportion of ATRX (but not ATRXt) is also found in nuclear speckles (called promyelocytic leukemia [PML] bodies) in human cells (Xue et al. 2003). One additional, striking finding in human metaphase preparations is that ATRX antibodies localize to the short arms of acrocentric chromosomes associated with the rDNA arrays (Gibbons et al. 2000). Genome-wide ChIP indicates that ATRX is most commonly associated with tandem repeats (G-rich rDNA, telomeres and interstitial tandem repeats, as well as A-T rich pericenric heterochromatin) and to a lesser extent CpG islands. Some G-rich repetitive sequences can take up non-B form DNA structures such as G-quadruplex structures in vitro and many ATRX targets are predicted to do so (Law et al. 2010). Furthermore, in a competition assay, recombinant ATRX protein binds to a G4 structure in preference to the unfolded DNA form. It is possible that ATRX plays a role to recognize and resolve non-B forms of DNA.

ATRX plays an important role in the incorporation of the histone variant H3.3 into telomeric and pericentromeric DNA (Goldberg et al. 2010; Wong et al. 2010). It does so with its partner protein Daxx, which directly binds H3.3 and acts as a histone chaperone (Drane

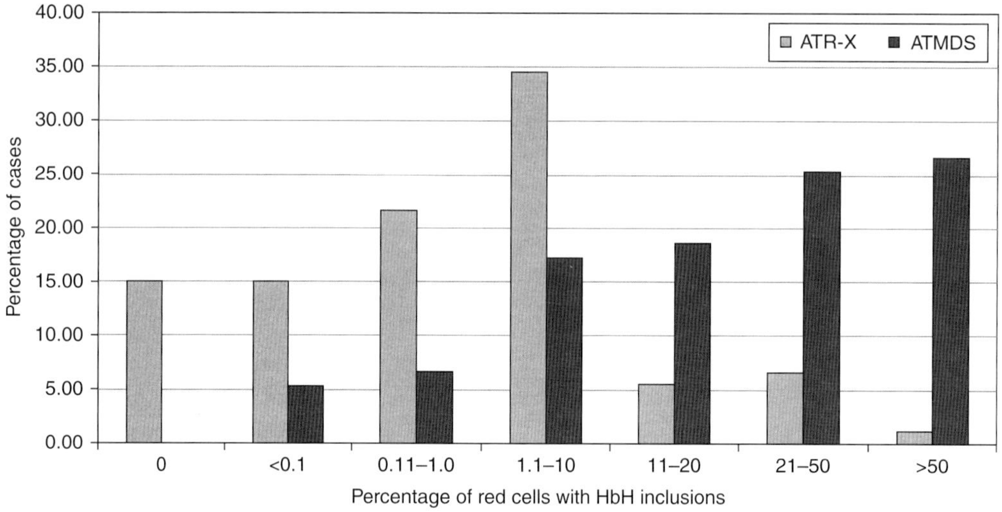

Figure 4. Percentage of cells with HbH inclusions in individuals with ATR-X syndrome or ATMDS.

et al. 2010; Goldberg et al. 2010); ATRX both directs where this occurs and facilitates the incorporation of H3.3 into chromatin. Tandem repetitive DNA is predicted to be an inherently unfriendly substrate for nucleosomes leading to rapid turnover (Kaplan et al. 2009). Unlike the canonical histones, H3.3 can be incorporated into chromatin in a replication-independent manner so is available for re-establishing the nucleosomal organization of tandem repetitive sequences (Ahmad and Henikoff 2002). An attractive, but as yet unproven, model is that ATRX acts at tandem repeats to prevent them taking up unusual structures by restablishing nucleosomal organization through the insertion of histone H3.3.

α-THALASSEMIA-MYELODYSPLASIA (ATMDS)

Occasionally, patients who previously showed normal erythropoiesis will develop an acquired form of α-thalassemia, which most commonly arises within the context of hematologic malignancy (Steensma et al. 2005). The first cases of this "acquired hemoglobin H (HbH) disease" were described in 1960 (White et al. 1960). Over the past half century, it has become clear that all cases of acquired α-thalassemia arise in the context of an underlying clonal disorder of hematopoiesis, most commonly a form of the myelodysplastic syndromes (MDS). Therefore, this condition is now commonly referred to as α-thalassemia-myelodysplasia (ATMDS, OMIM: 300448). In its most characteristic manifestation there is a severe form of HbH disease, characterized by striking hypochromic microcytic anemia, numerous HbH inclusions detectable by supravital staining of peripheral blood (Fig. 4), and measurable amounts of HbH in the hemolysate (1%–70%).

Other than the disease-defining red cell changes, there are no clinical features that clearly distinguish ATMDS from MDS more generally, although there may be a greater male predominance than is observed in chronic myeloid disorders overall (the male:female ratio in ATMDS is >5:1 whereas in MDS in general it is about 1.5:1) (Haas et al. 2009). The reason for this dramatic sex imbalance in ATMDS is unclear. Patients with ATMDS are diagnosed at similar ages to patients with chronic myeloid disorders who lack thalassemia (median age, 68 years), have similar marrow findings and karyotypic results, have a median survival typical for MDS overall (2–3 years), and die of the same complications—primarily death from infection and, in about 25% of cases, of complications of

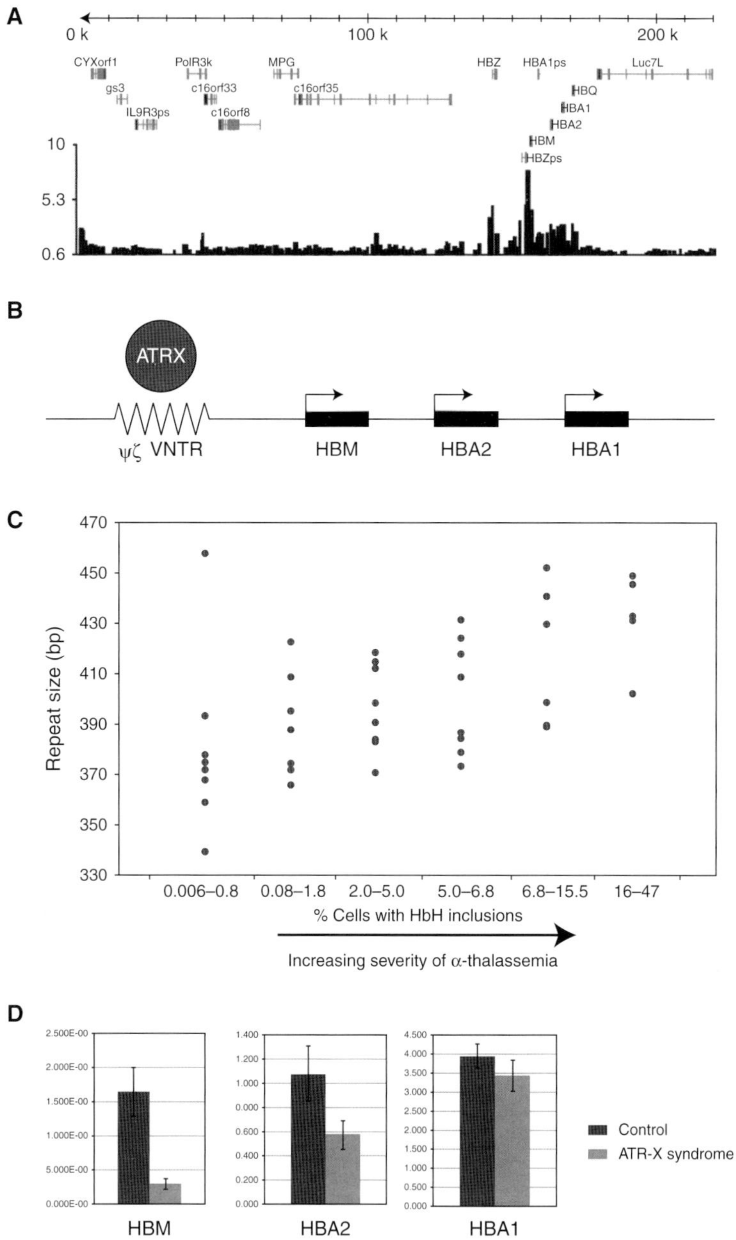

Figure 5. (*A*) Microarray analysis of ATRX ChIP DNA enrichment from human erythroblasts across the 220-kb terminal region of chromosome 16p containing the α-globin genes (HBM, HBA2, HBA1, HBQ). (*B*) A schematic showing the relationship of the ψζ VNTR (peak of ATRX binding) to the adjacent α-globin genes. (*C*) ψζ VNTR length was measured in ATR-X patients with α-thalassemia and the average length of the two alleles was plotted against the degree of α-thalassemia as measured by percent red cells showing HbH inclusions (from Law et al. 2010; adapted, with permission, from the author). (*D*) Comparison of α-globin gene expression between a cohort of normal controls ($n = 15$) and individuals with ATR-X syndrome ($n = 25$) showing that the degree of down-regulation is related to the distance from the ψζ VNTR.

progression to acute myeloid leukemia (AML) (Steensma et al. 2005). When ATMDS progresses to AML, and the ability of hematopoietic cells to differentiate is further impaired, some patients still continue to have detectable HbH, whereas in other cases the HbH inclusions disappear entirely.

The reticulocyte α/β-globin chain synthesis ratio (and α- to β-globin mRNA ratio) is severely reduced and in the most severely affected individuals, α-chain synthesis is almost completely abolished.

Patients with ATMDS are always anemic at presentation (mean hemoglobin 8.5 g/dL), with a reduced RBC count (mean $4.3 \times 10^{12}/L$) and markedly hypochromic with microcytic red cell indices (average MCH 22 pg and MCV 75 fL) (Steensma et al. 2005). These abnormalities are even more striking when compared with patients with MDS without thalassemia, who at presentation tend to have slightly higher than normal MCH and MCV values (average MCH 31 pg and MCV 97 fL).

ATRX Mutations in ATMDS

It is now clear that most patients with ATMDS have an acquired, somatic point mutation or an mRNA intron-exon splicing abnormality involving *ATRX* (Fig. 3). In one exception, the telomeric end of chromosome 16p, including the α-globin cluster, was deleted in a clonally restricted fashion as part of a complex MDS-associated karyotype with numerous large chromosomal rearrangements (Steensma et al. 2004b).

In general, patients with ATMDS have a more severe form of α-thalassemia than boys with inherited ATR-X syndrome. More than 30% of described ATMDS patients have had >50% of circulating erythrocytes containing HbH inclusions, whereas 90% of boys with ATR-X syndrome have less than 10% HbH-containing cells (Fig. 4) (Gibbons and Higgs 2000; Higgs 2004). This is not explained by the underlying ATRX mutation; even when an identical or similar *ATRX* mutation is detected in the germline in ATR-X syndrome and as a somatic change in the hematopoietic cells in ATMDS, the thalassemic phenotype is consistently more extensive in ATMDS than in ATR-X. For instance, a boy with germline c.576G > S, p.L192F ATRX mutation had 0.1% HbH, whereas a man with ATMDS who had the identical mutation as an acquired phenomenon had 50% HbH. The reason for this striking difference is unclear, and is suspected to be dependent on the cellular context, perhaps pointing to an interaction between *ATRX* mutations and a disturbed genetic or epigenetic background in MDS cells that is due to other mutations not present in boys with inherited ATR-X syndrome.

Role of ATRX Mutation in ATMDS and Other Neoplasms

It is debatable whether the ATRX mutations are driver or passenger changes. The loss of detectable HbH and loss of the ATRX mutation with time in an individual with ATMDS suggested the latter (Steensma et al. 2004a). The recent identification in a variety of cancers of frequent mutations in ATRX, its partner protein Daxx, or the histone variant H3.3 that ATRX/Daxx deposit into chromatin has reignited the issue (Heaphy et al. 2011; Jiao et al. 2011; Schwartzentruber et al. 2012). What links the affected cancers is that they appear to show a telomerase independent mechanism for maintaining their telomeres—alternative lengthening of telomeres—involving an aberrant from of homologous recombination. It is of course of considerable interest to determine whether ATMDS shares this characteristic and how the absence of ATRX is associated with this phenomenon.

CONCLUDING REMARKS

Both ATR-X syndrome and ATMDS have revealed the pivotal role of chromatin remodeling in the expression of the α-globin genes. This emphasizes the value of studying orphan diseases to reveal previously unknown biology. Furthermore, this pathway may provide a therapeutic target for manipulating α-globin expression with a view to reducing the toxic effects of the excess α-globin chains seen in β-thalassemia.

ACKNOWLEDGMENTS

Many thanks to Liz Rose for help with preparing the manuscript.

REFERENCES

Ahmad K, Henikoff S. 2002. Histone H3 variants specify modes of chromatin assembly. *Proc Natl Acad Sci* **99:** 16477–16484.

Argentaro A, Yang JC, Chapman L, Kowalczyk MS, Gibbons RJ, Higgs DR, Neuhaus D, Rhodes D. 2007. Structural consequences of disease-causing mutations in the ATRX-DNMT3-DNMT3L (ADD) domain of the chromatin-associated protein ATRX. *Proc Natl Acad Sci* **104:** 11939–11944.

Bezerra MA, Araujo AS, Phylipsen M, Balak D, Kimura EM, Oliveira DM, Costa FF, Sonati MF, Harteveld CL. 2008. The deletion of SOX8 is not associated with ATR-16 in an HbH family from Brazil. *Br J Haematol* **142:** 324–326.

Buckle VJ, Kearney L. 1994. New methods in cytogenetics. *Curr Opin Genet Dev* **4:** 374–382.

Cardoso C, Lutz Y, Mignon C, Compe E, Depetris D, Mattei MG, Fontes M, Colleaux L. 2000. ATR-X mutations cause impaired nuclear location and altered DNA binding properties of the XNP/ATR-X protein. *J Med Genet* **37:** 746–751.

Daniels RJ, Peden JF, Lloyd C, Horsley SW, Clark K, Tufarelli C, Kearney L, Buckle VJ, Doggett NA, Flint J, et al. 2001. Sequence, structure and pathology of the fully annotated terminal 2 Mb of the short arm of human chromosome 16. *Hum Mol Genet* **10:** 339–352.

Dhayalan A, Tamas R, Bock I, Tattermusch A, Dimitrova E, Kudithipudi S, Ragozin S, Jeltsch A. 2011. The ATRX-ADD domain binds to H3 tail peptides and reads the combined methylation state of K4 and K9. *Hum Mol Genet* **20:** 2195–2203.

Donnai D, Clayton-Smith J, Gibbons RJ, Higgs DR. 1991. The non-deletion a thalassaemia/mental retardation syndrome. Further support for X linkage. *J Med Genet* **28:** 742–745.

Drane P, Ouararhni K, Depaux A, Shuaib M, Hamiche A. 2010. The death-associated protein DAXX is a novel histone chaperone involved in the replication-independent deposition of H3.3. *Genes Dev* **24:** 1253–1265.

European Polycystic Kidney Disease Consortium. 1994. The polycystic kidney disease 1 gene encodes a 14 kb transcript and lies within a duplicated region on chromosome 16. *Cell* **77:** 881–894.

Eustermann S, Yang JC, Law MJ, Amos R, Chapman LM, Jelinska C, Garrick D, Clynes D, Gibbons RJ, Rhodes D, et al. 2011. Combinatorial readout of histone H3 modifications specifies localization of ATRX to heterochromatin. *Nat Struct Mol Biol* **18:** 777–782.

Fisher E, Scambler P. 1994. Human haploinsufficiency—one for sorrow, two for joy. *Nat Genet* **7:** 5–7.

Garrick D, Samara V, McDowell TL, Smith AJ, Dobbie L, Higgs DR, Gibbons RJ. 2004. A conserved truncated isoform of the ATR-X syndrome protein lacking the SWI/SNF-homology domain. *Gene* **326:** 23–34.

Garrick D, Sharpe JA, Arkell R, Dobbie L, Smith AJ, Wood WG, Higgs DR, Gibbons RJ. 2006. Loss of Atrx affects trophoblast development and the pattern of X-inactivation in extraembryonic tissues. *PLoS Genet* **2:** e58.

Gecz J, Pollard H, Conzalez G, Villard L, Stayton C, Millasseau P, Khrestchatisky M, Fontes M. 1994. Cloning and expression of the murine homologue of a putative human X-linked nuclear protein gene closely linked to PGK1 in Xq13.3. *Hum Mol Genet* **3:** 39–44.

Gibbons RJ, Higgs DR. 2000. The molecular-clinical spectrum of the ATR-X syndrome. *Am J Med Genet* **97:** 204–212.

Gibbons RJ, Wilkie AOM, Weatherall DJ, Higgs DR. 1991. A newly defined X linked mental retardation syndrome associated with a thalassaemia. *J Med Genet* **28:** 729–733.

Gibbons RJ, Suthers GK, Wilkie AOM, Buckle VJ, Higgs DR. 1992. X-linked α thalassemia/mental retardation (ATR-X) syndrome: Localisation to Xq12-21.31 by X-inactivation and linkage analysis. *Am J Hum Genet* **51:** 1136–1149.

Gibbons RJ, Brueton L, Buckle VJ, Burn J, Clayton-Smith J, Davison BCC, Gardner RJM, Homfray T, Kearney L, Kingston HM, et al. 1995a. The clinical and hematological features of the X-linked α thalassemia/mental retardation syndrome (ATR-X). *Am J Med Genet* **55:** 288–299.

Gibbons RJ, Picketts DJ, Villard L, Higgs DR. 1995b. Mutations in a putative global transcriptional regulator cause X-linked mental retardation with α-thalassemia (ATR-X syndrome). *Cell* **80:** 837–845.

Gibbons RJ, Bachoo S, Picketts DJ, Aftimos S, Asenbauer B, Bergoffen J, Berry SA, Dahl N, Fryer A, Keppler K, et al. 1997. Mutations in a transcriptional regulator (hATRX) establish the functional significance of a PHD-like domain. *Nat Genet* **17:** 146–148.

Gibbons RJ, McDowell TL, Raman S, O'Rourke DM, Garrick D, Ayyub H, Higgs DR. 2000. Mutations in ATRX, encoding a SWI/SNF-like protein, cause diverse changes in the pattern of DNA methylation. *Nat Genet* **24:** 368–371.

Gibbons RJ, Wada T, Fisher C, Malik N, Mitson M, Steensma D, Goudie D, Fryer A, Krantz I, Traeger-Synodinos J. 2008. Mutations in the chromatin associated protein ATRX. *Human Mutation* **29:** 796–802.

Gibson WT, Harvard C, Qiao Y, Somerville MJ, Lewis MES, Rajcan-Separovic E. 2008. Phenotype-genotype characterization of α-thalassemia mental retardation syndrome due to isolated monosomy of 16p13.3. *Am J Med Genet* **146A:** 225–232.

Goldberg AD, Banaszynski LA, Noh KM, Lewis PW, Elsaesser SJ, Stadler S, Dewell S, Law M, Guo X, Li X, et al. 2010. Distinct factors control histone variant H3.3 localization at specific genomic regions. *Cell* **140:** 678–691.

Haas PS, Roy NB, Gibbons RJ, Deville MA, Fisher C, Schwabe M, Bisse E, van Dorsselaer A, Higgs DR, Lubbert M. 2009. The role of X-inactivation in the gender bias of patients with acquired α-thalassaemia and myelodysplastic syndrome (ATMDS). *Br J Haematol* **144:** 538–545.

Hall JG. 1990. Genomic imprinting: Review and relevance to human diseases. *Am J Hum Genet* **46:** 857–873.

Harteveld CL, Voskamp A, Phylipsen M, Akkermans N, den Dunnen JT, White SJ, Giordano PC. 2005. Nine unknown rearrangements in 16p13.3 and 11p15.4 causing α- and β-thalassaemia characterised by high resolution multiplex ligation-dependent probe amplification. *J Med Genet* **42:** 922–931.

Harteveld CL, Kriek M, Bijlsma EK, Erjavec Z, Balak D, Phylipsen M, Voskamp A, Capua E, White SJ, Giordano PC. 2007. Refinement of the genetic cause of ATR-16. *Hum Genet* **122:** 283–292.

Heaphy CM, de Wilde RF, Jiao Y, Klein AP, Edil BH, Shi C, Bettegowda C, Rodriguez FJ, Eberhart CG, Hebbar S, et al. 2011. Altered telomeres in tumors with ATRX and DAXX mutations. *Science* **333:** 425.

Higgs DR. 2004. Ham-wasserman lecture. Gene regulation in hematopoiesis: New lessons from thalassemia. *Hematology* **2004:** 1–13.

Iwase S, Xiang B, Ghosh S, Ren T, Lewis PW, Cochrane JC, Allis CD, Picketts DJ, Patel DJ, Li H, et al. 2011. ATRX ADD domain links an atypical histone methylation recognition mechanism to human mental-retardation syndrome. *Nat Struct Mol Biol* **18:** 769–776.

Jiao Y, Shi C, Edil BH, de Wilde RF, Klimstra DS, Maitra A, Schulick RD, Tang LH, Wolfgang CL, Choti MA, et al. 2011. DAXX/ATRX, MEN1, and mTOR pathway genes are frequently altered in pancreatic neuroendocrine tumors. *Science* **331:** 1199–1203.

Kaplan N, Moore IK, Fondufe-Mittendorf Y, Gossett AJ, Tillo D, Field Y, LeProust EM, Hughes TR, Lieb JD, Widom J, et al. 2009. The DNA-encoded nucleosome organization of a eukaryotic genome. *Nature* **458:** 362–366.

Lamb J, Harris PC, Wilkie AOM, Wood WG, Dauwerse JG, Higgs DR. 1993. De novo truncation of chromosome 16p and healing with $(TTAGGG)_n$ in the α-thalassemia/mental retardation syndrome (ATR-16). *Am J Hum Genet* **52:** 668–676.

Law MJ, Lower KM, Voon HP, Hughes JR, Garrick D, Viprakasit V, Mitson M, De Gobbi M, Marra M, Morris A, et al. 2010. ATR-X syndrome protein targets tandem repeats and influences allele-specific expression in a size-dependent manner. *Cell* **143:** 367–378.

McDowell TL, Gibbons RJ, Sutherland H, O'Rourke DM, Bickmore WA, Pombo A, Turley H, Gatter K, Picketts DJ, Buckle VJ, et al. 1999. Localization of a putative transcriptional regulator (ATRX) at pericentromeric heterochromatin and the short arms of acrocentric chromosomes. *Proc Natl Acad Sci* **96:** 13983–13988.

Mitson M, Kelley LA, Sternberg MJ, Higgs DR, Gibbons RJ. 2011. Functional significance of mutations in the Snf2 domain of ATRX. *Hum Mol Genet* **20:** 2603–2610.

Muers MR, Sharpe JA, Garrick D, Sloane-Stanley J, Nolan PM, Hacker T, Wood WG, Higgs DR, Gibbons RJ. 2007. Defining the cause of skewed X-chromosome inactivation in X-linked mental retardation by use of a mouse model. *Am J Hum Genet* **80:** 1138–1149.

Pfeifer D, Poulat F, Holinski-Feder E, Kooy F, Scherer G. 2000. The SOX8 gene is located within 700 kb of the tip of chromosome 16p and is deleted in a patient with ATR-16 syndrome. *Genomics* **63:** 108–116.

Picketts DJ, Higgs DR, Bachoo S, Blake DJ, Quarrell OWJ, Gibbons RJ. 1996. *ATRX* encodes a novel member of the SNF2 family of proteins: Mutations point to a common mechanism underlying the ATR-X syndrome. *Hum Mol Genet* **5:** 1899–1907.

Picketts DJ, Tastan AO, Higgs DR, Gibbons RJ. 1998. Comparison of the human and murine ATRX gene identifies highly conserved, functionally important domains. *Mamm Genome* **9:** 400–403.

Price TS, Regan R, Mott R, Hedman A, Honey B, Daniels RJ, Smith L, Greenfield A, Tiganescu A, Buckle V, et al. 2005. SW-ARRAY: A dynamic programming solution for the identification of copy-number changes in genomic DNA using array comparative genome hybridization data. *Nucleic Acids Res* **33:** 3455–3464.

Redon R, Ishikawa S, Fitch KR, Feuk L, Perry GH, Andrews TD, Fiegler H, Shapero MH, Carson AR, Chen W, et al. 2006. Global variation in copy number in the human genome. *Nature* **444:** 444–454.

Schmickel RD. 1986. Contiguous gene syndromes: A component of recognizable syndromes. *J Pediatr* **109:** 231–241.

Schneider AS, Bischoff FZ, McCaskill C, Coady ML, Stopfer JE, Shaffer LG. 1996. Comprehensive 4-year follow-up on a case of maternal heterdisomy for chromosome 16. *Am J Med Genet* **66:** 204–208.

Schwartzentruber J, Korshunov A, Liu XY, Jones DT, Pfaff E, Jacob K, Sturm D, Fontebasso AM, Quang DA, Tonjes M, et al. 2012. Driver mutations in histone H3.3 and chromatin remodelling genes in paediatric glioblastoma. *Nature* **482:** 226–231.

Stayton CL, Dabovic B, Gulisano M, Gecz J, Broccoli V, Giovanazzi S, Bossolasco M, Monaco L, Rastan S, Boncinelli E, et al. 1994. Cloning and characterisation of a new human Xq13 gene, encoding a putative helicase. *Hum Mol Genet* **3:** 1957–1964.

Steensma DP, Higgs DR, Fisher CA, Gibbons RJ. 2004a. Acquired somatic ATRX mutations in myelodysplastic syndrome associated with α thalassemia (ATMDS) convey a more severe hematologic phenotype than germline ATRX mutations. *Blood* **103:** 2019–2026.

Steensma DP, Viprakasit V, Hendrick A, Goff DK, Leach J, Gibbons RJ, Higgs DR. 2004b. Deletion of the alpha-globin gene cluster as a cause of acquired α-thalassemia in myelodysplastic syndrome. *Blood* **103:** 1518–1520.

Steensma DP, Gibbons RJ, Higgs DR. 2005. Acquired α thalassemia in association with myelodysplastic syndrome and other hematologic malignancies. *Blood* **105:** 443–452.

Villard L, Gecz J, Mattéi JF, Fontés M, Saugier-Veber P, Munnich A, Lyonnet S. 1996a. XNP mutation in a large family with Juberg-Marsidi syndrome. *Nature Genetics* **12:** 359–360.

Villard L, Lacombe D, Fontés M. 1996b. A point mutation in the XNP gene, associated with an ATR-X phenotype without α-thalassemia. *Eur J Hum Genet* **4:** 316–320.

Villard L, Toutain A, Lossi A-M, Gecz J, Houdayer C, Moraine C, Fontès M. 1996c. Splicing mutation in the ATR-X gene can lead to a dysmorphic mental retardation phenotype without α-thalassemia. *Am J Hum Genet* **58:** 499–505.

Weatherall DJ, Higgs DR, Bunch C, Old JM, Hunt DM, Pressley L, Clegg JB, Bethlenfalvay NC, Sjolin S, Koler RD, et al. 1981. Hemoglobin H disease and mental retardation. A new syndrome or a remarkable coincidence? *N Engl J Med* **305:** 607–612.

White JC, Ellis M, Coleman PN, Beaven GH, Gratzer WB, Shooter EM, Skinner ER. 1960. An unstable haemoglobin associated with some cases of leukaemia. *Br J Haematol* **6:** 171–177.

Wilkie AOM, Buckle VJ, Harris PC, Lamb J, Barton NJ, Reeders ST, Lindenbaum RH, Nicholls RD, Barrow M, Bethlenfalvay NC, et al. 1990a. Clinical features and molecular analysis of the a thalassaemia/mental retardation syndromes. I: Cases due to deletions involving chromosome band 16p13.3. *Am J Hum Genet* **46:** 1112–1126.

Wilkie AOM, Zeitlin HC, Lindenbaum RH, Buckle VJ, Fischel-Ghodsian N, Chui DHK, Gardner-Medwin D, MacGillivray MH, Weatherall DJ, Higgs DR. 1990b. Clinical features and molecular analysis of the α thalassemia/mental retardation syndromes. II: Cases without detectable abnormality of the a globin complex. *Am J Hum Genet* **46:** 1127–1140.

Wilkie AOM, Pembrey ME, Gibbons RJ, Higgs DR, Porteous MEM, Burn J, Winter RM. 1991. The non-deletion type of a thalassaemia/mental retardation: A recognisable dysmorphic syndrome with X-linked inheritance. *J Med Genet* **28:** 724.

Wong LH, McGhie JD, Sim M, Anderson MA, Ahn S, Hannan RD, George AJ, Morgan KA, Mann JR, Choo KH. 2010. ATRX interacts with H3.3 in maintaining telomere structural integrity in pluripotent embryonic stem cells. *Genome Res* **20:** 351–360.

Xue Y, Gibbons R, Yan Z, Yang D, McDowell TL, Sechi S, Qin J, Zhou S, Higgs D, Wang W. 2003. The ATRX syndrome protein forms a chromatin-remodeling complex with Daxx and localizes in promyelocytic leukemia nuclear bodies. *Proc Natl Acad Sci* **100:** 10635–10640.

Management of the Thalassemias

Nancy F. Olivieri[1] and Gary M. Brittenham[2]

[1]Hemoglobinopathy Research University Health Network, Toronto, Ontario M5G 2C4, Canada
[2]Department of Pediatrics, Columbia University, New York, New York 10032
Correspondence: gmb31@columbia.edu

During the last 30 years, in addition to the considerable progress made in control and prevention of thalassemias[3], there have also been major advances in their symptomatic management, at least in wealthier countries where appropriate facilities are available. Remarkable improvements in survival in the severe forms of thalassemia have followed the more judicious use of blood transfusion and, in particular, the ability to manage the iron accumulation resulting from transfusion with its severe and ultimately lethal effects on endocrine and cardiac function.

Because the management of severe and milder intermittent forms of β- and α-thalassemia have been described in detail recently (Weatherall and Clegg 2001; Higgs 2009; Olivieri and Weatherall 2009; Taher et al. 2011, 2012), after a short introduction to the general principles of management of these conditions, this article focuses on the control of iron overload in the different forms of thalassemia.

GENERAL PRINCIPLES OF THE MANAGEMENT OF THALASSEMIAS

Although the more severe thalassemias are an extremely heterogeneous group of disorders, their general management follows the same principles. At first presentation, it is absolutely essential to obtain an accurate diagnosis of the form of the disease, ideally including its molecular basis. It is also important to perform a detailed family study to assess the pattern of inheritance at the same time. Once this information is available, the family requires well-informed counseling about the likely future course of the illness and, equally important, about the relative risks of having further affected children.

It is not uncommon for babies to first present with thalassemia with particularly low hemoglobin levels associated with intercurrent infection. Although they may need to receive transfusions until they recover from the presenting illness, it is important not to establish them on regular transfusion without a reasonable period of observation. Unless this precaution is taken, babies with various forms of thalassemia of intermediate severity may be placed on lifelong unnecessary transfusion. The important features to observe before long-term transfusion is considered include the patterns of growth and growth velocity, activity compared with infants of the same age, the occurrence of progressive splenomegaly, and early evidence of any skeletal

---

[3]See Cao and Kan (2013).

Copyright © 2013 Cold Spring Harbor Laboratory Press; all rights reserved
Cite this article as *Cold Spring Harb Perspect Med* doi: 10.1101/cshperspect.a011767

deformity. Although the hemoglobin level is also important, it should not be the only determinant of whether transfusion is required. In some forms of β-thalassemia intermedia, notably HbE β-thalassemia in particular, there is increasing evidence that patients may be able to adapt to relatively low hemoglobin levels and not require regular transfusion (Allen et al. 2010).

During later stages of development, these children—as well as continuing to be assessed regarding their growth, skeletal status, and spleen size—also need regular assessment of their hematologic findings together with their bone age, liver, endocrine and cardiac functions, as well as their iron status, which is considered in detail below.

Ideally, these patients should be managed in special centers with expertise in the management of thalassemia and with access to pediatricians in a variety of related specialties. Because so many of these patients now survive for much longer periods of time, it is also vital that these centers have staff with special knowledge of the manifestations of the thalassemias in adults, again with the appropriate specialist facilities in related fields.

Recent developments in more radical forms of treatment for the thalassemias—notably bone marrow transplantation, gene therapy, and stem cell therapy—have been reviewed elsewhere (Daley 2012; Lucarelli and Tisdale 2012; Nienhuis and Persons 2012).

IRON ACCUMULATION IN THALASSEMIA

Thalassemia patients who are not transfused may retain up to 75% of orally administered iron, accumulating between 3 and 9 mg of iron per day (Pippard et al. 1979). Because transfusions suppress erythroid expansion and reduce ineffective erythropoiesis, thereby reducing iron absorption (Pippard and Weatherall 1984), the goal of regular transfusions in thalassemia is to ameliorate these processes and correct anemia. The judicious application of a transfusion regimen in which pretransfusion hemoglobin concentrations do not exceed 9.5 g/dL achieves an optimal balance between accumulation of body iron and suppression of eryrthroid marrow activity (Cazzola et al. 1997). These regimens administer 100–200 mL RBCs/kg per year (Porter 2001). Transfusion requirements may be influenced by the severity of globin chain imbalance and by splenectomy, among other factors (Rebulla and Modell 1991), and daily accumulation of iron may therefore vary between 0.30 and 0.59 mg/kg (Cohen et al. 2008).

ASSESSMENT OF IRON IN VIVO

Serum Ferritin Concentration

Concentrations of plasma or serum ferritin, commonly used to estimate body iron, derive from ferritin synthesis as well as its release from damaged cells. If serum ferritin concentration exceeds 4000 μg/L, suggested as an upper physiological limit of the rate of synthesis, higher concentrations are believed to be due to the release of ferritin from damaged cells (Prieto et al. 1975; Worwood et al. 1980). Interpretation of values may be complicated by a variety of conditions, all common in thalassemia, that alter concentrations independently of changes in body iron burden, including ascorbate deficiency, fever, acute infection, chronic inflammation, acute and chronic hepatic damage, hemolysis, and ineffective erythropoiesis (Olivieri and Brittenham 1997). The 95% prediction intervals for hepatic iron concentration, given the plasma ferritin, are too broad to make the determination of ferritin a reliable predictor of body iron stores (Brittenham et al. 1993). A retrospective analysis in deferoxamine-treated patients found that during a period of 15 years, maintenance of serum ferritins of 2500 μg/L was associated with estimates of improved cardiac-disease-free survival (Olivieri et al. 1994). These findings were confirmed in subsequent studies, with lower ferritin concentration reported to predict even longer survivals (Telfer et al. 2000; Borgna-Pignatti et al. 2004a; Davis et al. 2004). These findings should not be interpreted to suggest, in the day-to-day management of individual patients, that serum ferritin is an accurate assessment of body iron burden,

 Cite this article as *Cold Spring Harb Perspect Med* doi: 10.1101/cshperspect.a011767

although in clinical practice serum ferritin continues to be measured, usually at 6-mo intervals. Documentation of changes may be useful to encourage patients and clinicians in compliance with treatment (Davis and Porter 2000) and, as described further below, regular measurements may also be useful in monitoring the risk of deferoxamine toxicity, especially in children (Porter and Davis 2002).

Liver Iron Concentration

The liver is the major repository of transfused iron. Hepatic parenchymal iron accumulation may rapidly progress to portal fibrosis in a significant percentage of patients, including children (Iancu and Neustein 1977). In thalassemia major, progressive liver disease itself is an important cause of death in young adults (Zurlo et al. 1989) and represents a risk factor for hepatocellular carcinoma (Borgna-Pignatti et al. 2004b).

The uncertainties with respect to serum ferritin concentration have led to recommendations for the application of quantitative measurements of body iron, especially in studies to evaluate the effectiveness of new chelating agents. Liver iron correlates closely with total body iron. In patients with transfusion-dependent thalassemia, the concentration of total body iron (in milligrams per kilogram of body weight) is equivalent to 10.6 times the hepatic iron concentration (in milligrams per gram of liver, dry weight [dw]). Using this equation, it is possible to reliably estimate total body iron stores that reach as high as 250 mg/kg of body weight, with a standard error of <7.9 (Angelucci et al. 2000). Elevated levels of liver iron are useful in predicting the risk of complications of iron loading in vivo (Brittenham 2011). Historically, assessment of liver iron was determined through biopsies of the liver but now generally relies largely on magnetic resonance imaging (MRI) exploiting the paramagnetic effects of tissue iron on surrounding tissues, which alters the relaxation time of molecules excited by application of a magnetic field (Porter and Shah 2010). One MRI method (R2; Ferriscan) that uses widely available equipment has been standardized and validated to be predictive over a clinically useful range of liver iron concentrations (St Pierre et al. 2005).

Assessment of Cardiac Iron

Even in transfused, nonchelated patients with thalassemia, cardiac disease related to iron overload does not usually develop during the second decade of life. As more extensively outlined below, this complication has become rare in patients in whom deferoxamine therapy is available and who are managed by attentive medical personnel.

Cardiovascular T2-star ($T2^*$) magnetic resonance (CMR $T2^*$) has been used to guide iron-chelating therapy (Anderson et al. 2001), but calibrated methods are not available for all patients (Brittenham 2011). Early, short-term cross-sectional studies involving CMR $T2^*$ (Anderson et al. 2002, 2006) suggested poor correlations between CMR $T2^*$ and liver iron (as well as serum ferritin) concentrations and proposed that CMR $T2^*$ measurements were superior to serial determinations of liver iron in the evaluation of responses to iron-chelating therapy. These conclusions have been challenged by the identification of a calibration error in $T2^*$ liver measurements conducted by the London investigators, leading in many studies to repeated underestimation of liver iron concentration (Pennell et al. 2006; Tanner et al. 2008). Even if iron within the heart is correctly imaged (Garbowski et al. 2008), simple cross-sectional associations between the two organs may have limited usefulness because of delays in loading and removal of cardiac iron, relative to these processes in the liver (Noetzli et al. 2008). The influences on CMR $T2^*$ of the rate of tissue iron loading, the age at the start of iron loading, the processes of iron absorption and ineffective erythropoiesis, and the particular iron-chelating therapy administered remain poorly understood. The clinical significance of measurements of CMR $T2^*$ was further questioned by a recent U.S. Food and Drug Administration inspection of a trial (Pennell et al. 2006) using this modality to evaluate the relative effectiveness of two iron-chelating therapies (Food and Drug Administration 2011).

In summary, no robust data appear to effectively contradict the long-established association of liver iron concentration and the risk of development of complications in thalassemia. Body iron burdens corresponding to hepatic iron concentrations exceeding ~15–20 mg iron/g liver, dw have been shown to place patients at heightened risk of cardiac disease and early death (Brittenham et al. 1994; Telfer et al. 2000), liver dysfunction (Jensen et al. 2003), and acceleration of hepatic portal fibrosis (Angelucci et al. 2002). Lack of certainty exists with respect to optimal hepatic iron concentrations that minimize the risk that hepatic fibrosis will progress to cirrhosis and its ultimate complication, hepatocellular carcinoma (Borgna-Pignatti et al. 2004a,b; Ko et al. 2007).

It is also of clinical relevance that the development of cardiac disease may be predicted from a liver iron concentration obtained as much as 13 years previously (Telfer et al. 2000). This suggests that measurements of liver iron may provide an opportunity to optimize therapy, before organ dysfunction develops, that is more extended than reportedly provided by measurements of CMR T2* (Kirk et al. 2009).

Extrahepatic Iron

The most common clinical problem in transfused patients with thalassemia is hypogonadotropic hypogonadism (Porter and Shah 2010), related to selective iron deposition in pituitary gonadotropes (Bergeron and Kovacs 1978) as body iron burden increases (Musallam et al. 2011). Following early studies observing that anterior pituitary iron imaged by magnetic resonance (Fujisawa et al. 1988) could be correlated with biochemical markers of hypogonadism (Chatterjee et al. 1998), techniques to image the anterior pituitary have improved (Argyropoulou et al. 2003; Christoforidis et al. 2007; Au et al. 2008b; Noetzli et al. 2011b) but are not widely used in clinical practice. Diabetes mellitus, attributed to several processes including insulin resistance as well as chronic pancreatic iron overload, is also a frequently observed complication in patients in whom body iron burden is poorly controlled (Olivieri and Brittenham 1997). During the last decade, there has been progress in the MRI of pancreatic iron (Midiri et al. 1999; Au et al. 2008a,b; Noetzli et al. 2009, 2011a); correlation of pancreatic MRI parameters with tests of pancreatic function is still awaited.

Nontransferrin-Bound Iron

In thalassemia, after plasma iron exceeds the transport capacity of circulating transferrin, a fraction of iron known as nontransferrin-bound iron (NTBI) appears in plasma. NTBI comprises a heterogeneous assortment of iron complexes that are believed to be the major mediators of extrahepatic tissue damage in transfusional iron overload (Breuer et al. 2000). NTBI enters hepatocytes, cardiomyocytes, anterior pituitary cells, and pancreatic β cells, leading to the generation of reactive oxygen species. Damage to lipids, proteins, DNA, and subcellular organelles may result in cellular dysfunction, apoptosis, and necrosis (Brittenham 2011). In addition to removing excess iron from cells, chelating therapy agents may reduce NTBI, but in clinical practice measurements they are not generally used to assess the response to therapy (Jacobs et al. 2005).

Other Means of Assessing Iron Loading

Serum iron, transferrin, transferrin saturation, transferrin receptor concentration, and urinary iron excretion do not quantitatively reflect body iron stores. A study of the iron-chelating agent deferasirox based the initial dose on a calculated rate of transfusional iron loading and then adjusted the dose according to measurements of serum ferritin levels and safety markers over a 1-yr period (Cappellini et al. 2010). As noted, the long-term efficacy and safety of this strategy are uncertain (Brittenham 2011).

CHELATION IN CLINICAL PRACTICE

Iron-chelating therapy should be considered in all patients with thalassemia who require long-term red-cell transfusion. Chelators should be avoided in patients who are pregnant or breast-feeding. Ideally, therapy should be initiated prophylactically, before clinically significant iron accumulation has occurred. Patients who have already undergone repeated transfusion without sufficient chelation can also be successfully treated, but they may require intensive regimens (Brittenham 2011).

Initiation of Iron-Chelating Therapy

A study in children with thalassemia major found that liver iron may exceed 7 mg/g dw between the 10th and 20th transfusion (Saxon et al. 1997). Before initiation (and any adjustment) of iron-chelating therapy, detailed documentation of the history of transfusion, previous iron-chelating therapy, if any, estimation of the rate of transfusional iron loading, and determination of the body iron load by measurement of the hepatic iron concentration should be undertaken. In clinical practice, body iron is also estimated regularly by serum ferritin concentration. The extent of any existing iron-induced hepatic, cardiac, or endocrine dysfunction should be established. In children and adolescents, height and weight velocity and percentiles and stage of sexual maturation should be documented. Pretherapy toxicity evaluations should be tailored to the chelating agent to be implemented, as detailed below.

The dose of an iron-chelating agent is determined by the presence or absence of cardiac iron overload, the rate of transfusional iron loading, and the body iron burden. In the absence of cardiac iron overload, the long-term objective is to maintain body iron at a level that permits safe storage while avoiding chelator toxicity. The greater the rate of transfusional iron loading, the greater the dose of an iron chelator that will be needed to control the accumulation of iron (as outlined in Table 1 [adapted from Brittenham 2011]).

Monitoring of Effectiveness

The toxic manifestations of iron depend not only on tissue concentrations but also the rate of accumulation, duration of exposure to increased iron, ascorbate status (which helps to determine the allocation of iron between macrophage and parenchymal cells), extent of internal redistribution of iron between macrophage and parenchymal sites, and noniron-related factors, such as alcohol and viral hepatitis (Olivieri and Brittenham 1997). If cardiac iron overload is present, eliminating excess iron from the heart becomes the primary therapeutic goal (see Table 1). Because chelators remove iron from the heart much more slowly than from the liver, prolonged, intensive iron-chelating therapy is usually needed. As a result, in chelated patients, the presence of cardiac iron cannot be predicted from the liver iron concentration. Iron-chelating therapy may have reduced the liver iron to an optimal range despite the persistence of severe cardiac iron overload. Nevertheless, cardiac iron loading can be prospectively predicted in the presence of sustained severe iron overload with elevated hepatic iron concentrations (Noetzli et al. 2008).

In the absence of cardiac iron overload, the primary therapeutic goal then becomes maintenance of the body iron burden at a level that permits safe storage while avoiding chelator toxicity from administering excessive amounts relative to the body iron. The best measure of the body iron load is the hepatic storage iron concentration, which reflects iron accumulations in both hepatocytes and reticuloendothelial macrophages (Kupffer cells). Therapy to maintain body iron at levels found in healthy, never-transfused individuals, corresponding to a hepatic iron of $\sim$0.2–1.6 mg iron/g liver, dw (Brittenham et al. 1982) may increase the probability of dose-related chelator toxicity. At the opposite extreme, body iron burdens corresponding to hepatic iron concentrations exceeding $\sim$15–20 mg iron/g liver, dw place patients at risk of serious complications of iron loading (Brittenham et al. 1994; Telfer et al. 2000; Angelucci et al. 2002; Jensen et al. 2003). In transfusion-dependent patients with thalassemia, hepatic iron concentrations of $\sim$3–7 mg/g of liver, dw are generally regarded as optimal, seeming to minimize the risks both of adverse drug effects and complications from iron overload. The safety and long-term effectiveness of off-label regimens of therapy during which body storage iron is reportedly normalized (Farmaki et al.

Table 1. Usual doses of deferoxamine or deferasirox for transfusional iron overload (to minimize interference with growth and skeletal development)

	No cardiac iron overload (T_2^*, ≥20 msec); daily transfusional iron intake			Cardiac iron overload (T_2^*, <20 msec)	
Hepatic iron concentration	<0.3 mg/kg of body weight	0.3–0.5 mg/kg of body weight	>0.5 mg/kg of body weight	Mild to moderate T_2^*, 10 to <20 msec	Severe T_2^*, <10 msec
≥15 mg/g, dw	Deferoxamin: 40–50 mg/kg/d, 8–10 h/d, 6 or 7 d/wk, by subcutaneous infusion; deferasirox: infusion, deferasirox: oral dose of 30–40 mg/kg daily	Deferoxamine: 40–50 mg/kg/d, 8–10 h/d, 6 or 7 d/wk by subcutaneous infusion; deferasirox: oral dose of 30–40 mg/kg daily	Deferoxamine: 40–50 mg/kg/d, 8–10 h/d, 6 or 7 d/wk by subcutaneous infusion; deferasirox: oral dose of 30–40 mg/kg daily	Deferoxamine: 50 mg/kg/d by continuous intravenous infusion; deferasirox: oral dose of 40 mg/kg daily, but uncertain efficacy in reducing cardiac iron	Deferoxamine: 50 mg/kg/day by continuous intravenous infusion, deferasirox: oral dose of 40 mg/kg daily, but uncertain efficacy in reducing cardiac iron
7 to 15 mg/g, dw	Deferoxamine: 30–40 mg/kg/d, 8–10 h/d, 5 d/wk by subcutaneous infusion; deferasirox: oral dose of 20–30 mg/kg daily	Deferoxamine: 40–50 mg/kg/d, 8–10 h/d, 6 or 7 d/wk by subcutaneous infusion; deferasirox: oral dose of 30–40 mg/kg daily	Deferoxamine 40–50 mg/kg/d, 8–10 h/d, 6 or 7 d/wk by subcutaneous infusion; deferasirox: oral dose of 30–40 mg/kg daily	Deferoxamine: 40–50 mg/kg/d, 8–10 h/d, 6 or 7 d/wk by subcutaneous infusion; deferasirox: oral dose of 30–40 mg/kg daily	Deferoxamine: 40–50 mg/kg/d by continuous infusion; deferasirox: oral dose of 30–40 mg/kg daily
3 to <7 mg/g, dw	Deferoxamine: 30–40 mg/kg/d, 8–10 h/d, 5 d/wk, by subcutaneous infusion; deferasirox: oral dose of 20–30 mg/kg	Deferoxamine: 30–40 mg/kg/d, 8–10 h/d, 5 d/wk, by subcutaneous infusion; deferasirox: oral dose of 20–30 mg/kg daily	Deferoxamine: 30–40 mg/kg/d, 8–10 h/d, 5 d/wk by subcutaneous infusion: deferasirox: oral dose of 20–30 mg/kg	Deferoxamine: 40–50 mg/kg/d, 8–10 h/d, 6 or 7 d/wk, by subcutaneous infusion; deferasirox: oral dose of 30–40 mg/kg daily	Deferoxamine: 40–50 mg/kg/d by continuous infusion; deferasirox: oral dose of 30–40 mg/kg daily
<3 mg/g, dw	Deferoxamine: suspend therapy; deferasirox: suspend therapy	Deferoxamine: suspend therapy; deferasirox: suspend therapy	Deferoxamine: suspend therapy; deferasirox: suspend therapy	Deferoxamine: adjust intravenous or subcutaneous dose according to therapeutic index 32; deferasirox: adjust oral dose, monitoring renal and hepatic function and blood count	Deferoxamine: adjust intravenous or subcutaneous dose according to therapeutic index 32; deferasirox adjust oral dose, monitoring renal and hepatic function and blood count

The dose of deferoxamine in young children should not exceed 25–30 mg/kg of body weight. The dose should be adjusted according to the therapeutic index (Pennell et al. 2006). The bioavailability may affect the response. T_2^*, Cardiac transverse relaxation time on magnetic resonance imaging; dw, dry weight.

2010) await confirmation. Patients with higher body iron burdens exceeding 7 up to ~15 mg iron/g liver, dw are at an increased risk of hepatic fibrosis, diabetes mellitus, and other complications and need more intensive iron-chelation therapy (Angelucci et al. 2000). Patients with still higher body iron burdens, recognized as having a greatly increased risk of early death, are candidates for special programs of management (Brittenham 2011).

In principle, the ineffectiveness of any drug may be related to dose prescribed, adherence to treatment, blood transfusion rate, variability in absorption, metabolism of the drug, or a combination of these and other factors. Administration of a chelator will be needed as long as transfusion is continued and will be lifelong in most patients; effectiveness needs to be sustained over time. But how is effectiveness defined?

Robust criteria to evaluate the effectiveness of iron-chelating therapy are derived from published results of prospective studies of patients with iron overload (Cartwright et al. 1979; Brittenham et al. 1982, 1994; Olivieri et al. 1995) and are based on the ability of the chelator to reduce and maintain body iron at levels associated with favorable clinical outcomes. These criteria should be applied to the clinical evaluation of patients, including in clinical trials of new agents. In brief, a chelating agent is judged to be effective if it reduces or maintains body iron, assessed by hepatic iron concentration, to concentrations associated with prolonged survival free of the complications of iron overload (hepatic iron 3.2–7.0 mg/g liver, dw). A chelating agent is judged to be ineffective if the body iron increases to or is maintained at levels associated with increased risk of the complications of iron overload (hepatic iron 7.1–14.9 mg/g liver, dw), including those of cardiac disease and premature death (hepatic iron $\geq$ 15.0 mg/g liver, dw).

Criteria that refer to "percentage changes" or "improving trends" from baseline hepatic iron may be fundamentally misleading in the interpretation of the effectiveness of an iron-chelating agent and may give rise to risks to individual patients. This is illustrated by contrasting the use of two criteria of effectiveness in two patients. Consider first a patient whose initial hepatic iron is 4 mg/g liver, dw and, after treatment with an iron-chelating agent, is shown to have hepatic iron of 6 mg/g liver, dw. Although the hepatic iron has increased by 30%, application of the criteria based on the reduction or maintenance of body iron at levels associated with favorable clinical outcomes would judge the chelator to be *effective*; the body iron has been maintained within optimal range. In contrast, the application of the criteria using "percentage changes" or "improving trends" would describe this therapy as *ineffective*. Next, consider another patient in whom initial hepatic iron was 40 mg/g liver, dw and, following iron-chelating treatment, has a hepatic iron of 36 mg/g liver, dw. Although hepatic iron has decreased by 10%, application of accepted criteria would judge the chelator to be *ineffective*, because body iron persists at a level associated with an increased risk of cardiac disease and premature death, but application of the criteria judging "percentage changes" or "improving trends" would describe this therapy as *effective*. Finally, the reporting of "mean changes" in tissue or serum parameters of iron loading, in the absence of provision of individual patient responses, does not adequately inform the clinician or patient about the expectation for effectiveness for a given agent. As above, changes or lack of changes in ferritin values, although potentially useful in retrospective evaluation of body iron, may be misleading and should not be used to define effectiveness in individual patients.

AGENTS USED FOR IRON CHELATION

Deferoxamine

Chemistry, Pharmacology, and Administration in Practice

Deferoxamine is a hexadentate chelator that binds iron at a 1:1 molar ratio, rendering it virtually inactive metabolically. The drug is poorly absorbed orally (Callender and Weatherall 1980) and rapidly metabolized in plasma (Summers et al. 1979), conferring a requirement for prolonged parenteral infusions during which plasma concentrations reach a plateau at 12 h.

Deferoxamine is administered subcutaneously or intravenously, usually using a portable pump, for 8–10 h/d, 5–7 d/wk (Propper et al. 1977; Pippard et al. 1978). Subcutaneous administration is preferred except in patients with severe cardiac iron deposition for whom continuous intravenous deferoxamine therapy is recommended (Davis and Porter 2000). Oral ascorbic acid at the equivalent of 2–3 mg/kg/d (Pippard et al. 1982) is usually prescribed at the beginning of an infusion.

Clinical Benefits

The evaluation of deferoxamine therapy antedates the common use of randomized, controlled trials (Brittenham 2011). For four decades following a randomized trial in which chronic intramuscular administration was shown to slow iron accumulation and arrest hepatic fibrosis in transfused patients (Barry et al. 1974), numerous observational studies provided evidence for deferoxamine-associated improvements in long-term survival related to improvements in cardiac disease. Cardiac disease even in transfused, nonchelated patients with thalassemia does not usually develop during the second decade of life, and if treatment with deferoxamine is available and supervised, it has become rare. Two trials of >10-yr duration confirmed unequivocally that long-term use of deferoxamine in thalassemia major is associated with long-term survival, free of the cardiac complications of iron overload (Brittenham et al. 1994; Olivieri et al. 1994). In one trial, patients with most serum ferritin concentrations <2500 μg/L had an estimated cardiac-disease-free survival of 91% after 15 years, whereas cardiac-disease-free survival in those in whom most ferritins had exceeded 2500 μg/L was <20% (Olivieri et al. 1994). The other trial classified patients as having received ineffective or effective chelation therapy using a threshold of body storage iron corresponding to a liver iron of ~15 mg iron/g liver, dw; the probability of survival to at least 25 yr was only 32% among patients above this threshold, whereas no deaths occurred among patients below the threshold (Brittenham et al. 1994). A subsequent report involving more than 250 patients reported that those who administered deferoxamine at least 4.7 times/wk enjoyed 95% survival at age 30 (Gabutti and Piga 1996). In the largest reported cohort in Italy, where deferoxamine was introduced in 1975, the prevalence of heart failure in patients aged ≥15 yr decreased from 5% in patients born between 1970 and 1974 to 2% in those born between 1980 and 1984 (Borgna-Pignatti et al. 1998). A follow-up report identified one death from cardiac disease in the 1980–1984 birth cohort and no such deaths in subsequent birth cohorts (Borgna-Pignatti et al. 2004a). These findings show that accumulation of iron in heart is not inevitable, provided that body storage iron is controlled (Porter 2007). These North American and Italian data are paralleled in expert programs in the United Kingdom (Porter and Davis 2002), although poorer results are reported in centers in that country where erratic compliance and other impediments to effective treatment are likely less well managed (Modell et al. 2000). It would therefore appear that cardiac disease develops almost exclusively in thalassemia patients in whom effective treatment is unavailable, who begin deferoxamine after irreversible damage has occurred and/or who are not expertly supervised during long-term treatment.

The most common complication of iron loading in thalassemia, hypogonadism, is also prevented by effective use of deferoxamine (Bronspiegel-Weintrob et al. 1990), although secondary amenorrhea in women, and secondary hypogonadism in men, may develop after age 21, even in those who have attained normal puberty (Gamberini et al. 2008).

Tolerability and Toxicity of Deferoxamine

Discomfort at the site of infusion may be mitigated with topical anesthetic or glucocorticoid creams (Brittenham 2011). Dose-related deferoxamine toxicities including visual changes, auditory toxicity, attenuation of linear growth, and skeletal dysplasia (Olivieri et al. 1986; Rodda et al. 1995; Chan et al. 2002; De Sanctis et al. 2006) may be minimized by maintaining doses of no more than 25–30 mg/kg in young children (Olivieri et al. 1992; Olivieri and Brittenham 1997). In both children and adults, as the hepatic iron concentration approaches optimal levels (see below), dose reduction is recommended; a useful recommendation is that the daily dose (mg/kg) divided by the serum ferritin (mg/L) should not exceed 0.025 (Porter and Huehns 1989; Porter et al. 1989). Allergy to deferoxamine is rare and most patients may be desensitized successfully (Porter 2001), with patients thereafter able to take deferoxamine. Monitoring of toxicity in patients who receive deferoxamine includes annual assessments of auditory function and vision and, in children, careful observations of linear growth including height velocity.

Deferasirox

Chemistry, Pharmacology, and Administration in Practice

The synthetic chelator deferasirox is a bis-hydroxyphenyl-triazole that, in contrast to deferoxamine, is well absorbed from the gastrointestinal tract and cleared from the circulation slowly with a plasma half-life of 11–19 h, supporting once-daily oral dosing (Galanello et al. 2003; Nisbet-Brown et al. 2003; Waldmeier et al. 2010). Because of low water solubility of the free ligand, deferasirox is administered as a suspension in water or fruit juice (Novartis Pharmaceuticals 2010). Hepatocytes readily take up deferasirox, which chelates hepatocellular iron. Deferasirox–iron complexes are excreted in the bile (Waldmeier et al. 2010).

The short-term effectiveness of deferasirox has been compared with deferoxamine in trials sponsored by Novartis (Cappellini et al. 2006; Piga et al. 2006; Vichinsky et al. 2007; Porter et al. 2008; Cappellini et al. 2010). In the largest trial in which 586 children with thalassemia were randomly assigned to either agent, with dosing according to the baseline hepatic iron concentration (Cappellini et al. 2006), the primary end point was the percentage of patients with either a maintained or reduced hepatic iron concentration at 1 yr. This end point was reached in 52.9% of patients assigned to deferasirox and in 66.4% of patients assigned to deferoxamine. This result, which did not meet a prespecified noninferiority target, was attributed to the relative underdosing of deferasirox (Brittenham 2011). No trial has established the long-term effectiveness of deferasirox in preventing organ toxicity or improving survival.

Tolerability and Unwanted Effects of Deferasirox

In the registration trial of deferasirox (Cappellini et al. 2006), gastrointestinal disturbances occurred in ~15% of patients, rash in 11%, and increases in serum creatinine levels in 38%. Similar rates have been observed in subsequent trials (Piga et al. 2006; Vichinsky et al. 2007; Porter et al. 2008). In 2010, on the basis of postmarketing studies, the Food and Drug Administration required a change in the prescribing information for deferasirox, stating that the drug could cause potentially fatal renal and hepatic impairment or failure as well as gastrointestinal hemorrhage. These adverse effects were reported to occur more frequently in older patients and in patients with high-risk myelodysplastic syndromes, thrombocytopenia, or underlying renal or hepatic impairment. In patients who receive deferasirox, serum creatinine, serum aminotransferases, and bilirubin levels should be assessed monthly (Novartis Pharmaceuticals 2010).

Deferiprone

In both the European Union and the United States, the approved use of the synthetic oral iron-chelating agent deferiprone is restricted to patients with transfusional iron overload owing to thalassemia syndromes when current chelation therapy is inadequate. Deferiprone is not approved for use in Canada. Because this agent is not recommended for primary iron-chelating therapy, it is considered here only briefly.

Chemistry, Pharmacology, and Administration in Practice

Deferiprone is a 3-hydroxypyridin-4-1 bidentate chelator that binds to iron in a 3:1 ratio

(Porter and Shah 2010). The drug has a plasma half-life of 1.5 h (Olivieri et al. 1990) and is usually given three times daily.

Clinical Benefits

In the United States, approval was based on a reduction in serum ferritin levels; sustained reduction in liver iron concentration during deferiprone treatment has not been shown. The FDA prescribing information states that there "are no controlled trials demonstrating a direct treatment benefit, such as improvement in disease-related symptoms, functioning, or increased survival" (Food and Drug Administration 2011).

Tolerability and Unwanted Effects

Common adverse effects include nausea, diarrhea, and arthropathy, including arthritis associated with clinically significant disability. The most serious adverse effects are agranulocytosis and neutropenia, increases in liver enzymes (Cohen et al. 2003), and progression of hepatic fibrosis associated with an increase in iron overload or hepatitis C. Weekly monitoring of the neutrophil count is recommended. Neurologic abnormalities have been reported during administration of higher-than-recommended doses (Beau-Salinas et al. 2009). In patients who receive deferiprone, weekly assessment of complete blood counts and monthly assessments of serum aminotransferases should be performed (European Medicines Agency 2009).

Approach to Therapy

An informed decision with respect to the choice of an iron chelator is best made with the patient or guardian. Despite the lack of data on long-term effectiveness, most patients now opt for deferasirox because of the ease of administration (Brittenham 2011). Patients, particularly those with severe iron overload with cardiac involvement, who are informed about the efficacy of deferoxamine in the reversal of iron-induced heart disease (Davis and Porter 2000) and improving long-term survival, may indicate a preference for the proven therapy. Deferasirox may be substituted for deferoxamine when cardiac function is improved, or it could also be a choice in patients who are unable to tolerate infusions of deferoxamine (Brittenham 2011).

Combination Therapy

The long-term effectiveness of a variety of binary combinations of chelating agents presently administered in off-label uses remains uncertain due to the absence of unequivocal evidence of the superiority of any specific combination over treatment with a single agent (Roberts et al. 2007; Brittenham 2011).

Monitoring

The control of iron-chelating therapy is best implemented by periodic measurements of hepatic iron and of cardiac function. Serum ferritin concentrations are usually measured twice a year. On an annual basis, hepatic, cardiac, and endocrine function should be evaluated and, in children and adolescents, growth and sexual maturation formally assessed. Monitoring of toxicity will depend on the potential adverse effects of the specific agent to be used as above.

Financial Costs of Iron-Chelating Therapy

In North America (Vichinsky et al. 2005) and Europe (Modell et al. 2001), the combined number of patients with transfusion-dependent thalassemia is $<$10,000. Globally, at least 100,000 patients with thalassemia require transfusions (Weatherall 2010). As previously discussed (Brittenham 2011), the annual per-patient costs of care for complications of iron overload are estimated to be \$15,000–\$20,000.

What is less well calculated, particularly in emerging countries, are costs associated with inadequate chelating therapy that may be collected from publications that report the costs in richer countries of failure of compliance with therapy (Delea et al. 2007). Full therapeutic doses of effective iron-chelating agents are affordable in only a minority of patients in countries worldwide, and regimens of treatment in these countries include the wider application of less effective drugs and combinations of drugs, or homeopathic doses of deferoxamine or deferasirox. As expected, heart and liver disease and endocrine dysfunction are, ultimately, not prevented by these regimens. The complications of iron overload that can be expected to develop result in increased costs in medical care, which, like chelating therapy itself, are generally borne primarily by the family of the individual in these countries. Unassisted by patient organizations that focus most efforts in North America and Europe, this situation is now being addressed through efforts by smaller charities.

CONCLUSIONS

Clearly, there have been major improvements in the symptomatic management of severe thalassemias, at least in the wealthier countries of the world. Although, as discussed by Weatherall (2010), owing to increasing international collaboration, at least some progress is also being made in the developing countries, but the situation is still far from satisfactory. As these countries go through the epidemiological transition resulting from improvements in public health, nutrition, and medical care, the number of babies who would previously have died before presenting for treatment with thalassemia is likely to decrease; hence, it is almost certain that overall, the number of patients with this condition will increase in the future. It may be some time before more radical forms of treatment such as gene therapy become available and, even then, they are likely to be very expensive. Hence, despite the increased development of public education, counseling, and prenatal diagnosis, it is likely that the symptomatic approach to the management of thalassemias, as outlined in this brief article, will remain of considerable importance for the foreseeable future.

REFERENCES

**Reference is also in this collection.*

Allen A, Fisher C, Premawardhena A, Peto T, Allen S, Arambepola M, Thayalsutha V, Olivieri N, Weatherall D. 2010. Adaptation to anemia in hemoglobin E-ss thalassemia. *Blood* **116:** 5368–5370.

Anderson LJ, Holden S, Davis B, Prescott E, Charrier CC, Bunce NH, Firmin DN, Wonke B, Porter J, Walker JM, et al. 2001. Cardiovascular T2-star (T2*) magnetic resonance for the early diagnosis of myocardial iron overload. *Eur Heart J* **22:** 2171–2179.

Anderson LJ, Wonke B, Prescott E, Holden S, Walker JM, Pennell DJ. 2002. Comparison of effects of oral deferiprone and subcutaneous desferrioxamine on myocardial iron concentrations and ventricular function in β-thalassaemia. *Lancet* **360:** 516–520.

Anderson LJ, Westwood MA, Prescott E, Walker JM, Pennell DJ, Wonke B. 2006. Development of thalassaemic iron overload cardiomyopathy despite low liver iron levels and meticulous compliance to desferrioxamine. *Acta Haematol* **115:** 106–108.

Angelucci E, Brittenham GM, McLaren CE, Ripalti M, Baronciani D, Giardini C, Galimberti M, Polchi P, Lucarelli G. 2000. Hepatic iron concentration and total body iron stores in thalassemia major. *N Engl J Med* **343:** 327–331.

Angelucci E, Muretto P, Nicolucci A, Baronciani D, Erer B, Gaziev J, Ripalti M, Sodani P, Tomassoni S, Visani G, et al. 2002. Effects of iron overload and hepatitis C virus positivity in determining progression of liver fibrosis in thalassemia following bone marrow transplantation. *Blood* **100:** 17–21.

Argyropoulou MI, Kiortsis DN, Efremidis SC. 2003. MRI of the liver and the pituitary gland in patients with β-thalassemia major: Does hepatic siderosis predict pituitary iron deposition? *Eur Radiol* **13:** 12–16.

* Arora N, Daley GQ. 2012. Pluripotent stem cells in research and treatment of hemoglobinopathies. *Cold Spring Harb Perpect Med* **2:** a011841.

Au WY, Lam WW, Chu W, Tam S, Wong WK, Liang R, Ha SY. 2008a. A T2* magnetic resonance imaging study of pancreatic iron overload in thalassemia major. *Haematologica* **93:** 116–119.

Au WY, Lam WW, Chu WW, Yuen HL, Ling AS, Li RC, Chan HM, Lee HK, Law MF, Liu HS, et al. 2008b. A cross-sectional magnetic resonance imaging assessment of organ specific hemosiderosis in 180 thalassemia major patients in Hong Kong. *Haematologica* **93:** 784–786.

Barry M, Flynn DM, Letsky EA, Risdon RA. 1974. Long-term chelation therapy in thalassaemia major: Effect on liver iron concentration, liver histology, and clinical progress. *Br Med J* **2:** 16–20.

Beau-Salinas F, Guitteny MA, Donadieu J, Jonville-Bera AP, Autret-Leca E. 2009. High doses of deferiprone may be associated with cerebellar syndrome. *BMJ* **338:** a2319.

Bergeron C, Kovacs K. 1978. Pituitary siderosis. A histologic, immunocytologic, and ultrastructural study. *Am J Pathol* **93:** 295–309.

Borgna-Pignatti C, Rugolotto S, De Stefano P, Piga A, Di Gregorio F, Gamberini MR, Sabato V, Melevendi C, Cappellini MD, Verlato G. 1998. Survival and disease complications in thalassemia major. *Ann NY Acad Sci* **850:** 227–231.

Borgna-Pignatti C, Rugolotto S, De Stefano P, Zhao H, Cappellini MD, Del Vecchio GC, Romeo MA, Forni GL, Gamberini MR, Ghilardi R, et al. 2004a. Survival and complications in patients with thalassemia major treated with transfusion and deferoxamine. *Haematologica* **89:** 1187–1193.

Borgna-Pignatti C, Vergine G, Lombardo T, Cappellini MD, Cianciulli P, Maggio A, Renda D, Lai ME, Mandas A, Forni G, et al. 2004b. Hepatocellular carcinoma in the thalassaemia syndromes. *Br J Haematol* **124:** 114–117.

Breuer W, Hershko C, Cabantchik ZI. 2000. The importance of non-transferrin bound iron in disorders of iron metabolism. *Transfus Sci* **23:** 185–192.

Brittenham GM. 2011. Iron-chelating therapy for transfusional iron overload. *N Engl J Med* **364:** 146–156.

Brittenham GM, Farrell DE, Harris JW, Feldman ES, Danish EH, Muir WA, Tripp JH, Bellon EM. 1982. Magnetic-susceptibility measurement of human iron stores. *N Engl J Med* **307:** 1671–1675.

Brittenham GM, Cohen AR, McLaren CE, Martin MB, Griffith PM, Nienhuis AW, Young NS, Allen CJ, Farrell DE, Harris JW. 1993. Hepatic iron stores and plasma ferritin concentration in patients with sickle cell anemia and thalassemia major. *Am J Hematol* **42:** 81–85.

Brittenham GM, Griffith PM, Nienhuis AW, McLaren CE, Young NS, Tucker EE, Allen CJ, Farrell DE, Harris JW. 1994. Efficacy of deferoxamine in preventing complications of iron overload in patients with thalassemia major. *N Engl J Med* **331:** 567–573.

Bronspiegel-Weintrob N, Olivieri NF, Tyler B, Andrews DF, Freedman MH, Holland FJ. 1990. Effect of age at the start of iron chelation therapy on gonadal function in β-thalassemia major. *N Engl J Med* **323:** 713–719.

Callender ST, Weatherall DJ. 1980. Iron chelation with oral desferrioxamine. *Lancet* **316:** 689.

* Cao A, Kan YW. 2013. The prevention of thalassemia. *Cold Spring Harb Perpect Med* **3:** a011775.

Cappellini MD, Cohen A, Piga A, Bejaoui M, Perrotta S, Agaoglu L, Aydinok Y, Kattamis A, Kilinc Y, Porter J, et al. 2006. A phase 3 study of deferasirox (ICL670), a once-daily oral iron chelator, in patients with β-thalassemia. *Blood* **107:** 3455–3462.

Cappellini MD, Porter J, El-Beshlawy A, Li CK, Seymour JF, Elalfy M, Gattermann N, Giraudier S, Lee JW, Chan LL, et al. 2010. Tailoring iron chelation by iron intake and serum ferritin: The prospective EPIC study of deferasirox in 1744 patients with transfusion-dependent anemias. *Haematologica* **95:** 557–566.

Cartwright GE, Edwards CQ, Kravitz K, Skolnick M, Amos DB, Johnson A, Buskjaer L. 1979. Hereditary hemochromatosis. Phenotypic expression of the disease. *N Engl J Med* **301:** 175–179.

Cazzola M, Borgna-Pignatti C, Locatelli F, Ponchio L, Beguin Y, De Stefano P. 1997. A moderate transfusion regimen may reduce iron loading in β-thalassemia major without producing excessive expansion of erythropoiesis. *Transfusion* **37:** 135–140.

Chan YL, Pang LM, Chik KW, Cheng JC, Li CK. 2002. Patterns of bone diseases in transfusion-dependent homozygous thalassaemia major: Predominance of osteoporosis and desferrioxamine-induced bone dysplasia. *Pediatr Radiol* **32:** 492–497.

Chatterjee R, Katz M, Oatridge A, Bydder GM, Porter JB. 1998. Selective loss of anterior pituitary volume with severe pituitary-gonadal insufficiency in poorly compliant male thalassemic patients with pubertal arrest. *Ann NY Acad Sci* **850:** 479–482.

Christoforidis A, Haritandi A, Perifanis V, Tsatra I, Athanassiou-Metaxa M, Dimitriadis AS. 2007. MRI for the determination of pituitary iron overload in children and young adults with β-thalassaemia major. *Eur J Radiol* **62:** 138–142.

Cohen AR, Galanello R, Piga A, De Sanctis V, Tricta F. 2003. Safety and effectiveness of long-term therapy with the oral iron chelator deferiprone. *Blood* **102:** 1583–1587.

Cohen AR, Glimm E, Porter JB. 2008. Effect of transfusional iron intake on response to chelation therapy in β-thalassemia major. *Blood* **111:** 583–587.

Davis BA, Porter JB. 2000. Long-term outcome of continuous 24-hour deferoxamine infusion via indwelling intravenous catheters in high-risk β-thalassemia. *Blood* **95:** 1229–1236.

Davis BA, O'Sullivan C, Jarritt PH, Porter JB. 2004. Value of sequential monitoring of left ventricular ejection fraction in the management of thalassemia major. *Blood* **104:** 263–269.

Delea TE, Edelsberg J, Sofrygin O, Thomas SK, Baladi JF, Phatak PD, Coates TD. 2007. Consequences and costs of noncompliance with iron chelation therapy in patients with transfusion-dependent thalassemia: A literature review. *Transfusion* **47:** 1919–1929.

De Sanctis V, Roos M, Gasser T, Fortini M, Raiola G, Galati MC. 2006. Impact of long-term iron chelation therapy on growth and endocrine functions in thalassemia. *J Pediatr Endocrinol Metab* **19:** 471–480.

European Medicines Agency. 2009. Ferriprox (deferiprone): Summary of product characteristics. European Public Assessment Report, rev. ed. 14 (http://wwwemaeuropaeu/humandocs/Humans/EPAR/ferriprox/ferriproxhtm).

Farmaki K, Tzoumari I, Pappa C, Chouliaras G, Berdoukas V. 2010. Normalisation of total body iron load with very intensive combined chelation reverses cardiac and endocrine complications of thalassaemia major. *Br J Haematol* **148:** 466–475.

Food and Drug Administration. 2011. Briefing document for the Oncologic Drugs Advisory Committee Meeting, September 14 (http://www.fda.gov/downloads/AdvisoryCommittees/CommitteesMeetingMaterials/Drugs/OncologicDrugsAdvisoryCommittee/UCM271537.pdf).

Fujisawa I, Morikawa M, Nakano Y, Konishi J. 1988. Hemochromatosis of the pituitary gland: MR imaging. *Radiology* **168:** 213–214.

Gabutti V, Piga A. 1996. Results of long-term iron-chelating therapy. *Acta Haematol* **95:** 26–36.

Galanello R, Piga A, Alberti D, Rouan MC, Bigler H, Sechaud R. 2003. Safety, tolerability, and pharmacokinetics of ICL670, a new orally active iron-chelating agent in patients with transfusion-dependent iron overload due to β-thalassemia. *J Clin Pharmacol* **43:** 565–572.

Gamberini MR, De Sanctis V, Gilli G. 2008. Hypogonadism, diabetes mellitus, hypothyroidism, hypoparathyroidism: Incidence and prevalence related to iron overload and chelation therapy in patients with thalassaemia major followed from 1980 to 2007 in the Ferrara Centre. *Pediatr Endocrinol Rev* **6:** 158–169.

Garbowski M, Eleftheriou P, Pennell D Tanner M Porter JB. 2008. Impact of compliance, ferritin and LIC on long-term trends in myocardial T2* with deferasirox. *Blood* **112:** 116.

Higgs DR. 2009. The pathophysiology and clinical features of a thalassemia. In *Disorders of hemoglobin* (ed. Steinberg MH, Forget BG, Higgs DR, Weatherall DJ), pp. 266–295. Cambridge University Press, Cambridge.

Iancu TC, Neustein HB. 1977. Ferritin in human liver cells of homozygous β-thalassaemia: Ultrastructural observations. *Br J Haematol* **37:** 527–535.

Jacobs EM, Hendriks JC, van Tits BL, Evans PJ, Breuer W, Liu DY, Jansen EH, Jauhiainen K, Sturm B, Porter JB, et al. 2005. Results of an international round robin for the quantification of serum non-transferrin-bound iron: Need for defining standardization and a clinically relevant isoform. *Anal Biochem* **341:** 241–250.

Jensen PD, Jensen FT, Christensen T, Eiskjaer H, Baandrup U, Nielsen JL. 2003. Evaluation of myocardial iron by magnetic resonance imaging during iron chelation therapy with deferrioxamine: Indication of close relation between myocardial iron content and chelatable iron pool. *Blood* **101:** 4632–4639.

Kirk P, Roughton M, Porter JB, Walker JM, Tanner MA, Patel J, Wu D, Taylor J, Westwood MA, Anderson LJ, et al. 2009. Cardiac T2* magnetic resonance for prediction of cardiac complications in thalassemia major. *Circulation* **120:** 1961–1968.

Ko C, Siddaiah N, Berger J, Gish R, Brandhagen D, Sterling RK, Cotler SJ, Fontana RJ, McCashland TM, Han SH, et al. 2007. Prevalence of hepatic iron overload and association with hepatocellular cancer in end-stage liver disease: Results from the National Hemochromatosis Transplant Registry. *Liver Int* **27:** 1394–1401.

* Lucarelli G, Isgrò A, Sodani P, Gaziev J. 2012. Hematopoietic stem cell transplantation in thalassemia and sickle cell anemia. *Cold Spring Harb Perpect Med* **2:** a011825.

Midiri M, Lo Casto A, Sparacia G, D'Angelo P, Malizia R, Finazzo M, Montalto G, Solbiati L, Lagalla R, De Maria M. 1999. MR imaging of pancreatic changes in patients with transfusion-dependent β-thalassemia major. *AJR Am J Roentgenol* **173:** 187–192.

Modell B, Khan M, Darlison M. 2000. Survival in β-thalassaemia major in the UK: Data from the UK Thalassaemia Register. *Lancet* **355:** 2051–2052.

Modell B, Khan M, Darlison M, King A, Layton M, Old J, Petrou M, Varnavides L. 2001. A national register for surveillance of inherited disorders: β-Thalassaemia in the United Kingdom. *Bull World Health Org* **79:** 1006–1013

Musallam KM, Cappellini MD, Wood JC, Motta I, Graziadei G, Tamim H, Taher AT. 2011. Elevated liver iron concentration is a marker of increased morbidity in patients with β thalassemia intermedia. *Haematologica* **96:** 1605–1612.

* Nienhuis AW, Persons DA. 2012. Development of gene therapy for thalassemia. *Cold Spring Harb Perpect Med* **2:** a011833.

Nisbet-Brown E, Olivieri NF, Giardina PJ, Grady RW, Neufeld EJ, Sechaud R, Krebs-Brown AJ, Anderson JR, Alberti D, Sizer KC, et al. 2003. Effectiveness and safety of ICL670 in iron-loaded patients with thalassaemia: A randomised, double-blind, placebo-controlled, dose-escalation trial. *Lancet* **361:** 1597–1602.

Noetzli LJ, Carson SM, Nord AS, Coates TD, Wood JC. 2008. Longitudinal analysis of heart and liver iron in thalassemia major. *Blood* **112:** 2973–2978.

Noetzli LJ, Papudesi J, Coates TD, Wood JC. 2009. Pancreatic iron loading predicts cardiac iron loading in thalassemia major. *Blood* **114:** 4021–4026.

Noetzli LJ, Mittelman SD, Watanabe RM, Coates TD, Wood JC. 2011a. Pancreatic iron and glucose dysregulation in thalassemia major. *Am J Hematol* doi: 10.1002/ajh.22223.

Noetzli LJ, Panigrahy A, Mittelman SD, Hyderi A, Dongelyan A, Coates TD, Wood JC. 2011b. Pituitary iron and volume predict hypogonadism in transfusional iron overload. *Am J Hematol* doi: 10.1002/ajh.22247.

Novartis Pharmaceuticals. 2010. Exjade (deferasirox) prescribing information (http://www.exjade.com).

Olivieri NF, Brittenham GM. 1997. Iron-chelating therapy and the treatment of thalassemia. *Blood* **89:** 739–761.

Olivieri NF, Weatherall DJ. 2009. Clinical aspects of β thalassemia and related disorders. In *Disorders of hemoglobin* (ed. Steinberg MH, Forget BG, Higgs DR, Weatherall DJ), pp. 357–416. Cambridge University Press, Cambridge.

Olivieri NF, Buncic JR, Chew E, Gallant T, Harrison RV, Keenan N, Logan W, Mitchell D, Ricci G, Skarf B, et al. 1986. Visual and auditory neurotoxicity in patients receiving subcutaneous deferoxamine infusions. *N Engl J Med* **314:** 869–873.

Olivieri NF, Koren G, Hermann C, Bentur Y, Chung D, Klein J, St Louis P, Freedman MH, McClelland RA, Templeton DM. 1990. Comparison of oral iron chelator L1 and desferrioxamine in iron-loaded patients. *Lancet* **336:** 1275–1279.

Olivieri NF, Koren G, Harris J, Khattak S, Freedman MH, Templeton DM, Bailey JD, Reilly BJ. 1992. Growth failure and bony changes induced by deferoxamine. *Am J Pediatr Hematol Oncol* **14:** 48–56.

Olivieri NF, Nathan DG, MacMillan JH, Wayne AS, Liu PP, McGee A, Martin M, Koren G, Cohen AR. 1994. Survival in medically treated patients with homozygous β-thalassemia. *N Engl J Med* **331:** 574–578.

Olivieri NF, Brittenham GM, Matsui D, Berkovitch M, Blendis LM, Cameron RG, McClelland RA, Liu PP, Templeton DM, Koren G. 1995. Iron-chelation therapy with oral deferipronein patients with thalassemia major. *N Engl J Med* **332:** 918–922.

Pennell DJ, Berdoukas V, Karagiorga M, Ladis V, Piga A, Aessopos A, Gotsis ED, Tanner MA, Smith GC, Westwood MA, et al. 2006. Randomized controlled trial of deferiprone or deferoxamine in β-thalassemia major patients with asymptomatic myocardial siderosis. *Blood* **107:** 3738–3744.

Piga A, Galanello R, Forni GL, Cappellini MD, Origa R, Zappu A, Donato G, Bordone E, Lavagetto A, Zanaboni L, et al. 2006. Randomized phase II trial of deferasirox (Exjade, ICL670), a once-daily, orally-administered iron chelator, in comparison to deferoxamine in thalassemia patients with transfusional iron overload. *Haematologica* **91:** 873–880.

Pippard MJ, Weatherall DJ. 1984. Iron absorption in non-transfused iron loading anaemias: Prediction of risk for iron loading, and response to iron chelation treatment, in β thalassaemia intermedia and congenital sideroblastic anaemias. *Haematologia (Budap)* **17:** 17–24.

Pippard MJ, Letsky EA, Callender ST, Weatherall DJ. 1978. Prevention of iron loading in transfusion-dependent thalassaemia. *Lancet* **1:** 1178–1181.

Pippard MJ, Callender ST, Warner GT, Weatherall DJ. 1979. Iron absorption and loading in β-thalassaemia intermedia. *Lancet* **2:** 819–821.

Pippard MJ, Callender ST, Finch CA. 1982. Ferrioxamine excretion in iron-loaded man. *Blood* **60:** 288–294.

Porter JB. 2001. Practical management of iron overload. *Br J Haematol* **115:** 239–252.

Porter JB. 2007. Concepts and goals in the management of transfusional iron overload. *Am J Hematol* **82:** 1136–1139.

Porter JB, Davis BA. 2002. Monitoring chelation therapy to achieve optimal outcome in the treatment of thalassaemia. *Best Pract Res Clin Haematol* **15:** 329–368.

Porter JB, Huehns ER. 1989. The toxic effects of desferrioxamine. *Baillieres Clin Haematol* **2:** 459–474.

Porter JB, Shah FT. 2010. Iron overload in thalassemia and related conditions: Therapeutic goals and assessment of response to chelation therapies. *Hematol Oncol Clin North Am* **24:** 1109–1130.

Porter JB, Jaswon MS, Huehns ER, East CA, Hazell JW. 1989. Desferrioxamine ototoxicity: Evaluation of risk factors in thalassaemic patients and guidelines for safe dosage. *Br J Haematol* **73:** 403–409.

Porter J, Galanello R, Saglio G, Neufeld EJ, Vichinsky E, Cappellini MD, Olivieri N, Piga A, Cunningham MJ, Soulieres D, et al. 2008. Relative response of patients with myelodysplastic syndromes and other transfusion-dependent anaemias to deferasirox (ICL670): A 1-yr prospective study. *Eur J Haematol* **80:** 168–176.

Prieto J, Barry M, Sherlock S. 1975. Serum ferritin in patients with iron overload and with acute and chronic liver diseases. *Gastroenterology* **68:** 525–533.

Propper RD, Cooper B, Rufo RR, Nienhuis AW, Anderson WF, Bunn HF, Rosenthal A, Nathan DG. 1977. Continuous subcutaenous administration of deferoxamine in patients with iron overload. *N Engl J Med* **297:** 418–423.

Rebulla P, Modell B. 1991. Transfusion requirements and effects in patients with thalassaemia major. Cooleycare Programme. *Lancet* **337:** 277–280.

Roberts DJ, Brunskill SJ, Doree C, Williams S, Howard J, Hyde CJ. 2007. Oral deferiprone for iron chelation in people with thalassaemia. *Cochrane Database Syst Rev* No. 3: CD004839.

Rodda CP, Reid ED, Johnson S, Doery J, Matthews R, Bowden DK. 1995. Short stature in homozygous β-thalassaemia is due to disproportionate truncal shortening. *Clin Endocrinol (Oxf)* **42:** 587–592.

Saxon B, Brittenham G, Nisbet-Brown E, Chait P, Klein N, Muraca M, Olivieri N. 1997. Liver biopsy is safe and provides quantitative guidelines for initiation of chelating therapy in children with thalassemia major. *Blood* **90:** 130a.

St Pierre TG, Clark PR, Chua-anusorn W, Fleming AJ, Jeffrey GP, Olynyk JK, Pootrakul P, Robins E, Lindeman R. 2005. Noninvasive measurement and imaging of liver iron concentrations using proton magnetic resonance. *Blood* **105:** 855–861.

Summers MR, Jacobs A, Tudway D, Perera P, Ricketts C. 1979. Studies in desferrioxamine and ferrioxamine metabolism in normal and iron-loaded subjects. *Br J Haematol* **42:** 547–555.

Taher AT, Musallam KM, Cappellini MD, Weatherall DJ. 2011. Optimal management of β thalassaemia intermedia. *Br J Haematol* **152:** 512–523.

Taher AT, Musallam KM, Karimi M, Cappellini MD. 2012. Contemporary approaches to treatment of β-thalassemia intermedia. *Blood Rev* **26:** S24–S27.

Tanner MA, Galanello R, Dessi C, Smith GC, Westwood MA, Agus A, Pibiri M, Nair SV, Walker JM, Pennell DJ. 2008. Combined chelation therapy in thalassemia major for the treatment of severe myocardial siderosis with left ventricular dysfunction. *J Cardiovasc Magn Reson* **10:** 12.

Telfer PT, Prestcott E, Holden S, Walker M, Hoffbrand AV, Wonke B. 2000. Hepatic iron concentration combined with long-term monitoring of serum ferritin to predict complications of iron overload in thalassaemia major. *Br J Haematol* **110:** 971–977.

Vichinsky EP, MacKlin EA, Waye JS, Lorey F, Olivieri NF. 2005. Changes in the epidemiology of thalassemia in North America: A new minority disease. *Pediatrics* **116:** e818–e825.

Vichinsky E, Onyekwere O, Porter J, Swerdlow P, Eckman J, Lane P, Files B, Hassell K, Kelly P, Wilson F, et al. 2007. A randomised comparison of deferasirox versus deferoxamine for the treatment of transfusional iron overload in sickle cell disease. *Br J Haematol* **136:** 501–508.

Waldmeier F, Bruin GJ, Glaenzel U, Hazell K, Sechaud R, Warrington S, Porter JB. 2010. Pharmacokinetics, metabolism, and disposition of deferasirox in β-thalassemic patients with transfusion-dependent iron overload who are at pharmacokinetic steady state. *Drug Metab Dispos* **38:** 808–816.

Weatherall DJ. 2010. The inherited diseases of hemoglobin are an emerging global health burden. *Blood* **115:** 4331–4336.

Weatherall D, Clegg JB. 2001. *The thalassemia syndromes*. Blackwell Science, Oxford, England.

Worwood M, Cragg SJ, Jacobs A, McLaren C, Ricketts C, Economidou J. 1980. Binding of serum ferritin to concanavalin A: Patients with homozygous β thalassaemia and transfusional iron overload. *Br J Haematol* **46:** 409–416.

Zurlo MG, De Stefano P, Borgna-Pignatti C, Di Palma A, Piga A, Melevendi C, Di Gregorio F, Burattini MG, Terzoli S. 1989. Survival and causes of death in thalassaemia major. *Lancet* **2:** 27–30.

The Prevention of Thalassemia

Antonio Cao[1,3] and Yuet Wai Kan[2]

[1]Dipartimento di Scienze Biomediche e Biotecnologie, Università degli Studi di Cagliari, Via Jenner snc 09121 Cagliari, Sardinia, Italy

[2]Department of Medicine, University of California, San Francisco, California 94143-0793

Correspondence: yw.kan@ucsf.edu

The thalassemias are among the most common inherited diseases worldwide, affecting individuals originating from the Mediterranean area, Middle East, Transcaucasia, Central Asia, Indian subcontinent, and Southeast Asia. As the diseases require long-term care, prevention of the homozygous state constitutes a major armament in the management. This article discusses the major prevention programs that are set up in many countries in Europe, Asia, and Australia, often drawing from the experience in Sardinia. These comprehensive programs involve carrier detections, molecular diagnostics, genetic counseling, and prenatal diagnosis. Variability of clinical severity can be attributable to interactions with α-thalassemia and mutations that increase fetal productions. Special methods that are currently quite expensive and not widely applicable are preimplantation and preconception diagnosis. The recent successful studies of fetal DNA in maternal plasma may allow future prenatal diagnosis that is noninvasive for the fetus.

The thalassemias are a group of autosomal recessive disorders caused by reduction or absent production of one or more of the globin chains that make up the hemoglobin (Hb) tetramers. According to the type of globin chain involved, two main types (i.e., the α- and β-thalassemias) can be distinguished. In addition, complex thalassemias resulting from defective production of two to four different globin chains (δβ-, γδβ-, and εγδβ-thalassemia) are recognized (Weatherall and Clegg 2001). The clinical types of thalassemias that are targets of prevention are β-thalassemia major resulting from homozygosity for β-thalassemia and Hb Bart's fetal hydrops syndrome caused by deletion or dysfunction of all four α-globin genes (Weatherall and Clegg 2001).

EPIDEMIOLOGY AND GLOBAL BURDEN OF THALASSEMIA DISORDERS

Thalassemias are among the commonest autosomal recessive disorders worldwide (Modell and Darlison 2008; Weatherall 2010a,b) and are prevalent in populations in the Mediterranean area, the Middle East, Transcaucasus, Central Asia, the Indian subcontinent, and the Far East. However, they are also quite common

[3]Dr. Kan, the editors, and publisher deeply regret that Professor Antonio Cao, a remarkable contributor to the study and alleviation of thalassemia, passed away on June 21, 2012.

Copyright © 2013 Cold Spring Harbor Laboratory Press; all rights reserved

Cite this article as *Cold Spring Harb Perspect Med* doi: 10.1101/cshperspect.a011775

in populations of African heritage. Because of population migration, nowadays thalassemias are also common in Northern Europe, North Central and South America, and Australia.

Worldwide 56,000 conceptions have a major thalassemia disorder of which approximately 30,000 are affected by β-thalassemia major and 3500 succumb perinatally from the hydrops fetalis syndrome. Most (or many) of these patients are born in developing and low-income countries where they create an enormous health burden. Together with sickle cell anemia, it has also been estimated that, worldwide, 9 million carriers become pregnant annually and 1.33 million pregnancies are at risk for a thalassemia major condition (Modell and Darlison 2008; Weatherall 2010b).

POPULATION CONTROL OF THALASSEMIAS: PROGRAMS FOR PREVENTION

In the late 1970s pilot population programs directed to prevent β-thalassemia major by carrier screening, counseling, and prenatal diagnosis started in several at-risk populations in the Mediterranean area [Cyprus, Sardinia, several regions of Continental Italy (Delta Po area, Sicily), and Greece] (Angastiniotis and Hadjiminas 1981; Cao et al. 1996; Loukopoulos 1996). At present, several countries have set up comprehensive national prevention programs, which include public awareness and education, carrier screening, and counseling, as well as information on prenatal diagnosis and preimplantation diagnosis. These countries are Italy, Greece, Cyprus, UK, France, Iran, Thailand, Australia, Singapore, Taiwan, Hong Kong, and Cuba (A Eleftheiou and M Angastionitis, pers. comm.). Partial programs including antenatal screening according to ethnic origin are available in several countries in Northern Europe (Netherlands, Belgium, and Germany) (Modell et al. 2007). In many developing countries, prenatal diagnosis service has also been introduced (Canatan et al. 2006; Chern et al. 2006; Weatherall 2010b).

In most programs, carrier screening and counseling are being performed on a voluntary basis. In Cyprus, however, the Orthodox church requires a certificate proving that screening for β-thalassemia has been performed before marriage, but allows the final decision on marriage and a reproduction option to be left to the couple (Angastiniotis and Hadjiminas 1981; Kalokairinou 2007). When Cyprus became a member of the European Union in 2004, the requirement for this certification became unnecessary for civil weddings (Kolnagou and Kontoghiorghes 2009). In a number of Muslim countries including Lebanon, Iran, Saudi Arabia, Tunisia, United Arab Emirates, Bahrain, Qatar, and Gaza Strip (Samavat and Modell 2004; Inati et al. 2006), the national premarital programs are mandatory and aimed at limiting carrier marriage. In most of these countries, nowadays, interruption of the pregnancy is permitted for severe fetal disorders but only within the fourth month of gestation. In China, prenatal screening was compulsory to obtain marriage permission by the marriage law office (Li 2009). Because of worldwide criticism, the requirement for a certificate of premarital screening was withdrawn.

One of the obstacles for introducing prospective screening in Continental Europe is the lack of awareness among policymakers, health professionals, and groups at risk, of hemoglobin disorders and the failure to recognize that hemoglobin disorders nowadays present a significant health problem (Modell et al. 2007). In developing countries, it has been suggested that the ideal place where these preventive measures could be organized is through existing health care services. Intensive education of health professionals as well as the population at large in the field of preventive genetics is needed (reviewed in Alwan and Modell 2003) and major health organizations and funding agencies must support these initiatives (Weatherall 2010b).

In α-thalassemia, screening is recommended only to detect couples at risk for hydrops fetalis syndrome because severe toxemia complications (hypertension, preeclampsia) in the mother may occur (Chui 2005). So far, screening has been mostly retrospective. But prospective screening has recently been introduced in several South Asian at-risk countries such as

Hong Kong, Southern China, Thailand, Taiwan, Malaysia, Singapore, Maldives, and Sri Lanka (Premawardhena et al. 2004; A Eleftheiou and M Angastionitis, pers. comm.). Because of the mild phenotype, prevention of HbH disease is not justified and should be discouraged.

PROGRAMS FOR CONTROL OF β-THALASSEMIA: EDUCATION

The successful programs in the Mediterranean area are all characterized by intensive education directed toward the health personnel and to the population at large. Periodic meetings are held among physicians and especially pediatricians and obstetricians, family planning workers, nurses, and social workers, to discuss the clinical characteristics, the natural history, the principles of genetic counseling, and the methodologies for preventing the birth of affected children (Fig. 1). Population education makes use of mass media, posters, and informational booklets, which are left at various key sites such as family planning clinics, marriage registries, and counseling rooms. The essential messages should include the clinical manifestation, natural history, available therapies, and life expectancy for the homozygous state of thalassemia major. It is important to emphasize that the carrier state can be easily identified at the thalassemia centers and is not associated with stigmatization, whereas couples who are both carriers have several reproductive options, including prenatal diagnosis. Contact with parents' and patients' associations should be regularly maintained to update them on the content of the messages to avoid potential misinformation. Furthermore, parents' associations can play an invaluable role as an influential group by raising economical support for all aspects of thalassemia management and providing psychological assistance to parents. Furthermore, in several countries, an annual thalassemia meeting or thalassemia day is organized and a specific news magazine is regularly issued.

Formal education on thalassemias was also introduced in the educational curriculum at the end of secondary school (Cao et al. 2008; Kalokairinou 2008; Amato et al. 2009). A critical evaluation of the information channels by which the population was informed, was performed in Sardinia and showed that a large majority of the target population (70%) was informed via physicians (family doctors, obstetricians, and

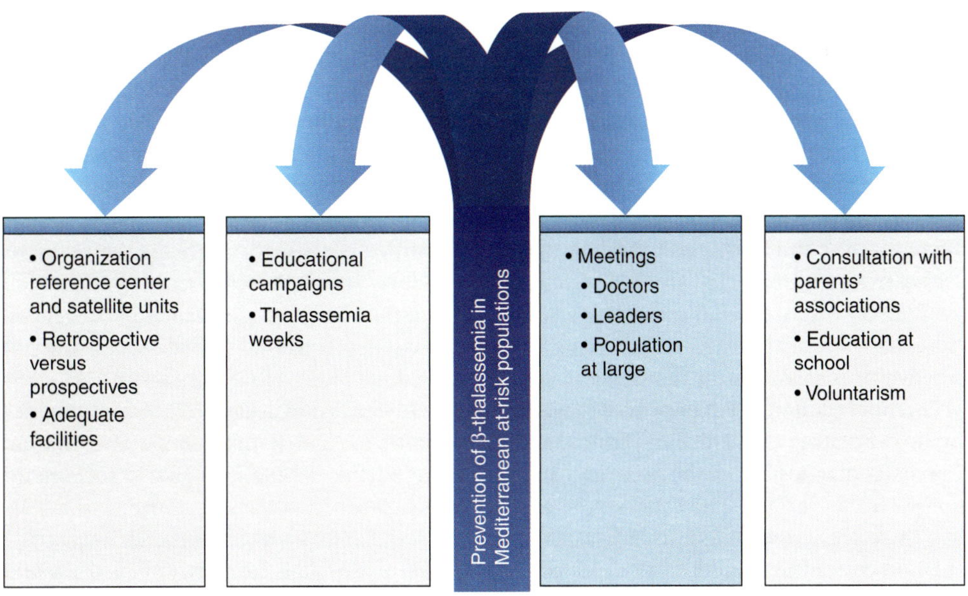

Figure 1. Strategy for prevention of β-thalassemias in Mediterranean at-risk populations.

genetic counselors). These programs were initially directed to couples with a previous affected child (retrospective diagnosis). Although very useful for individual families, retrospective diagnosis has a limited effect on the control of homozygous β-thalassemia at the population level. Nowadays, the target populations for screenings are couples before marriage, at preconception, and early following marriage (prospective diagnosis). Both in Cyprus and in Italy, the percentage of young unmarried couples requesting screening is increasing, indicating a continuous improvement of the knowledge on thalassemia. In all these programs, carrier screening has been performed on a voluntary basis. Information counseling preceded testing but written informed consent, which should be performed according to accepted principles, is usually not requested.

Another critical prerequisite for a screening program is the organization of adequate facilities to meet the demand both for screening and prenatal diagnosis. In all the successful programs, a large majority of the adults counseled are already fully informed about thalassemia and the methodology for its prevention, and are capable of making an informed decision about reproduction. Furthermore, as seen later, these programs have resulted in a consistent decline in the birth rate of homozygous β-thalassemia.

In Sardinia, a hub and spokes network with a main center and several satellite units was instrumental in achieving such success. A similar organization has been set up in Cyprus. In Sardinia, the distribution of the population in small villages facilitates the spread of information and extends the counseling and testing to families of carriers and parents. The introduction of teaching on thalassemia in secondary school has also played a critical role in enforcing the knowledge of thalassemia in the family members. Potential economic barriers were prevented by incorporating the screening, counseling, and prenatal diagnosis into the network of the National Health Service. The experience gathered in these preventive programs indicates that population education should be modulated according to the cultural and religious characteristics of the target population.

CARRIER DETECTION

Heterozygotes for β-thalassemias, both of the β^0 or the β^+ variety are characterized hematologically by high red blood cell count, microcytosis, hypochromia, increased HbA_2, and unbalanced globin chain synthesis. Several procedures have been proposed for β-thalassemia carrier screening (Cao et al. 1998). The methodology most often used in the preliminary selection of individuals at risk of being carriers is the determination of the mean corpuscular volume (MCV) and mean corpuscular Hb concentration (MCH) (Fig. 2). An alternative for this approach proposed for the South Asian population (especially for subjects living in rural areas) is the one-tube fragility test (OF) associated with the dichlorophenol indophenol (DCIP) dye test to detect the presence of HbE, which is very common in this region and when interacting with β-thalassemia leads to β-thalassemia major or intermedia (Fucharoen et al. 2004). Those individuals resulting positive at this screening test are further investigated by quantitative determination of HbA_2, which is the most important feature for identifying heterozygous β-thalassemia (Weatherall and Clegg 2001). However, both MCV–MCH determination and especially the one-tube fragility test, may miss those β-thalassemia heterozygotes who coinherited α-thalassemia (Melis et al. 1983; Rosatelli et al. 1984). In carriers of β^0-thalassemia, the α-thalassemia determinants giving rise to this effect are the deletion of two α-globin genes or one of the nondeletion mutations affecting the major α_2-globin gene. These carriers may be still easily identified because they have the typical increase of HbA_2 levels. Another cause for normal or borderline MCV–MCH values in carriers of β-thalassemia is a mild β-thalassemia mutation, including HbE, Hb Knossos, and Hb Malay (Weatherall and Clegg 2001). A high frequency of both α- and β-thalassemia is common in many of the populations at risk such as those living in the Mediterranean area, and the Middle and Far East. In Sardinia, for instance, because of the high frequency of α-thalassemia, 13.5% of β-thalassemia carriers have normal–borderline MCV–MCH values. The major

 Cite this article as *Cold Spring Harb Perspect Med* doi: 10.1101/cshperspect.a011775

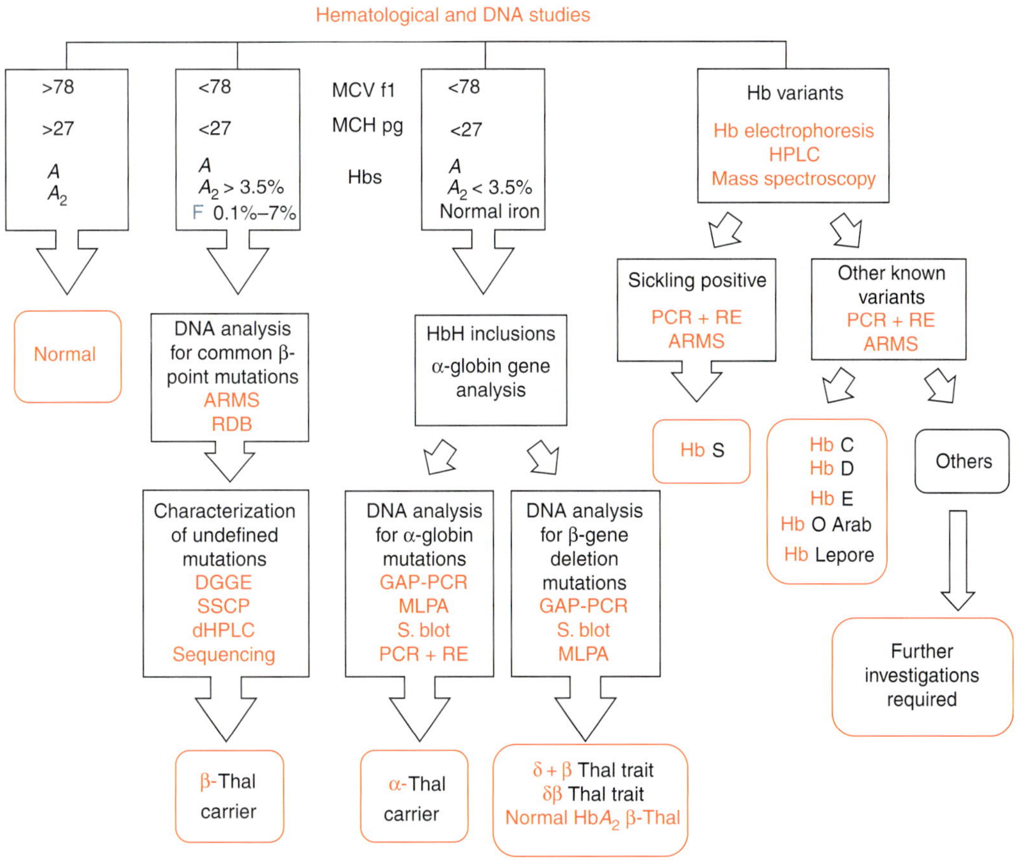

Figure 2. Flow chart for β-thalassemia carrier screening applied in Sardinia. Thal, thalassemia; S. blot, Southern blotting.

hematologic characteristic for identifying the β-thalassemia carrier state is the increase of HbA_2 levels (3.8%–6% in our laboratory). However, a variable but substantial number (in the different populations at risk) of β-thalassemia heterozygotes may have a normal level of HbA_2 (normal HbA_2 β-thalassemia). The first mechanism determining this phenotype is the presence of heterozygosity for a β-thalassemia mutation associated with a significant residual output of β-globin chains from the affected locus (mild β-thalassemia), such as the common Mediterranean β^+-thalassemia IVS nt 6 mutations (HBB c. 92 + 6 T → C) or some promoter mutation, i.e., the −92 C → T (HBB c. 142 C → T). However, normal HbA_2 β-thalassemia may also result from coinheritance with β-thalassemia, of δ-thalassemia mutations in *cis* or in *trans*, which determine a decrease in δ-chain production, thereby resulting in normalization of HbA_2 (normal HbA_2 β-thalassemia type 2) (Kattamis et al. 1979; Moi et al. 1988). Double heterozygotes for β- and δ-thalassemia may be diagnosed by family studies, globin chain synthesis analysis, and/or α-, δ-, and β-gene analysis. Many different mutations have been so far defined. δ-Thalassemia is a relatively common genetic condition (Loudianos et al. 1992; Moi et al. 1992). The presence of a δ-chain variant may also be associated with normal HbA_2 levels (Waye et al. 2003; Phylipsen et al. 2010; So et al. 2011). This condition may be recognized from the presence of an additional band in the electrophoretic pattern or a peak on the chromatogram in the region where HbA_2 also migrates. In β-thalassemia carriers, the sum of normal plus variant HbA_2 is, however, increased. Normal HbA_2 β-thalassemia heterozygotes should be differentiated from α-thalassemia carriers who show the same hematological features. The rare εγδβ-, γδβ-, and δβ-thalassemias are also characterized by microcytosis and normal HbA_2 levels. δβ-Thalassemia may be easily defined by a marked increase of HbF. Associated iron deficiency may result in decreased levels of HbA_2. This effect, however, is only observed in the presence of severe anemia (Galanello et al. 1981). Iron studies can easily exclude this association.

Finally, a critical problem in β-thalassemia carrier screening is the identification of silent β-thalassemia or the triple-quadruple α-gene arrangement, which by interacting with typical β-thalassemia may result in the clinical features of mild or severe β-thalassemia (thalassemia intermedia or major) (Galanello et al. 1983; Thein et al. 1984; Kulozik et al. 1987; Ristaldi et al. 1990; Harteveld et al. 2008; Sollaino et al. 2009; Faa et al. 2010). Silent β-thalassemia (normal HbA_2 β-thalassemia type 1) shows normal hematological features and HbA_2 level and may be identified solely by globin chain synthesis analysis or β-globin gene analysis (Kattamis et al. 1979; Gonzalez-Redondo et al. 1989; Galanello et al. 1994). The triple or quadruple α-globin gene arrangement, which also displays the phenotype of normal HbA_2 β-thalassemia type 1, may show a slight unbalance of $\alpha/\beta + \gamma$-globin chain synthesis but may also have a balanced synthesis (Harteveld et al. 2008; Faa et al. 2010). The presence of these atypical carriers is generally suspected in families that already had a child affected by mild β-thalassemia, despite having one apparently normal parent, the other being a typical β-thalassemia carrier.

The finding of individuals with borderline HbA_2 (level between 3.3% and 3.8% in our laboratory) and normal MCV and MCH creates a relevant problem in the process of thalassemia carrier identification. Known determinants causing this phenotype are the triple α-globin gene arrangement, some β-promoter mutation (HBB c. 151 C $\rightarrow$ T), and rare HBB or HBD gene variant (Table 1) (Galanello et al. 1983, 1994; Gasperini et al. 1993). Recently, we performed in the Sardinian population molecular analysis of the KLF1 gene (which codes for a zinc-finger transcriptional factor) in a large group of individuals with borderline HbA_2. KLF1 is able to activate in the perinatal period the β-globin gene and at the same time the BCL11A gene (Borg et al. 2010; Zhou et al. 2010), which is capable in turn to suppress the γ-chain production (Borg et al. 2010; Zhou et al. 2010). KLF1 gene mutation has been associated with many different phenotypes including hereditary persistence of fetal hemoglobin (HPFH), congenital dyserythropoietic anemia, and the (Lu) blood group phenotype (Singleton et al. 2008; Arnaud et al. 2010; Borg et al. 2010; Satta et al. 2011).

Table 1. Causes of borderline HbA2 levels

Some promoter mutation of the β-globin gene
Some silent β-thalassemia
Some HBB or HBD variants
Triple or quadruple α-globin arrangement
KLF1 mutations

In our study, we show that a consistent proportion of individuals (35%) with borderline HbA_2 have pathogenic mutation in KLF1, resulting in the activation of the HbD gene and thereby in increased HbA_2 levels (Perseu et al. 2011). This finding may facilitate the carrier detection and counseling thereby contributing to the success of the program.

In our opinion, a carrier detection procedure should be designed to avoid missing any couple at risk. For this reason we have included in the first group of examination, in addition to determination of MCV–MCH, the quantitative evaluation of HbA_2, which may be obtained by electrophoresis or by high-pressure liquid chromatography (HPLC). HPLC has the additional advantage to quantitate also HbF and to detect clinical relevant Hb variants including Hb Knossos (a mild β-thalassemia allele), HbS, HbC, HbD Punjab, HbO Arab, and HbE, all of which may interact with β-thalassemia heterozygosity leading to thalassemia major or intermedia or sickle cell β-thalassemia. By this approach only silent β-thalassemia or the triple or quadruple α-globin gene arrangement may escape detection.

The critical issue in the screening process is the differential diagnosis of individuals with the phenotype of normal HbA_2 β-thalassemia type 2 (low MCV, low MCH, normal, or borderline HbA_2), which may result from iron deficiency, α-thalassemia, γδβ-thalassemia, double heterozygosity for β- and δ-thalassemia or mild β-thalassemia. After exclusion of iron deficiency by erythrocyte ZnPP and transferrin saturation determination, we differentiate the different thalassemia genotypes by α-/β-globin chain synthesis or α-, δ-, and β-gene analysis.

A hematological phenotype characterized by low MCV and MCH, normal or reduced HbA_2, and high HbF levels indicate the presence of δβ-thalassemia. δβ-Thalassemia may be confused with HPFH. The differentiation is based on globin gene synthesis analysis or β-gene cluster studies; in addition, HPFH is characterized by lack of microcytosis and hypochromia.

MOLECULAR DIAGNOSIS

Molecular diagnosis of α-, β-, and δ-thalassemia when necessary may be performed by one of the PCR-based available procedures. β-Thalassemias display a marked heterogeneity at the molecular level. So far, >200 different mutations have been described. A complete updated list is available at the Globin Gene Server Web site—http://globin.cse.psu.edu. The large majority of these mutations are point mutation or the addition or deletion of one or a few nucleotides causing a frameshift. More rarely, β-thalassemias are produced by globin gene deletion. Despite this heterogeneity a limited number of molecular defects are present in each population at risk. This information may be useful for designing the appropriate primers or probes for each at-risk population (Fig. 3). Nowadays, known mutations are detected by a PCR-based procedure among which the most commonly used are commercially available RDB (reverse oligonucleotide hybridization) analysis (Saiki et al. 1989) and allele-specific amplification (amplification refractory mutation system [ARMS]) (Newton et al. 1989). Both methods, especially the RDB, can screen in one step up to 20–30 different mutations. Methodologies currently used for identification of unknown mutations by scanning the entire gene were denaturing gradient gel electrophoresis (DDGE) (Myers et al. 1985; Cai and Kan 1990; Rosatelli et al. 1992), denaturing high-pressure liquid chromatography (dHPLC) (Sanger et al. 1977), and single-strand conformation polymorphism (SSCP) analysis (Orita et al. 1989). When mutation is not detected by these procedures, polyacrylamide gel electrophoresis of PCR-amplified product (gapPCR) may lead to the detection of a small deletion of the β-globin gene, which may be suspected by finding unusually high HbA_2. The presence of rare large deletions, including deletion of LCR, is finally investigated by multiple ligation-dependent probe amplification (MLPA). These methods are now usually replaced by DNA sequence analysis.

Detection of the most frequent mutations/deletions of the α-globin gene, as for the β-globin gene detection, is also available using the PCR–RDB-based kits. Known deletions may be detected by PCR-based methods using primers flanking the deletion breakpoint. Primer panels targeted to the most common deletion found in the area of the geographic origin of the proband can be used (Galanello et al. 1998; Chong et al. 2000; Old et al. 2001). MLPA (Harteveld et al. 2005) or quantitative real-time PCR assay may also be used to detect less common mutations or novel deletion (Fallah et al. 2010). α-Thalassemia mental retardation 16 (ATR16), a contiguous gene deletion, may also be detected by MLPA or array comparative genome hybridization (CGH) (Gibson et al. 2008).

PREDICTION OF A MILD PHENOTYPE

The severity of homozygous β-thalassemia depends on the extent of the imbalance between α-globin and non-α-globin (including β- and γ-chains) chains. Any conditions able to reduce this unbalance results in a mild clinical phenotype referred to as thalassemia intermedia. The first molecular mechanism is related to severity of the β-thalassemia mutation present in each case (Higgs et al. 2011). Mild β^+-thalassemia (which also includes HbE) and silent mutations are associated with adequate residual output of

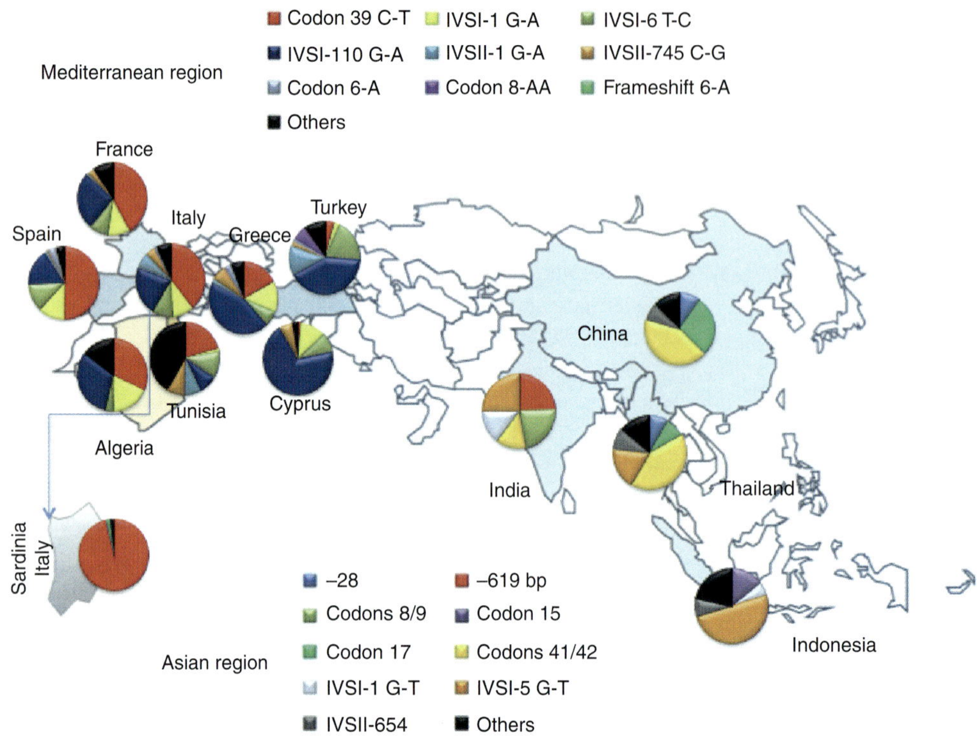

Figure 3. Population distribution of common β-thalassemia mutations.

the β-globin gene from the affected β-globin locus and thereby usually result in attenuated phenotypes in the homozygosity or compound heterozygous state. In contrast, compound heterozygotes for a mild β^+-thalassemia or silent mutation and a severe mutation produce a variable phenotype ranging from thalassemia major to thalassemia intermedia. Therefore, the presence of this genotype does not consent to consistently predict a mild phenotype.

The second recognized mechanism for producing a mild phenotype is the coinheritance with homozygous β-thalassemia of α-thalassemia, which, by reducing the output of α-chains, decreases the α-/non-α-chain imbalance (Wainscoat et al. 1983; Galanello et al. 1989). A consistent ameliorating effect is observed when the α-thalassemia determinant consists of the deletion of two α-globin structural genes or of an inactivating mutation of the major α2-globin gene (Wainscoat et al. 1983; Galanello et al. 1989; Thein 2008). However, because this effect is not constantly seen, the coinheritance of α-thalassemia cannot be recommended to predict the phenotype.

Finally, the coinheritance of genetic determinants able to sustain a continuous high production of γ-globin chains in adult life, may ameliorate homozygous β-thalassemia by reducing the α-/non-α-chain imbalance.

Genetic conditions associated with increased HbF in adult life and therefore able to ameriolate the phenotype of β-thalassemia are listed in Table 2. Recent studies using genome-wide association studies (GWAS) have identified two quantitative trait loci (QTL) (BCL11A on chromosome 2p16 and HBS1L-MYB intergenic region most likely involving MYB on chromosome 6q23) that account for ~20%–30% of the common variation in HbF levels in healthy adults as well as in persons with β-thalassemia and sickle cell disease (Uda et al. 2008; Thein and Menzel 2009). In Sardinians, the alleles of BCL11A associated with high HbF were found

Table 2. Genetic conditions with increased HbF in adult life

Linked to the β-cluster
HPFH
δβ-Thalassemias
Deletions of the 5′ region of the HBB promoter
Point mutation of the Gγ (HBG2) (−158 C → T Gγ mutation)
Point mutation of the Aγ (HBG1) (−196 G → T Aγ mutation)
QTL 11q (HS3 locus); locus downstream from HBG1 (Galarneau et al. 2010)
Unlinked to the β-cluster
BCL11A
HBSA1L-MYB region
Others (chromosome 13)

HPFH, hereditary persistence of fetal hemoglobin.

more frequently in patients with thalassemia intermedia as compared to those affected by thalassemia major, indicating that those loci exert, by increasing HbF, an ameliorating effect on homozygous β^0-thalassemia (Uda et al. 2008; Galanello et al. 2009). These results were confirmed in patients with sickle cell anemia from the United States and Brazil (Lettre et al. 2008) and β-thalassemias in France (Badens et al. 2011). The two loci, in addition to coinherited α-thalassemia, act in additive fashion, accounting for ~75% of the variation in clinical severity in Sardinian β^0-thalassemias (Galanello et al. 2009). Genotyping at these loci in prospective parents may have a practical utility as a prognostic indicator for the severity of β-thalassemia and sickle cell anemia, thereby improving the genetic counseling. It is likely that many other HbF-associated QTL exist such as that recently defined on chromosome 13 (Sankaran and Nathan 2010).

GENETIC COUNSELING

Limited information is available in the literature concerning the approach for counseling by thalassemia centers. A working party several years ago discussed the problem and presented some recommendations (WHO 1994). I would like therefore to concentrate on our own experience in Sardinia. According to international rules, counseling has been performed in a nondirective way and was based on a private interview. A pediatrician trained in genetics is responsible for counseling the couples identified to be at risk, whereas a trained nurse deals with individuals not yet married. The information provided is aimed at giving an informed basis on which to make a decision about reproduction. The discussion includes various options available such as birth control, mate selection, adoption, fetal testing including prenatal diagnosis and preimplantation diagnosis, or artificial insemination by normal donors as well as the details of the natural history of the disease. When the genotypes indicate the possible development of a mild phenotype, we also prudently discuss this point, emphasizing that the development of this phenotype should be considered a probability and not a certainty.

I strongly believe that the counseling according to international guidelines should be based on truth, complete honesty, and frankness. Other experts, before the recent discovery on HbF QTL, were of the opinion that this phenotype prediction should not be discussed (Weatherall and Clegg 2001).

In cases with a previous nonaffected child, we propose HLA typing on fetal DNA, to check whether the sibling could be HLA identical and therefore a suitable bone marrow donor (Orofino et al. 2003). This information could be useful for deciding on noninterruption of the pregnancy in the case of an affected fetus. In Sardinia, as well in other Mediterranean countries (Cyprus, Continental Italy, and Greece) and from recently published papers also in Muslim countries such as Iran, the option of prenatal diagnosis is followed by the majority of the couples counseled. Last, we inform the counselee about the risk for their relatives and recommend informing them about the result. This inductive screening is a very efficient way to multiply the efficacy of the screening as reported also in Pakistani populations (Ahmed et al. 2002).

PRENATAL DIAGNOSIS

The possibility of prenatal diagnosis of inherited hemoglobinopathies was first suggested by the discovery that β-globin chain synthesis in cord blood of β-thalassemia heterozygotes was significantly lower than normal (Kan and Nathan 1968) and that adult hemoglobin synthesis could be detected in fetuses at midtrimester (Hollenberg et al. 1971). Later on, the β^S-chain was detected in the blood of fetuses (Kan et al. 1972). Fetal blood sampling by fetoscopy or placentocentesis was then set up (Hobbins and Mahoney 1974; Kan et al. 1974) and paved the way to introduce prenatal diagnosis by globin chain synthesis on fetal blood (Kan et al. 1975). These techniques were used for 5 years. The first molecular diagnosis was performed for the diagnosis of α-thalassemia by the technique of molecular hybridization (Kan et al. 1976). For a short period diagnosis of β-thalassemia by DNA analysis was obtained indirectly by analysis of DNA polymorphisms linked to the β-globin gene (Kan et al. 1980) or directly by mutation-specific oligonucleotide hybridization on electrophoretically separated DNA fragments (Pirastu et al. 1983).

Nowadays, thalassemias are detected directly by the analysis of amplified DNA from fetal trophoblast or amniotic fluid cells. Fetal DNA is obtained transabdominally by chorionic villus sampling. This technique has been chosen by many centers because of the advantage of being used within the first trimester and because it is associated with a low risk (~1%) of fetal mortality. Fetal DNA is analyzed with one of the methods described for the detection of known mutations in the process of carrier identification. Prenatal diagnosis by DNA analysis is today available in many at-risk populations, including those from Hong Kong, Taiwan, China, and India, in addition to those from the Mediterranean area, Northern Europe, and Northern and South America (Eleftheiou and Angastionitis 2011).

Misdiagnosis may occur for several reasons: i.e., failure to amplify the DNA fragment, mispaternity, maternal contamination, and sample exchange. To limit the possibility of misdiagnosis, we analyze the fetal DNA with two different techniques (ARMS and RDB) and carry out DNA polymorphism analysis to detect maternal contamination and mispaternity. The PCR-based technique of prenatal diagnosis turned out to be very reliable. In our center we had no misdiagnoses in ~6000 cases performed so far. Similar results have been reported by others (Angastiniotis et al. 1995).

Fetal anemia may be diagnosed at mid-trimester gestation by quantitation of middle cerebral artery peak systolic velocity. This technique has been applied successfully to prenatal identification of Hb Bart's fetal hydrops syndrome (Lam and Tang 2002; Srisupundit et al. 2009). However, false negatives have been recorded using this technology. Mutation analysis of α-globin genes may be used to detect deletion and nondeletion α-thalassemia.

PREIMPLANTATION AND PRECONCEPTIONAL GENETIC DIAGNOSIS

The progress in molecular genetics to analyze the genotype of a single cell, together with advances in assisted reproduction techniques, has paved the way for developing preimplantation and preconceptional diagnosis (Handyside et al. 1990; Monk and Holding 1990). These techniques are nowadays widely available, including countries where β-thalassemia is prevalent such as Cyprus (Kuliev et al. 2011), Continental Greece (Zachaki et al. 2011), and China (Jiao et al. 2003; Xu et al. 2009). Preimplantation genetic diagnosis is performed either by biopsy of one to two blastomeres in eight-cell embryos after in vitro fertilization (by intracytoplasmatic sperm injection) or by biopsy of trophectoderm cells from blastocyst (Kuliev et al. 1998; Kokkali et al. 2007).

Preconceptional diagnosis is based on the analysis of the first polar body of unfertilized eggs followed by analysis of the second polar bodies after fertilization, which is performed to avoid misdiagnosis resulting from recombination during the first meiosis (Verlinsky et al. 1990). Diagnosis was obtained by multiple nested PCR analysis to detect the mutations as well as polymorphic alleles at the β-globin cluster (Kuliev et al. 2011; Zachaki et al. 2011). HLA typing of the embryo to select a nonaffected fetus HLA compatible with a previous affected sibling was recently proposed (Kuliev et al. 2011).

The biopsy of blastomeres from an eight-cell embryo is nowadays the most frequently used

 Cite this article as *Cold Spring Harb Perspect Med* doi: 10.1101/cshperspect.a011775

procedure worldwide. This approach minimizes the risk of misdiagnosis that could result from PCR failure, contamination, and allelic dropout that is a definite risk in single-cell analysis. In a large series, misdiagnosis was rare and the pregnancy rate high (32%) (Kuliev et al. 2011). The motivations for couples following counseling to opt for preimplantation genetic diagnosis (PGD) are problems of infertility, an unsuccessful reproductive history owing to repeated terminations of affected fetuses, moral or religious attitude against pregnancy termination, and the desire to have a nonaffected fetus HLA compatible with a previously homozygously affected sibling for future bone marrow transplantation. Genetic counseling for couples considering PGD should include a discussion on the risk associated with in vitro fertilization and embryo biopsy, the risk of failure and/or misdiagnosis, and the need of subsequent prenatal diagnosis for confirming the results. At present, however, the technical demand for these procedures, the difficulty in organizing the service, and the high cost preclude its use for monitoring worldwide pregnancies at risk.

PRENATAL DIAGNOSIS BY NONINVASIVE PROCEDURES

Several attempts have been made to make fetal diagnosis by analysis of fetal cells in maternal circulation (Simpson and Elias 1993; Adinolfi 1995; Cheung et al. 1996; Bianchi et al. 1997). However, the complexity of the cell isolation process and the lack of reproducibility preclude the use of this approach in clinical practice. Likewise, many studies have been performed to make the diagnosis of β-thalassemia or α-thalassemia by analysis of fetal DNA in maternal plasma (Lo et al. 1997; Lun et al. 2008). These noninvasive diagnostic methods, while gradually being introduced for chromosomal defects, require more refinement for globin and other single gene disorders. It should, however, be mentioned that a diagnosis of β-thalassemia was recently realized by the Lo group with next-generation sequencing (paired and massively, parallel sequencing) of the fetal and maternal DNA genomes present in maternal plasma (Lo et al. 2010).

OUTCOME OF THE PROGRAMS

The primary effect of the programs for preventing β-thalassemia in the Mediterranean populations has been the acquisition by the couples at risk of the appropriate knowledge to make informed decisions on the reproductive options available to them. This awareness of the genetic risk and methods of prevention led to a marked acceptance of fetal testing leading to a consistent decline in the birth rate of thalassemia major in

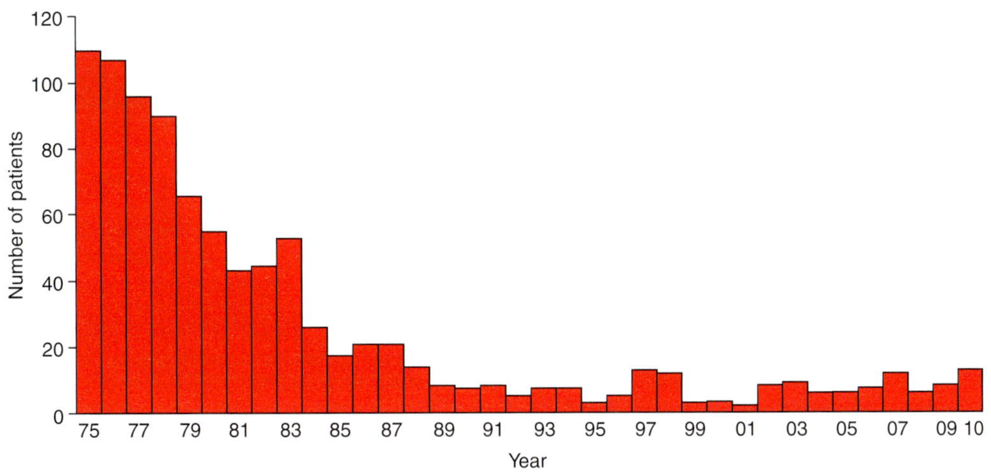

Figure 4. Decline of birth rate of thalassemia major in Sardinia.

these populations. In Sardinia, for instance, the number of thalassemia major children born per year predicted on the basis of the carrier rate, assuming a random mating, shows a reduction from 1:250 live births to 1:1660 in 2009 with an effective prevention of 85% of the cases (Fig. 4). Similar results have been reported in Cyprus (Angastiniotis and Modell 1998). The dissolution of the marriage or the decision not to have children is less commonly followed, except in Muslim countries, where the breakup of marriages seems to occur more often.

Despite well-organized prevention programs, reasons for the birth of affected children are negative attitudes against fetal testing following counseling and decision to continue the pregnancy despite the documented presence of an affected fetus for religious or ethical reasons. Less common motivations are misinformation, lack of information, mispaternity, and mistakes in fetal testing. Although research on this topic is not available, it seems that the programs had no adverse effects.

FUTURE PROSPECTIVES

Education on thalassemias may be implemented by using informatic channels. The continuous reduction in the cost of DNA analysis may lead to its application as the initial step in the screening process. Using microchip technology targeted to detect all the known α- and β-thalassemias as well as DNA polymorphisms is useful to predict the phenotype, thereby making it possible to skip the preliminary hematological analysis.

Technical advances in the methodologies of fetal DNA or fetal cells analysis in maternal plasma may allow in the future the introduction of this important noninvasive technique in clinical practice.

Finally, the most important challenge for the future is the organization of national preventive programs in populations in which thalassemias are prevalent such as the Middle East, the Indian subcontinent, and the Far East, where several factors previously discussed preclude at the present time the realization of such programs.

REFERENCES

Adinolfi M. 1995. Non- or minimally invasive prenatal diagnostic tests on maternal blood samples or transcervical cells. *Prenat Diagn* **15:** 889–896.

Ahmed S, Saleem M, Modell B, Petrou M. 2002. Screening extended families for genetic hemoglobin disorders in Pakistan. *N Engl J Med* **347:** 1162–1168.

Alwan A, Modell B. 2003. Recommendations for introducing genetics services in developing countries. *Nat Rev Genet* **4:** 61–68.

Amato A, Grisanti P, Lerone M, Ponzini D, Di Biagio P, Cappabianca MP, Giordano PC. 2009. Prevention strategies for severe hemoglobinopathies in endemic and non-endemic immigration countries: The Latium example. *Prenat Diagn* **29:** 1171–1174.

Angastiniotis MA, Hadjiminas MG. 1981. Prevention of thalassaemia in Cyprus. *Lancet* **1:** 369–371.

Angastiniotis M, Modell B. 1998. Global epidemiology of hemoglobin disorders. *Ann NY Acad Sci* **850:** 251–269.

Angastiniotis M, Modell B, Englezos P, Boulyjenkov V. 1995. Prevention and control of haemoglobinopathies. *Bull World Health Organ* **73:** 375–386.

Arnaud L, Saison C, Helias V, Lucien N, Steschenko D, Giarratana MC, Prehu C, Foliguet B, Montout L, de Brevern AG, et al. 2010. A dominant mutation in the gene encoding the erythroid transcription factor KLF1 causes a congenital dyserythropoietic anemia. *Am J Hum Genet* **87:** 721–727.

Badens C, Joly P, Agouti I, Thuret I, Gonnet K, Fattoum S, Francina A, Simeoni MC, Loundou A, Pissard S. 2011. Variants in genetic modifiers of β-thalassemia can help to predict the major or intermedia type of the disease. *Haematologica* **96:** 1712–1714.

Bianchi DW, Williams JM, Sullivan LM, Hanson FW, Klinger KW, Shuber AP. 1997. PCR quantitation of fetal cells in maternal blood in normal and aneuploid pregnancies. *Am J Hum Genet* **61:** 822–829.

Borg J, Papadopoulos P, Georgitsi M, Gutierrez L, Grech G, Fanis P, Phylactides M, Verkerk AJ, van der Spek PJ, Scerri CA, et al. 2010. Haploinsufficiency for the erythroid transcription factor KLF1 causes hereditary persistence of fetal hemoglobin. *Nat Genet* **42:** 801–805.

Cai SP, Kan YW. 1990. Identification of the multiple β-thalassemia mutations by denaturing gradient gel electrophoresis. *J Clin Invest* **85:** 550–553.

Canatan D, Kose MR, Ustundag M, Haznedaroglu D, Ozbas S. 2006. Hemoglobinopathy control program in Turkey. *Community Genet* **9:** 124–126.

Cao A, Rosatelli MC, Galanello R. 1996. Control of β-thalassaemia by carrier screening, genetic counselling and prenatal diagnosis: The Sardinian experience. *Ciba Found Symp* **197:** 137–151; discussion 151–135.

Cao A, Galanello R, Rosatelli MC. 1998. Prenatal diagnosis and screening of the haemoglobinopathies. *Baillieres Clin Haematol* **11:** 215–238.

Cao A, Congiu R, Sollaino MC, Desogus MF, Demartis FR, Loi D, Cau M, Galanello R. 2008. Thalassaemia and glucose-6-phosphate dehydrogenase screening in 13- to 14-year-old students of the Sardinian population: Preliminary findings. *Community Genet* **11:** 121–128.

Chern JP, Lin KH, Su YN, Lu MY, Jou ST, Lin DT, Wang SC, Lin KS. 2006. Impact of a national β-thalassemia carrier screening program on the birth rate of thalassemia major. *Pediatr Blood Cancer* **46:** 72–76.

Cheung MC, Goldberg JD, Kan YW. 1996. Prenatal diagnosis of sickle cell anaemia and thalassaemia by analysis of fetal cells in maternal blood. *Nat Genet* **14:** 264–268.

Chong SS, Boehm CD, Higgs DR, Cutting GR. 2000. Single-tube multiplex-PCR screen for common deletional determinants of α-thalassemia. *Blood* **95:** 360–362.

Chui DH. 2005. α-Thalassemia: Hb H disease and Hb Barts hydrops fetalis. *Ann NY Acad Sci* **1054:** 25–32.

Faa V, Masala M, Cao A, Rosatelli MC. 2010. α Globin gene duplications in β thalassemia patients with intact β globin gene. *Blood Cells Mol Dis* **44:** 156–158.

Fallah MS, Mahdian R, Aleyasin SA, Jamali S, Hayat-Nosaeid M, Karimipour M, Raeisi M, Zeinali S. 2010. Development of a quantitative real-time PCR assay for detection of unknown α-globin gene deletions. *Blood Cells Mol Dis* **45:** 58–64.

Fucharoen G, Sanchaisuriya K, Sae-ung N, Dangwibul S, Fucharoen S. 2004. A simplified screening strategy for thalassaemia and haemoglobin E in rural communities in south-east Asia. *Bull World Health Organ* **82:** 364–372.

Galanello R, Ruggeri R, Addis M, Paglietti E, Cao A. 1981. Hemoglobin A2 in iron deficient β-thalassemia heterozygotes. *Hemoglobin* **5:** 613–618.

Galanello R, Ruggeri R, Paglietti E, Addis M, Melis MA, Cao A. 1983. A family with segregating triplicated α globin loci and β thalassemia. *Blood* **62:** 1035–1040.

Galanello R, Dessi E, Melis MA, Addis M, Sanna MA, Rosatelli C, Argiolu F, Giagu N, Turco MP, Cacace E, et al. 1989. Molecular analysis of β^0-thalassemia intermedia in Sardinia. *Blood* **74:** 823–827.

Galanello R, Barella S, Ideo A, Gasperini D, Rosatelli C, Paderi L, Paglietti E, Sollaino C, Perseu L, Loi D, et al. 1994. Genotype of subjects with borderline hemoglobin A2 levels: Implication for β-thalassemia carrier screening. *Am J Hematol* **46:** 79–81.

Galanello R, Sollaino C, Paglietti E, Barella S, Perra C, Doneddu I, Pirroni MG, Maccioni L, Cao A. 1998. α-Thalassemia carrier identification by DNA analysis in the screening for thalassemia. *Am J Hematol* **59:** 273–278.

Galanello R, Sanna S, Perseu L, Sollaino MC, Satta S, Lai ME, Barella S, Uda M, Usala G, Abecasis GR, et al. 2009. Amelioration of Sardinian β^0 thalassemia by genetic modifiers. *Blood* **114:** 3935–3937.

Galarneau G, Palmer CD, Sankaran VG, Orkin SH, Hirschhorn JN, Lettre G. 2010. Fine-mapping at three loci known to affect fetal hemoglobin levels explains additional genetic variation. *Nat Genet* **42:** 1049–1051.

Gasperini D, Cao A, Paderi L, Barella S, Paglietti E, Perseu L, Loi D, Galanello R. 1993. Normal individuals with high Hb A2 levels. *Br J Haematol* **84:** 166–168.

Gibson WT, Harvard C, Qiao Y, Somerville MJ, Lewis ME, Rajcan-Separovic E. 2008. Phenotype-genotype characterization of α-thalassemia mental retardation syndrome due to isolated monosomy of 16p13.3. *Am J Med Genet A* **146A:** 225–232.

Gonzalez-Redondo JM, Stoming TA, Kutlar A, Kutlar F, Lanclos KD, Howard EF, Fei YJ, Aksoy M, Altay C, Gurgey A, et al. 1989. A C→T substitution at nt −101 in a conserved DNA sequence of the promotor region of the β-globin gene is associated with "silent" β-thalassemia. *Blood* **73:** 1705–1711.

Handyside AH, Kontogianni EH, Hardy K, Winston RM. 1990. Pregnancies from biopsied human preimplantation embryos sexed by Y-specific DNA amplification. *Nature* **344:** 768–770.

Harteveld CL, Voskamp A, Phylipsen M, Akkermans N, den Dunnen JT, White SJ, Giordano PC. 2005. Nine unknown rearrangements in 16p13.3 and 11p15.4 causing α- and β-thalassaemia characterised by high resolution multiplex ligation-dependent probe amplification. *J Med Genet* **42:** 922–931.

Harteveld CL, Refaldi C, Cassinerio E, Cappellini MD, Giordano PC. 2008. Segmental duplications involving the α-globin gene cluster are causing β-thalassemia intermedia phenotypes in β-thalassemia heterozygous patients. *Blood Cells Mol Dis* **40:** 312–316.

Higgs DR, Engel JD, Stamatoyannopoulos G. 2011. Thalassaemia. *Lancet* **379:** 373–383.

Hobbins JC, Mahoney MJ. 1974. In utero diagnosis of hemoglobinopathies. Technic for obtaining fetal blood. *N Engl J Med* **290:** 1065–1067.

Hollenberg MD, Kaback MM, Kazazian HH Jr, 1971. Adult hemoglobin synthesis by reticulocytes from the human fetus at midtrimester. *Science* **174:** 698–702.

Inati A, Zeineh N, Isma'eel H, Koussa S, Gharzuddine W, Taher A. 2006. β-thalassemia: The Lebanese experience. *Clin Lab Haematol* **28:** 217–227.

Jiao ZX, Zhuang GL, Zhou CQ, Shu YM, Li J, Liang XY, Zhang MF, Deng MF. 2003. Successful preimplantation genetic diagnosis for β-thalassemia using primer extension preamplification. *Zhonghua Fu Chan Ke Za Zhi* **38:** 143–146.

Kalokairinou EM. 2007. The experience of β-thalassaemia and its prevention in Cyprus. *Med Law* **26:** 291–307.

Kalokairinou EM. 2008. The experience of β-thalassaemia and its prevention in Cyprus. *Med Law* **27:** 825–841.

Kan YW, Nathan DG. 1968. β Thalassemia trait: Detection at birth. *Science* **161:** 589–590.

Kan YW, Dozy AM, Alter BP, Frigoletto FD, Nathan DG. 1972. Detection of the sickle gene in the human fetus. Potential for intrauterine diagnosis of sickle-cell anemia. *N Engl J Med* **287:** 1–5.

Kan YW, Valenti C, Carnazza V, Guidotti R, Rieder RF. 1974. Fetal blood-sampling in utero. *Lancet* **1:** 79–80.

Kan YW, Golbus MS, Klein P, Dozy AM. 1975. Successful application of prenatal diagnosis in a pregnancy at risk for homozygous β-thalassemia. *N Engl J Med* **292:** 1096–1099.

Kan YW, Golbus MS, Dozy AM. 1976. Prenatal diagnosis of α-thalassemia. Clinical application of molecular hybridization. *N Engl J Med* **295:** 1165–1167.

Kan YW, Lee KY, Furbetta M, Angius A, Cao A. 1980. Polymorphism of DNA sequence in the β-globin gene region. Application to prenatal diagnosis of β^0 thalassemia in Sardinia. *N Engl J Med* **302:** 185–188.

Kattamis C, Metaxotou-Mavromati A, Wood WG, Nash JR, Weatherall DJ. 1979. The heterogeneity of normal Hb A2-β thalassaemia in Greece. *Br J Haematol* **42:** 109–123.

Kokkali G, Traeger-Synodinos J, Vrettou C, Stavrou D, Jones GM, Cram DS, Makrakis E, Trounson AO, Kanavakis E, Pantos K. 2007. Blastocyst biopsy versus cleavage stage biopsy and blastocyst transfer for preimplantation genetic diagnosis of β-thalassaemia: A pilot study. *Hum Reprod* **22:** 1443–1449.

Kolnagou A, Kontoghiorghes GJ. 2009. Advances in the prevention and treatment are changing thalassemia from a fatal to a chronic disease. Experience from a Cyprus model and its use as a paradigm for future applications. *Hemoglobin* **33:** 287–295.

Kuliev A, Rechitsky S, Verlinsky O, Ivakhnenko V, Evsikov S, Wolf G, Angastiniotis M, Georghiou D, Kukharenko V, Strom C, et al. 1998. Preimplantation diagnosis of thalassemias. *J Assist Reprod Genet* **15:** 219–225.

Kuliev A, Pakhalchuk T, Verlinsky O, Rechitsky S. 2011. Preimplantation genetic diagnosis for hemoglobinopathies. *Hemoglobin* **31:** 273–277.

Kulozik AE, Thein SL, Wainscoat JS, Gale R, Kay LA, Wood JK, Weatherall DJ, Huehns ER. 1987. Thalassaemia intermedia: Interaction of the triple α-globin gene arrangement and heterozygous β-thalassaemia. *Br J Haematol* **66:** 109–112.

Lam YH, Tang MH. 2002. Middle cerebral artery Doppler study in fetuses with homozygous α-thalassaemia-1 at 12–13 weeks of gestation. *Prenat Diagn* **22:** 56–58.

Lettre G, Sankaran VG, Bezerra MA, Araujo AS, Uda M, Sanna S, Cao A, Schlessinger D, Costa FF, Hirschhorn JN, et al. 2008. DNA polymorphisms at the BCL11A, HBS1L-MYB, and β-globin loci associate with fetal hemoglobin levels and pain crises in sickle cell disease. *Proc Natl Acad Sci* **105:** 11869–11874.

Li DZ. 2009. Premarital screening for thalassemia in mainland China. *Prenat Diagn* **29:** 637–638.

Lo YM, Corbetta N, Chamberlain PF, Rai V, Sargent IL, Redman CW, Wainscoat JS. 1997. Presence of fetal DNA in maternal plasma and serum. *Lancet* **350:** 485–487.

Lo YM, Chan KC, Sun H, Chen EZ, Jiang P, Lun FM, Zheng YW, Leung TY, Lau TK, Cantor CR, et al. 2010. Maternal plasma DNA sequencing reveals the genome-wide genetic and mutational profile of the fetus. *Sci Transl Med* **2:** 61ra91.

Loudianos G, Murru S, Ristaldi MS, Cossu P, Pilia G, Porcu S, Sciarratta GV, Parodi MI, Cao A, Pirastu M. 1992. A novel δ-thalassemia mutation A G→C substitution at codon 30 of the δ-globin gene in a person of southern Italian origin. *Hum Mutat* **1:** 169–171.

Loukopoulos D. 1996. Current status of thalassemia and the sickle cell syndromes in Greece. *Semin Hematol* **33:** 76–86.

Lun FM, Tsui NB, Chan KC, Leung TY, Lau TK, Charoenkwan P, Chow KC, Lo WY, Wanapirak C, Sanguansermsri T, et al. 2008. Noninvasive prenatal diagnosis of monogenic diseases by digital size selection and relative mutation dosage on DNA in maternal plasma. *Proc Natl Acad Sci* **105:** 19920–19925.

Melis MA, Galanello R, Cao A. 1983. α Globin gene analysis in a Sardinian family with interacting α and β thalassaemia genes. *Br J Haematol* **53:** 667–671.

Modell B, Darlison M. 2008. Global epidemiology of haemoglobin disorders and derived service indicators. *Bull World Health Organ* **86:** 480–487.

Modell B, Darlison M, Birgens H, Cario H, Faustino P, Giordano PC, Gulbis B, Hopmeier P, Lena-Russo D, Romao L, et al. 2007. Epidemiology of haemoglobin disorders in Europe: An overview. *Scand J Clin Lab Invest* **67:** 39–69.

Moi P, Paglietti E, Sanna A, Brancati C, Tagarelli A, Galanello R, Cao A, Pirastu M. 1988. Delineation of the molecular basis of δ- and normal HbA2 β-thalassemia. *Blood* **72:** 530–533.

Moi P, Loudianos G, Lavinha J, Murru S, Cossu P, Casu R, Oggiano L, Longinotti M, Cao A, Pirastu M. 1992. δ-Thalassemia due to a mutation in an erythroid-specific binding protein sequence 3′ to the δ-globin gene. *Blood* **79:** 512–516.

Monk M, Holding C. 1990. Amplification of a β-haemoglobin sequence in individual human oocytes and polar bodies. *Lancet* **335:** 985–988.

Myers RM, Fischer SG, Lerman LS, Maniatis T. 1985. Nearly all single base substitutions in DNA fragments joined to a GC-clamp can be detected by denaturing gradient gel electrophoresis. *Nucleic Acids Res* **13:** 3131–3145.

Newton CR, Graham A, Heptinstall LE, Powell SJ, Summers C, Kalsheker N, Smith JC, Markham AF. 1989. Analysis of any point mutation in DNA. The amplification refractory mutation system (ARMS). *Nucleic Acids Res* **17:** 2503–2516.

Old JM, Khan SN, Verma I, Fucharoen S, Kleanthous M, Ioannou P, Kotea N, Fisher C, Riazuddin S, Saxena R, et al. 2001. A multi-center study in order to further define the molecular basis of β-thalassemia in Thailand, Pakistan, Sri Lanka, Mauritius, Syria, and India, and to develop a simple molecular diagnostic strategy by amplification refractory mutation system-polymerase chain reaction. *Hemoglobin* **25:** 397–407.

Orita M, Iwahana H, Kanazawa H, Hayashi K, Sekiya T. 1989. Detection of polymorphisms of human DNA by gel electrophoresis as single-strand conformation polymorphisms. *Proc Natl Acad Sci* **86:** 2766–2770.

Orofino MG, Argiolu F, Sanna MA, Rosatelli MC, Tuveri T, Scalas MT, Badiali M, Cossu P, Puddu R, Lai ME, et al. 2003. Fetal HLA typing in β thalassaemia: Implications for haemopoietic stem-cell transplantation. *Lancet* **362:** 41–42.

Perseu L, Satta S, Moi P, Demartis FR, Manunza L, Sollaino MC, Barella S, Cao A, Galanello R. 2011. KLF1 gene mutations cause borderline HbA2. *Blood* **118:** 4454–4458.

Phylipsen M, Harteveld CL, de Metz M, Gallivan MV, Arkesteijn SG, Luo HY, Chui DH, Giordano PC. 2010. New and known β-thalassemia determinants masked by known and new δ gene defects [Hb A(2)-Ramallah or delta6(A3)Glu→Gln, GAG≫CAG]. *Hemoglobin* **34:** 445–450.

Pirastu M, Kan YW, Cao A, Conner BJ, Teplitz RL, Wallace RB. 1983. Prenatal diagnosis of β-thalassemia. Detection of a single nucleotide mutation in DNA. *N Engl J Med* **309:** 284–287.

Premawardhena A, De Silva S, Arambepola M, Olivieri N, Merson L, Muraco J, Allen A, Fisher C, Peto T,

Vichinsky E, et al. 2004. Thalassemia in Sri Lanka: A progress report. *Hum Mol Genet* **13:** R203–R206.

Ristaldi MS, Murru S, Loudianos G, Casula L, Porcu S, Pigheddu D, Fanni B, Sciarratta GV, Agosti S, Parodi MI, et al. 1990. The C-T substitution in the distal CACCC box of the β-globin gene promoter is a common cause of silent β thalassaemia in the Italian population. *Br J Haematol* **74:** 480–486.

Rosatelli C, Falchi AM, Scalas MT, Tuveri T, Furbetta M, Cao A. 1984. Hematological phenotype of the double heterozygous state for α and β thalassemia. *Hemoglobin* **8:** 25–35.

Rosatelli MC, Dozy A, Faa V, Meloni A, Sardu R, Saba L, Kan YW, Cao A. 1992. Molecular characterization of β-thalassemia in the Sardinian population. *Am J Hum Genet* **50:** 422–426.

Saiki RK, Walsh PS, Levenson CH, Erlich HA. 1989. Genetic analysis of amplified DNA with immobilized sequence-specific oligonucleotide probes. *Proc Natl Acad Sci* **86:** 6230–6234.

Samavat A, Modell B. 2004. Iranian national thalassaemia screening programme. *BMJ* **329:** 1134–1137.

Sanger F, Nicklen S, Coulson AR. 1977. DNA sequencing with chain-terminating inhibitors. *Proc Natl Acad Sci* **74:** 5463–5467.

Sankaran VG, Nathan DG. 2010. Reversing the hemoglobin switch. *N Engl J Med* **363:** 2258–2260.

Satta S, Perseu L, Moi P, Asunis I, Cabriolu A, Maccioni L, Demartis FR, Manunza L, Cao A, Galanello R. 2011. Compound heterozygosity for KLF1 mutations associated with remarkable increase of fetal hemoglobin and red cell protoporphyrin. *Haematologica* **96:** 767–770.

Simpson JL, Elias S. 1993. Isolating fetal cells from maternal blood. Advances in prenatal diagnosis through molecular technology. *JAMA* **270:** 2357–2361.

Singleton BK, Burton NM, Green C, Brady RL, Anstee DJ. 2008. Mutations in EKLF/KLF1 form the molecular basis of the rare blood group In(Lu) phenotype. *Blood* **112:** 2081–2088.

So CC, Chan AY, Luo HY, Verhovsek M, Chui DH, Ling SC, Chan LC. 2011. Hb A2 Hong Kong—A novel δ-globin variant in a Chinese family masks the diagnosis of β-thalassemia trait. *Hemoglobin* **35:** 162–165.

Sollaino MC, Paglietti ME, Perseu L, Giagu N, Loi D, Galanello R. 2009. Association of α globin gene quadruplication and heterozygous β thalassemia in patients with thalassemia intermedia. *Haematologica* **94:** 1445–1448.

Srisupundit K, Piyamongkol W, Tongsong T. 2009. Identification of fetuses with hemoglobin Bart's disease using middle cerebral artery peak systolic velocity. *Ultrasound Obstet Gynecol* **33:** 694–697.

Thein SL. 2008. Genetic modifiers of the β-haemoglobinopathies. *Br J Haematol* **141:** 357–366.

Thein SL, Menzel S. 2009. Discovering the genetics underlying foetal haemoglobin production in adults. *Br J Haematol* **145:** 455–467.

Thein SL, Al-Hakim I, Hoffbrand AV. 1984. Thalassaemia intermedia: A new molecular basis. *Br J Haematol* **56:** 333–337.

Uda M, Galanello R, Sanna S, Lettre G, Sankaran VG, Chen W, Usala G, Busonero F, Maschio A, Albai G, et al. 2008. Genome-wide association study shows BCL11A associated with persistent fetal hemoglobin and amelioration of the phenotype of β-thalassemia. *Proc Natl Acad Sci* **105:** 1620–1625.

Verlinsky Y, Ginsberg N, Lifchez A, Valle J, Moise J, Strom CM. 1990. Analysis of the first polar body: Preconception genetic diagnosis. *Hum Reprod* **5:** 826–829.

Wainscoat JS, Kanavakis E, Wood WG, Letsky EA, Huehns ER, Marsh GW, Higgs DR, Clegg JB, Weatherall DJ. 1983. Thalassaemia intermedia in Cyprus: The interaction of α and β thalassaemia. *Br J Haematol* **53:** 411–416.

Waye JS, Patterson M, Walker L, Eng B, Nakamura LM, Lafferty JD, Yong SL, Wu JK, Chui DH. 2003. β-Thalassemia in association with a new δ-chain hemoglobin variant [δ116(g18)Arg→Leu]: Implications for carrier screening and prenatal diagnosis. *Am J Hematol* **74:** 179–181.

Weatherall DJ. 2010a. The inherited diseases of hemoglobin are an emerging global health burden. *Blood* **115:** 4331–4336.

Weatherall DJ. 2010b. Thalassemia as a global health problem: Recent progress toward its control in the developing countries. *Ann NY Acad Sci* **1202:** 17–23.

Weatherall DJ, Clegg JB. 2001. *The thalassaemia syndromes.* Blackwell Science, Oxford.

WHO. 1994. Guidelines for control of haemoglobin disorders. In *WHO/HDP/HB/GL/941.* WHO, Geneva.

Xu YW, Zeng YH, Deng J, Liu Y, Gao L, Zhou CQ, Zhuang GL. 2009. Preimplantation genetic diagnosis for α-thalassaemia in China. *J Assist Reprod Genet* **26:** 399–403.

Zachaki S, Vrettou C, Destouni A, Kokkali G, Traeger-Synodinos J, Kanavakis E. 2011. Novel and known microsatellite markers within the β-globin cluster to support robust preimplantation genetic diagnosis of β-thalassemia and sickle cell syndromes. *Hemoglobin* **35:** 56–66.

Zhou D, Liu K, Sun CW, Pawlik KM, Townes TM. 2010. KLF1 regulates BCL11A expression and γ- to β-globin gene switching. *Nat Genet* **42:** 742–744.

The Natural History of Sickle Cell Disease

Graham R. Serjeant

Sickle Cell Trust (Jamaica), Kingston 6, Jamaica

Correspondence: grserjeant@cwjamaica.com

The term sickle cell disease embraces a group of genetic conditions in which pathology results from the inheritance of the sickle cell gene either homozygously or as a double heterozygote with another interacting gene. The spectrum of resulting conditions is therefore influenced by the geography of individual hemoglobin genes, but in most populations, the commonest genotype at birth is homozygous sickle cell (SS) disease. Because this genotype generally manifests a greater mortality, the relative proportion of sickle cell genotypes is influenced by age as well as the geographical distribution of individual genes.

Inheritance of the abnormal sickle cell gene from one parent and a gene for normal HbA from the other parent results in the sickle cell trait. Persons with the sickle cell trait have a relative resistance to falciparum malaria, are less likely to get the disease, run lower parasite counts, and are less likely to die. This survival advantage is most marked during a window in early childhood between the loss of passively acquired maternal immunity and the development of active immunity, but the length and timing of this window may vary between communities and be influenced by the pattern of malarial transmission. The survival advantage relative to normal individuals with HbA has contributed to high frequencies of the sickle cell trait in areas with a history of malaria. The prevalence of the sickle cell trait varies markedly between different regions but reaches levels as high as 40% in some areas of sub-Saharan Africa, eastern Saudi Arabia, and central India. Persons with the sickle cell trait have both HbA and HbS, with HbS levels between 20% and 45%, which are largely genetically determined. The levels of HbS in the sickle cell trait are not enough under normal physiological conditions to cause the problems relating to sickling except in the renal medulla where vascular damage causes an inability to concentrate the urine and a tendency to painless hematuria. Other features possibly associated with the trait include pulmonary embolism and an increased risk of sudden death under unusual physiological conditions, but these events are rare and the sickle cell trait, under normal conditions, is considered generally harmless and must remain outside the definition of sickle cell disease.

GEOGRAPHY OF SICKLE CELL DISEASE

Populations of African Origin

The sickle cell trait is widespread throughout Africa with low frequencies (<1%–2%) in the north and south of the continent and high but variable frequencies throughout much of

Copyright © 2013 Cold Spring Harbor Laboratory Press; all rights reserved
Cite this article as *Cold Spring Harb Perspect Med* doi: 10.1101/cshperspect.a011783

equatorial Africa. Characterization of the DNA structure flanking the β-globin locus of HbS suggests that the mutation has arisen on at least three independent occasions in the African continent, referred to as β-globin haplotypes and named after the areas where they were first described: Benin, Senegal, and Central African Republic or Bantu (Pagnier et al. 1984; Nagel et al 1985; Chebloune et al. 1988). The HbC trait is believed to be a relatively recent mutation limited to West Africa where it occurs at high frequencies (>20%) in central Ghana and Burkina Faso, in only 2% in Nigeria, and does not occur, except in peoples of West African origin, in East and Central Africa. Only limited data are available on the type and distribution of α- and β-thalassemia genes in the African continent.

The most comprehensive data on populations of West African origin derive from the Americas and particularly from the United States and from the Caribbean. Screening of 100,000 consecutive nonoperative deliveries in the Jamaica Cohort Study found the sickle cell trait in 10.0% and the HbC trait in 3.6% (Serjeant et al. 1986) and the relative frequencies among Afro-Americans were 8% and 2%. In South America, these genes are largely confined to Brazil where they may occur in up to 7% of peoples of African origin.

The β-thalassemia trait derives from many different molecular mutations but their clinical significance depends on the degree of β-chain synthesis and, hence, HbA. Most common among Jamaicans are the mild β^+-thalassemia mutations, which occur in about 1%, and are usually owing to mutations in the promoter region (−88 C > T, −29 A > G). The more severe β^0-thalassemia mutations occur in ~0.5%, and are usually owing to mutations in the intervening sequence (IVSII-849 A > G). When inherited with the HbS gene, sickle cell-β^+-thalassemia generally results in 15%–25% HbA and a mild clinical course, whereas sickle cell-β^0-thalassemia has no HbA and is often clinically indistinguishable from homozygous sickle cell (SS) disease.

The frequencies of the genotypes of sickle cell disease at birth in Jamaica are one in 300 for SS disease, one in 500 for sickle cell-hemoglobin C (SC) disease, one in 3000 for sickle cell-β^+-thalassemia, and one in 7000 for sickle cell-β^0-thalassemia; slightly lower figures occur in the United States. Rare forms of sickle cell disease may result from the interaction of HbS with HbD Punjab, HbO Arab, and Hb Lepore.

Populations in India and the Arabian Gulf

In Saudi Arabia, the sickle cell trait is widespread but reaches its highest prevalence in the eastern Province. In India, the trait occurs most commonly among the tribal peoples in central India (southeastern Gujarat, Maharastra, Madhya Pradesh, Chhattisgarh, western Odisha) with a smaller focus in the south of the country (northern Tamil Nadu and Kerala), and trait frequencies as high as 40% have been described in some groups. In these areas, the DNA structure flanking the β-globin locus differs from that in African peoples, suggesting that this is a fourth independent occurrence of the HbS mutation and is referred to as the Asian haplotype. This haplotype is often associated with high levels of fetal hemoglobin and frequent α-thalassemia, both of which tend to inhibit sickling and ameliorate the clinical course. Sickle cell disease in the Arabian Gulf and India is predominantly SS disease.

COMPARISON OF GENOTYPES

SS disease and sickle cell-β^0-thalassemia are most severe clinically and hematologically although both genotypes may vary markedly in severity. SC disease is generally mild but more prone to proliferative sickle retinopathy (PSR). Sickle cell-β^+-thalassemia presents a very wide clinical spectrum depending on the molecular mutation for the β-thalassemia gene and the amount of HbA produced. In peoples of West African origin, the sickle cell-β^+-thalassemia gene is associated with a mild reduction of normal β-chain synthesis, HbA levels of 15%–25%, and generally a mild clinical course, although this genotype is also more prone to PSR. The increased risk of PSR in the generally milder forms of sickle cell disease, SC and $S\beta^+$-thalassemia, remains largely unexplained. In Indians, the common β^+-thalassemia gene occurring in association with HbS produces only 3%–5% HbA and a generally severe clinical course.

The frequency and severity of clinical complications varies between the genotypes but there is a pattern for symptoms to be more marked in the generally severe SS and $S\beta^0$-thalassemia and more mild in SC disease and the West African form of $S\beta^+$-thalassemia with the notable exception of PSR. Much of the subsequent presentation will focus on the features of SS disease and be presented by patient age because this is a major determinant of clinical features.

CLINICAL FEATURES

The First Year

Symptoms develop as the γ-chain synthesis, which results in 60%–80% fetal hemoglobin (HbF) at birth, is replaced by β-chain synthesis and increasing levels of the abnormal HbS. This transition occurs at different rates in different individuals but in some, pathological levels of HbS may be reached within 8–10 wk of birth and potentially life-threatening complications may occur from that age. Early complications include dactylitis and death may result from the acute chest syndrome and abnormal splenic function rendering the infant prone to overwhelming septicemia and to acute splenic sequestration.

The onset of significant mortality between 6 and 12 mo implies that detection of the disease at birth is vital to implement family education and prophylactic measures; this may be a particular challenge because the mother has an apparently normal, asymptomatic child and may not understand the urgency. Overwhelming blood infections may be largely prevented by pneumococcal prophylaxis, which includes the conjugate pneumococcal vaccine at 2, 4, and 6 mo (Halasa et al. 2007), and regular penicillin (Gaston et al. 1986) although there are concerns that the conjugate vaccine may select nonvaccine serotypes (McCavit et al. 2011). Acute splenic sequestration may commence as early as 3 mo of age but is most common in the second 6 mo of life; two-thirds of attacks occur before 2 yr of age, and events become rare after 6 yr of age. Splenectomy has been advocated after two attacks, and instructing mothers in the detection of splenic enlargement and when to seek urgent attention has significantly reduced mortality from this complication (Emond et al. 1985). The mother also receives an important psychological message that although her child has a currently incurable disease, much can be done by close observation and simple interventions to improve its outcome.

Dactylitis, a complication caused by avascular necrosis of the bone marrow in the small bones of the hands and feet starts as early as 10–12 wk of age and affected 50% of children by the age of 2 yr in the Jamaican Cohort Study (Stevens et al. 1981). Attacks usually resolve over 5–7 d with simple analgesia, but commonly recur, and become rare after the age of 5 yr when the active bone marrow withdraws from the small bones. Although most attacks resolve completely, clinically and radiologically, infection superimposed on the avascular tissue may sometimes cause premature fusion and a permanent shortening of the affected small bones. Dactylitis must be carefully explained to the mother and medical attendants because the occasional misdiagnosis of child abuse may cause great distress to the family. A history of dactylitis may predict a more severe clinical course (Miller et al. 2000).

Early Years

Acute chest syndrome is a term that has become widely used in sickle cell disease because of the complex pathology which may include components from infection, infarction, fat embolism, and pulmonary sequestration. Characterized by new pulmonary infiltrates and a very variable clinical picture including chest pain, cough, and dyspnea, this syndrome should be monitored closely clinically and by pulse oximetry. Exchange transfusion can be lifesaving when rapid deterioration signals acute pulmonary sequestration. This syndrome may also follow hypoventilation associated with the pleuritic pain of avascular necrosis of the ribs or sternum and may be prevented by incentive spirometry in such cases (Bellet et al. 1995). The acute chest syndrome is one of the commonest causes of death after the age of 2 yr.

Stroke is also a feature of early childhood with a median age of onset of 6 yr, some cases occurring as early as the second year of life, with an incidence of 8% by the age of 14 yr in the Jamaican cohort (Balkaran et al. 1992). The principal pathology at this age is cerebral infarction secondary to stenosis of major cerebral vessels, even as large as the internal carotid artery. The mechanism of this pathology is poorly understood and clearly different from the small vessel occlusion characterizing other manifestations of the disease. Recurrent strokes occur within 3 yr of the initial event in 50%–70% and many recurrent episodes may be prevented with chronic transfusion programs (Sarnaik et al. 1979; Pegelow et al. 1995). Risk factors include narrowing of cerebral vessels, which may be detected by transcranial Doppler, and in these, chronic transfusion has prevented the initial stroke from occurring (Adams et al. 1998).

Hypersplenism, characterized by persistent splenic enlargement, significant red cell sequestration, marked bone marrow expansion, and a new hematological equilibrium, also emerges around this age and affected 5% of children in the Jamaican cohort. Some events resolve spontaneously; others are treated by splenectomy or chronic transfusion, but if unresolved, this complication may lead to significant morbidity and mortality. The relationship between acute splenic sequestration and hypersplenism is complex and ill understood; some episodes of acute splenic sequestration evolve into hypersplenism and episodes of acute splenic sequestration may become superimposed on hypersplenism, but the two pathologies tend to occur at different ages and have different risk factors.

The aplastic crisis also emerges as a common problem at this age and is an avoidable cause of the occasional death. Almost all clinically defined aplastic crises are owing to human parvovirus B19, which destroys the red cell precursors in the bone marrow for 8–10 d until neutralized by the production of parvovirus-specific IgG. In hematologically normal people, such infection reduces the reticulocyte count to zero but with a mean red cell survival of 120 d, the hemoglobin decreases by only 0.5 g/dL. However, with red cell survival of 10–12 d typical of African SS disease, events in the bone marrow are rapidly reflected in the peripheral blood and the hemoglobin falls by ~1 g/d to life-threatening levels unless treated by transfusion. The outcome is entirely predictable and benign as long as oxygen delivery is maintained by a single unit transfusion, and immunity to parvovirus seems to be lifelong. Seroconversion, assessed by parvovirus-specific IgG, occurred in 70% by age 20 yr in both Jamaica (Serjeant et al. 2001) and Uganda (Ndugwa et al. 2004), and most cases present clinically between 2 and 10 yr of age.

Later Childhood and Early Adolescence

Many problems emerge in adolescence, which unfortunately coincides with the transition from pediatric to adult care when many patients are lost to regular follow-up. These problems include nocturnal enuresis, an increasing prevalence of bone pain crisis, avascular necrosis of the femoral head, leg ulceration in some communities, priapism, problems associated with delayed growth and puberty, pregnancy, and the need for effective contraception.

Nocturnal enuresis has received little attention because it is not life threatening but it may have a profound effect on behavior and social performance. Bedwetting more than or equal to twice weekly at the age of 8 yr occurred in 45% of children in the Jamaican cohort compared with 19% among normal (AA) controls (Readett et al. 1990a). Enuretic patients tended to have higher urinary volumes and lower bladder capacity (Readett et al. 1990b). Enuresis always resolves eventually but may persist in some patients up to age 18–20 yr. The deleterious effects are more social and psychological rather than medical because of the obvious stress caused to patient and family.

The bone pain crisis, generally resulting from avascular necrosis of bone marrow, is the adult counterpart of dactylitis. Although commonly believed to result from vaso-occlusion and often referred to as the vaso-occlusive crisis (VOC), the bilateral, symmetrical involvement (Serjeant et al. 1994) and the common precipitation by skin cooling are difficult to explain on this basis. Bone pain crises increase in adolescence and early adult life, especially in males (Baum et al. 1987; Platt et al. 1991), and decline in both severity and frequency after the age of 25–30 yr, in most patients. Risk factors include a high hemoglobin level, low HbF level, and in the third trimester of pregnancy precipitating factors include cold exposure, infections, and stress. Although this complication is the cause of ~90% of hospital admissions in the United States (Yang et al. 1995) and the United Kingdom (Brozovic et al. 1987), it is not an intrinsically severe pathology and death is unusual. The mechanism of the progressive amelioration of bone pains after 30 yr of age is unclear but may reassure patients.

Avascular necrosis of the femoral head also results from avascular necrosis of bone marrow but because of its weight bearing in walking, secondary damage may occur to the articular surface of the femoral head. The prevalence in unselected series is uncertain although probably ~10% and most commonly occurs in late adolescence. The outcome is determined by age at occurrence; infarction of the immature capital epiphysis allows remodeling and reasonable function, whereas involvement of the mature femoral head typically results in segmental damage with painful limitation of movement, which may require prosthetic surgery. Early diagnosis, for which magnetic resonance imaging (MRI) seems most sensitive, and avoidance of weight bearing may allow the bone to heal without deformity. Core decompression, presumably by relieving the intramedullary pressure, may relieve pain and improve movement (Mukisi-Mukaza et al. 2009) and possibly delay the use of total hip replacement.

Chronic leg ulceration is a major problem in adolescence but its frequency varies widely in different communities, affecting over 50% of Jamaican patients (Cumming et al. 2008), but only 5%–10% of those in the United States (Koshy et al. 1989). Although rarely a cause of death, leg ulceration has a profound impact on education and subsequent employment potential, on social relationships, and may cause depression. Although characterized by a chronic healing/relapsing course, ulcers eventually heal in most patients by the age of 30 yr. The cause of the geographic variability is not currently understood.

Physical growth and sexual development are commonly delayed from a variety of mechanisms and it is necessary to reassure the families that this is characteristic of the disease and generally does not require special therapy because families may use expensive tonics to "fatten" the child with little or no effect. Body habitus may be permanently affected with the appearance of long thin limbs, narrow shoulders and hips, and a relative absence of body fat. Puberty, assessed by menarche, is delayed by an average of 2.5 yr but there is no difference in the interval between menarche and first pregnancy. Pregnancy is associated with an increased risk of bone pain crisis, acute chest syndrome, and maternal death, especially in the last trimester and immediate postpartum period, and close antenatal monitoring with hospital delivery is advised. If contraception is requested, there is no evidence that any form of contraception is contraindicated in SS disease.

Priapism is a distressing complication affecting ~40% of postpubertal males in Jamaica (Emond et al. 1980). Defined as painful erections unassociated with sexual desire, there are two clinical forms. Stuttering priapism, which is nocturnal, lasts 2–4 h, with normal intervening sexual function and major attacks lasting >12 h and usually followed by permanent vascular damage and impotence. Stuttering attacks interfere with sleep, leading to daytime somnolence but more importantly may be a prodrome for major events. Stuttering attacks are usually underreported because of embarrassment and the lack of realization that it is owing to sickle cell disease, so direct questioning may be necessary in its detection. The natural history is poorly documented but stuttering events generally resolve with time.

Later Adult Life

A second peak of stroke occurs around 25 yr of age and these are predominantly hemorrhagic including subarachnoid hemorrhage often secondary to berry aneurysms or direct intracerebral or intraventricular hemorrhage, which have a poor prognosis.

The acute chest syndrome continues to be a major clinical problem and increasing pulmonary fibrosis may be associated with pulmonary hypertension, placing further stress on cardiac function. Congestive cardiac failure becomes more common, possibly secondary to pulmonary hypertension, but also to a reduced compliance of the left ventricular wall secondary to the sustained hypertrophy and intraventricular fibrosis. Hemoglobin levels commonly decline over the age of 40 yr compromising cardiac function further.

Deteriorating renal function is a major contributor to morbidity beyond the age of 40 yr, progressive glomerular fibrosis being associated with declining glomerular filtration rates (GFR), falling erythropoietin levels, and a gradual decline in total hemoglobin (Morris et al. 1991). Often clinically silent, patients do not present with the renal abnormalities but the gradually decreasing hemoglobin level impairs cardiac function. Renal failure in SS disease should be defined on the basis of lower serum creatinine levels than in the general population, and an upper limit of normal of 70–80 μmol/L was associated with significantly lowered GFR in the Jamaican cohort (Thompson et al. 2007).

PERSPECTIVE

From the foregoing summary of major clinical events, these are clearly influenced by patient age. The disease is asymptomatic in the first 1–2 mo of life but between 3–6 mo, dactylitis and splenic problems emerge. The second 6 mo of the first year hold serious dangers and the greatest mortality rate for the patient from acute splenic sequestration, overwhelming septicemias, and acute chest syndrome, most of which can be prevented or more effectively treated, but only if the underlying diagnosis of sickle cell disease is known, which is a major justification for newborn screening. From 1–5 yr of age, symptoms continue to be common with dactylitis and acute splenic sequestration (both of which may recur), acute chest syndrome continues as an important problem, and stroke emerges. After age 5 yr, dactylitis becomes rare but is replaced by the adult counterpart of the bone pain crisis as active bone marrow disappears from the small bones of the hands and feet and becomes focused in the juxta-articular areas of the long bones, the vertebra, ribs, and sternum.

Adolescence is a serious time for many patients faced with enuresis, leg ulceration, increasing bone pain crisis, delayed growth and sexual development, priapism, and hip disease. Major clinical and social support is needed at this time.

In early adult life, many of these problems continue and may be amplified by the difficulties of management of pregnancy. After 25–30 yr of age, bone pain crises wane in frequency and severity and there may be a long period free of bone pains when the quality of life is better than previously and patients can engage in full-time employment and social activities and commonly default from clinic follow-up. After the age of 40 yr, renal impairment with the associated decrease in hemoglobin level becomes common and may lead to cardiac and other problems.

Overall, survival in SS disease in developed societies now shows a median survival of 45–55 yr in the United States and Jamaica (Platt et al. 1994; Wierenga et al. 2001). However, the absence of survival data from earlier periods makes it difficult to assess whether this improvement is real or apparent because of less symptomatic bias in the observed populations. This bias, which was undoubtedly strong in the past, has been reduced by widespread population screening and especially neonatal detection, although even in the relatively short time frame of the Jamaican cohort, significantly improved survival was shown when comparing the first with subsequent terciles (Lee et al. 1995). Furthermore, there should be continued improvement because many of the interventions developed in the past 30 yr address problems in early childhood and the beneficial effect of these is yet to filter through to greater survival.

The outcome of the disease in developing societies where services are less organized and resources more limited remains unknown and in sub-Saharan Africa, median survival may be <5 yr.

DETERMINANTS OF SEVERITY

It is likely that many factors, both genetic and environmental, determine expression and final outcome of the disease.

Fetal Hemoglobin

High levels of HbF, present at birth in virtually all patients, protect against vaso-occlusive and hemolytic features of the disease in the first few months. The level at which HbF persists varies widely between patients and between geographic areas but patients maintaining higher levels tend to have less hemolysis and more mild clinical courses. In African SS disease hemolysis was well established with lower hemoglobin and higher reticulocyte counts from 1 mo of age in the Jamaican cohort (Serjeant et al. 1981). In the Asian haplotype, higher HbF levels persist and are associated with a milder clinical course and probably greater survival but may render patients more prone to hypersplenism. The genetics of the persistence of γ-chain synthesis allowing HbF to persist is complex, but patients with higher HbF levels tend to have one or both parents with the sickle cell trait and a modest elevation of HbF suggesting a genetic mechanism that is not yet clearly defined.

α-Thalassemia

Deletion of one of a pair of closely linked α-thalassemia genes (heterozygous α^+-thalassemia) in SS disease tends to lower the mean cell hemoglobin concentration (MCHC), which as a determinant of HbS polymerization (Seakins et al. 1973), is likely to reduce intravascular sickling. Deletion of both of the closely linked α-thalassemia genes (homozygous α^+-thalassemia) has an even greater effect in lowering the MCHC but any consequent reduction in sickling may be offset by an increase in total hemoglobin. α-Thalassemia therefore may provide a fascinating model to distinguish the effects of inhibiting sickling in small vessels and the potentially deleterious effect of a higher hematocrit impairing blood flow in larger vessels. In African SS disease in Jamaica, heterozygous α^+-thalassemia occurred in 35% and homozygous α^+-thalassemia patients in 4% (Higgs et al. 1982), whereas in the Asian haplotype even higher levels have been found (Kar et al. 1986; Padmos et al. 1991).

MCHC

Greece is the only population so far described in which SS disease is not associated with high frequencies of α-thalassemia, yet Greek patients were found to have significantly lower MCHC and features of milder disease compared with Jamaicans (Christakis et al. 1990). The lowered MCHC in this population remains to be confirmed and the mechanism explored.

Environmental Factors

The importance of environmental factors in the outcome of sickle cell disease is undoubted, whether it relates to climatic factors or skin cooling precipitating bone pain crises or apparently random events such as contact with *Streptococcus pneumonia* or parvovirus B19. Malaria continues to be a major determinant of morbidity and mortality in affected areas such as sub-Saharan Africa and central India. Susceptibility to other potentially serious infections by *Salmonella* spp. may reflect carriage rates in the general population. Other important factors are nutrition, access to public health measures such as immunization, and socioeconomic status, which may determine the families' access to communication, transport, and medical care. The importance of environmental influences and their mechanisms hold great potential for increasing the understanding and further amelioration of sickle cell disease.

Genetic versus Environmental Factors Modifying SS Disease

One of the classic methods for distinguishing genetic from environmental influences is the study of identical twins. Identical or monozygotic twins occur in 0.3%–0.4% of births, but to be useful for studies of hematology and clinical features, both twins must survive long enough to provide adequate data. These conditions are not frequently met and the largest available series contained six pairs of identical twins in Jamaica (Weatherall et al. 2005). A comparison of twin pairs showed that genetic factors influenced growth and hematological indices but that clinical events in the twin pairs were frequently discordant, suggesting the importance of nongenetic influences.

GEOGRAPHIC VARIABILITY

Reflecting both genetic and environmental factors is the variability of SS disease in different geographic areas. The β-globin haplotypes probably represent separate occurrences of the HbS mutation and do show minor differences in disease features depending on the DNA structures on which such mutations occurred. Comparison of features in the three principal African haplotypes, Benin, Bantu, and Senegal, have been difficult to conduct because of the dominance of the Benin haplotype in the United States and the Caribbean but the approximately equal prevalence in Brazil should allow comparison of Benin and Bantu haplotypes. Most studies currently available have been small and reached conflicting conclusions. In the Eastern Province of Saudi Arabia and central India, the Asian haplotype appears to have less hemolysis, greater persistence of splenomegaly, generally more mild clinical features, and probably greater survival, although the extent to which this is owing to the haplotype or to the commonly co-inherited genes for α-thalassemia and persistence of HbF remains unclear.

CAUSES OF DEATH

The causes of death are strongly influenced by the prevalence of malaria and other infections, and almost certainly by the availability and sophistication of medical and other services. No systematic data are available in sub-Saharan Africa but survival is likely to be markedly shortened and median duration may be as short as 5 yr.

African SS disease in the United States, where malaria is absent, has a much better prognosis, although even under the most sophisticated conditions, median survival is shortened by at least 20 yr relative to the general population. Mortality patterns are undoubtedly changing because of diagnosis by newborn screening and more effective interventions, but an earlier review in Jamaica indicated that the highest mortality occurred in the second 6 mo of the first year of life due principally to acute splenic sequestration, the acute chest syndrome, and septicemias (Thomas et al. 1982). At later ages, acute chest syndrome persists as a major cause of death and the pattern of septicemias may be changing with the selection of penicillin-resistant and/or nonvaccine serotypes of *Streptococcus pneumoniae* and the emergence of *Salmonella* as an important organism (Wright et al. 1997). After the age of 30 yr, cumulative organ damage impairs function especially in the lungs with pulmonary fibrosis and pulmonary hypertension and in the kidney with a progressive glomerular fibrosis.

In SS disease with the Asian haplotype, almost no data are available on survival or on the causes of death although there are suggestions that cumulative end organ damage may be less common than in African SS disease.

Ascertaining the cause of death is vital to defining strategies to prevent the responsible pathologies and the dearth of available data, lack of blood and viral cultures, and of autopsies are major impediments in Africa, Arabia, and India. However, even in Jamaica, with autopsy rates as high as 90%, detailed histology and blood cultures sometimes failed to detect a plausible cause of death. Furthermore, the effects on survival of recent long-term interventions such as hydroxyurea and chronic transfusion programs will need to be assessed and the current data on mortality and hydroxyurea (Steinberg et al. 2003; Bakanay et al. 2005) illustrate the complexity of interpreting results.

 Cite this article as *Cold Spring Harb Perspect Med* doi: 10.1101/cshperspect.a011783

PREVENTION OF SICKLE CELL DISEASE

It is now widely recognized that prevention of disease is more cost effective than trying to treat established pathology. Nowhere is this challenge more relevant than sub-Saharan Africa where figures based on population size, birth rates, and sickle cell trait frequency estimate that more than 250,000 babies with SS disease are born each year. The available resources cannot provide adequate clinical care to a problem of this scale and prevention of affected births must be one approach to the future control of the disease. Because the disease results from the inheritance of abnormal hemoglobin genes from both parents, affected children could be avoided if one parent had a normal hemoglobin (AA) genotype. Genotype identification with education and counseling on its significance would allow persons to make informed decisions on their choice of partner and should they wish, avoid the risks of a child with sickle cell disease. In Bahrain, a voluntary premarital screening program halved the births with SS disease and screening became mandatory in 2004 (Al-Arrayed 2005). The Kingdom of Saudi Arabia implemented a similar program of premarital screening in February 2004, and a review of the first 6 yr of the program (2004–2009) indicates that out of 1.5 million marriage proposals, there had been a steadily increasing rate of marriage cancellations among at-risk couples (Memish and Saeed 2011). Among at-risk couples who continue to marry, other potential options include prenatal diagnosis and preimplantation diagnosis (Xu et al. 1999).

It remains unknown whether, in societies without traditions of arranged marriages, genotype identification and counseling at younger ages will influence reproductive decisions and such studies should have a high priority at this time. A project in Orchomenos, Greece in 1966 examined 2300 families with a sickle cell trait frequency of 23%, but assessment of genotypes in 101 marriages occurring over the subsequent 7 yr showed no significant deviation from random (Stammatoyannopoulos 1973). Factors militating against success in this study may have included the small numbers and lack of awareness of the disease among this relatively uneducated farming community. The Manchester Project, set in the Parish of Manchester in central Jamaica, has advantages of scale and of a population with greater awareness of sickle cell disease. Screening for the detection of HbS, HbC, β-thalassemia trait, and other abnormal hemoglobins has been performed in approximately 16,000 senior school children aged 16–19 yr over the last 6 yr. Instead of looking at the genotypes of selected partners, this study will be based on the offspring of screened students and whether the distribution of genotypes differs from that predicted by the Hardy-Weinberg hypothesis. Newborn screening of more than 32,000 births in the parishes serving this area identified 770 babies so far born to mothers screened as school girls but much greater numbers will be needed to test the hypothesis.

Great advances have been made in the management of sickle cell disease but many of these are expensive and dependent on medical infrastructure, which may take a long time to develop in high-risk societies. In the meantime prevention, where possible, will be an essential component to overall management of the problem of sickle cell disease.

REFERENCES

Adams RJ, McKie VC, Hsu L, Files B, Vichinsky E, Pegelow C, Abboud M, Gallagher D, Kutlar A, Nichols FT, et al. 1998. Prevention of a first stroke by transfusions in children with sickle cell anemia and abnormal results on transcranial Doppler ultrasonography. *N Engl J Med* **339:** 5–11.

Al Arrayed S. 2005. Campaign to control genetic blood diseases in Bahrain. *Community Genet* **8:** 52–55.

Bakanay SM, Dainer E, Clair B, Adekile A, Daitch L, Wells L, Holley L, Smith D, Kutlar A. 2005. Mortality in sickle cell patients on hydroxyurea therapy. *Blood* **105:** 545–547.

Balkaran B, Char G, Morris JS, Serjeant BE, Serjeant GR. 1992. Stroke in a cohort study of patients with homozygous sickle cell disease. *J Pediatr* **120:** 360–366.

Baum KF, Dunn DT, Maude GH, Serjeant GR. 1987. The painful crisis of homozygous sickle cell disease: A study of risk factors. *Arch Int Med* **147:** 1231–1234.

Bellet PS, Kalinyak KA, Shukla R, Gelfand MJ, Rucknagel DL. 1995. Incentive spirometry to prevent acute pulmonary complications in sickle cell diseases. *N Engl J Med* **333:** 699–703.

Brozović M, Davies SC, Brownell AI. 1987. Acute admissions of patients with sickle cell disease who live in Britain. *Br Med J* **294:** 1206–1208.

Chebloune Y, Pagnier J, Trabuchet G, Faure C, Verdier G, Labie D, Nigon V. 1988. Structural analysis of the 5′ flanking region of the β-globin gene in African sickle cell anemia patients: Further evidence for three origins of the sickle cell mutation in Africa. *Proc Natl Acad Sci* **85:** 4431–4435.

Christakis J, Vavatsi N, Hassapopoulou H, Papadopoulou M, Mandraveli K, Loukopoulos D, Morris J, Serjeant BE, Serjeant GR. 1990. Comparison of homozygous sickle cell disease in Northern Greece and Jamaica. *Lancet* **335:** 637–640.

Cumming V, King L, Fraser R, Serjeant G, Reid M. 2008. Venous incompetence, poverty and lactate dehydrogenase in Jamaica are important predictors of leg ulceration in sickle cell anaemia. *Br J Haematol* **142:** 119–125.

Emond A, Holman R, Hayes RJ, Serjeant GR. 1980. Priapism and impotence in homozygous sickle cell disease. *Arch Intern Med* **140:** 1434–1437.

Emond AM, Collis R, Darvill D, Higgs DR, Maude GH, Serjeant GR. 1985. Acute splenic sequestration in homozygous sickle cell disease: Natural history and management. *J Pediatr* **107:** 201–206.

Gaston MH, Verter JI, Woods G, Pegelow C, Kelleher J, Presbury G, Zarkowsky H, Vichinsky E, Iyer R, Lobel JS, et al. 1986. Prophylaxis with oral penicillin in children with sickle cell anemia. A randomized trial. *N Engl J Med* **314:** 1593–1599.

Halasa NB, Shankar SM, Talbot TR, Arbogast PG, Mitchel EF, Wang WC, Schaffner W, Craig AS, Griffin MR. 2007. Incidence of invasive pneumococcal disease among individuals with sickle cell disease before and after the introduction of the pneumococcal conjugate vaccine. *Clin Infect Dis* **44:** 1428–1433.

Higgs DR, Aldridge BE, Lamb J, Clegg JB, Weatherall DJ, Hayes RJ, Grandison Y, Lowrie Y, Mason KP, Serjeant BE, Serjeant GR. 1982. The interaction of α-thalassemia and homozygous sickle-cell disease. *N Engl J Med* **306:** 1441–1446.

Kar BC, Satapathy RK, Kulozik AE, Kulozik M, Sirr S, Serjeant BE, Serjeant GR. 1986. Sickle cell disease in Orissa State, India. *Lancet* **328:** 1198–1201.

Koshy M, Entsuah R, Koranda A, Kraus AP, Johnson R, Bellvue R, Flournoy-Gill Z, Levy P. 1989. Leg ulcers in patients with sickle cell disease. *Blood* **74:** 1403–1408.

Lee A, Thomas P, Cupidore L, Serjeant B, Serjeant G. 1995. Improved survival in homozygous sickle cell disease: Lessons from a cohort study. *Br Med J* **311:** 160–162.

McCavit TL, Quinn CT, Techasaensiri C, Rogers ZR. 2011. Increase in invasive *Streptococcus pneumonia* infections in children with sickle cell disease since pneumococcal conjugate vaccine licensure. *J Pediatr* **158:** 505–507.

Memish ZA, Saeed MY. 2011. Six-year outcome of the national premarital screening and genetic counseling program for sickle cell disease and β-thalassemia in Saudi Arabia. *Ann Saudi Med* **31:** 229–235.

Miller ST, Sleeper LA, Pegelow CH, Enos LE, Wang WC, Weiner SJ, Wethers DL, Smith J, Kinney TR. 2000. Prediction of adverse outcomes in children with sickle cell disease. *N Engl J Med* **342:** 83–89.

Morris J, Dunn D, Beckford M, Grandison Y, Mason K, Higgs DR, De Ceulaer K, Serjeant BE, Serjeant GR. 1991. The haematology of homozygous sickle cell disease after the age of forty years. *Br J Haematol* **77:** 382–385.

Mukisi-Mukaza M, Manicom O, Alexis C, Bashoun K, Donkerwolcke M, Burny F. 2009. Treatment of sickle cell disease's hip necrosis by core decompression: A prospective case-control study. *Orthop Traumatol Surg Res* **95:** 498–504.

Nagel RL, Fabry ME, Pagnier J, Zohoun I, Wajcman H, Baudin V, Labie D. 1985. Hematologically and genetically distinct forms of sickle cell anemia in Africa. The Senegal type and the Benin type. *N Engl J Med* **312:** 880–884.

Ndugwa C, Kiggundu M, Cohen B, Serjeant GR. 2004. Aplastic crises and acute anaemic episodes in homozygous sickle cell disease. *E Afr Med J* **81:** 493–494.

Padmos MA, Roberts GT, Sackey K, Kulozik A, Bail S, Morris JS, Serjeant BE, Serjeant GR. 1991. Two different forms of homozygous sickle cell disease occur in Saudi Arabia. *Br J Haematol* **79:** 93–98.

Pagnier J, Mears JG, Dunda-Belkhodja O, Schaefer-Rego KE, Beldjord C, Nagel RL, Labie D. 1984. Evidence for the multicentric origin of the sickle cell hemoglobin gene in Africa. *Proc Natl Acad Sci* **81:** 1771–1773.

Pegelow CH, Adams RJ, McKie V, Abboud M, Berman B, Miller ST, Olivieri N, Vichinsky E, Wang W, Brambilla D. 1995. Risk of recurrent stroke in patients with sickle cell disease treated with erythrocyte transfusions. *J Pediatr* **126:** 896–899.

Platt OS, Thorington BD, Brambilla DJ, Milner PF, Rosse WF, Vichinsky E, Kinney TR. 1991. Pasin in sickle cell disease. Rates and risk factors. *N Engl J Med* **325:** 11–16.

Platt OS, Brambilla DJ, Rosse WF, Milner PF, Castro O, Steinberg MH, Klug PP. 1994. Mortality in sickle cell disease. Life expectancy and risk factors for early death. *N Engl J Med* **330:** 1639–1644.

Readett DRJ, Morris JS, Serjeant GR. 1990a. Nocturnal enuresis in sickle haemoglobinopathies. *Arch Dis Child* **65:** 290–293.

Readett DRJ, Morris J, Serjeant GR. 1990b. Determinants of nocturnal enuresis in homozygous sickle cell disease. *Arch Dis Child* **65:** 615–618.

Sarnaik S, Soorya D, Kim J, Ravindranath Y, Lusher J. 1979. Periodic transfusions for sickle cell anemia and CNS infarction. *Am J Dis Child* **133:** 1254–1257.

Seakins M, Gibbs WN, Milner PF, Bertles JF. 1973. Erythrocyte Hb-S concentration. An important factor in the low oxygen affinity of blood in sickle cell anemia. *J Clin Invest* **52:** 422–432.

Serjeant GR, Grandison Y, Lowrie Y, Mason K, Phillips J, Serjeant BE, Vaidya S. 1981. The development of haematological changes in homozygous sickle cell disease: A cohort study from birth to six years. *Br J Haematol* **48:** 533–543.

Serjeant GR, Serjeant BE, Forbes M, Hayes RJ, Higgs DR, Lehmann H. 1986. Haemoglobin gene frequencies in the Jamaican population: A study of 100,000 newborns. *Br J Haematol* **64:** 253–262.

Serjeant GR, De Ceulaer C, Lethbridge R, Morris JS, Singhal A, Thomas PW. 1994. The painful crisis of homozygous sickle cell disease—Clinical features. *Br J Haematol* **87:** 586–591.

Serjeant BE, Hambleton IR, Kerr S, Kilty CG, Serjeant GR. 2001. Haematological response to parvovirus B19 infection in sickle-cell disease. *Lancet* **358:** 1779–1780.

Stammatoyannopoulos G. 1973. Problems of screening and counseling in the hemoglobinopathies. In *Proceedings of the IVth International Conference on Birth Defects*, Vienna, pp. 268–276.

Steinberg MH, Barton F, Castro O, Pegelow CH, Ballas SK, Kutlar A, Orringer E, Bellevue R, Olivieri N, Eckman J, et al. 2003. Effect of hydroxyurea on mortality and morbidity in adult sickle cell anemia: Risks and benefits up to 9 years of treatment. *JAMA* **289:** 1645–1651.

Stevens MCG, Padwick M, Serjeant GR. 1981. Observations on the natural history of dactylitis in homozygous sickle cell disease. *Clin Pediatr* **20:** 311–317.

Thomas AN, Pattison C, Serjeant GR. 1982. Causes of death in sickle-cell disease in Jamaica. *Br Med J* **285:** 633–635.

Thompson J, Hambleton IR, Reid M, Serjeant GR. 2007. Albuminuria and renal function in homozygous sickle cell disease: Observations from a cohort study. *Arch Intern Med* **167:** 701–708.

Weatherall MW, Higgs DR, Weiss H, Weatherall DJ, Serjeant GR. 2005. Genotype/phenotype relationships in sickle cell disease: A pilot twin study. *Clin Lab Haematol* **27:** 384–390.

Wierenga KJ, Hambleton IR, Lewis NA. 2001. Survival estimates for patients with homozygous sickle-cell disease in Jamaica: A clinic-based population study. *Lancet* **357:** 680–683.

Wright J, Thomas P, Serjeant GR. 1997. Septicemia caused by salmonella infection; an overlooked complication of sickle cell disease. *J Pediatr* **130:** 394–399.

Xu K, Shi ZM, Veeck LL, Hughes MR, Rosenwaks Z. 1999. First unaffected pregnancy using preimplantation genetic diagnosis for sickle cell anemia. *JAMA* **281:** 1701–1706.

Yang Y-M, Shah AK, Watson M, Mankad V. 1995. Comparison of costs to the health sector of comprehensive and episodic health care for sickle cell disease patients. *Public Health Rep* **110:** 80–86.

The Search for Genetic Modifiers of Disease Severity in the β-Hemoglobinopathies

Guillaume Lettre

Montreal Heart Institute and Université Montréal, Montréal, Québec H1T 1C8, Canada

Correspondence: guillaume.lettre@mhi-humangenetics.org

Sickle cell disease (SCD) and β-thalassemia, two monogenic diseases caused by mutations in the *β-globin* gene, affect millions of individuals worldwide. These hemoglobin disorders are characterized by extreme clinical heterogeneity, complicating patient management and treatment. A better understanding of this patient-to-patient clinical variability would dramatically improve care and might also guide the development of novel therapies. Studies of the natural history of these β-hemoglobinopathies have identified fetal hemoglobin levels and concomitant α-thalassemia as important modifiers of disease severity. Several small-scale studies have attempted to identify additional genetic modifiers of SCD and β-thalassemia, without much success. Fortunately, improved knowledge of the human genome and the development of new genomic tools, such as genome-wide genotyping arrays and next-generation DNA sequencers, offer new opportunities to use genetics to better understand the causes of the many complications observed in β-hemoglobinopathy patients. Here I discuss the most important factors to consider when planning an experiment to find associations between β-hemoglobinopathy-related complications and DNA sequence variants, with a focus on how to successfully perform a genome-wide association study. I also review the literature and explain why most published findings in the field of SCD modifier genetics are likely to be false-positive reports, with the goal to draw lessons allowing investigators to design better genetic experiments.

Two of the promises of the Human Genome Project were to give the medical community the resources to better grasp interindividual variation in disease and treatment and to uncover previously unknown biology (Lander et al. 2001; Venter et al. 2001). In a recent review article to celebrate the 10-year anniversary of our genome's sequence, the genomicist Eric Lander highlighted many of the outstanding accomplishments made toward achieving these promises: from evolutionary conservation and non-coding RNAs to landscape of epigenetic marks and maps of genetic variation (Lander 2011). As anticipated, the field of human genetics has prospered on the foundation set by the Human Genome Project, in particular through the discovery and characterization of genetic variation by flagship projects such as the HapMap Project (Altshuler et al. 2005, 2010; Frazer et al. 2007) and the 1000 Genomes Project (1000 Genomes Project Consortium 2010). This has led to novel insights into the selective and recombination forces that have shaped our genome, but most and foremost this has had a

Copyright © 2013 Cold Spring Harbor Laboratory Press; all rights reserved
Cite this article as *Cold Spring Harb Perspect Med* doi: 10.1101/cshperspect.a015032

dramatic impact on our ability to identify genetic risk factors for complex human diseases. Progress in our theoretical understanding of the genetic variation present in our genome—both in terms of its frequency spectrum and correlated nature—was paved by advances in the development of high-throughput DNA genotyping and sequencing technologies. Together, these new genetic knowledge and technologic capabilities led to the discovery of more than 1000 common DNA sequence variants (so-called single-nucleotide polymorphisms, SNPs) associated with common human diseases and other complex traits by genome-wide association studies (GWAS) (Manolio et al. 2009).

Although most of the attention following the release of the human genome sequence has focused on the genetics of common diseases—which is understandable after decades of unfruitful research—our DNA sequence also helped us make significant progress in identifying genes responsible for rare genetic diseases. More recently, we have seen many studies on various human syndromes that used whole-exome DNA resequencing to find causal genes, often after many years of unsuccessful positional cloning and linkage study efforts (Teer and Mullikin 2010). These successes, together with the convergence of tools initially developed to study complex human diseases, have made us better appreciate that many *simple* Mendelian genetic diseases are, in fact, all but simple. This is particularly true for the β-hemoglobinopathies sickle cell disease (SCD) and β-thalassemia, which are the most common monogenic diseases in the world (Weatherall and Clegg 2001; Weatherall 2010) and are characterized by a very heterogeneous clinical course between patients with the same mutations, ranging from nearly asymptomatic to life-threatening conditions (Steinberg et al. 2009). The interpatient clinical variability in the β-hemoglobinopathies has sparked a vivid interest in identifying genetic modifiers of severity for these diseases. The rationale for this quest is that such genetic modifiers, if they exist, might lead to more precise (or personalized) prognostic tests, but also guide the development of more specific and effective therapies.

Fetal hemoglobin (HbF) levels and concomitant α-thalassemia are the two best-characterized modifiers of severity in SCD and β-thalassemia. Because these two topics are reviewed in other work (Nienhuis and Nathan 2012), they are not extensively discussed here. Rather, I focus on the main practical questions that one needs to address before undertaking an experiment to find new genetic modifiers for SCD or β-thalassemia. What are the main considerations in terms of phenotype definition, genotype acquisition, data analysis, and result validation for such genetic experiments? Finally, with a focus on SCD, I review the existing literature on genetic modifiers and assess the robustness of previously identified genetic associations, highlight the lessons learned from these earlier studies, and discuss how we might build on these results to design future experiments.

PHENOTYPE CONSIDERATIONS

Phenotype Definitions

Until a few years ago, the main preoccupation of human geneticists was to collect sufficient genetic information in their patient samples to begin the search for genetic risk factors. Nowadays, the emergence of genome-wide genotyping platforms and next-generation DNA sequencers has streamlined what used to be an incredible task: One can now sequence a complete human genome at high coverage in 2 wk for less than $5000. This problem solved, scientists are now redirecting their attention to the importance of the quality and harmonization of the phenotypes that are being studied. Although many phenotypes are straightforward to measure (e.g., HbF levels, white blood cell counts), others are more difficult to ascertain (e.g., pulmonary hypertension), and phenotype definitions are not always consistent, even between clinicians at the same clinic. As one example in the field of complex trait genetics, phenotype heterogeneity has been proposed as one possible reason why it has proven so difficult to understand the genetics of psychiatric diseases, despite a very large genetic contribution (Rasetti and

Weinberger 2011). Generally, simpler phenotypes that are reproducible and easy to collect (e.g., height) or endo-phenotypes, which are often quantitative traits related to disease etiology (e.g., cholesterol levels for atherosclerosis), have been studied more successfully than complex clinical end points using genetic strategies. There are also important practical considerations, especially in the context of the β-hemoglobinopathies, that influence the large-scale clinical availability of certain phenotypes when attempting to find genetic associations. Transcranial Doppler (TCD) ultrasonography and magnetic resonance imaging (MRI) results provide robust prognostic tools for cerebral vascular accidents (strokes) in children with SCD (Adams et al. 1992; DeBaun et al. 1995). They would thus appear as ideal phenotypes to identify modifiers of stroke risk in SCD. But these exquisite imaging measurements are often impossible to collect in populations in which the prevalence of SCD is high (e.g., Africa), limiting these studies to a small number of patients, which often leads to false-positive reports (Lohmueller et al. 2003) (see below).

The choice of phenotypes and ascertainment scheme, for instance, selection of a case-control sample as defined by the presence or absence of a particular clinical SCD complication, will also determine how analyses of associations of phenotypes with genotype information must be performed. Both dichotomous and continuous traits can be analyzed using simple statistics (e.g., χ^2 or Fisher's exact test, linear or logistic regression), but these analyses need to control for relevant confounding factors or matching variables such as age, sex, or medications. They also need to take into account family structure if related individuals are recruited. For continuous traits, the investigator could focus on patients with the most extreme phenotypes. For example, analyzing the bottom and top fifth percentile of the phenotypic distribution would reduce genotyping or sequencing costs by 90%; this strategy was successfully used by one of the two groups that found the association between the *BCL11A* locus and HbF levels (Menzel et al. 2007). This is a promising strategy as long as there is evidence that the genetic architecture of the trait is the same in the extremes as in the rest of the distribution. In other words, this approach works if the same genes/alleles control phenotypic variation across the phenotypic distribution, which is usually unknown, although recent work on adult height suggests that it is not always the case (Chan et al. 2011). Finally, although retrospective case-control samples are often easier to assemble, one cannot sufficiently emphasize the virtues of prospective studies. They are more expensive and time-consuming but eliminate many caveats and offer unique advantages to identify and test predictive risk factors. Longitudinal data are also amenable to powerful statistical methodologies, which can become important when the phenotypes are rare. This is certainly one of the main reasons why the large prospective Cooperative Study of Sickle Cell Disease (CSSCD), which was initiated in 1977 in the United States to determine the natural history of SCD from birth to death, has had such a tremendous impact on how we understand and treat SCD (Farber et al. 1985; Gaston et al. 1987).

Heritability

Most importantly, for genetic association studies to be successful, the phenotypes need to have a genetic component. Heritability is the key concept here, but it is often misunderstood (Visscher et al. 2008). It is defined, in the case of narrow-sense heritability (h^2), as the ratio of the phenotypic variation explained by additive genetic factors over the total phenotypic variation observed in the population. A low heritability estimate indicates that a large fraction of the phenotypic variation (or disease risk on a liability scale) is due to environmental factors (plus other non-additive genetic effects) and that, consequently, the phenotype is harder to track genetically. Heritability is usually estimated by comparing phenotype concordance among relatives, although new genomic-based methods have been developed that use only distantly related individuals (Visscher et al. 2006; Yang et al. 2011).

There is surprisingly little information available on the heritability of the phenotypes and complications relevant to the severity of the β-hemoglobinopathies. The estimate for the heritability of HbF levels is high ($h^2 \sim 0.6$–0.9), but this was obtained in healthy individuals of European descent (Garner et al. 2000; Pilia et al. 2006). In Jamaica, a study with nine identical pairs of twins with sickle cell anemia showed good concordance for many hematological traits, whereas the clinical complications were often discordant (Weatherall et al. 2005). With a focus on stroke, two groups also found evidence of familial aggregation in SCD patients (Driscoll et al. 2003; Kwiatkowski et al. 2003). Finally, in our analysis of genome-wide genotypes in the CSSCD, we identified 104 pairs of full sibs in which we analyzed phenotypic correlation (Table 1). Correlation coefficients were high for HbF levels, pain crisis rate, and stroke; moderate for acute chest syndrome, leg ulcer, osteonecrosis, and priapism; and low for survival. As the size and clinical quality of the β-hemoglobinopathy cohorts increase, estimating heritability of the complications before trying to find their genetic modifiers is certainly an avenue of research that should be pursued.

Table 1. Correlation of eight phenotypes in 104 pairs of full sibs from the Cooperative Study of Sickle Cell Disease (CSSCD)

Phenotype	Number of pairs with phenotype available (affected siblings/unaffected siblings)	Correlation coefficient
Acute chest syndrome rate	104	0.166
Fetal hemoglobin levels	84	0.579
Leg ulcer	84 (26/142)	0.183
Osteonecrosis	104 (37/171)	0.265
Pain crisis rate	104	0.683
Priapism (male only)	25 (9/41)	0.218
Stroke	104 (16/192)	0.361
Survival	104 (7/201)	0.024

These estimates must be considered carefully because the sample size is small and they do not take into account important confounding factors, such as socioeconomic status.

GENOTYPE CONSIDERATIONS

GWAS

Before the development of genome-wide genotyping arrays, genetic searches for complex diseases and traits used a combination of linkage- and candidate-gene-based approaches. For reasons that fall outside the scope of this article, linkage is inappropriate to find common genetic variants of small effect on phenotypic variation (Risch and Merikangas 1996), and candidate-gene studies are limited to the known biology. Except for the identification of the association between HbF levels and genetic variation at the *HBS1L–MYB* intergenic region, first by linkage (Craig et al. 1996; Garner et al. 1998) and subsequently by association study (Thein et al. 2007), these strategies have not proven very useful to find genetic modifiers of severity in the β-hemoglobinopathies.

Genetic association analysis consists of testing whether a specific allele, for example, the A allele at an A/G biallelic SNP, is found more often in cases than controls or whether it is correlated with a continuous trait (Hirschhorn and Daly 2005). Because alleles at different markers may be correlated in the genome, a phenomenon called linkage disequilibrium (LD) (Slatkin 2008), it is possible to test association with a large number of variants by genotyping only a subset of them. Because of LD, GWAS test associations systematically across the whole genome, but for markers that are common in the population of interest (usually SNPs with a minor allele frequency [MAF] $>5\%$; for rarer markers, DNA sequencing is required because of weak LD) (see below).

The first successful GWAS was performed in 2005 and identified the complement factor H (*CFH*) gene as a risk factor for age-related macular degeneration among a cohort of 96 cases and 50 controls, and genotypes at ~100,000 SNPs (Klein et al. 2005). Since then, GWAS have increased in size both in terms of the number of participants genotyped and markers tested. For example, in a recent study on body mass index, about 2.5 million SNPs were analyzed in about 250,000 individuals (Speliotes et al.

2010). For the β-hemoglobinopathies, GWAS have so far been limited to the discovery of *BCL11A* and its role on HbF levels (Menzel et al. 2007; Uda et al. 2008), SCD pain crises (Lettre et al. 2008), and β-thalassemia transfusion dependency (Uda et al. 2008; Nuinoon et al. 2009). In the next sections, the key concepts required for a successful GWAS are reviewed.

Statistical Power

For genetic association analysis, the statistical power is defined as the probability of rejecting the null hypothesis of no association when the null hypothesis is indeed false. In other words, if there is a true association between genotype and phenotype, what is the probability of finding it given your study design? Power depends on sample size, marker allele frequency, linkage disequilibrium (LD) strength between the genotyped markers and the causal variants, effect size of the genetic variants on the phenotype, and the number of independent markers tested. This latter factor, also referred to as the "multiple hypothesis burden," is particularly important in GWAS because hundreds of thousands or even millions of SNPs are usually tested. To limit the number of false-positive associations (type I errors), we use very stringent significance criteria to declare significance, but this has the disadvantage of reducing power. For GWAS, a generally accepted criterion to claim genome-wide significance is a p-value $< 5 \times 10^{-8}$ (Altshuler et al. 2005). Power is a critical concept to estimate before embarking on any GWAS project but also essential to interpret published GWAS results correctly. Fortunately, user-friendly tools exist to compute the power of association studies (Purcell et al. 2003; Gauderman and Morrison 2006).

Population: Coverage, Stratification, and Admixture

Genetic variation differs between populations because of mutations, genetic drift, migration, and natural selection. As a consequence, not all genetic variants are present in all populations, or the allele frequency of a given variant may differ across populations. For example, populations of African ancestry are genetically more ancient and carry more genetic variation (both rare and common) than populations of European or Asian descent (1000 Genomes Project Consortium 2010). Because the content of commercial genome-wide genotyping arrays is fixed, they cover more exhaustively the common genetic variations in non-African populations (Bhangale et al. 2005). This can have a major impact on genetic discovery efforts, especially for diseases that are more prevalent in Africans or individuals of African ancestry such as SCD. Novel genotyping arrays that were designed using empirical data from the 1000 Genomes Project, as well as imputation methods (Li et al. 2009) that can probabilistically infer genotypes at ungenotyped markers using genotypes from reference sets such as the HapMap or 1000 Genomes Projects, can, however, mitigate the bias introduced by genotyping the more limited markers present in commercial genotyping arrays.

Allele frequency differences between populations can also result in false-positive genetic association results when the prevalence of a disease (or the distribution of a quantitative trait) differs across these populations. The classic example of this important confounder, called population stratification (Price et al. 2010), is the spurious association between height and the DNA polymorphisms responsible for lactase persistence in populations of European descent because both height and lactose tolerance follow a North–West/South–East cline within Europe (Campbell et al. 2005). Using genome-wide genotypes or a limited set of ancestry informative markers, several methods now exist to detect and correct for the effect of population stratification on association results (Price et al. 2006; Kang et al. 2010). It is also worth mentioning that family-based association tests can be protected from the effect of population stratification, providing additional value to this study design.

Whereas population stratification applies to subpopulations from the same continent (e.g., European Americans), admixture describes individuals whose ancestors are from different continents, such as African Americans or Hispanic Americans (Seldin et al. 2011). The chromosomes in admixed populations are mosaics with chromosomal segments from each ancestral populations: This can be useful for gene mapping but also presents several analytical challenges that must be appropriately accounted for when performing GWAS (Price et al. 2009). Correction for admixture is particularly important in the search for genetic modifiers of the β-hemoglobinopathies because many studied populations are, in fact, admixed.

Replication

As mentioned above, the challenge in GWAS is to identify the few authentic genetic associations among millions of markers tested. We can increase the stringency of the significance threshold to account for the number of tests performed, but this has the disadvantage of decreasing power. To compensate for the power loss, we can analyze more clinical or population samples by GWAS, but this increases costs. Most successful GWAS have used staged designs: All markers are first tested in an initial "discovery" population, and then the most promising markers (based on statistical evidence or biological candidacy) are analyzed in replication cohorts. In fact, the replication of genetic associations in independent samples—ideally, a replication study would test the same phenotype, allele, direction of effect, and genetic model with a different genotyping platform—is now the gold standard and is essentially mandatory before publication because it has been shown to be key to avoid false-positive reports (Lohmueller et al. 2003; Chanock et al. 2007). Given the importance of replicating GWAS findings, any proposal to identify β-hemoglobinopathy modifiers should include a plan to replicate association results in independent samples. It is critical to consider accessibility to replication cohorts (independent samples, same phenotypes) early in the design of any GWAS experiments.

Next-Generation DNA Sequencing

The GWAS framework is extremely efficient to capture the role of common genetic variation (MAF $> 5\%$) in phenotypic variation. Rare (MAF $< 0.1\%$) and low-frequency (MAF $0.1\% - 5\%$) variants are usually poorly correlated with common variants surveyed by GWAS and have therefore not been tested comprehensively for their role in disease risk. With the development of affordable next-generation DNA sequencing technologies and data release from the 1000 Genomes Project (1000 Genomes Project Consortium 2010), we now have the tools to test the role of rare and low-frequency variants in complex human phenotypes. Genetic results for HbF, implicating rare familial mutations in the *β-globin* locus and *KLF1* (Borg et al. 2010) and low-frequency variants in *MYB* (Galarneau et al. 2010), strongly hint that rare genetic variation might also modify disease severity in the β-hemoglobinopathies. With many ongoing whole-exome sequencing projects, we should have a better idea of the role of rare and low-frequency genetic variants in SCD and β-thalassemia complications in the coming years.

REVIEW OF PUBLISHED GENETIC MODIFIERS IN SCD

Several studies have been published describing genetic associations between DNA polymorphisms in strong candidate genes and many complications specific to SCD (for review, see Fertrin and Costa 2010). Table 2 provides the details of most of the published genetic associations with acute chest syndrome, leg ulcer, osteonecrosis, pain crisis, and priapism. Based on these data, we can make the following observations: (1) Most sample sizes are small. (2) Reported odds ratios (OR) are high in comparison to what we now regard as realistic effect sizes in genetic association studies (usually OR < 1.2). (3) For many studies, the significance threshold was not adjusted for the number of tests performed. (4) Few results have been replicated in independent cohorts (Table 2). These are all characteristics usually linked to false-positive reports (Lohmueller et al. 2003), which is reflected by the low discovery power that any of these studies had to find genetic associations even under an ideal scenario (OR = 1.5; last column in Table 2).

Table 2. Published genetic associations with sickle cell disease–related complications

Complication	References	Sample size (cases/controls)	Gene (variant)	Reported odds ratio	Reported *p*-value	Appropriate significance threshold[a]	Replication study (reference)	Power (%)[b]
Acute chest syndrome	Sharan et al. 2004	18/27	*NOS3* (rs2070744)	6.5	0.005	0.05	No	10
Leg ulcer	Nolan et al. 2006	243/516	*KL* (rs516306)	1.8	0.0076	0.0002 (0.05/215 tests)	No	26
			MAP3K7 (rs157702)	1.6	0.008		No	
			SMAD7 (rs736839)	2.0	0.0004		No	
	Ofosu et al. 1987	9/29	*HLA* (B35/Cw4)	17.0	<0.005	0.05	No	10
Osteonecrosis	Baldwin et al. 2005	442/455	*KL* (rs211239)	2.6	0.001	0.0008 (0.05/66 tests)	Yes, but association did not replicate (Ulug et al. 2009).	47
			BMP6 (rs267196)	1.9	0.001		Yes, but association did not replicate; see text (Ulug et al. 2009).	
			ANXA2 (rs7170421)	3.4	<0.001		Yes, but association did not replicate (Ulug et al. 2009).	
Pain crisis	Mendonca et al. 2010	48/39	MBL2 (exon 1 0/A and −221 X/Y)	3.2	0.02	0.05	No	29
	Al-Subaie et al. 2009	127/130	*GP1BA* (rs6065)	1.8	0.004	0.02 (0.05/3 tests)	No	45
			ITGA2B (rs5911)	2.0	0.0008		No	
			ITGA2 (rs1801106)	1.9	0.002		No	
Priapism	Nolan et al. 2005	148/529	*KL* (rs2249358)	2.6	Not reported	0.001 (0.05/44 tests)	Yes, but association did not replicate (Elliott et al. 2007).	40
	Elliott et al. 2007	83/116	*TGFBR3* (rs7526590)	2.7	0.00058	0.001 (0.05/49 tests)	No	5
			AQP1 (rs10244884)	2.1	0.00068		No	
			ITGAV (rs3768780)	2.0	0.0009		No	

This is not an exhaustive list. Meeting abstracts were not considered because of the difficulty to assess the quality of the results.

[a]The appropriate significance threshold corresponds to a Bonferroni correction for the number of genes/tests performed in each study.

[b]Discovery power for each study was calculated under the following assumptions: odds ratio = 1.5, effect allele frequency = 20%, sample size = number of cases and controls available in the study, α level = appropriate significance threshold, prevalence as determined from the Cooperative Study of Sickle Cell Disease (CSSCD) phase I clinical data set.

The reported association between SNPs at the *BMP6* locus and osteonecrosis risk is a good example of the inconsistency that exists in the literature. The association was first identified in 442 cases and 455 controls from the CSSCD (Baldwin et al. 2005) with a replication attempt in another SCD cohort that included 39 osteonecrosis cases and 205 controls (Ulug et al. 2009). Although the investigators of the replication study claimed that they could replicate the *BMP6*–osteonecrosis association, careful analysis of the articles shows that the two studies have marginal associations with different *BMP6* SNPs (rs267196 and rs3812163) that are 124 kb away from each other and not in LD ($r^2 \sim 0$). By definition, because a different SNP was tested in the replication study, the original association between *BMP6* and osteonecrosis has not been replicated yet. Thus, none of the associations presented in Table 2 appear robust.

Stroke

Because of its devastating consequences and apparent heritability (Table 1) (Driscoll et al. 2003; Kwiatkowski et al. 2003), many investigators have attempted to identify genetic risk factors for stroke in SCD (Hoppe et al. 2004, 2007; Sebastiani et al. 2005; Voetsch et al. 2008). In a recent replication study, Flanagan and colleagues attempted to confirm these previous stroke association results in an independent SCD cohort of 130 stroke cases and 103 controls (Flanagan et al. 2011). This study had 44% power to replicate an association with stroke (using the same assumptions reported in Table 2). The researchers could replicate the known protective effect of α-thalassemia on stroke risk, and also reported nominally significant association results for SNPs in four genes: *ADYC9*, *ANXA2*, *TEK*, and *TGFBR3* (Flanagan et al. 2011). These results may seem encouraging, but once again they must be interpreted with caution. SNPs in *ADYC9*, *ANXA2*, *TEK*, and *TGFBR3* were initially related to stroke risk in SCD through a Bayesian network (Sebastiani et al. 2005). Bayesian networks integrate many variables, in this case, laboratory measures such as HbF levels but also genotypes, to predict disease outcome. Although they are theoretically interesting as methods to analyze large data sets with many variable dependencies, the biological and/or clinical interpretation of the different nodes in the network is difficult. In this specific example, although the Bayesian network was validated in an independent cohort of SCD patients with or without stroke, the marginal effect of the SNPs in the network on stroke risk remains unclear. In fact, we tested by logistic regression in the CSSCD the association between stroke and genotypes at *ADCY9*-rs2238432, *ANXA2*-rs11853426, *TEK*-rs489347, and *TGFBR3*-rs284875, and observed no significant results ($p > 0.15$) (KS Lo and G Lettre, unpubl.).

Two Convincing Associations

Besides HbF levels and α-thalassemia, there are two other convincing genetic modifiers of complications in SCD. Importantly, these associations are more compelling because they were first very robustly confirmed with phenotypes in non-SCD populations. Polymorphisms in the promoter of the *UGT1A1* gene are associated with serum bilirubin levels in many studies and different populations, including patients with SCD; the same polymorphisms are also associated with the risk of formation of gallstones in SCD patients (Passon et al. 2001; Fertrin et al. 2003; Chaar et al. 2005, 2006; Haverfield et al. 2005; Carpenter et al. 2008; Milton et al. 2012). Renal failure is one of the major complications of SCD and is a strong predictor of mortality in SCD (Platt et al. 1994). Admixture mapping and association studies have identified the *MYH9–APOL1* locus as an important genetic risk factor for kidney disease in populations of African ancestry (Kao et al. 2008; Kopp et al. 2008; Genovese et al. 2010). More recently, risk alleles at the *MYH9–APOL1* locus were also associated in SCD patients with proteinuria and decreased glomerular filtration rate, both hallmarks of SCD nephropathy (Ashley-Koch et al. 2011). The *UGT1A1* and *MYH9–APOL1* results are reminders of the importance of large sample size, access to replication cohorts, and clearly defined or well-harmonized phenotypes to find true-positive genetic associations.

CONCLUSIONS

The identification of genetic modifiers of disease severity for the β-hemoglobinopathies promises to improve patient management and treatment. As highlighted in this article, many of the previous associations between SCD complications and genetic variation are false-positive reports. This result is not specific to SCD because most of the association studies performed before the GWAS era were underpowered. Fortunately, robust associations for many complex human phenotypes, including some relevant to hemoglobin disorders (e.g., HbF levels, kidney disease), now exist and allow us to identify the key factors that lead to successful genetic searches: clear phenotype definition and heritability estimate, large sample size and knowledge of statistical power, control for the multiple hypothesis burden and other confounders such as population stratification and admixture, and access to independent samples for replication. Because the importance of these factors are well appreciated and with large collaborations of geneticists and clinicians working on the β-hemoglobinopathies already in place, we are finally in a position to ask convincingly important questions, test clinically meaningful hypotheses, and, hopefully, gain insight into the pathophysiology of SCD and β-thalassemia. This is clearly an exciting time for β-hemoglobinopathy patients, their families, and physicians.

ACKNOWLEDGMENTS

I thank Alexander P. Reiner, Aikaterini Kritikou, and Geneviève Galarneau for comments on earlier versions of this article, and Samuel Lessard for identifying related individuals in the CSSCD. Dr. Martin H. Steinberg kindly provided access to the CSSCD GWAS data set (funded by the NHLBI STAMPEED). Work on SCD in my laboratory is supported by an Innovations in Clinical Research Award grant from the Doris Duke Charitable Foundation, as well as by the Canada Research Chair program, the Canadian Institute of Health Research, and the Fonds de Recherche Santé Québec.

REFERENCES

*Reference is also in this collection.

1000 Genomes Project Consortium. 2010. A map of human genome variation from population-scale sequencing. *Nature* **467:** 1061–1073.

Adams R, McKie V, Nichols F, Carl E, Zhang DL, McKie K, Figueroa R, Litaker M, Thompson W, Hess D. 1992. The use of transcranial ultrasonography to predict stroke in sickle cell disease. *N Engl J Med* **326:** 605–610.

Al-Subaie AM, Fawaz NA, Mahdi N, Al-Absi IK, Al-Ola K, Ameen G, Almawi WY. 2009. Human platelet alloantigens (HPA) 1, HPA2, HPA3, HPA4, and HPA5 polymorphisms in sickle cell anemia patients with vaso-occlusive crisis. *Eur J Haematol* **83:** 579–585.

Altshuler D, Brooks LD, Chakravarti A, Collins FS, Daly MJ, Donnelly P. 2005. A haplotype map of the human genome. *Nature* **437:** 1299–1320.

Altshuler DM, Gibbs RA, Peltonen L, Altshuler DM, Gibbs RA, Peltonen L, Dermitzakis E, Schaffner SF, Yu F, Peltonen L, et al. 2010. Integrating common and rare genetic variation in diverse human populations. *Nature* **467:** 52–58.

Ashley-Koch AE, Okocha EC, Garrett ME, Soldano K, De Castro LM, Jonassaint JC, Orringer EP, Eckman JR, Telen MJ. 2011. MYH9 and APOL1 are both associated with sickle cell disease nephropathy. *Br J Haematol* **155:** 386–394.

Baldwin C, Nolan VG, Wyszynski DF, Ma QL, Sebastiani P, Embury SH, Bisbee A, Farrell J, Farrer L, Steinberg MH. 2005. Association of klotho, bone morphogenic protein 6, and annexin A2 polymorphisms with sickle cell osteonecrosis. *Blood* **106:** 372–375.

Bhangale TR, Rieder MJ, Livingston RJ, Nickerson DA. 2005. Comprehensive identification and characterization of diallelic insertion-deletion polymorphisms in 330 human candidate genes. *Hum Mol Genet* **14:** 59–69.

Borg J, Papadopoulos P, Georgitsi M, Gutierrez L, Grech G, Fanis P, Phylactides M, Verkerk AJ, van der Spek PJ, Scerri CA, et al. 2010. Haploinsufficiency for the erythroid transcription factor KLF1 causes hereditary persistence of fetal hemoglobin. *Nat Genet* **42:** 801–805.

Campbell CD, Ogburn EL, Lunetta KL, Lyon HN, Freedman ML, Groop LC, Altshuler D, Ardlie KG, Hirschhorn JN. 2005. Demonstrating stratification in a European American population. *Nat Genet* **37:** 868–872.

Carpenter SL, Lieff S, Howard TA, Eggleston B, Ware RE. 2008. UGT1A1 promoter polymorphisms and the development of hyperbilirubinemia and gallbladder disease in children with sickle cell anemia. *Am J Hematol* **83:** 800–803.

Chaar V, Keclard L, Diara JP, Leturdu C, Elion J, Krishnamoorthy R, Clayton J, Romana M. 2005. Association of UGT1A1 polymorphism with prevalence and age at onset of cholelithiasis in sickle cell anemia. *Haematologica* **90:** 188–199.

Chaar V, Keclard L, Etienne-Julan M, Diara JP, Elion J, Krishnamoorthy R, Romana M. 2006. UGT1A1 polymorphism outweighs the modest effect of deletional (−3.7 kb) α-thalassemia on cholelithogenesis in sickle cell anemia. *Am J Hematol* **81:** 377–379.

Chan Y, Holmen OL, Dauber A, Vatten L, Havulinna AS, Skorpen F, Kvaloy K, Silander K, Nguyen TT, Willer C, et al. 2011. Common variants show predicted polygenic effects on height in the tails of the distribution, except in extremely short individuals. *PLoS Genet* **7:** e1002439.

Chanock SJ, Manolio T, Boehnke M, Boerwinkle E, Hunter DJ, Thomas G, Hirschhorn JN, Abecasis G, Altshuler D, Bailey-Wilson JE, et al. 2007. Replicating genotype–phenotype associations. *Nature* **447:** 655–660.

Craig JE, Rochette J, Fisher CA, Weatherall DJ, Marc S, Lathrop GM, Demenais F, Thein S. 1996. Dissecting the loci controlling fetal haemoglobin production on chromosomes 11p and 6q by the regressive approach. *Nat Genet* **12:** 58–64.

DeBaun MR, Glauser TA, Siegel M, Borders J, Lee B. 1995. Noninvasive central nervous system imaging in sickle cell anemia. A preliminary study comparing transcranial Doppler with magnetic resonance angiography. *J Pediatr Hematol Oncol* **17:** 29–33.

Driscoll MC, Hurlet A, Styles L, McKie V, Files B, Olivieri N, Pegelow C, Berman B, Drachtman R, Patel K, et al. 2003. Stroke risk in siblings with sickle cell anemia. *Blood* **101:** 2401–2404.

Elliott L, Ashley-Koch AE, De Castro L, Jonassaint J, Price J, Ataga KI, Levesque MC, Brice Weinberg J, Eckman JR, Orringer EP, et al. 2007. Genetic polymorphisms associated with priapism in sickle cell disease. *Br J Haematol* **137:** 262–267.

Farber MD, Koshy M, Kinney TR. 1985. Cooperative Study of Sickle Cell Disease: Demographic and socioeconomic characteristics of patients and families with sickle cell disease. *J Chronic Dis* **38:** 495–505.

Fertrin KY, Costa FF. 2010. Genomic polymorphisms in sickle cell disease: Implications for clinical diversity and treatment. *Exp Rev Hematol* **3:** 443–458.

Fertrin KY, Melo MB, Assis AM, Saad ST, Costa FF. 2003. UDP-glucuronosyltransferase 1 gene promoter polymorphism is associated with increased serum bilirubin levels and cholecystectomy in patients with sickle cell anemia. *Clin Genet* **64:** 160–162.

Flanagan JM, Frohlich DM, Howard TA, Schultz WH, Driscoll C, Nagasubramanian R, Mortier NA, Kimble AC, Aygun B, Adams RJ, et al. 2011. Genetic predictors for stroke in children with sickle cell anemia. *Blood* **117:** 6681–6684.

Frazer KA, Ballinger DG, Cox DR, Hinds DA, Stuve LL, Gibbs RA, Belmont JW, Boudreau A, Hardenbol P, Leal SM, et al. 2007. A second generation human haplotype map of over 3.1 million SNPs. *Nature* **449:** 851–861.

Galarneau G, Palmer CD, Sankaran VG, Orkin SH, Hirschhorn JN, Lettre G. 2010. Fine-mapping at three loci known to affect fetal hemoglobin levels explains additional genetic variation. *Nat Genet* **42:** 1049–1051.

Garner C, Mitchell J, Hatzis T, Reittie J, Farrall M, Thein SL. 1998. Haplotype mapping of a major quantitative-trait locus for fetal hemoglobin production, on chromosome 6q23. *Am J Hum Genet* **62:** 1468–1474.

Garner C, Tatu T, Reittie JE, Littlewood T, Darley J, Cervino S, Farrall M, Kelly P, Spector TD, Thein SL. 2000. Genetic influences on F cells and other hematologic variables: A twin heritability study. *Blood* **95:** 342–346.

Gaston M, Smith J, Gallagher D, Flournoy-Gill Z, West S, Bellevue R, Farber M, Grover R, Koshy M, Ritchey AK, et al. 1987. Recruitment in the Cooperative Study of Sickle Cell Disease (CSSCD). *Control Clin Trials* **8:** 131S–140S.

Gauderman WJ, Morrison JM. 2006. QUANTO 1.1: A computer program for power and sample size calculations for genetic-epidemiology studies. http://hydra.usc.edu/gxe.

Genovese G, Friedman DJ, Ross MD, Lecordier L, Uzureau P, Freedman BI, Bowden DW, Langefeld CD, Oleksyk TK, Uscinski Knob AL, et al. 2010. Association of trypanolytic ApoL1 variants with kidney disease in African Americans. *Science* **329:** 841–845.

Haverfield EV, McKenzie CA, Forrester T, Bouzekri N, Harding R, Serjeant G, Walker T, Peto TE, Ward R, Weatherall DJ. 2005. UGT1A1 variation and gallstone formation in sickle cell disease. *Blood* **105:** 968–972.

Hirschhorn JN, Daly MJ. 2005. Genome-wide association studies for common diseases and complex traits. *Nat Rev Genet* **6:** 95–108.

Hoppe C, Klitz W, Cheng S, Apple R, Steiner L, Robles L, Girard T, Vichinsky E, Styles L. 2004. Gene interactions and stroke risk in children with sickle cell anemia. *Blood* **103:** 2391–2396.

Hoppe C, Klitz W, D'Harlingue K, Cheng S, Grow M, Steiner L, Noble J, Adams R, Styles L. 2007. Confirmation of an association between the TNF(−308) promoter polymorphism and stroke risk in children with sickle cell anemia. *Stroke* **38:** 2241–2246.

Kang HM, Sul JH, Service SK, Zaitlen NA, Kong SY, Freimer NB, Sabatti C, Eskin E. 2010. Variance component model to account for sample structure in genome-wide association studies. *Nat Genet* **42:** 348–354.

Kao WH, Klag MJ, Meoni LA, Reich D, Berthier-Schaad Y, Li M, Coresh J, Patterson N, Tandon A, Powe NR, et al. 2008. MYH9 is associated with nondiabetic end-stage renal disease in African Americans. *Nat Genet* **40:** 1185–1192.

Klein RJ, Zeiss C, Chew EY, Tsai JY, Sackler RS, Haynes C, Henning AK, SanGiovanni JP, Mane SM, Mayne ST, et al. 2005. Complement factor H polymorphism in age-related macular degeneration. *Science* **308:** 385–389.

Kopp JB, Smith MW, Nelson GW, Johnson RC, Freedman BI, Bowden DW, Oleksyk T, McKenzie LM, Kajiyama H, Ahuja TS, et al. 2008. MYH9 is a major-effect risk gene for focal segmental glomerulosclerosis. *Nat Genet* **40:** 1175–1184.

Kwiatkowski JL, Hunter JV, Smith-Whitley K, Katz ML, Shults J, Ohene-Frempong K. 2003. Transcranial Doppler ultrasonography in siblings with sickle cell disease. *Br J Haematol* **121:** 932–937.

Lander ES. 2011. Initial impact of the sequencing of the human genome. *Nature* **470:** 187–197.

Lander ES, Linton LM, Birren B, Nusbaum C, Zody MC, Baldwin J, Devon K, Dewar K, Doyle M, FitzHugh W, et al. 2001. Initial sequencing and analysis of the human genome. *Nature* **409:** 860–921.

Lettre G, Sankaran VG, Bezerra MA, Araujo AS, Uda M, Sanna S, Cao A, Schlessinger D, Costa FF, Hirschhorn JN, et al. 2008. DNA polymorphisms at the *BCL11A*, *HBS1L-MYB*, and *β-globin* loci associate with fetal hemoglobin levels and pain crises in sickle cell disease. *Proc Natl Acad Sci* **105:** 11869–11874.

Li Y, Willer C, Sanna S, Abecasis G. 2009. Genotype imputation. *Annu Rev Genomics Hum Genet* **10:** 387–406.

Lohmueller KE, Pearce CL, Pike M, Lander ES, Hirschhorn JN. 2003. Meta-analysis of genetic association studies supports a contribution of common variants to susceptibility to common disease. *Nat Genet* **33:** 177–182.

Manolio TA, Collins FS, Cox NJ, Goldstein DB, Hindorff LA, Hunter DJ, McCarthy MI, Ramos EM, Cardon LR, Chakravarti A, et al. 2009. Finding the missing heritability of complex diseases. *Nature* **461:** 747–753.

Mendonca TF, Oliveira MC, Vasconcelos LR, Pereira LM, Moura P, Bezerra MA, Santos MN, Araujo AS, Cavalcanti MS. 2010. Association of variant alleles of *MBL2* gene with vasoocclusive crisis in children with sickle cell anemia. *Blood Cells Mol Dis* **44:** 224–228.

Menzel S, Garner C, Gut I, Matsuda F, Yamaguchi M, Heath S, Foglio M, Zelenika D, Boland A, Rooks H, et al. 2007. A QTL influencing F cell production maps to a gene encoding a zinc-finger protein on chromosome 2p15. *Nat Genet* **39:** 1197–1199.

Milton JN, Sebastiani P, Solovieff N, Hartley SW, Bhatnagar P, Arking DE, Dworkis DA, Casella JF, Barron-Casella E, Bean CJ, et al. 2012. A genome-wide association study of total bilirubin and cholelithiasis risk in sickle cell anemia. *PLoS ONE* **7:** e34741.

* Nienhuis AW, Nathan DG. 2012. Pathophysiology and clinical manifestations of the β-thalassemias. *Cold Spring Harb Perspect Med* **2:** a011726.

Nolan VG, Baldwin C, Ma Q, Wyszynski DF, Amirault Y, Farrell JJ, Bisbee A, Embury SH, Farrer LA, Steinberg MH. 2005. Association of single nucleotide polymorphisms in klotho with priapism in sickle cell anaemia. *Br J Haematol* **128:** 266–272.

Nolan VG, Adewoye A, Baldwin C, Wang L, Ma Q, Wyszynski DF, Farrell JJ, Sebastiani P, Farrer LA, Steinberg MH. 2006. Sickle cell leg ulcers: Associations with haemolysis and SNPs in Klotho, TEK and genes of the TGF-β/BMP pathway. *Br J Haematol* **133:** 570–578.

Nuinoon M, Makarasara W, Mushiroda T, Setianingsih I, Wahidiyat PA, Sripichai O, Kumasaka N, Takahashi A, Svasti S, Munkongdee T, et al. 2009. A genome-wide association identified the common genetic variants influence disease severity in β^0-thalassemia/hemoglobin E. *Hum Genet* **127:** 303–314.

Ofosu MD, Castro O, Alarif L. 1987. Sickle cell leg ulcers are associated with HLA-B35 and Cw4. *Arch Dermatol* **123:** 482–484.

Passon RG, Howard TA, Zimmerman SA, Schultz WH, Ware RE. 2001. Influence of bilirubin uridine diphosphate-glucuronosyltransferase 1A promoter polymorphisms on serum bilirubin levels and cholelithiasis in children with sickle cell anemia. *J Pediatr Hematol Oncol* **23:** 448–451.

Pilia G, Chen WM, Scuteri A, Orru M, Albai G, Dei M, Lai S, Usala G, Lai M, Loi P, et al. 2006. Heritability of cardiovascular and personality traits in 6,148 Sardinians. *PLoS Genet* **2:** e132.

Platt OS, Brambilla DJ, Rosse WF, Milner PF, Castro O, Steinberg MH, Klug PP. 1994. Mortality in sickle cell disease. Life expectancy and risk factors for early death. *N Engl J Med* **330:** 1639–1644.

Price AL, Patterson NJ, Plenge RM, Weinblatt ME, Shadick NA, Reich D. 2006. Principal Components Analysis corrects for stratification in genome-wide association studies. *Nat Genet* **38:** 904–909.

Price AL, Tandon A, Patterson N, Barnes KC, Rafaels N, Ruczinski I, Beaty TH, Mathias R, Reich D, Myers S. 2009. Sensitive detection of chromosomal segments of distinct ancestry in admixed populations. *PLoS Genet* **5:** e1000519.

Price AL, Zaitlen NA, Reich D, Patterson N. 2010. New approaches to population stratification in genome-wide association studies. *Nat Rev Genet* **11:** 459–463.

Purcell S, Cherny SS, Sham PC. 2003. Genetic power calculator: Design of linkage and association genetic mapping studies of complex traits. *Bioinformatics* **19:** 149–150.

Rasetti R, Weinberger DR. 2011. Intermediate phenotypes in psychiatric disorders. *Curr Opin Genet Dev* **21:** 340–348.

Risch N, Merikangas K. 1996. The future of genetic studies of complex human diseases. *Science* **273:** 1516–1517.

Sebastiani P, Ramoni MF, Nolan V, Baldwin CT, Steinberg MH. 2005. Genetic dissection and prognostic modeling of overt stroke in sickle cell anemia. *Nat Genet* **37:** 435–440.

Seldin MF, Pasaniuc B, Price AL. 2011. New approaches to disease mapping in admixed populations. *Nat Rev Genet* **12:** 523–528.

Sharan K, Surrey S, Ballas S, Borowski M, Devoto M, Wang KF, Sandler E, Keller M. 2004. Association of T-786C eNOS gene polymorphism with increased susceptibility to acute chest syndrome in females with sickle cell disease. *Br J Haematol* **124:** 240–243.

Slatkin M. 2008. Linkage disequilibrium—Understanding the evolutionary past and mapping the medical future. *Nat Rev Genet* **9:** 477–485.

Speliotes EK, Willer CJ, Berndt SI, Monda KL, Thorleifsson G, Jackson AU, Allen HL, Lindgren CM, Luan J, Magi R, et al. 2010. Association analyses of 249,796 individuals reveal 18 new loci associated with body mass index. *Nat Genet* **42:** 937–948.

Steinberg MH, Forget BG, Higgs DR, Weatherall D. 2009. *Disorders of hemoglobin: Genetics, pathophysiology, and clinical management*, 2nd ed. Cambridge University Press, Cambridge.

Teer JK, Mullikin JC. 2010. Exome sequencing: The sweet spot before whole genomes. *Human Mol Genet* **19:** R145–R151.

Thein SL, Menzel S, Peng X, Best S, Jiang J, Close J, Silver N, Gerovasilli A, Ping C, Yamaguchi M, et al. 2007. Intergenic variants of *HBS1L–MYB* are responsible for a major quantitative trait locus on chromosome 6q23 influencing fetal hemoglobin levels in adults. *Proc Natl Acad Sci* **104:** 11346–11351.

Uda M, Galanello R, Sanna S, Lettre G, Sankaran VG, Chen W, Usala G, Busonero F, Maschio A, Albai G, et al. 2008. Genome-wide association study shows BCL11A associated with persistent fetal hemoglobin and amelioration of the phenotype of β-thalassemia. *Proc Natl Acad Sci* **105:** 1620–1625.

Ulug P, Vasavda N, Awogbade M, Cunningham J, Menzel S, Thein SL. 2009. Association of sickle avascular necrosis with bone morphogenic protein 6. *Ann Hematol* **88:** 803–805.

Venter JC, Adams MD, Myers EW, Li PW, Mural RJ, Sutton GG, Smith HO, Yandell M, Evans CA, Holt RA, et al. 2001. The sequence of the human genome. *Science* **291:** 1304–1351.

Visscher PM, Medland SE, Ferreira MA, Morley KI, Zhu G, Cornes BK, Montgomery GW, Martin NG. 2006. Assumption-free estimation of heritability from genome-wide identity-by-descent sharing between full siblings. *PLoS Genet* **2:** e41.

Visscher PM, Hill WG, Wray NR. 2008. Heritability in the genomics era—Concepts and misconceptions. *Nat Rev Genet* **9:** 255–266.

Voetsch B, Jin RC, Bierl C, Deus-Silva L, Camargo EC, Annichino-Bizacchi JM, Handy DE, Loscalzo J. 2008. Role of promoter polymorphisms in the plasma glutathione peroxidase (GPx-3) gene as a risk factor for cerebral venous thrombosis. *Stroke* **39:** 303–307.

Weatherall DJ. 2010. The inherited diseases of hemoglobin are an emerging global health burden. *Blood* **115:** 4331–4336.

Weatherall DJ, Clegg JB. 2001. Inherited haemoglobin disorders: An increasing global health problem. *Bull World Health Organ* **79:** 704–712.

Weatherall MW, Higgs DR, Weiss H, Weatherall DJ, Serjeant GR. 2005. Phenotype/genotype relationships in sickle cell disease: A pilot twin study. *Clin Lab Haematol* **27:** 384–390.

Yang J, Lee SH, Goddard ME, Visscher PM. 2011. GCTA: A tool for genome-wide complex trait analysis. *Am J Hum Genet* **88:** 76–82.

Current Management of Sickle Cell Anemia

Patrick T. McGann[1,2], Alecia C. Nero[3], and Russell E. Ware[1,2]

[1]Texas Children's Center for Global Health, Houston, Texas 77030

[2]Texas Children's Hematology Center, Baylor College of Medicine, Houston, Texas 77030

[3]University of Texas Southwestern Medical Center, Dallas, Texas 75390

Correspondence: reware@bcm.edu

Proper management of sickle cell anemia (SCA) begins with establishing the correct diagnosis early in life, ideally during the newborn period. The identification of affected infants by neonatal screening programs allows early initiation of prophylactic penicillin and pneumococcal immunizations, which help prevent overwhelming sepsis. Ongoing education of families promotes the early recognition of disease-released complications, which allows prompt and appropriate medical evaluation and therapeutic intervention. Periodic evaluation by trained specialists helps provide comprehensive care, including transcranial Doppler examinations to identify children at risk for primary stroke, plus assessments for other parenchymal organ damage as patients become teens and adults. Treatment approaches that previously highlighted acute vaso-occlusive events are now evolving to the concept of preventive therapy. Liberalized use of blood transfusions and early consideration of hydroxyurea treatment represent a new treatment paradigm for SCA management.

The natural history of untreated sickle cell anemia (SCA) is well described and documents serious morbidity and early mortality (Powars 1975; Platt et al. 1994; Serjeant 1995; Powars et al. 2005). Hemolytic anemia, acute vaso-occlusive events (VOEs), and chronic end-organ damage begin early in life, and complications accumulate throughout childhood. Without early identification or specific interventions, many patients with SCA have poor quality of life, and most die as young adults of SCA-related complications (Diggs 1973; Rogers et al. 1978).

Fortunately great strides have occurred over the past 40 years, and better management strategies have altered this previously bleak outlook. Despite the complexity and multifactorial pathophysiology of vaso-occlusion (Ware 2010a), relatively straightforward measures have greatly improved outcomes for children with SCA: (1) early identification by neonatal screening programs; (2) education of parents and patients about medical complications and early recognition; (3) preventive measures with prophylactic penicillin and pneumococcal immunizations; (4) aggressive treatment of acute VOEs including hydration, analgesics, antibiotics, and transfusions; (5) screening programs for early signs of organ damage, especially primary stroke risk using transcranial Doppler (TCD) examinations; and (6) therapeutic intervention with transfusions, hydroxyurea, or stem cell transplantation. For children receiving medical care

Copyright © 2013 Cold Spring Harbor Laboratory Press; all rights reserved
Cite this article as *Cold Spring Harb Perspect Med* doi: 10.1101/cshperspect.a011817

at comprehensive care programs, 95%–99% survival rates into adulthood are documented (Telfer et al. 2007; Quinn et al. 2010). For adults with SCA, screening programs and anticipatory guidance are less standardized but still critical, and the benefits of preventive therapy using hydroxyurea are even more compelling.

Here we will focus primarily on the management of SCA (HbSS or HbS/β^0-thalassemia). We emphasize and summarize general principles of care and management, rather than discussing details of pathophysiology or mechanisms of disease. The management of specific examples of acute VOEs will be highlighted.

EARLY IDENTIFICATION

Perhaps the most critical aspect of optimizing SCA management is early identification of affected patients, before the onset of signs and symptoms of disease. Without early diagnosis and intervention, SCA often acts as a swift and invisible killer, with many infants dying suddenly of bacterial sepsis or acute splenic sequestration crisis (ASSC) within the first few years of life (Pearson et al. 1969, 1977; Rogers et al. 1978; Powars et al. 1981). Sometimes fatal complications occur even before families or medical providers are aware the infants have SCA.

With the recognition that infants with SCA have greatly increased risk of bacterial sepsis, the landmark multicenter double-blinded placebo-controlled PROPS trial proved that penicillin prophylaxis significantly lowered the risk of bacteremia and death (Gaston et al. 1986). This simple intervention provided the justification needed for newborn screening of SCA, to identify affected infants soon after birth and to allow lifesaving prophylactic antibiotic therapy. Although a 1987 NIH Consensus Conference recommended newborn screening for SCA, universal screening was not accomplished in all U.S. states and territories until 2006.

Newborn screening programs in the United States, Jamaica, and Europe have documented the utility of early identification of SCA, with a marked reduction in morbidity and mortality, especially in the first 5 years of life (Rogers et al. 1978; Vichinsky et al. 1988; Almeida et al. 2001; Bardakdjian-Michau et al. 2001). Figure 1 illustrates that early identification of SCA through neonatal screening programs has contributed to the improved survival rates (Quinn et al. 2010).

Testing of newborns in the United States for SCA began with targeted screening, which involves selecting at-risk populations to screen, such as babies whose parents are African-American. Such an approach is problematic in several ways, and has evolved now to universal screening for all newborns. In contrast, most European countries still perform targeted screening for infants most likely to be affected, such as those of African ancestry. Although potentially cost-effective, targeted screening almost certainly misses some babies with SCA, and presents difficulties related to both equity and logistics (Grosse et al. 2005). Despite the high burden of disease, newborn screening has yet to be implemented

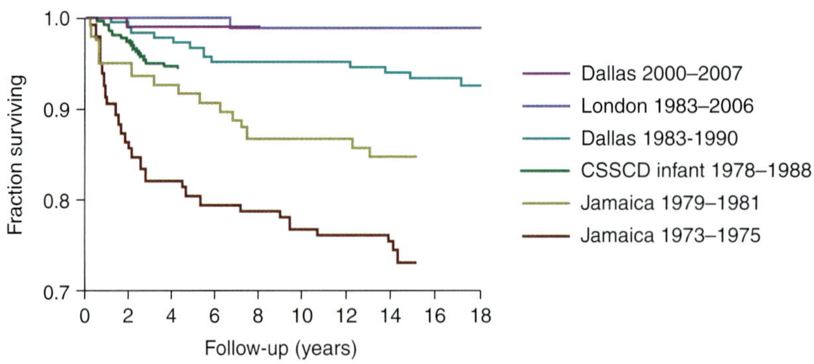

Figure 1. Survival curves of infants with SCA in the United States and Jamaica, by era. This research was originally published in *Blood*. (From Quinn et al. 2010; reprinted, with permission, © American Society of Hematology.)

systematically in Africa, although pilot studies document a high incidence of trait and disease.

Newborn screening for SCA requires a small dried blood spot (DBS) for analysis, collected from cord blood or the infant's heel/toe. Collection technique is important; DBS specimens have variable quality by the amount and distribution of blood on the filter paper. Specimens are most easily collected in the neonatal period for babies born in the hospital, or during initial immunization visits for babies born at home. Testing in the United States is most commonly performed by hemoglobin electrophoresis using isoelectric focusing (IEF), which easily distinguishes abnormal sickle hemoglobin (Hemoglobin S, HbS) from normal hemoglobin (HbA) and fetal hemoglobin (HbF), as illustrated in Figure 2. High-performance liquid chromatography (HPLC), capillary electrophoresis (CE) techniques, and even DNA-based laboratory diagnosis also can be used for accurate diagnosis. When possible, parental studies should also be performed to confirm the diagnosis of SCA.

EDUCATION

Parental and family sickle cell education should begin once the diagnosis is made and continue throughout childhood. Given variable parental education and literacy, education should be provided in written and spoken form, and should be repeated with each visit to ensure information is comprehended. When possible, both parents should receive education, plus extended family members and other caregivers, to become knowledgeable about SCA.

Education in the early newborn period should focus on the basics of SCA, including its genetics and inheritance, need for penicillin prophylaxis, and benefits of protein-conjugated pneumococcal immunizations. At each visit these key points should be repeated to parents and caregivers. The importance of regular medical care should be emphasized, especially the need for prompt medical evaluation for fever. Antipyretics should never be given for fever at home, because this treatment can mask a serious infection. Education for young patients should also include signs and symptoms of ASSC: pallor, fussiness or irritability, and tender splenomegaly. Teaching parents to palpate their baby's spleen regularly, ideally several times a day during routine diaper changes, allows early diagnosis of ASSC and potentially prompt and lifesaving intervention.

As the child grows, education should focus more on recognition and early medical intervention for acute vaso-occlusive complications

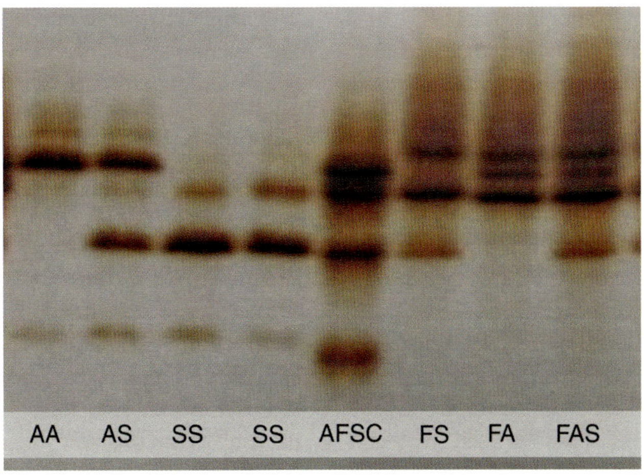

Figure 2. Isoelectric focusing (IEF) electrophoresis technique for identification of SCA. Blood specimens from AA, AS, and SS patient controls are shown on the *left*, along with a manufactured Hb AFSC control in the *center*. Newborn samples, typically obtained from dried blood spots, are tested for the abnormal FS pattern indicating SCA, versus the normal FA pattern or FAS pattern indicating sickle cell trait.

such as dactylitis and other painful events, respiratory distress, acute chest syndrome, and stroke. Parents should learn to manage mild pain at home with oral hydration and analgesia. At an early age, families should be introduced to possible treatment options including hydroxyurea and transfusions, and even stem cell transplantation, if available. As affected children grow up and enter adolescence, it is critical to provide ongoing education about SCA and its complications, to provide young patients with the skills necessary to understand and advocate for their own medical care. Such self-awareness and investment in their medical care becomes critically important on transition from pediatric to adult hematology care. Unfortunately, evidence suggests increased morbidity and mortality in late adolescence and early adulthood following this transition of care (Brousseau et al. 2010; Quinn et al. 2010).

PREVENTIVE MEASURES

As the result of vascular congestion, intraparenchymal sickling, and hypoxic injury to the spleen, infants with SCA have early loss of filtrative splenic function and are susceptible to acute life-threatening infections, particularly from encapsulated bacteria such as *Streptococcus pneumoniae* and *Haemophilus influenzae* type b (Winkelstein and Drachman 1968; Pearson et al. 1969; Pearson 1977). This risk remains increased throughout life, but is most significantly increased in the first 5 years, when bacteremia incidence is the highest (Overturf et al. 1977; Powars et al. 1981; Gill et al. 1995).

The PROPS trial showed that prophylactic oral penicillin reduced the frequency of bacterial infection by 84% among young children (age 3–36 mo) with SCA (Gaston et al. 1986). However, the follow-up PROPS 2 study was unable to show benefit from penicillin after age 5 yr, primarily owing to their lower incidence of bacteremia (Falletta et al. 1995). After introduction of protein-conjugated pneumococcal vaccines, the incidence of invasive pneumococcal disease decreased by 93.4% among young children with SCA (Halasa et al. 2007). In Africa the dangers of pneumococcal sepsis for children with SCA have been questioned, but recent data confirm its prevalence and lethality (Williams et al. 2009).

Based on overwhelming evidence, early pneumococcal prophylaxis is recommended for all infants with SCA. Penicillin 125 mg by mouth twice daily should begin by 3–4 mo of age, as a liquid formulation or crushed tablet. The penicillin dose should be increased to 250 mg by mouth twice a day as the child grows, typically at 3 yr of age. Some international programs recommend monthly IM penicillin to help ensure compliance. Oral erythromycin can be used as a substitute if penicillin allergy or rash develops, but this is uncommon.

Pneumococcal immunization should begin with locally available pneumococcal conjugate vaccines (7, 10, or 13 valency) and supplemented at age 2 and 5–7 yr with the 23-valent pneumococcal polysaccharide vaccine (Pneumovax). Additional recommended vaccinations include the *H. influenzae* type b series, meningococcal conjugate vaccine (Menactra), and yearly influenza. When locally feasible, the published immunization schedule for high-risk children as recommended by the American Academy of Pediatrics (www2.aap.org/immunization/IZSchedule.html) should be followed.

Penicillin prophylaxis should continue through age 5, when the risk of invasive bacterial disease is lower. Once the immunization series is up to date, including the Pneumovax booster, children with SCA may discontinue penicillin prophylaxis. However, penicillin should be continued indefinitely if a child has had culture-positive sepsis or a surgical splenectomy.

The dramatically increased risk of overwhelming and rapidly fatal infection among young patients with SCA must be understood by all caregivers and medical providers. Fever >38.5°C is a medical emergency requiring prompt medical evaluation, including physical examination with vital signs and splenic palpation, blood culture, complete blood count, reticulocytes, urinalysis, and chest X-ray if clinically warranted. Type and crossmatch should be obtained if there is extreme pallor, splenomegaly, clinical instability, or acute respiratory or neurologic symptoms. After obtaining the blood culture, broad-spectrum antibiotics (e.g., ceftriaxone) should be administered intravenously. Addition of another broad-spectrum antibiotic (e.g., vancomycin) should be considered if the child appears toxic, has high fever, or suspicion of central nervous system (CNS) infection.

Hospitalization is recommended if clinical or laboratory indicators suggest sepsis, such as hemodynamic compromise including hypotension, child <1 yr of age, prior history of sepsis, temperature $>40°C$, WBC $>30 \times 10^9/L$, or $<3 \times 10^9/L$, concurrent symptoms such as pain or acute anemia, or if close follow-up is not reliable (Lane et al. 2001).

ACUTE VASO-OCCLUSION

Pain Events

The sudden onset of pain that occurs frequently in patients with SCA results from acute intravascular sickling, so is often referred to as painful VOE or vaso-occlusive "crisis" (VOC). Although many providers and patients use the simple phrase "pain crisis," VOE is preferable because it broadly defines the process and avoids stigma about pain perception and management.

The VOE results from erythrocyte sickling, microvascular occlusion, and tissue ischemia/reperfusion, and is a hallmark clinical feature of SCA. Pain is the most common cause of acute morbidity of SCA, and is associated with severity of disease and early mortality among young adults (Platt et al. 1991). Pain often accompanies acute chest syndrome (ACS), a serious and potentially life-threatening complication of SCA, in 72% of cases (Platt et al. 1994; Vichinsky et al. 2000). This association usually follows sternal or truncal pain, which leads to splinting and poor inspiratory effort, and lack of active and complete inspiration following opioid-induced sedation. The frequency and severity of pain in SCA is more than just episodic and acute, however; pain in SCA is often chronic, underrecognized and underreported, and therefore undertreated (Solomon 2008). Pain diaries of 232 adult patients showed that SCA pain is common and often chronic; pain was present on 54.5% of days and 29.3% of patients reported pain on $>95\%$ of days (Smith et al. 2008).

The pathophysiology of vaso-occlusive pain is multifactorial and complex, and includes various blood cells including reticulocytes and neutrophils, plus plasma factors and vascular endothelium (Ware 2010a). Several factors have been identified as triggers of painful VOE, with individual patients often recognizing their own specific triggers. The most commonly described triggers include cold temperatures and especially cold water, as well as dehydration, overexertion, and menses (Redwood et al. 1976; Resar and Oski 1991; Yoong and Tuck 2002).

A combination approach of nonpharmacologic and pharmacologic agents should be used for acute management of vaso-occlusive painful events. Nonpharmacologic interventions with shown effectiveness include oral hydration, heat, massage, and various cognitive-behavioral and self-relaxation techniques (Rees et al. 2003; Dampier et al. 2004). Cold packs can increase local sickling and may exacerbate pain, so should be avoided. Pharmacologic interventions should begin at home with nonopioid analgesics, including acetaminophen and nonsteroidal anti-inflammatory drugs (NSAIDs); ibuprofen (10 mg/kg or 800 mg for adults >40 kg every 6–8 h) is an effective oral agent given its potent analgesic and anti-inflammatory properties. Corticosteroids may reduce the duration of painful VOE but on discontinuation, are associated with an increased frequency of rebound painful episodes requiring readmission (Griffin et al. 1994), and so are relatively contraindicated for routine pain management. If pain is not controlled with increased hydration, oral analgesia, and other conservative measures, opioids should be used. Oral narcotic therapy such as codeine and its derivatives can often be used effectively at home, in selected settings and patients.

In the event of severe vaso-occlusive pain requiring formal medical evaluation, aggressive pain management should be implemented promptly with intravenous hydration and opioid (morphine or hydromorphone) analgesia, and adjuvant intravenous NSAID therapy. Historically, the most painful VOEs have been evaluated and treated in the local emergency room, but given the relative lack of sickle cell knowledge and familiarity among some health-care providers, such evaluations have delays in initiating appropriate analgesia, poor pain control, and a high rate of hospital admission. A sickle cell day hospital approach, which features staffing by experienced sickle cell providers, is increasingly used and results in improved pain management, better patient satisfaction, and decreased rates of hospitalization for both adults and children with SCA (Benjamin et al. 2000; Raphael et al. 2008).

During treatment of VOE, frequent evaluation of pain is important to assess the degree of relief and potential side effects of narcotic analgesia. When hospitalization is required, continuous opioid infusion by patient-controlled analgesia (PCA) is recommended (van Beers et al. 2007; Jacob et al. 2008). Figure 3 illustrates a convenient algorithm to consider for management of mild to severe painful VOE, but flexibility should exist for individual patient preferences. When opioids are used, an aggressive bowel regimen should be used concurrently to reduce gastrointestinal complications, especially

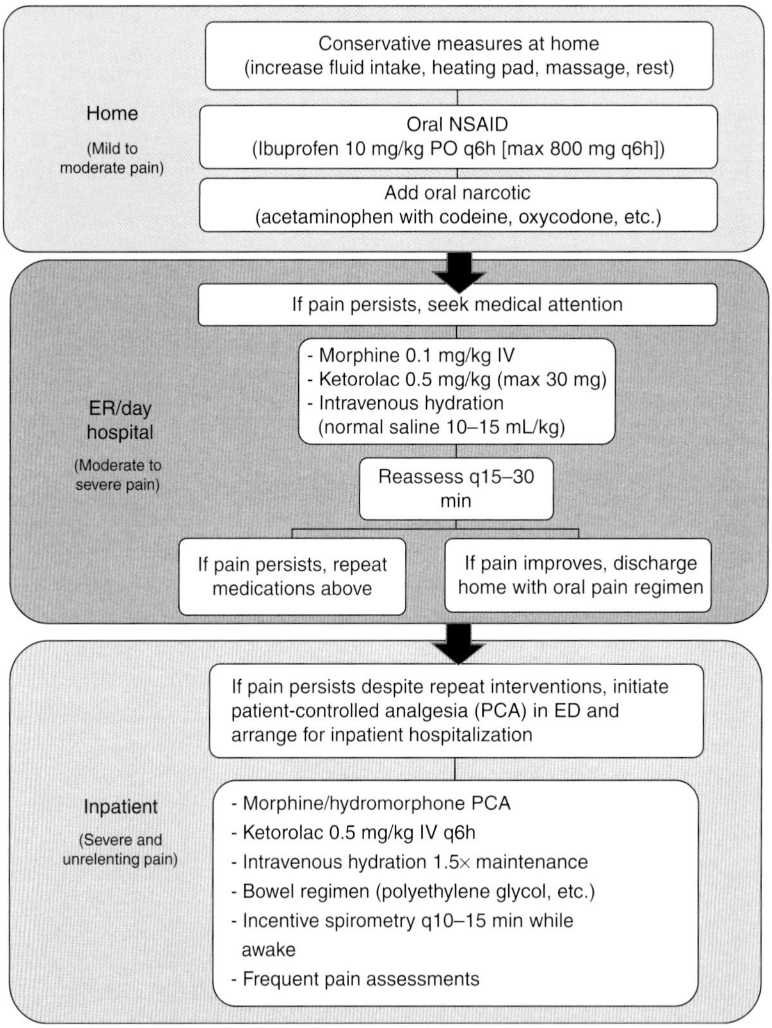

Figure 3. Algorithm for the management of painful vaso-occlusive events.

hypomotility (O'Brien et al. 2010). Teaching and encouraging frequent incentive spirometry with ambulation can reduce the risk of developing complications including ACS (Bellet et al. 1995; Ahmad et al. 2011).

Acute Splenic Sequestration Crisis

ASSC remains an important cause of morbidity and mortality for young children with SCA. Most ASSC events occur in infants or toddlers before age 2 yr. In some cases, ASSC may be the first clinical manifestation of SCA, and hence should be emphasized in the education of families during the first year of life.

The pathophysiology of ASSC involves erythrocyte sickling and rapid accumulation within the spleen. ASSC is clinically defined as a decrease in baseline hemoglobin concentration by ≥ 2 g/dL, in the presence of active reticulocytosis and splenomegaly; mild thrombocytopenia is common. The acute sequestration event can result in severe anemia, occasionally with hypovolemia, and even can evolve to circulatory shock or death (Topley et al. 1981; Emond et al. 1985; Powell et al. 1992).

Medical management of ASSC begins with early recognition and diagnosis; parents and caregivers must be educated about early signs and symptoms and the need to seek urgent medical evaluation. After initial assessment including vital signs and physical examination, laboratory studies should include complete blood count with reticulocytes, blood culture if febrile, and type and crossmatch. Transfusion volumes should not be excessive because a transfusion "overshoot" phenomenon can occur when the spleen abruptly unloads trapped erythrocytes, raising the hemoglobin level above the target goal. Small aliquots (from the same unit) should be administered every 12–24 h to treat anemia and hypovolemia, while avoiding hyperviscosity following splenic release.

Recurrent ASSC events are common, occurring in about half of the children who survive the first episode. Chronic transfusions can be implemented after the first episode, but their benefits on reducing recurrent events or avoiding splenectomy are limited (Kinney et al. 1990). The benefits of splenectomy for ASSC must be compared to its infectious and other postoperative risks; surgery is usually recommended only after one severe or life-threatening ASSC event, or after several recurrent ASSC events. Pneumococcal immunizations should be completed before surgery, and then lifelong penicillin prophylaxis is recommended (Ammann et al. 1977; Deodhar et al. 1993). Partial or subtotal splenectomy could potentially preserve some filtrative and immunological splenic function, but published reports in SCA are sparse and anecdotal experiences have been unsuccessful (Rice et al. 2003).

Acute Chest Syndrome

ACS is a common cause of morbidity and a leading cause of death among adults with SCA (Castro et al. 1994; Gill et al. 1995; Vichinsky et al. 2000; Powars et al. 2005). First described over 30 years ago (Charache et al. 1979), ACS has a complex pathophysiology that remains poorly defined. Numerous etiologies have been proposed including typical and atypical bacterial pathogens (Miller et al. 1991; Vichinsky et al. 2000; Neumayr et al. 2003), viral infection (Lowenthal et al. 1996), fat embolism (Vichinsky et al. 1994), intrapulmonary sequestration of erythrocytes (Vichinsky et al. 1994), and nitric oxide–hemoglobin interactions (Gladwin et al. 1999).

ACS is defined as a constellation of signs and symptoms including respiratory distress with tachypnea and dyspnea, hypoxemia, fever, elevated WBC count, mild anemia, and new infiltrate on chest X-ray (Castro et al. 1994; Vichinsky et al. 2000; Ballas et al. 2010). ACS is often characterized by rapid clinical decline, so a high index of suspicion is needed for early identification and intervention. The onset of ACS may be insidious, often including pain, with nearly 50% of patients admitted with a different diagnosis. Sternal and rib pain often leads to splinting and poor inspiration, which coupled with mild respiratory depression from opioids, can quickly deteriorate into a serious condition requiring aggressive respiratory and hematological support. To avoid this sequence of events, incentive spirometry can reduce the incidence of ACS and should be mandatory for all hospitalized patients with SCA (Bellet et al 1995; Ahmad et al. 2011).

Worsening hypoxemia or tachypnea, and early radiographic changes should lead to aggressive incentive spirometry, and oxygen therapy to correct hypoxemia. Close management of fluid status is warranted to prevent fluid overload. Although hyperhydration is typically recommended for painful VOE to improve blood flow, intravenous fluids should be limited to 50%–75% maintenance in the setting of evolving ACS, to reduce the risks of developing pulmonary edema or pleural effusions and worsening respiratory distress. Despite the lack of randomized clinical trials investigating the efficacy of antibiotics for ACS (Martí-Carvajal et al. 2007), coverage is provided for typical community-acquired and atypical pathogens, such as a broad-spectrum third-generation cephalosporin and macrolide (Lottenberg and Hassell 2005). Bronchodilators are not effective for all patients but are recommended for patients with ACS and concurrent reactive airways disease or asthma (Knight-Madden and Hambleton 2003; Knight-Madden et al. 2005).

Transfusions are often used for ACS (Lanzkowsky et al. 1978; Vichinsky et al. 2000), improving both anemia and oxygen-carrying capacity. Early simple transfusions are beneficial and can avert clinical deterioration that might warrant later exchange transfusion. Significant respiratory distress or clinical decompensation, hemoglobin $\geq$2 g/dL below baseline, and oxygen saturation $>$5% below baseline are all indications for packed red blood cell (PRBC) transfusion (Miller 2011). Automated erythrocytapheresis should be used for severe ACS associated with significant respiratory distress or hypoxia (Kleinman et al. 1984; Velasquez et al. 2009). There is no current evidence that inhaled nitric oxide (NO) has a beneficial role in the current management of ACS (Gladwin et al. 1999; Al Hajeri et al. 2008), but several clinical trials are ongoing. For recurrent ACS, both chronic transfusion programs (Miller et al. 2001a; Hankins et al. 2005a) and hydroxyurea (Steinberg et al. 2003; Hankins et al. 2005b; Wang et al. 2011) can reduce the frequency and severity of additional ACS events.

Stroke

Cerebrovascular accidents are a relatively common and devastating complication of SCA, with an overt stroke incidence rate of 11% by age 20 years and 24% by age 45 (Ohene-Frempong et al. 1998). Clinical stroke events represent only a fraction of the cerebrovascular complications of SCA, which include silent cerebral infarctions (Miller et al. 2001b; Pegelow et al. 2001; DeBaun et al. 2012) and other neurocognitive deficits (Schatz et al. 2001; Thompson et al. 2002).

In the setting of acute clinical stroke, quickly reestablishing cerebral blood flow is crucial; new-onset weakness or aphasia suggests stroke and intervention should never depend on confirmatory radiological imaging. Modest IV hydration can help acutely, and should be provided while blood is being crossmatched for transfusion. Particularly for children with severe anemia, a simple PRBC transfusion can rapidly reduce intravascular sludging and help improve cerebral blood flow, which is critical to help reverse acute symptoms and prevent stroke progression. When available, exchange transfusion should be performed promptly to reduce HbS $<$30%, with a target hemoglobin concentration of $\sim$10 g/dL (Swerdlow 2006). A retrospective analysis of children with overt stroke suggested children receiving exchange transfusion had a significantly lower risk of recurrent stroke, compared to children receiving only simple transfusion (Hulbert et al. 2006).

After an initial stroke, the risk of recurrent stroke events is 47%–93% without specific treatment (Powars et al. 1978; Balkaran et al. 1992; Pegelow et al. 1995). Chronic transfusions provided every 3–4 wk to maintain HbS of $\sim$30% are recommended to prevent recurrent events (Lusher et al. 1976; Pegelow et al. 1995; Strater et al. 2002; Platt 2006). Once initiated, transfusions should be continued indefinitely, because discontinuation of transfusions is associated with increased risk of recurrent events (Wang et al. 1991; Adams and Bramilla 2005).

 Cite this article as *Cold Spring Harb Perspect Med* doi: 10.1101/cshperspect.a011817

Although efficacious, 10%–20% of patients will develop a second stroke despite transfusions (Pegelow et al. 1995; Scothorn et al 2002). Recent evidence further shows progression of vasculopathy and silent cerebral infarctions among chronically transfused patients (Hulbert et al. 2011). Hydroxyurea for the prevention of recurrent stroke, coupled with phlebotomy to remove iron overload, has shown efficacy (Ware et al. 1999, 2004), but in a phase III randomized clinical trial was inferior to transfusions and chelation therapy (Ware and Helms 2012). In certain clinical settings in which chronic transfusions are unsafe or otherwise not feasible, however, hydroxyurea may be a viable treatment option (Ali et al. 2011).

SCREENING PROGRAMS

Prevention of complications is the ideal way to reduce the morbidity of acute VOE and potentially limit chronic organ damage; this is especially true when considering devastating neurologic events like stroke. Because blood flow velocity is inversely related to vessel diameter, the measurement of intracerebral arterial blood flow by transcranial Doppler ultrasonography (TCD) allows easy and noninvasive identification of large vessel stenosis and can accurately identify children at increased risk of developing primary stroke (Adams et al. 1992). Children with SCA have increased mean flow velocities when compared to children without anemia (Adams et al. 1989, 1992), and markedly elevated TCD values represent a biomarker of cerebrovascular disease and a significant risk factor for primary stroke.

Based on its utility of screening for stroke risk, both with efficacy in clinical trials (Adams et al. 1998) and effectiveness in clinical practice (McCarville et al. 2008; Enninful-Eghan et al. 2010; Bernaudin et al. 2011; Kwiatkowski et al. 2011), TCD screening is recommended annually for children with SCA starting at the age of 2–3 years. Figure 4 illustrates the recommended screening regimen and algorithm for management of TCD results (Platt 2006; McCarville et al. 2008). The risk of first stroke can be nearly eliminated by the initiation of a chronic transfusion program to decrease HbS <30% for time-averaged maximum blood flow velocities >200 cm/sec in the internal carotid or middle cerebral artery (Adams et al. 1998; Lee et al. 2006). Despite its clear efficacy for prevention of primary stroke, the risks of chronic transfusion therapy are significant and include iron overload, alloimmunization, and cost (Rosse et al. 1990; Harmatz et al. 2000; Wayne et al. 2000). Hydroxyurea therapy at maximum tolerated dose (MTD) also can reduce TCD velocities (Zimmerman et al. 2007) and the ongoing TWiTCH trial (ClinicalTrials.gov NCT00122980)

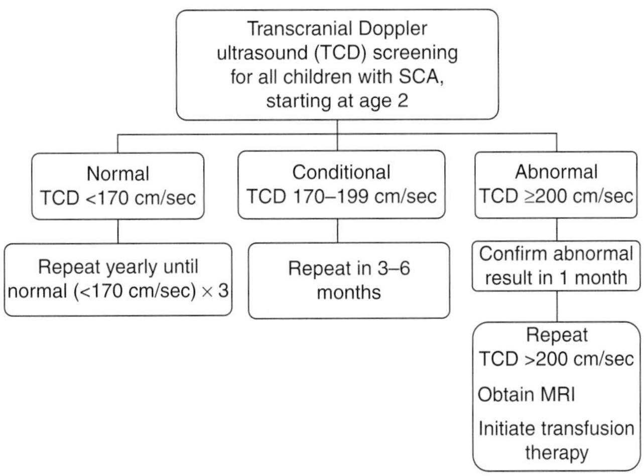

Figure 4. Transcranial Doppler screening algorithm for children with SCA.

compares the efficacy of hydroxyurea versus transfusions for prevention of primary stroke in children with abnormal TCD velocities.

ADULT CARE

Improved care of the child with SCA has led to increased survival rates into adulthood, and now attention must be focused on appropriate management of this older population. The success of neonatal and pediatric care leads to the formidable task of transitioning patients to the adult healthcare arena, in which providers often are neither trained adequately nor prepared to manage their unique needs. Written guidelines exist for managing adults with SCA (Lottenberg and Hassell 2005) but most of these recommendations do not derive from evidence-based research. The upcoming National Heart, Lung, and Blood Institute (NHLBI) guidelines document for SCA care, anticipated for release and distribution in 2013, will provide current evidence-based guidelines.

Adults with SCD should receive immunization boosters including pneumococcal, meningococcal, and even varicella vaccines based on age-specific recommendations. Management of acute VOEs such as pain, ACS, stroke, and priapism are similar for adults as for pediatric patients, but new issues may emerge related to chronic lung disease, especially restrictive pattern; leg ulcers that are difficult to manage and heal; and hepatic, endocrine, and even cardiac damage related to transfusional iron overload.

Additional screening for specific organ damage is recommended for teens and adults. Baseline pulse oximetry readings will help establish a baseline value and identify chronic hypoxemia. Blood pressure and serum creatinine are typically lower than age-related published values; hence, values at the upper limit of normal may indicate renal dysfunction. Dipstick urinalysis identifies gross proteinuria, but quantitative testing for microalbuminuria, using a 24-h urine collection if warranted, detects subclinical renal disease. Screening for hepatitis B, C, and HIV is warranted for patients who receive chronic blood transfusions; serum ferritin with review of transfusion history may identify patients who need iron chelation. Periodic screening for tricuspid regurgitation jet (TR jet) velocities can identify patients with pulmonary hypertension and possible early mortality (Gladwin et al. 2004). Ophthalmological screening for retinal disease is warranted every 1–3 yr with specialist referral for patients with abnormalities. Additional screening tests for the general adult population should also be provided, including strategies to prevent or identify cancer, hyperlipidemia, bone density loss, and even diabetes. (In the latter case, the lack of HbA makes screening with HbA1c innacurate, but serum fructosamine can be used.)

THERAPEUTIC INTERVENTION

Urgent erythrocyte transfusions are indicated for many acute complications of SCA including ACS, ASSC, transient aplastic crisis owing to parvovirus B19 infection, and acute stroke. In these settings, transfused blood helps to alleviate anemia, improve circulating blood volume, increase oxygen-carrying capacity, and provide erythrocytes that cannot sickle. If the posttransfusion target is high enough, transfusions also help suppress endogenous sickle erythropoiesis. Elective transfusions are often given for preoperative management, to prevent perioperative sickle-related complications. Chronic transfusions given on a monthly basis are also highly efficacious for primary and secondary stroke prevention. In contrast, transfusions are not indicated for acute painful events or anemia per se (recognizing that almost all patients have a baseline steady-state partially compensated hemolytic anemia), and have little role in the management of standard VOE (Smith-Whitley and Thompson 2012).

In most acute settings, simple transfusions with packed erythrocytes (PRBC) should be administered. PRBC are readily available across the United States, and are routinely tested for HIV as well as hepatitis B and C. As a general principle, simple transfusions should be given with a target of alleviating anemia or treating the underlying condition; whole units (or half units for small pediatric patients) should be administered whenever possible, instead of fixed volumes (e.g., 10 mL/kg), to help limit foreign antigen exposure. The posttransfusion target hemoglobin concentration should not exceed 10–11 g/dL in the untreated patient because hyperviscosity can occur; in chronically transfused patients with low %HbS, however, the posttransfusion target can be raised to help suppress endogenous erythrocyte production. It is also important for the Blood Bank to be aware that the patient has SCA, because extending red blood cell (RBC) phenotype matching for minor blood group antigens is recommended to help prevent alloimmunization (Yazdanbakhsh et al. 2012).

For patients with neurological indications for chronic transfusion therapy such as abnormal TCD velocities or stroke, repeated simple transfusions are effective in preventing primary and secondary stroke, respectively, but ultimately result in transfusional iron overload. For this reason, partial exchange transfusions or isovolemic erythrocytapheresis is recommended to minimize iron accumulation. In most patients, intravenous access for exchange transfusions is facilitated by the placement of an implantable device. With chronic transfusions, the goal is typically HbS $\leq$30% as a pretransfusion value, which typically requires transfusion every 3–5 weeks depending on the type and volume of each transfusion, the patient's own erythropoietic drive, and the response to transfusion therapy. Chelation therapy for transfusional iron overload should be considered for all patients on chronic transfusions, but also for teens and adults who have a large cumulative number of episodic or sporadic transfusions.

HYDROXYUREA

Increased fetal hemoglobin (HbF) levels have been associated with a less severe phenotype of SCA (Conley et al. 1963; Diggs 1973; Platt et al. 1991, 1994) and HbF induction has become a desired pharmacologic end point for SCA therapy (Charache 1990; Charache et al. 1992). Hydroxyurea has been shown to potently increase HbF and is currently the most effective disease-modifying therapy for both adults and children with SCA (Platt et al. 1984; Steinberg et al. 2003; Zimmerman et al. 2004; Hankins et al. 2005b). The first clinical experience with hydroxyurea for SCA was reported nearly 30 years ago in seminal proof-of-principles studies (Platt et al. 1984). Subsequently, a multicenter phase II study documented laboratory efficacy (increased Hb, %HbF, and MCV; decreased WBC, ANC, ARC, and platelets) of hydroxyurea using a dose escalation schedule to MTD (Charache et al. 1992). The Multi-Center Study of Hydroxyurea (MSH) double-blinded, placebo-controlled randomized clinical trial showed clinical efficacy of hydroxyurea for adults with severe SCA, with significantly reduced time to first painful event, plus fewer episodes of ACS, transfusions, and hospitalizations (Charache et al. 1995).

In children with SCA, similar laboratory and clinical efficacy have been shown in open-label trials (Kinney et al. 1999; Wang et al. 2001; Zimmerman et al. 2004; Hankins et al. 2005b; Thornburg et al. 2009). In hydroxyurea study of long-term effects (HUSTLE), all pediatric patients with medication adherence had HbF responses, although responses were variable and possibly related to differences in drug absorption, pharmacokinetics, and pharmacogenetics (Ware et al. 2011). The results from the double-blinded, placebo-controlled multicenter randomized BABY HUG study show the safety and clinical efficacy of hydroxyurea for young infants with SCA, regardless of previous clinical severity (Wang et al. 2011). The primary end point of BABY HUG was the ability of hydroxyurea to prevent chronic organ damage (kidney, spleen), and the short-term study results were equivocal. Anecdotal reports suggest prevention and even reversal of chronic organ damage with hydroxyurea therapy (Zimmerman et al. 2007; Hankins et al. 2008; Thornburg et al. 2009), so further investigation of the BABY HUG cohort is necessary.

Long-term follow up from MSH and the Greek Laikon Study of Hydroxyurea in Sickle Cell Syndromes documented reduced mortality for adult patients with SCA on hydroxyurea (Steinberg et al. 2010; Voskaridou et al. 2010). There is now indisputable evidence that hydroxyurea has laboratory and clinical efficacy for all ages; a growing body of evidence also supports the long-term safety of hydroxyurea and the ability of hydroxyurea to prevent chronic organ damage and reduce mortality (McGann et al. 2011). Whereas hydroxyurea previously has been reserved for older patients with a severe clinical course, hydroxyurea use should be liberalized and offered to all adults with SCA. An increasing number of pediatric hematologists believe hydroxyurea should now be considered as a treatment option for all children with SCA, regardless of age or previous clinical course.

Hydroxyurea should be initiated by an experienced clinician familiar with laboratory monitoring and appropriate dose escalation to MTD. Hydroxyurea treatment should commence at ~20 mg/kg/d by mouth, once daily. Complete blood count (CBC) should be checked every 4 wk to monitor for myelosuppression, which is typically mild and dose dependent, and always reversible by holding the hydroxyurea dose temporarily (Heeney and Ware 2008). Dose adequacy and medication compliance can be assessed by reviewing changes in CBC parameters and reviewing the peripheral blood smear (Fig. 5). To reach MTD, the daily dose should be escalated by ~5 mg/kg every 8 wk until MTD is mild neutropenia (e.g., ANC of 1500–3000 × 10^6/L) or reticulocytopenia (ARC of 100–150 × 10^9/L) is reached on a stable dose. Drug toxicity is usually defined by cytopenias such as ANC <1.0 × 10^9/L, hemoglobin <7.0 g/dL with low reticulocyte count, ARC <80 × 10^9/L, and platelets <80 × 10^9/L (Heeney and Ware 2008). Given the unlikelihood of true hydroxyurea "nonresponders," efforts must be made to encourage medication compliance to reach and maintain a stable and efficacious hydroxyurea MTD (Ware 2010b).

OTHER TREATMENTS

HbF induction can be achieved by a group of short-chain fatty acids that inhibit the enzyme histone deacetylase; such HDAC inhibitors, primarily butyrate, can alter chromatin structure and induce HbF production by altering the transcription of the γ-globin gene (McCaffrey et al. 1997; Ataga 2009; Bradner et al. 2010). Clinical experience with HDAC inhibitors for SCA is limited but anecdotal reports suggest robust HbF induction in some patients with SCA (Atweh et al. 1999; Hines et al. 2008).

Decitabine is a nucleoside analog that induces HbF induction via epigenetic modulation, specifically hypomethylation of the γ-globin gene promoter. Experience with decitabine for SCA is also relatively limited, but several reports suggest clinical and laboratory efficacy of subcutaneously administered decitabine in adults who were not responsive to hydroxyurea (Creusot et al. 1982; DeSimone et al. 2002; Saunthararajah et al. 2003, 2008). Prospective trials of decitabine are warranted to determine if it has efficacy for a broad spectrum of patients with SCA.

Additional treatments that target specific pathways of the pathophysiology of SCA are just entering into clinical trials. One new promising inhibitor of the Gardos channel was found to have favorable effects on hemolysis and RBC

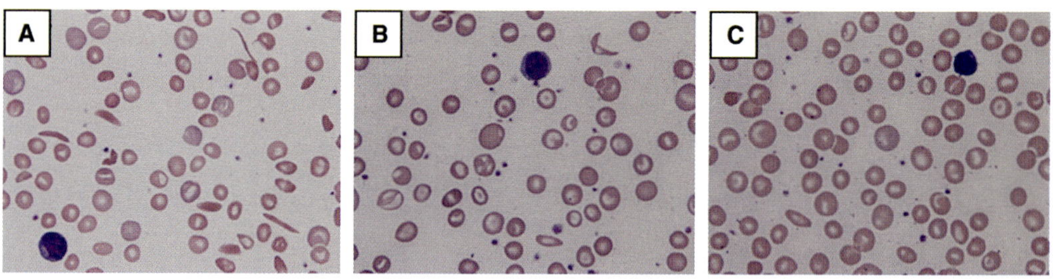

Figure 5. Blood smear changes with hydroxyurea therapy. *A* illustrates the untreated patient with anemia and numerous sickled forms; *B* is after initiation of hydroxyurea treatment with macrocytosis and more target cells; and *C* is after reaching a stable hydroxyurea MTD with less anemia and fewer sickled forms. (From Ware 2010b; reprinted, with permission.)

survival, yet did not have clinical efficacy in a phase III randomized clinical trial (Ataga et al. 2011).

CONCLUDING REMARKS

Decades of observational studies and therapeutic trials have contributed to a greater understanding of the pathophysiology and management of SCA. Based on these results, relatively simple interventions can substantially improve the survival of SCA, especially among children. Newborn screening, early preventive treatments, education about complications, and screening programs improve both the morbidity and mortality of SCA. Going forward, attention must focus on the care and management of teens and adults with SCA, and address quality of life as well as medical complications. More aggressive treatment of SCA is supported by current evidence, and therapeutic options with hydroxyurea should be considered early in life.

REFERENCES

Adams RJ, Bramilla D. 2005. Discontinuing prophylactic transfusions used to prevent stroke in sickle cell disease. *N Engl J Med* **353:** 2769–2778.

Adams RJ, Nichols FT, McKie VC, McKie KM, Stephens S, Carl E, Thompson WO. 1989. Transcranial Doppler: Influence of hematocrit in children with sickle cell anemia without stroke. *J Cardiovasc Technol* **8:** 97–101.

Adams R, McKie V, Nichols F, Carl E, Zhang DL, McKie K, Figueroa R, Litaker M, Thompson W, Hess D. 1992. The use of transcranial ultrasoundography to predict stroke in sickle cell disease. *N Engl J Med* **326:** 605–610.

Adams RJ, McKie VC, Hsu L, Files B, Vichinsky E, Pegelow C, Abboud M, Gallagher D, Kutlar A, Nichols FT, et al. 1998. Prevention of a first stroke by transfusions in children with sickle cell anemia and abnormal results on transcranial Doppler ultrasonography. *N Engl J Med* **339:** 5–11.

Ahmad FA, Macias CG, Allen JY. 2011. The use of incentive spirometry in pediatric patients with sickle cell disease to reduce the incidence of acute chest syndrome. *J Pediatr Hematol Oncol* **33:** 415–420.

Al Hajeri A, Serjeant GR, Fedorowicz Z. 2008. Inhaled nitric oxide for acute chest syndrome in people with sickle cell disease. *Cochrane Database Syst Rev* **1:** CD006957.

Ali SB, Moosang M, King L, Knight-Madden J, Reid M. 2011. Stroke recurrence in children with sickle cell disease treated with hydroxyurea following first clinical stroke. *Am J Hematol* **86:** 846–850.

Almeida AM, Henthorn JS, Davies SC. 2001. Neonatal screening for haemoglobinopathies: The results of a 10-year programme in an English Health Region. *Br J Hematol* **112:** 32–35.

Ammann AJ, Addiego J, Wara DW, Lubin B, Smith WB, Mentzer WC. 1977. Polyvalent pneumococcal-polysaccharide immunization of patients with sickle-cell anemia and patients with splenectomy. *N Engl J Med* **297:** 897–900.

Ataga KI. 2009. Novel therapies in sickle cell disease. *Hematology Am Soc Hematol Educ Program* **2006:** 54–61.

Ataga KI, Reid M, Ballas SK, Yasin Z, Bigelow C, James LS, Smith WR, Galaceros F, Kutlar A, Hull JH, et al. 2011. Improvements in haemolysis and indicators of erythrocyte survival do not correlate with acute vaso-occlusive crises in patients with sickle cell disease: A phase III randomized placebo-controlled double-blind study of the Gardos channel blocker senicapoc (ICA-17043). *Br J Hematol* **153:** 92–104.

Atweh GF, Sutton M, Nassif I, Boosalis V, Dover GJ, Wallenstein S, Wright E, McMahon L, Stamatoyannopoulos G, Faller DV, et al. 1999. Sustained induction of fetal hemoglobin by pulse butyrate therapy in sickle cell disease. *Blood* **93:** 1790–1797.

Balkaran B, Char G, Morris JS, Thomas TW, Serjeant BE, Serjeant GR. 1992. Stroke in a cohort of patients with homozygous sickle cell disease. *J Pediatr* **120:** 360–366.

Ballas SK, Lieff S, Benjamin LJ, Dampler CD, Heeney MM, Hoppe C, Johnson CS, Rogers ZR, Smith-Whitley K, Wang WC, et al. 2010. Definitions of the phenotypic manifestations of sickle cell disease. *Am J Hematol* **85:** 6–13.

Bardakdjian-Michau J, Guilloud-Batailie M, Maier-Redelsperger M, Elion J, Girot R, Feingold J, Galactéros F, de Montalembert M. 2001. Decreased morbidity in homozygous sickle cell disease detected at birth. *Hemoglobin* **26:** 211–217.

Bellet PS, Kalinyak KA, Shukla R, Gelfand MJ, Rucknagel DL. 1995. Incentive spirometry to prevent acute pulmonary complications in sickle cell diseases. *N Engl J Med* **333:** 699–703.

Benjamin LR, Swinson GI, Nagel RL. 2000. Sickle cell anemia day hospital: An approach for the management of uncomplicated painful crises. *Blood* **95:** 1130–1136.

Bernaudin F, Verlhac S, Arnaud C, Kamdem A, Chevret S, Hau I, Coïc L, Leveillé E, Lemarchand E, Lesprit E, et al. 2011. Impact of early transcranial Doppler screening and intensive therapy on cerebral vasculopathy outcome in a newborn sickle cell anemia cohort. *Blood* **117:** 1130–1140.

Bradner JE, Mak R, Tanguturi SK, Mazitschek R, Haggarty SJ, Ross K, Chang CY, Bosco J, West N, Morse E, et al. 2010. Chemical genetic strategy identifies histone deacetylase 1 (HDAC1) and HDAC2 as therapeutic targets in sickle cell disease. *Proc Natl Acad Sci* **107:** 12617–12622.

Brousseau DC, Owens PL, Mosso AL, Panepinto JA, Steiner CA. 2010. Acute care utilization and rehospitalization for sickle cell disease. *JAMA* **303:** 1288–1294.

Castro O, Brambilla DJ, Thorington B, Reindorf CA, Scott RB, Gillette P, Vera JC, Levy PS. 1994. The acute chest syndrome in sickle cell disease: Incidence and risk factors. *Blood* **84:** 643–649.

Charache S. 1990. Fetal hemoglobin, sickling, and sickle cell disease. *Adv Pediatr* **37:** 1–31.

Charache S, Scott JC, Charache P. 1979. "Acute chest syndrome" in adults with sickle cell anemia. *Arch Intern Med* **139:** 67–69.

Charache S, Dover GJ, Moore RD, Eckert S, Ballas SK, Koshy M, Milner PF, Orringer EP, Phillips G Jr, Platt OS. 1992. Hydroxyurea: Effects on hemoglobin F production in patients with sickle cell anemia. *Blood* **79:** 2555–2565.

Charache S, Terrin ML, Moore RD, Dover GJ, Barton FB, Eckert SV, McMahon RP, Bonds DR, and the Investigators of the Multicenter Study of Hydroxyurea in Sickle Cell Anemia. 1995. Effect of hydroxyurea on the frequency of painful crises in sickle cell anemia. *N Engl J Med* **332:** 1317–1322.

Conley CL, Weatherall DJ, Richardson SN, Shepard MK, Charache S. 1963. Hereditary persistence of fetal hemoglobin: A study of 79 affected persons in 15 Negro families in Baltimore. *Blood* **21:** 261–281.

Creusot F, Acs G, Christman JK. 1982. Inhibition of DNA methyltransferase and induction of Friend erythroleukemia cell differentiation by 5-azacytidine and 5-aza2′-deoxycytidine. *J Biol Chem* **257:** 2041–2048.

Dampier C, Ely E, Eggleston B, Brodecki D, O'Neal P. 2004. Physical and cognitive-behavioral activities used in the home management of sickle pain: A daily diary study in children and adolescents. *Pediatr Blood Cancer* **43:** 674–678.

DeBaun MR, Armstrong FD, McKinstry RC, Ware RE, Vichinsky E, Kirkham FJ. 2012. Silent cerebral infarcts: A review on a prevalent and progressive cause of neurologic injury in sickle cell anemia. *Blood* **119:** 4587–4596.

Deodhar HA, Marshall RJ, Barnes JN. 1993. Increased risk of sepsis after splenectomy. *BMJ* **307:** 1408–1409.

DeSimone J, Koshy M, Dorn L, Lavelle D, Bressler L, Molokie R, Talischy N. 2002. Maintenance of elevated fetal hemoglobin levels by decitabine during dose interval treatment of sickle cell anemia. *Blood* **99:** 3905–3908.

Diggs LW. 1973. Anatomic lesions in sickle cell disease. In *Sickle cell disease: Diagnosis, management, education, and research* (ed. Abramson H, et al.), pp. 189–229. Mosby, St. Louis.

Emond AM, Collis R, Darvill D, Higgs DR, Maude GH, Serjeant GR. 1985. Acute splenic sequestration in homozygous sickle cell disease: Natural history and management. *J Pediatr* **107:** 201–206.

Enninful-Eghan H, Moore RH, Ichord R, Smith-Whitley K, Kwiatkowski JL. 2010. Transcranial Doppler ultrasonography and prophylactic transfusion program is effective in preventing overt stroke in children with sickle cell disease. *J Pediatr* **157:** 479–484.

Falletta JM, Woods GM, Verter JI, Buchanan GR, Pegelow CH, Iyer RV, Miller ST, Holbrook CT, Kinney TR, Vichinsky E, et al. 1995. Discontinuing penicillin prophylaxis in children with sickle cell anemia. Prophylactic Penicillin Study II. *J Pediatr* **127:** 685–690.

Gaston MH, Verter JI, Woods G, Pegelow C, Kelleher J, Presbury G, Zarkowsky H, Vichinsky E, Iyer R, Lobel JS, et al. 1986. Prophylaxis with oral penicillin in children with sickle cell anemia. *N Engl J Med* **314:** 1593–1599.

Gill FM, Sleeper LA, Weiner SJ, Brown AK, Bellevue R, Grover R, Pegelow CH, Vichinsky E. 1995. Clinical events in the first decade in a cohort of infants with sickle cell disease. Cooperative Study of Sickle Cell Disease. *Blood* **86:** 776–783.

Gladwin MT, Schechter AN, Shelhamer JH, Ognibene FP. 1999. The acute chest syndrome in sickle cell disease: Possible role of nitric oxide in its pathophysiology and treatment. *Am J Resp Crit Care Med* **159:** 1368–1376.

Gladwin MT, Sachdev V, Jison ML, Shizukuda Y, Plehn JF, Minter K, Brown B, Coles WA, Nichols JS, Ernst I, et al. 2004. Pulmonary hypertension as a risk factor for death in patients with sickle cell disease. *N Engl J Med* **350:** 886–895.

Griffin TC, McIntire D, Buchanan GR. 1994. High-dose intravenous methylprednisolone therapy for pain in children and adolescents with sickle cell disease. *N Engl J Med* **330:** 733–737.

Grosse SD, Olney RS, Baily MA. 2005. The cost effectiveness of universal versus selective newborn screening for sickle cell disease in the US and the UK: A critique. *Appl Health Econ Health Policy* **4:** 239–247.

Halasa NB, Shankar SM, Talbot TR, Arbogast PG, Mitchel EF, Wang WC, Schaffner W, Craig AS, Griffin MR. 2007. Incidence of invasive pneumococcal disease among individuals with sickle cell disease before and after the introduction of the pneumococcal conjugate vaccine. *Clin Infect Dis* **44:** 1428–1433.

Hankins J, Jeng M, Harris S, Li CS, Liu T, Wang W. 2005a. Chronic transfusion therapy for children with sickle cell disease and recurrent acute chest syndrome. *J Pediatr Hematol Oncol* **27:** 158–161.

Hankins JS, Ware RE, Rogers ZR, Wynn LW, Lane PA, Scott JP, Wang WC. 2005b. Long-term hydroxyurea therapy for infants with sickle cell anemia: The HUSOFT extension study. *Blood* **106:** 2269–2275.

Hankins JS, Helton KJ, McCarville MB, Li CS, Wang WC, Ware RE. 2008. Preservation of spleen and brain function in children with sickle cell anemia treated with hydroxyurea. *Pediatr Blood Cancer* **50:** 293–297.

Harmatz P, Butensky E, Quirolo K, Williams R, Ferrell L, Moyer T, Golden D, Neumayr L, Vichinsky E. 2000. Severity of iron overload in patients with sickle cell disease receiving chronic red blood cell transfusion therapy. *Blood* **96:** 76–79.

Heeney MM, Ware RE. 2008. Hydroxyurea for children with sickle cell disease. *Pediatr Clin N Am* **55:** 483–501.

Hines P, Dover GJ, Resar LM. 2008. Pulsed-dosing with oral sodium phenylbutyrate increases hemoglobin F in a patient with sickle cell anemia. *Pediatr Blood Cancer* **50:** 357–359.

Hulbert ML, Scothorn DJ, Panepinto JA, Scott JP, Buchanan GR, Sarnaik S, Fallon R, Chu JY, Wang W, Casella JF, et al. 2006. Exchange blood transfusion compared with simple transfusion for first overt stroke is associated with a lower risk of subsequent stroke: A retrospective cohort study of 137 children with sickle cell anemia. *J Pediatr* **149:** 710–712.

Hulbert ML, McKinstry RC, Lacey JL, Moran CJ, Panepinto JA, Thompson AA, Sarnaik SA, Woods GM, Casella JF, Inusa B, et al. 2011. Silent cerebral infarcts occur despite regular blood transfusion therapy after first strokes in children with sickle cell disease. *Blood* **117:** 772–779.

Jacob E, Hockenberry M, Mueller BU. 2008. Effects of patient controlled analgesia hydromorphone during acute painful episodes in adolescents with sickle cell disease: A pilot study. *J Pain Manag* **1:** 173–178.

Kinney TR, Ware RE, Schultz WH, Filston HC. 1990. Long-term management of splenic sequestration in children with sickle cell disease. *J Pediatr* **117:** 194–199.

Kinney TR, Helms RW, O'Branski EE, Ohene-Frempong K, Wang W, Daeschner C, Vichinsky E, Redding-Lallinger R, Gee B, Platt OS, et al. 1999. Safety of hydroxyurea in children with sickle cell anemia: Results of the HUG-KIDS study, a phase I/II trial. Pediatric Hydroxyurea Group. *Blood* **94:** 1550–1554.

Kleinman SH, Hurvitz CG, Goldfinger D. 1984. Use of erythrocytapheresis in the treatment of patients with sickle cell anemia. *J Clin Apher* **2:** 170–176.

Knight-Madden JA, Hambleton IR. 2003. Inhaled bronchodilators for acute chest syndrome in people with sickle cell disease. *Cochrane Database Syst Rev* **3:** CD003733.

Knight-Madden JM, Forrester TS, Lewis NA, Greenough A. 2005. Asthma in children with sickle cell disease and its association with acute chest syndrome. *Thorax* **60:** 206–210.

Kwiatkowski JL, Yim Eunsil, Miller S, Adams RJ. 2011. Effect of transfusion therapy on transcranial doppler ultrasonography velocities in children with sickle cell disease. *Pediatr Blood Cancer* **56:** 777–782.

Lane PA, Buchanan GR, Hutter JJ, Austin RF, Britton HA, Rogers ZR, Eckman JR, DeBaun MR, Wang WC, Matthew P, et al. 2001. Sickle cell disease in children and adolescents: Diagnosis, guidelines for comprehensive care and care paths and protocols for management of acute and chronic complications. Health Resources and Services Administration, Department of Health and Human Services, http://www.scinfo.org/protchild-index.htm.

Lanzkowsky P, Shende A, Karayalcin G, Kim YJ, Aballi AJ. 1978. Partial exchange transfusion in sickle cell anemia: Use in children with serious complications. *Am J Dis Child* **132:** 1206–1208.

Lee MT, Piomelli S, Granger S, Miller ST, Harkness S, Brambilla DJ, Adams RJ. 2006. Stroke prevention trial in sickle cell anemia (STOP): Extended follow-up and final results. *Blood* **108:** 847–852.

Lottenberg R, Hassell KL. 2005. An evidence-based approach to the treatment of adults with sickle cell disease. *Hematol Am Soc Hematol Educ Program* 2006: 58–65.

Lowenthal EA, Wells A, Emanuel PD, Player R, Prchal JT. 1996. Sickle cell acute chest syndrome associated with parvovirus B19 infection: Case series and review. *Am J Hematol* **51:** 207–213.

Lusher JM, Haghighat H, Khalifa AS. 1976. A prophylactic transfusion program for children with sickle cell anemia complicated by CNS infarction. *Am J Hematol* **1:** 265–273.

Martí-Carvajal AJ, Conterno LO, Knight-Madden JM. 2007. Antibiotics for treating acute chest syndrome in people with sickle cell disease. *Cochrane Database Syst Rev* **2:** 1–14.

McCaffrey PG, Newsome DA, Fibach E, Yoshida M, Su MS. 1997. Induction of γ-globin by histone decetylase inhibitors. *Blood* **90:** 2075–2083.

McCarville MB, Goodin GS, Fortner G, Chin-Shang L, Smeltzer MP, Adams R, Wang W. 2008. Evaluation of a comprehensive transcranial doppler screening program for children with sickle cell anemia. *Pediatr Blood Cancer* **50:** 818–821.

McGann PT, Ware RE. 2011. Hydroxyurea for sickle cell anemia: What have we learned and what questions still remain? *Curr Opinion Hematol* **18:** 158–165.

Miller ST. 2011. How I treat acute chest syndrome in children with sickle cell disease. *Blood* **117:** 5297–5305.

Miller ST, Hammerschlag MR, Chirgwin K, Rao SP, Roblin P, Gelling M, Stilerman T, Schachter J, Cassell G. 1991. Role of *Chlamydia pneumoniae* in acute chest syndrome of sickle cell disease. *J Pediatr* **118:** 30–33.

Miller ST, Wright E, Abboud M, Berman B, Files B, Scher CD, Styles L, Adams RJ. 2001a. Impact of chronic transfusion on incidence of pain and acute chest syndrome during the Stroke Prevention Trial (STOP) in sickle-cell anemia. *J Pediatr* **139:** 785–789.

Miller ST, Macklin EA, Pegelow CH, Kinney TR, Sleeper LA, Bello JA, DeWitt LD, Gallagher DM, Guarini L, Moser FG, et al. 2001b. Silent infarction as a risk factor for overt stroke in children with sickle cell anemia: A report from the Cooperative Study of Sickle Cell Disease. *J Pediatr* **139:** 385–390.

Neumayr L, Lennette E, Kelly D, Earles A, Embury S, Groncy P, Grossi M, Grover R, McMahon L, Swerdlow P, et al. 2003. Mycoplasma disease and acute chest syndrome in sickle cell disease. *Pediatrics* **112:** 87–95.

O'Brien SH, Fan L, Kelleher KJ. 2010. Inpatient use of laxatives during opioid administration in children with sickle cell disease. *Pediatr Blood Cancer* **54:** 559–562.

Ohene-Frempong K, Weiner SJ, Sleeper LA, Miller ST, Embury S, Moohr JW, Wethers DL, Pegelow CH, Gill FM. 1998. Cerebrovascular accidents in sickle cell disease: Rates and risk factors. *Blood* **91:** 288–294.

Overturf GD, Powars D, Barraf LJ. 1977. Bacterial meningitis and septicemia in sickle cell disease. *Am J Dis Child* **131:** 784–787.

Pearson HA. 1977. Sickle cell anemia and severe infections due to encapsulated bacteria. *J Infect Dis* **136:** S25–S30.

Pearson HA, Spencer RP, Cornelius EA. 1969. Functional asplenia in sickle-cell anemia. *N Engl J Med* **281:** 923–926.

Pegelow CH, Adams RJ, McKie V, Abboud M, Berman B, Miller ST, Olivieri N, Vichinsky E, Wang W, Brambilla D. 1995. Risk of recurrent stroke in patients with sickle cell disease treated with erythrocyte transfusions. *J Pediatr* **126:** 896–899.

Pegelow CH, Macklin EA, Moser FG, Wang WC, Bello JA, Miller ST, Vichinsky EP, DeBaun MR, Guarini L, Zimmerman RA, et al. 2001. Longitudinal changes in brain magnetic resonance imaging findings in children with sickle cell disease. *Blood* **99:** 3014–3018.

Platt OS. 2006. Prevention and management of stroke in sickle cell anemia. *Hematology Am Soc Hematol Educ Program* **2006:** 54–57.

Platt OS, Orkin SH, Dover G, Beardsley GP, Miller B, Nathan DG. 1984. Hydroxyurea enhances fetal hemoglobin production in sickle cell anemia. *J Clin Invest* **74:** 652–656.

Platt OS, Thorington BD, Brambilla DJ, Milner PF, Rosse WF, Vichinsky E, Kinney TR. 1991. Pain in sickle cell disease. Rates and risk factors. *N Engl J Med* **325:** 11–16.

Platt OS, Brambilla DJ, Rosse WF, Milner PF, Castro O, Steinberg MH, Klug PP. 1994. Mortality in sickle cell disease—Life expectancy and risk factors for early death. *N Engl J Med* **330:** 1639–1644.

Powars DR. 1975. Natural history of sickle cell disease—The first ten years. *Semin Hematol* **12:** 267–285.

Powars D, Wilson B, Imbus C, Pegelow C, Allen J. 1978. The natural history of stroke in sickle cell disease. *Am J Med* **65:** 461–71.

Powars D, Overturf G, Weiss J, Lee S, Chan L. 1981. Pneumococcal septicemia in children with sickle cell anemia. Changing trend of survival. *JAMA* **245:** 1839–1842.

Powars DR, Chan LS, Hiti A, Ramicone E, Johnson C. 2005. Outcome of sickle cell anemia: A 4-decade observational study of 1056 patients. *Medicine* **84:** 363–376.

Powell RW, Levine GL, Yang YM, Mankad VN. 1992. Acute splenic sequestration crisis in sickle cell disease: Early detection and treatment. *J Pediatr Surg* **27:** 215–219.

Quinn CT, Rogers ZR, McCavit TL, Buchanan GR. 2010. Improved survival of children and adolescents with sickle cell disease. *Blood* **115:** 3447–3452.

Raphael JL, Kamdar A, Wang T, Liu H, Mahoney DH, Mueller BU. 2008. Day hospital versus inpatient management of uncomplicated vaso-occlusive crises in children with sickle cell disease. *Pediatr Blood Cancer* **51:** 398–401.

Redwood AM, Williams EM, Desal P, Serjeant GR. 1976. Climate and painful crisis of sickle-cell disease in Jamaica. *Br Med J* **1:** 66.

Rees DC, Olujohungbe AD, Parker NE, Stephens AD, Telfer P, Wright J. 2003. Guidelines for the management of the acute painful crisis in sickle cell disease. *Br J Hematol* **120:** 744–752.

Resar LM, Oski FA. 1991. Cold water exposure and vaso-occlusive crises in sickle cell anemia. *J Pediatr* **118:** 407.

Rice HE, Oldham KT, Hillery CA, Skinner MA, O'Hara SM, Ware RE. 2003. Clinical and hematologic benefits of partial splenectomy for congenital hemolytic anemias in children. *Ann Surg* **237:** 281–288.

Rogers DW, Clarke JM, Cupidore L, Ramlal AM, Sparke BR, Serjeant GR. 1978. Early deaths in Jamaican children with sickle cell disease. *Br Med J* **1:** 1515–1516.

Rosse WF, Gallagher D, Kinney TR, Castro O, Dosik H, Moohr J, Wang W, Levy PS. 1990. Transfusion and alloimmunization in sickle cell disease. The Cooperative Study of Sickle Cell Disease. *Blood* **76:** 1431–1437.

Saunthararajah Y, Hillery CA, Lavelle D, Molokie R, Dorn L, Bressler L, Gavazova S, Chen YH, Hoffman R, DeSimone J. 2003. Effects of 5-aza-2′-deoxycytidine on fetal hemoglobin levels, red cell adhesion, and hematopoietic differentiation in patients with sickle cell disease. *Blood* **102:** 3865–3870.

Saunthararajah Y, Molokie R, Saraf S, Sidhwani S, Gowhari M, Vara S, Lavelle D, DeSimone J. 2008. Clinical effectiveness of decitabine in severe sickle cell disease. *Br J Haematol* **141:** 126–129.

Schatz J, Brown RT, Pascual JM, Hsu L, DeBaun MR. 2001. Poor school and cognitive functioning with silent cerebral infarcts and sickle cell disease. *Neurology* **56:** 1109–1111.

Scothorn DJ, Price C, Schwartz D, Terrill C, Buchanan GR, Shurney W, Sarniak I, Fallon R, Chu JY, Pegelow CH, et al. 2002. Risk of recurrent stroke in children with sickle cell disease receiving blood transfusion therapy for at least five years after initial stroke. *J Pediatr* **140:** 348.

Serjeant GR. 1995. Natural history and determinants of clinical severity of sickle cell disease. *Curr Opin Hematol* **2:** 103–108.

Smith WR, Penberthy LT, Bovbjerg VE, McClish DK, Roberts JD, Dahman B, Aisiku IP, Levenson JL, Roseff SD. 2008. Daily assessment of pain in adults with sickle cell disease. *Ann Int Med* **148:** 94–101.

Smith-Whitley K, Thompson AA. 2012. Indications and complications of transfusions in sickle cell disease. *Pediatr Blood Cancer* **59:** 358–364.

Solomon LR. 2008. Treatment and prevention of pain due to vaso-occlusive crises in adults with sickle cell disease: An educational void. *Blood* **111:** 997–1003.

Steinberg MH, Barton F, Castro O, Pegelow CH, Ballas SK, Kutlar A, Orringer E, Bellevue R, Olivieri N, Eckman J, et al. 2003. Effect of hydroxyurea on mortality and morbidity in adult sickle cell anemia: Risks and benefits up to 9 years of treatment. *J Am Med Assoc* **289:** 1645–1651.

Steinberg MH, McCarthy WF, Castro O, Ballas SK, Armstrong FD, Smith W, Ataga K, Swerdlow P, Kutlar A, DeCastro L, et al. 2010. The risks and benefits of long-term use of hydroxyurea in sickle cell anemia: A 17.5 year follow-up. *Am J Hematol* **85:** 403–408.

Sträter R, Becker S, von Eckardstein A, Heinecke A, Gutsche S, Junker R, Kurnik K, Schobess R, Nowak-Göttl U. 2002. Prospective assessment of risk factors for recurrent stroke during childhood-a 5-year follow-up study. *Lancet* **360:** 1540–1545.

Swerdlow PS. 2006. Red cell exchange in sickle cell disease. *Am Soc Hematol Educ Program* **2006:** 48–53.

Telfer P, Coen P, Chakravorty S, Wilkey O, Evans J, Newell H, Smalling B, Amos R, Stephens A, Rogers D, et al. 2007. Clinical outcomes in children with sickle cell disease living in England: A neonatal cohort in East London. *Haematologica* **92:** 905–912.

Thompson RJ, Gustafson KE, Bonner MJ, Ware RE. 2002. Neurocognitive development of young children with sickle cell disease through three years of age. *J Pediatr Psychol* **27:** 235–244.

Thornburg CD, Dixon N, Burgett S, Mortier NA, Schultz WH, Zimmerman SA, Bonner M, Hardy KK, Calatroni A, Ware RE. 2009. A pilot study of hydroxyurea to prevent chronic organ damage in young children with sickle cell anemia. *Pediatr Blood Cancer* **52:** 609–615.

Topley JM, Rogers DW, Stevens MCG, Serjeant GR. 1981. Acute splenic sequestration and hypersplenism first five years in homozygous sickle cell disease. *Arch Dis Child* **56:** 765–769.

van Beers EJ, van Tuijn CF, Nieuwkerk PT, Friederich PW, Vranken JH, Biemond BJ. 2007. Patient-controlled analgesia versus continuous infusion of morphine during vaso-occlusive crisis in sickle cell disease, a randomized controlled trial. *Am J Hematol* **82:** 955–960.

Velasquez MP, Mariscalco MM, Goldstein SL, Airewele GE. 2009. Erythrocytapheresis in children with sickle cell disease and acute chest syndrome. *Pediatr Blood Cancer* **53:** 1060–1063.

Vichinsky E, Hurst D, Earles A, Kleman K, Lubin B. 1988. Newborn screening for sickle cell disease: Effect on mortality. *Pediatrics* **81:** 749–755.

Vichinsky E, Williams R, Das M, Earles AN, Lewis N, Adler A, McQuitty J. 1994. Pulmonary fat embolism: A distinct cause of severe acute chest syndrome in sickle cell anemia. *Blood* **83:** 3107–3112.

Vichinsky EP, Neumayr LD, Earles AN, Williams R, Lennette ET, Dean D, Nickerson B, Orringer E, McKie V, Bellevue R, et al. 2000. Causes and outcomes of the acute chest syndrome in sickle cell disease. National Acute Chest Syndrome Study Group. *N Engl J Med* **342:** 1855–1865.

Voskaridou E, Christoulas D, Bilalis A, Plata E, Varvagiannis K, Stamatopoulos G, Sinopoulou K, Balassopoulou A, Loukopoulos D, Terpos E. 2010. The effect of prolonged administration of hydroxyurea on morbidity and mortality in adult patients with sickle cell syndromes: Results of a 17-year, single-center trial (LaSHS). *Blood* **115:** 2354–2363.

Wang WC, Kovnar EH, Tonkin IL, Mulhern RK, Langston JW, Day SW, Schell MJ, Wilimas JA. 1991. High risk of recurrent stroke after discontinuance of five to twelve years of transfusion therapy in patients with sickle cell disease. *J Pediatr* **118:** 377–382.

Wang WC, Wynn LW, Rogers ZR, Scott JP, Lane PA, Ware RE. 2001. A two-year pilot trial of hydroxyurea in very young children with sickle-cell anemia. *J Pediatr* **139:** 790–796.

Wang WC, Ware RE, Miller ST, Iyer RV, Casella JF, Minniti CP, Rana S, Thornburg CD, Rogers ZR, Kalpatthi RV, et al. 2011. Hydroxycarbamide in very young children with sickle-cell anaemia: A multicentre, randomised, controlled trial (BABY HUG). *Lancet* **377:** 1663–1672.

Ware RE. 2010a. Acute vaso-occlusion in sickle cell anemia. In *European Haematology Association Educational Handbook*. Hague, The Netherlands.

Ware RE. 2010b. How I use hydroxyurea to treat young patients with sickle cell anemia. *Blood* **115:** 5300–5311.

Ware RE, Zimmerman SA, Schultz WH. 1999. Hydroxyurea as an alternative to blood transfusions for the prevention of recurrent stroke in children with sickle cell disease. *Blood* **94:** 3022–3026.

Ware RE, Zimmerman SA, Sylvestre PB, Mortier NA, Davis JS, Treem WR, Schultz WH. 2004. Prevention of secondary stroke and resolution of transfusional iron overload in children with sickle cell anemia using hydroxyurea and phlebotomy. *J Pediatr* **145:** 346–352.

Ware RE, Despotovic JM, Mortier NA, Flanagan JM, He J, Smeltzer MP, Kimble AC, Aygun B, Wu S, Howard T, et al. 2011. Pharmacokinetics, pharmacodynamics, and pharmacogenetics of hydroxyurea treatment for children with sickle cell anemia. *Blood* **118:** 4985–4991.

Ware RE, Helms RW, for the SWiTCH Investigators. 2012. Stroke with transfusions changing to hydroxyurea (SWiTCH). *Blood* **119:** 3925–3932.

Wayne AS, Schoenike SE, Pegelow CH. 2000. Financial analysis of chronic transfusion for stroke prevention in sickle cell disease. *Blood* **96:** 2369–2372.

Williams TN, Uyoga S, Macharia A, Ndila C, McAuley CF, Opi DH, Mwarumba S, Makani J, Komba A, Ndiritu MN, et al. 2009. Bacteraemia in Kenyan children with sickle-cell anaemia: A retrospective cohort and case-control study. *Lancet* **374:** 1364–1370.

Winkelstein JA, Drachman RH. 1968. Deficiency of pneumococcal serum opsonizing activity in sickle cell disease. *N Engl J Med* **279:** 459–466.

Yazdanbakhsh K, Ware RE, Noizat-Pirenne F. 2012. Red blood cell alloimmunization in sickle cell disease: Pathophysiology, risk factors, and transfusion management. *Blood* **120:** 528–537.

Yoong WC, Tuck SM. 2002. Menstrual pattern in women with sickle cell anaemia and its association with sickling crises. *J Obstet Gynaecol* **22:** 399–401.

Zimmerman SA, Schultz WH, Davis JS, Pickens CV, Mortier NA, Howard TA, Ware RE. 2004. Sustained long-term hematologic efficacy of hydroxyurea at maximum tolerated dose in children with sickle cell disease. *Blood* **103:** 2039–2045.

Zimmerman SA, Schultz WH, Burgett S, Mortier NA, Ware RE. 2007. Hydroxyurea therapy lowers transcranial Doppler flow velocities in children with sickle cell anemia. *Blood* **110:** 1043–1047.

Hematopoietic Stem Cell Transplantation in Thalassemia and Sickle Cell Anemia

Guido Lucarelli, Antonella Isgrò, Pietro Sodani, and Javid Gaziev

International Center for Transplantation in Thalassemia and Sickle Cell Anemia–Mediterranean Institute of Hematology, Policlinic of the University of Rome Tor Vergata, Tor Vergata, Italy

Correspondence: g.lucarelli@fondazioneime.org

The globally widespread single-gene disorders β-thalassemia and sickle cell anemia (SCA) can only be cured by allogeneic hematopoietic stem cell transplantation (HSCT). HSCT treatment of thalassemia has substantially improved over the last two decades, with advancements in preventive strategies, control of transplant-related complications, and preparative regimens. A risk class–based transplantation approach results in disease-free survival probabilities of 90%, 84%, and 78% for class 1, 2, and 3 thalassemia patients, respectively. Because of disease advancement, adult thalassemia patients have a higher risk for transplant-related toxicity and a 65% cure rate. Patients without matched donors could benefit from haploidentical mother-to-child transplantation. There is a high cure rate for children with SCA who receive HSCT following myeloablative conditioning protocols. Novel non-myeloablative transplantation protocols could make HSCT available to adult SCA patients who were previously excluded from allogeneic stem cell transplantation.

Thalassemia and sickle cell anemia (SCA) are the two most widely diffused hereditary hemoglobinopathies in the world. Thalassemia originated in Mediterranean, Middle Eastern, and Asian countries, and sickle cell anemia originated throughout African countries; both soon became globally spread because of spontaneous migration and migration imposed through slavery. Modern migration is currently causing both diseases, particularly SCA, to further invade European countries (see Williams and Weatherall 2012).

β-Thalassemia involves deficient or absent synthesis of the β-globin chains that constitute adult hemoglobin molecules. This genetic defect affects erythropoiesis through the entire process of maturation and proliferation of the erythron, with severe apoptosis due to accumulation of α2 chains in the erythroid precursors. Early death of the red blood cells continues in the blood circulation. Patients without transfusions experience an enormous expansion of the erythroid and severe hemolytic anemia.

In SCA, a single nucleotide substitution—valine replacing glutamic acid at the sixth position of the β-globin chain of hemoglobin A—forms a pathological hemoglobin called hemoglobin S (HbS). The propensity of HbS polymerization causes the red blood cell structure to change to a stable sickle shape. Resulting episodic anemia and acute and chronic vaso-occlusion of small and large vessels in virtually all organs cause the polymorphic clinical features

Copyright © 2013 Cold Spring Harbor Laboratory Press; all rights reserved
Cite this article as *Cold Spring Harb Perspect Med* doi: 10.1101/cshperspect.a011825

of the sickle cell disease. Intramedullary apoptosis is also present but moderate, never at the levels seen in thalassemia.

Many thalassemic patients need continuous red blood cell replacement through a sequential program of red blood cell transfusions; the SCA patient does not. Without transfusions, thalassemic patients have tremendous skeleton deformities, as well as hepatomegaly and splenomegaly due to expansion of the hematopoietic system with extramedullary hematopoiesis. The SCA patient can suffer from severe organ damage due to the sequence of HbS polymerization, formation of microemboli, adhesion to the intima of the vessels, and organ infarction. The pathogeneses and clinical aspects of thalassemia and sickle cell anemia are described in greater detail in other articles in this collection.

Because both thalassemia and SCA are genetic diseases in which the genetic defect is expressed in the hematopoietic system, they are both curable by allogeneic cellular gene therapy through hematopoietic stem cell transplantation. On the other hand, the separate pathogeneses of these two disorders explain why there must also be some differences between the transplantation procedures.

HEMATOPOIETIC STEM CELL TRANSPLANTATION FOR THALASSEMIA

HSCT for Thalassemia: Risk Class Approach

Allogeneic HSCT is used to treat hemoglobinopathies; after conditioning to overcome the immunological barrier, allogeneic stem cells are used as vectors to correct the basic genetic defect by replanting genes that are essential for normal hematopoiesis. Essentially, allogeneic HSCT treatment of these diseases can be considered as allogeneic stem cell gene therapy. It may some day be possible to use autologous stem cells transformed by the insertion of normal genes as vectors, but there is no indication that this approach will be a clinical option in the foreseeable future.

For treating diseases other than aplastic anemia, preparatory regimens for HSCT must achieve two objectives: elimination of the disordered marrow and establishment of a tolerant environment that will permit transplanted marrow to survive and thrive. There is a considerable body of experience with the use of busulfan (BU) and its derivatives for ablating marrow in patients undergoing HSCT for the treatment of non-malignant conditions (Parkman et al. 1978; Kapoor et al. 1981; Hobbs et al. 1986). Cyclophosphamide (CY) is a well-established agent for providing adequate immunosuppression for allogeneic engraftment (Thomas et al. 1972; Storb et al. 1991). The combination of BU and CY can eradicate thalassemia and facilitate sustained allogeneic engraftment. The two fundamental treatments in pretransplant preparation include eradication of the hematopoietic system by administration of a total dose of 14–16 mg/kg busulfan and suppression of the immune system by administration of a total dose of 200 mg/kg cyclophosphamide. Because of the overlap of the anti-proliferative activity of busulfan on the immune cell system and of cyclophosphamide on the hematopoietic stem cell system, the summed toxicity of these dosages of busulfan and cyclophosphamide reaches the tolerability threshold and cannot be increased.

When the allogeneic graft starts to proliferate in the recipient, an immunological graft-versus-host reaction (graft-versus-host disease [GVHD]) may occur. GVHD can affect one or more organs to varying degrees, with the most frequently affected being the skin, gastrointestinal tract, and liver. GVHD is a serious complication of bone marrow transplantation and can be fatal; therefore, the prophylactic administration of cyclosporine (an immunosuppressive drug) is an important part of the pretransplant and posttransplant treatment. Three or four doses of methotrexate may also be given in addition to cyclosporine during the first 15 d after transplant.

The influence of pretransplant characteristics on the outcome of transplantation was analyzed in 161 patients <17 yr of age, who were all treated with exactly the same regimen (Storb et al. 1977; Lucarelli et al. 1990). Multivariate analysis showed that hepatomegaly of more than 2 cm, portal fibrosis, and irregular chelation history were associated with a significantly reduced probability of survival. The quality of chelation was characterized as regular when deferoxamine therapy was initiated no later than 18 mo after the first transfusion and was administered subcutaneously for 8–10 h continuously for at least 5 d/wk. Any deviation from this regimen was defined as irregular chelation. On the basis of these risk factors, patients were categorized into three risk classes: Class 1 patients had none of these adverse risk factors, class 2 patients had one or two adverse risk factors, and class 3 patients had all three.

HSCT from HLA-Matched Related Donors

The results of transplants from HLA-matched related donors to class 1, class 2, and class 3 patients, obtained by our group, are reported in Table 1 (Lucarelli et al. 2008; Isgrò et al. 2010).

Adult Patients

Adult thalassemia patients have more advanced disease, with both disease- and treatment-related organ complications that are mainly due to prolonged exposure to iron overload. From November 1988 through September 1996, 107 >16-yr-old patients received transplants from matched donors at the Hospital of Pesaro. The median age for this population was 20 yr, with a range of 17–35 yr. The probabilities of survival, thalassemia-free survival, rejection mortality, and non-rejection mortality for this entire group were 66%, 62%, 4%, and 37%, respectively (Lucarelli et al. 1992, 1999).

From April 1997, 15 high-risk adult patients were prepared for transplantation with a reduced total dose of 90 mg/kg cyclophosphamide. The probabilities of survival, thalassemia-free survival, rejection mortality, and non-rejection mortality were 65%, 65%, 7%, and 28%, respectively (Gaziev et al. 2005).

HSCT FROM ALTERNATIVE DONORS FOR THALASSEMIA

A major obstacle to successful transplantation is the limited number of HLA-matched related donors within families. Approximately 60% of patients lack a suitable family donor. Some of these patients could benefit from HSCT from matched unrelated donors. However, the chance of finding a matched unrelated donor is strongly dependent on the patient's ethnic background.

One study reported the outcome of bone marrow transplant (BMT) from matched unrelated donors who were prospectively selected

Table 1. Outcomes of hematopoietic cell transplantation from HLA-matched donors in cases of β-thalassemia

Patients	Stem cell source	OS (%)	TFS (%)	Reference
<17 yr: 515 (class 1–2)	MSD/MRD	Class 1–2: 88%	Class 1–2: 85%	Lucarelli et al. 2008
73 (class 3)		Class 3: 87%	Class 3: 82%	
Adult: 107		Adult: 66%	Adult: 62%	
68	MUD	79.3%	40%	La Nasa et al. 2005a
33	Cord blood	100%	79%	Locatelli et al. 2003
70	MSD/MRD	83%	73%	Ghavamzadeh et al. 1997
49	Related: 28	89% (all patients)	Related: 82%	Hongeng et al. 2006
	Unrelated: 21		Unrelated: 71%	
37 (class 1–2)	MSD	Class 1–2: 97%	Class 1–2: 89%	Isgrò et al. 2010
35 (class 3)		Class 3: 87%	Class 3: 80%	
3 (class 1)	MSD	Class 1: NA	Class 1: NA	Sabloff et al. 2011
75 (class 2)		Class 2: 91%	Class 2: 88%	
64 (class 3)		Class 3: 64%	Class 3: 62%	
37 (NA)				

Abbreviations: OS, overall survival; TFR, thalassemia-free survival; NA, not available; MSD, matched sibling donor; MRD, matched-related donor; MUD, matched unrelated donor; SCD, sickle cell disease.

using high-resolution molecular typing for HLA class I and class II loci; this study included 68 patients with thalassemia major who received BU/CY or BU/FLU (fludarabine) and/or thiotepa (TT) as a conditioning regimen (La Nasa et al. 2005a). The study included 14 class 1, 16 class 2, and 38 class 3 patients. Overall rates of survival, thalassemia-free survival, rejection, and transplant-related mortality (TRM) for the entire cohort were 79.3%, 65.8%, 14.4%, and 20.7%, respectively. The incidences of acute grade II–IV or grade III–IV GVHD were 40% and 17%, respectively. Ten of 56 evaluable patients (18%) developed chronic GVHD. Class 1 and class 2 patients had much better overall and disease-free survival (96.7% and 80%, respectively) than class 3 patients (65.5% and 54.5%, respectively). These investigators have also recently reported encouraging results in adult thalassemia patients who received bone marrow from matched unrelated donors, with rates of survival, thalassemia-free survival, TRM, and rejection of 70%, 70%, 30%, and 4%, respectively (La Nasa et al. 2005b). Similar results were reported in another study from Asia that included 21 children who received transplants from matched unrelated donors, as well as 28 patients who received transplants from matched related donors (Hongeng et al. 2006). The 2-yr thalassemia-free survival for patients who received transplants from matched unrelated donors was 71%, compared with 82% for patients who received transplants from matched related donors.

Together, these data strongly suggest that improvements in donor selection and transplantation preparation have improved the safety of unrelated donor HSCT for thalassemia treatment. This appears to be a viable treatment option for selected patients when there is no suitable sibling donor.

Results of Transplant from Haploidentical Donors

Haploidentical transplantation may extend the use of this treatment option to the 50%–60% of patients who lack a suitably matched familial donor or an HLA-identical unrelated donor. The presence of fetal cells in maternal blood and of maternal cells in fetal blood (fetomaternal microchimerism) suggests that immunological tolerance may exist between mother and offspring. The combination of a megadose of purified $CD34^+$ cells and a highly immuno-myeloablative conditioning regimen is crucial for overcoming the barrier of residual anti-donor cytotoxic T-lymphocyte precursors in T-cell-depleted mismatched transplants, and the addition of BMMCs (including NK cells, mesenchymal stem cells, and T-cells) to a T-cell-depleted allograft may help engraftment and control GVHD. Recently, we reported the outcomes of 31 children with thalassemia who received transplants from haploidentical donors; 27 were from mothers, two from brothers, and two from fathers (Sodani et al. 2011). Eight patients received T-cell-depleted peripheral blood progenitor cells ($CD34^+$ immunoselection) and $CD3^+$- and $CD19^+$-depleted bone marrow stem cells. Twenty-three patients received $CD34^+$ mobilized peripheral and bone marrow progenitor cells. Positive selection was performed using the CliniMACS procedure. The $CD34^+$ grafts contained a median of 14.2 $\times$ 10^6/kg $CD34^+$ cells (range, 5.4 $\times$ 10^6/kg to 39 $\times$ 10^6/kg), 2 $\times$ 10^5/kg $CD3^+$ cells, and 0.19 $\times$ 10^6/kg $CD19^+$ cells.

Between Day −59 and Day −11 before the transplantation, 60 mg/kg daily hydroxyurea and 3 mg/kg daily azathioprine were administered to eradicate marrow; and growth factors, granulocyte colony-stimulating factor, and erythropoietin were given twice weekly to maintain stem cell proliferation in the face of hypertransfusion, thereby facilitating the effect of the hydroxyurea. Fludarabine was administered at a dosage of 30 mg/m^2 per day from Day −17 through Day −13. Starting on Day −10, the first 17 patients received 14 doses of 1 mg/kg oral busulfan (BU), three times daily over 4 d (total dose of 14 mg/kg over 4 d). The following 14 patients received a corresponding dose of BU, given intravenously, followed by intravenous CY at a dose of 50 mg/kg daily on each of the next 4 d (total dose of 200 mg/kg); 10 mg/kg thiotepa and 12.5 mg/kg anti-thymocyte globulin (ATG-Fresenius S) were given daily from Days

−5 to −2. All patients received cyclosporine for GVHD prophylaxis for the first 2 mo posttransplantation.

All patients showed donor chimerism by Day 14 after HSCT. A granulocyte count of >500/mL occurred after a median time of 13 d (range of 11–17). Seven patients rejected their grafts and survived with thalassemia; three patients showed early mixed chimerism, which was deemed persistent when observed at 14, 38, and 42 mo after the transplants. In 19 cases, the transplantation was successful with complete allogeneic reconstitution. Two patients died from transplantation-related causes; one died on Day +114 of Epstein-Barr virus cerebral lymphoma, and the other died on Day +92 from CMV pneumonia. All of the 22 cured children are no longer transfusion-dependent, with hemoglobin levels ranging from 10.3 g/dL to 13.8 g/dL (Sodani et al. 2010, 2011).

Mixed Chimerism following HSCT for Thalassemia

Mixed hematopoietic chimerism (MC) is an interesting phenomenon that sometimes occurs after HSCT for thalassemia. The incidence of MC was evaluated in 335 patients who received BMT from HLA-matched family donors for thalassemia; it was found to be 32.2% at 2 mo after transplant (Lucarelli et al. 2002). Of the 227 patients with complete donor chimerism, none rejected their grafts, whereas graft loss occurred in 35 of 108 patients (32.4%) with MC, indicating that MC is a risk factor for graft rejection in thalassemia patients. The percentage of residual host hematopoietic cells (RHCs) determined at 2 mo after transplant was predictive for graft rejection, with nearly all patients experiencing graft rejection when RHCs exceeded 25%. The risk of graft rejection was only 13% in patients with <10% RHCs and was 41% in patients with 10%–25% RHCs (Andreani et al. 1996, 2000). Unlike hematological malignancies in which residual host cells are predictive of relapse, patients with thalassemia can have lifelong stable mixed chimerism without rejection following transplantation. Of patients receiving BMT for thalassemia following myeloablative conditioning, 10% became persistent mixed chimeras and became transfusion independent, suggesting that once donor-host tolerance is established, a limited number of engrafted donor cells might be sufficient to provide significant improvement of disease phenotype in patients with thalassemia major.

We recently published a paper describing four long-term transplanted patients affected by hemoglobinopathies (Andreani et al. 2011a). These patients are characterized by the presence of few donor-engrafted nucleated cells, both in the peripheral blood and in the bone marrow; the majority of the erythrocytes were of donor origin. Moreover, by analyzing singularly picked-up burst-forming unit erythroid colonies, we showed that the proportion of donor-derived erythroid precursors was equivalent to that observed in the mature nucleated cells, rather than that of the red blood cells. These results suggest that in patients characterized by the presence of PMC after HSCT, a selective advantage of the maturation of donor erythroid precursors might successfully contrast the problems associated with the recipient ineffective erythropoiesis (Andreani et al. 2011b).

HEMATOPOIETIC STEM CELL TRANSPLANTATION FOR SICKLE CELL DISEASE

Sickle cell disease (SCD) is associated with considerable morbidity and premature mortality, and HSCT is the only available therapy with curative intent. Outcomes of HSCT from matched sibling donors in pediatric patients have been excellent; however, the applicability of HSCT has been limited by lack of matched sibling donors and concerns regarding regimen-related toxicity. A reduced-intensity conditioning regimen has the potential to decrease regimen-related morbidity and mortality. Improved understanding of the typical sequence of complications and the effects of treatments such as hydroxyurea and blood transfusions, as well as the impact of transplantation on organ damage, are likely to influence the timing and indications for transplantation.

Following a demonstration of successful use of the combination of busulfan and cyclophosphamide in transplant for thalassemia, this protocol was applied to the transplant preparation for treating the other genetic hemoglobinopathy, sickle cell anemia, in 1986 (Vermylen et al. 1993). We first performed a transplant in a 16-yr-old African girl with SCA in 1993 (Giardini et al. 1993). Since then, there has been slow progress in stem cell transplantation for SCA, with only approximately 250 transplants performed to date worldwide, mainly reported in multicenter studies (Table 2). A major reason for this very conservative approach to using transplantation for SCA treatment is the absence of universally accepted indications for transplantation.

Transplantation has been offered primarily to younger patients with overt symptomatology. Most patients have received an HSCT following a preparative regimen with 14–16 mg/kg busulfan and 200 mg/kg cyclophosphamide. Additional immunosuppressive agents sometimes include equine antithymocyte globulin, rabbit antithymocyte globulin, antilymphocyte globulin, or total lymphoid irradiation. Cyclosporine A, alone or in combination with methylprednisolone or methotrexate, has been used for post-transplant graft-versus-host disease (GVHD) prophylaxis. The results of HSCT after myeloablative conditioning in children have been very encouraging, with disease-free survival in most studies of ~85% and a TRM rate of <10% (Table 2). The results of transplantation might be even better if HSCT were performed in children before the development of irreversible sickle vasculopathy-related complications (Vermylen et al. 1998; Lucarelli et al. 2011).

The experience of cord blood transplantation (CBT) has been limited in patients with SCD, with a total of 33 cases reported, two-thirds of whom were in the related donor setting (Hsieh et al. 2011). The majority of related CBT have been 6/6 HLA-matched. In unrelated CBT, although more feasible for pediatric patients on the basis of the availability of 4/6 HLA-matched cord blood unit, it appears to be associated with a great risk of graft rejection and GVHD. The largest study consisting of seven patients reported one death, an EFS of 43%, five patients developed acute GVHD, and one patient chronic GVHD (Adamkiewicz et al. 2007). Moreover, CBT for adults is limited by the necessary total nucleated cell count per kilogram of body weight for engraftment to occur. At this time, transplant should be safely extended to patients with SCD using HLA identical donors, minimizing morbidity and mortality. The attainment of tolerance should allow extension of this curative approach to alternative donor sources.

The Rome Experience in HSCT for Sickle Cell Disease

A total of 11 SCA patients (Lucarelli et al. 2011) with a median age of 12 yr (range of 2–16) received hematopoietic stem cell transplantations from HLA-identical, related donors, following a myeloablative conditioning regimen. Indications for transplantation were vaso-occlusive crisis, acute chest syndrome, avascular bone necrosis, chronic red blood cell transfusions, or hemorrhagic stroke. Five patients received preconditioning cytoreductive/immunosuppressive chemotherapy with hydroxyurea,

Table 2. Worldwide reported outcomes of myeloablative HSCT for sickle cell disease

	n = 50 (Vermylen et al. 1998)	*n* = 50 (Walters et al. 2000)	*n* = 11 (Locatelli et al. 2003)	*n* = 67 (Panepinto et al. 2007)	*n* = 87 (Bernaudin et al. 2007)	*n* = 11 (Lucarelli et al. 2011)
OS	93%	94%	100%	97%	92%	90%
DFS	85%	84%	90%	85%	86%	90%
Rejection	10%	10%	9%	13%	5%	10%

Abbreviations: OS, overall survival; DFS, disease-free survival.

azathioprine, and fludarabine, as described elsewhere (Lucarelli et al. 2011). The remaining six patients were prepared for the transplant with busulfan, cyclophosphamide, and anti-thymocyte globulin. They received cyclosporine (CSA), low-dose methylprednisolone, and a short course of MTX as GVHD prophylaxis. All patients had sustained engraftment with donor chimerism of 100%, except for one patient, who showed a persistent MC with 25% of donor cells. Starting 2 mo after transplantation, this patient showed a progressive decrease in donor chimerism until 25%; she had stable mixed chimerism for more than 4 yr without SCD-related events and was transfusion independent (Andreani et al. 2011a). All of the remaining patients had full donor chimerism in lymphoid and myeloid lineages. Seven patients developed grade 2 acute GVHD. The only patient who developed grade 3 GVHD had received a peripheral blood stem cell transplantation. All of the patients responded promptly to the steroid treatment given to control the acute GVHD. Four out of 11 patients had mild skin chronic GVHD that resolved completely following steroid use. Ten out of 11 patients survived without sickle cell disease, with Lansky/Karnofsky scores of 100. Median follow-up for survivors was 43 mo (range of 12–68 mo). One patient died 1 yr after transplantation in her home country; she was on tapering immunosuppressive treatment for chronic skin GVHD and died of pneumonia. The probabilities of survival, SCA-free survival, and transplant-related mortality 5 yr after transplant were 90%, 90%, and 10%, respectively. After transplantation, none of the patients had complications typical of SCA, such as pains, stroke, or acute chest syndrome.

In sickle cell trait carriers, the use of G-CSF for peripheral stem cell mobilization may induce HbS polymerization and red blood cell sickling, and therefore transform a relatively stable steady state into a cascade of events resulting in a sickle cell crisis and, in severe cases, multi-organ dysfunction. We suggest the use of BM-derived HSCs versus G-CSF-mobilized HSCs especially when the donor is an HbS trait carrier.

Mixed Chimerism following Myeloablative HSCT for SCA

A subgroup of patients who undergo HSCT for SCA develop lifelong stable MC once donor–host tolerance is established (Andreani et al. 1996, 2000). Walters et al. (2001) observed stable MC in 13 out of 50 patients (26%) who showed SCD-free survival for a median duration of 6.9 yr (range of 4.2–13 yr). Of note, among these patients, five had mixed donor chimerism of <75% (range of 11%–74%), and none of them developed sickle cell-related complications during a follow-up period that ranged from 22 to 70.2 mo (median of 36.3 mo).

Recently we reported the clinical course of a 6-yr-old girl who received a bone marrow transplant for sickle cell anemia (M Marziali, A Isgrò, D Fraboni, et al., in prep.). Four years after transplant, the patient showed stable mixed chimerism, as evidenced by a 39% bone marrow (BM) level of donor-origin erythroid precursors, and 80% donor-origin red blood cells (RBCs) in peripheral blood (PB). The patient received BM from her HLA-matched sister (Hb AA) after a conditioning regimen of 14 mg/kg BU, 200 mg/kg CY, and 10 mg/kg anti-thymocyte globulin (ATG). For prophylaxis against graft-versus-host disease (GVHD), the patient received cyclosporine (starting on Day −1) and a short course of methotrexate (MTX; 10 mg/m^2 on posttransplant Days 1, 3, and 6 with folinic acid rescue). The course after allogeneic hematopoietic stem cell transplantation was uneventful, with rapid hematologic engraftment and no signs of acute or chronic GVHD. Molecular analysis of sorted cell subgroups revealed MC in nucleated cells, CD34$^+$ progenitors, and RBCs in the PB and BM. Four years after transplantation, the level of donor NCs was 39% in PB and 36% at the BM level, in parallel to a very high proportion of donor-derived RBCs (80%) in PB. The proportions of donor-derived RBCs and BFU-E at the BM level were 40% and 46%, respectively, indicating the presence of quantitatively different red cell/nucleated cell chimerism. A possible explanation for the presence of a greater proportion of donor-derived RBC may be found in the improved survival of donor erythroid precursors, compared with the host counterparts that might be destroyed during ineffective erythropoiesis.

New Approach to Overcome Current Limitations: Reduced-Intensity Conditioning Regimens and Non-Myeloablative Regimens

A frequently encountered barrier to marrow transplantation for SCD treatment is lack of an HLA-identical donor. Another major limitation of current approaches is the high regimen-related toxicity that has stopped HSCT from being offered to patients older than 16 yr of age or with advanced organ damage. Attempts to reduce the intensity of preparative regimens for SCD patients have been based on two approaches. One is to produce myeloablation, which requires donor marrow infusion for hematopoietic recovery. The second approach is to not eradicate host hematopoiesis, which would allow hematopoietic recovery even without donor stem cell infusion. Sustained engraftment and alleviation of clinical phenotype have been reported in SCD patients following a reduced-intensity conditioning regimen; However, HSCT using non-myeloablative regimens, although well tolerated, has generally not resulted in stable engraftment.

In a recent study, seven patients (18 yr or younger) with high-risk SCD were given a bone marrow graft from an HLA-matched sibling donor (Krishnamurti et al. 2008). The conditioning regimen consisted of a total dose of 6.4 mg/kg intravenous or 8 mg/kg oral BU, a total dose of 175 mg/m^2 FLU, a total dose of 130 mg/kg equine ATG, and TLI 500 cGy. Regimen-related toxicity was minimal. All but one patient had sustained engraftment. None of the patients developed greater than grade 2 acute GVHD or extensive chronic GVHD. Five out of seven patients developed MC. At a follow-up of 2–8.5 yr after HSCT, all patients were alive, and six of the seven patients had no laboratory or clinical evidence of disease. These data are encouraging in terms of sustained engraftment with minimal toxicity. On the other hand, unfavorable outcomes have been reported from using a reduced-intensity conditioning regimen to treat adults with SCD (Van Beisen et al. 2000).

The rationale for lower-intensity conditioning regimens in hemoglobinopathies stems from historical experience indicating that stable MC is sufficient for cure (Nesci et al. 1998; Andreani et al. 2000). Thus, myeloablation may not be necessary for a successful outcome. In a retrospective multicenter series, Iannone et al. (2003) described a non-myeloablative HSCT approach in six pediatric patients with SCD and one with thalassemia. This very-low-intensity conditioning regimen consisted of minimal intensity treatment with FLU/2 Gy TBI. Two patients also received ATG. Regimen-related toxicity was minimal, but the treatment resulted in only transient donor engraftment in six of seven patients, suggesting that more intensive conditioning is required in individuals with pre-HSCT transfusion exposures.

More recently, a novel non-myeloablative conditioning regimen was reported to be successful in adults with SCD (Hsieh et al. 2009). Ten adults with a median age of 26 yr (range of 16–45 yr) and severe SCD underwent non-myeloablative transplantation with PBSC from HLA-matched siblings. The conditioning regimen included a total dose of 1 mg/kg alemtuzumab on Days 7–3 and TBI of 300 cGy on Day 2. As GVHD prophylaxis, recipients received sirolimus initiated on Day 1 at a dose of 5 mg every 4 h for three doses, then 5 mg daily. Patients received 5.51×10^6 to 23.8×10^6 $CD34^+$ cells/kg. Nine of 10 patients had sustained engraftment, and one had rejection. No patient died. None of the patients had full donor chimerism in both lymphoid and myeloid lines. Acute or chronic GVHD did not develop in any patient. If studies using this method can show sustained mixed donor chimerism with low regimen-related toxicity, the paradigm that only severely affected patients with SCD be offered HSCT may shift, thus allowing more low-risk patients the option of cure without long-term morbidity.

Sustained and Full Fetal Hemoglobin Production after Failure of Bone Marrow Transplant

HbF increase has been reported in three transplanted SCD patients after graft failure (Ferster et al. 1995). Recent clinical observations performed by our group strongly support such novel approaches. The reactivation of HbF synthesis has been documented after bone marrow transplant (BMT) failure and autologous reconstitution in patients affected by β-hemoglobin disorders (Paciaroni et al. 2009).

It is possible that after myeloablative treatment, the vast majority of committed autologous erythroid progenitors were destroyed, and only autologous hematopoietic stem cells survived in marrow niches, restoring the ontological HbF regulation mechanisms (Bank 2006). Alternatively, the increased HbF may be due to a new status, induced by modulations and influences operating in the setting of the BMT (microenvironment and stage-specific cytokines), that reprograms globin-gene expression or selects "fetal clone" proliferation. These cases strongly support further research efforts to find ways to reverse the hemoglobin switch and induce HbF production in adults in order to treat β-hemoglobin disorders.

CONCLUSIONS

The only radical cure for homozygous thalassemia is to transplant bone marrow from an HLA-identical donor who is normal or heterozygous for thalassemia, which is capable of producing and maintaining a normal hemoglobin level in the recipient. All thalassemic patients, together with their parents and siblings, should be HLA typed for this purpose. When an HLA donor is available, we believe that bone marrow transplantation is mandatory in thalassemic patients of class 1 and 2, as well as for those of class 3 who are aged <17 yr; adult patients should also be offered this possibility of a cure, but 30% will die of transplant-related mortality. Patients without matched family or unrelated donors could benefit from haploidentical mother-to-child transplantation, which has shown encouraging results, although it is still in the experimental phase.

A high cure rate has also been achieved by performing HSCT on children with SCA following current myeloablative conditioning protocols. Again, we think that HLA typing should be performed for all family members of an SCA child. If a genotypically identical sibling or phenotypically identical parent is identified, hematopoietic stem cell transplantation should be performed in all SCA patients aged <17 yr, before the major complications of the disease affect the child. The novel non-myeloablative transplantation protocol for SCD patients could serve as a platform for its wide application, not only in adults but also in high-risk pediatric patients with SCD.

REFERENCES

**Reference is also in this collection.*

Adamkiewicz TV, Szabolcs P, Haight A, Baker KS, Staba S, Kedar A, Chiange KY, Krishnamurti L, Boyer MW, Kurtzberg J, et al. 2007. Unrelated cord blood transplantation in children with sickle cell disease: Review of four-center experience. *Pediatr Transplant* **11:** 641–644.

Andreani M, Manna M, Lucarelli G, Tonucci P, Agostinelli F, Ripalti M, Rapa S, Talevi N, Galimberti M, Nesci S. 1996. Persistence of mixed chimerism in patients transplanted for the treatment of thalassemia. *Blood* **87:** 3494–3499.

Andreani M, Nesci S, Lucarelli G, Tonucci P, Rapa S, Angelucci E, Persini B, Agostinelli F, Donati M, Manna M. 2000. Long-term survival of ex-thalassemic patients with persistent mixed cjimerism after marrow transplantation. *Bone Marrow Transplant* **25:** 401–404.

Andreani M, Testi M, Battarra M, Lucarelli G. 2011a. Split chimerism between nucleated and red blood cells after bone marrow transplantation for haemoglobinopathies. *Chimerism* **2:** 21–22.

Andreani M, Testi M, Gaziev J, Condello R, Bontadini A, Tazzari PL, Ricci F, De Felice L, Agostini F, Fraboni D, et al. 2011b. Quantitatively different red cell/nucleated cell chimerism in patients with long-term, persistent hematopoietic mixed chimerism after bone marrow transplantation for thalassemia major or sickle cell disease. *Haematologica* **96:** 128–133.

Bank A. 2006. Regulation of human fetal haemoglobin: New players, new complexities. *Blood* **107:** 435–443.

Bernaudin F, Socie G, Kuentz M, Chevret S, Duval M, Bertrand Y, Vannier JP, Yakouben K, Thuret I, Bordigoni P, et al. 2007. Long-term results of related myeloablative stem-cell transplantation to cure sickle cell disease. *Blood* **110:** 2749–2756.

Ferster A, Corazza F, Vertongen F, Bujan W, Devalck C, Fondu P, Cochaux P, Lambermont M, Khaladji Z, Sariban E. 1995. Transplanted sickle-cell disease patients with autologous bone marrow recovery after graft failure develop increased levels of fetal haemoglobin which corrects disease severity. *Brit J Haematol* **90:** 804–808.

Gaziev J, Sodani P, Polchi P, Andreani M, Lucarelli G. 2005. Bone marrow transplantation in adults with thalassemia. Treatment and long-term follow-up. *Ann NY Acad Sci* **1054:** 196–205.

Ghavamzadeh A, Nasseri P, Eshraghian MR, Jahani M, Baybordi I, Nateghi J, Khodabandeh A, Sadjadi AR, Mohyeddin M, Khademi Y. 1997. Prognostic factors in bone marrow transplantation for β thalassemia major: Experiences from Iran. *Bone Marrow Transplant* **22:** 1167–1169.

Giardini C, Galimberti M, Lucarelli G, Polchi P, Angelucci E, Baronciani F, Agostinelli F, Giorni G, Muretto P. 1993. Bone marrow transplantation in sickle-cell anemia in Pesaro. *Bone Marrow Transplant* **12:** 122–123.

Hobbs JR, Hugh-Jones K, Shaw PJ, Downie CJC, Williamson S. 1986. Engraftment rates related to busulphan and cyclophosphamide dosages for displacement bone marrow transplants in fifty children. *Bone Marrow Transplant* **1:** 201–208.

Hongeng S, Pakakasama S, Chuansumrit A, Sirachainan N, Kitpoka P, Udomsubpayakul U, Ungkanont A, Jootar S. 2006. Outcomes of transplantation with related and unrelated-donor stem cells in children with severe thalassemia. *Biol Blood Marrow Transplant* **12:** 683–687.

Hsieh MM, Kang EM, Fitzhugh CD, et al. 2009. Allogeneic hematopoietic stem-cell transplantation for sickle cell disease. *N Engl J Med* **361:** 2309–2317.

Hsieh MM, Fitzhugh CD, Tisdale JF. 2011. Allogeneic hematopoietic stem cell transplantation for sickle cell disease: The time is now. *Blood* **118:** 1197–1207.

Iannone R, Casella JF, Fuchs EJ, Chen AR, Jones RJ, Woolfrey A, Amylon M, Sullivan KM, Storb RF, Walters MC. 2003. Results of minimally toxic nonmyeloablative transplantation in patients with sicule cell anemia and β-thalassemia. *Biol Blood Marrow Transplant* **9:** 519–528.

Isgrò A, Gaziev J, Sodani P, Lucarelli G. 2010. Progress in hematopoietic stem cell transplantation as allogeneic cellular gene therapy in thalassemia. *Ann NY Acad Sci* **1202:** 149–154.

Kapoor N, Kirkpatrick D, Blaese RM, Oleske J, Hilgartner MH, Changanti RS, Good RA, O'Reilly RJ. 1981. Reconstitution of normal megakaryocytopoiesis and immunologic functions in Wiskott-Aldrich syndrome by marrow transplantation following myeloablation and immunosuppression with busulfan and cyclophosphamide. *Blood* **57:** 692–696.

Krishnamurti L, Kharbanda S, Biernacki M, Zhang W, Baker KS, Wagner JE, Wu CJ. 2008. Stable long-term donor engraftment following reduced-intensity hematopoietic cell transplantation for sickle cell disease. *Biol Blood Marrow Transplant* **14:** 1270–1278.

La Nasa G, Argiolu F, Giardini C, Pession A, Fagioli F, Caocci G, Vacca A, De Stefano P, Piras E, Ledda A, et al. 2005a. Unrelated bone marrow transplantation for β-thalassemia patients: The experience of the Italian Bone Marrow Tranplant Group. *Ann NY Acad Sci* **1054:** 186–195.

La Nasa G, Caocci G, Argiolu F, Giardini C, Locatelli F, Vacca A, Ororfino MG, Piras E, Addari MC, Ledda A, et al. 2005b. Unrelated donor stem cell transplantation in adult patients with thalassemia. *Bone Marrow Transplant* **36:** 971–975.

Locatelli F, Rocha V, Reed W, Bernaudin F, Ertem M, Grafakos S, Brichard B, Li X, Nagler A, Giorgiani G, et al. 2003. Related umbilical cord blood transplantation in patients with thalassemia and sickle cell disease. *Blood* **101:** 2137–2143.

Lucarelli G, Gaziev J. 2008. Advances in the allogeneic transplantation for thalassemia. *Blood Rev* **22:** 53–63.

Lucarelli G, Clift RA, Galimberti M, Angelucci E, Giardini C, Baronciani D, Polchi P, Andreani M, Gazlev D, Erer B, et al. 1990. Bone marrow transplantation in patients with thalassemia. *N Engl J Med* **322:** 417–421.

Lucarelli G, Galimberti M, Polchi P, Angelucci E, Baronciani D, Durazzi SM, Giardini C, Albertini F, Clift RA. 1992. Bone marrow transplantation in adult thalassemia. *Blood* **80:** 1603–1607.

Lucarelli G, Clift RA, Galimberti M, Angelucci E, Giardini C, Baronciani D, Polchi P, Andreani M, Gazlev D, Erer B, et al. 1999. Bone marrow transplantation in adult thalassemic patients. *Blood* **93:** 1164–1167.

Lucarelli G, Andreani M, Angelucci E. 2002. The cure of thalassemia by bone marrow transplantation. *Blood Rev* **16:** 81–85.

Lucarelli G, Gaziev J, Isgrò A, Sodani P, Paciaroni K, Alfieri C, De Angelis G, Marziali M, Simone MD, Gallucci C, et al. 2011. Allogeneic cellular gene therapy in hemoglobinopathies—evaluation of hematopoietic SCT in sickle cell anemia. *Bone Marrow Transplant* doi: 10.1038/bmt.2011.79.

Nesci S, Manna M, Lucarelli G, Tonucci P, Donati M, Buffi O, Agostinelli F, Andreani M. 1998. Mixed chimerism after bone marrow transplantation in thalassemia. *Ann NY Acad Sci* **850:** 495–497.

Paciaroni K, Gallucci C, De Angelis G, Alfieri C, Roveda A, Lucarelli G. 2009. Sustained and full fetal hemoglobin production after failure of bone marrow transplant in a patient homozygous for β^0-thalassemia: A clinical remission despite genetic disease and transplant rejection. *Am J Hematol* **84:** 372–373.

Panepinto JA, Walters MC, Carreras J, Marsh J, Bredeson CN, Gale RP, Hale GA, Horan J, Hows JM, Klein JP, et al. 2007. Matched-related donor transplantation for sickle cell disease: Report from the Center for International Blood and Transplant Research. *Br J Haematol* **137:** 479–485.

Parkman R, Rappeport J, Geha R, Belli J, Cassady R, Levey R, Nathan DG, Rosen FS. 1978. Complete correction of the Wiskott-Aldrich syndrome by allogeneic bone-marrow transplantation. *N Engl J Med* **298:** 921–927.

Sabloff M, Chandy M, Wang Z, Logan BR, Ghavamzadeh A, Li CK, Irfan SM, Bredeson CN, Cowan MJ, Gale RP, et al. 2011. HLA-matched sibling bone marrow transplantation for β-thalassemia major. *Blood* **117:** 1745–1750.

Sodani P, Isgrò A, Gaziev J, Polchi P, Paciaroni K, Marziali M, Simone MD, Roveda A, Montuoro A, Alfieri C, et al. 2010. Purified T-depleted, CD34+ peripheral blood and bone marrow cell transplantation from haploidentical mother to child with thalassemia. *Blood* **115:** 1296–1302 (Erratum in *Blood* **116:** 1627).

Sodani P, Isgrò A, Gaziev J, Paciaroni K, Marziali M, Simone MD, Roveda A, De Angelis G, Gallucci C, Torelli F, et al. 2011. T cell-depleted HLA-haploidentical stem cell transplantation in thalassemia young patients. *Pediatr Rep* **22:** e13.

Storb R, Champlin RE. 1991. Bone marrow transplantation for severe aplastic anemia. *Bone Marrow Transplant* **8:** 69–72.

Storb R, Weiden PL, Graham TC, Lerner KG, Nelson N, Thomas ED. 1977. Hemopoietic grafts between DLA-identical canine littermates following dimethyl myleran. Evidence for resistance to grafts not associated with DLA and abrogated by antithymocyte serum. *Transplantation* **24:** 349–357.

Thomas ED, Storb R, Fefer A, Slichter SJ, Bryant JI, Buckner CD, Neiman PE, Clift RA, Funk DD, Lerner KE. 1972. Aplastic anaemia treated by marrow transplantation. *Lancet* **1:** 284–289.

van Beisen K, Bartholomew A, Stock W, Peace D, Devine S, Sher D, Sosman J, Chen YH, Koshy M, Hoffman R. 2000. Fludarabine-based conditioning regimen for allogeneic transplantation in adult with sickle cell disease. *Bone Marrow Transplant* **26:** 445–449.

Vermylen C, Cornu G. 1993. Bone marrow transplantation in sickle cell anemia. *Blood Rev* **7:** 1–3.

Vermylen C, Cornu G, Ferster A, Brichard B, Ninane J, Ferrant A, Zenebergh A, Maes P, Dhooge C, Benoit Y, et al. 1998. Haematopoietic stem cell transplantation for sickle cell anaemia: The first 50 patients transplanted in Belgium. *Bone Marrow Transplant* **22:** 1–6.

Walters MC, Storb R, Patience M, Leisenring W, Taylor T, Sanders JE, Buchanan GE, Rogers ZR, Dinndorf P, Davies SC, et al. 2000. Impact of bone marrow transplantation for symptomatic sickle cell disease: An interim report. Multicenter investigation of bone marrow transplantation for sickle cell disease. *Blood* **95:** 1918–1924.

Walters MC, Patience M, Leisenring W, Rogers ZR, Aquino VM, Buchanan GR, Roberts IA, Yeager AM, Hsu L, Adamkiewicz T, et al. 2001. Stable mixed hematopoietic chimerism after bone marrow transplantation for sickle cell anemia. *Biol Blood Marrow Transplant* **7:** 665–673.

* Williams TN, Weatherall DJ. 2012. World distribution, population genetics, and burden of thalassemic and sickle cell syndromes. *Cold Spring Harb Perspect Med* **2:** a011692.

Development of Gene Therapy for Thalassemia

Arthur W. Nienhuis and Derek A. Persons

Department of Hematology, Division of Experimental Hematology, St. Jude Children's Research Hospital, Memphis, Tennessee 38105

Correspondence: arthur.nienhuis@stjude.org

Retroviral vector–mediated gene transfer into hematopoietic stem cells provides a potentially curative therapy for severe β-thalassemia. Lentiviral vectors based on human immunodeficiency virus have been developed for this purpose and have been shown to be effective in curing thalassemia in mouse models. One participant in an ongoing clinical trial has achieved transfusion independence after gene transfer into bone marrow stem cells owing, in part, to a genetically modified, dominant clone. Ongoing efforts are focused on improving the efficiency of lentiviral vector–mediated gene transfer into stem cells so that the curative potential of gene transfer can be consistently achieved.

The development of gene transfer for the treatment of thalassemia has been a goal for more than three decades (Mercola and Cline 1980). The anticipated targets for gene transfer are hematopoietic stem cells (Naldini 2011; Rivière et al. 2012). Stem and progenitor cells are to be harvested from patient bone marrow or peripheral blood following cytokine mobilization, purified by immunoselection, transduced, and returned to the patient (Fig. 1). Various types of vectors have been considered for gene transfer into stem cells, with integrating retroviruses being the leading candidate. Recently, one participant in an ongoing gene transfer trial for severe β-thalassemia has achieved clinical benefit with elimination of his transfusion requirement (Cavazzana-Calvo et al. 2010).

Overall, the field of gene therapy evolved slowly over the last three decades (Weatherall 1988). In 1995, the NIH Director convened a panel of experts to evaluate the field (Orkin and Motulsky 1995). Numerous, early clinical trials had been unsuccessful and uninformative. The panel, while noting that well-designed clinical trials are necessary to advance a new therapy, concluded that the methodology of gene transfer had not yet evolved to justify widespread attempts at clinical trials. Proof of efficacy in relevant preclinical models was judged to be necessary before attempting additional clinical trials. Although some academic investigators remained committed to preclinical research and ultimately to clinical trials, commercial interest declined, and there was a general sense of pessimism that pervaded the field.

SUCCESSFUL GENE THERAPY FOR IMMUNODEFICIENCIES

Success initially came with the report of ex vivo gene therapy trials of gene transfer into hematopoietic stem cells performed in patients with

Copyright © 2013 Cold Spring Harbor Laboratory Press; all rights reserved
Cite this article as *Cold Spring Harb Perspect Med* doi: 10.1101/cshperspect.a011833

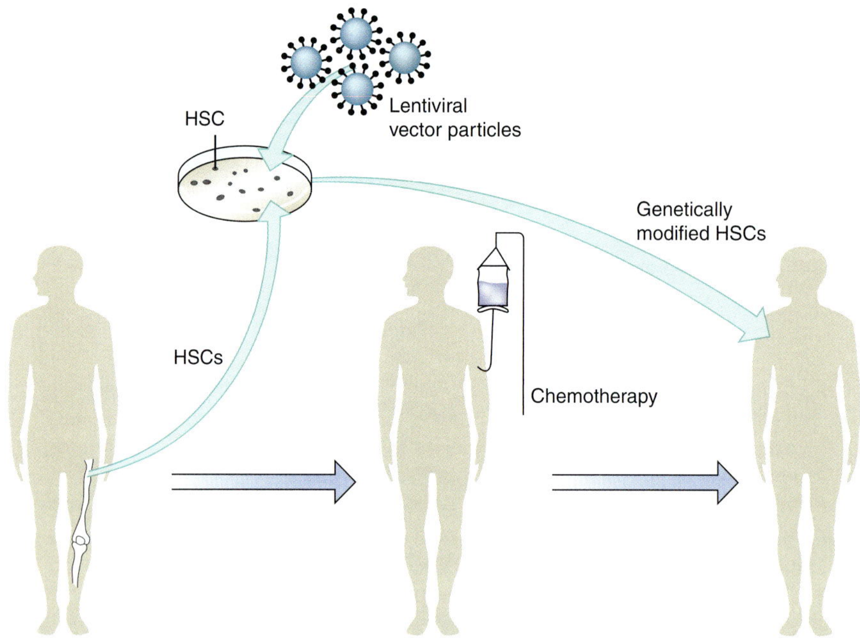

Figure 1. Gene therapy procedures: Cells were collected from the bone marrow of a patient with β-thalassemia (Cavazzano-Calvo et al. 2010). After purification, the hematopoietic stem cells (HSCs) were maintained in culture for 34 h. Lentiviral vector particles containing a functional β-globin gene were added, and culture was continued for 18 h. The patient underwent chemotherapy to eradicate the remaining HSCs and make room for the genetically modified cells. The transduced HSCs were then transplanted into the patient by infusion into a peripheral vein. (Adapted from Persons 2010; reprinted, with permission, from the author.)

severe combined immunodeficiency (SCID) secondary to deficiency in the common γ-chain (X-SCID) (Hacein-Bey-Abina et al. 2002, 2010; Gaspar et al. 2004, 2011b) or adenosine deaminase deficiency (ADA-SCID) (Aiuti et al. 2002, 2009; Gaspar et al. 2011b). Gene-corrected lymphoid cells in these disorders have a growth advantage compared with uncorrected cells and fully repopulate the lymphoid tissues. Gammaretroviral vectors, based on a mouse tumor virus, were used with the coding sequences for the missing protein encoded by a cDNA under the control of the enhancer/promoter elements in the long terminal repeats (LTRs). This initial experience with successful gene therapy has recently been reviewed (Naldini 2011; Rivière et al. 2012). Twenty patients with X-SCID and 40 patients with ADA deficiency have undergone the gene transfer procedure. Generally, the procedure has been successful with recovery of lymphoid elements and correction of the immunodeficiency, although the procedure has failed in a few older patients with ADA deficiency (Thrasher et al. 2005; Gaspar et al. 2011a). A similar vector was used to treat three individuals with chronic granulomatous disease (CGD) (Ott et al. 2006). Although clinical improvement occurred in two patients initially, complications—to be described below—limited the overall outcome. Recently, a successful gene therapy trial has also been reported for Wiskott–Aldrich syndrome (WAS), with 10 participants showing evidence of clinical improvement (Boztug et al. 2010; Braun et al. 2011; C Klein, pers. comm.). These trials provided proof-of-principle that stem cell–targeted gene transfer can achieve a therapeutic benefit.

The gene therapy trials that used a gammaretroviral vector with intact LTRs have been complicated by insertional mutagenesis and proto-oncogene activation (Hacein-Bey-Abina et al. 2003, 2008; Howe et al. 2008). Five of the

20 participants in the trials for X-SCID have developed T-cell leukemia. Recently, four of the 10 participants in the trial for WAS have also developed leukemia (Braun et al. 2011; C Klein, pers. comm.). Three patients with CGD developed myelodysplastic syndrome, and one has died of acute myeloid leukemia (Stein et al. 2010). Although the risk of insertional mutagenesis with proto-oncogene activation was recognized when the trials were initiated, this complication was not observed in murine models or in large animal studies over the years (Kiem et al. 2004; Nienhuis et al. 2006). Only one monkey developed myeloid leukemia 5 yr after the gene transfer procedure, thought secondary to vector-mediated inactivation of an anti-apoptotic gene (Seggewiss et al. 2006).

Studies of the pattern of vector integration in patients from the gene therapy trials has uncovered a skewed pattern with preferential integration near or within transcriptionally active genes (Cattoglio et al. 2007; Deichmann et al. 2007, 2011; Schwarzwaelder et al. 2007; Biasco et al. 2012). Remarkably, the majority of patients with leukemia in both the X-SCID and WAS trials had an insertion near or within the LM02 gene, suggesting that its activation was the initiating event in the development of leukemia (Hacein-Bey-Abina et al. 2003, 2008; Schwarzwaelder et al. 2007; Braun et al. 2011; C Klein, pers. comm.). Integrations in the CGD patients were found in the MDS-EVI1, PRDM1, and SETB1 loci (Stein et al. 2010). None of the patients with ADA deficiency has developed leukemia despite vector integration near or within proto-oncogenes including the LM02 gene (Aiuti et al. 2007). The basis for the different outcomes for X-SCID versus ADA-SCID is not understood. A recent review discusses the various possibilities (Shaw and Kohn 2011).

POTENTIALLY SAFER INTEGRATING RETROVIRAL VECTORS

The occurrence of leukemia in the trials for X-SCID, WAS, and CGD has prompted a search for safer vectors, as recently reviewed comprehensively (Wu and Dunbar 2011; Cooray et al. 2012). A self-inactivating (SIN) design involves removal of the enhancer/promoter from the 3′ LTR. During retroviral integration, a portion of the 5′ LTR is copied from the 3′ LTR so that both LTRs lack the enhancer/promoter in the integrated SIN vector genome, thereby reducing the risk for proto-oncogene activation (Yu et al. 1986; Zufferey et al. 1998). The SIN design requires the use of an internal promoter, either retroviral or cellular, to achieve expression of the therapeutic transgene. A SIN design has been applied to deriving a gammaretroviral vector that is now in clinical trials for X-SCID in the United States and Europe (Thornhill et al. 2008). Success in the first patient treated with the new vector (Pai et al. 2011) was reported at the most recent meeting of the American Society of Hematology.

Another modification designed to improve safety is the incorporation of an insulator element into the vector. The insulator element is usually placed in the 3′ LTR of a SIN vector so that it is copied over into the 5′ LTR upon vector integration, resulting in flanking of the transgene with insulator elements (Emery 2011; Cooray et al. 2012). The most well studied of insulator elements is that located in the 5′ hypersensitive site 4 (HS4) of the chicken β-globin gene locus (Fig. 2) (Chung et al. 1993, 1997; Giles et al. 2010). This insulator has two separable activities; it protects integrated transgenes from silencing by nearby heterochromatin, a so-called barrier activity, and it also blocks interaction between linked enhancer and promoter elements when located between the two (Recillas-Targa et al. 2002). As a barrier element, the insulator prevents spread of silencing histone modifications into the transcribed locus and CpG methylation of the protected promoter sequences (Pikaart et al. 1998). Experimentally, the erythroid-specific LCR elements included in globin gene lentiviral vectors did not activate the proto-oncogene LM02 in human T-cells as expected (Ryu et al. 2008), and the cHS4 insulator reduced gene activation by LCR elements in human erythroleukemia cells (Ryu et al. 2007).

The activities of HS4 have been mapped to a 275-bp core element, which includes five protein-binding sites as determined by DNase I

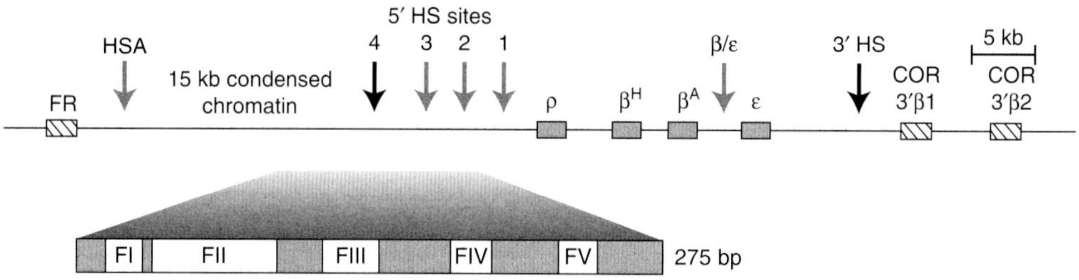

Figure 2. Schematic representation of the chicken *β-globin* cluster and surrounding loci. Boxes represent the folate receptor (FR), *β-globin* (ρ, $β^H$, $β^A$, and ε), and chicken olfactory receptor (COR) genes (not to scale). (Arrows) DNase I hypersensitive sites. The core of the HS4 element is expanded to show the positions of the five in vitro footprinted sequences (Chung et al. 1997).

footprinting (Fig. 2) (Chung et al. 1997; Giles et al. 2010). The sites have been designated footprint I through footprint V (FI, FII, FIII, FIV, and FV). FII binds the protein CTCF, and this protein is sufficient for enhancer blocking through its interaction with other proteins and intracellular membranes (Bell et al. 1999; Giles et al. 2010; Ghirlando et al. 2012). CTCF and associated proteins anchor the insulator to nuclear structures including the nucleolus, creating a chromatin loop that is thought to inhibit enhancer–promoter interactions. FIV is required for barrier activity and has been shown to bind USF proteins that recruit histone-modifying enzymes associated with transcriptionally active chromatin (West et al. 2004). Such recruitment appears to block encroachment of transcriptionally repressive chromatin into the transcribed region. The remaining protein-binding sites (F1, FIII, and FV) have all been shown to bind the protein VEZF1, which, in the context of recruiting other proteins, protects downstream DNA within the transcribed region from DNA methylation, particularly of promoter elements (Dickson et al. 2010). Overall, insulators are involved in the organization of chromatin domains and in maintaining boundaries, specifically those between repressed chromatin and transcriptionally active chromatin structure (Ghirlando et al. 2012; Kolovos et al. 2012). Other insulators have been considered for vector development, and a limited number have been tested (Emery 2011). For example, a combination of FII from the chicken HS4 and a homologous region from the T-cell receptor, α/δ BEAD-1 insulator greatly reduced the transforming potential of retroviral vectors (Ramezani et al. 2008).

DEVELOPMENT OF GLOBIN GENE VECTORS

Gammaretroviral Vectors

Early studies explored the use of DNA viruses including SV40 (Mulligan et al. 1979), adenovirus (Karlsson et al. 1986), and adeno-associated viral vectors (Walsh et al. 1992; Maina et al. 2008) for transfer of globin-encoding sequences. However, integrating gammaretroviral vectors was judged to be potentially more suitable for achieving permanent genetic modification of hematopoietic cells. With evolving knowledge of the β-globin gene locus, retroviral vectors were constructed that incorporated the natural promoter sequences, the introns, and the 3′-untranslated sequences in an effort to enhance expression (Cone et al. 1987). Transfer of the complete globin gene into murine stem cells could be accomplished, but the level of expression remained relatively low (Cone et al. 1987; Karlsson et al. 1987; Dzierzak et al. 1988; Sadelain et al. 1995).

Locus Control Region

Only with the discovery of the locus control region (LCR) upstream from the β-globin gene cluster did it become feasible to design vectors that achieved higher levels of expression.

 Cite this article as *Cold Spring Harb Perspect Med* doi: 10.1101/cshperspect.a011833

The LCR was discovered by virtue of its functional activity in providing copy number–dependent, integration site–independent globin gene expression in erythroid cells in transgenic mice (Grosveld et al. 1987). Four DNase I hypersensitive (HS) sites are distributed over ~20 kb upstream from the human β-globin gene locus (Tuan and London 1984; Tuan et al. 1985; Forrester et al. 1987). Each HS site marks a 200- to 400-bp core element of exposed DNA that includes binding sites for many erythroid or ubiquitously expressed transcriptional factors. The LCR is thought to function by forming a folded structure whereby multiple proteins bound to HS sites interact in forming a holocomplex, which by virtue of loop formation is brought into contact with a promoter within a transcription factory within the nucleus, thereby activating that promoter (Palstra et al. 2008; Schoenfelder et al. 2010; Edelman and Fraser 2012).

Gammaretroviral Vectors Continued

The series of HS sites upstream from the ε-globin gene has been tailored to minimize the size of vectors by focusing on the core, HS elements for use in gene transfer vectors (Plavec et al. 1993; Emery et al. 1998). Unfortunately, these elements often resulted in gammaretroviral vector instability with low titer. Various strategies were attempted to overcome these limitations, but none was fully successful (Novak et al. 1990; Chang et al. 1992; Takekoshi et al. 1995; Ren et al. 1996; Sabatino et al. 2000; Fragkos et al. 2005). Extensive mutagenesis was required to achieve adequate titers with vectors containing multiple HS sites from the β-globin locus to remove potential splice or polyadenylation sites and to eliminate sequences in the β-globin gene that were unnecessary and reduced titer (Leboulch et al. 1994). Another approach that gave results that were somewhat encouraging was the screening of many vectors to identify one that gave a fairly high titer despite the presence of multiple HS fragments. However, highly variable expression was found in mouse erythroleukemia cell clones containing the vector that had been selected based on titer (Sadelain et al. 1995). Addition of an insulator improved expression (Emery et al. 2000) as did other modifications designed to increase globin gene expression (Li et al. 1999). Efforts continued through 2002 to adapt gammaretroviral vectors for transfer of a globin gene with high expression (Nicolini et al. 2002; Emory et al. 2002). Such vectors were also tested for their ability either to correct β-thalassemia in a mouse model (Nishino et al. 2006) or to express an antisickling β-globin in human cells from patients with sickle cell disease (Oh et al. 2004). Overall, the extensive efforts over nearly two decades to develop gammaretroviral vectors for globin gene transfer had limited success (Ellis and Pannell 2001), and ultimately committed investigators migrated to the use of lentiviral vectors because of the advantages discussed in the next section.

Lentiviral Vectors

Lentiviral vectors based on the genome of human immunodeficiency virus (HIV) have several inherent advantages for stem cell–targeted gene transfer (Naldini et al. 1996). HIV-based vectors are able to transduce non-mitotic cells because of the ability of the preintegration complex (PIC) to traverse the intact nuclear membrane, a property that is dependent on the inclusion of specific viral sequences within the gene transfer vector (Follenzi et al. 2000; Sirven et al. 2000; Zennou et al. 2000; Cooray et al. 2012). In addition, the PIC of HIV-based vectors persists in the nucleus much longer than the PIC of gammaretroviral vectors, thereby extending the window during which vector integration may occur (Hass et al. 2000). Activation of the quiescent hematopoietic cells to the G_0/G_1 phase of the cell cycle is required for lentiviral vector transduction (Uchida et al. 1998; Case et al. 1999; Sutton et al. 1999). Lentiviral vectors can be packaged with a variety of envelope proteins (Naldini and Verma 2000). Historically, the envelope protein from vesicular stomatitis virus (VSV-G) has been used to pseudotype lentiviral vector particles for experimental use. A recent study confirms that particles having the VSV-G envelope protein are superior at transducing $CD34^+$ cells relative to the envelope proteins of several gammaretroviruses (Kim et al. 2010).

The chicken HS4 insulator has also been tested within the LTRs of a SIN lentiviral globin vector. Its incorporation into the vector may reduce titer (Urbinati et al 2009), but optimization has been found to improve titer and transgene expression (Arumugam et al. 2007; Hanawa et al. 2009). The core 250-bp element of the cHS4 insulator has been extended with sequences immediately downstream (Aker et al. 2007) or with sequences from the 3′ end of the cHS4 fragment (Arumugam et al. 2009) in efforts to improve vector expression. Overall results suggest that the use of chromatin insulators may improve the expression and safety of integrating gene transfer vectors as summarized in a recent review (Emery 2011).

Although gammaretroviral vectors have a predilection to integrate near transcriptional start sites (Wu et al. 2003), HIV-based lentiviral vector genomes are more uniformly distributed throughout transcribed genes (Schröder et al. 2002). Preferential intragenic integration has also been shown for a β-globin gene vector in human and murine hematopoietic cells (Imren et al. 2004; Ronen et al. 2011). This is only a relative advantage because inactivation of a tumor suppressor gene may occur with intragenic integration. In addition, some lentiviral vector integrations do indeed occur near a transcription start site. Nonetheless, direct comparison in a tumor-prone mouse model suggests that SIN lentiviral vectors with an internal promoter are significantly less genotoxic than standard gammaretroviral vectors with intact LTRs (Montini et al. 2006). Aberrant splicing events are much reduced by use of SIN lentiviral vector design compared with vectors having intact LTRs (Cesana et al. 2012). Recent data also suggest that the lentiviral vector integration profile in preclinical models and in a clinical trial is inherently less oncogenic than that of gammaretroviral vectors (Biffi et al. 2011). However, insertional mutagenesis and gene activation have been documented with lentiviral vectors (Hargrove et al. 2008; Modlich et al. 2009; Montini et al. 2009), which has led to the development of leukemia in a murine model (Heckl et al. 2012). Thus, use of lentiviral vectors will likely mitigate, but not eliminate, the risk of insertional mutagenesis.

HIV-based vectors having a SIN design are in clinical trials for metachromatic leukodystrophy (MLD) (Biffi and Naldini 2007), adrenoleukodystrophy (ALD) (Cartier et al. 2009, 2012), WAS (Galy and Thrasher 2011; Scaramuzza et al. 2012), and indeed for β-thalassemia (Cavazzana-Calvo et al. 2010). The results in four patients with MLD are particularly striking in that expression of the therapeutic transgene occurred in >50% of hematopoietic cells following stem cell transduction and autologous transplantation with major clinical benefit (A Biffi and L Naldini, pers. comm.). Thus, the greater efficiency of gene transfer into stem cells that had been predicted for lentiviral vectors has apparently been achieved in these early clinical trials.

Another advantage of lentiviral vectors is their well-developed mechanism for nuclear-to-cytoplasmic transfer of unspliced mRNA, which facilitated the inclusion of globin genes with intact introns, LCR fragments, and insulator elements (May et al. 2000; Pawliuk et al. 2001). Globin vectors were envisioned as potentially being therapeutically useful both for thalassemia and sickle cell disease. Accordingly, an anti-sickling variant (Oh et al. 2004) or the natural γ-globin gene (Pestina et al. 2009) has been used to facilitate their ultimate use in treatment of sickle cell disease.

Correction of Mouse Models

Transgenic mouse strains having either severe β-thalassemia (Ciavatta et al. 1995; Yang et al. 1995) or sickle cell disease (Pászty et al. 1997) have been derived by the standard transgenic and/or homologous recombination methodologies. Globin vectors have been tested for their ability to correct these phenotypes. Quantitative studies using various transgenic mouse strains had suggested that expression of a transgene in 20% of the erythroid precursors along with achievement of expression at 20% of the level of the normal β-globin gene was likely to be sufficient to correct the phenotype of β-thalassemia intermedia in the mouse model

(Persons et al. 2001). Similar studies were performed in the sickle mouse model and suggested that 9%–16% HbF significantly improved the sickle phenotype (Blouin et al. 2000). Collectively, these studies supported the therapeutic potential of available globin gene vectors for correcting hemoglobin disorders.

Several studies have shown that lentiviral vector–mediated globin gene transfer will ameliorate the phenotype of murine thalassemia intermedia (Imren et al. 2002; May et al. 2002; Persons et al. 2003a; Hanawa et al. 2004; Negre et al. 2011). A mouse model of thalassemia major was developed by transplanting adult mice, following lethal irradiation, with fetal liver cells taken from homozygous, β-globin-null mice. These animals developed severe anemia and massive splenomegaly as well as extramedullary hematopoiesis by 60 d of age (Rivella et al. 2003). The phenotype could be rescued by lentiviral vector–mediated, globin gene transfer into the fetal liver cells before transplantation. Significant correction of the erythroid phenotype occurred in animals in which >95% of the hemoglobin contained a human β-globin chain (Rivella et al. 2003). Several studies have also documented the ability of lentiviral vectors to ameliorate or correct the phenotype of sickle cell disease in the mouse model (Pawliuk et al. 2001; Levasseru et al. 2003; Perumbeti et al. 2009; Pestina et al. 2009).

Correction of Human Thalassemic Erythroid Cells

Lentiviral vectors will also correct the phenotype of β-thalassemia major in cultured erythroid cells derived from patients with this disorder. Production of normal amounts of β-globin has been achieved in erythroid cultures with correction of ineffective erythropoiesis and a reduction in apoptosis (Puthenveetil et al. 2004). Gene-corrected primitive cells were transplanted into immune-deficient mice, and effective human erythropoiesis was documented 3–4 mo after transplantation (Puthenveetil et al. 2004). A recent study that used cells from a large number of patients provided additional evidence that the β-thalassemia major phenotype may be corrected by gene transfer (Roselli et al. 2010). Therapeutic levels of fetal hemoglobin have also been achieved in erythroid progeny of β-thalassemic $CD34^+$ cells after lentiviral vector–mediated γ-globin gene transfer in a two-stage culture system (Fig. 3) (Wilber et al. 2011a).

In Vivo Selection

Of concern with regard to the development of future clinical trials for treatment of severe β-thalassemia is the level of stem cell transduction that can be achieved. Strategies have been explored to amplify the genetically modified stem cell pool in vivo to increase its contribution to hematopoiesis relative to the contribution of non-transduced stem cells. A dual gene vector encoding γ-globin with associated LCR elements and a variant methylguanine methyltransferase, driven by a constitutive promoter, has been used to transduce repopulating cells in the mouse model. On initial reconstitution, animals had subtherapeutic levels of γ-globin-expressing cells, but in vivo drug selection amplified the genetically modified population, leading to correction of the β-thalassemia intermedia phenotype in the murine model (Persons et al. 2003b; Zhao et al. 2009). Selective amplification of genetically modified cells by drug selection has also been achieved in nonhuman primate models (Larochelle et al. 2009; Beard et al. 2010).

CLINICAL TRIALS FOR SEVERE β-THALASSEMIA

Plans for the first clinical trial using a lentiviral vector to transfer a globin gene into bone marrow cells from patients with β-thalassemia were announced in 2005 (Bank et al. 2005). The lentiviral vector that was used is a self-inactivating variant that included large segments of the β-locus control region as well as 250-bp core elements from the chicken HS4 insulator in the LTRs (Fig. 4). To date, three participants have been enrolled on the clinical trial, which is open in Paris (Cavazzana-Calvo et al. 2010; P Leboulch, pers. comm.). The vector encodes a mutated adult β-globin [β^A(T87Q)] that has

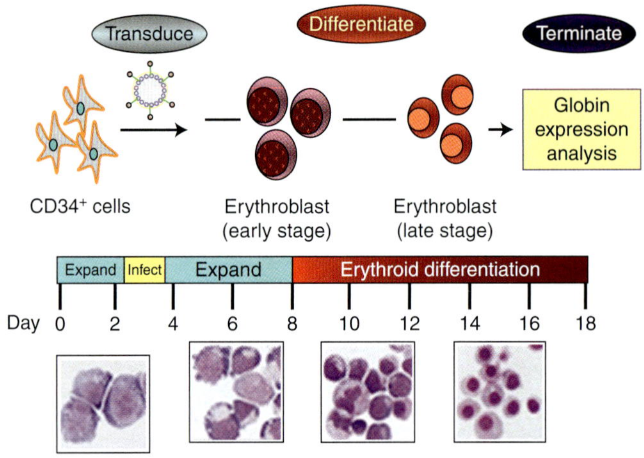

Figure 3. Two-stage culture model of human erythropoiesis. $CD34^+$ cells from normal or thalassemic individuals are initiated into culture in medium designed to expand erythroid progenitors. The $CD34^+$ cells are transduced with a retroviral vector between days 2 and 4 of the expansion phase of the culture. After 8 d, the culture medium is changed to one that favors erythroid differentiation. Full hemoglobinization and often enucleation of the red cells can be achieved, indicating significant maturation. Aliquots were taken at various intervals and at the end of culture to analyze the pattern of hemoglobin production.

anti-sickling properties and can be distinguished from normal adult β-globin by high performance liquid chromatography. The first patient failed to engraft because the purified $CD34^+$ cells had been compromised technically without relation to the gene therapy vector. The second participant has now been followed for 5 yr after the gene transfer procedure. His clinical course is described in detail below. The third patient was transplanted ~2 mo ago without complication. Hematopoietic reconstitution is ongoing, and the variant β-globin can be detected in blood reticulocytes.

The second participant in the clinical trial was 18 yr old at the time of treatment and has now been followed for nearly 5 yr following the

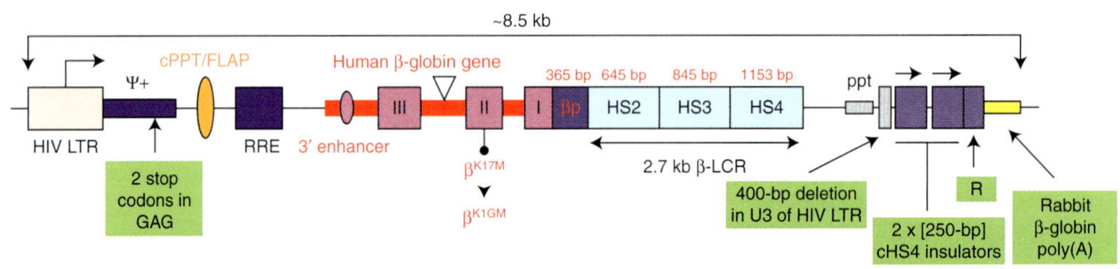

Figure 4. Diagram of the lentiglobin vector showing the 3′ β-globin enhancer, the 372-bp IVS2 deletion, the $\beta^{A\text{-}T87Q}$ mutation [ACA(Thr) to CAG(Gln)], and the DNase I hypersensitive sites (HS) 2, HS3, and HS4 of the human β-globin locus control region (LCR). Safety modifications including the two stop codons in the Ψ+ signal, the 400-bp deletion in the U3 of the right HIV LTR, the rabbit β-globin poly(A) signal, and the two 250-bp cHS4 chromatin insulator core fragments are indicated. (HIV LTR) Human immunodeficiency type-1 virus long terminal repeat; (Ψ+) packaging signal; (cPPT/FLAP) central polypurine tract/DNA FLAP; (RRE) Rev-responsive element; (βp) human β-globin promoter; (ppt) polypurine tract. Southern blot analysis of transduced cells indicates that up to about one-third of integrated vector copies contain a deletion of one of the cHS4 cores within the right U3, and about two-thirds of integrated vector copies contain a deletion of one of the cHS4 cores within the left U3 (Cavazzana-Calvo et al. 2010).

gene transfer procedure (Cavazzana-Calvo et al. 2010). He has severe β^E/β^0-thalassemia and began on regular transfusion therapy at age 3 because of severe and poorly tolerated anemia and significant enlargement of the liver and spleen. Splenectomy was performed at age 6. The participant was conditioned with intravenous Busulfex (3.2 mg/kg per day). His transduced bone marrow $CD34^+$ cells were found to have 0.6 vector genomes/transduced cell. The patient received the transduced bone marrow cells in early June of 2007 and was last transfused in early June of 2008. His Hb level has remained between 9 and 10 g/dL. Approximately one-third of the total hemoglobin is composed of the therapeutic hemoglobin Hb-β^A(T87Q), one-third of which is HbE and one-third is HbF. The ratio of γ to β^E globin did not change following the transplant procedure, suggesting that enhancement of HbF production was not a significant factor in the patient's clinical improvement. The patient has undergone frequent phlebotomies to accelerate the clearance of iron that had accumulated during the several years that he was transfusion dependent.

Insertion site analysis detected a dominant clone in which the vector was integrated into the high-mobility group AT-Hook2 (HMGA2) locus in granulocytes, monocytes, and erythroblasts (Cavazzana-Calvo et al. 2010). This integration site was not found in lymphoid cells. The proportion of genetically modified cells in whole blood increased gradually over time, but the clone having the HMGA2 integration site remained stable at ~2% total blood cells. Molecular analysis indicated that the HMGA2 gene that had been disrupted by vector integration was actively expressed. All three exons were contained within a truncated RNA product that arose by alternate splicing of the third exon into a cryptic 3′ splice signal located within the insulator core and cleavage/polyadenylation within the adjacent LTR. The disrupted locus encodes a chimeric protein in that the first 83 amino acids are encoded by the three HMGA2 exons fused to 52 amino acids from the read-through transcript. Missing from this chimeric RNA is the normal 3′ untranslated region, which includes two binding sites for LET7 microRNA, which is thought to decrease stability of the transcript within the cytoplasm. Analysis of erythroid cells suggested that the HMGA2 integration site was not particularly favorable for high expression of the transferred globin gene.

A multicenter phase I clinical trial to evaluate globin gene transfer in β-thalassemia is being developed and led by Michel Sadelain of Memorial Sloan-Kettering Cancer Center (Sadelain et al. 2010). A collaboration between investigators in Greece and in Seattle has also been established to begin preparation for a clinical trial. These investigators have compared mobilization of stem cells with G-CSF alone, mobilization with G-CSF after 1 mo of pretreatment with hydroxyurea to prevent hyperleukocytosis and splenic enlargement, and mobilization with plerixafor. Plerixafor resulted in rapid and effective mobilization in both splenectomized and unsplectomized patients and was well tolerated (Yannaki et al. 2012). We are also planning a trial at St. Jude Children's Research Hospital.

In summary, gene therapy trials for β-thalassemia are underway or soon to begin. One patient clearly benefited from the gene transfer procedure but in the context of clonal dominance secondary to proto-oncogene activation. This outcome is reminiscent of experience with gammaretroviral gene transfer for the treatment of CGD (Grez et al. 2011). In general, although a clinical benefit has been achieved in several CGD patients, the presence of genetically modified cells has been transient and nearly undetectable in most after a few months following the transplant procedure. Of the 12 patients with CGD who have been treated by gene therapy, only three engrafted with high levels of gene-modified cells long term, and in all three cases, clonal expansion was triggered by insertional activation of the proto-oncogene EVI1 (Grez et al. 2011). On the other hand, the outcomes with lentiviral vectors for metachromatic leukodystrophy (A Biffi and L Naldini, pers. comm.) and adrenoleukodystrophy (Cartier et al. 2009, 2012) have generally been more favorable, with persistence of high levels of genetically modified cells for months to years following gene transfer. Only further clinical experience will determine which outcome is most likely for most patients with β-thalassemia.

STRATEGIES TO IMPROVE STEM CELL–TARGETED GENE TRANSFER

Compared in Table 1 are the transduction conditions and outcomes in the lentiviral clinical trials conducted to date. Incorporated into the conditions used for transduction for the metachromatic leukodystrophy trial are strategies to reduce proteasome activity (Santoni de Sio et al. 2008, 2009). Prior studies had shown that lentiviral vector–mediated gene transfer into mobilized peripheral blood stem cells is restricted by proteasome activity and that such activity is down-regulated by early active cytokines that enhance transduction. A period of 12 h rest was incorporated into the transduction protocol to allow cells to recover with reduced proteasome activity, and a proteasome inhibitor was included in the culture medium.

Various other strategies have also been proposed to increase the proliferative potential of primitive cells and to enhance their engraftment. These include transient silencing of PTEN (phosphatase tensin homolog deleted on chromosome 10) (Kim et al. 2012); inhibition of aldehyde dehydrogenase, which increased radioprotective capacity in the murine model (Muramoto et al. 2010); and addition of nicotinamide, an inhibitor of various ribosylases,

Table 1. Transduction conditions and outcome in lentiviral clinical trials

Disease	Metachromatic leukodystrophy	Adrenoleukodystrophy	β-Thalassemia
Age of patient		7 or 7.5 yr	19 yr
Source of CD34$^+$ cells	G-CSF mobilization Peripheral blood	G-CSF mobilization Peripheral blood	Bone marrow
Medium	X-vivo 15	X-vivo 20	X-vivo 20
Cell density/viral density	1×10^6 mL	2.0×10^6 mL^{-1}/ 5.0×10^7 mL^{-1}	2.0×10^6 mL^{-1} / 1.1×10^8 mL^{-1}
Cytokines (ng/mL)	SCF, 300 TPO, 100 FlT3L, 300 IL3, 60	SCF, 100 TPO, 100 FlT-3, 100 IL3, 60	SCF, 300 TPO, 10 FlT3L, 300 IL3, 100
Protamine/ RetroNectin		4 μg/mL / 50 μg/mL	4 μg/mL / 50 μg/mL
Prestimulation	24 h	19 h/no retro	34 h
Transduction	12 h wash/rest 12 h/12 h proteasome inhibitor MG132	16 h	18 h
Cells recovered		4.6×10^6 or 7.2×10^6 cells/kg	3×10^8 total
Labeling—bulk—CFU		50% or 33%	0.6 vg/cell
In vivo labeling	>50%	23%–13%–10%[a] 25%–17%–10%[a]	70% RBCs 11% Erythroid 19% granulocyte 9% B-lymphoid
Myeloablation		Cytoxan + busulfan	Busulfan
Clinical outcome	Polyclonal/prevention of neurological deterioration	Sustained polyclonal hematopoiesis/clinical improvement	Polyclonal with dominant clone/transfusion independent

G-CSF, granulocyte-colony stimulating factor; CFU, colony-forming unit; RBCs, red blood cells.
[a]Values obtained over 16 mo following gene transfer procedure.

which enhances hematopoietic stem cell activity in the context of ex vivo expansion of functional $CD34^+$ cells (Peled et al. 2012). Whether any of these strategies will ultimately be applicable to the in vitro transduction of hematopoietic stem cells with therapeutic vectors remains to be determined. Efforts are also focused on understanding homing and engraftment with the ultimate goal of improving the outcome of the gene transfer procedure (Larochelle et al. 2012).

Rather than transferring a globin gene, another approach would involve the activation of the endogenous γ-globin genes to compensate for the lack of β-globin synthesis (Wilber et al. 2011b). Future prospects include the development of targeted vectors with improved efficiency of gene transfer into stem cells (Verhoeyen et al. 2012). Ongoing efforts are also directed toward achieving targeted correction of the causative mutation in repopulating stem cells by homologous recombination (Lombardo et al. 2011). Recently, alternative targets for gene transfer including induced pluripotent stem cells derived from patients' somatic cells have been proposed, although the development of this strategy is far more distant than stem cell–targeted gene transfer (Papapetrou et al. 2011).

CONCLUDING REMARKS

Three decades of effort to achieve therapeutic globin gene transfer has ultimately proved productive despite periods in which success seemed quite distant. Basic discoveries such as the development of lentiviral vectors and the characterization of the β-globin LCR were essential milestones on the road to potential success. One patient with a severe form of β-thalassemia has clearly benefitted, with achievement of transfusion independence albeit with clonal dominance. Ongoing trials for metachromatic leukodystrophy and adrenoleukodystrophy suggest that therapeutic levels of gene transfer into repopulating stem cells can be achieved. Further refinements in the methodology for gene transfer is likely to bring more definite success for the treatment of β-thalassemia.

ACKNOWLEDGMENTS

We thank Pat Streich for expert assistance in preparation of the manuscript.

REFERENCES

Aiuti A, Slavin S, Aker M, Ficara F, Deola S, Mortellaro A, Morecki S, Andolfi G, Tabucchi A, Carlucci F, et al. 2002. Correction of ADA-SCID by stem cell gene therapy combined with nonmyeloablative conditioning. *Science* **296:** 2410–2413.

Aiuti A, Cassani B, Andolfi G, Mirolo M, Biasco L, Recchia A, Urbinati F, Valacca C, Scaramuzza S, Aker M, et al. 2007. Multilineage hematopoietic reconstitution without clonal selection in ADA-SCID patients treated with stem cell gene therapy. *J Clin Invest* **117:** 2233–2240.

Aiuti A, Cattaneo F, Galimberti S, Benninghoff U, Cassani B, Callegaro L, Scaramuzza S, Andolfi G, Mirolo M, Brigida I, et al. 2009. Gene therapy for immunodeficiency due to adenosine deaminase deficiency. *N Engl J Med* **360:** 447–458.

Aker M, Tubb J, Groth AC, Bukovsky AA, Bell AC, Felsenfeld G, Kiem HP, Stamatoyannopoulos G, Emery DW. 2007. Extended core sequences from the cHS4 insulator are necessary for protecting retroviral vectors from silencing position effects. *Hum Gene Ther* **18:** 333–343.

Arumugam PI, Scholes J, Perelman N, Xia P, Yee JK, Malik P. 2007. Improved human β-globin expression from self-inactivating lentiviral vectors carrying the chicken hypersensitive site-4 (cHS4) insulator element. *Mol Ther* **15:** 1863–1871.

Arumugam PI, Urbinati F, Velu CS, Higashimoto T, Grimes HL, Malik P. 2009. The 3′ region of the chicken hypersensitive site-4 insulator has properties similar to its core and is required for full insulator activity. *PLoS ONE* **4:** e6995.

Bank A, Dorazio R, Leboulch P. 2005. A phase I/II clinical trial of β-globin gene therapy for β-thalassemia. *Ann NY Acad Sci* **1054:** 308–316.

Beard BC, Trobridge GD, Ironside C, McCune JS, Adair JE, Kiem HP. 2010. Efficient and stable MGMT-mediated selection of long-term repopulating stem cells in nonhuman primates. *J Clin Invest* **120:** 2345–2354.

Bell AC, West AG, Felsenfeld G. 1999. The protein CTCF is required for the enhancer blocking activity of vertebrate insulators. *Cell* **98:** 387–396.

Biasco L, Baricordi C, Aiuti A. 2012. Retroviral integrations in gene therapy trials. *Mol Ther* **20:** 709–716.

Biffi A, Naldini L. 2007. Novel candidate disease for gene therapy: Metachromatic leukodystrophy. *Expert Opin Biol Ther* **7:** 1193–1205.

Biffi A, Bartolomae CC, Cesana D, Cartier N, Aubourg P, Ranzani M, Cesani M, Benedicenti F, Plati T, Rubagotti E, et al. 2011. Lentiviral vector common integration sites in preclinical models and a clinical trial reflect a benign integration bias and not oncogenic selection. *Blood* **117:** 5332–5339.

Blouin MJ, Beauchemin H, Wright A, De Paepe M, Sorette M, Bleau AM, Nakamoto B, Ou CN, Stamatoyannopoulos G,

Trudel M. 2000. Genetic correction of sickle cell disease: Insights using transgenic mouse models. *Nat Med* **6:** 177–182.

Boztug K, Schmidt M, Schwarzer A, Banerjee PP, Díez IA, Dewey RA, Böhm M, Nowrouzi A, Ball CR, Glimm H, et al. 2010. Stem-cell gene therapy for the Wiskott–Aldrich syndrome. *N Engl J Med* **363:** 1918–1927.

Braun CJ, Boztug K, Schmidt M, Albert MH, Schwarzer A, Paruzynski A, Modlich U, Beier R, Gobring G, Naundorf S, et al. 2011. Efficacy of gene therapy for Wiskott–Aldrich syndrome. *Blood* **118:** 81.

Cartier N, Hacein-Bey-Abina S, Bartholomae CC, Veres G, Schmidt M, Kutschera I, Vidaud M, Abel U, Dal-Cortivo L, Caccavelli L, et al. 2009. Hematopoietic stem cell gene therapy with a lentiviral vector in X-linked adrenoleukodystrophy. *Science* **326:** 818–823.

Cartier N, Hacein-Bey-Abina S, Bartholomae CC, Bougnères P, Schmidt M, Kalle CV, Fischer A, Cavazzana-Calvo M, Aubourg P. 2012. Lentiviral hematopoietic cell gene therapy for X-linked adrenoleukodystrophy. *Methods Enzymol* **507:** 187–198.

Case SS, Price MA, Jordan CT, Yu XJ, Wang L, Bauer G, Haas DL, Xu D, Stripecke R, Naldini L, et al. 1999. Stable transduction of quiescent $CD34^+CD38^-$ human hematopoietic cells by HIV-1-based lentiviral vectors. *Proc Natl Acad Sci* **96:** 2988–2993.

Cattoglio C, Facchini G, Sartori D, Antonelli A, Miccio A, Cassani B, Schmidt M, von Kalle C, Howe S, Thrasher AJ, et al. 2007. Hot spots of retroviral integration in human $CD34^+$ hematopoietic cells. *Blood* **110:** 1770–1778.

Cavazzana-Calvo M, Payen E, Negre O, Wang G, Hehir K, Fusil F, Down J, Denaro M, Brady T, Westerman K, et al. 2010. Transfusion independence and HMGA2 activation after gene therapy of human β-thalassaemia. *Nature* **467:** 318–322.

Cesana D, Sgualdino J, Rudilosso L, Merella S, Naldini L, Montini E. 2012. Whole transcriptome characterization of aberrant splicing events induced by lentiviral vector integrations. *J Clin Invest* **122:** 1667–1676.

Chang JC, Liu D, Kan YW. 1992. A 36-base-pair core sequence of locus control region enhances retrovirally transferred human β-globin gene expression. *Proc Natl Acad Sci* **89:** 3107–3110.

Chung JH, Whiteley M, Felsenfeld G. 1993. A 5′ element of the chicken β-globin domain serves as an insulator in human erythroid cells and protects against position effect in *Drosophila*. *Cell* **74:** 505–514.

Chung JH, Bell AC, Felsenfeld G. 1997. Characterization of the chicken β-globin insulator. *Proc Natl Acad Sci* **94:** 575–580.

Ciavatta DJ, Ryan TM, Farmer SC, Townes TM. 1995. Mouse model of human β^0 thalassemia: Targeted deletion of the mouse β^{maj}- and β^{min}-globin genes in embryonic stem cells. *Proc Natl Acad Sci* **92:** 9259–9263.

Cone RD, Weber-Benarous A, Baorto D, Mulligan RC. 1987. Regulated expression of a complete human β-globin gene encoded by a transmissible retrovirus vector. *Mol Cell Biol* **7:** 887–897.

Cooray S, Howe SJ, Thrasher AJ. 2012. Retrovirus and lentivirus vector design and methods of cell conditioning. *Methods Enzymol* **507:** 29–57.

Deichmann A, Hacein-Bey-Abina S, Schmidt M, Garrigue A, Brugman MH, Hu J, Glimm H, Gyapay G, Prum B, Fraser CC, et al. 2007. Vector integration is nonrandom and clustered and influences the fate of lymphopoiesis in SCID-X1 gene therapy. *J Clin Invest* **117:** 2225–2232.

Deichmann A, Brugman MH, Bartholomae CC, Schwarzwaelder K, Verstegen MM, Howe SJ, Arens A, Ott MG, Hoelzer D, Seger R, et al. 2011. Insertion sites in engrafted cells cluster within a limited repertoire of genomic areas after gammaretroviral vector gene therapy. *Mol Ther* **19:** 2031–2039.

Dickson J, Gowher H, Strogantsev R, Gaszner M, Hair A, Felsenfeld G, West AG. 2010. VEZF1 elements mediate protection from DNA methylation. *PLoS Genet* **6:** e1000804.

Dzierzak EA, Papayannopoulou T, Mulligan RC. 1988. Lineage-specific expression of a human β-globin gene in murine bone marrow transplant recipients reconstituted with retrovirus-transduced stem cells. *Nature* **331:** 35–41.

Edelman LB, Fraser P. 2012. Transcription factories: Genetic programming in three dimensions. *Curr Opin Genet Dev* **22:** 110–114.

Ellis J, Pannell D. 2001. The β-globin locus control region versus gene therapy vectors: A struggle for expression. *Clin Genet* **59:** 17–24.

Emery DW. 2011. The use of chromatin insulators to improve the expression and safety of integrating gene transfer vectors. *Hum Gene Ther* **22:** 761–774.

Emery DW, Chen H, Li Q, Stamatoyannopoulos G. 1998. Development of a condensed locus control region cassette and testing in retrovirus vectors for $^A\gamma$-globin. *Blood Cells Mol Dis* **24:** 322–339.

Emery DW, Yannaki E, Tubb J, Stamatoyannopoulos G. 2000. A chromatin insulator protects retrovirus vectors from chromosomal position effects. *Proc Natl Acad Sci* **97:** 9150–9155.

Emery DW, Yannaki E, Tubb J, Nishino T, Li Q, Stamatoyannopoulos G. 2002. Development of virus vectors for gene therapy of β chain hemoglobinopathies: Flanking with a chromatin insulator reduces γ-globin gene silencing in vivo. *Blood* **100:** 2012–2019.

Follenzi A, Ailles LE, Bakovic S, Geuna M, Naldini L. 2000. Gene transfer by lentiviral vectors is limited by nuclear translocation and rescued by HIV-1 pol sequences. *Nat Genet* **25:** 217–222.

Forrester WC, Takegawa S, Papayannopoulou T, Stamatoyannopoulos G, Groudine M. 1987. Evidence for a locus activation region: The formation of developmentally stable hypersensitive sites in globin-expressing hybrids. *Nucleic Acids Res* **15:** 10159–10177.

Fragkos M, Anagnou NP, Tubb J, Emery DW. 2005. Use of the hereditary persistence of fetal hemoglobin 2 enhancer to increase the expression of oncoretrovirus vectors for human γ-globin. *Gene Ther* **12:** 1591–1600.

Galy A, Thrasher AJ. 2011. Gene therapy for the Wiskott–Aldrich syndrome. *Curr Opin Allergy Clin Immunol* **11:** 545–550.

Gaspar HB, Parsley KL, Howe S, King D, Gilmour KC, Sinclair J, Brouns G, Schmidt M, Von Kalle C, Barington T, et al. 2004. Gene therapy of X-linked severe combined immunodeficiency by use of a pseudotyped gammaretroviral vector. *Lancet* **364:** 2181–2187.

Gaspar HB, Cooray S, Gilmour KC, Parsley KL, Adams S, Howe SJ, Al Ghonaium A, Bayford J, Brown L, Davies EG, et al. 2011a. Long-term persistence of a polyclonal T cell repertoire after gene therapy for X-linked severe combined immunodeficiency. *Sci Transl Med* **3:** 97ra79.

Gaspar HB, Cooray S, Gilmour KC, Parsley KL, Zhang F, Adams S, Bjorkegren E, Bayford J, Brown L, Davies EG, et al. 2011b. Hematopoietic stem cell gene therapy for adenosine deaminase-deficient severe combined immunodeficiency leads to long-term immunological recovery and metabolic correction. *Sci Transl Med* **3:** 97ra80.

Ghirlando R, Giles K, Gowher H, Xiao T, Xu Z, Yao H, Felsenfeld G. 2012. Chromatin domains, insulators, and the regulation of gene expression. *Biochim Biophys Acta* **1819**: 644–651.

Giles KE, Gowher H, Ghirlando R, Jin C, Felsenfeld G. 2010. Chromatin boundaries, insulators, and long-range interactions in the nucleus. *Cold Spring Harb Symp Quant Biol* **75:** 79–85.

Grez M, Reichenbach J, Schwäble J, Seger R, Dinauer MC, Thrasher AJ. 2011. Gene therapy of chronic granulomatous disease: The engraftment dilemma. *Mol Ther* **19:** 28–35.

Grosveld F, van Assendelft GB, Greaves DR, Kollias G. 1987. Position-independent, high-level expression of the human β-globin gene in transgenic mice. *Cell* **51:** 975–985.

Hacein-Bey-Abina S, Le Deist F, Carlier F, Bouneaud C, Hue C, De Villartay JP, Thrasher AJ, Wulffraat N, Sorensen R, Dupuis-Girod S, et al. 2002. Sustained correction of X-linked severe combined immunodeficiency by ex vivo gene therapy. *N Engl J Med* **346:** 1185–1193.

Hacein-Bey-Abina S, Von Kalle C, Schmidt M, McCormack MP, Wulffraat N, Leboulch P, Lim A, Osborne CS, Pawliuk R, Morillon E, et al. 2003. LMO2-associated clonal T cell proliferation in two patients after gene therapy for SCID-X1. *Science* **302:** 415–419.

Hacein-Bey-Abina S, Garrigue A, Wang GP, Soulier J, Lim A, Morillon E, Clappier E, Caccavelli L, Delabesse E, Beldjord K, et al. 2008. Insertional oncogenesis in 4 patients after retrovirus-mediated gene therapy of SCID-X1. *J Clin Invest* **118:** 3132–3142.

Hacein-Bey-Abina S, Hauer J, Lim A, Picard C, Wang GP, Berry CC, Martinache C, Rieux-Laucat F, Latour S, Belohradsky BH, et al. 2010. Efficacy of gene therapy for X-linked severe combined immunodeficiency. *N Engl J Med* **363:** 355–364.

Hanawa H, Hargrove PW, Kepes S, Srivastava DK, Nienhuis AW, Persons DA. 2004. Extended β-globin locus control region elements promote consistent therapeutic expression of a γ-globin lentiviral vector in murine β-thalassemia. *Blood* **104:** 2281–2290.

Hanawa H, Yamamoto M, Zhao H, Shimada T, Persons DA. 2009. Optimized lentiviral vector design improves titer and transgene expression of vectors containing the chicken β-globin locus HS4 insulator element. *Mol Ther* **17:** 667–674.

Hargrove PW, Kepes S, Hanawa H, Obenauer JC, Pei D, Cheng C, Gray JT, Neale G, Persons DA. 2008. Globin lentiviral vector insertions can perturb the expression of endogenous genes in β-thalassemic hematopoietic cells. *Mol Ther* **16:** 525–533.

Hass DL, Case SS, Crooks GM, Kohn DB. 2000. Critical factors influencing stable transduction of human $CD34^+$ cells with HIV-1-derived lentiviral vectors. *Mol Ther* **2:** 71–80.

Heckl D, Schwarzer A, Haemmerle R, Steinemann D, Rudolph C, Skawran B, Knoess S, Krause J, Li Z, Schlegelberger B, et al. 2012. Lentiviral vector induced insertional haploinsufficiency of Ebf1 causes murine leukemia. *Mol Ther* **20:** 1187–1195.

Howe SJ, Mansour MR, Schwarzwaelder K, Bartholomae C, Hubank M, Kempski H, Brugman MH, Pike-Overzet K, Chatters SJ, de Ridder D, et al. 2008. Insertional mutagenesis combined with acquired somatic mutations causes leukemogenesis following gene therapy of SCID-X1 patients. *J Clin Invest* **118:** 3143–3150.

Imren S, Payen E, Westerman KA, Pawliuk R, Fabry ME, Eaves CJ, Cavilla B, Wadsworth LD, Beuzard Y, Bouhassira EE, et al. 2002. Permanent and panerythroid correction of murine β thalassemia by multiple lentiviral integration in hematopoietic stem cells. *Proc Natl Acad Sci* **99:** 14380–14385.

Imren S, Fabry ME, Westerman KA, Pawliuk R, Tang P, Rosten PM, Nagel RL, Leboulch P, Eaves CJ, Humphries RK. 2004. High-level β-globin expression and preferred intragenic integration after lentiviral transduction of human cord blood stem cells. *J Clin Invest* **114:** 953–962.

Karlsson S, Van Doren K, Schweiger SG, Nienhuis AW, Gluzman Y. 1986. Stable gene transfer and tissue-specific expression of a human globin gene using adenoviral vectors. *EMBO J* **5:** 2377–2385.

Karlsson S, Papayannopoulou T, Schweiger SG, Stamatoyannopoulos G, Nienhuis AW. 1987. Retroviral-mediated transfer of genomic globin genes leads to regulated production of RNA and protein. *Proc Natl Acad Sci* **84:** 2411–2415.

Kiem HP, Sellers S, Thomasson B, Morris JC, Tisdale JF, Horn PA, Hematti P, Adler R, Kuramoto K, Calmels B, et al. 2004. Long-term clinical and molecular follow-up of large animals receiving retrovirally transduced stem and progenitor cells: No progression to clonal hematopoiesis or leukemia. *Mol Ther* **9:** 389–395.

Kim YS, Wielgosz MM, Hargrove P, Kepes S, Gray J, Persons DA, Nienhuis AW. 2010. Transduction of human primitive repopulating hematopoietic cells with lentiviral vectors pseudotyped with various envelope proteins. *Mol Ther* **18:** 1310–1317.

Kim I, Kim YJ, Métais JY, Dunbar CE, Larochelle A. 2012. Transient silencing of PTEN in human $CD34^+$ cells enhances their proliferative potential and ability to engraft immunodeficient mice. *Exp Hematol* **40:** 84–91.

Kolovos P, Knoch TA, Grosveld FG, Cook PR, Papantonis A. 2012. Enhancers and silencers: An integrated and simple model for their function. *Epigenetics Chromatin* **5:** 1.

Larochelle A, Choi U, Shou Y, Naumann N, Loktionova NA, Clevenger JR, Krouse A, Metzger M, Donahue RE, Kang E, et al. 2009. In vivo selection of hematopoietic progenitor cells and temozolomide dose intensification in rhesus macaques through lentiviral transduction with a drug resistance gene. *J Clin Invest* **119:** 1952–1963.

Larochelle A, Gillette JM, Desmond R, Ichwan B, Cantilena A, Cerf A, Barrett AJ, Wayne AS, Lippincott-Schwartz J, Dunbar CE. 2012. Bone marrow homing and engraftment of human hematopoietic stem and progenitor cells is mediated by a polarized membrane domain. *Blood* **119:** 1848–1855.

Leboulch P, Huang GM, Humphries RK, Oh YH, Eaves CJ, Tuan DY, London IM. 1994. Mutagenesis of retroviral vectors transducing human β-globin gene and β-globin locus control region derivatives results in stable transmission of an active transcriptional structure. *EMBO J* **13:** 3065–3076.

Levasseur DN, Ryan TM, Pawlik KM, Townes TM. 2003. Correction of a mouse model of sickle cell disease: Lentiviral/antisickling β-globin gene transduction of unmobilized, purified hematopoietic stem cells. *Blood* **102:** 4312–4319.

Li Q, Emery DW, Fernandez M, Han H, Stamatoyannopoulos G. 1999. Development of viral vectors for gene therapy of β-chain hemoglobinopathies: Optimization of a γ-globin gene expression cassette. *Blood* **93:** 2208–2216.

Lombardo A, Cesana D, Genovese P, Di Stefano B, Provasi E, Colombo DF, Neri M, Magnani Z, Cantore A, Lo Riso P, et al. 2011. Site-specific integration and tailoring of cassette design for sustainable gene transfer. *Nat Methods* **8:** 861–869.

Maina N, Zhong L, Li X, Zhao W, Han Z, Bischof D, Aslanidi G, Zolotukhin S, Weigel-Van Aken KA, Rivers AE, et al. 2008. Optimization of recombinant adeno-associated viral vectors for human β-globin gene transfer and transgene expression. *Hum Gene Ther* **19:** 365–375.

May C, Rivella S, Callegari J, Heller G, Gaensler KM, Luzzatto L, Sadelain M. 2000. Therapeutic haemoglobin synthesis in β-thalassaemic mice expressing lentivirus-encoded human β-globin. *Nature* **406:** 82–86.

May C, Rivella S, Chadburn A, Sadelain M. 2002. Successful treatment of murine β-thalassemia intermedia by transfer of the human β-globin gene. *Blood* **99:** 1902–1908.

Mercola KE, Cline MJ. 1980. The potentials of inserting new genetic information. *N Engl J Med* **303:** 1297–1300.

Modlich U, Navarro S, Zychlinski D, Maetzig T, Knoess S, Brugman MH, Schambach A, Charrier S, Galy A, Thrasher AJ, et al. 2009. Insertional transformation of hematopoietic cells by self-inactivating lentiviral and gammaretroviral vectors. *Mol Ther* **17:** 1919–1928.

Montini E, Cesana D, Schmidt M, Sanvito F, Ponzoni M, Bartholomae C, Sergi Sergi L, Benedicenti F, Ambrosi A, Di Serio C, et al. 2006. Hematopoietic stem cell gene transfer in a tumor-prone mouse model uncovers low genotoxicity of lentiviral vector integration. *Nat Biotechnol* **24:** 687–696.

Montini E, Cesana D, Schmidt M, Sanvito F, Bartholomae CC, Ranzani M, Benedicenti F, Sergi LS, Ambrosi A, Ponzoni M, et al. 2009. The genotoxic potential of retroviral vectors is strongly modulated by vector design and integration site selection in a mouse model of HSC gene therapy. *J Clin Invest* **119:** 964–975.

Mulligan RC, Howard BH, Berg P. 1979. Synthesis of rabbit β-globin in cultured monkey kidney cells following infection with a SV40 β-globin recombinant genome. *Nature* **277:** 108–114.

Muramoto GG, Russell JL, Safi R, Salter AB, Himburg HA, Daher P, Meadows SK, Doan P, Storms RW, Chao NJ, et al. 2010. Inhibition of aldehyde dehydrogenase expands hematopoietic stem cells with radioprotective capacity. *Stem Cells* **28:** 523–534.

Naldini L. 2011. Ex vivo gene transfer and correction for cell-based therapies. *Nat Rev Genet* **12:** 301–315.

Naldini L, Verma IM. 2000. Lentiviral vectors. *Adv Virus Res* **55:** 599–609.

Naldini L, Blömer U, Gallay P, Ory D, Mulligan R, Gage FH, Verma IM, Trono D. 1996. In vivo gene delivery and stable transduction of nondividing cells by a lentiviral vector. *Science* **272:** 263–267.

Negre O, Fusil F, Colomb C, Roth S, Gillet-Legrand B, Henri A, Beuzard Y, Bushman F, Leboulch P, Payen E. 2011. Correction of murine β-thalassemia after minimal lentiviral gene transfer and homeostatic in vivo erythroid expansion. *Blood* **117:** 5321–5331.

Nicolini FE, Imren S, Oh IH, Humphries RK, Leboulch P, Fabry ME, Nagel RL, Eaves CJ. 2002. Expression of a human β-globin transgene in erythroid cells derived from retrovirally transduced transplantable human fetal liver and cord blood cells. *Blood* **100:** 1257–1264.

Nienhuis AW, Dunbar CE, Sorrentino BP. 2006. Genotoxicity of retroviral integration in hematopoietic cells. *Mol Ther* **13:** 1031–1049.

Nishino T, Tubb J, Emery DW. 2006. Partial correction of murine β-thalassemia with a gammaretrovirus vector for human γ-globin. *Blood Cells, Mol Dis* **37:** 1–7.

Novak U, Harris EA, Forrester W, Groudine M, Gelinas R. 1990. High-level β-globin expression after retroviral transfer of locus activation region-containing human β-globin gene derivatives into murine erythroleukemia cells. *Proc Natl Acad Sci* **87:** 3386–3390.

Oh IH, Fabry ME, Humphries RK, Pawliuk R, Leboulch P, Hoffman R, Nagel RL, Eaves C. 2004. Expression of an anti-sickling β-globin in human erythroblasts derived from retrovirally transduced primitive normal and sickle cell disease hematopoietic cells. *Exp Hematol* **32:** 461–469.

Orkin SH, Motulsky AG. 1995. Report and recommendations of the panel to assess the NIH investment in research on gene therapy. http://www.nih.gov/news/panelrep.html.

Ott MG, Schmidt M, Schwarzwaelder K, Stein S, Siler U, Koehl U, Glimm H, Kühlcke K, Schilz A, Kunkel H, et al. 2006. Correction of X-linked chronic granulomatous disease by gene therapy, augmented by insertional activation of MDS1-EVI1, PRDM16 or SETBP1. *Nat Med* **12:** 401–409.

Pai S-Y, Notarangelo LD, Harris C, Cattaneo F, Wladkowski M, Armant M, Kao GS, Wielgosz M, Nienhuis AW, Modlich U, et al. 2011. Somatic gene therapy for X-linked severe combined immunodeficiency using a self-inactivating modified gammaretroviral vector results in an improved preclinical safety profile and early clinical efficacy in a human patient. *Blood* **118:** 80.

Palstra RJ, de Laat W, Grosveld F. 2008. β-globin regulation and long-range interactions. *Adv Genet* **61:** 107–142.

Papapetrou EP, Lee G, Malani N, Setty M, Riviere I, Tirunagari LM, Kadota K, Roth SL, Giardina P, Viale A, et al. 2011. Genomic safe harbors permit high β-globin transgene expression in thalassemia induced pluripotent stem cells. *Nat Biotechnol* **29:** 73–78.

Pászty C, Brion CM, Manci E, Witkowska HE, Stevens ME, Mohandas N, Rubin EM. 1997. Transgenic knockout mice with exclusively human sickle hemoglobin and sickle cell disease. *Science* **278:** 876–878.

Pawliuk R, Westerman KA, Fabry ME, Payen E, Tighe R, Bouhassira EE, Acharya SA, Ellis J, London IM, Eaves CJ, et al. 2001. Correction of sickle cell disease in transgenic mouse models by gene therapy. *Science* **294:** 2368–2371.

Peled T, Shoham H, Aschengrau D, Yackoubov D, Frei G, Rosenheimer GN, Lerrer B, Cohen HY, Nagler A, Fibach E, et al. 2012. Nicotinamide, a SIRT1 inhibitor, inhibits differentiation and facilitates expansion of hematopoietic progenitor cells with enhanced bone marrow homing and engraftment. *Exp Hematol* **40:** 342–355.

Persons DA. 2010. Targeting β-thalassaemia. *Nature* **467:** 277–278.

Persons DA, Allay ER, Sabatino DE, Kelly P, Bodine DM, Nienhuis AW. 2001. Functional requirements for phenotypic correction of murine β-thalassemia: Implications for human gene therapy. *Blood* **97:** 3275–3282.

Persons DA, Hargrove PW, Allay ER, Hanawa H, Nienhuis AW. 2003a. The degree of phenotypic correction of murine β-thalassemia intermedia following lentiviral-mediated transfer of a human γ-globin gene is influenced by chromosomal position effects and vector copy number. *Blood* **101:** 2175–2183.

Persons DA, Allay ER, Sawai N, Hargrove PW, Brent TP, Hanawa H, Nienhuis AW, Sorrentino BP. 2003b. Successful treatment of murine β-thalassemia using in vivo selection of genetically modified, drug-resistant hematopoietic stem cells. *Blood* **102:** 506–513.

Perumbeti A, Higashimoto T, Urbinati F, Franco R, Meiselman HJ, Witte D, Malik P. 2009. A novel human γ-globin gene vector for genetic correction of sickle cell anemia in a humanized sickle mouse model: Critical determinants for successful correction. *Blood* **114:** 1174–1185.

Pestina TI, Hargrove PW, Jay D, Gray JT, Boyd KM, Persons DA. 2009. Correction of murine sickle cell disease using γ-globin lentiviral vectors to mediate high-level expression of fetal hemoglobin. *Mol Ther* **17:** 245–252.

Pikaart MJ, Recillas-Targa F, Felsenfeld G. 1998. Loss of transcriptional activity of a transgene is accompanied by DNA methylation and histone deacetylation and is prevented by insulators. *Genes Dev* **12:** 2852–2862.

Plavec I, Papayannopoulou T, Maury C, Meyer F. 1993. A human β-globin gene fused to the human β-globin locus control region is expressed at high levels in erythroid cells of mice engrafted with retrovirus-transduced hematopoietic stem cells. *Blood* **81:** 1384–1392.

Puthenveetil G, Scholes J, Carbonell D, Qureshi N, Xia P, Zeng L, Li S, Yu Y, Hiti AL, Yee JK, et al. 2004. Successful correction of the human β-thalassemia major phenotype using a lentiviral vector. *Blood* **104:** 3445–3453.

Ramezani A, Hawley TS, Hawley RG. 2008. Combinatiorial incorporation of enhancer-blocking components of chicken β-globin 5′HS4 and human T-cell receptor α/δ BEAD-1 Insulators in self-inactivating retroviral vectors reduces their genotoxic potential. *Stem Cells* **26:** 3257–3266.

Recillas-Targa F, Pikaart MJ, Burgess-Beusse B, Bell AC, Litt MD, West AG, Gaszner M, Felsenfeld G. 2002. Position-effect protection and enhancer blocking by the chicken β-globin insulator are separable activities. *Proc Natl Acad Sci* **99:** 6883–6888.

Ren S, Wong BY, Li J, Luo XN, Wong PM, Atweh GF. 1996. Production of genetically stable high-titer retroviral vectors that carry a human γ-globin gene under the control of the α-globin locus control region. *Blood* **87:** 2518–2524.

Rivella S, May C, Chadburn A, Rivière I, Sadelain M. 2003. A novel murine model of Cooley anemia and its rescue by lentiviral-mediated human β-globin gene transfer. *Blood* **101:** 2932–2939.

Rivière I, Dunbar CE, Sadelain M. 2012. Hematopoietic stem cell engineering at a crossroads. *Blood* **119:** 1107–1116.

Ronen K, Negre O, Roth S, Colomb C, Malani N, Denaro M, Brady T, Fusil F, Gillet-Legrand B, Hehir K, et al. 2011. Distribution of lentiviral vector integration sites in mice following therapeutic gene transfer to treat β-thalassemia. *Mol Ther* **19:** 1273–1286.

Roselli EA, Mezzadra R, Frittoli MC, Maruggi G, Biral E, Mavilio F, Mastropietro F, Amato A, Tonon G, Refaldi C, et al. 2010. Correction of β-thalassemia major by gene transfer in haematopoietic progenitors of pediatric patients. *EMBO Mol Med* **2:** 315–328.

Ryu BY, Persons DA, Evans-Galea MV, Gray JT, Nienhuis AW. 2007. A chromatin insulator blocks interactions between globin regulatory elements and cellular promoters in erythroid cells. *Blood Cells Mol Dis* **39:** 221–228.

Ryu BY, Evans-Galea MV, Gray JT, Bodine DM, Persons DA, Nienhuis AW. 2008. An experimental system for the evaluation of retroviral vector design to diminish the risk for proto-oncogene activation. *Blood* **111:** 1866–1875.

Sabatino DE, Seidel NE, Aviles-Mendoza GJ, Cline AP, Anderson SM, Gallagher PG, Bodine DM. 2000. Long-term expression of γ-globin mRNA in mouse erythrocytes from retrovirus vectors containing the human γ-globin gene fused to the ankyrin-1 promoter. *Proc Natl Acad Sci* **97:** 13294–13299.

Sadelain M, Wang CH, Antoniou M, Grosveld F, Mulligan RC. 1995. Generation of a high-titer retroviral vector capable of expressing high levels of the human β-globin gene. *Proc Natl Acad Sci* **92:** 6728–6732.

Sadelain M, Rivière I, Wang X, Boulad F, Prockop S, Giardina P, Maggio A, Galanello R, Locatelli F, Yannaki E. 2010. Strategy for a multicenter phase I clinical trial to evaluate globin gene transfer in β-thalassemia. *Ann NY Acad Sci* **1202:** 52–58.

Santoni de Sio F, Naldini L. 2009. Short-term culture of human $CD34^+$ cells for lentiviral gene transfer. *Methods Mol Biol* **506:** 59–70.

Santoni de Sio FR, Gritti A, Cascio P, Neri M, Sampaolesi M, Galli C, Luban J, Naldini L. 2008. Lentiviral vector gene transfer is limited by the proteasome at postentry steps in various types of stem cells. *Stem Cells* **26:** 2142–2152.

Scaramuzza S, Biasco L, Ripamonti A, Castiello MC, Loperfido M, Draghici E, Hernandez RJ, Benedicenti F, Radrizzani M, Salomoni M, et al. 2012. Preclinical safety and efficacy of human CD34$^+$ cells transduced with lentiviral vector for the treatment of Wiskott–Aldrich syndrome. *Mol Ther* doi: 10.1038/mt.2012.23.

Schoenfelder S, Sexton T, Chakalova L, Cope NF, Horton A, Andrews S, Kurukuti S, Mitchell JA, Umlauf D, Dimitrova DS, et al. 2010. Preferential associations between co-regulated genes reveal a transcriptional interactome in erythroid cells. *Nat Genet* **42:** 53–61.

Schröder AR, Shinn P, Chen H, Berry C, Ecker JR, Bushman F. 2002. HIV-1 integration in the human genome favors active genes and local hotspots. *Cell* **110:** 521–529.

Schwarzwaelder K, Howe SJ, Schmidt M, Brugman MH, Deichmann A, Glimm H, Schmidt S, Prinz C, Wissler M, King DJ, et al. 2007. Gammaretrovirus-mediated correction of SCID-X1 is associated with skewed vector integration site distribution in vivo. *J Clin Invest* **117:** 2241–2249.

Seggewiss R, Pittaluga S, Adler RL, Guenaga FJ, Ferguson C, Pilz IH, Ryu B, Sorrentino BP, Young WS III, Donahue RE, et al. 2006. Acute myeloid leukemia is associated with retroviral gene transfer to hematopoietic progenitor cells in a rhesus macaque. *Blood* **107:** 3865–3867.

Shaw KL, Kohn DB. 2011. A tale of two SCIDs. *Sci Transl Med* **3:** 97ps36.

Sirven A, Pflumio F, Zennou V, Titeux M, Vainchenker W, Coulombel L, Dubart-Kupperschmitt A, Charneau P. 2000. The human immunodeficiency virus type-1 central DNA flap is a crucial determinant for lentiviral vector nuclear import and gene transduction of human hematopoietic stem cells. *Blood* **96:** 4103–4110.

Stein S, Ott MG, Schultze-Strasser S, Jauch A, Burwinkel B, Kinner A, Schmidt M, Krämer A, Schwäble J, Glimm H, et al. 2010. Genomic instability and myelodysplasia with monosomy 7 consequent to EVI1 activation after gene therapy for chronic granulomatous disease. *Nat Med* **16:** 198–204.

Sutton RE, Reitsma MJ, Uchida N, Brown PO. 1999. Transduction of human progenitor hematopoietic stem cells by human immunodeficiency virus type 1-based vectors is cell cycle dependent. *J Virol* **73:** 649–660.

Takekoshi KJ, Oh YH, Westerman KW, London IM, Leboulch P. 1995. Retroviral transfer of a human β-globin/δ-globin hybrid gene linked to β locus control region hypersensitive site 2 aimed at the gene therapy of sickle cell disease. *Proc Natl Acad Sci* **92:** 3014–3018.

Thornhill SI, Schambach A, Howe SJ, Ulaganathan M, Grassman E, Williams D, Schiedlmeier B, Sebire NJ, Gaspar HB, Kinnon C, et al. 2008. Self-inactivating gammaretroviral vectors for gene therapy of X-linked severe combined immunodeficiency. *Mol Ther* **16:** 590–598.

Thrasher AJ, Hacein-Bey-Abina S, Gaspar HB, Blanche S, Davies EG, Parsley K, Gilmour K, King D, Howe S, Sinclair J, et al. 2005. Failure of SCID-X1 gene therapy in older patients. *Blood* **105:** 4255–4257.

Tuan D, London IM. 1984. Mapping of DNase I-hypersensitive sites in the upstream DNA of human embryonic ε-globin gene in K562 leukemia cells. *Proc Natl Acad Sci* **81:** 2718–2722.

Tuan D, Solomon W, Li Q, London IM. 1985. The "β-like-globin" gene domain in human erythroid cells. *Proc Natl Acad Sci* **82:** 6384–6388.

Uchida N, Sutton RE, Friera AM, He D, Reitsma MJ, Chang WC, Veres G, Scollay R, Weissman IL. 1998. HIV, but not murine leukemia virus, vectors mediate high efficiency gene transfer into freshly isolated G_0/G_1 human hematopoietic stem cells. *Proc Natl Acad Sci* **95:** 11939–11944.

Urbinati F, Arumugam P, Higashimoto T, Perumbeti A, Mitts K, Xia P, Malik P. 2009. Mechanism of reduction in titers from lentivirus vectors carrying large inserts in the 3′LTR. *Mol Ther* **17:** 1527–1536.

Verhoeyen E, Relouzat F, Cambot M, Costa C, Nègre D, Legrand F, Joubert C, Le Grand R, Cosset FL, Leboulch P, et al. 2012. Stem cell factor–displaying SIV vectors together with a low conditioning regimen allow for long-term engraftment of gene marked autologous hematopoietic stem cells in macaques. *Hum Gene Ther* **23:** 754–768.

Walsh CE, Liu JM, Xiao X, Young NS, Nienhuis AW, Samulski RJ. 1992. Regulated high level expression of a human γ-globin gene introduced into erythroid cells by an adeno-associated virus vector. *Proc Natl Acad Sci* **89:** 7257–7261.

Weatherall DJ. 1988. The slow road to gene therapy. *Nature* **331:** 13–14.

West AG, Huang S, Gaszner M, Litt MD, Felsenfeld G. 2004. Recruitment of histone modifications by USF proteins at a vertebrate barrier element. *Mol Cell* **16:** 453–463.

Wilber A, Hargrove PW, Kim YS, Riberdy JM, Sankaran VG, Papanikolaou E, Georgomanoli M, Anagnou NP, Orkin SH, Nienhuis AW, et al. 2011a. Therapeutic levels of fetal hemoglobin in erythroid progeny of β-thalassemic CD34$^+$ cells after lentiviral vector-mediated gene transfer. *Blood* **117:** 2817–2826.

Wilber A, Nienhuis AW, Persons DA. 2011b. Transcriptional regulation of fetal to adult hemoglobin switching: New therapeutic opportunities. *Blood* **117:** 3945–3953.

Wu C, Dunbar CE. 2011. Stem cell gene therapy: The risks of insertional mutagenesis and approaches to minimize genotoxicity. *Front Med* **5:** 356–371.

Wu X, Li Y, Crise B, Burgess SM. 2003. Transcription start regions in the human genome are favored targets for MLV integration. *Science* **300:** 1749–1751.

Yang B, Kirby S, Lewis J, Detloff PJ, Maeda N, Smithies O. 1995. A mouse model for $β^0$-thalassemia. *Proc Natl Acad Sci* **92:** 11608–11612.

Yannaki E, Papayannopoulou T, Jonlin E, Zervou F, Karponi G, Xagorari A, Becker P, Psatha N, Batsis I, Kaloyannidis P, et al. 2012. Hematopoietic stem cell mobilization for gene therapy of adult patients with severe β-thalassemia: Results of clinical trials using G-CSF or plerixafor in splenectomized and nonsplenectomized subjects. *Mol Ther* **20:** 230–238.

Yu SF, von Rüden T, Kantoff PW, Garber C, Seiberg M, Rüther U, Anderson WF, Wagner EF, Gilboa E. 1986. Self-inactivating retroviral vectors designed for transfer of whole genes into mammalian cells. *Proc Natl Acad Sci* **83:** 3194–3198.

Zennou V, Petit C, Guetard D, Nerhbass U, Montagnier L, Charneau P. 2000. HIV-1 genome nuclear import is mediated by a central DNA flap. *Cell* **101:** 173–185.

Zhao H, Pestina TI, Nasimuzzaman M, Mehta P, Hargrove PW, Persons DA. 2009. Amelioration of murine β-thalassemia through drug selection of hematopoietic stem cells transduced with a lentiviral vector encoding both γ-globin and the MGMT drug-resistance gene. *Blood* **113:** 5747–5756.

Zufferey R, Dull T, Mandel RJ, Bukovsky A, Quiroz D, Naldini L, Trono D. 1998. Self-inactivating lentivirus vector for safe and efficient in vivo gene delivery. *J Virol* **72:** 9873–9880.

Pluripotent Stem Cells in Research and Treatment of Hemoglobinopathies

Natasha Arora and George Q. Daley

Department of Biological Chemistry and Molecular Pharmacology, Harvard Medical School, Boston, Massachusetts 02115; Harvard Stem Cell Institute, Cambridge, Massachusetts 02138; Division of Pediatric Hematology/Oncology, Children's Hospital Boston, Boston, Massachusetts 02115

Correspondence: george.daley@childrens.harvard.edu

Pluripotent stem cells (PSCs) hold great promise for research and treatment of hemoglobinopathies. In principle, patient-specific induced pluripotent stem cells could be derived from a blood sample, genetically corrected to repair the disease-causing mutation, differentiated into hematopoietic stem cells (HSCs), and returned to the patient to provide a cure through autologous gene and cell therapy. However, there are many challenges at each step of this complex treatment paradigm. Gene repair is currently inefficient in stem cells, but use of zinc finger nucleases and transcription activator-like effector nucleases appear to be a major advance. To date, no successful protocol exists for differentiating PSCs into definitive HSCs. PSCs can be directly differentiated into primitive red blood cells, but not yet in sufficient numbers to enable treating patients, and the cost of clinical scale differentiation is prohibitively expensive with current differentiation methods and efficiencies. Here we review the progress, promise, and remaining hurdles in realizing the potential of PSCs for cell therapy.

Pluripotent stem cells (PSCs), whether embryonic stem cells (ESCs) or induced pluripotent stem cells (iPSCs), have the potential to form any cell type in the body. ESCs are derived from the inner cell mass of the mammalian blastocyst (Evans and Kaufman 1981; Martin 1981; Thomson et al. 1998), whereas iPSCs are somatic cells reprogrammed to a pluripotent state by exogenous expression of transcription factors responsible for conferring pluripotency on ESCs (Takahashi et al. 2007; Yu et al. 2007; Park et al. 2008b). Many diverse types of somatic cells that are readily obtained from patients—like skin fibroblasts, blood, and keratinocytes—when transduced with Oct4, Sox2, Klf4, and c-Myc, will revert to a pluripotent state, albeit inefficiently (Aasen et al. 2008; Loh et al. 2009). Although the early reprogramming experiments were performed with retroviruses and lentiviruses, safer transgene-free reprogramming methods have since been developed that use synthetic mRNA or non-integrating episomes to express the reprogramming factors (Yu et al. 2009; Warren et al. 2010). In addition, although iPSCs are remarkably similar to ESCs in pluripotent function, many studies have identified subtle but potentially significant molecular differences, showing the need to continue to study and improve reprogramming methods to achieve the closest facsimile to the naturally

Copyright © 2013 Cold Spring Harbor Laboratory Press; all rights reserved
Cite this article as *Cold Spring Harb Perspect Med* doi: 10.1101/cshperspect.a011841

derived ESCs, which remain the gold standard (International Stem Cell Initiative et al. 2007; Kim et al. 2011). With regard to applications for hematologic disease, there appear to be differences in the differentiation potential of human PSC lines to form various tissues of clinical interest, particularly blood (Osafune et al. 2008; Choi et al. 2009). Lanza's group examined the blood differentiation potential of 22 human PSC lines generated at multiple institutions. The 14 human ESC lines tested showed higher hematopoietic colony forming unit (CFU) activity and a 1000-fold increase in erythroid differentiation efficiency, whereas the eight human iPSC lines displayed limited growth potential and decreased CFU activity (Feng et al. 2010).

DISEASE MODELING

PSCs are a powerful tool for modeling disease. Disease-specific human ESCs have been generated from embryos with genetic mutations causing diseases like thalassemia, Fanconi anemia, and cystic fibrosis (Pickering et al. 2005; Verlinsky et al. 2005; Mateizel et al. 2006; Ben-Yosef et al. 2008). However, affected embryos diagnosed through preimplantation testing are the prerequisite for deriving disease-relevant human ESCs, and because only a subset of predominantly Mendelian genetic disorders are subject to such testing, it is not feasible to obtain human ESCs for any specific disease of interest. On the other hand, to generate disease-specific human iPSCs, only a simple skin biopsy or blood draw from an affected patient is necessary (Park et al. 2008a; Soldner et al. 2009). Fibroblasts can be grown out of the skin biopsy or mononuclear cells collected from the peripheral blood sample and reprogrammed with the four reprogramming factors.

Disease-specific human iPSCs have already been generated for numerous diseases and are being used to model diseases in vitro. Human iPSCs generated from a patient with type 1 diabetes were differentiated into β-like cells that produced insulin and responded to glucose (Maehr et al. 2009). Parkinson's disease iPSCs were differentiated toward dopaminergic neurons, which were functionally tested in a rat model of Parkinson's disease (Hargus et al. 2010). These studies and numerous others have shown the power of disease-specific human iPSCs in modeling diseases in vitro and represent compelling proof-of-principle for the potential of cell therapies.

HEMOGLOBINOPATHIES

Since the first demonstration of mutant hemoglobin more than a half century ago, hundreds of mutations in the α-globin and β-globin loci have been identified. Many mutations are silent or do not present a clinical phenotype, but a few are common and cause severe phenotypes. Two of the most common hemoglobinopathies caused by these mutations are sickle cell anemia (SCA) and thalassemia.

Sickle cell anemia, caused by an adenine-to-thymine point mutation in the sixth codon of the β-globin gene, changes a polar glutamic acid to a nonpolar valine (Ingram 1956, 1957). This subtle change predisposes hemoglobin to polymerization, which induces rigidity and structural deformations of red blood cells (RBCs), recognized by a characteristic sickling. Sickled RBCs cause vaso-occlusion, which leads to painful crises, organ damage, and premature death (Hofrichter et al. 1974; Crepeau et al. 1978; Dykes et al. 1979; Noguchi and Schechter 1981; Brittenham et al. 1985; Eaton and Hofrichter 1987).

Current treatments for SCA are limited. Hydroxyurea reduces the frequency of painful crises by lowering white blood cell counts, thereby reducing cellular sludging, and increasing production of γ-globin, a fetal hemoglobin, which reduces RBC sickling (Letvin et al. 1984; Platt et al. 1984). In the setting of disease complications, RBC transfusions can be used to temporarily increase the number of circulating normal RBCs, but frequent RBC transfusions result in sickle antibody sensitization and iron accumulation, which itself causes organ damage (Orlina et al. 1978; Coles et al. 1981; Davies et al. 1986; Reisner et al. 1987; Cox et al. 1988; Rosse et al. 1990; Vichinsky et al. 1990). Bone marrow transplants are a treatment option but are riskier than RBC transfusions, given the potential for graft failure and immunological complications like graft-versus-host disease. Marrow transplant can provide a cure but remains an underutilized approach because of the considerable morbidity and mortality associated with such a heroic intervention.

Thalassemias are disorders related to imbalances in the production of the hemoglobin chains that can be caused by several different mutations. In α-thalassemia, production of α-globin chains is impaired, resulting in free β-globin polypeptides, which form an unstable hemoglobin molecule (Ingram and Stretton 1959; Ottolenghi et al. 1974; Taylor et al. 1974). In β-thalassemia, production of β-globin chains is defective, resulting in free unstable α-globin chains that are toxic to erythroid precursors (Fessas 1963; Fessas et al. 1965). As with SCA, treatment options are limited. In severe thalassemia (major), patients require regular transfusion, which comes with the risk of antibody sensitization and the complications of iron overload. Treatments like hydroxyurea, which increase fetal hemoglobin (HbF) levels, reduce the severity of β-thalassemia because they decrease the number of unpaired α-globin chains (Platt et al. 1984; Weatherall 2001; Koren et al. 2008). However, hydroxyurea is not effective in β-thalassemia patients because they need higher levels of HbF than SCA patients to ameliorate their symptoms (Alebouyeh et al. 2004; Yavarian et al. 2004). As with SCA, bone marrow transplant is a treatment option and the only chance for a cure; despite the risks of graft failure and graft-versus-host disease, it is frequently used for patients with thalassemia major.

HEMOGLOBINOPATHY DISEASE MODELS—IN VIVO AND IN VITRO

Two mouse models of β-thalassemia and one mouse model of α-thalassemia have been reported (Skow et al. 1983; Yang et al. 1995; Chang et al. 1996). These models have been used for numerous gene transfer studies to test the efficacy of lentiviruses containing various components of the human β-globin locus control region (LCR), promoter elements, and γ-globin or β-globin gene constructs (May et al. 2000, 2002; Imren et al. 2002; Nishino et al. 2006; Han et al. 2007; Lisowski and Sadelain 2007). The studies show hemoglobin tetramers containing human β-globin at levels high enough to ameliorate the disease phenotype.

There are two humanized mouse models of SCA, the "Berkeley" mouse and the "Townes" mouse, which have proven valuable in studying disease mechanisms and testing therapeutic strategies, such as gene therapy (Pawliuk et al. 2001; Xu et al. 2011). The Berkeley mouse is a transgenic strain engineered to express normal human α-globin and γ-globin as well as β^S-globin, the sickle cell mutant (Paszty et al. 1997). The Townes mouse is a knock-in model, in which both the mouse α-globin and β-globin genes were replaced with the human genes. The β-globin locus contains $^A\gamma$-globin and β^S-globin (Wu et al. 2006).

In 2002, the first proof-of-principle experiment for combined gene and cell therapy of blood disorders in a murine model was performed using an immunodeficient $Rag2^{-/-}$ mouse strain (Rideout et al. 2002). In this case, $Rag2^{-/-}$ ESCs were generated through somatic cell nuclear transfer, a method whereby somatic cells are reprogrammed to pluripotency via nuclear transplantation, creation of a cloned mouse embryo, and extraction of ESCs from the blastocyst stage. One allele of the Rag2 gene was repaired by homologous recombination in the ESCs, which were then differentiated toward hematopoietic precursors and transplanted back into the $Rag2^{-/-}$ mouse. Upon analysis, the engrafted mice showed mature lymphoid cells. In 2006, a similar experiment was performed with the Townes humanized SCA mouse model (Wu et al. 2006). Autologous ESCs were genetically corrected by homologous recombination, differentiated toward the hematopoietic lineage, and transplanted into the Townes mice. Analysis showed high levels of adult human HbA, reduced RBC sickling, and corrected pathology in the animals. In 2007 with the advent of iPSCs, the same experiment was performed with autologous iPSCs made from the Townes mice (Hanna et al. 2007). This was indeed an impressive demonstration of the potential for harnessing personalized stem cells for the treatment and potential cure of SCA. In a tour de force of modern cell and molecular biology, the Jaenisch group derived iPSCs from the murine model of human SCA, restored a normal allele of human β-globin through homologous recombination, then derived engraftable hematopoietic precursors through directed in vitro differentiation, and treated mice through hematopoietic transplant. Although to date the engraftment of mice with hematopoietic progenitors derived from ESCs/iPSCs remains less robust than from marrow sources, and in mice has required genetic modification of the cells, the therapeutic effect of even partial hematopoietic chimerism is quite dramatic for the hemoglobinopathies, owing to the longer half-life of the repaired cells relative to their diseased counterparts. Consequently, the peripheral blood of engrafted mice shows remarkable restoration of normal erythroid indices, compelling evidence for the potential therapeutic use of iPSCs in hemoglobinopathies.

Two groups have reported the generation of human iPSC models of SCA (Mali et al. 2008; Sebastiano et al. 2011). Zinc finger nucleases (ZFNs) have been used to correct the sickle cell mutation (Sebastiano et al. 2011). However, neither report differentiated the cells and functionally tested them to show sickling or a reduction in sickling after correction.

A handful of human PSC models for thalassemia have also been reported. Human iPSCs derived from β-thalassemia fibroblasts were used to identify genomic safe harbors for integration of gene therapy vectors (Papapetrou et al. 2010). The study reported human β-globin levels similar to those of previously reported studies that used mouse models. β-Thalassemia human ESCs have been generated through nuclear transfer using the nucleus from a fibroblast of a β-thalassemia patient, which could also be used for cell or gene therapy studies (Fan et al. 2011). Thalassemia human ESCs have also been derived from preimplantation genetic diagnosis embryos (Verlinsky et al. 2005).

CURRENT STATE OF DIRECTED DIFFERENTIATION

Before using human PSCs for cell therapy, it must first be possible to differentiate human PSCs into the desired cell type for transplantation. In the case of hemoglobinopathies, the desired cells are hematopoietic stem cells (HSCs) or mature RBCs. Second, the issue of autologous cells versus HLA matched cells from a cell bank must be addressed. If a patient's cells are to be used for autologous treatment, then gene correction becomes an extra step.

Deriving large numbers of HSCs from human PSCs is a challenge in the field. Many groups have developed directed differentiation protocols, but most reports analyze $CD34^+$ cells for gene expression, presence of hematopoietic cell surface markers, and CFU potential (Kaufman 2009). High hematopoietic activity in vitro does not, however, correlate with high in vivo engraftment potential when the cells are tested in functional transplant assays (Ledran et al. 2008). One of the biggest challenges in generating HSCs from human PSCs is that HSCs are difficult to maintain in culture, making it difficult to capture, sustain, and expand to clinical scale even if a substantial number of HSCs could be generated in vitro.

Existing directed differentiation protocols use coculture with stromal cell lines or formation of embryoid bodies (EBs), disorganized tissue masses that resemble the early gastrulating mammalian embryo. In all but one report, the percent chimerism of in vitro–derived hematopoietic cells is <2%, which is significantly less than the chimerism seen from xenotransplants with cord blood cells (Wang et al. 2005; Tian et al. 2006; Ledran et al. 2008; Chicha et al. 2011). Lako's group assessed the effect of primary stromal cells and stromal cell lines from the mouse AGM and fetal liver on hematopoietic differentiation (Ledran et al. 2008). Conditioned medium or extracellular matrix from the stromal lines alone was not sufficient to induce strong hematopoietic differentiation. They reported a down-regulation of pluripotency genes followed by activation of mesodermal and hematopoietic genes. Peak hematopoietic CFU activity and accumulation of $CD34^+$ cells occurred on Days 18-21 of differentiation and were enhanced by TGF-β signaling. Xenotransplants resulted in 2%–16% multilineage donor chimerism, which also engrafted secondary recipients. The highest engraftment was seen from cells cocultured with the AGM stromal line. These data, which represent the highest standard to date, are promising yet still significantly short of what would be required before considering scale-up to a clinical stage for human transplantation.

Directly differentiating human PSCs into definitive RBCs has proven equally challenging (Chang et al. 2011). It is possible to generate large numbers of primitive immature RBCs, but generating definitive adult-like RBCs has proven elusive for the field. One group has cultured whole human PSC colonies on immortalized human fetal liver cells until $CD34^+$ cells arose, then treated the $CD34^+$ cells with hematopoietic differentiation cytokines and cocultured with mouse stromal cells (Olivier et al. 2006; Qiu et al. 2008). A second group cultured whole human PSC colonies on mouse fetal liver stromal cells with hematopoietic differentiation cytokines (Ma et al. 2008). A third group differentiated human PSCs as EBs, which were then directly plated on Matrigel with hematopoietic differentiation cytokines (Chang et al. 2006). A fourth group differentiated human PSCs as EBs, which were then dissociated to single cells and plated in methylcellulose with hematopoietic differentiation cytokines (Lu et al. 2008). Although there are subtle differences among the RBCs generated under each method, they all result in similarly primitive cells. The cells accumulate hemoglobin, appear red in color, and bind oxygen. They mature appropriately as measured by cell surface markers CD71 and CD235a, but remain mostly nucleated and have a primitive hemoglobin expression pattern, expressing ε-globin and γ-globin with little or no β-globin. An unanswered question in the field is what mechanism will drive the switch from primitive to definitive RBCs in vitro.

GENE CORRECTION

The chief advantage of PSCs as targets for combined gene repair and cell therapy is that gene correction is performed in immortalized cells that can be characterized extensively for the fidelity and precision of gene repair and multiple other safety parameters, and the cells can be expanded from a master cell bank to clinical scale before being differentiated into the desired transplantable cell type. A disadvantage is that gene correction is currently inefficient in human PSCs. However, recent work with ZFNs and transcription activator-like effector nucleases (TALENs) has made homologous recombination 200-fold to 1400-fold more efficient than standard transfection-based methods (Hockemeyer et al. 2009, 2011; Zou et al. 2009). As a proof-of-principle, the point mutation causing α_1-antitrypsin deficiency has been corrected in human iPSCs with ZFNs (Yusa et al. 2011). Of the colonies that incorporated the correction vector, 54% were targeted at one allele, and 4% were targeted at both alleles. The corrected clones differentiated into normal hepatocyte-like cells that did not show the disease phenotype.

Once disease-specific human PSCs are corrected, they must be differentiated into HSCs. However, large numbers of HSCs would need to be generated and maintained long enough to transplant them into the patient, and both directed differentiation and scale-up of cell culture are current challenges for the field. Alternatively, corrected human PSCs could theoretically be differentiated into RBCs for transfusion, but this would provide only a temporary amelioration of disease, in contrast to HSC transplantation, which offers promise of a cure. For patients with antibody sensitization due to multiple prior transfusions, autologous human PSCs might provide a source of compatible RBCs. Because RBCs are enucleate and have a limited life span, such cells could be irradiated before transfusion, eliminating many of the major safety concerns associated with transplantation of stem cells that would otherwise persist in the patient. Indeed, RBC transfusions from autologous human PSCs represent one of the most appealing near-term strategies for cell-based therapies derived from autologous human iPSCs.

PROHIBITIVE COST

Another major barrier to the widespread application of human PSCs in disease treatment is the high cost of goods for derivation, maintenance, quality control, banking, expansion, and differentiation of cells, made even more cumbersome if one envisions personalized therapies. There are $\sim 2 \times 10^{12}$ RBCs per unit for transfusion, but the best of current differentiation methods generates only 10^{10}–10^{11} RBCs per six-well plate of human PSCs, meaning that the production of one transfusable unit would require 20 to 200 plates of human PSCs (Lu et al. 2008). For RBC production on a clinical scale, $\sim 2.6 \times 10^{9}$ plates of human PSCs would be needed to meet the transfusion needs of only the United States.

The highly specialized maintenance and differentiation media for PSCs is prohibitively expensive, and manual passaging and cell processing are labor intensive. Moreover, for individualized therapies, the cost of the extensive quality analyses that would be required to ensure safety adds considerably to treatment costs. The advantages of immunological identity would have to outweigh the complications of using immune suppression for cells produced as an "off-the-shelf" therapy from a master cell bank, as discussed below. Because human PSCs are adherent cells, opportunities for bioreactor production of large numbers of cells for clinical application are limited.

Currently, cytokines are used to direct the differentiation of human PSCs toward hematopoietic cells, and the costs of cytokines to differentiate human PSCs to RBCs would be prohibitively expensive when considering large-scale production from a cell bank. From an individual's autologous iPSCs, it would be less expensive because of the smaller scale, but the cost would fall on one patient, making patient-specific treatment also prohibitively expensive.

AUTOLOGOUS CELLS VERSUS CELL BANKING

For cell therapies to become financially feasible, one option is to envision mass production of cells from master cell banks that could be delivered to patients "off-the-shelf," such that economies of scale could be realized for costs of production, maintenance, quality control, and differentiation; and costs could be distributed across a large number of patients. Such an approach would necessarily compromise the chief advantage of personalized, individualized cell therapies—that of immunological identity—and would thus require the use of immune suppression, as currently used for organ transplantation. A major question then becomes, what number of PSCs harboring what extent of genetic diversity would be required to constitute a cell bank that might provide a suitable immunological match for most of the human population? Two groups have attempted to model the number of cell lines needed for such a cell bank, based on known immunological heterogeneity of patient populations in the United Kingdom and Japan. The approximations were made using HLA information from registries of kidney transplant recipients, assuming that HLA matching requirements and the use of immune suppression would be similar for solid organ transplants and cell transplants of human PSC derivatives. It is well known that the function and durability of renal allografts is directly proportional to the degree of immunological match, and when considering immunological identity at the human histocompatibility loci HLA-A, HLA-B, and HLA-DR, a single antigen match is considered minimally advantageous. Consequently, in the United Kingdom, a cell bank of 150 randomly chosen human PSC lines would provide a full 6/6 antigen match at HLA-A, HLA-B, and HLA-DR for <20% of the population, and 37.9% would have a beneficial match (only one HLA-A or HLA-B mismatch or better), whereas 84.9% of the population would have a minimally advantageous match at a single HLA-DR locus, or better. However, if the bank is constructed from highly selected human PSCs that harbor homozygous HLA haplotypes (the same A, B, and DR loci on both chromosomes), then only 10 cell lines would provide matches for similar percentages of the population (Taylor et al. 2005). In the Japanese population, a cell bank of 170 human PSC lines would provide a single mismatch or better for 80% of patients (Nakajima et al. 2007), whereas a cell bank of 30 highly selected homozygous human PSC lines would provide a full match for 82.2% of the Japanese population, and 50 lines would provide a full match for 73%-90.7% (Nakatsuji et al. 2008; Okita and

Yamanaka 2011). The suspected reason for the difference in beneficial matches is the low ethnic diversity in Japan compared with the United Kingdom, suggesting that the cell bank strategy would be a more reasonable option in less diverse regions and more difficult in regions with high ethnic diversity like the United States.

If a cell bank strategy were to be implemented, depending on the country or region, there would still be a significant percentage of the population for which there would be no beneficial match in the bank. Because of the considerable degree of polymorphism in the HLA loci in human populations, increasing the number of human PSC lines in the cell bank meets a diminishing return and has only a minimal effect on the percentage of the population with a beneficial match. In such cases and for severe disease, the extra effort and expense of personalized cell therapies would have to be considered.

CONCLUDING REMARKS

We have come a long way from identifying the genetic mutations that cause hemoglobinopathies, but we are still far from a cell therapy treatment or cure. Although the advent of PSCs offers an appealing theoretical platform for treating not only hemoglobinopathies but any of dozens of genetic and malignant bone marrow disorders for which current approaches of bone marrow transplant remain infeasible because of lack of suitable donors or immunological barriers, many hurdles must first be overcome. Advances are needed to improve the efficiency of transgene-free human iPSC derivation, to enhance directed differentiation toward engraftable HSCs or transfusable RBCs, and to reduce the cost of maintaining and differentiating human PSCs in order to make cell therapy treatments feasibly affordable. A hybrid effort, with the majority of the population served by off-the-shelf treatments manufactured from master cell banks, accepting the trade-offs of immune suppression, whereas others are treated with individualized strategies, may ultimately prove the only sustainable mode of delivering innovative cell therapies based on PSCs.

ACKNOWLEDGMENTS

G.Q.D. is an investigator of the Howard Hughes Medical Institute and the Manton Center for Orphan Disease Research. Research was funded by grants from the U.S. National Institutes of Health (NIH) to G.Q.D. (NHLBI U01-HL100-001), the Doris Duke Medical Foundation, and the CHB Stem Cell Program.

REFERENCES

Aasen T, Raya A, Barrero MJ, Garreta E, Consiglio A, Gonzalez F, Vassena R, Bilic J, Pekarik V, Tiscornia G, et al. 2008. Efficient and rapid generation of induced pluripotent stem cells from human keratinocytes. *Nat Biotechnol* **26:** 1276–1284.

Alebouyeh M, Moussavi F, Haddad-Deylami H, Vossough P. 2004. Hydroxyurea in the treatment of major β-thalassemia and importance of genetic screening. *Ann Hematol* **83:** 430–433.

Ben-Yosef D, Malcov M, Eiges R. 2008. PGD-derived human embryonic stem cell lines as a powerful tool for the study of human genetic disorders. *Mol Cell Endocrinol* **282:** 153–158.

Brittenham GM, Schechter AN, Noguchi CT. 1985. Hemoglobin S polymerization: Primary determinant of the hemolytic and clinical severity of the sickling syndromes. *Blood* **65:** 183–189.

Chang J, Lu RH, Xu SM, Meneses J, Chan K, Pedersen R, Kan YW. 1996. Inactivation of mouse α-globin gene by homologous recombination: Mouse model of hemoglobin H disease. *Blood* **88:** 1846–1851.

Chang KH, Nelson AM, Cao H, Wang L, Nakamoto B, Ware CB, Papayannopoulou T. 2006. Definitive-like erythroid cells derived from human embryonic stem cells coexpress high levels of embryonic and fetal globins with little or no adult globin. *Blood* **108:** 1515–1523.

Chang KH, Bonig H, Papayannopoulou T. 2011. Generation and characterization of erythroid cells from human embryonic stem cells and induced pluripotent stem cells: An overview. *Stem Cells Int* **2011:** 791604.

Chicha L, Feki A, Boni A, Irion O, Hovatta O, Jaconi M. 2011. Human pluripotent stem cells differentiated in fully defined medium generate hematopoietic $CD34^-$ and $CD34^+$ progenitors with distinct characteristics. *PLoS ONE* **6:** e14733.

Choi KD, Yu J, Smuga-Otto K, Salvagiotto G, Rehrauer W, Vodyanik M, Thomson J, Slukvin I. 2009. Hematopoietic and endothelial differentiation of human induced pluripotent stem cells. *Stem Cells* **27:** 559–567.

Coles SM, Klein HG, Holland PV. 1981. Alloimmunization in two multitransfused patient populations. *Transfusion* **21:** 462–466.

Cox JV, Steane E, Cunningham G, Frenkel EP. 1988. Risk of alloimmunization and delayed hemolytic transfusion reactions in patients with sickle cell disease. *Arch Intern Med* **148:** 2485–2489.

Crepeau RH, Dykes G, Garrell R, Edelstein SJ. 1978. Diameter of haemoglobin S fibres in sickled cells. *Nature* **274:** 616–617.

Davies SC, McWilliam AC, Hewitt PE, Devenish A, Brozovic M. 1986. Red cell alloimmunization in sickle cell disease. *Br J Haematol* **63:** 241–245.

Dykes GW, Crepeau RH, Edelstein SJ. 1979. Three-dimensional reconstruction of the 14-filament fibers of hemoglobin S. *J Mol Biol* **130:** 451–472.

Eaton WA, Hofrichter J. 1987. Hemoglobin S gelation and sickle cell disease. *Blood* **70:** 1245–1266.

Evans MJ, Kaufman MH. 1981. Establishment in culture of pluripotential cells from mouse embryos. *Nature* **292:** 154–156.

Fan Y, Jiang Y, Chen X, Ou Z, Yin Y, Huang S, Kou Z, Li Q, Long X, Liu J, et al. 2011. Derivation of cloned human blastocysts by histone deacetylase inhibitor treatment after somatic cell nuclear transfer with β-thalassemia fibroblasts. *Stem Cells Dev* **20:** 1951–1959.

Feng Q, Lu SJ, Klimanskaya I, Gomes I, Kim D, Chung Y, Honig GR, Kim KS, Lanza R. 2010. Hemangioblastic derivatives from human induced pluripotent stem cells exhibit limited expansion and early senescence. *Stem Cells* **28:** 704–712.

Fessas P. 1963. Inclusions of hemoglobin erythroblasts and erythrocytes of thalassemia. *Blood* **21:** 21–32.

Fessas P, Loukopoulos D, Thorell B. 1965. Absorption spectra of inclusion bodies in β-thalassemia. *Blood* **25:** 105–109.

Han XD, Lin C, Chang J, Sadelain M, Kan YW. 2007. Fetal gene therapy of α-thalassemia in a mouse model. *Proc Natl Acad Sci* **104:** 9007–9011.

Hanna J, Wernig M, Markoulaki S, Sun CW, Meissner A, Cassady JP, Beard C, Brambrink T, Wu LC, Townes TM, et al. 2007. Treatment of sickle cell anemia mouse model with iPS cells generated from autologous skin. *Science* **318:** 1920–1923.

Hargus G, Cooper O, Deleidi M, Levy A, Lee K, Marlow E, Yow A, Soldner F, Hockemeyer D, Hallett PJ, et al. 2010. Differentiated Parkinson patient-derived induced pluripotent stem cells grow in the adult rodent brain and reduce motor asymmetry in Parkinsonian rats. *Proc Natl Acad Sci* **107:** 15921–15926.

Hockemeyer D, Soldner F, Beard C, Gao Q, Mitalipova M, DeKelver RC, Katibah GE, Amora R, Boydston EA, Zeitler B, et al. 2009. Efficient targeting of expressed and silent genes in human ESCs and iPSCs using zinc-finger nucleases. *Nat Biotechnol* **27:** 851–857.

Hockemeyer D, Wang H, Kiani S, Lai CS, Gao Q, Cassady JP, Cost GJ, Zhang L, Santiago Y, Miller JC, et al. 2011. Genetic engineering of human pluripotent cells using TALE nucleases. *Nat Biotechnol* **29:** 731–734.

Hofrichter J, Ross PD, Eaton WA. 1974. Kinetics and mechanism of deoxyhemoglobin S gelation: A new approach to understanding sickle cell disease. *Proc Natl Acad Sci* **71:** 4864–4868.

Imren S, Payen E, Westerman KA, Pawliuk R, Fabry ME, Eaves CJ, Cavilla B, Wadsworth LD, Beuzard Y, Bouhassira EE, et al. 2002. Permanent and panerythroid correction of murine β thalassemia by multiple lentiviral integration in hematopoietic stem cells. *Proc Natl Acad Sci* **99:** 14380–14385.

Ingram VM. 1956. A specific chemical difference between the globins of normal human and sickle-cell anaemia haemoglobin. *Nature* **178:** 792–794.

Ingram VM. 1957. Gene mutations in human haemoglobin: The chemical difference between normal and sickle cell haemoglobin. *Nature* **180:** 326–328.

Ingram VM, Stretton AO. 1959. Genetic basis of the thalassaemia diseases. *Nature* **184:** 1903–1909.

International Stem Cell Initiative , Adewumi O, Aflatoonian B, Ahrlund-Richter L, Amit M, Andrews PW, Beighton G, Bello PA, Benvenisty N, Berry LS, et al. 2007. Characterization of human embryonic stem cell lines by the International Stem Cell Initiative. *Nat Biotechnol* **25:** 803–816.

Kaufman DS. 2009. Toward clinical therapies using hematopoietic cells derived from human pluripotent stem cells. *Blood* **114:** 3513–3523.

Kim K, Zhao R, Doi A, Ng K, Unternaehrer J, Cahan P, Hongguang H, Loh YH, Aryee MJ, Lensch MW, et al. 2011. Donor cell type can influence the epigenome and differentiation potential of human induced pluripotent stem cells. *Nat Biotechnol* **29:** 1117–1119.

Koren A, Levin C, Dgany O, Kransnov T, Elhasid R, Zalman L, Palmor H, Tamary H. 2008. Response to hydroxyurea therapy in beta-thalassemia. *Am J Hematol* **83**(5)**:** 366–370.

Ledran MH, Krassowska A, Armstrong L, Dimmick I, Renstrom J, Lang R, Yung S, Santibanez-Coref M, Dzierzak E, Stojkovic M, et al. 2008. Efficient hematopoietic differentiation of human embryonic stem cells on stromal cells derived from hematopoietic niches. *Cell Stem Cell* **3:** 85–98.

Letvin NL, Linch DC, Beardsley GP, McIntyre KW, Nathan DG. 1984. Augmentation of fetal-hemoglobin production in anemic monkeys by hydroxyurea. *N Engl J Med* **310:** 869–873.

Lisowski L, Sadelain M. 2007. Locus control region elements HS1 and HS4 enhance the therapeutic efficacy of globin gene transfer in β-thalassemic mice. *Blood* **110:** 4175–4178.

Loh YH, Agarwal S, Park IH, Urbach A, Huo H, Heffner GC, Kim K, Miller JD, Ng K, Daley GQ. 2009. Generation of induced pluripotent stem cells from human blood. *Blood* **113:** 5476–5479.

Lu SJ, Feng Q, Park JS, Vida L, Lee BS, Strausbauch M, Wettstein PJ, Honig GR, Lanza R. 2008. Biologic properties and enucleation of red blood cells from human embryonic stem cells. *Blood* **112:** 4475–4484.

Ma F, Ebihara Y, Umeda K, Sakai H, Hanada S, Zhang H, Zaike Y, Tsuchida E, Nakahata T, Nakauchi H, et al. 2008. Generation of functional erythrocytes from human embryonic stem cell-derived definitive hematopoiesis. *Proc Natl Acad Sci* **105:** 13087–13092.

Maehr R, Chen S, Snitow M, Ludwig T, Yagasaki L, Goland R, Leibel RL, Melton DA. 2009. Generation of pluripotent stem cells from patients with type 1 diabetes. *Proc Natl Acad Sci* **106:** 15768–15773.

Mali P, Ye Z, Hommond HH, Yu X, Lin J, Chen G, Zou J, Cheng L. 2008. Improved efficiency and pace of generating induced pluripotent stem cells from human adult and fetal fibroblasts. *Stem Cells* **26:** 1998–2005.

Martin GR. 1981. Isolation of a pluripotent cell line from early mouse embryos cultured in medium conditioned by teratocarcinoma stem cells. *Proc Natl Acad Sci* **78**(12): 7634–7638.

Mateizel I, De Temmerman N, Ullmann U, Cauffman G, Sermon K, Van de Velde H, De Rycke M, Degreef E, Devroey P, Liebaers I, et al. 2006. Derivation of human embryonic stem cell lines from embryos obtained after IVF and after PGD for monogenic disorders. *Hum Reprod* **21:** 503–511.

May C, Rivella S, Callegari J, Heller G, Gaensler KM, Luzzatto L, Sadelain M. 2000. Therapeutic haemoglobin synthesis in β-thalassaemic mice expressing lentivirus-encoded human β-globin. *Nature* **406:** 82–86.

May C, Rivella S, Chadburn A, Sadelain M. 2002. Successful treatment of murine β-thalassemia intermedia by transfer of the human β-globin gene. *Blood* **99:** 1902–1908.

Nakajima F, Tokunaga K, Nakatsuji N. 2007. Human leukocyte antigen matching estimations in a hypothetical bank of human embryonic stem cell lines in the Japanese population for use in cell transplantation therapy. *Stem Cells* **25:** 983–985.

Nakatsuji N, Nakajima F, Tokunaga K. 2008. HLA-haplotype banking and iPS cells. *Nat Biotechnol* **26:** 739–740.

Nishino T, Cao H, Stamatoyannopoulos G, Emery DW. 2006. Effects of human γ-globin in murine β-thalassaemia. *Br J Haematol* **134:** 100–108.

Noguchi CT, Schechter AN. 1981. The intracellular polymerization of sickle hemoglobin and its relevance to sickle cell disease. *Blood* **58:** 1057–1068.

Okita K, Yamanaka S. 2011. Induced pluripotent stem cells: Opportunities and challenges. *Philos Trans R Soc Lond B Biol Sci* **366:** 2198–2207.

Olivier EN, Qiu C, Velho M, Hirsch RE, Bouhassira EE. 2006. Large-scale production of embryonic red blood cells from human embryonic stem cells. *Exp Hematol* **34:** 1635–1642.

Orlina AR, Unger PJ, Koshy M. 1978. Post-transfusion alloimmunization in patients with sickle cell disease. *Am J Hematol* **5:** 101–106.

Osafune K, Caron L, Borowiak M, Martinez RJ, Fitz-Gerald CS, Sato Y, Cowan CA, Chien KR, Melton DA. 2008. Marked differences in differentiation propensity among human embryonic stem cell lines. *Nat Biotechnol* **26:** 313–315.

Ottolenghi S, Lanyon WG, Paul J, Williamson R, Weatherall DJ, Clegg JB, Pritchard J, Pootrakul S, Boon WH. 1974. The severe form of α thalassaemia is caused by a haemoglobin gene deletion. *Nature* **251:** 389–392.

Papapetrou EP, Lee G, Malani N, Setty M, Riviere I, Tirunagari LM, Kadota K, Roth SL, Giardina P, Viale A, et al. 2010. Genomic safe harbors permit high β-globin transgene expression in thalassemia induced pluripotent stem cells. *Nat Biotechnol* **29:** 73–78.

Park IH, Arora N, Huo H, Maherali N, Ahfeldt T, Shimamura A, Lensch MW, Cowan C, Hochedlinger K, Daley GQ. 2008a. Disease-specific induced pluripotent stem cells. *Cell* **134:** 877–886.

Park IH, Lerou PH, Zhao R, Huo H, Daley GQ. 2008b. Generation of human-induced pluripotent stem cells. *Nat Protoc* **3:** 1180–1186.

Paszty C, Brion CM, Manci E, Witkowska HE, Stevens ME, Mohandas N, Rubin EM. 1997. Transgenic knockout mice with exclusively human sickle hemoglobin and sickle cell disease. *Science* **278:** 876–878.

Pawliuk R, Westerman KA, Fabry ME, Payen E, Tighe R, Bouhassira EE, Acharya SA, Ellis J, London IM, Eaves CJ, et al. 2001. Correction of sickle cell disease in transgenic mouse models by gene therapy. *Science* **294:** 2368–2371.

Pickering SJ, Minger SL, Patel M, Taylor H, Black C, Burns CJ, Ekonomou A, Braude PR. 2005. Generation of a human embryonic stem cell line encoding the cystic fibrosis mutation ΔF508, using preimplantation genetic diagnosis. *Reprod Biomed Online* **10:** 390–397.

Platt OS, Orkin SH, Dover G, Beardsley GP, Miller B, Nathan DG. 1984. Hydroxyurea enhances fetal hemoglobin production in sickle cell anemia. *J Clin Invest* **74:** 652–656.

Qiu C, Olivier EN, Velho M, Bouhassira EE. 2008. Globin switches in yolk sac-like primitive and fetal-like definitive red blood cells produced from human embryonic stem cells. *Blood* **111:** 2400–2408.

Reisner EG, Kostyu DD, Phillips G, Walker C, Dawson DV. 1987. Alloantibody responses in multiply transfused sickle cell patients. *Tissue Antigens* **30:** 161–166.

Rideout WM III, Hochedlinger K, Kyba M, Daley GQ, Jaenisch R. 2002. Correction of a genetic defect by nuclear transplantation and combined cell and gene therapy. *Cell* **109:** 17–27.

Rosse WF, Gallagher D, Kinney TR, Castro O, Dosik H, Moohr J, Wang W, Levy PS. 1990. Transfusion and alloimmunization in sickle cell disease. The Cooperative Study of Sickle Cell Disease. *Blood* **76:** 1431–1437.

Sebastiano V, Maeder ML, Angstman JF, Haddad B, Khayter C, Yeo DT, Goodwin MJ, Hawkins JS, Ramirez CL, Batista LF, et al. 2011. In situ genetic correction of the sickle cell anemia mutation in human induced pluripotent stem cells using engineered zinc finger nucleases. *Stem Cells* **29:** 1717–1726.

Skow LC, Burkhart BA, Johnson FM, Popp RA, Popp DM, Goldberg SZ, Anderson WF, Barnett LB, Lewis SE. 1983. A mouse model for β-thalassemia. *Cell* **34:** 1043–1052.

Soldner F, Hockemeyer D, Beard C, Gao Q, Bell GW, Cook EG, Hargus G, Blak A, Cooper O, Mitalipova M, et al. 2009. Parkinson's disease patient-derived induced pluripotent stem cells free of viral reprogramming factors. *Cell* **136:** 964–977.

Takahashi K, Tanabe K, Ohnuki M, Narita M, Ichisaka T, Tomoda K, Yamanaka S. 2007. Induction of pluripotent stem cells from adult human fibroblasts by defined factors. *Cell* **131:** 861–872.

Taylor JM, Dozy A, Kan YW, Varmus HE, Lie-Injo LE, Ganesan J, Todd D. 1974. Genetic lesion in homozygous α thalassaemia (hydrops fetalis). *Nature* **251:** 392–393.

Taylor CJ, Bolton EM, Pocock S, Sharples LD, Pedersen RA, Bradley JA. 2005. Banking on human embryonic stem cells: Estimating the number of donor cell lines needed for HLA matching. *Lancet* **366:** 2019–2025.

Thomson JA, Itskovitz-Eldor J, Shapiro SS, Waknitz MA, Swiergiel JJ, Marshall VS, Jones JM. 1998. Embryonic stem cell lines derived from human blastocysts. *Science* **282:** 1145–1147.

Tian X, Woll PS, Morris JK, Linehan JL, Kaufman DS. 2006. Hematopoietic engraftment of human embryonic stem cell-derived cells is regulated by recipient innate immunity. *Stem Cells* **24:** 1370–1380.

Verlinsky Y, Strelchenko N, Kukharenko V, Rechitsky S, Verlinsky O, Galat V, Kuliev A. 2005. Human embryonic stem cell lines with genetic disorders. *Reprod Biomed Online* **10:** 105–110.

Vichinsky EP, Earles A, Johnson RA, Hoag MS, Williams A, Lubin B. 1990. Alloimmunization in sickle cell anemia and transfusion of racially unmatched blood. *N Engl J Med* **322:** 1617–1621.

Wang L, Menendez P, Shojaei F, Li L, Mazurier F, Dick JE, Cerdan C, Levac K, Bhatia M. 2005. Generation of hematopoietic repopulating cells from human embryonic stem cells independent of ectopic HOXB4 expression. *J Exp Med* **201:** 1603–1614.

Warren L, Manos PD, Ahfeldt T, Loh YH, Li H, Lau F, Ebina W, Mandal PK, Smith ZD, Meissner A, et al. 2010. Highly efficient reprogramming to pluripotency and directed differentiation of human cells with synthetic modified mRNA. *Cell Stem Cell* **7:** 618–630.

Weatherall DJ. 2001. Phenotype–genotype relationships in monogenic disease: Lessons from the thalassaemias. *Nat Rev Genet* **2:** 245–255.

Wu LC, Sun CW, Ryan TM, Pawlik KM, Ren J, Townes TM. 2006. Correction of sickle cell disease by homologous recombination in embryonic stem cells. *Blood* **108:** 1183–1188.

Xu J, Peng C, Sankaran VG, Shao Z, Esrick EB, Chong BG, Ippolito GC, Fujiwara Y, Ebert BL, Tucker PW, et al. 2011. Correction of sickle cell disease in adult mice by interference with fetal hemoglobin silencing. *Science* **334:** 993–996.

Yang B, Kirby S, Lewis J, Detloff PJ, Maeda N, Smithies O. 1995. A mouse model for β^0-thalassemia. *Proc Natl Acad Sci* **92:** 11608–11612.

Yavarian M, Karimi M, Bakker E, Harteveld CL, Giordano PC. 2004. Response to hydroxyurea treatment in Iranian transfusion-dependent β-thalassemia patients. *Haematologica* **89:** 1172–1178.

Yu J, Vodyanik MA, Smuga-Otto K, Antosiewicz-Bourget J, Frane JL, Tian S, Nie J, Jonsdottir GA, Ruotti V, Stewart R, et al. 2007. Induced pluripotent stem cell lines derived from human somatic cells. *Science* **318:** 1917–1920.

Yu J, Hu K, Smuga-Otto K, Tian S, Stewart R, Slukvin II, Thomson JA. 2009. Human induced pluripotent stem cells free of vector and transgene sequences. *Science* **324:** 797–801.

Yusa K, Rashid ST, Strick-Marchand H, Varela I, Liu PQ, Paschon DE, Miranda E, Ordonez A, Hannan NR, Rouhani FJ, et al. 2011. Targeted gene correction of α1-antitrypsin deficiency in induced pluripotent stem cells. *Nature* **478:** 391–394.

Zou J, Maeder ML, Mali P, Pruett-Miller SM, Thibodeau-Beganny S, Chou BK, Chen G, Ye Z, Park IH, Daley GQ, et al. 2009. Gene targeting of a disease-related gene in human induced pluripotent stem and embryonic stem cells. *Cell Stem Cell* **5:** 97–110.

Hemoglobin Variants: Biochemical Properties and Clinical Correlates

Christopher S. Thom[1,2], Claire F. Dickson[3], David A. Gell[3], and Mitchell J. Weiss[2]

[1]Cell and Molecular Biology Graduate Group, University of Pennsylvania School of Medicine, Philadelphia, Pennsylvania 19104

[2]Hematology Department, Children's Hospital of Philadelphia, Philadelphia, Pennsylvania 19104

[3]Menzies Research Institute, University of Tasmania, Hobart, Australia

Correspondence: weissmi@email.chop.edu

Diseases affecting hemoglobin synthesis and function are extremely common worldwide. More than 1000 naturally occurring human hemoglobin variants with single amino acid substitutions throughout the molecule have been discovered, mainly through their clinical and/or laboratory manifestations. These variants alter hemoglobin structure and biochemical properties with physiological effects ranging from insignificant to severe. Studies of these mutations in patients and in the laboratory have produced a wealth of information on hemoglobin biochemistry and biology with significant implications for hematology practice. More generally, landmark studies of hemoglobin performed over the past 60 years have established important paradigms for the disciplines of structural biology, genetics, biochemistry, and medicine. Here we review the major classes of hemoglobin variants, emphasizing general concepts and illustrative examples.

Globin gene mutations affecting hemoglobin (Hb), the major blood oxygen (O_2) carrier, are common, affecting an estimated 7% of the world's population (Weatherall and Clegg 2001; Kohne 2011). These mutations are broadly subdivided into those that impair globin protein subunit production (thalassemias) and those that produce structurally abnormal globin proteins (Hb variants). The latter class is mainly composed of missense mutations that cause single amino acid substitutions in the globin protein, resulting in an abnormal, or "variant" Hb tetramer. Less commonly, Hb variants are associated with deletions, multiple amino acid substitutions, antitermination mutations, and altered posttranslational processing (Table 1).

Naturally occurring Hb mutations cause a range of biochemical abnormalities, some of which produce clinically significant symptoms. The most common and medically important Hb variants include HbS (Lettre 2012; Williams and Weatherall 2012; Cao and Kan 2013; Serjeant 2013), HbC (Lettre 2012; Williams and Weatherall 2012; Cao and Kan 2013; Serjeant 2013), HbE (see the sections on Selected Variants that Illustrate Important Aspects of Hemoglobin

Copyright © 2013 Cold Spring Harbor Laboratory Press; all rights reserved
Cite this article as *Cold Spring Harb Perspect Med* doi: 10.1101/cshperspect.a011858

Table 1. Hb variants that are discussed in this article

Name	Globin site (fold)	Amino acid substitution	Molecular mechanism	Clinical phenotype	Other biochemical and laboratory findings
			Unstable mutants		
Brockton	β138 (H16)	Ala > Pro	Altered secondary structure	Hemolytic anemia, reticulocytosis	
Philly	β35 (C1)	Tyr > Phe	Altered α1β1 interface	Hemolytic anemia, reticulocytosis	Decreased cooperativity, increased oxygen affinity
Peterborough	β111 (G13)	Val > Phe	Altered α1β1 interface	Hemolytic anemia, reticulocytosis	Decreased oxygen affinity
Stanmore	β111 (G13)	Val > Ala	Altered α1β1 interface	Hemolytic anemia	Decreased oxygen affinity
J-Guantanamo	β128 (H6)	Ala > Asp	Altered α1β1 interface	Hemolytic anemia	Target cells
Khartoum	β124 (H2)	Pro > Arg	Altered α1β1 interface	Normal	
Prato	α1 or α2 31 (B12)	Arg > Ser	Altered α1β1 interface	Anisocytosis, hypochromia	Mildly unstable in isopropanol
Lombard	α2 103 (G10)	His > Tyr	Altered α1β1 interface	Anemia	
Contaldo	α1 or α2 103 (G10)	His > Arg	Altered α1β1 interface	Hemolytic anemia	
Foggia	α2 117 (GH5)	Phe > Ser	Altered α1β1 interface	Microcytosis	Rapidly degraded α chains
Groene Hart	α1 119 (H2)	Pro > Ser	Altered α1β1 interface, disrupted AHSP binding	Hemolytic anemia, microcytosis	
Turriff	α1 or α2 99 (G6)	Lys > Glu	Altered α1β1 interface, disrupted AHSP binding	Normal	Comigrates with HbA_{1C}, rapidly degraded α chains
Beziers	α1 99 (G6)	Lys > Asn	Altered α1β1 interface, disrupted AHSP binding	Normal	Comigrates with HbA_{1C}
Hirosaki	α2 43 (CE1)	Phe > Leu	Altered heme pocket	Heinz body hemolytic anemia	Hyperunstable
Terre Haute	β106 (G8)	Leu > Arg	Altered heme pocket	Heinz body hemolytic anemia, dominant inclusion body thalassemia	Hyperunstable
			High affinity variants		
Kempsey	β99 (G1)	Asp > Asn	Unstable T state	Erythrocytosis	Decreased cooperativity
Hiroshima	β146 (HC3)	His > Asp	Mutated Bohr proton donor	Erythrocytosis	Decreased cooperativity, decreased Bohr effect
York	β146 (HC3)	His > Pro	Mutated Bohr proton donor	Erythrocytosis	Decreased cooperativity, decreased Bohr effect
Cowtown	β146 (HC3)	His > Leu	Mutated Bohr proton donor	Erythrocytosis	Decreased Bohr effect

Continued

Table 1. *Continued*

Name	Globin site (fold)	Amino acid substitution	Molecular mechanism	Clinical phenotype	Other biochemical and laboratory findings
Rahere	β82 (EF6)	Lys > Thr	Altered 2,3DPG binding site	Erythrocytosis	
Providence	β82 (EF6)	Lys > Asn	Altered 2,3DPG binding site	Erythrocytosis	Low oxygen affinity
Helsinki	β82 (EF6)	Lys > Met	Altered 2,3DPG binding site	Erythrocytosis	Decreased Bohr effect
			Low affinity variants		
Kansas	β102 (G4)	Asn > Thr	Unstable R state	Cyanosis	Decreased cooperativity
Beth Israel	β102 (G4)	Asn > Ser	Unstable R state	Cyanosis	Decreased cooperativity
St. Mandé	β102 (G4)	Asn > Tyr	Unstable R state	Cyanosis	
			Methemoglobin variants		
M-Iwate	α1 or α2 (F8)	His > Tyr	Oxidized heme	Pseudocyanosis (Methemoglobinemia)	Abnormal visible spectrum
M-Saskatoon	β63 (E7)	His > Tyr	Oxidized heme	Pseudocyanosis (Methemoglobinemia)	Abnormal visible spectrum
			Globin chain elongation variants		
Constant Spring	α2 142 (HC3)	Stop > Gln	Antitermination mutant	Microcytosis	Decreased mRNA stability
Cranston	β145 (HC3)	+CT	Frameshift, elongated globin	Hemolytic anemia	Increased oxygen affinity, decreased cooperativity
			Variants with multiple effects		
HbE	β26 (B8)	Glu > Lys	Unstable, reduced synthesis	Microcytosis	
Bruxelles	β41 (C7) or β42 (CD1)	Phe > 0	Altered heme pocket	Hemolytic anemia, cyanosis, splenomegaly, reticulocytosis	Heinz bodies, decreased cooperativity
Warsaw	β42 (CD1)	Phe > Val	Altered heme pocket	Hemolytic anemia, cyanosis	Heinz bodies, decreased cooperativity
Hammersmith	β42 (CD1)	Phe > Ser	Altered heme pocket	Hemolytic anemia, cyanosis	Heinz bodies, decreased cooperativity
Buccuresti-Louisville	β42 (CD1)	Phe > Leu	Altered heme pocket	Hemolytic anemia, cyanosis	Heinz bodies, decreased cooperativity
Zurich	β63 (E7)	His > Arg	Altered heme pocket	Normal, but hypersensitive to oxidative stress	Decreased cooperativity, increased oxygen affinity, increased CO affinity
Jamaica Plain	β6 (A3) and β68 (E12)	Glu > Val and Leu > Phe	Altered secondary structure	Hemolytic anemia, cyanosis, splenomegaly, splenic sequestration	Unstable, Heinz bodies, sickle cell phenotype, decreased oxygen affinity

Continued

Table 1. *Continued*

Name	Globin site (fold)	Amino acid substitution	Molecular mechanism	Clinical phenotype	Other biochemical and laboratory findings
Quebec-Chori	β87 (F3)	Thr > Ile	Altered interaction with HbS polymer	Normal	Promotes HbS polymerization
D-Ibadan	β87 (F3)	Thr > Lys	Altered interaction with HbS polymer	Normal	Inhibits HbS polymerization
Bristol-Alesha	β67 (E11)	Val > Met	Altered heme pocket	Hemolytic anemia, reticulocytosis	Heinz bodies, decreased cooperativity, decreased Bohr effect, decreased oxygen affinity
Toms River	γ67 (E11)	Val > Met	Altered heme pocket	Anemia, cyanosis	Unstable, low oxygen affinity

For a full listing of hemoglobin variants, see The Globin Gene Server (http://globin.bx.psu.edu; Hardison et al. 2002; Giardine et al. 2011).

Biology and Variants that Affect Multiple Hemoglobin Functions; see also Musallam et al. 2012), and some thalassemias (e.g., "thalassemic hemoglobinopathies"), all of which are under positive genetic selection because they confer survival advantages in areas where malaria is endemic (Weatherall and Clegg 2001). In addition to these prevalent mutant proteins, there are also >1000 other known naturally occurring Hb variants, which are rare individually but common collectively. Most Hb mutants are cataloged on the Globin Gene database (HbVar, http://globin.bx.psu.edu; Hardison et al. 2002; Giardine et al. 2011). By convention, these variants are named after the geographic origin of the affected individual. Although many Hb variants are clinically silent, some produce clinical manifestations of varying severity. Analyses of these variants, which can be considered to be "experiments of nature" (Garrod 1928), have generated valuable insights into structure–function relationships within the Hb molecule, with interesting and important clinical consequences.

The goal of this work is to provide a succinct conceptual framework for understanding the biology and clinical implications of Hb variants. We explore major concepts of Hb biology followed by a discussion of selected Hb variants that reinforce basic principles. For more extensive reviews of this topic, see *Nathan and Oski's Hematology of Infancy and Childhood* (Nathan et al. 2009); *Hemoglobin: Molecular, Genetic, and Clinical Effects* (Bunn and Forget 1986); *Disorders of Hemoglobin: Genetics, Pathophysiology, and Clinical Management* (Steinberg et al. 2001); and the Globin Gene database (Hardison et al. 2002; Giardine et al. 2011).

BASIC PRINCIPLES

Hemoglobin Synthesis, Structure, and Function

Hemoglobin is a heterotetramer composed of α-like and β-like globin subunits, each bound to a heme prosthetic group. The major functions of Hb are to transport oxygen (O_2) from the lungs to peripheral tissues and carbon dioxide (CO_2) from the tissues to the lungs. The kinetics of Hb-O_2 binding and release are fine-tuned for this purpose and adaptable according to developmental ontogeny and metabolic perturbations. Moreover, the Hb molecule must limit potential problems caused by its associated iron and free O_2, reactive molecules capable of inflicting damage through the production of reactive oxygen species. Efforts to understand how Hb structure imparts these critical functions have been ongoing for more than 50 years. Hemoglobin was one of the first proteins to be sequenced (Konigsberg et al. 1961; Schroeder et al. 1961; Watson and Kendrew 1961) and the globin genes were among the earliest to be cloned (Rabbitts 1976). In the late 1950s, Perutz and colleagues determined the three-dimensional structure of Hb through X-ray crystallography (Perutz 1960; Perutz et al. 1960). More recent studies refined this structure to high resolution (Paoli et al. 1996; Park et al. 2006). In addition, O_2 and other ligand-binding properties have been measured in detail for native Hb and many naturally occurring mutants. All of this work provides a substantial framework for defining the O_2 delivery properties of Hb, as well as the molecular consequences of variant mutations (see, for example, Lehmann 1957; Konigsberg and Lehmann 1965; Shimizu et al. 1965; Perutz and Lehmann 1968; Perutz 1970; Bunn and Forget 1986).

Globin polypeptides are synthesized from separate α-like and β-like globin gene clusters located on human chromosomes 16 and 11, respectively. Nascent globin chains rapidly incorporate heme, which stabilizes their native folding into Hb subunits composed of seven or eight α helices named A–H, which fold together into a globular structure (Fig. 1A). Hb subunits bind O_2 and other ligands via the heme iron buried within an evolutionarily conserved hydrophobic pocket that faces the outside of HbA tetramers (Fig. 1B). Heme iron is axially coordinated to globin proteins by an invariant histidine residue in helix F8, termed the "proximal" histidine (Fig. 1C). The opposite axial position binds O_2, which is stabilized by interaction with the conserved "distal" histidine in helix E7 (Fig. 1C). Multiple additional amino acids within the globin proteins stabilize heme binding through noncovalent interactions (Fig. 1C). Iron must be in its reduced (ferrous, Fe^{2+}) state for Hb to bind O_2. Oxidized or "met" Hb (ferric, Fe^{3+}) cannot bind O_2 and is relatively unstable, tending to lose hemin and denature. Thus, red blood cells have evolved elaborate mechanisms to maintain Hb in its reduced state (Bunn and Forget 1986; Ganz and Nemeth 2012). For example, the methemoglobin reductase system converts methemoglobin (metHb) to its reduced form. Not surprisingly, globin mutations that alter amino acids within the ligand pocket frequently produce strong functional effects, including destabilization, altered affinity for O_2, and increased rates of metHb formation and heme loss (see the section on Selected Variants that Illustrate Important Aspects of Hemoglobin Biology). Variants promoting autoxidation are termed "M-Hbs" (see the sections on Selected Variants that Illustrate Important Aspects of Hemoglobin Biology and Methemoglobin ("M-Type") Variants). Because Hb gains its distinctive color from the heme group, alterations that affect the environment of the heme iron, including changes in the surrounding amino acids, different gas ligands, or redox state, produce characteristic changes in visible light absorption. These color changes are used clinically to assess Hb-O_2 saturation, metHb formation and the effects of Hb variants.

Within the Hb tetramer, each globin subunit binds the unlike chain through two distinct interfaces, termed α1β1 and α1β2 (Fig. 1D). Individual globin subunits form dimers through the extremely high affinity α1β1 interaction. Globin chain monomers are relatively unstable compared with dimers, with a tendency to form intracellular precipitates that damage erythrocytes, causing hemolytic anemia. Thus, mutations that impair the α1β1 interaction can cause erythrotoxicity by favoring the accumulation of monomeric subunits (Fig. 2C). The α1β2 interaction, which is lower affinity, mediates tetramerization. Oxygen binding destabilizes the α1β2 interaction, resulting in a transition of the quaternary structure from the "T" (tense, low affinity, deoxygenated) to "R" (relaxed, high affinity, oxygenated) state, which facilitates O_2 binding to additional subunits (Fig. 2A). This process causes cooperative O_2 binding, which is illustrated by the characteristic sigmoidal shape of the Hb-O_2 equilibrium curve (Fig. 2B). Cooperativity allows maximal O_2 release over relatively small drops in O_2 tension. Mutations within the α1β2 interaction region can alter the functional properties of Hb, mainly by perturbing O_2-binding characteristics (Fig. 2C, cyan spheres; Table 1).

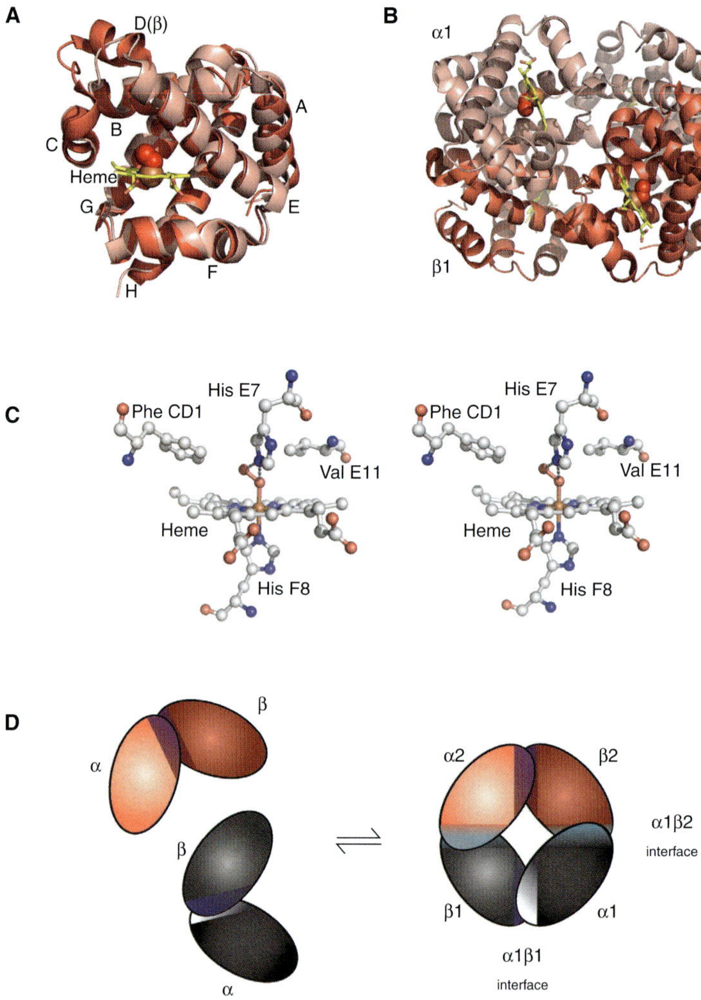

Figure 1. The structure of Hb. (*A*) The α (pink) and β (red) Hb subunits have conserved α-helical folds. Helices are labeled A–H from the amino terminus. The α subunit lacks helix D. (*B*) The high O_2 affinity R state quaternary structure of Hb with O_2 (red spheres) bound at all four heme sites (protoporphyrin-IX as yellow sticks, with central iron atom as orange sphere). (*C*) Stereo (wall-eye) diagram of the heme pocket of β showing the proximal (F8) and distal (E7) histidines and selected residues in the distal heme pocket that influence ligand binding and autoxidation. (*D*) Hb tetramer is assembled from two identical αβ dimers (shown in red and gray for clarity). In the tetramer, each subunit makes contact with the unlike chain through a high affinity dimerization α1β1 interface and a lower affinity α1β2 dimer–tetramer interface (cyan).

Cooperativity represents a general phenomenon termed allosteric regulation, in which effector molecules control the properties of enzymes or other proteins by binding to regions that are distinct from the active sites. Several allosteric regulators, in addition to O_2, influence the properties of Hb. For example, H^+ binds Hb to promote O_2 release in a process termed the Bohr effect (Perutz et al. 1984). In peripheral tissues, abundant CO_2 is taken up by red blood cells and metabolized by carbonic anhydrase to carbonic acid, resulting in the production of H^+ and consequent O_2 release. In contrast, relatively low CO_2 and high pH in the lung favors O_2 binding to Hb. Additionally, CO_2 forms carbamino adducts with the amino termini of both α and β

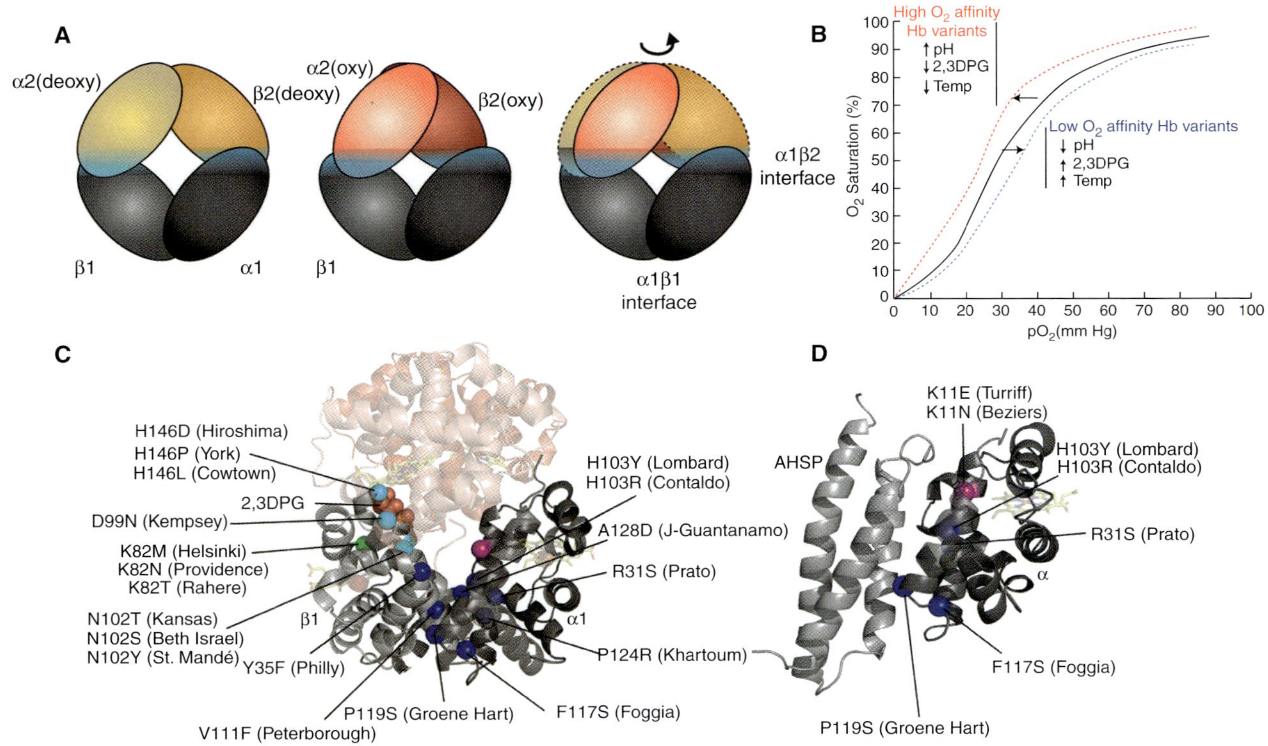

Figure 2. Hb variants with altered subunit interactions. (*A*) Conversion from the low O_2 affinity (deoxy, T state) to high O_2 affinity (oxy, R state) involves a relative rotation of the α1β1 and α2β2 dimers, with changes in contacts across the α1β2 (and α2β1) interface (cyan). In this cartoon, the α1β1 dimer is held stationary to reveal the relative motion of the α2β2 dimer in going from the deoxy (orange) to the oxy (red) states. (*B*) The sigmoidal shape of the Hb-O_2 saturation curve shows allosteric regulation by changes in pH, temperature, and 2,3DPG. These regulators, as well as Hb variants, influence the shape of the curve. High oxygen affinity variants, high pH, low 2,3DPG, or low temperature induce a "left shift" in the saturation curve (red line). Conversely, low oxygen affinity variants, low pH, high 2,3DPG, or high temperature induce a "right shift" (blue line). (*C*) Hb sequence variants at the allosteric α1β2 interface (cyan spheres) show an impaired response to O_2 binding. Some sequence variants disrupt binding to other allosteric regulators, e.g., substitutions at βK82 (green) disrupt interactions with 2,3DPG that normally stabilize the low O_2 affinity T state. Mutations that disrupt α1β1 (and α2β2) dimerization (blue spheres) increase the concentration of free monomers, which are unstable. (*D*) Some α mutations that disrupt binding to β may also disturb binding to the chaperone, AHSP. Other α variants, such as Turriff and Beziers (pink sphere), may inhibit only AHSP binding.

globins in vivo. These interactions produce minor effects on whole blood O_2 affinity compared to the impact of CO_2 on pH (Bunn and Forget 1986). The compound 2,3-diphosphoglycerate (2,3DPG), formed as a by-product of glycolysis and present at relatively high concentrations in red blood cells, is another important allosteric regulator of Hb. 2,3DPG binds and stabilizes T state Hb to facilitate O_2 release. Hemoglobin binding sites for H^+ and 2,3DPG have been identified (Perutz et al. 1969, 1984; Perutz 1970; Arnone 1972). Together, these interactions allow Hb to sense metabolic activity and modulate O_2 binding accordingly. This environmental sensory function can be altered by mutations that affect the affinity of Hb for its allosteric regulators (Fig. 2C, green sphere; Table 1).

Identification of Hb Variants and Their Clinical Implications

Many Hb variants are readily ascertained through physical examination and/or routine laboratory testing, which explains why so many have been discovered. Uncommonly, some variants are identified through evaluation of ill patients with severe anemia or clinically significant cyanosis. Many amino acid substitutions alter surface charge and are thus detected by electrophoresis or chromatography, techniques which are routinely performed on neonates in North America and Europe. Interestingly, a few Hb variants migrate similarly to HbA_{1c}, a glycosylated form of Hb that reflects long-term control of blood glucose levels in diabetic patients (Table 1) (reviewed in Little and Roberts 2009). In this way, specific Hb amino acid substitutions can artificially elevate HbA_{1c} measurement and interfere with diabetic management. Other Hb variants are clinically benign but produce obvious changes in skin color. For example, mutations that increase Hb-O_2 affinity typically stimulate erythropoietic drive by inhibiting O_2 tissue delivery, causing erythrocytosis associated with a ruddy complexion (Nathan et al. 2009). Mutations that reduce O_2 affinity produce a bluish hue to the skin (cyanosis) caused by abnormally high levels of deoxyHb. Mutations that favor oxidation of Hb iron to the met form (M-Hbs), also cause blue-tinged skin, whereas the blood itself appears brown. Studies of one family with congenital cyanosis led Horlein and Weber to describe the first known hemoglobinopathy, caused by the variant Hb-M Saskatoon (Horlein and Weber 1948). The M-Hb variant Iwate, which causes "black blood" or "hereditary nigremia," was first described more than 200 years ago in Japan (Shibata et al. 1960).

It is important to note that many Hb variants affecting skin color are not clinically damaging beyond their cosmetic effects. However, these conditions can mimic more life-threatening problems such as cardiopulmonary and myeloproliferative disorders, which must be excluded. Thus, patients with cyanotic or polycythemic Hb variants may mistakenly undergo unnecessary and potentially dangerous medical procedures. Historical examples include cyanotic patients undergoing surgery or catheterization for presumed heart defects and polycythemic patients receiving radioactive ^{32}P for presumed polycythemia vera (Steinberg et al. 2001; Nathan et al. 2009). As stated, "the primary reason for establishing the diagnosis [of M Hbs] is to prevent iatrogenic misadventures that might arise under the mistaken impression that the patient has a cardiac or pulmonary disorder" (Bunn and Forget 1986). Most Hb variants with altered O_2 affinity are relatively simple to diagnose through history, physical examination, and laboratory testing (Wajcman et al. 2001; Wajcman and Moradkhani 2011).

Globin Gene Synthesis Is Developmentally Regulated

Developmental regulation of the α-like and β-like globin gene families is of great medical significance (Sankaran and Orkin 2013). Hemoglobin F ($\alpha2\gamma2$) is the most highly expressed form during late fetal gestation. After birth, expression gradually switches to HbA ($\alpha2\beta2$) over several months. Thus, symptomatic mutations affecting α or γ globins are present prenatally or at birth, whereas the manifestations of β-globin mutations are typically delayed until a few months after birth. Interestingly, γ-globin gene mutations that are apparent at birth, most typically reflected by cyanosis or hemolytic anemia (e.g., Hb-F Toms River, sections on Selected Variants that Illustrate Important Aspects of Hemoglobin Biology and Variants that Affect Multiple Hemoglobin Functions, and Table 1), fade over a few weeks as Hb production switches from F (α2γ2) to A (α2β2).

Laboratory Testing for Hb Variants

In many countries, routine testing of all newborns is performed to identify common hemoglobinopathies such as some thalassemias and HbS. Isoelectric focusing or high-performance liquid chromatography (HPLC), the most commonly used tests, identify most structurally abnormal Hbs (Wajcman et al. 2001). In this way, many benign Hb variants are discovered incidentally. Clinical indications for laboratory testing to investigate potential Hb variants are listed in Table 2.

Specific laboratory tests to investigate Hb variants include:

a. *Physical methods of Hb separation.* These include electrophoretic and chromatographic techniques that examine the physical properties of α1β1 dimers or individual globin subunits. Specific standards, such as HbA, HbS, HbC, HbF, and HbA_2, are typically examined as controls. Hb variants may show altered migration in these assays. Historically, cellulose acetate and citrate agar electrophoresis were most commonly used to detect variant Hbs. Current clinical testing more typically uses isoelectric focusing and HPLC, which are more sensitive.

b. *Hb-O_2 binding curve.* This test, performed on whole red blood cells or hemolysate, indicates the percent (%) oxygenated Hb at a given O_2 partial pressure (Fig. 2B). Hemoglobin variants with an abnormally high O_2 affinity (see sections on Selected Variants that Illustrate Important Aspects of Hemoglobin Biology and High Oxygen Affinity Variants) will become saturated at lower O_2 pressures producing a "left-shifted" O_2 equilibrium curve, whereas mutations that reduce O_2 affinity (see sections on Selected Variants that Illustrate Important Aspects of Hemoglobin Biology and Low Oxygen Affinity Variants) will cause the opposite "right shift." Determination of Hb-O_2 affinity responses to allosteric regulators, particularly 2,3DPG or H^+ (pH changes), can provide insight into the structural causes of the observed phenotypes. Unfortunately, few clinical laboratories currently offer this assay.

c. *Visible wavelength spectroscopy.* Hemoglobin variants with amino acid substitutions in the heme pocket affect visible light absorbance. For example, M-type Hbs show characteristic spectra that can distinguish them from methemoglobinemia caused by an enzyme deficiency in the metHb reductase system (Bunn and Forget 1986; Ganz and Nemeth 2012; Dailey and Meissner 2013). Pulse oximetry is a noninvasive spectrophotometric test that measures absorbance ratios at specific wavelengths for oxygenated (660 nm) and deoxygenated (940 nm) blood (Tremper and Barker 1989). This can produce confusing and sometimes misleading results in patients with variant Hbs that show unique light absorbance properties (Verhovsek et al. 2010). In these cases, analysis of arterial blood O_2 concentration may be required to rule out hypoxia. Analyzing these variant Hbs using light absorbance throughout the full visible wavelength spectrum can provide useful diagnostic information.

Table 2. Clinical indications for laboratory testing to diagnose Hb variants

Indications for hemoglobin testing
Routine newborn testing for common hemoglobinopathies (i.e., HbS, HbC, thalassemias)
Cyanosis with adequate arterial oxygenation and no apparent cardiopulmonary disease
Erythrocytosis with normal or elevated erythropoietin levels
Unexplained hemolytic anemia
Unexplained thalassemia phenotype
Family history consistent with an Hb variant

d. *Hemoglobin stability testing.* Typically, Hb stability is impaired in variants that are associated with hemolytic anemia. Hemoglobin stability tests measure the propensity for Hb to denature on exposure to various stresses including heat (Carrell and Kay 1972), isopropanol (Bender et al. 1981), mechanical agitation (Asakura et al. 1975), and zinc acetate (Roth et al. 1976). The Heinz body test uses supravital stains, such as methylene blue or crystal violet, to detect aggregated globins within erythrocytes (Eisinger et al. 1985).

e. *Specialized testing.* Mass spectrometry analysis of patient hemolysate and DNA sequencing of globin genes are specialized confirmatory tests to identify amino acid and nucleotide alterations associated with suspected Hb variants (Wajcman et al. 2001; Wajcman and Moradkhani 2011). DNA sequencing may readily elucidate the presence of an Hb variant. However, additional biochemical and structural studies, including those discussed in this section, are required to determine how the variant affects Hb function. Crystallographic analysis is the highest resolution approach to determine the effects of globin amino acid substitutions on molecular structure. Crystallography has been used historically as a research tool to assess the effects of some interesting Hb variants (see, for example, Pulsinelli et al. 1973; Perutz et al. 1984).

SELECTED VARIANTS THAT ILLUSTRATE IMPORTANT ASPECTS OF HEMOGLOBIN BIOLOGY

Unstable Variants

Unstable variants frequently cause congenital Heinz body hemolytic anemia detected by laboratory screening and clinical symptoms (see sections on Basic Principles and Laboratory Testing for Hb Variants and Table 2). Mutations that alter any step in globin processing, including subunit folding, heme interaction, dimerization, or tetramerization, can destabilize Hb. Bunn and Forget note five general mechanisms that destabilize Hbs: amino acid substitutions within the heme pocket, disruption of secondary structure, substitution in the hydrophobic interior of the subunit, amino acid deletions, and elongation of the subunit (Bunn and Forget 1986).

More than 75% of Hb is α helical (Perutz et al. 1960; Park et al. 2006). This structure is particularly susceptible to disruptions by proline substitutions (Levitt 1981). For example, in Hb Brockton (β138 [H16] Ala > Pro) the substituted proline disrupts intermolecular hydrogen bonding between β138Ala and β134Val in helix H (Russu and Ho 1986; Moo-Penn et al. 1988). This produces an unstable variant with a propensity to precipitate and aggregate, thereby damaging erythrocytes and predisposing to hemolysis. Hb Brockton does not show altered O_2 binding affinity or electrophoretic mobility shifts. This variant was identified by HPLC analysis of patient globin chains and its altered X-ray crystallography diffraction pattern shows local disruption of helix H (Moo-Penn et al. 1988).

Mutations at the α1β1 interface can cause hemolytic anemia by inhibiting heterodimer formation, favoring the accumulation of free globin subunits, which themselves are unstable, particularly α chains (Fig. 2C, blue spheres). Examples include Hb Philly (β35 [C1] Tyr > Phe) (Rieder et al. 1969), Hb Peterborough (β111 [G13] Val > Phe) (King et al. 1972), Hb Stanmore (β111 [G13] Val > Ala) (Como et al. 1991), and Hb J-Guantanamo (β128 [H6] Ala > Asp) (Martínez et al. 1977). Hb Khartoum (β124 [H2] Pro > Arg) contains a substitution at the α1β1 interface that is mildly destabilizing in vitro, but does not cause clinical symptoms (Clegg et al. 1969; Argos et al. 1979).

Interestingly, some α-globin (*HBA*) gene mutations affecting the α1β1 interface may also destabilize free α chains by inhibiting their binding to α-hemoglobin stabilizing protein (AHSP), an erythroid molecular chaperone that facilitates Hb assembly (Fig. 2D, blue spheres) (reviewed in Weiss et al. 2005; Favero and Costa 2011). These α-globin variants include Hbs Prato (α1 or α2 31 [B12] Arg > Ser) (Marinucci et al. 1979), Lombard (α2 103 [G10] His > Tyr) (Hoyer et al. 2002), Contaldo (α1 or α2 103 [G10] His > Arg) (Sciarratta et al. 1984), Foggia (α2 117 [GH5] Phe > Ser) (Lacerra et al. 2008), Groene Hart (α1 119 [H2] Pro > Ser) (Harteveld et al. 2002; Vasseur-Godbillon et al. 2006; Giordano et al. 2007; Vasseur et al. 2009), and others (Wajcman et al. 2008; Yu et al. 2009). Naturally occurring α-globin variants with amino acid substitutions at position 99 including Hb Turriff (α1 or α2 99 [G6] Lys > Glu) (Langdown et al. 1992) and Hb Beziers (α1 99 [G6] Lys > Asn) (Lacan et al. 2004) bind β globin normally but show impaired interaction with AHSP and may be mildly destabilizing (Fig. 2D, violet spheres) (see also Yu et al. 2009; Khandros et al. 2012; Mollan et al. 2012). Antitermination mutations can also destabilize α globin in part by impairing its binding to AHSP (Turbpaiboon et al. 2006).

Hyperunstable Hb variants precipitate shortly after synthesis and are not incorporated into Hb tetramers. These ephemeral proteins are difficult to isolate. In this case, electrophoresis may be falsely negative due to the rapid turnover of these variants, making DNA sequencing a critical diagnostic test. Affected patients show a dominantly inherited "inclusion body thalassemia" resulting from both the precipitated variant globin and consequent chain imbalance with accumulation of the unaffected globin, which is unstable in its free form (Stamatoyannopoulos et al. 1974). Patient erythrocytes typically display abnormal morphology with microcytosis, hypochromia, moderate to severe anisopoikilocytosis, basophilic stippling, and inclusions that may be become particularly prominent following splenectomy (Steinberg et al. 2001). Weatherall, Thein, and colleagues characterized several hyperunstable mutations in exon 3 of the β-globin gene (Thein et al. 1990). All of the mutations were frameshifts or nonsense codons that produced relatively long (>120 amino acid) proteins with carboxy-terminal truncations. The investigators proposed that truncated globins causing dominantly inherited thalassemia are long enough to bind heme posttranslationally, which rendered them relatively resistant to proteolytic degradation, allowing for subsequent precipitation of heme-containing aggregates detected as Heinz bodies. Missense mutations also cause hyperunstable Hb variants. Hb Hirosaki (α2 43 [CE1] Phe > Leu) was discovered in a family with hemolytic anemia (Ohba et al. 1975; Tanaka et al. 2005). After several tests failed to identify a soluble variant Hb within erythrocytes, DNA sequencing was used to characterize the mutation. Hb Terre Haute (β106 [G8] Leu > Arg) is another hyperunstable variant associated with a severe Heinz body hemolytic anemia and globin chain imbalance (Coleman et al. 1991). In initial studies of patient erythroid cells, performed in 1979, abnormal Hb tetramers were not detected and peptide mapping of radiolabeled nascent globins identified a β112 (G14) Cys > Arg substitution, originally termed Hb Indianapolis (Adams et al. 1979). However, the β112 Cys > Arg mutation was subsequently identified in unrelated individuals with much less severe disease. In 1991, reevaluation of the original pedigree by DNA analysis discovered a β106 Leu > Arg mutation, which was renamed Hb Terre Haute (Coleman et al. 1991). Most likely, incomplete tryptic cleavage of the abnormal β-globin peptide in the earlier studies led to misidentification of the causative mutation. This work reflects the interesting historical point that many Hb variants were identified through laborious and technically challenging protein studies performed some time ago, before DNA sequence analysis of patient globin genes was feasible. Reevaluation of these mutations through genetic testing has yielded some interesting surprises (see also the discussion on Hb Bristol-Alesha and the section on Variants that Affect Multiple Hemoglobin Functions).

High Oxygen Affinity Variants

Hemoglobin variants with increased O_2 affinity cause erythrocytosis by stimulating erythropoietic drive (see Hebbel et al. 1978 for a description of the associated physiology). This commonly results from amino acid substitutions that stabilize the R (high O_2 affinity) state relative to the T (low O_2 affinity) state and/or inhibit responses to environmental allosteric regulators that stimulate O_2 release, including H^+ (Bohr effect) or 2,3DPG (see the sections on

Basic Principles and Hemoglobin Synthesis, Structure, and Function). Because T to R state transitions are mediated largely through α1β2 interactions, high affinity variants frequently result from substitutions that alter this interface (Fig. 2C, cyan spheres). For example, the amino acid replacement in Hb Kempsey (β99 [G1] Asp > Asn) perturbs α1β2 interactions by preventing the formation of a hydrogen bond between β99 Asp and α42 Tyr, which normally stabilizes the deoxygenated low O_2 affinity T state (Fig. 3) (Reed et al. 1968; Lindstrom et al. 1973; Bunn et al. 1974). This structural change shifts quaternary equilibrium toward the oxygenated R form, which impairs O_2 release to peripheral tissues and thus increases erythropoietic drive. The carboxyl termini of globin chains are also involved in α1β2 interactions that stabilize the low O_2 affinity T state and numerous substitutions within these regions cause high O_2 affinity variants. In addition, β146 His at the carboxyl terminus contributes significantly to the Bohr effect by forming a salt bridge with β94 Asp (Perutz et al. 1984). This interaction is disrupted in several high O_2 affinity variants with β146 substitutions: Hb Hiroshima (β146 [HC3] His > Asp) (Hamilton et al. 1969; Perutz et al. 1971; Imai et al. 1972; Olson et al. 1972), Hb York (β146 [HC3] His > Pro) (Bare et al. 1976), and Hb Cowtown (β146 [HC3] His > Leu) (Fig. 2C) (Schneider et al. 1979; Perutz et al. 1984). These variants show reduced Bohr effect with impaired release of O_2 under acidic conditions.

Several high affinity Hb variants are caused by substitutions that inhibit interaction with 2,3DPG, which normally binds globin chains to stimulate O_2 release (Fig. 2C). For example, Hb Rahere (β82 [EF6] Lys > Thr) replaces an

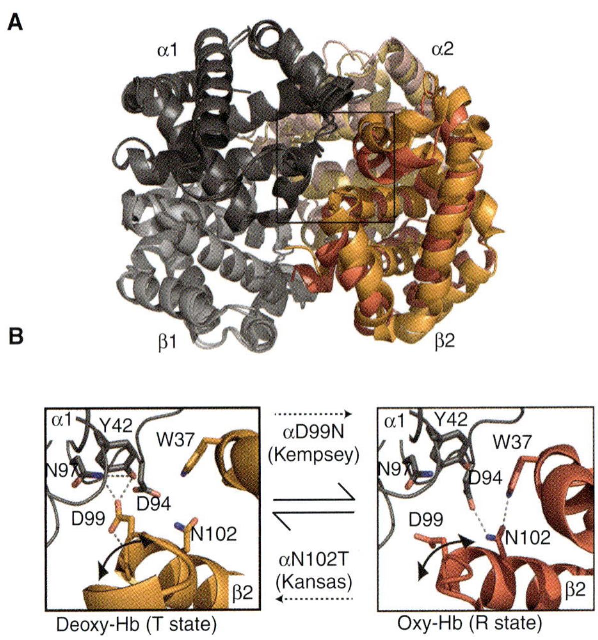

Figure 3. Hb variants that affect allosteric regulation. (*A*) The α2 and β2 subunits indicated on the *right* side of the figure show an overlay of the R-state (red/pink) and T-state (orange/light orange) quaternary structures of Hb. The allosteric α1β2 interface is boxed. (*B*) Detail of the α1β2 interface in the deoxy T state (β2 chain in orange, PDB 2DN2) and oxy R state (β2 chain in red, PDB 2DN1) showing selected H-bonding interactions. Note that the two H-bonding networks use nonoverlapping sets of side chains, hence mutations in these residues affect only one state. Mutation of Asp99 to Asn (Hb Kempsey) compromises electrostatic interactions in the deoxygenated state, thereby favoring the R state and causing impaired O_2 release. Mutation of Asn102 to Thr (Hb Kansas) abrogates interactions with Asp94, favoring the T state and O_2 binding inhibition.

invariant lysine in the 2,3DPG binding site of β globin, thereby reducing the affinity for this allosteric regulator (Lorkin et al. 1975; Sugihara et al. 1985). Consequently, Hb Rahere shows blunted O_2 release in response to added 2,3DPG in vitro. In vivo, O_2 release in peripheral tissues is inhibited, resulting in elevated blood Hb levels. Similarly, Hb Providence (β82 [EF6] Lys > Asn) (Bonaventura et al. 1976; Moo-Penn et al. 1976) and Hb Helsinki (β82 [EF6] Lys > Met) (Ikkala et al. 1976; Charache et al. 1977) are high O_2 variants caused by different amino acid substitutions at the 2,3DPG binding site on β82.

Low Oxygen Affinity Variants

Low O_2 affinity Hb variants typically present with cyanosis. In general, these variants are caused by globin amino acid substitutions that tip the quaternary equilibrium of Hb tetramers from the high affinity oxygenated R state to the low affinity deoxygenated T state; more or less the opposite of what occurs for high O_2 affinity variants (see the sections on Selected Variants that Illustrate Important Aspects of Hemoglobin Biology and High Oxygen Affinity Variants). This does not inhibit Hb-O_2 release in tissue capillaries, but rather, interferes with Hb-O_2 uptake if the P_{50} has increased to ≥ 50 mm Hg. Paradoxically, low O_2 affinity Hb variants can be associated with mild anemia thought to be caused by increased O_2 tissue delivery with reduced erythropoietic drive (Stamatoyannopoulos et al. 1969). In addition, some low O_2 affinity mutants are unstable and therefore associated with not only cyanosis but also Heinz body hemolytic anemia (see the sections on Selected Variants that Illustrate Important Aspects of Hemoglobin Biology and Unstable Variants).

Several low affinity variants involve replacements at the α1β2 interface, which plays an important role in Hb cooperativity. Hb Kansas (β102 [G4] Asn > Thr) is a well-studied low O_2 affinity variant (Fig. 3B) (Reissmann et al. 1961; Bonaventura and Riggs 1968; Gibson et al. 1973; Riggs and Gibson 1973). Affected individuals are markedly cyanotic, although clinically well. Replacement of Asn102 at the α1β2 interface inhibits the formation of a hydrogen bond with Asp94 that normally stabilizes the oxygenated R structure. A similar mechanism causes low O_2 affinity in two other Hb variants through different substitutions of the same amino acid residue (B102 [G4] Asn) in Hb Beth Israel (β102 [G4] Asn > Ser) (Nagel et al. 1976) and Hb St. Mandé (β102 [G4] Asn > Tyr) (Arous et al. 1981; Poyart et al. 1990).

Methemoglobin ("M-Type") Variants

Hemoglobin iron must be in its reduced (Fe^{2+}, ferrous) state to bind O_2. Moreover, oxidized (Fe^{3+}, ferric, met) Hb is intrinsically unstable with a tendency to release heme. Hemoglobin reduction is maintained through intrinsic features of the Hb protein and extrinsic antioxidant pathways within red blood cells (see the sections on Basic Principles and Hemoglobin Synthesis, Structure, and Function). Exposure to oxidant drugs or toxins, genetic alterations in erythroid metHb reductase enzyme systems (Ganz and Nemeth 2012), or globin chain variants can predispose to methemoglobinemia. These disorders present as "pseudocyanosis," (i.e., low Hb-O_2 saturation), despite adequate arterial oxygenation. Detailed in vitro analyses of the red cell and isolated Hb samples can usually distinguish wild-type metHb resulting from toxins or defective reductase systems and M-type Hb variants that are predisposed to spontaneous oxidation (Bunn and Forget 1986; Steinberg et al. 2001; Nathan et al. 2009). For example, various M-Hbs show characteristic visible absorbance spectra.

Globin variants associated with metHb formation are typically caused by amino acid substitutions within the heme pocket. For example, four different M-Hbs occur when tyrosine replaces the α- or β-globin proximal or distal histidine residues that interact with heme (Fig. 1C) (reviewed in Adachi et al. 2011). In Hb M-Iwate (α1 or α2 87 [F8] His > Tyr), the proximal histidine is replaced by tyrosine (Fig. 4A), which deprotonates and coordinates to the heme iron (Fig. 4B) (Konigsberg and Lehmann 1965; Shimizu et al. 1965). Ferric heme, bound through the native His F8, is readily reduced by metHb reductase (Fig. 4C). Tyrosine (F8) coordination stabilizes the oxidized ferric state and decreases reactivity with metHb reductases. This interaction also distorts the position of heme and helix F within the altered α subunits. In native Hb, movement of the proximal His F8 and F helix away from the heme group stabilizes the deoxygenated T state and reduces the O_2 affinity of the native β subunit partners. Hence, substitution of the normal His F8 side chain for a longer Tyr F8 (Fig. 4A) also stabilizes the deoxygenated T state and reduces O_2 affinity of the native β subunit in Hb M-Iwate (Nagai et al. 2000; Jin et al. 2004). These biochemical and structural alterations underlie the lack of cooperativity and severe cyanosis in patients with Hb M-Iwate, in which metHb levels can exceed 20% (normal <2%) (Ameri et al. 1999). In contrast, Hb M-Saskatoon (β63 [E7] His > Tyr) replaces the distal His with Tyr (Fig. 4D) (Horlein and Weber 1948; Hayashi et al. 1966). In this variant, the protonated form of mutant Tyr can bind ferric heme iron to generate a hexacoordinate structure that is relatively easily reduced by cellular metHb reductases (Fig. 4E). Consequently, patients with Hb M-Saskatoon have lower levels of circulating oxidized Hb compared with those with Hb M-Iwate. Comparative studies of these patients and their variant M-Hbs have contributed greatly to understanding the biochemical properties of the heme iron, including its interactions with various ligands and nearby amino acids such as the proximal and distal histidines.

Globin Chain Elongation Mutants

Antitermination and frameshift mutations that add irrelevant amino acids to the carboxyl terminus of globin proteins produce interesting variants that can damage erythrocytes (Nathan et al. 2009). The most clinically significant

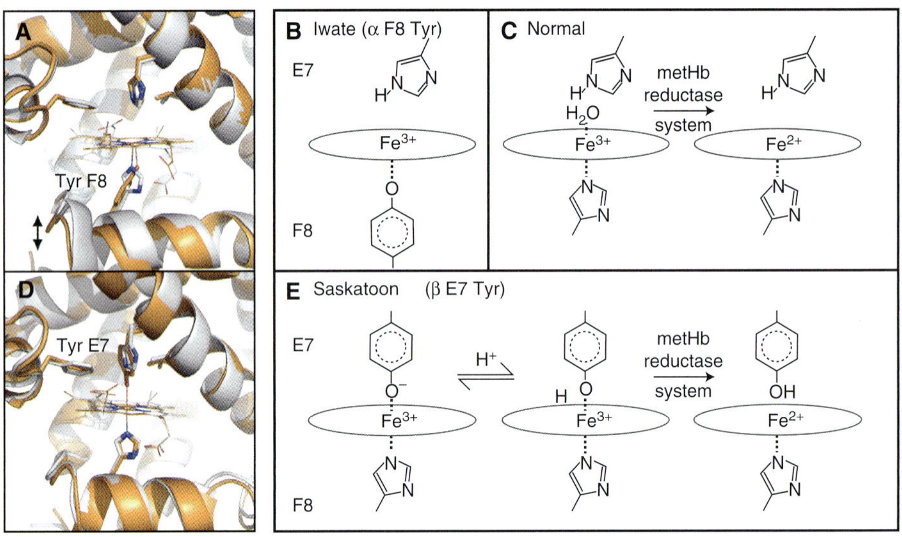

Figure 4. Examples of M-type hemoglobins. (*A*) The heme pocket of wild-type Mb (gray) and the F8 His > Tyr mutation (orange, PDB 1HRM), which serves as a model for Hb M-Iwate. Effects on the protein fold are to increase the distance from the heme to the F helix, recapitulating features of the deoxy-α (T state) structure. (*B*) In Hb M-Iwate, α 87 Tyr F8 is deprotonated and favors Fe^{3+} oxidation state, resulting in rapid autoxidation. This ferric form is not reduced by metHb reductase. (*C*) With the normal His F8 present, ferric heme is readily reduced. (*D*) Substituting the distal His E7 side chain in Mb for a larger Tyr E7 (orange, PDB 1MGN), as also occurs in Hb Saskatoon, brings the Tyr hydroxyl group within binding distance of the iron, forming a hexacoordinate iron site. (*E*) In Hb Saskatoon the hexacoordinate ferric iron in the effected β chains *can* be reduced by metHb reductase, possibly owing to a transient protonation of Tyr E7. Note that Mb is used as a model for Hb subunits in *A* and *D*, whereas *B* and *E* are based on Hb spectroscopy.

example is Hb Constant Spring (α2 142 [HC3] Stop > Gln), caused by an antitermination mutation at the α2 stop codon (Clegg et al. 1971; Efremov et al. 1971; Milner et al. 1971; Clegg and Weatherall 1974). This elongates the protein by 31 amino acids, generating an unstable protein that is relatively underrepresented in hemolysates, but can be detected by physical methods (see sections on Basic Principles, Laboratory Testing for Hb Variants, and *Physical methods of Hb separation*). In addition, Hb Constant Spring mRNA is rapidly degraded in developing erythrocytes, owing to ribosomal entry into the 3′UTR, causing displacement of RNA-bound stabilizing proteins with a resultant thalassemia syndrome (Hunt et al. 1982; Derry et al. 1984; Weiss and Liebhaber 1994; Morales et al. 1997).

Hb Constant Spring contributes to α-thalassemia syndromes, particularly when combined with two α-globin deletional alleles ($-/\alpha^{CS}\alpha$), which produces a severe form of HbH disease (Viprakasit and Tanphaichitr 2002). Isolated Hb Constant Spring, in its heterozygous ($\alpha\alpha/\alpha^{CS}\alpha$) or homozygous ($\alpha^{CS}\alpha/\alpha^{CS}\alpha$) forms, results in more severe anemia than occurs when the same α alleles are deleted ($\alpha\alpha/-\alpha$) or ($-\alpha/-\alpha$) (Schrier et al. 1997). This is due to the cytotoxic effects of the unstable Constant Spring protein. Although most common in Southeast Asia, Hb Constant Spring is increasingly identified in other geographic regions, largely through global migration (Lal et al. 2011). In fact, it was first discovered in a Chinese family living in Constant Spring, Jamaica.

One example of a β-globin chain elongation mutant is Hb Cranston (β 145 [HC3] +CT) (Bunn et al. 1975). This mutation introduces a frameshift at the normal stop codon to generate a β chain that is extended by 11 amino acids. This results in an unstable Hb tetramer with high O_2 affinity and diminished cooperativity (McDonald et al. 1980; Shaeffer et al. 1980). Affected patients show compensated hemolytic anemia with the variants accounting for 30% of total Hb in the hemolysate. Interestingly, the structure of Hb Cranston was investigated simultaneously with studies to determine the β-globin mRNA 3′ untranslated sequence (Forget et al. 1975). Cross-comparison of the protein and mRNA sequencing data allowed Bunn, Forget, and colleagues to more rapidly define normal β-globin gene structures and ascertain that the Hb Cranston mutation likely arose by nonhomologous crossover of two normal β-globin genes.

Variants That Affect Multiple Hemoglobin Functions

Not surprisingly, amino acid substitutions within critical regions of globin proteins can produce multiple effects. For example, HbE (β26 [B8] Glu > Lys), a common variant in Southeast Asia, contains an amino acid substitution that renders β chains mildly unstable in vitro with minimal clinical significance (Frischer and Bowman 1975; Huisman 1997; Rees et al. 1998; see also Musallam et al. 2012). However, this mutation also creates an alternate splice site in the β-globin mRNA, leading to reduced synthesis of productive transcripts with resultant thalassemia (Orkin et al. 1982). HbE is particularly deleterious when coinherited with a more severe β-thalassemic allele, which happens commonly in Southeast Asia.

Mutations that alter the heme pocket commonly produce multiple biochemical effects. For example, deletion or substitution of the conserved Phe residue at the CD1 helical region in the heme pocket markedly destabilizes the affected globin and also alters its O_2 affinity. Thus, Hb Bruxelles (β42 [CD1] Phe > 0) (Blouquit et al. 1989; Griffon et al. 1996), Hb Warsaw (β42 [CD1] Phe > Val) (Honig et al. 1990), Hb Hammersmith (β42 [CD1] Phe > Ser) (Dacie et al. 1967), and Hb Buccuresti-Louisville (β42 [CD1] Phe > Leu) (Bratu et al. 1971; Keeling et al. 1971) cause both congenital Heinz body hemolytic anemia and cyanosis. These combined effects arise from severely reduced cooperativity, rapid rates of autoxidation and hemin loss, and unfolding of these unstable globin variants (Griffon et al. 1996).

Another interesting heme pocket variant is Hb Zurich (β63 [E7] His > Arg) in which the distal His is replaced by Arg (Huisman et al. 1961; Bachmann and Martihr 1962; Frick et al.

1962; Tucker et al. 1978; Phillips et al. 1981; Springer et al. 1989). The large, highly polar variant His side chain swings out of the distal heme pocket, and the positively charged guanidino group forms a salt bridge with a heme propionate (Fig. 5A). This results in an enlarged ligand-binding pocket, destabilizing O_2 binding and causing iron autoxidation via exposure to water. Affected individuals show increased sensitivity to oxidant agents, including sulfa drugs, which more easily enter the expanded heme pocket. Loss of the distal histidine markedly decreases O_2 affinity but has little effect on carbon monoxide (CO) binding. As a result, individuals with Hb Zurich tend to have supranormal levels of CO-Hb, which ironically protects the heme iron from oxidation and the globin from denaturation. Affected subjects who are cigarette smokers accumulate even higher levels of CO-Hb, which tends to protect against hemolysis. Thus, "the pathology of a mutant protein is ameliorated by a normally toxic pollutant" (Bunn and Forget 1986).

Two other recently identified Hb variants illustrate how multiple biochemical defects produce unique phenotypes. Hb Jamaica Plain

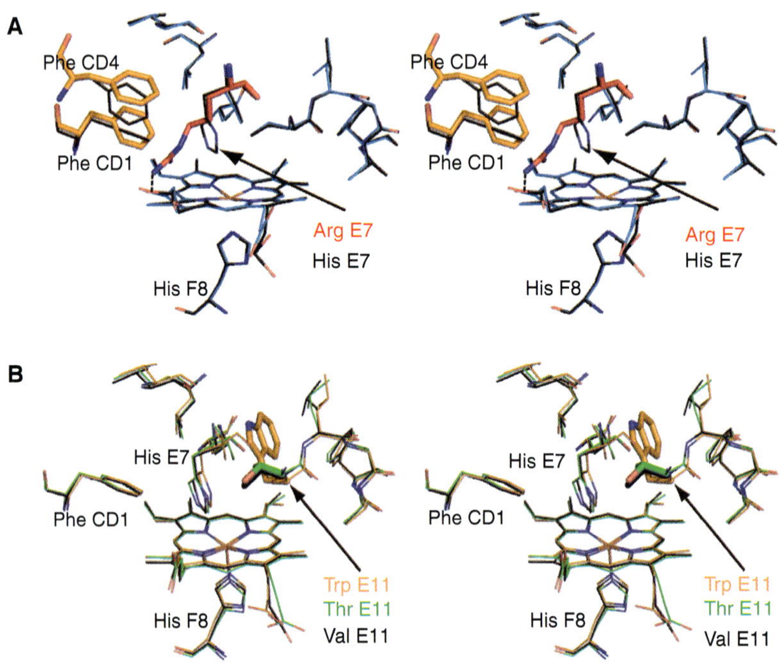

Figure 5. Hb variants with amino acid changes in the heme pocket. (*A*) Stereo diagram of a model of the deoxy Hb Zurich β heme pocket (blue) overlaid with the wild-type β heme pocket (black, PDB 2DN2). This model of Hb Zurich is based on the structures solved by Phillips et al. (1981) and Tucker et al. (1978). It was generated with the macromolecular modeling program Coot (Emsley et al. 2010) by mutating the distal His of deoxy-β (PDB 2DN2) to Arg. The distal Arg E7 (red) is oriented toward the CD corner, disturbing the position of Phe CD1 and Phe CD4 (orange). The heme pocket entrance is much wider allowing increased access to ligands. However, unlike normal His E7, mutant Arg is unable to stabilize bound O_2 via hydrogen bonding. (*B*) Stereo diagram showing the structural changes associated with substitution of β Val E11. The wild-type structure carrying the branched hydrophobic side chain Val (black bonds, PDB 2DN2) is overlaid with structures carrying the largest aromatic side chain Trp E11 (orange, PDB 101K) or a polar side chain Thr E11 (green PDB, 1HDB). It is clear that no major backbone or side chain repacking occurs. Thus, mutations in this position are likely to have specific effects in changing the volume of the heme pocket accessible to solvent or diatomic ligands and the electrostatic properties of the pocket. These changes will manifest as differences in O_2 binding, ligand selectivity, autoxidation, and heme loss.

 Cite this article as *Cold Spring Harb Perspect Med* doi: 10.1101/cshperspect.a011858

(β6 [A3] Glu > Val and β68 [E12] Leu > Phe) contains two defects in the same β chain, a β6 Glu to Val substitution that causes sickle cell anemia (Serjeant 2013) and a β68 amino acid substitution that reduces O_2 affinity, probably by destabilizing the oxygenated conformation through steric effects introduced into helix E (Geva et al. 2004). The affected patient, who was heterozygous for the mutant allele, showed symptoms of sickle cell anemia that were precipitated by infection or airplane travel. Thus, an amino acid substitution that reduces O_2 affinity exacerbates the effects of a sickling mutation in the same globin chain.

Several other Hb variants modulate the severity of sickle cell anemia (see also Cao and Kan 2013; Serjeant 2013). For example, γ globin inhibits polymerization of HbS (Nagel et al. 1979). This effect is attributable to differences in several amino acid residues compared with the corresponding β chain, including γ80 and γ87 (Adachi and Asakura 1979; Nagel et al. 1979; Adachi et al. 1996). Hb D-Ibadan (β87 [F3] Thr > Lys), which introduces a lysine residue at the β87 position, is presumed to have decreased interaction with the mutant Val residue at HbS β6 (Watson-Williams et al. 1965; Nagel et al. 1979). Thus, Hb D-Ibadan inhibits HbS polymerization. In contrast, Hb Quebec-Chori (β87 [F3] Thr > Ile) was identified in a compound heterozygous patient with mild to moderately severe sickle cell anemia (Witkowska et al. 1991). The introduction of an isoleucine at β87 causes Hb Quebec-Chori to promote deoxygenated HbS polymerization.

Hb Bristol-Alesha (β67 [E11] Val > Met) (Molchanova et al. 1993; Rees et al. 1996) and Hb Toms River (γ67 [E11] Val > Met) (Crowley et al. 2011), which contain analogous amino acid substitutions in β and γ chains, respectively, represent interesting globin variants with multiple biochemical defects. Hb Bristol-Alesha was initially observed in patients with moderately severe hemolytic anemia. Studies of the mutant protein in patient erythrocytes revealed a β67 Val > Asp substitution, predicted to render the protein unstable by introducing a highly charged polar residue into the hydrophobic heme pocket. However, subsequent DNA analysis of the same patient identified a Val > Met codon substitution (Rees et al. 1996). The investigators concluded that the mutant Met residue was converted to Asp posttranslationally, probably through oxidative mechanisms. More recently, the analogous variant was identified in fetal (γ) globin (Hb Toms River) (Crowley et al. 2011). The affected patient was a newborn presenting with both cyanosis and anemia. DNA testing revealed the codon change (Val > Met at E11). Mass spectrometry of patient hemolysate indicated a mixture of variant γ globins containing either Met or Asp at position E11. Although no structural studies have been performed with Hb chains carrying Met or Asp E11, structures with polar (Thr) or large aromatic (Trp) substitutions are available. These indicate that a range of amino acids can be accommodated without gross changes in the heme pocket structure (Fig. 5B). Instead, altered steric and electrostatic interactions with the distal His and diatomic ligands entering the heme pocket are likely to be functionally significant. Biochemical studies indicated that the Hb Toms River Met substitution produced a stable, low O_2 affinity variant γ globin, causing cyanosis. Its gradual posttranslational conversion to Asp destabilized the molecule, causing hemolytic anemia. This provides an example of how posttranslational modifications of variant globins can modify phenotypes. The reason that Hb Bristol-Alesha causes predominantly hemolytic anemia whereas Hb Toms River causes mainly cyanosis probably reflects different rates of Met conversion to Asp in the variant globin chains.

CONCLUDING REMARKS

More than 1000 Hb variants are known to exist (Globin Gene Server; Hardison et al. 2002; Giardine et al. 2011). These are mainly missense mutations that destabilize Hb, alter Hb-O_2 affinity, or most commonly, alter Hb function minimally. Moreover, variants that do alter Hb biochemistry are rarely life threatening or health compromising. Nonetheless, studies of these variants have been of great benefit to science and medicine for two main reasons. First, identification of Hb gene mutations as a cause for cyanosis, erythrocytosis, or mild hemolysis in otherwise healthy patients provides reassurance and minimizes additional diagnostic procedures, sparing expense and risk. Second, efforts to understand how Hb variants produce their structural, biochemical, and clinical effects has generated important insights into red blood cell function and also created general paradigms for the study of protein biology. Despite hundreds of studies over more than 50 years, new Hb variants continue to emerge, yielding new insights into this important molecule.

ACKNOWLEDGMENTS

We thank Drs. Franklin Bunn, Kazuhiko Adachi, and John Olson for comments on the manuscript. Hemoglobin research in M.J.W.'s laboratory is funded through National Institutes of Health (NIH) grants DK061692, HL087427, and P30DK090969. The authors declare no competing financial interests.

REFERENCES

**Reference is also in this collection.*

Adachi K, Asakura T. 1979. The solubility of sickle and non-sickle hemoglobins in concentrated phosphate buffer. *J. Biol Chem* **254:** 4079–4084.

Adachi K, Pang J, Konitzer P, Surrey S. 1996. Polymerization of recombinant hemoglobin F γE6V and hemoglobin F γE6V, γQ87T alone, and in mixtures with hemoglobin S. *Blood* **87:** 1617–1624.

Adachi K, Surrey S, Nagai M. 2011. Hemoglobinopathies due to amino acid mutation/deletion: HbS and HbM. In *Hemoglobin: Recent developments and topics*, pp. 179–210. Research Signpost, Kerala, India.

Adams J, Boxer L, Baehner R, Forget B, Tsistrakis G, Steinberg M. 1979. Hemoglobin Indianapolis (β112 [G14] arginine). An unstable β-chain variant producing the phenotype of severe β-thalassemia. *J Clin Invest* **63:** 931–938.

Ameri A, Fairbanks V, Yanik G, Mahdi F, Thibodeau S, McCormick D, Boxer L, McDonagh K. 1999. Identification of the molecular genetic defect of patients with methemoglobin M-Kankakee (M-Iwate), α87 (F8) His→Tyr: Evidence for an electrostatic model of αM hemoglobin assembly. *Blood* **94:** 1825–1826.

Argos P, Rossman MG, Grau UM, Zuber H, Frank G, Tratschin JD. 1979. Thermal stability and protein structure. *Biochemistry* **18:** 5698–5703.

Arnone A. 1972. X-ray diffraction study of binding of 2,3-diphosphoglycerate to human deoxyhaemoglobin. *Nature* **237:** 146–149.

Arous N, Braconnier F, Thillet J, Blouquit Y, Galacteros F, Chevrier M, Bordahandy C, Rosa J. 1981. Hemoglobin Saint Mandé [β102 (G4) Asn→Tyr]: A new low oxygen affinity variant. *FEBS Lett* **126:** 114–116.

Asakura T, Adachi K, Shapiro M, Friedman S, Schwartz E. 1975. Mechanical precipitation of hemoglobin köln. *Biochim Biophys Acta* **412:** 197–201.

Bachmann F, Marti HR. 1962. Hemoglobin Zürich. II. Physicochemical properties of the abnormal hemoglobin. *Blood* **20:** 272–86.

Bare GH, Bromberg PA, Alben JO, Brimhall B, Jones RT, Mintz S, Rother I. 1976. Altered C-terminal salt bridges in haemoglobin York cause high oxygen affinity. *Nature* **259:** 155–156.

Bender JW, Adachi K, Asakura T. 1981. Precipitation of oxyhemoglobins A and S by isopropanol. *Hemoglobin* **5:** 463–474.

Blouquit Y, Bardakdjian J, Lena-Russo D, Arous N, Perrimond H, Orsini A, Rosa J, Galacteros F. 1989. Hb Bruxelles: α 2A β (2)41 or 42(C7 or CD1)Phe deleted. *Hemoglobin* **13:** 465–474.

Bonaventura J, Riggs A. 1968. Hemoglobin Kansas, a human hemoglobin with a neutral amino acid substitution and an abnormal oxygen equilibrium. *J. Biol Chem* **243:** 980–991.

Bonaventura J, Bonaventura C, Sullivan B, Ferruzzi G, McCurdy PR, Fox J, Moo-Penn WF. 1976. Hemoglobin providence. Functional consequences of two alterations of the 2,3-diphosphoglycerate binding site at position β82. *J. Biol Chem* **251:** 7563–7571.

Bratu V, Lorkin PA, Lehmann H, Predescu C. 1971. Haemoglobin Buccureşti 42(CD1) Phe-Leu, a cause of unstable haemoglobin haemolytic anaemia. *Biochim Biophys Acta* **251:** 1–6.

Bunn HF, Forget BG. 1986. *Hemoglobin: Molecular, genetic and clinical aspects*. W.B. Saunders, Philadelphia.

Bunn HF, Wohl RC, Bradley TB, Cooley M, Gibson QH. 1974. Functional properties of hemoglobin Kempsey. *J Biol Chem* **249:** 7402–7409.

Bunn HF, Schmidt GJ, Haney DN, Dluhy RG. 1975. Hemoglobin Cranston, an unstable variant having an elongated β chain due to nonhomologous crossover between two normal β chain genes. *Proc Natl Acad Sci* **72:** 3609–3613.

* Cao A, Kan YW. 2013. The prevention of thalassemia. *Cold Spring Harb Perspect Med* **3:** a011775.

Carrell RW, Kay R. 1972. A simple method for the detection of unstable haemoglobins. *Br J Haematol* **23:** 615–619.

Charache S, Fox J, McCurdy P, Kazazian H Jr, Winslow R, Hathaway P, van Beneden R, Jessop M. 1977. Postsynthetic deamidation of hemoglobin Providence (β82 Lys replaced by Asn, Asp) and its effect on oxygen transport. *J. Clin Invest* **59:** 652–658.

Clegg JB, Weatherall DJ. 1974. Hemoglobin Constant Spring, and unusual α-chain variant involved in the etiology of hemoglobin H disease. *Ann NY Acad Sci* **232:** 168–178.

Clegg J, Weatherall D, Boon W, Mustafa D. 1969. Two new haemoglobin variants involving proline substitutions. *Nature* **222:** 379–380.

Clegg JB, Weatherall DJ, Milner PF. 1971. Haemoglobin Constant Spring—A chain termination mutant? *Nature* **234:** 337–340.

Coleman MB, Steinberg MH, Adams JG 3rd. 1991. Hemoglobin Terre Haute arginine β106. A posthumous correction to the original structure of hemoglobin Indianapolis. *J Biol Chem* **266:** 5798–5800.

Como PF, Wylie BR, Trent RJ, Bruce D, Volpato F, Wilkinson T, Kronenberg H, Holland RA, Tibben EA. 1991. A new unstable and low oxygen affinity hemoglobin variant: Hb Stanmore (β111[G13]Val→Ala). *Hemoglobin* **15:** 53–65.

Crowley M, Mollan T, Abdulmalik O, Butler AD, Goodwin E, Sarkar A, Stolle C, Gow A, Olson J, Weiss M. 2011. A hemoglobin variant associated with neonatal cyanosis and anemia. *N Engl J Med* **364:** 1837–1843.

Dacie JV, Shinton NK, Gaffney PJ Jr, Lehmann H. 1967. Haemoglobin Hammersmith (β-42 [CDI] Phe replaced by ser). *Nature* **216:** 663–665.

* Dailey HA, Meissner PN. 2013. Erythroid heme biosynthesis and its disorders. *Cold Spring Harb Perspect Med* **3:** a011676.

Derry S, Wood WG, Pippard M, Clegg JB, Weatherall DJ, Wickramasinghe SN, Darley J, Fucharoen S, Wasi P. 1984. Hematologic and biosynthetic studies in homozygous hemoglobin Constant Spring. *J. Clin Invest* **73:** 1673–1682.

Efremov GD, Wrightstone RN, Huisman TH, Schroeder WA, Hyman C, Ortega J, Williams K. 1971. An unusual hemoglobin anomaly and its relation to α-thalassemia and hemoglobin-H disease. *J. Clin Invest* **50:** 1628–1636.

Eisinger J, Flores J, Tyson JA, Shohet SB. 1985. Fluorescent cytoplasm and Heinz bodies of hemoglobin Köln erythrocytes: Evidence for intracellular heme catabolism. *Blood* **65:** 886–893.

Emsley P, Lohkamp B, Scott WG, Cowtan K. 2010. Features and development of Coot. *Acta Cryst* **66:** 486–501.

Favero ME, Costa FF. 2011. α-Hemoglobin-stabilizing protein: An erythroid molecular chaperone. *Biochem Res Int* **2011:** 373859.

Forget BG, Marotta CA, Weissman SM, Cohen-Solal M. 1975. Nucleotide sequences of the 3′-terminal untranslated region of messenger RNA for human β globin chain. *Proc Natl Acad Sci* **72:** 3614–3618.

Frick PG, Hitzig WH, Betke K. 1962. Hemoglobin Zurich. I. A new hemoglobin anomaly associated with acute hemolytic episodes with inclusion bodies after sulfonamide therapy. *Blood* **20:** 261–271.

Frischer H, Bowman J. 1975. Hemoglobin E, an oxidatively unstable mutation. *J Lab Clin Med* **85:** 531–539.

* Ganz T, Nemeth E. 2012. Iron metabolism: Interactions with normal and disordered erythropoiesis. *Cold Spring Harb Perspect Med* **2:** a011668.

Garrod A. 1928. The lessons of rare maladies: Annual oration before the medical society of London. *Lancet* **1:** 914–915.

Geva A, Clark JJ, Zhang Y, Popowicz A, Manning JM, Neufeld EJ. 2004. Hemoglobin Jamaica Plain—A sickling hemoglobin with reduced oxygen affinity. *N Engl J Med* **351:** 1532–1538.

Giardine B, Borg J, Higgs D, Peterson K, Philipsen S, Maglott D, Singleton B, Anstee D, Basak A, Clark B, et al. 2011. Systematic documentation and analysis of human genetic variation in hemoglobinopathies using the microattribution approach. *Nat Genet* **43:** 295–301.

Gibson QH, Riggs A, Imamura T. 1973. Kinetic and equilibrium properties of hemoglobin Kansas. *J. Biol Chem* **248:** 5976–5986.

Giordano PC, Zweegman S, Akkermans N, Arkesteijn SGJ, van Delft Peter, Versteegh FGA, Wajcman Henri, Harteveld CL. 2007. The first case of Hb Groene Hart (α119[H2]Pro→Ser, CCT→TCT [α1]) homozygosity confirms that a thalassemia phenotype is associated with this abnormal hemoglobin variant. *Hemoglobin* **31:** 179–182.

Griffon N, Badens C, Lena-Russo D, Kister J, Bardakdjian J, Wajcman H, Marden MC, Poyart C. 1996. Hb Bruxelles, deletion of $Phe^{\beta42}$, shows a low oxygen affinity and low cooperativity of ligand binding. *J. Biol Chem* **271:** 25916–25920.

Hamilton HB, Iuchi I, Miyaji T, Shibata S. 1969. Hemoglobin Hiroshima (β^{143} histidine→aspartic acid): A newly identified fast moving β chain variant associated with increased oxygen affinity and compensatory erythremia. *J. Clin Invest* **48:** 525–535.

Hardison R, Chui D, Giardine B, Riemer C, Patrinos G, Anagnou N, Miller W, Wajcman H. 2002. HbVar: A relational database of human hemoglobin variants and thalassemia mutations at the globin gene server. *Hum Mutat* **19:** 225–233.

Harteveld C, van Delft P, Plug R, Versteegh F, Hagen B, van Rooijen I, Kok P, Wajcman H, Kister J, Giordano PC. 2002. Hb Groene Hart: A new Pro→Ser amino acid substitution at position 119 of the α1-globin chain is associated with a mild α-thalassemia phenotype. *Hemoglobin* **26:** 255–60.

Hayashi A, Shimizu A, Yamamura Y, Watari H. 1966. Hemoglobins M: Identification of Iwate, Boston, and Saskatoon variants. *Science* **152:** 207–208.

Hebbel RP, Eaton JW, Kronenberg RS, Zanjani ED, Moore LG, Berger EM. 1978. Human llamas: Adaptation to altitude in subjects with high hemoglobin oxygen affinity. *J Clin Invest* **62:** 593–600.

Honig GR, Vida LN, Rosenblum BB, Perutz MF, Fermi G. 1990. Hemoglobin Warsaw ($Phe^{\beta42(CD1)}$→Val), an unstable variant with decreased oxygen affinity. Characterization of its synthesis, functional properties, and structure. *J. Biol Chem* **265:** 126–132.

Horlein H, Weber G. 1948. Ueber chronische familiäre methämoglobinämie und eine neue modifikation des methämoglobins. *Dtsch Med Wochenschr* **73:** 476–478.

Hoyer JD, McCormick DJ, Snow K, Kwon JH, Booth D, Duarte M, Grayson G, Kubik KS, Holmes MW, Fairbanks VF. 2002. Four new variants of the α2-globin gene without clinical or hematologic effects: Hb Park Ridge (α9[α7]Asn→Lys [α2]), Hb Norton (α72[EF1]-His→Asp [α2]), Hb Lombard (α103[G10]His→Tyr [α2]), and Hb San Antonio (A113[GH2]Leu→Arg [A2]). *Hemoglobin* **26:** 175–179.

Huisman TH. 1997. Hb E and α-thalassemia; variability in the assembly of β^E chain containing tetramers. *Hemoglobin* **21:** 227–236.

Huisman TH, Horton B, Bridges MT, Betke K, Hitzig WH. 1961. A new abnormal human hemoglobin-Hb: Zurich. *Clin Chim Acta* **6:** 347–355.

Hunt DM, Higgs DR, Winichagoon P, Clegg JB, Weatherall DJ. 1982. Haemoglobin Constant Spring has an unstable α chain messenger RNA. *Br J Haematol* **51:** 405–413.

Ikkala E, Koskela J, Pikkarainen P, Rahiala EL, El-Hazmi MA, Nagai K, Lang A, Lehmann H. 1976. Hb Helsinki: A variant with a high oxygen affinity and a substitution at a 2,3-DPG binding site (β82[EF6] Lys replaced by Met). *Acta Haematol* **56:** 257–275.

Imai K, Hamilton HB, Miyaji T, Shibata S. 1972. Physicochemical studies of the relation between structure and function in hemoglobin Hiroshima (HC3, histidine leads to aspartate). *Biochemistry* **11:** 114–121.

Jin Y, Nagai M, Nagai Y, Nagatomo S, Kitagawa T. 2004. Heme structures of five variants of hemoglobin M probed by resonance Raman spectroscopy. *Biochemistry* **43:** 8517–8527.

Keeling MM, Ogden LL, Wrightstone RN, Wilson JB, Reynolds CA, Kitchens JL, Huisman TH. 1971. Hemoglobin Louisville (β-42 [CD1] phe-leu): An unstable variant causing mild hemolytic anemia. *J Clin Invest* **50:** 2395–2402.

Khandros E, Mollan TL, Yu X, Wang X, Yao Y, D'Souza J, Gell DA, Olson JS, Weiss MJ. 2012. Insights into hemoglobin assembly through in vivo mutagenesis of α-hemoglobin stabilizing protein. *J Biol Chem* **287:** 11325–11327.

King MA, Wiltshire BG, Lehmann H, Morimoto H. 1972. An unstable haemoglobin with reduced oxygen affinity: Haemoglobin Peterborough, 3 (GI3) Valine lead to Phenylalanine, its interaction with normal haemoglobin and with haemoglobin Lepore. *Br J Haematol* **22:** 125–134.

Kohne E. 2011. Hemoglobinopathies: Clinical manifestations, diagnosis, and treatment. *Dtsch Ärztebl Int* **108:** 532.

Konigsberg W, Lehmann H. 1965. The amino acid substitution in hemoglobin M-Iwate. *Biochim Biophys Acta* **107:** 266–269.

Konigsberg W, Guidotti G, Hill RJ. 1961. The amino acid sequence of the α chain of human hemoglobin. *J. Biol Chem* **236:** PC55–PC56.

Lacan P, Aubry M, Couprie N, Francina A. 2004. Two new α chain variants: Hb Die (α93[FG5]Val→Ala [α1]) and Hb Beziers (α99[G6]Lys→Asn [α1]). *Hemoglobin* **28:** 59–63.

Lacerra G, Scarano C, Musollino G, Flagiello A, Pucci P, Carestia C. 2008. Hb Foggia or α117(GH5)Phe→Ser: A new α2 globin allele affecting the αHb-AHSP interaction. *Haematologica* **93:** 141–142.

Lal A, Goldrich ML, Haines DA, Azimi M, Singer ST, Vichinsky EP. 2011. Heterogeneity of hemoglobin H disease in childhood. *N. Engl J Med* **364:** 710–718.

Langdown JV, Davidson RJ, Williamson D. 1992. A new α chain variant, Hb Turriff (α99[G6]Lys→Glu): The interference of abnormal hemoglobins in Hb A1c determination. *Hemoglobin* **16:** 11–17.

Lehmann H. 1957. Haemoglobin and its abnormalities. *Practitioner* **178:** 198–214.

* Lettre G. 2012. The search for genetic modifiers of disease severity in the β-hemoglobinopathies. *Cold Spring Harb Perspect Med* **2:** a015032.

Levitt M. 1981. Effect of proline residues on protein folding. *J. Mol Biol* **145:** 251–263.

Lindstrom TR, Baldassare JJ, Bunn HF, Ho C. 1973. Nuclear magnetic resonance and spin-label studies of hemoglobin Kempsey. *Biochemistry* **12:** 4212–4217.

Little R, Roberts W. 2009. A review of variant hemoglobins interfering with hemoglobin A1c measurement. *J Diabetes Sci Technol* **3:** 446–451.

Lorkin P, Stephens A, Beard M, Wrigley P, Adams L, Lehmann H. 1975. Haemoglobin Rahere (β Lys-Thr): A new high affinity haemoglobin associated with decreased 2,3-diphosphoglycerate binding and relative polycythaemia. *Br Med J* **4:** 200–202.

Marinucci M, Mavilio F, Massa A, Gabbianelli M, Fontanarosa PP, Camagna A, Ignesti C, Tentori L. 1979. A new abnormal human hemoglobin: Hb Prato (α2 31 [B12] Arg leads to Ser β2). *Biochim Biophys Acta* **578:** 534–540.

Martínez G, Lima F, Colombo B. 1977. Haemoglobin J Guantanamo (α 2 β 2 128 [H6] Ala replaced by Asp). A new fast unstable haemoglobin found in a Cuban family. *Biochim Biophys Acta* **491:** 1–6.

McDonald MJ, Lund DP, Bleichman M, Bunn HF, De Young A, Noble RW, Foster B, Arnone A. 1980. Equilibrium, kinetic and structural properties of hemoglobin Cranston, an elongated β chain variant. *J. Mol Biol* **140:** 357–375.

Milner PF, Clegg JB, Weatherall DJ. 1971. Haemoglobin-H disease due to a unique haemoglobin variant with an elongated α-chain. *Lancet* **1:** 729–732.

Molchanova TP, Postnikov YuV, Pobedimskaya DD, Smetanina NS, Moschan AA, Kazanetz EG, Tokarev YuN, Huisman TH. 1993. Hb Alesha or α 2 β (2)67 (E11)Val→Met: A new unstable hemoglobin variant identified through sequencing of amplified DNA. *Hemoglobin* **17:** 217–225.

Mollan TL, Khandros E, Weiss MJ, Olson JS. 2012. The kinetics of α-globin binding to α hemoglobin stabilizing protein (AHSP) indicate preferential stabilization of a hemichrome folding intermediate. *J Biol Chem* **287:** 11338–11350

Moo-Penn WF, Jue DL, Bechtel KC, Johnson MH, Schmidt RM. 1976. Hemoglobin Providence. A human hemoglobin variant occurring in two forms in vivo. *J. Biol Chem* **251:** 7557–7562.

Moo-Penn W, Jue D, Johnson M, Olsen K, Shih D, Jones R, Lux S, Rodgers P, Arnone A. 1988. Hemoglobin Brockton (β 138 [H16] Ala→Pro): An unstable variant near the C-terminus of the β-subunits with normal oxygen-binding properties. *Biochemistry* **27:** 7614–7619.

Morales J, Russell JE, Liebhaber SA. 1997. Destabilization of human α-globin mRNA by translation anti-termination is controlled during erythroid differentiation and is paralleled by phased shortening of the poly(A) tail. *J Biol Chem* **272:** 6607–6613.

* Musallam KM, Taher AT, Rachmilewitz EA. 2012. β-Thalassemia intermedia: A clinical perspective. *Cold Spring Harb Perspect Med* **2:** a013482.

Nagai M, Aki M, Li R, Jin Y, Sakai H, Nagatomo S, Kitagawa T. 2000. Heme structure of hemoglobin M Iwate (α87[F8]His→Tyr): A UV and visible resonance Raman study. *Biochemistry* **39:** 13093–13105.

Nagel RL, Lynfield J, Johnson J, Landau L, Bookchin RM, Harris MB. 1976. Hemoglobin Beth Israel. A mutant causing clinically apparent cyanosis. *N. Engl J Med* **295:** 125–130.

Nagel RL, Bookchin RM, Johnson J, Labie D, Wajcman H, Isaac-Sodeye WA, Honig GR, Schilirò G, Crookston JH, Matsutomo K. 1979. Structural bases of the inhibitory effects of hemoglobin F and hemoglobin A2 on the polymerization of hemoglobin S. *Proc Natl Acad Sci* **76:** 670–672.

Nathan D, Orkin S, Ginsburg D, Look A, Fisher D, Lux S. 2009. *Nathan and Oski's hematology of infancy and childhood*, 7th ed. Saunders Elsevier, Philadelphia.

Ohba Y, Miyaji T, Matsuoka M, Yokoyama M, Numakura H. 1975. Hemoglobin Hirosaki (α43 [CE 1] Phe replaced by Leu), a new unstable variant. *Biochim Biophys Acta* **405:** 155–160.

Olson J, Gibson Q, Nagel R, Hamilton H. 1972. The ligand-binding properties of hemoglobin Hiroshima ($\alpha_2\beta_2^{146\ asp}$). *J Biol Chem* **247:** 7485–7493.

Orkin S, Kazazian HJ, Antonarakis S, Ostrer H, Goff S, Sexton J. 1982. Abnormal RNA processing due to the exon mutation of β^E-globin gene. *Nature* **300:** 768–769.

Paoli M, Liddington R, Tame J, Wilkinson A, Dodson G. 1996. Crystal structure of T state haemoglobin with oxygen bound at all four haems. *J Mol Biol* **256:** 775–792.

Park S, Yokoyama T, Shibayama N, Shiro Y, Tame JR. 2006. 1.25 A resolution crystal structures of human haemoglobin in the oxy, deoxy and carbonmonoxy forms. *J Mol Biol* **360:** 690–701.

Perutz M. 1960. Structure of hemoglobin. *Brookhaven Symp Biol* **13:** 165–183.

Perutz M. 1970. Stereochemistry of cooperative effects in haemoglobin. *Nature* **228:** 726–739.

Perutz M, Lehmann H. 1968. Molecular pathology of human haemoglobin. *Nature* **219:** 902–909.

Perutz M, Rossman M, Cullis A, Muirhead H, Will G, North A. 1960. Structure of haemoglobin: Three-dimensional Fourier synthesis at 5.5-A resolution, obtained by X-ray analysis. *Nature* **185:** 416–422.

Perutz MF, Muirhead H, Mazzarella L, Crowther RA, Greer J, Kilmartin JV. 1969. Identification of residues responsible for the alkaline Bohr effect in haemoglobin. *Nature* **222:** 1240–1243.

Perutz MF, Pulsinelli P, Eyck LT, Kilmartin JV, Shibata S, Iuchi I, Miyaji T, Hamilton HB. 1971. Haemoglobin Hiroshima and the mechanism of the alkaline Bohr effect. *Nat New Biol* **232:** 147–149.

Perutz M, Fermi G, Shih TB. 1984. Structure of deoxyhemoglobin Cowtown (His HC3[146] β→Leu): Origin of the alkaline Bohr effect and electrostatic interactions in hemoglobin. *Proc Natl Acad Sci* **81:** 4781–4784.

Phillips SE, Hall D, Perutz MF. 1981. Structure of deoxyhaemoglobin Zürich (HisE7[63 β]—greater than Arg). *J Mol Biol* **150:** 137–141.

Poyart C, Schaad O, Kister J, Galacteros F, Edelstein SJ, Blouquit Y, Arous N. 1990. Hemoglobin Saint Mandé [β102 (G4) Asn→Tyr]. Functional studies and structural modeling reveal an altered T state. *Eur J Biochem* **194:** 343–348.

Pulsinelli PD, Perutz M, Nagel R. 1973. Structure of hemoglobin M Boston, a variant with a five-coordinated ferric heme. *Proc Natl Acad Sci* **70:** 3870–3874.

Rabbitts TH. 1976. Bacterial cloning of plasmids carrying copies of rabbit globin messenger RNA. *Nature* **260:** 221–225.

Reed CS, Hampson R, Gordon S, Jones RT, Novy MJ, Brimhall B, Edwards MJ, Koler RD. 1968. Erythrocytosis secondary to increased oxygen affinity of a mutant hemoglobin, hemoglobin Kempsey. *Blood* **31:** 623–632.

Rees DC, Rochette J, Schofield C, Green B, Morris M, Parker NE, Sasaki H, Tanaka A, Ohba Y, Clegg JB. 1996. A novel silent posttranslational mechanism converts methionine to aspartate in hemoglobin Bristol (β67[E11] Val-Met→Asp). *Blood* **88:** 341–348.

Rees DC, Clegg JB, Weatherall DJ. 1998. Is hemoglobin instability important in the interaction between hemoglobin E and β thalassemia? *Blood* **92:** 2141–2146.

Reissmann K, Ruth W, Nomura T. 1961. A human hemoglobin with lowered oxygen affinity and impaired heme-heme interactions. *J Clin Invest* **40:** 1826–1833.

Rieder RF, Oski FA, Clegg JB. 1969. Hemoglobin Philly (β35 tyrosine phenylalanine): Studies in the molecular pathology of hemoglobin. *J. Clin Invest* **48:** 1627–1642.

Riggs A, Gibson QH. 1973. Oxygen equilibrium and kinetics of isolated subunits from hemoglobin Kansas. *Proc Natl Acad Sci* **70:** 1718–1720.

Roth EF Jr, Elbaum D, Bookchin RM, Nagel RL. 1976. The conformational requirements for the mechanical precipitation of hemoglobin S and other mutants. *Blood* **48:** 265–271.

Russu IM, Ho C. 1986. Assessment of role of β 146-histidyl and other histidyl residues in the Bohr effect of human normal adult hemoglobin. *Biochemistry* **25:** 1706–1716.

* Sankaran VG, Orkin SH. 2013. The switch from fetal to adult hemoglobin. *Cold Spring Harb Perspect Med* **3:** a011643.

Schneider RG, Bremner JE, Brimhall B, Jones RT, Shih TB. 1979. Hemoglobin Cowtown (β 146 HC3 His-Leu): A mutant with high oxygen affinity and erythrocytosis. *Am J Clin Pathol* **72:** 1028–1032.

Schrier SL, Bunyaratvej A, Khuhapinant A, Fucharoen S, Aljurf M, Snyder LM, Keifer CR, Ma L, Mohandas N. 1997. The unusual pathobiology of hemoglobin constant spring red blood cells. *Blood* **89:** 1762–1769.

Schroeder WA, Jones RT, Shelton JR, Shelton JB, Cormick J, McCalla K. 1961. A partial sequence of the amino acid residues in the γ chain of human hemoglobin F. *Proc Natl Acad Sci* **47:** 811–818.

Sciarratta GV, Ivaldi G, Molaro GL, Sansone G, Salkie ML, Wilson JB, Reese AL, Huisman TH. 1984. The characterization of hemoglobin Manitoba or $\alpha_2$102(G9)Ser→Arg β_2 and hemoglobin Contaldo or $\alpha_2$103(G10)His→Arg β_2 by high performance liquid chromatography. *Hemoglobin* **8:** 169–181.

* Serjeant G. 2013. The natural history of sickle cell disease. *Cold Spring Harb Perspect Med* **3:** a011783.

Shaeffer JR, Schmidt GJ, Kingston RE, Bunn HF. 1980. Synthesis of hemoglobin Cranston, and elongated β chain variant. *J Mol Biol* **140:** 377–389.

Shibata S, Tanuira A, Iuchi I. 1960. Hemoglobin M1 demonstration of a new abnormal hemoglobin in hereditary nigremia. *Acta Haematol Jpn* **23:** 96–105.

Shimizu A, Tsugita A, Hayashi A, Yamamura Y. 1965. The primary structure of hemoglobin M-Iwate. *Biochim Biophys Acta* **107:** 270–277.

Springer BA, Egeberg KD, Sligar SG, Rohlfs RJ, Mathews AJ, Olson JS. 1989. Discrimination between oxygen and carbon monoxide and inhibition of autooxidation by myoglobin. Site-directed mutagenesis of the distal histidine. *J Biol Chem* **264:** 3057–3060.

Stamatoyannopoulos G, Parer JT, Finch CA. 1969. Physiologic implications of a hemoglobin with decreased oxygen affinity (hemoglobin seattle). *N Engl J Med* **281:** 916–919.

Stamatoyannopoulos G, Woodson R, Papayannopoulou T, Heywood D, Kurachi S. 1974. Inclusion-body β-thalassemia trait. A form of β thalassemia producing clinical manifestations in simple heterozygotes. *N Engl J Med* **290:** 939–943.

Steinberg M, Forget B, Higgs D, Nagel R. 2001. *Disorders of hemoglobin: Genetics, pathophysiology, and clinical management.* Cambridge University Press, Cambridge.

Sugihara J, Imamura T, Nagafuchi S, Bonaventura J, Bonaventura C, Cashon R. 1985. Hemoglobin Rahere, a human hemoglobin variant with amino acid substitution at the 2,3-diphosphoglycerate binding site. Functional consequences of the alteration and effects of bezafibrate on the oxygen bindings. *J Clin Invest* **76:** 1169–73.

Tanaka Y, Matsui K, Matsuda K, Shinohara K, Haranob K. 2005. A family with hemoglobin Hirosaki. *Int J Hematol* **82:** 124–126.

Thein SL, Hesketh C, Taylor P, Temperley IJ, Hutchinson RM, Old JM, Wood WG, Clegg JB, Weatherall DJ. 1990. Molecular basis for dominantly inherited inclusion body β-thalassemia. *Proc Natl Acad Sci* **87:** 3924–3928.

Tremper K, Barker S. 1989. Pulse oximetry. *Anesthesiology* **70:** 98–108.

Tucker PW, Phillips SE, Perutz MF, Houtchens R, Caughey WS. 1978. Structure of hemoglobins Zürich (His E7[63]β replaced by Arg) and Sydney (Val E11[67]β replaced by Ala) and role of the distal residues in ligand binding. *Proc Natl Acad Sci* **75:** 1076–1080.

Turbpaiboon C, Limjindaporn T, Wongwiwat W, U-Pratya Y, Siritanaratkul N, Yenchitsomanus P-thai, Jitrapakdee S, Wilairat P. 2006. Impaired interaction of α-haemoglobin-stabilising protein with α-globin termination mutant in a yeast two-hybrid system. *Br J Haematol* **132:** 370–373.

Vasseur C, Domingues-Hamdi E, Brillet T, Marden Michael C, Baudin-Creuza V. 2009. The α-hemoglobin stabilizing protein and expression of unstable α-Hb variants. *Clin Biochem* **42:** 1818–1823.

Vasseur-Godbillon C, Marden M, Giordano P, Wajcman H, Baudin-Creuza V. 2006. Impaired binding of AHSP to α chain variants: Hb Groene Hart illustrates a mechanism leading to unstable hemoglobins with α thalassemic like syndrome. *Blood Cells Mol Dis* **37:** 173–179.

Verhovsek M, Henderson M, Cox G, Luo H, Steinberg M, Chui D. 2010. Unexpectedly low pulse oximetry measurements associated with variant hemoglobins: A systematic review. *Am J Hematol* **85:** 882–885.

Viprakasit V, Tanphaichitr VS. 2002. Compound heterozygosity for α^0-thalassemia (- -THAI) and Hb constant spring causes severe Hb H disease. *Hemoglobin* **26:** 155–162.

Wajcman H, Moradkhani K. 2011. Abnormal haemoglobins: Detection and characterization. *Indian J Med Res* **134:** 538–546.

Wajcman H, Préhu C, Bardakdjian-Michau J, Promé D, Riou J, Godart C, Mathis M, Hurtrel D, Galactéros F. 2001. Abnormal hemoglobins: Laboratory methods. *Hemoglobin* **25:** 169–181.

Wajcman H, Traeger-Synodinos J, Papassotiriou I, Giordano PC, Harteveld CL, Baudin-Creuza V, Old J. 2008. Unstable and thalassemic α chain hemoglobin variants: A cause of Hb H disease and thalassemia intermedia. *Hemoglobin* **32:** 327–349.

Watson HC, Kendrew JC. 1961. The amino-acid sequence of sperm whale myoglobin. Comparison between the amino-acid sequences of sperm whale myoglobin and of human hemoglobin. *Nature* **190:** 670–672.

Watson-Williams EJ, Beale D, Irvine D, Lehmann H. 1965. A new haemoglobin, D-Ibadan (β-87 threonine–lysine), producing no sickle-cell haemoglobin D disease with haemoglobin S. *Nature* **205:** 1273–1276.

Weatherall D, Clegg J. 2001. Inherited haemoglobin disorders: An increasing global health problem. *Bull World Health Organ* **79:** 704–712.

Weiss IM, Liebhaber SA. 1994. Erythroid cell-specific determinants of α-globin mRNA stability. *Mol Cell Biol* **14:** 8123–8132.

Weiss MJ, Zhou S, Feng L, Gell DA, Mackay JP, Shi Y, Gow AJ. 2005. Role of α-hemoglobin-stabilizing protein in normal erythropoiesis and β-thalassemia. *Ann NY Acad Sci* **1054:** 103–117.

* Williams TN, Weatherall DJ. 2012. World distribution, population genetics, and health burden of the hemoglobinopathies. *Cold Spring Harb Perspect Med* **2:** a011692.

Witkowska HE, Lubin BH, Beuzard Y, Baruchel S, Esseltine DW, Vichinsky EP, Kleman KM, Bardakdjian-Michau J, Pinkoski L, Cahn S. 1991. Sickle cell disease in a patient with sickle cell trait and compound heterozygosity for hemoglobin S and hemoglobin Quebec-Chori. *N Engl J Med* **325:** 1150–1154.

Yu X, Mollan TL, Butler Andrew, Gow AJ, Olson JS, Weiss MJ. 2009. Analysis of human α globin gene mutations that impair binding to the α hemoglobin stabilizing protein. *Blood* **113:** 5961–5969.

Cell-Free Hemoglobin and Its Scavenger Proteins: New Disease Models Leading the Way to Targeted Therapies

Dominik J. Schaer[1,2,3] and Paul W. Buehler[4]

[1]Division of Internal Medicine, University Hospital, Zurich CH-8091, Switzerland

[2]Center of Integrative Human Physiology University of Zurich, Zurich CH-8057, Switzerland

[3]Center of Evolutionary Medicine, University of Zurich, Zurich CH-8057, Switzerland

[4]Laboratory of Biochemistry and Vascular Biology, Division of Hematology, Center for Biologics Evaluation and Research, Food and Drug Administration, Bethesda, Maryland 20892

Correspondence: dominik.schaer@usz.ch; paul.buehler@fda.hhs.gov

Hemoglobin (Hb) has multiple pathophysiologic effects when released into the intravascular space during hemolysis. The extracellular effects of Hb have resulted in novel models of toxicity, which help to explain endothelial dysfunction and cardiovascular complications that accompany genetic hemolytic anemias, malaria, blood transfusion, and atherosclerosis. The majority of models focus on nitric oxide (NO) depletion; however, in local tissue environments, Hb can also act as a pro-oxidant and inflammatory agent. This can alter cellular differentiation with the potential to deviate immune responses. The understanding of these mechanisms set in the context of natural scavenger and detoxification systems may accelerate the development of novel treatment strategies.

The capability of hemoglobin (Hb) to efficiently load and transport oxygen is inevitably linked to a broad reactivity pattern with alternative ligands, such as carbon monoxide (CO), nitric oxide (NO), hydrogen peroxide (H_2O_2), and many others. The biochemistry of these reactions has been the focus of Hb research for decades. More recently, however, these Hb reactions were examined as a potential cause of adverse pathophysiologic processes that accompany red blood cell (RBC) destruction (i.e., hemolysis) and the accumulation of extracellular free Hb (Baek et al. 2012; Gladwin et al. 2012). Other biologic activities of extracellular free Hb can be traced back either to direct interactions of its globin or heme components with specific cellular receptors and signaling pathways, or to the secondary effects of heme breakdown by the heme oxygenases. The evolving picture suggests that Hb as a toxin can adversely affect the outcome of diverse conditions, including the hemolytic anemias, sepsis, malaria, blood transfusion, and atherosclerosis, in which local accumulation of extracellular Hb causes oxidative stress and changes macrophage polarization in the atherosclerotic plaque microenvironment. The recognition that Hb is a disease-modifying compound, and concurrent research on protec-

Copyright © 2013 Cold Spring Harbor Laboratory Press; all rights reserved
Cite this article as *Cold Spring Harb Perspect Med* doi: 10.1101/cshperspect.a013433

tive Hb scavenger proteins have provided a framework for novel pathophysiologic models and may lead the way toward a new era of targeted treatment strategies. The toxic effects of free Hb appear to depend on the amounts in the extracellular space, anatomic location, and the activity of scavenger and detoxification pathways. These factors may vary considerably among different disease states and, therefore, extrapolations or application of a general Hb toxicity model to heterogeneous conditions must be considered cautiously. In this article we will summarize current evidence that supports Hb's role as a disease modifier in hemolytic anemias, malaria, blood transfusion, and atherosclerosis, and how scavenger protein-based therapeutics could be used to attenuate the underlying pathophysiologic processes.

DISEASE STATES THAT ARE MODULATED BY THE TOXICITY OF EXTRACELLULAR HEMOGLOBIN

Sickle Cell Disease and Hemoglobin-Based Oxygen Carriers

Two areas of research, sickle cell disease (SCD), which represents a condition of chronic low-level plasma Hb exposure (3–10 μM plasma heme), and Hb-based oxygen carrier (HBOC) therapy, which represents a condition of acute high-level Hb exposure (>500 μM plasma heme), have driven the evaluation of pathophysiologic models to better understand the roles of extracellular Hb toxicity as a general disease process (Buehler et al. 2010). A pronounced systemic, and in some animal models a pulmonary hypertensive, response is observed within seconds of exposure to cell-free Hb or HBOCs (Buehler et al. 2010). This acute response is likely related to the interaction of Hb with NO and is suspected to be a cause of acute myocardial infarction and stroke in certain subjects receiving HBOCs (Natanson et al. 2008; Silverman and Weiskopf 2009). HBOCs are typically transfused in large quantities reaching millimolar plasma concentrations of extracellular Hb. These dosing levels are required to meet O_2 delivery and volume replacement needs in patients with severe hemorrhage.

Sickle cell anemia is a chronic low-level hemolytic disease; however, some of the sequelae mimic those of HBOC administration. The typical complications of SCD are vasculopathies, stroke, pulmonary hypertension (PH), and renal failure, which suggest a pathophysiology of unopposed constriction within the vasculature and, therefore, may be related to an Hb-induced reduction in NO bioavailability. The NO depletion hypothesis is based on the findings that plasma from patients with SCD had elevated levels of free Hb, and accordingly, that the plasma from these patients had higher ex vivo NO-depleting activity (Reiter et al. 2002). In other studies, positive correlations were found between surrogate markers of hemolysis, PH, and disease-related mortality (Gladwin et al. 2004). PH, measured by Doppler echocardiography, was estimated to occur in up to 30% of patients with SCD and was therefore hypothesized to be a paradigmatic effect of NO depletion that could accompany chronic hemolytic diseases in general (Rother et al. 2005). The NO depletion hypothesis, however, has been challenged by other studies that found a lower, but still relevant, prevalence of PH when assessed by pulmonary artery catheterization (the gold standard for measuring blood pressure within the pulmonary circulation) (Parent et al. 2011). Even fewer patients in this cohort were found to have precapillary PH, which would be expected if Hb-mediated NO depletion within the pulmonary vasculature were highly relevant. Further concerns with the NO depletion hypothesis in SCD are related to the validity of hemolytic surrogate markers, such as plasma lactic acid dehydrogenase (LDH), and to the generally low plasma Hb concentrations (in the low micromolar range) detected in SCD patients, even during vasoocclusive crisis, when free Hb concentrations remain orders of magnitude below those associated with some hyperhemolytic states or HBOC administration (Bunn et al. 2010). Other proposed mechanisms of Hb toxicity during SCD or HBOC administration are related to the pro-oxidative and proinflammatory properties of Hb/heme, which can damage the endothelium or activate components of the coagulation and innate immune systems (Jeney et al.

 Cite this article as *Cold Spring Harb Perspect Med* doi: 10.1101/cshperspect.a013433

2002; Belcher et al. 2003). The history of research on Hb toxicity in SCD and HBOC therapy illustrates that multiple pathways of toxicity for free Hb can theoretically coexist and that emphasis on one pathway may not fully define the extent and complexity of Hb-induced pathophysiology in conditions where it may contribute to disease processes.

Transfusion of Stored Blood

RBC transfusion with acute blood loss or chronic anemia is one of the most common therapeutic interventions in medicine. Over the last two decades, however, more restrictive transfusion practices have been implemented, and transfusion thresholds have been lowered by limiting unnecessary transfusions (Barr and Bailie 2011). Accumulating evidence suggests that the adverse consequences of some transfusions are related to the storage period between blood donation and transfusion (Wang et al. 2012). The maximum allowed storage time is now 42 d with most storage solutions, and current blood bank practices following a "first in, first out" system that favors the distribution of older units first. A 2011 meta-analysis of retrospective data from 409,966 patients estimated that a significant (odds ratio, 1.16; 95% confidence interval, 1.07–1.24) risk of death exists when relatively old, rather than fresh, blood is transfused. The definition of older stored blood in the analyzed studies ranged from 9 d up to 42 d, with a range of 1–10 units of blood transfused per patient (Wang et al. 2012). The RBC transfusion-associated risk of death may be increased in certain high-risk populations, such as critically ill or massively transfused (>10 units of blood within 24 h) patients. It is important to point out that storage time-related morbidity/mortality and potential mechanisms associated with transfusion-related toxicity remains controversial and not all studies support the concept that older blood is detrimental (Middelburg et al. 2012). RBCs may undergo complex biochemical and structural changes, collectively referred to as the "RBC storage lesion." Damaged or aged RBCs can accumulate over time as a low-quality population within stored blood bags. It is therefore possible that some degree of acute hemolysis occurs when older stored blood is transfused. Current regulations only require that 75% of transfused RBCs survive within the circulation for 24 h after transfusion. This means that up to 25% of RBCs may be lost within the first 24 h after transfusion, owing to either intravascular hemolysis or the trapping of RBCs by macrophages in the spleen or liver. In the most extreme situation, this implies an acute burden of >10 g extra Hb per unit of transfused older RBCs that are either acutely released into the circulation or metabolized by spleen and liver macrophages after erythrophagocytosis.

Small studies in healthy volunteers suggest that extravascular hemolysis and increased RBC-Hb catabolism occur when as little as one unit of blood is transfused after 40 days of storage (Hod et al. 2011). Following single-unit transfusion, other investigators found that intravascular Hb exposure is associated with concomitant elevation in plasma nitrite levels and NO depletion in humans following small volume blood transfusion (Berra et al. 2012).

Animal models of RBC transfusion have been designed to explore the possible roles of intravascular hemolysis and cell-free Hb in the pathophysiologic consequences of the massive transfusion of older stored blood in guinea pigs (Baek et al. 2012). These studies suggest that high-volume transfusions of older blood, but not fresh blood, result in posttransfusion intravascular and extravascular hemolysis that approaches a 15%–20% loss of transfused RBCs within 24 h. Transfusion of older, rather than fresh, blood caused an acute hypertensive response, hemoglobinuria, acute renal failure, and vascular injury (Fig. 1) (Baek et al. 2012). Infusion of transfused old blood with the specific Hb scavenger haptoglobin (Hp) attenuated several of the transfusion-related adverse effects. These antagonist experiments proved that free Hb was a contributor to the observed adverse effects. Additionally, animal studies focused on the acute hemodynamic effects of stored blood have suggested that NO donors could be a therapeutic option to limit transfusion-related and hemolysis-induced pulmonary and/or systemic hypertensive responses in mice, rats, and sheep

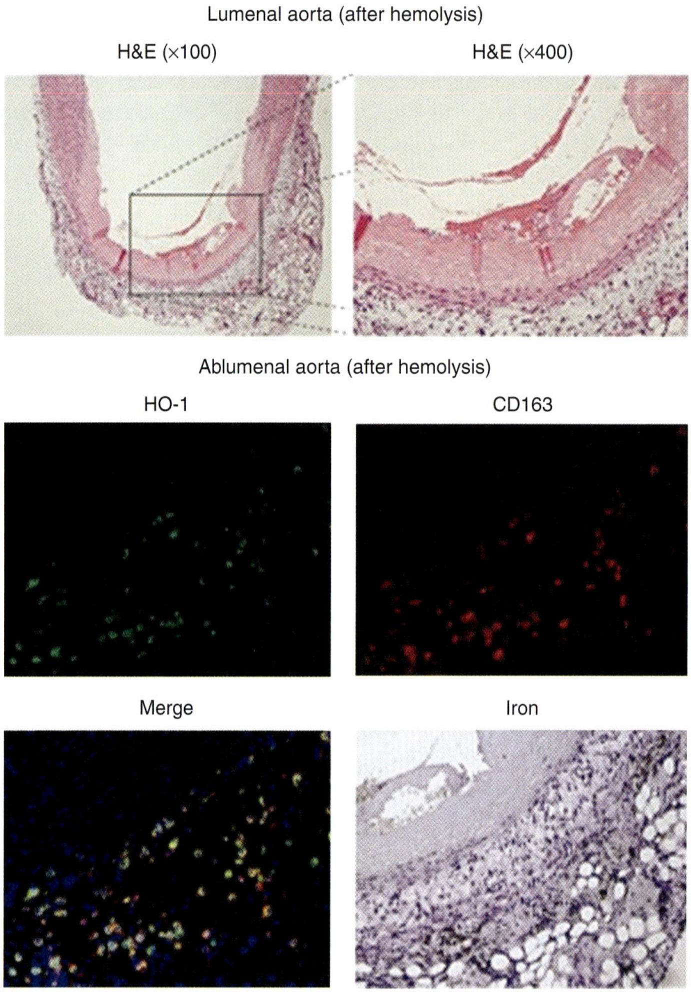

Figure 1. Vascular effects of old blood transfusion-associated hemolysis. The *top* panel shows regions of vascular necrosis appearing to initiate within the vascular lumen and progress to the tunica media. These observations were captured 24 h after massive transfusion to guinea pigs at 100× and 400× magnification. The injury in isolated regions of the aorta in transfused animals was consistent with coagulative necrosis and may be the result of hemoglobin and/or damaged red blood cells. The *lower* panels indicate macrophage accumulation to hemoglobin-rich regions within the adventitia consistent with HO-1, CD163, and nonheme iron accumulation.

(Yu et al. 2008, 2009, 2012; Donadee et al. 2011; Baron et al. 2012). One of these studies identified Hb and Hb-containing RBC membrane microparticles released in vitro before transfusion as an additional factor that may enhance the adverse posttransfusion effects (Donadee et al. 2011). Blood transfusion leading to stimulation of the heme breakdown pathway and accumulation of heme catabolites was associated with anti-inflammatory and immunosuppressive effects in multiple animal models, and such effects could contribute to the immune-compromised state of severely ill and massively transfused patients (Yazdanbakhsh et al. 2011). On the other hand, proinflammatory effects may be triggered by free heme through the stimulation of toll-like receptor 4 (TLR-4), by reaction intermediates emanating from Hb/heme peroxidative reactions with phospholipids, or by Hb/heme breakdown products

(D'Agnillo and Alayash 2001; Jeney et al. 2002; Jia et al. 2007; Lin et al. 2010; Yazdanbakhsh et al. 2011). In a mouse blood transfusion model, older stored RBCs induced an acute phase cytokine release response that accompanied the clearance of transfused RBCs in the spleen (Hod et al. 2010).

In summary, intravascular release and enhanced turnover of Hb after transfusion of older stored blood could contribute to cardiovascular and renal dysfunction, as well as inflammation and enhanced susceptibility to infection in severely ill patients. However, similar to interpretation of retrospective human data, the translation of animal-based transfusion model data to the human situation is nearly impossible. In animal studies, RBCs are often maximally stressed and RBCs of animals and humans behave with varying degrees of difference both under storage and following transfusion. As a result it is unlikely that any model will provide a definitive understanding of the potential complications associated with human blood storage. The best that animal studies can be expected to provide are suggestions for pathophysiologic mechanisms that may help to explain poor outcomes previously reported in clinical settings.

Malaria

Malaria is a paradigmatic hemolytic condition and considered a driving force in human evolution. Several common hemoglobinopathies, such as SCD or the thalassemias, have been proposed to result, in part, from the coevolutionary interactions between the plasmodium parasite and its human host (Rosenthal 2011). Animal studies in mice have been used to construct hypothetic models of the pathophysiologic role of the Hb molecule in malarial disease progression and tolerance. In the acute phase of severe malaria, hemolysis appears to exaggerate inflammatory processes and endothelial damage through NO depletion, oxidation, and inflammation (Gramaglia et al. 2006; Pamplona et al. 2007). The appearance of Hb and heme in plasma has been linked to the development of cerebral malaria, which remains the most severe and difficult to treat complication of the infection. In contrast, chronic hemolysis (before malaria infection) in SCD induces a state of malaria tolerance by enhancing the basal activity of heme defense systems, such as heme oxygenase 1 (HO-1) up-regulation, which drives the generation of metabolites with anti-inflammatory effects (i.e., CO, bilirubin, and ferritin) (Ferreira et al. 2011). Thus, active hemolysis before infection does not alter the course of parasitemia per se, but may prevent progression to fatal disease complications. However, similar mechanisms that confer a state of relative tolerance to severe malaria may also enhance host susceptibility to concurrent nonplasmodium infections. Infection with nontyphoid *Salmonella* during the acute hemolytic phase of plasmodium infection was associated with the appearance of an immature granulocyte population with reduced oxidative burst capacity and, consequently, a reduced capacity to kill phagocytosed *Salmonella* (Cunnington et al. 2011). This effect was mimicked experimentally by heme injection and attenuated when HO-1 was inhibited. As a result, animals that were coinfected with malaria were much more susceptible to fatal *Salmonella* infection than control animals that were not heme stressed.

It is overly simplistic to reduce the pathophysiology of a complex parasite infection with diverse host–pathogen interaction pathways to the level of the Hb molecule as a single mechanism of pathology. However, based on the discussed animal studies, the impact of hemolysis in malaria could be a relevant component. Low-level hemolysis before plasmodium infection may induce a state of tolerance, attenuating the deleterious effects of Hb/heme in the acute phase. Acute Hb and heme exposure exaggerate inflammatory and oxidative tissue damage in nontolerant individuals and may enhance susceptibility to superinfection by nonmalaria pathogens. An important lesson to take away from these studies is that Hb and heme may have diverse pathophysiology in patients with inflammatory diseases and infection. The net impact of extracellular Hb in a given disease state can be protective or deleterious, and the effect is likely determined by the anatomic site, timing, and degree of hemolysis on one hand, and availability of scavenger systems on the other hand.

Atherosclerosis—A Paradigm for Hb as a Locally Active Disease Modifier

As exemplified in the sections above, the pathophysiology of extracellular Hb has largely been studied on a systemic level. It is likely, however, that Hb also acts as an important modulator of local disease processes, where RBCs extravasate and are lysed within tissues. Under such conditions, very high extracellular concentrations of Hb and heme (in the millimolar range) could affect cell viability, activation state, and differentiation (Boyle et al. 2009).

One of the paradigmatic conditions in which local Hb/heme effects have been studied is atherosclerosis. The principal findings of these investigations may similarly apply to other clinical situations, such as wound healing or hemorrhagic stroke. Microscopic hemorrhage from fragile neovessels is a hallmark of advanced plaques, and a predictor of catastrophic clinical events such as myocardial infarction and stroke (Kolodgie et al. 2003). Besides the cholesterol and phospholipid-rich RBC membrane components that are deposited in the areas of microbleeding, Hb is the most abundant and biologically reactive compound within these lesions (Schaer et al. 2006).

The potential effects of Hb are diverse. Heme can oxidize low-density lipoprotein (LDL) and other lipids, and transform these compounds into proinflammatory and cytotoxic species (e.g., oxidized LDL) that have an established impact on atherosclerotic plaque progression (Nagy et al. 2010). Hb-derived free heme, as well as heme- and globin-degradation products, have a cytotoxic impact on endothelial and vascular smooth muscle cells, and may act as proinflammatory mediators that perpetuate inflammatory cell recruitment through the stimulation of chemokine (e.g., IL-8) expression (Boyle et al. 2009; Kaempfer et al. 2011). Free Hb in the extracellular environment may also polarize the differentiation plasticity of certain cell types. The observation that phenotypically distinct macrophage subpopulations cluster within hemorrhagic and nonhemorrhagic plaque areas has stimulated research exploring whether Hb and/or heme are the factors reshaping monocyte-macrophage lineage differentiation in these lesions (Boyle et al. 2009).

The specific interest in macrophages is twofold. On the one hand, macrophages are the principal inflammatory cell type within atherosclerotic plaques, and primary mediators of disease progression. On the other hand, they are the only cell type that expresses the Hb scavenger receptor CD163, and possess a highly specialized physiology to handle and detoxify extracellular Hb. Because of this unique function, macrophages in Hb-rich environments have very high expression levels of HO-1, ferritin, and the iron exporter ferroportin. In vitro and in the coronary arteries of patients, Hb/heme-exposed macrophages were described as a distinct phenotype with high expression of the Hb detoxification pathway ($CD163^{high}$ $HO\text{-}1^{high}$), high mannose receptor expression, and profoundly suppressed expression of HLA class 2 (Boyle et al. 2009; Kaempfer et al. 2011; Finn et al. 2012). This nonfoamy macrophage subtype localizes to the postbleeding plaque areas, has enhanced antioxidative and iron export functions, and is resistant to cholesterol loading as a result of up-regulated ATP-binding cassette transporters (ABCA1 and ABCG1) and down-regulated scavenger receptors (SR-A/B and CD36). The available data suggest that these Hb/heme-driven macrophages constitute a unique phenotype (referred to as M-hem or M-Hb), with distinct and nonoverlapping features, within the known spectrum of macrophage phenotypes extending from proinflammatory (M1) to anti-inflammatory (M2) macrophages. The Hb/heme-driven macrophage phenotype polarization appears to be triggered by increased intracellular heme levels via a pathway involving signaling by Nrf-2 and the activating transcription factor 1 (ATF1), with subsequent coinduction of HO-1 and liver-X-receptor β (LXRβ) (Kaempfer et al. 2011; Boyle et al. 2012). The functional significance of the profoundly suppressed levels of HLA class 2 in these macrophages has not been evaluated, but the spectrum of functional consequences of this intriguing Hb/heme-induced condition may include enhanced control of autoimmunity and/or increased susceptibility to infection.

Accumulating evidence now suggests that Hb/heme can impact multiple levels of innate and acquired immunity. Hb should therefore be considered a novel intrinsic alarm molecule that serves to signal bleeding and tissue destruction.

In summary, Hb may play a dual role in atherosclerosis. The proinflammatory and oxidative effects of heme within the extracellular milieu initiate and promote plaque formation, whereas Hb/heme drives macrophage polarization toward a protective, nonfoamy, and antioxidative macrophage phenotype that halts lesion progression at later stages of the disease. It remains to be explored how these opposing processes could be shifted favorably by pharmacological interventions.

MECHANISMS OF HEMOGLOBIN-MEDIATED ADVERSE EFFECTS

The primary acute pathophysiologic responses to extracellular Hb in plasma are blood pressure elevation (Rother et al. 2005) and pro-oxidative toxicity occurring in vascular and renal tissues (Boretti et al. 2009; Baek et al. 2012). The processes driving Hb-induced toxicity are depicted in Figure 2 and likely involve a complex interaction between NO balance, oxidation, and inflammation.

Nitric Oxide Consumption—Systemic and PH

The most accepted hypothesis for Hb-induced hypertension involves the interactions of NO with both oxy- and deoxyhemoglobin (Doherty et al. 1998; Olson et al. 2004). The pathophysiologic response to Hb and its consumption of NO is observed primarily as an increase in systolic/diastolic and mean arterial blood pressure (MAP) with cardiac output either unchanged (Boretti et al. 2009) or decreased (Minneci et al. 2005). This consistently causes an increase in systemic vascular resistance, or vasoconstriction (Reiter et al. 2002; Minneci et al. 2005; Boretti et al. 2009). In addition, heart rate and stroke volume are often deceased with the aggregate hemodynamic response resulting in decreased perfusion to some organ systems, particularly the kidneys.

Following infusion of a bolus of Hb into the vascular space, the MAP response appears to be related to the pharmacokinetics of the protein in circulation (Fig. 3), with rapid blood pressure normalization and parallel renal clearance of free Hb. However, in animal models of RBC transfusion-related hemolysis and cases in which an HBOC with a long intravascular circulatory time is administered, there is a clear disconnect between pharmacodynamics (blood pressure response and duration) and pharmacokinetics (maximal concentration and total exposure) (Buehler et al. 2007). This suggests that some degree of tolerance to extracellular Hb in circulation occurs with longer-term exposures. If analyses of Hb-mediated vascular wall injury are considered, however, the duration of sustained intravascular Hb exposure appears to be a critical determinant of the cumulative toxicity (Baek et al. 2012).

Within the context of potential pathology, Hb access to NO progresses via two primary reactions: (1) NO dioxygenation, and (2) iron nitrosylation of NO deoxy-Hb:

$$(1)\quad \text{Hb-Fe}^{2+}(O_2) + \text{NO} \rightarrow \{\text{Hb}(\text{Fe}^{3+}\text{OONO}^{\bullet})\} \rightarrow \text{Hb-Fe}^{3+} + \text{NO}_3^{-},$$

$$(2)\quad \text{Hb-Fe}^{2+} + \text{NO} \rightarrow \text{Hb(NO)}.$$

Our understanding of the factors controlling these processes is largely owing to the extensive study of HBOCs and their effects on systemic and pulmonary hemodynamics (Doherty et al. 1998; Olson et al. 2004). Targeted mutagenesis to limit interactions with NO or chemical modification of human Hb attenuated the systemic vascular response, presumably via reduced NO consumption within sites of NO generation or decreased access to sites of NO production. Hb exposure also stimulates other vasoconstrictive physiologic systems and effectors such as angiotensin, endothelin, and prostaglandins (Buehler et al. 2010).

Pulmonary arterial pressure (PAP) is regulated by NO, prostacyclin, endothelin, as well as thromboxane; several of these regulators have been studied in association with Hb (Heller et al. 1998; Farber and Loscalzo 2004; Yu et al. 2009). The primary cause for PAP elevation

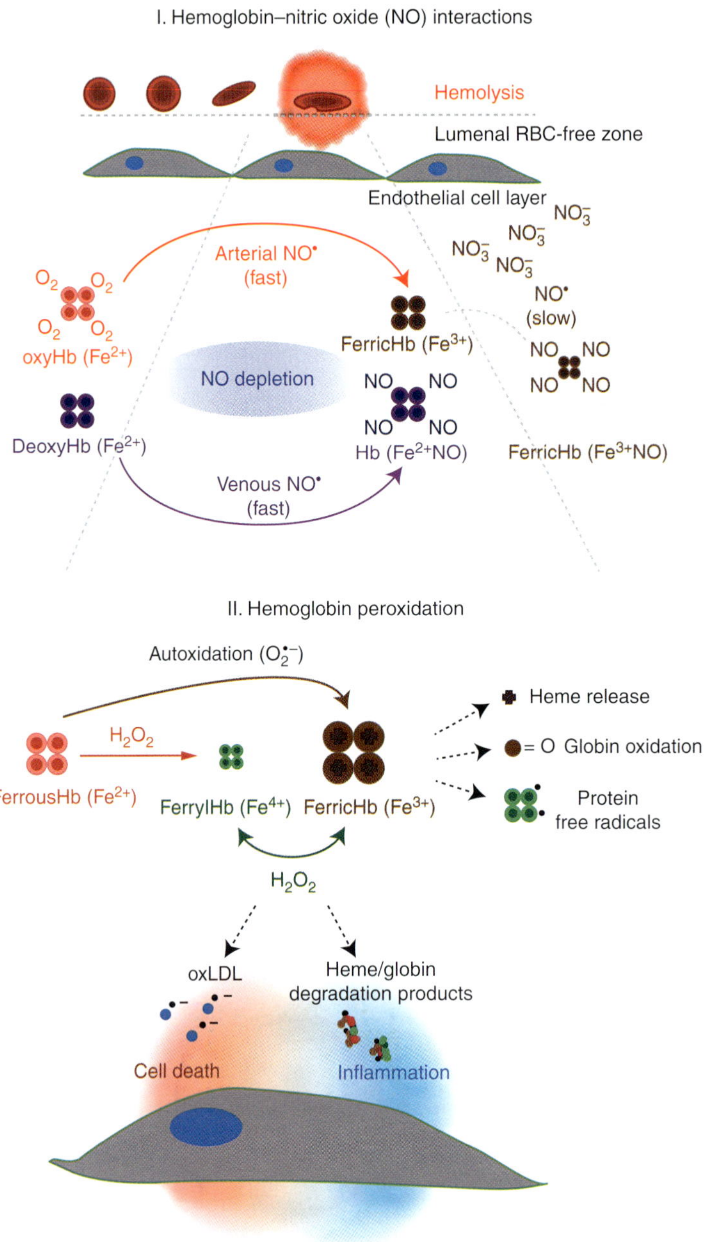

Figure 2. Schematic of hemoglobin-mediated toxicity initiated by intravascular hemolysis. (*I*) NO reactions with hemoglobin—Red blood cells (RBC) do not enter the vascular luminal RBC-free zone; however, free hemoglobin can enter the RBC-free zone and gain access to endothelial and perivascular spaces. When free oxyhemoglobin (red) enters the perivascular space it can react with NO rapidly via dioxygenation forming ferric hemoglobin (brown) and nitrate (NO_3^-). Additionally, deoxyhemoglobin (purple) can react with NO to form iron nitrosyl hemoglobin. Additionally, ferric hemoglobin can react with NO, but at much slower rates. It remains unknown whether this reaction is of any biologic significance. (*II*) Hemoglobin peroxidative reactions—Oxyhemoglobin (red) undergoes peroxidation to ferryl hemoglobin (green) followed by a redox cycle that is driven by hydrogen peroxide. (*Legend continues on following page.*)

induced by extracellular Hb has also been attributed to NO consumption. Elevated pulmonary vascular resistance was shown in dogs (Minneci et al. 2005) and PH was shown in rodents and sheep following exposure to Hb (Yu et al. 2008). These studies suggest a causal relationship between Hb and NO, based on the use of NO synthase inhibitors and reversal of PH via NO supplementation before Hb exposure. In other acute Hb exposures, such as transfusion-related hemolysis, PH is reported to occur more readily in animals with endothelial dysfunction compared with normal animals (Yu et al. 2012). Pretransfusion NO inhalation was shown to attenuate this pathology (Baron et al. 2012). Certain animal species show PH in response to HBOCs; however, it is unknown whether this effect regularly occurs following HBOC infusion in humans and whether a short-term increase in PAP could contribute to adverse events following extracellular Hb exposure or HBOC administration. Publicly available data on the incidence of PH from HBOC clinical trials show one event among 504 subjects treated with diaspirin-cross-linked Hb (DCLHb, HemAssist) (no events among 505 control subjects) and three events among 708 subjects treated with HBOC-201 (no events among 618 control subjects). It is unknown whether subjects who experienced acute transient PAP elevation developed clinically relevant events.

In Vivo Oxidation of Hemoglobin

Hydrogen peroxide and lipid hydroperoxides in tissue compartments are important components influencing the oxidative toxicity of heme proteins (Moore et al. 1998). Intracellular hydrogen peroxide (H_2O_2) is generated by the dismutation of superoxide ($O_2^{\bullet -}$), either spontaneously or by superoxide dismutase, and is involved in cellular signaling. Its primary fate is intracellular conversion to H_2O and O_2 by catalases and peroxidases. The remaining H_2O_2 can be converted to reactive oxygen species or secreted into the extracellular environment, typically in low (1–15 μM) quantities (D'Agnillo et al. 2001). During inflammation, certain cell types, such as granulocytes and macrophages, increase the production and release of H_2O_2. The production, from H_2O_2 and chloride (Cl^-), and secretion of hypochlorous acid are also increased. The increased levels of pro-oxidants could interact with extracellular Hb within localized tissue compartments and certain vascular sites. The reactions of Hb with peroxides have been extensively studied under "test-tube" conditions, and can be summarized as follows: Deoxy-Hb reacts with H_2O_2 via the following steps: (1) oxo-ferryl Hb [$Hb(Fe^{4+}=O)$] generation, (2) ferric Hb [$Hb(Fe^{3+})$] generation, and (3) protein radical generation [$^{\bullet}Hb(Fe^{4+}=O)$].

$$\text{(1)} \quad Hb(Fe^{2+}) + H_2O_2 \rightarrow Hb(Fe^{4+}=O) + H_2O,$$

$$\text{(2)} \quad Hb(Fe^{4+}=O) + H^+ \rightarrow Hb(Fe^{3+})\ OH^-,$$

$$\text{(3)} \quad Hb(Fe^{3+}) + H_2O_2 \rightarrow {}^{\bullet}Hb(Fe^{4+}=O) + H_2O.$$

Ultimately, the globin chain-free radical shown in reaction (3) is available to participate in localized amino acid oxidations (e.g., within Hb) or transfer to LDL (Miller et al. 1997; Vallelian et al. 2008; Pimenova et al. 2010). Peroxidative Hb reactions with production of ferryl Hb and free radicals have been considered an important component of Hb and HBOC toxicity in the past (Alayash 2004). However, despite

Figure 2. (*Continued*) In the presence of large quantities of reducing agents ferryl stability in this reaction is reduced to an extent that the species cannot accumulate, and in most cases cannot be directly detected. The most prevalent oxidized Hb species in vivo is therefore ferric Hb, which can result from peroxidation but also from autoxidation and NO dioxygenation depicted above. In vitro, the result of redox cycling mediated by H_2O_2 can be release of heme, globin chain radicals, and oxidation of amino acids within the hemoglobin molecule. The protein may also degrade to precipitate formed by globin-heme adducts and cross-links that physically damage endothelial cells leading to inflammation. Additionally, free radical transfer from ferryl radical oxidizes lipid, leading to potential tissue oxidation as well as inflammation.

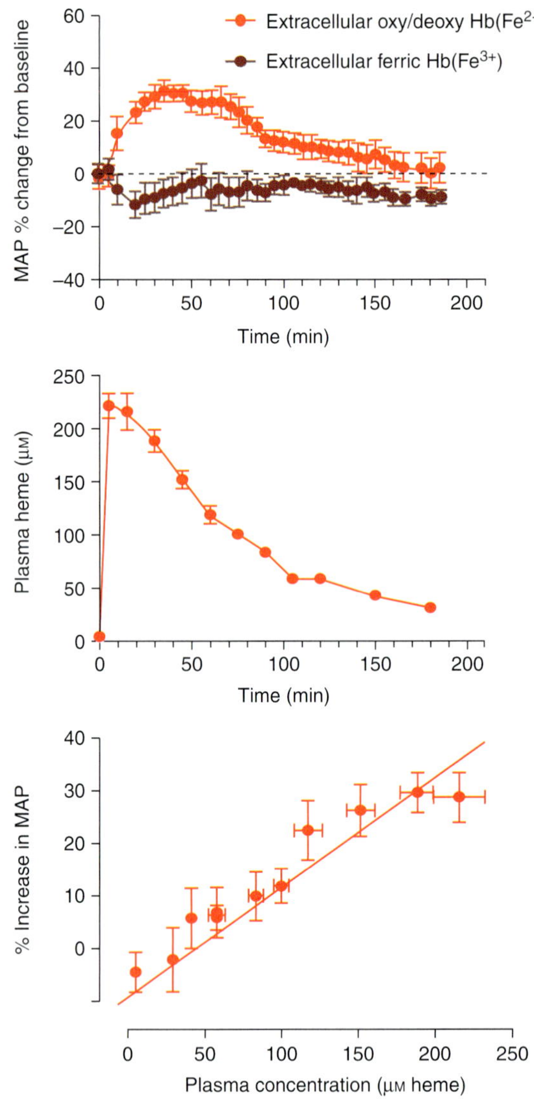

Figure 3. Plasma concentration blood pressure relationships. (*Top*) The blood pressure response to ferrous (red) and ferric (brown) hemoglobin following a 20% top load to conscious guinea pigs. Ferric hemoglobin shows no response, whereas ferrous hemoglobin shows a rapid and transient increase in mean arterial pressure. This is consistent with the hemoglobin plasma concentration time course shown (*middle*). These two time effects can be better visualized by plotting plasma concentration versus percent mean arterial blood pressure change from baseline (*bottom*).

the extensive biochemical knowledge of oxidative Hb reactions in vitro, it is far less clear to what extent these reactions do occur in vivo and how they could contribute to Hb toxicity. For example, it is not consistently possible to trap ferryl Hb or Hb-associated free radicals in vivo. An explanation for this discrepancy might be that the reactions occur in vivo in a complex environment with different oxidants and in the presence of large quantities of reductants (e.g., ascorbate, urate, and enzymatic reductants). These conditions likely prevent ferryl Hb accumulation above a certain threshold of detection. In vivo evidence for these oxidative

 Cite this article as *Cold Spring Harb Perspect Med* doi: 10.1101/cshperspect.a013433

Hb processes is therefore largely based on indirect markers such as intramolecular Hb cross-links and globin amino acid oxidations that have been detected in some studies (Moore et al. 1998; Boutaud et al. 2010).

The kidneys are heavily exposed to heme and globin components owing to the extensive and rapid filtration of extracellular Hb. The mechanism of Hb toxicity in the kidneys may be related to redox cycling and ferryl Hb formation (Zager and Gamelin 1989; Moore et al. 1998; Boutaud et al. 2010). During rhabdomyolysis, redox cycling of ferric and ferryl myoglobin, driven by lipid hydroperoxides, enhances tissue oxidation and F_2-isoprostane release (Moore et al. 1998; Boutaud et al. 2010). Observations of the excretion of covalently cross-linked dimeric, tetrameric, and polymeric Hb, as well as iron deposition and oxidative tissue modifications, in the kidneys of Hb-infused animals were interpreted as indicators of the peroxidative activity of Hb in vivo (Boretti et al. 2009; Butt et al. 2010). Within the gastrointestinal circulation of rats, an increase in microvascular permeability to fluorescently labeled albumin was reported to occur with several HBOCs (Baldwin et al. 2002). These effects appeared to be causally linked to the rates of oxidation of the individual HBOCs (Baldwin et al. 2002). Coadministration of the antioxidant sodium selenite reduced mesenteric vascular leakage and mucosal epithelial damage (Baldwin et al. 2003).

PHYSIOLOGIC HEMOGLOBIN SCAVENGERS, ANTIOXIDANTS, AND DETOXIFICATION SYSTEMS

The Hb scavenger system is a multiple-component pathway made up of several soluble plasma proteins (e.g., haptoglobin [Hp], hemopexin [Hpx], α1-microglobulin, and albumin) (Schaer and Alayash 2012). Cell-based receptors bind the Hb, or heme complexes of these proteins, and protect against systemic Hb and heme toxicity, until heme is degraded to metabolites by the heme oxygenases. The most extensively studied pathways are those involving the plasma Hb scavenger Hp and the monocyte/macrophage Hb–Hp scavenger receptor CD163 (Kristiansen et al. 2001; Schaer et al. 2006b). Additionally, the plasma heme-binding protein Hpx and its interaction with the LDL receptor-related protein 1 (LRP-1) appear to be critical when heme is released from Hb and other proteins (Nielsen et al. 2010). Hp remains the most important plasma Hb-binding protein, whereas the role of Hpx is secondary and generally required only when heme is released from Hb following oxidation. Figure 4 depicts the process of Hb/heme clearance and iron recycling within the plasma, macrophages, and erythroid cellular compartments.

Haptoglobin

The primary Hb scavenger in plasma is Hp (Levy et al. 2010). Each Hp β chain binds one Hb αβ dimer in a 1:1 stoichiometry. In mammalian species, the primary phenotype of Hp is 1-1, meaning that the protein contains two identical β chains and two α1 chains assembled as βα1-α1β (90 kDa). The disulfide bond between the α1 chains occurs at only one cysteine residue. In humans, certain evolutionary changes have driven the expression of two additional phenotypes, Hp 2-1 and Hp 2-2, in which the α2 chain has two exposed cysteine residues available for disulfide bonding (Wicher and Fries 2010). Hp 2-1 contains an α2 chain, two α1 chains, and three β chains (135 kDa). Hp 2-2 contains only α2 chains and β chains that can form a multitude of self-polymerizing proteins with a molecular mass ranging from ~135 kDa to 800 kDa. Additional mutations exist in humans, but remain rare and isolated. The formation of Hb–Hp creates a stable and irreversible complex that circulates in the plasma until it is cleared by CD163-expressing peripheral blood monocytes or tissue-resident macrophages (Kristiansen et al. 2001).

Hemopexin

This protein is primarily expressed in the liver and, like Hp, is classified as a class I acute phase protein induced by proinflammatory cytokines (i.e., IL-1 and IL-6) (Tolosano et al. 1999, 2010).

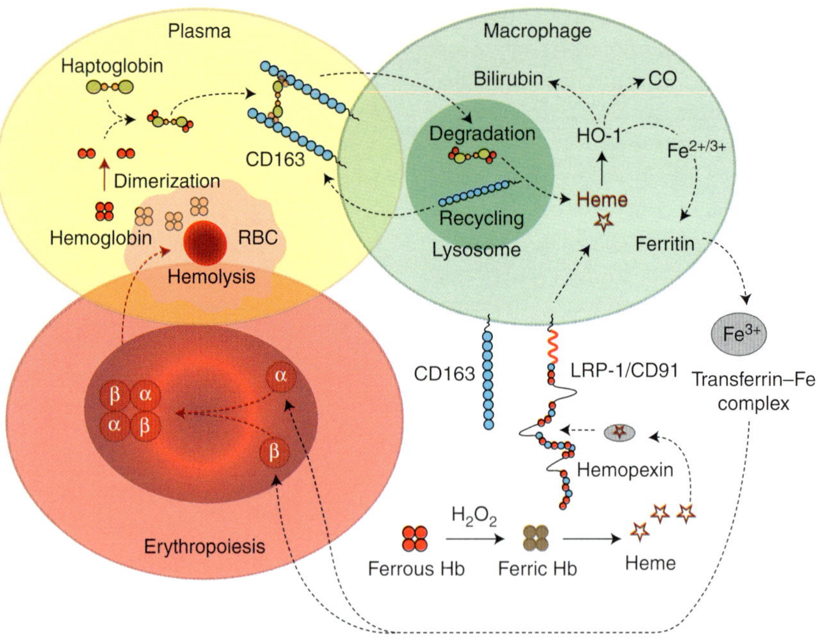

Figure 4. Physiologic hemoglobin scavengers and detoxification systems. Hb can be released from red blood cells (RBC) into the plasma and extracellular environment during hemolysis or tissue injury. In the extracellular compartment, Hb reacts with peroxides (H_2O_2 and lipid hydroperoxides) and promotes oxidative tissue damage. In the plasma or extracellular space, Hb is sequestered in the Hb–Hp complex. Complex formation prevents Hb-induced hypertensive and oxidative reactions. The Hb–Hp complex is subsequently endocytosed by the macrophage Hb scavenger receptor CD163. Within the macrophage, heme is released from globin and degraded by HO-1 into bilirubin and carbon monoxide (CO). The released iron induces ferritin synthesis. Iron can either be exported for iron recycling or it can be stored in a ferritin complex. Backup systems such as hemopexin (Hpx) and the Hpx–heme complex receptor (LDL receptor-related protein, LRP) pathway can bind and detoxify free heme that is released from oxidized Hb.

The interaction of free heme with Hpx is reported to be the strongest plasma heme–protein interaction with a $K_d < 10^{-13}$ M (Tolosano et al. 2010). Following inflammation, Hpx is synthesized in the liver, generating mean plasma concentrations of ~10 μM (0.6 mg/mL) (Tolosano et al. 1999, 2010). In sickle cell anemia and thalassemia, Hpx is typically depleted in the plasma. Once heme-Hpx binding occurs, interaction and uptake by the macrophage-associated LRP-1 receptor removes the heme from the circulation (Nielsen et al. 2010).

Monocyte/Macrophage CD163

In the circulation and tissues, the Hb–Hp complex is cleared by the Hb scavenger receptor CD163, a 130-kDa scavenger receptor tethered to peripheral blood monocytes/macrophages and tissue-resident macrophages (Kristiansen et al. 2001; Schaer et al. 2006b, 2007). CD163 is the primary receptor responsible for removing free Hb (primarily as Hb–Hp) from the circulation. The binding of the Hb–Hp complex to the scavenger receptor cysteine-rich domain three of CD163 and its subsequent uncoupling following uptake are calcium-dependent. CD163 is cleaved from monocyte and macrophage cell surfaces by a TNF-α-converting enzyme/A Disintigrin and Metalloproteinase 17 mechanism under conditions of inflammation (Etzerodt et al. 2010). The inflammation-linked cleavage and plasma levels of CD163 have been shown to be elevated during general tissue inflammation; the highest concentrations were found in patients with macrophage-activation syndromes (Schaer et al. 2005).

Endocytosis of the Hb–Hp complex results in cellular delivery of heme and up-regulation of HO-1. This provides a link to enhanced macrophage HO-1 clearance of heme (Schaer et al. 2006a; Vallelian et al. 2010). The end result of this process is the up-regulation of the synthesis of heme metabolites (i.e., bilirubin, CO, and ferritin) that may function as antioxidants (bilirubin and ferritin) and signaling molecules (CO) (Otterbein et al. 2003).

Hemoxygenase

The loss of heme oxygenase (HO) function, or HO-1 deficiency, in humans was first reported in 1999 (Yachie et al. 1999). The disease was described in a 6-yr-old boy, who suffered from severe growth retardation, increased RBC fragility, chronic hemolysis with paradoxically elevated Hp levels, elevated ferritin levels, iron deposition in renal and hepatic tissues, and marked kidney injury (Yachie et al. 1999). This very rare condition provided important insight into the functional role and importance of this class of enzymes. HO is the primary enzyme of heme catabolism. Two predominant isoforms are involved in heme metabolism, HO-1 and HO-2, and both enzymatically degrade heme to biliverdin, bilirubin, iron, and CO (Abraham and Kappas 2008). HO-2 is constitutively expressed; however, HO-1 is induced by various stimuli including heme, lipopolysaccharide, hypoxia, and heavy metals (Abraham and Kappas 2008). However, heme is the most potent physiologic inducer of HO-1 (Balla et al. 1993). Although it appears evident that HOs are essential to keep intracellular heme homeostasis within a physiologic (e.g., nontoxic range), many effects that are associated with under- or overexpression of these enzymes are poorly understood. For example, the mechanisms and pathways of cellular heme toxicity (e.g., in cases of inappropriately low HO-1 expression or excessive heme load) have not yet been systematically explored. Cellular heme toxicity is mostly considered to be caused by nonspecific oxidative reactions that may damage cellular proteins and lipid components. However, it is likely that more complex pharmacologic effects of the protoporphyrin exist and could disturb essential metabolic pathways. Likewise, it is uncertain to what extent the protection against heme toxicity by HO is dependent on biologic activities of its metabolic end products (CO, iron, and bilirubin) in addition to the primary catabolic heme "removal" activity (Otterbein et al. 2003).

SCAVENGER PROTEIN-BASED THERAPEUTICS FOR HEMOLYTIC CONDITIONS

Several studies have explored the protective activities of Hp and Hpx against Hb/heme-driven disease processes in vitro and in vivo, and novel therapeutics based on these proteins have been proposed. Hp can prevent heme release from met-Hb (Fe^{3+}) and, in the presence of H_2O_2, attenuate the peroxidation of globin chain amino acids and the oxidative modification of external molecules (i.e., other proteins and lipids) (Miller et al. 1997; Buehler et al. 2009; Pimenova et al. 2010). These protective functions appear to be related to the ability of Hp to firmly stabilize the structure of the bound Hb $\alpha\beta$ dimers, and to a less understood mechanism whereby Hp shields peroxidative reactions within the Hb–Hp complex from external substrates. The Hb–Hp complex can still react with many ligands such as O_2, NO, H_2O_2, and nitrite (Boretti et al. 2009; Buehler et al. 2009; Roche et al. 2012). It is unknown whether these reactions have any biological function. Some hypotheses suggest that the H_2O_2 reaction may attenuate oxidative stress in inflamed tissues, whereas the enhanced nitrite-reductase activity of the complex may help to control the adverse vascular effects of unopposed NO consumption within the circulation. Another essential function of Hp in vivo is the capture of the relatively small Hb dimers within the circulation by a protein complex too large for renal clearance and extravasation (Boretti et al. 2009; Baek et al. 2012). This intravascular sequestration limits toxic Hb exposure in the kidneys and other sensitive tissues, and may also prevent access of the NO scavenger Hb to the vascular wall (Fig. 5). Proof of concept

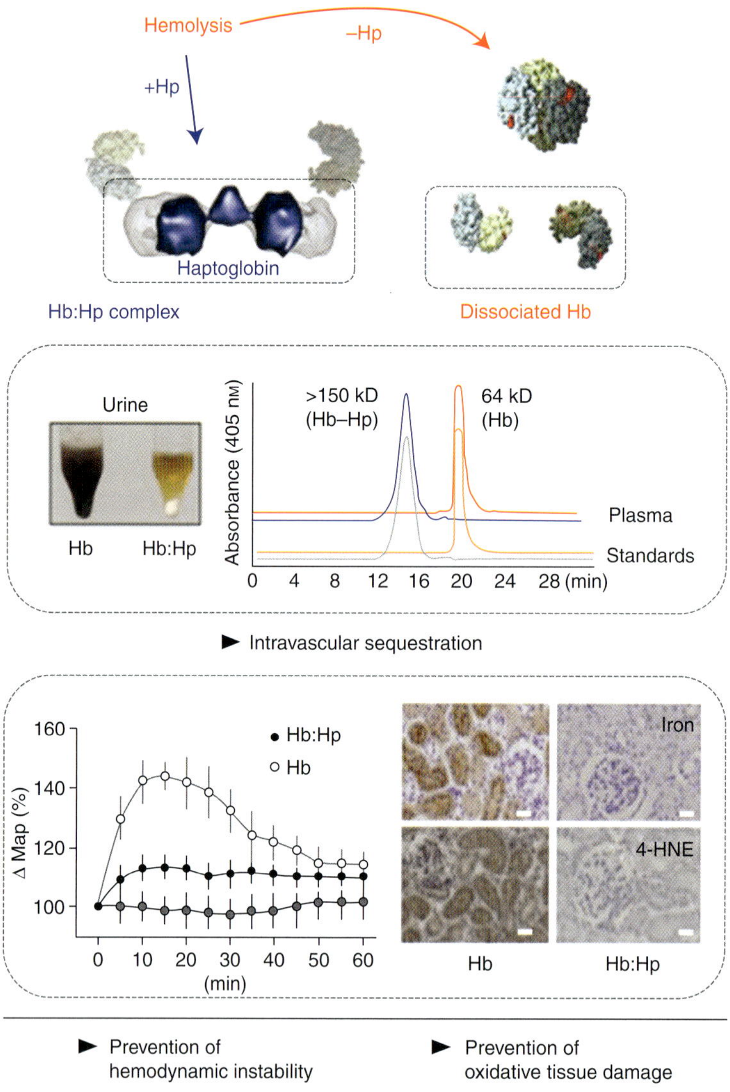

Figure 5. The haptoglobin paradigm of damage prevention during hemolysis. After hemolysis free Hb can dissociate into αβ dimers with a relatively small molecule size of ~32 kD (eluting at 19.8 min in the shown plasma size exclusion chromatography [SEC] profile). Susceptible tissues are exposed to toxic Hb. The results are hemoglobinuria, hemodynamic instability with acute hypertension, and oxidative tissue damage. With haptoglobin (Hp) treatment 100% of the free Hb is captured within the large sized (>150 kD) Hb:Hp complex that elutes at 14.7 min in plasma SEC (note that there remains no free Hb in the Hp-treated plasma sample). As a result, free Hb remains compartmentalized within the circulation. Hemoglobinuria, hemodynamic instability, and oxidative tissue damage are prevented. The image in the *left middle* panel shows urine collections of Hb and Hb:Hp infused guinea pigs. The *lower left* panel shows blood pressure recordings of Hb, Hb:Hp, or starch (control) infused animals. The kidney sections in the *lower right* panel were stained for 4-HNE as a marker of oxidative tissue damage (brown color) in Hb-infused (but not in Hb transfused +Hp-treated) animals.

studies of the therapeutic potential of Hp have been conducted in animal models of Hb infusion and blood transfusion (Boretti et al. 2009; Baek et al. 2012). Application of human plasma-derived Hp in these models attenuated the Hb-mediated acute hypertensive response, oxidative renal damage, and vascular injury. Human plasma-derived Hp is approved as a drug for the treatment of hemolytic conditions in Japan (haptoglobin injection, Benesis, Osaka, Japan). Unfortunately, this product is not available outside of Japan and there are insufficient data available to estimate the clinical efficacy of Hp, but there are reports of therapeutic benefits (Imaizumi et al. 1994). Other Hp products are expected to be developed by European and U.S.-based plasma fractionation companies, and further preclinical and clinical studies will likely be conducted in the near future to further explore the efficacy and safety of Hp for the treatment of diverse hemolytic conditions.

The rationale for the use of Hpx as a therapeutic scavenger protein in hemolytic conditions is based on the idea that free heme could be an ultimate mediator of Hb toxicity. This idea is supported by multiple in vitro observations and by findings in the Hpx-knockout mouse, which is susceptible to renal damage after acute intravascular hemolysis (Tolosano et al. 1999). Severe sepsis can be accompanied by hemolysis and Hpx depletion. In a mouse model of severe sepsis, plasma-derived Hpx was protective against organ damage and death (Larsen et al. 2011). The protective effect in this model may be related to the ability of Hpx to prevent the interaction of free heme with innate immune receptors (e.g., toll-like receptors), interrupting the synergistic proinflammatory effects of heme and the sepsis-inducing pathogen. Currently, there is no Hpx product available for clinical application and, therefore, no safety and/or efficacy data exist for its therapeutic use.

Several functions of Hpx have been described that are not directly related to its heme scavenging activity (Tolosano et al. 2010). These functions involve enzymatic activities as well as direct anti-inflammatory effects. Nonheme-binding effects of Hpx might be responsible for the intriguing observation that, in another mouse model of sepsis, Hpx inhibited neutrophil migration to the focus of infection. In this model, lethality was paradoxically increased in septic wild-type mice treated with the heme-binding protein, and reduced in Hpx-knockout animals (Spiller et al. 2011). It was suggested that many of the reported nonheme-binding functions of Hpx were related to the experimental conditions or to impurities and degradation products in some experimental Hpx preparations (Mauk et al. 2011); careful examination of such effects may be critical in the preclinical evaluation of Hpx-based therapeutics. Also, no comparative experimental data currently exist to determine which of the plasma-scavenging proteins within the Hb/heme-binding cascade might be the most protective against hemoglobin-driven pathologies.

ACKNOWLEDGMENTS

This work is supported by the Swiss National Science Foundation (grants 310030/120658 and 31003A/138500), the University of Zurich Research Priority Program "Integrative Human Physiology," the Swiss Federal Commission for Technology and Innovation (CTI), and FDA Internal Funding.

REFERENCES

Abraham NG, Kappas A. 2008. Pharmacological and clinical aspects of heme oxygenase. *Pharmacol Rev* **60:** 79–127.

Alayash AI. 2004. Oxygen therapeutics: Can we tame haemoglobin? *Nat Rev Drug Discov* **3:** 152–159.

Baek JH, D'Agnillo F, Vallelian F, Pereira CP, Williams MC, Jia Y, Schaer DJ, Buehler PW. 2012. Hemoglobin-driven pathophysiology is an in vivo consequence of the red blood cell storage lesion that can be attenuated in guinea pigs by haptoglobin therapy. *J Clin Invest* **122:** 1444–1458.

Baldwin AL, Wiley EB, Alayash AI. 2002. Comparison of effects of two hemoglobin-based O(2) carriers on intestinal integrity and microvascular leakage. *Am J Physiol Heart Circ Physiol* **283:** H1292–H1301.

Baldwin AL, Wiley EB, Summers AG, Alayash AI. 2003. Sodium selenite reduces hemoglobin-induced venular leakage in the rat mesentery. *Am J Physiol Heart Circ Physiol* **284:** H81–H91.

Balla J, Jacob HS, Balla G, Nath K, Eaton JW, Vercellotti GM. 1993. Endothelial-cell heme uptake from heme proteins: Induction of sensitization and desensitization to oxidant damage. *Proc Natl Acad Sci* **90:** 9285–9289.

Baron DM, Yu B, Lei C, Bagchi A, Beloiartsev A, Stowell CP, Steinbicker AU, Malhotra R, Bloch KD, Zapol WM. 2012. Pulmonary hypertension in lambs transfused with stored blood is prevented by breathing nitric oxide. *Anesthesiology* **116:** 637–647.

Barr PJ, Bailie KE. 2011. Transfusion thresholds in FOCUS. *N Engl J Med* **365:** 2532–2533.

Belcher JD, Bryant CJ, Nguyen J, Bowlin PR, Kielbik MC, Bischof JC, Hebbel RP, Vercellotti GM. 2003. Transgenic sickle mice have vascular inflammation. *Blood* **101:** 3953–3959.

Berra L, Coppadoro A, Yu B, Lei C, Spagnolli E, Steinbicker AU, Bloch KD, Lin T, Sammy FY, Warren HS, et al. 2012. Transfusion of stored blood does not alter reactive hyperemia in healthy volunteers. *Anesthesiology* **117:** 56–63.

Boretti FS, Buehler PW, D'Agnillo F, Kluge K, Glaus T, Butt OI, Jia Y, Goede J, Pereira CP, Maggiorini M, et al. 2009. Sequestration of extracellular hemoglobin within a haptoglobin complex decreases its hypertensive and oxidative effects in dogs and guinea pigs. *J Clin Invest* **119:** 2271–2280.

Boutaud O, Moore KP, Reeder BJ, Harry D, Howie AJ, Wang S, Carney CK, Masterson TS, Amin T, Wright DW, et al. 2010. Acetaminophen inhibits hemoprotein-catalyzed lipid peroxidation and attenuates rhabdomyolysis-induced renal failure. *Proc Natl Acad Sci* **107:** 2699–2704.

Boyle JJ, Harrington HA, Piper E, Elderfield K, Stark J, Landis RC, Haskard DO. 2009. Coronary intraplaque hemorrhage evokes a novel atheroprotective macrophage phenotype. *Am J Pathol* **174:** 1097–1108.

Boyle JJ, Johns M, Kampfer T, Nguyen AT, Game L, Schaer DJ, Mason JC, Haskard DO. 2012. Activating transcription factor 1 directs Mhem atheroprotective macrophages through coordinated iron handling and foam cell protection. *Circ Res* **110:** 20–33.

Buehler PW, D'Agnillo F, Hoffman V, Alayash AI. 2007. Effects of endogenous ascorbate on oxidation, oxygenation, and toxicokinetics of cell-free modified hemoglobin after exchange transfusion in rat and guinea pig. *J Pharmacol Exp Ther* **323:** 49–60.

Buehler PW, Abraham B, Vallelian F, Linnemayr C, Pereira CP, Cipollo JF, Jia Y, Mikolajczyk M, Boretti FS, Schoedon G, et al. 2009. Haptoglobin preserves the CD163 hemoglobin scavenger pathway by shielding hemoglobin from peroxidative modification. *Blood* **113:** 2578–2586.

Buehler PW, D'Agnillo F, Schaer DJ. 2010. Hemoglobin-based oxygen carriers: From mechanisms of toxicity and clearance to rational drug design. *Trends Mol Med* **16:** 447–457.

Bunn HF, Nathan DG, Dover GJ, Hebbel RP, Platt OS, Rosse WF, Ware RE. 2010. Pulmonary hypertension and nitric oxide depletion in sickle cell disease. *Blood* **116:** 687–692.

Butt OI, Buehler PW, D'Agnillo F. 2010. Differential induction of renal heme oxygenase and ferritin in ascorbate and nonascorbate producing species transfused with modified cell-free hemoglobin. *Antioxid Redox Signal* **12:** 199–208.

Cunnington AJ, de Souza JB, Walther M, Riley EM. 2011. Malaria impairs resistance to *Salmonella* through heme- and heme oxygenase-dependent dysfunctional granulocyte mobilization. *Nat Med* **18:** 120–127.

D'Agnillo F, Alayash AI. 2001. Redox cycling of diaspirin cross-linked hemoglobin induces G2/M arrest and apoptosis in cultured endothelial cells. *Blood* **98:** 3315–3323.

Doherty DH, Doyle MP, Curry SR, Vali RJ, Fattor TJ, Olson JS, Lemon DD. 1998. Rate of reaction with nitric oxide determines the hypertensive effect of cell-free hemoglobin. *Nat Biotechnol* **16:** 672–676.

Donadee C, Raat NJ, Kanias T, Tejero J, Lee JS, Kelley EE, Zhao X, Liu C, Reynolds H, Azarov I, et al. 2011. Nitric oxide scavenging by red blood cell microparticles and cell-free hemoglobin as a mechanism for the red cell storage lesion. *Circulation* **124:** 465–476.

Etzerodt A, Maniecki MB, Moller K, Moller HJ, Moestrup SK. 2010. Tumor necrosis factor α-converting enzyme (TACE/ADAM17) mediates ectodomain shedding of the scavenger receptor CD163. *J Leukoc Biol* **88:** 1201–1205.

Farber HW, Loscalzo J. 2004. Pulmonary arterial hypertension. *N Engl J Med* **351:** 1655–1665.

Ferreira A, Marguti I, Bechmann I, Jeney V, Chora A, Palha NR, Rebelo S, Henri A, Beuzard Y, Soares MP. 2011. Sickle hemoglobin confers tolerance to Plasmodium infection. *Cell* **145:** 398–409.

Finn AV, Nakano M, Polavarapu R, Karmali V, Saeed O, Zhao X, Yazdani S, Otsuka F, Davis T, Habib A, et al. 2012. Hemoglobin directs macrophage differentiation and prevents foam cell formation in human atherosclerotic plaques. *J Am Coll Cardiol* **59:** 166–177.

Gladwin MT, Sachdev V, Jison ML, Shizukuda Y, Plehn JF, Minter K, Brown B, Coles WA, Nichols JS, Ernst I, et al. 2004. Pulmonary hypertension as a risk factor for death in patients with sickle cell disease. *N Engl J Med* **350:** 886–895.

Gladwin MT, Kanias T, Kim-Shapiro DB. 2012. Hemolysis and cell-free hemoglobin drive an intrinsic mechanism for human disease. *J Clin Invest* **122:** 1205–1208.

Gramaglia I, Sobolewski P, Meays D, Contreras R, Nolan JP, Frangos JA, Intaglietta M, van der Heyde HC. 2006. Low nitric oxide bioavailability contributes to the genesis of experimental cerebral malaria. *Nat Med* **12:** 1417–1422.

Heller A, Ragaller M, Schmeck J, Fluth H, Muller M, Albrecht DM, Koch T. 1998. Role of NO and endothelin in hemoglobin-induced pulmonary vasoconstriction. *Shock* **10:** 401–406.

Hod EA, Zhang N, Sokol SA, Wojczyk BS, Francis RO, Ansaldi D, Francis KP, Della-Latta P, Whittier S, Sheth S, et al. 2010. Transfusion of red blood cells after prolonged storage produces harmful effects that are mediated by iron and inflammation. *Blood* **115:** 4284–4292.

Hod EA, Brittenham GM, Billote GB, Francis RO, Ginzburg YZ, Hendrickson JE, Jhang J, Schwartz J, Sharma S, Sheth S, et al. 2011. Transfusion of human volunteers with older, stored red blood cells produces extravascular hemolysis and circulating non-transferrin-bound iron. *Blood* **118:** 6675–6682.

Imaizumi H, Tsunoda K, Ichimiya N, Okamoto T, Namiki A. 1994. Repeated large-dose haptoglobin therapy in an extensively burned patient: Case report. *J Emerg Med* **12:** 33–37.

Jeney V, Balla J, Yachie A, Varga Z, Vercellotti GM, Eaton JW, Balla G. 2002. Pro-oxidant and cytotoxic effects of circulating heme. *Blood* **100:** 879–887.

Jia Y, Buehler PW, Boykins RA, Venable RM, Alayash AI. 2007. Structural basis of peroxide-mediated changes in human hemoglobin: A novel oxidative pathway. *J Biol Chem* **282:** 4894–4907.

Kaempfer T, Duerst E, Gehrig P, Roschitzki B, Rutishauser D, Grossmann J, Schoedon G, Vallelian F, Schaer DJ. 2011. Extracellular hemoglobin polarizes the macrophage proteome toward Hb-clearance, enhanced antioxidant capacity and suppressed HLA class 2 expression. *J Proteome Res* **10:** 2397–2408.

Kolodgie FD, Gold HK, Burke AP, Fowler DR, Kruth HS, Weber DK, Farb A, Guerrero LJ, Hayase M, Kutys R, et al. 2003. Intraplaque hemorrhage and progression of coronary atheroma. *N Engl J Med* **349:** 2316–2325.

Kristiansen M, Graversen JH, Jacobsen C, Sonne O, Hoffman HJ, Law SK, Moestrup SK. 2001. Identification of the haemoglobin scavenger receptor. *Nature* **409:** 198–201.

Larsen R, Gozzelino R, Jeney V, Tokaji L, Bozza FA, Japiassu AM, Bonaparte D, Cavalcante MM, Chora A, Ferreira A, et al. 2011. A central role for free heme in the pathogenesis of severe sepsis. *Sci Transl Med* **2:** 51ra71.

Levy AP, Asleh R, Blum S, Levy NS, Miller-Lotan R, Kalet-Litman S, Anbinder Y, Lache O, Nakhoul FM, Asaf R, et al. 2010. Haptoglobin: Basic and clinical aspects. *Antioxid Redox Signal* **12:** 293–304.

Lin T, Kwak YH, Sammy F, He P, Thundivalappil S, Sun G, Chao W, Warren HS. 2010. Synergistic inflammation is induced by blood degradation products with microbial Toll-like receptor agonists and is blocked by hemopexin. *J Infect Dis* **202:** 624–632.

Lin S, Zhong Q, Lv FL, Zhou Y, Li JQ, Wang JZ, Yang QW, Yin Q. 2012. Heme activates TLR4-mediated inflammatory injury via MyD88/TRIF signaling pathway in intracerebral hemorrhage. *J Neuroinflammation* **9:** 46.

Mauk MR, Smith A, Mauk AG. 2011. An alternative view of the proposed alternative activities of hemopexin. *Protein Sci* **20:** 791–805.

Middelburg RA, van den Watering LM, Briet E, van der Bom JG. 2012. Storage time of red blood cells and mortality of transfusion recipients. *Transfus Med Rev* **27:** 36–43.

Miller YI, Altamentova SM, Shaklai N. 1997. Oxidation of low-density lipoprotein by hemoglobin stems from a heme-initiated globin radical: Antioxidant role of haptoglobin. *Biochemistry* **36:** 12189–12198.

Minneci PC, Deans KJ, Zhi H, Yuen PS, Star RA, Banks SM, Schechter AN, Natanson C, Gladwin MT, Solomon SB. 2005. Hemolysis-associated endothelial dysfunction mediated by accelerated NO inactivation by decompartmentalized oxyhemoglobin. *J Clin Invest* **115:** 3409–3417.

Moore KP, Holt SG, Patel RP, Svistunenko DA, Zackert W, Goodier D, Reeder BJ, Clozel M, Anand R, Cooper CE, et al. 1998. A causative role for redox cycling of myoglobin and its inhibition by alkalinization in the pathogenesis and treatment of rhabdomyolysis-induced renal failure. *J Biol Chem* **273:** 31731–31737.

Nagy E, Eaton JW, Jeney V, Soares MP, Varga Z, Galajda Z, Szentmiklosi J, Mehes G, Csonka T, Smith A, et al. 2010. Red cells, hemoglobin, heme, iron, and atherogenesis. *Arterioscler Thromb Vasc Biol* **30:** 1347–1353.

Natanson C, Kern SJ, Lurie P, Banks SM, Wolfe SM. 2008. Cell-free hemoglobin-based blood substitutes and risk of myocardial infarction and death: A meta-analysis. *JAMA* **299:** 2304–2312.

Nielsen MJ, Moller HJ, Moestrup SK. 2010. Hemoglobin and heme scavenger receptors. *Antioxid Redox Signal* **12:** 261–273.

Olson JS, Foley EW, Rogge C, Tsai AL, Doyle MP, Lemon DD. 2004. NO scavenging and the hypertensive effect of hemoglobin-based blood substitutes. *Free Radic Biol Med* **36:** 685–697.

Otterbein LE, Soares MP, Yamashita K, Bach FH. 2003. Heme oxygenase-1: Unleashing the protective properties of heme. *Trends Immunol* **24:** 449–455.

Pamplona A, Ferreira A, Balla J, Jeney V, Balla G, Epiphanio S, Chora A, Rodrigues CD, Gregoire IP, Cunha-Rodrigues M, et al. 2007. Heme oxygenase-1 and carbon monoxide suppress the pathogenesis of experimental cerebral malaria. *Nat Med* **13:** 703–710.

Parent F, Bachir D, Inamo J, Lionnet F, Driss F, Loko G, Habibi A, Bennani S, Savale L, Adnot S, et al. 2011. A hemodynamic study of pulmonary hypertension in sickle cell disease. *N Engl J Med* **365:** 44–53.

Pimenova T, Pereira CP, Gehrig P, Buehler PW, Schaer DJ, Zenobi R. 2010. Quantitative mass spectrometry defines an oxidative hotspot in hemoglobin that is specifically protected by haptoglobin. *J Proteome Res* **9:** 4061–4070.

Reiter CD, Wang X, Tanus-Santos JE, Hogg N, Cannon RO III, Schechter AN, Gladwin MT. 2002. Cell-free hemoglobin limits nitric oxide bioavailability in sickle-cell disease. *Nat Med* **8:** 1383–1389.

Roche C, Dantsker D, Alayash AI, Friedman JM. 2012. Enhanced nitrite reductase activity associated with the haptoglobin complexed hemoglobin dimer: Functional and antioxidative implications. *Nitric Oxide* **27:** 32–39.

Rosenthal PJ. 2011. Lessons from sickle cell disease in the treatment and control of malaria. *N Engl J Med* **364:** 2549–2551.

Rother RP, Bell L, Hillmen P, Gladwin MT. 2005. The clinical sequelae of intravascular hemolysis and extracellular plasma hemoglobin: A novel mechanism of human disease. *JAMA* **293:** 1653–1662.

Schaer DJ, Alayash AI. 2012. Clearance and control mechanisms of hemoglobin from cradle to grave. *Antioxid Redox Signal* **12:** 181–184.

Schaer DJ, Schleiffenbaum B, Kurrer M, Imhof A, Bachli E, Fehr J, Moller HJ, Moestrup SK, Schaffner A. 2005. Soluble hemoglobin-haptoglobin scavenger receptor CD163 as a lineage-specific marker in the reactive hemophagocytic syndrome. *Eur J Haematol* **74:** 6–10.

Schaer CA, Schoedon G, Imhof A, Kurrer MO, Schaer DJ. 2006a. Constitutive endocytosis of CD163 mediates hemoglobin-heme uptake and determines the noninflammatory and protective transcriptional response of macrophages to hemoglobin. *Circ Res* **99:** 943–950.

Schaer DJ, Schaer CA, Buehler PW, Boykins RA, Schoedon G, Alayash AI, Schaffner A. 2006b. CD163 is the macrophage scavenger receptor for native and chemically modified hemoglobins in the absence of haptoglobin. *Blood* **107:** 373–380.

Schaer CA, Vallelian F, Imhof A, Schoedon G, Schaer DJ. 2007. CD163-expressing monocytes constitute an endotoxin-sensitive Hb clearance compartment within the vascular system. *J Leukoc Biol* **82:** 106–110.

Silverman TA, Weiskopf RB. 2009. Hemoglobin-based oxygen carriers: Current status and future directions. *Anesthesiology* **111:** 946–963.

Spiller F, Costa C, Souto FO, Vinchi F, Mestriner FL, Laure HJ, Alves-Filho JC, Freitas A, Rosa JC, Ferreira SH, et al. 2011. Inhibition of neutrophil migration by hemopexin leads to increased mortality due to sepsis in mice. *Am J Respir Crit Care Med* **183:** 922–931.

Tolosano E, Hirsch E, Patrucco E, Camaschella C, Navone R, Silengo L, Altruda F. 1999. Defective recovery and severe renal damage after acute hemolysis in hemopexin-deficient mice. *Blood* **94:** 3906–3914.

Tolosano E, Fagoonee S, Morello N, Vinchi F, Fiorito V. 2010. Heme scavenging and the other facets of hemopexin. *Antioxid Redox Signal* **12:** 305–320.

Vallelian F, Pimenova T, Pereira CP, Abraham B, Mikolajczyk MG, Schoedon G, Zenobi R, Alayash AI, Buehler PW, Schaer DJ. 2008. The reaction of hydrogen peroxide with hemoglobin induces extensive α-globin crosslinking and impairs the interaction of hemoglobin with endogenous scavenger pathways. *Free Radic Biol Med* **45:** 1150–1158.

Vallelian F, Schaer CA, Kaempfer T, Gehrig P, Duerst E, Schoedon G, Schaer DJ. 2010. Glucocorticoid treatment skews human monocyte differentiation into a hemoglobin-clearance phenotype with enhanced heme-iron recycling and antioxidant capacity. *Blood* **116:** 5347–5356.

Wang D, Sun J, Solomon SB, Klein HG, Natanson C. 2012. Transfusion of older stored blood and risk of death: A meta-analysis. *Transfusion* **52:** 1184–1195.

Wicher KB, Fries E. 2010. Evolutionary aspects of hemoglobin scavengers. *Antioxid Redox Signal* **12:** 249–259.

Yachie A, Niida Y, Wada T, Igarashi N, Kaneda H, Toma T, Ohta K, Kasahara Y, Koizumi S. 1999. Oxidative stress causes enhanced endothelial cell injury in human heme oxygenase-1 deficiency. *J Clin Invest* **103:** 129–135.

Yazdanbakhsh K, Bao W, Zhong H. 2011. Immunoregulatory effects of stored red blood cells. *Hematology Am Soc Hematol Educ Program* **2011:** 466–469.

Yu B, Raher MJ, Volpato GP, Bloch KD, Ichinose F, Zapol WM. 2008. Inhaled nitric oxide enables artificial blood transfusion without hypertension. *Circulation* **117:** 1982–1990.

Yu B, Volpato GP, Chang K, Bloch KD, Zapol WM. 2009. Prevention of the pulmonary vasoconstrictor effects of HBOC-201 in awake lambs by continuously breathing nitric oxide. *Anesthesiology* **110:** 113–122.

Yu B, Lei C, Baron DM, Steinbicker AU, Bloch KD, Zapol WM. 2012. Diabetes augments and inhaled nitric oxide prevents the adverse hemodynamic effects of transfusing syngeneic stored blood in mice. *Transfusion* **52:** 1410–1422.

Zager RA, Gamelin LM. 1989. Pathogenetic mechanisms in experimental hemoglobinuric acute renal failure. *Am J Physiol* **256:** F446–F455.

Iron Deficiency Anemia: A Common and Curable Disease

Jeffery L. Miller

Molecular Medicine Branch, National Institute of Diabetes and Digestive and Kidney Diseases, National Institutes of Health, Bethesda, Maryland 20892

Correspondence: jm7f@nih.gov

Iron deficiency anemia arises when the balance of iron intake, iron stores, and the body's loss of iron are insufficient to fully support production of erythrocytes. Iron deficiency anemia rarely causes death, but the impact on human health is significant. In the developed world, this disease is easily identified and treated, but frequently overlooked by physicians. In contrast, it is a health problem that affects major portions of the population in underdeveloped countries. Overall, the prevention and successful treatment for iron deficiency anemia remains woefully insufficient worldwide, especially among underprivileged women and children. Here, clinical and laboratory features of the disease are discussed, and then focus is placed on relevant economic, environmental, infectious, and genetic factors that converge among global populations.

Iron comprises 5% of the earth's crust. Its redox states make iron useful for evolving biological processes. Growing lists of biomolecules that bind or incorporate iron are being catalogued according to their structural similarities. An excellent source of information is provided at http://metallo.scripps.edu/PROMISE/PROMISE_LIST.html.

Four general categories of proteins contain iron: (1) mononuclear iron proteins (e.g., superoxide dismutase), (2) diiron-carboxylate proteins (e.g., ribonucleotide reductase, ferritin), (3) iron-sulfur proteins (e.g., aconitase), and (4) heme proteins (e.g., hemoglobin). Among these four categories, the first three protein groups are detected at lower levels, but they are functionally important. Hemoglobin is the most abundant iron-containing protein in humans. More than one-half of total-body iron is contained within hemoglobin. Based on the location of hemoglobin in erythrocytes, anemia is a characteristic trait of iron deficiency.

Despite iron's plentifulness on earth, iron deficiency is extremely common in humans, and is the most prevalent cause of anemia worldwide. To more fully understand iron deficiency anemia, consideration must be directed toward concepts of iron supply and demand for the production of erythrocytes. Erythropoiesis-related demands for iron are created by three variables: tissue oxygenation, erythrocyte turnover, and erythrocyte loss from hemorrhage. Tissue oxygenation requirements and erythrocyte production generally remain stable during adulthood in the absence of hemorrhage, disease, or altered physical activity. As such, iron

Copyright © 2013 Cold Spring Harbor Laboratory Press; all rights reserved
Cite this article as *Cold Spring Harb Perspect Med* doi: 10.1101/cshperspect.a011866

homeostasis (Fig. 1) also remains stable. Approximately 20 mL of senescent erythrocytes are cleared daily, and the 20 mg of iron in those cells is recycled for the production of new erythrocytes. Owing to a shorter half-life of circulating erythrocytes in iron deficiency anemia, iron is recovered sooner in those patients, but the amount of iron in each microcytic erythrocyte is reduced (Macdougall et al. 1970). In the event of hemorrhage, additional iron must be absorbed from the diet to meet the steady-state demands of the host.

Erythrocytes and their precursors require large amounts of iron for the production of heme (Fig. 2) and hemoglobin. Iron is central to hemoglobin structure and function (Perutz 1982). The most immediate source of iron for erythroblasts is mono- or diferric transferrin, found in high concentrations in the plasma. Iron deficiency anemia is typically associated with low iron saturation of available transferrin. Iron is loaded onto diferric transferrin from three sources: the gut (diet), macrophages (recycled iron), and the liver (stored ferritin iron). In general, iron stores are reduced or lost before the host develops anemia. Therefore, dietary and erythrocyte-recycled iron must meet the demands for erythrocyte production. If iron losses continue, the newly produced erythrocytes will have decreased hemoglobin, causing the amount of iron provided by the same number of senescent erythrocytes to be reduced. Unlike thalassemia trait, increased numbers of erythrocytes are not produced in the iron-deficient state to compensate for the reduction in intracellular hemoglobin content. For this reason, reticulocytosis is usually not present. In the absence of major hemorrhage, iron deficiency anemia generally develops slowly over the course of months or years. Resolution of iron deficiency

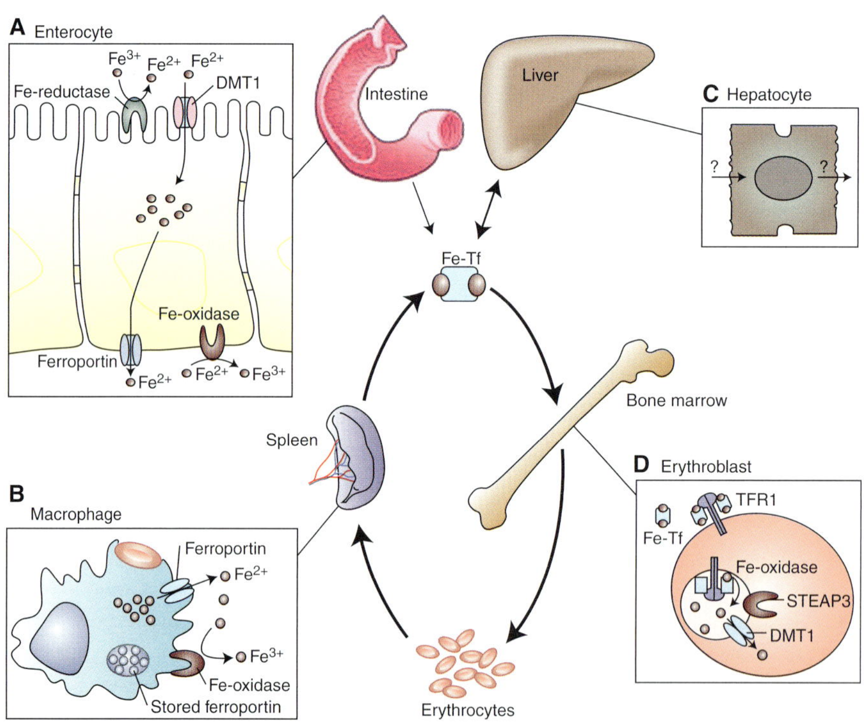

Figure 1. Iron homeostasis in humans. Each day, 20 mg of iron is recycled between circulating transferrin (Fe-Tf) and erythrocytes. This recycling pathway is supported by (*A*) intestinal iron absorption, (*B*) erythrophagocytosis, (*C*) hepatic iron stores, and (*D*) iron incorporation into hemoglobin. For more details, see Andrews 2008. (From Andrews 2008; reprinted, with permission, © American Society of Hematology.)

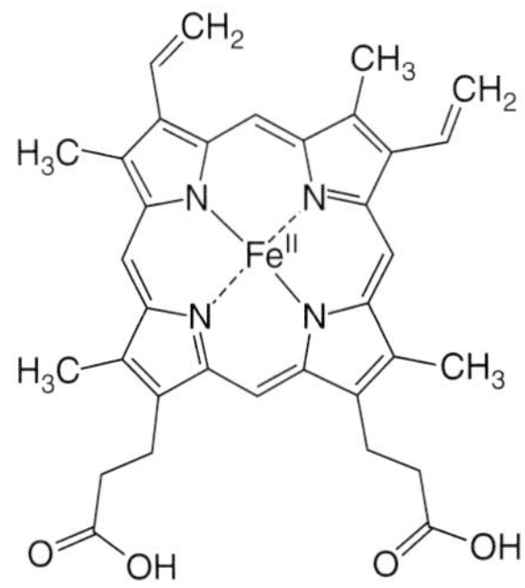

Figure 2. The heme B molecule. (The image is a public domain illustration from Wikipedia and can be found at http://commons.wikimedia.org/wiki/File:Heme_b.svg.)

anemia may be equally slow depending on the availability of iron in the diet as well as the adequacy of gastrointestinal function.

For decades, it has been possible to diagnose and fully reverse the anemia of iron deficiency at a relatively low cost. Unfortunately, iron deficiency has maintained itself as the most common anemia and nutritional disorder worldwide. This seemingly inexplicable paradox of high prevalence despite effective treatment represents a major challenge to public health efforts. Multiple obstacles involving economics, cultural barriers, and infectious diseases converge and make eradication of this disease more difficult. The additional challenges that are encountered by certain human subpopulations in select geographies need to be overcome to achieve therapeutic success in the global community.

THE MAGNITUDE OF THE PROBLEM

30%–50% of anemia in children and other groups is caused by iron deficiency (World Health Organization 2007). Because 1.6 billion people are anemic (McLean et al. 2009), several hundred million manifest iron deficiency anemia. As such, iron deficiency is the most common cause of anemia worldwide. Iron deficiency anemia afflicts a subset of the two billion people worldwide who are nutritionally iron deficient (Viteri 1998). Therefore, the health burden of iron deficiency may be extrapolated from the global prevalence of anemia. Anemia is not distributed evenly throughout the world, as there is a fivefold increase in underdeveloped geographies. In some global regions, the prevalence of anemia among young children is >50% (Table 1) and even approaches 100% in some locales (Lutter 2008). In the same regions, 40%–50% of the population remains anemic at all ages with the exception of nonelderly men (McLean et al. 2009). By comparison, the disease burden is far less in developed nations. In the United States, the prevalence of anemia as a result of iron deficiency is reduced among all age and gender groups (Clark 2008). However, approximately 10 million people are iron deficient in the United States, including 5 million who have iron deficiency anemia.

RECOGNIZING IRON DEFICIENCY ANEMIA

The clinical constellation of signs and symptoms for this disease depends largely on the magnitude of the anemia. Hemoglobin values used for the definition of anemia vary according to age, sex, race, and whether the blood was sampled from the capillary (finger stick) or venous (phlebotomy) source (Newman 2006; Cable et al. 2011a). In the absence of automated testing, portable devices or visual matching of hemoglobin color have been proven to be 95% accurate for identifying the hemoglobin level within 1–2 g/dL of reference values (Ingram and Lewis 2000; Lewis and Emmanuel 2001). Patients may complain of poor mental performance or cold intolerance (Rosenzweig and Volpe 1999). Fatigue and exercise-associated

Table 1. Prevalence of anemia in infants and young children (birth to 5 years of age)

Global region	Prevalence (%)
Africa	64.6
Asia	47.7
Europe	16.7
Latin America	39.5
North America	3.4
Oceania	28.0

Data adapted from McLean et al. 2009.

dyspnea are regularly reported. Although rare, glossitis or dysphagia may be identified at presentation (Cook 2005; Novacek 2006). Recognition of these features may trigger appropriate laboratory tests and therapy.

Iron deficiency anemia is associated with some rather striking neurological sequelae. Some subjects possess the compulsion to move their lower extremities while at rest. The restless leg syndrome is now recognized as a reversible symptom of reduced brain iron levels that is particularly prevalent during pregnancy (Vivarelli et al. 1976; Goodman et al. 1988). Pica is another associated neurological comorbidity. Pica is defined as dietary compulsions for materials that may not usually be consumed in the diet of humans without iron deficiency. Pica, specifically geophagia, has been reported in a majority of pregnant African women living in regions where iron deficiency anemia is extremely common (Njiru et al. 2011). The molecular basis for this unusual behavior is unknown. Iron deficiency is also known to cause cognitive dysfunction. Neurological damage is particularly relevant during infancy brain development. Long-lasting cognitive challenges occur despite therapy later in life (Lozoff et al. 1991). Therefore, iron deficiency anemia during infancy should be aggressively treated to avoid the potential for cognitive problems.

Laboratory evaluation reveals characteristic changes in blood parameters for iron regulation storage, transport, and utilization. Hepcidin is the main regulator of iron in humans. Unfortunately, interlaboratory variation and lack of adequate standards have muted the advance of hepcidin assays in clinical care (Kroot et al. 2009). Research studies suggest that low or absent hepcidin levels may be diagnostic of simple iron deficiency (Tanno et al. 2010).

Currently, the central parameter for determination of significant iron deficiency as well as therapeutic response is ferritin (Mei et al. 2005; O'Meara et al. 2011). The ferritin protein complex structure acts as a cage to contain up to 4500 iron molecules (Fig. 3) (Harrison and Arosio 1996). A serum ferritin level of 15 μg/L or less is diagnostic of iron deficiency, and correlates specifically with the absence of stainable

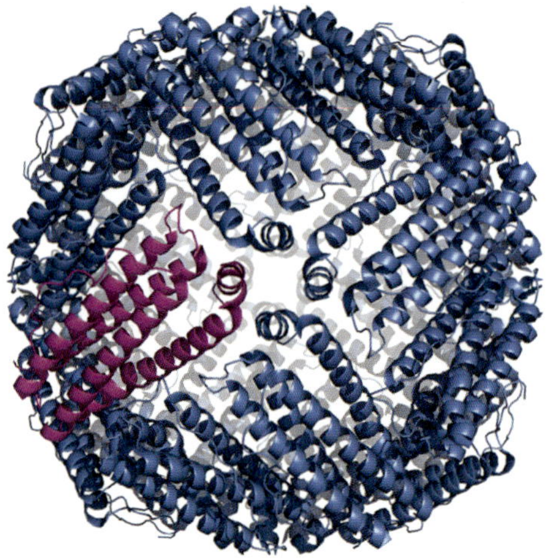

Figure 3. Three-dimensional ribbon representation of the ferritin protein cage. (The image is a public domain illustration from Wikipedia and can be found at http://commons.wikimedia.org/wiki/File:Ferritin.png.)

bone marrow iron (Baker 2000). Even at higher ferritin levels, ~40 μg/L, erythropoiesis may be affected. Iron deficiency also causes increased release of soluble transferrin from erythroblasts. Therefore, ratios of soluble transferrin receptor and ferritin are used to detect iron-deficient erythropoiesis (Cable et al. 2011b). When significant inflammation is present, serum ferritin levels may not reflect accurate iron stores. In patients with chronic renal disease, ferritin levels of >400 μg/L are regularly detected in the absence of significant marrow iron stores (Rocha et al. 2009).

As expected, serum iron levels are reduced in iron deficiency anemia. Transferrin protein expression increases in iron-deficient states, so the iron saturation on transferrin is reduced to values of <15%. Notably, apoferric and monoferric transferrin are the dominant species, and relatively minor amounts of diferric iron are present in the plasma (Finch and Huebers 1982). The reduction of diferric iron is pertinent for erythropoiesis owing to the relative levels and saturation kinetics of transferrin receptors on erythroblasts (Cazzola et al. 1987).

In addition to iron parameters, a lack of iron has measurable effects on erythropoiesis and erythroid cells. Low reticulocyte hemoglobin

 Cite this article as *Cold Spring Harb Perspect Med* doi: 10.1101/cshperspect.a011866

content is derived from reduced hemoglobin production, and may be useful for screening infants and children for iron deficiency (Brugnara et al. 1999; Ullrich et al. 2005). However, thalassemia mutations may confound the interpretation of reticulocyte hemoglobin quantitation (Mast et al. 2002). Iron deficiency accounts for increased production of zinc protoporphyrin in association with decreased heme production during erythroblast maturation. Ratios of these two parameters may be quantified using washed erythrocytes. Although values >40 moles of zinc protoporphyrin per mole heme denote an iron-deficient state (Hastka et al. 1992), the presence of a recent illness may reduce the sensitivity of this test (Crowell et al. 2006).

The manifestation of anemia or microcytosis is usually somewhat delayed relative to the loss of body iron stores. About 1% of erythrocytes are replaced daily, and the recycling of iron from the senescent cells continues to support the production of new cells. Eventually, the complete blood count (CBC) will reflect the effects on erythropoiesis. A combination of increased red cell distribution width (RDW), decreased red blood cell (RBC) count, decreased RBC hemoglobin, and decreased mean cell volume may be manifested. Unless the iron deficiency is reversed, the hemoglobin and hematocrit levels decrease to sufficiently low levels to be classified as anemia.

MAJOR CAUSES OF IRON DEFICIENCY ANEMIA

Blood Loss

Each milliliter of packed RBCs (~2.5 mL of whole blood) contains ~1.0 mg of iron. Each day, 1.0 mg of iron is absorbed from the diet and 20 mg of iron from senescent erythrocytes are available to support erythropoiesis. Once iron stores are depleted, dietary and recycled erythrocyte iron are not usually sufficient to compensate for acute blood loss.

In all cases of iron deficiency anemia, blood loss should be considered. Hemorrhage itself is by far the most common mechanism for acute iron loss and anemia. Hemorrhage decreases the host's red cell mass, decreases the supply of iron for erythropoiesis, and increases the iron demand for erythropoiesis. Chronic blood loss from menstruation or hookworm infection (see below) has the greatest impact worldwide. Less than 2 mL of blood is lost daily in the stool of healthy adults (Ahlquist et al. 1985). Detection of occult blood losses of up to 60 mL/d may be difficult without specialized stool tests (Rockey 1999). Bleeding may occur from multiple sites along the intestinal tract, with an increased incidence of bleeding from the colon (Lanas et al. 2009). Sometimes overlooked causes of blood loss include blood donation and nosebleeds. Intravascular hemolysis with hemoglobinuria such as occurs in malaria results in iron loss in the absence of hemorrhage. After chronic physical exertion, significant iron is lost in sweat and may contribute to the deficient state (Reinke et al. 2010). A full history should be queried in all new cases of iron deficiency anemia.

The Maternal–Fetal Bridge of Iron Deficiency

Requirements for iron are greatest around the time of birth. Iron demand is high in menstruating as well as pregnant females. During pregnancy, it is estimated that ~1200 mg of iron are required from conception through delivery (Lee and Okam 2011). Iron intake and stores in the mother must satisfy fetal development, and blood loss at delivery. Additionally, the maternal erythrocyte mass should increase from 350 to 450 mL. By comparison, pregnant women without iron supplements only increase their red cell mass by 180–250 mL (Pedersen and Milman 2003). One interpretation of this difference is that fetal iron demands are prioritized over the red cell mass of the mother. Postpartum, iron is lost as lactoferrin in breast milk. Those losses are balanced by the absence of menstruation in the lactating female.

Maternal iron deficiency anemia during pregnancy and the perinatal period have devastating effects on both the mother and child. In addition to the direct effects of anemia, reduced fetal brain maturation, pediatric cognitive defects, and maternal depression are associated with iron deficiency anemia (Black et al. 2011). The reversibility of cognitive defects caused at an early age by iron deficiency is unclear. Importantly, untreated iron deficiency in pregnant females will be passed to the infant. If left untreated during infancy, childhood, and adolescence, anemia and iron-associated cognitive defects may conceivably be passed between generations much like genetic traits. Unless the iron deficiency is treated at some stage of life, the cycle of iron deficiency from mother to child may remain unbroken for several generations.

In the fetus and during infancy, iron is required for the growth and development of all tissues. The human growth rate is almost logarithmic during this period (Anderson and Holford 2008). The hemoglobin and myoglobin requirements must be met before the accumulation of storage iron. Although serum ferritin levels are elevated in the fetus compared with the mother, the fetal iron stores correlate well with the iron stores of the mother (Milman et al. 1987). In anemic mothers with iron deficiency, the fetal ferritin levels remain 10 times higher than the maternal ferritin at the time of delivery (Erdem et al. 2002). At birth, the fetal red cell mass is ~50 mL/kg (Phillips et al. 1986), compared with 25–30 mL/kg in adults (Fairbanks et al. 1996). During the first year of postnatal life, total-body iron increases by ~240 mg (Oski 1993). Around 80% of that iron is used for expanded hemoglobin production (50%) and iron stores (30%). Beyond the first year, iron intake or stores must remain sufficient to support the ongoing growth and increased red cell mass (Moser et al. 2011). In children and young adults, iron deficiency remains one of the top three contributors to the overall disease burden in those populations (Gore et al. 2011).

Malaria

Iron deficiency anemia and malaria coexist in most tropical regions of the world. Malaria contributes to iron deficiency anemia by causing intravascular hemolysis with subsequent loss of hemoglobin iron in the urine. This clinical feature was described in 1898 as blackwater fever (Connolly 1898). Malaria also causes an immune response that suppresses erythropoietin (Burgmann et al. 1996) as well as direct effects on erythropoiesis (Skorokhod et al. 2010). The host may also increase hepcidin expression for protection from liver-stage malaria (Portugal et al. 2011). Of course, increased hepcidin restricts iron and might delay erythroid recovery.

It is essential to understand the complex interplay between iron, hepcidin, and malaria when considering efforts to eradicate iron deficiency in malaria-endemic regions. If iron redistribution by hepcidin is beneficial for malaria, restricted iron could benefit the infected host. This hypothesis may help explain the recent report of potential harm caused by iron supplementation among preschool children in malaria-endemic areas (Sazawal et al. 2006). In that study, iron and folic acid supplements were stopped owing to an increased risk for hospitalization or death from malaria in Zanzibar, Tanzania. Increased caution for iron supplementation in malaria-endemic regions was proposed (World Health Organization 2007). However, a recent Cochrane paper recommended, "iron supplementation should not be withheld from children living in malaria-endemic countries" (Okebe et al. 2011). Treatment of iron deficiency anemia is less clear in areas where access to proper malarial prevention and treatment are suboptimal. Further studies and resolution of this critical, but complex issue are awaited.

Hookworm

Like iron deficiency anemia, hookworm infection affects several hundred million humans worldwide (Bungiro and Cappello 2011). Amazingly, a recent study reported that there is a considerable overlap between malaria and hookworm in sub-Saharan Africa (Fig. 4) (Brooker et al. 2006). Worldwide, there are two hookworm species that infect humans. Both are found in tropical regions based on the requirement of moist soil for survival. The worm is introduced to the soil by fecal matter in regions where sanitation is not present. From the soil, the parasite accesses the duodenum of a new human host directly by mouth, or indirectly via the skin.

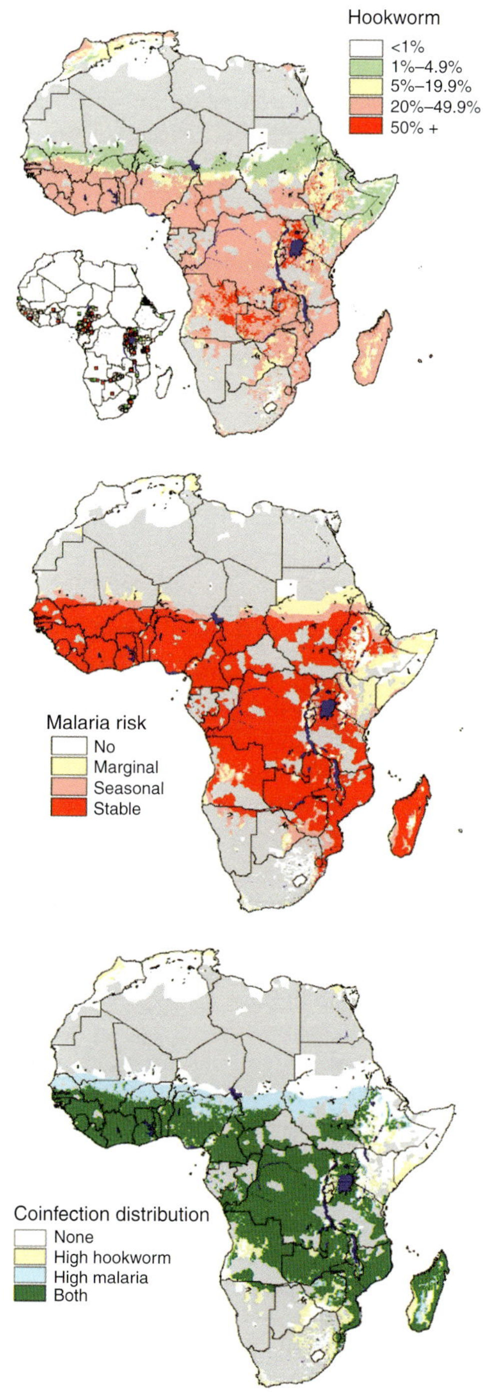

Figure 4. Map of hookworm (*top*), *Plasmodium falciparum* (*middle*), and overlapping (*bottom*) prevalence. (From Brooker et al. 2006; reprinted, with express permission, from BioMed Central Ltd., © 2006.)

Once in the gut, the worm may be retained for several years as it releases eggs in the stool. A hookworm infection should be suspected in cases of travel or habitation in the tropics, iron deficiency anemia, and mild eosinophilia. Owing to their location in the small bowel, capsule endoscopy is helpful for diagnosis if eggs are not present in the stool (Li et al. 2007). Hookworms live on 0.3–0.5 mL of blood extravasated daily from the intestinal mucosa. Heavily infected patients are simply unable to maintain adequate iron stores and become anemic (Smith and Brooker 2010). Even without additional iron supplements, antihelminthic drugs can cause reversal of iron deficiency anemia (Radhika et al. 2011). It is currently unknown whether hookworm infection causes increased production of hepcidin in the host.

Diet and Malabsorption of Iron

Local economics generally dictate the level of nutrition worldwide. The diet, by itself, infrequently causes iron deficiency anemia in the absence of severe malnourishment or a comorbidity. A vegan diet is usually sufficient to prevent anemia even though the iron stores of the host may be low (Craig 1994). The diet becomes far more relevant when the iron stores are lost, or anemia has already developed, and the host requires additional iron absorption from the gut for recovery. This occurs in the multiple settings described earlier in this work: blood loss, rapid growth during infancy, malaria, and hookworm. In these settings, the diet and iron supplements become critical for maintaining iron availability. Supplemental dietary iron may be needed, because the average Western diet is not sufficient to meet the needs of pregnancy (Lee and Okam 2011). In addition to the iron content itself, the bioavailability of iron for absorption depends largely on the dietary components (Sharp 2010). Iron in the form of heme is especially bioavailable, and meat-containing diets are also beneficial (López and Martos 2004). Vitamin C improves dietary availability of iron as well as avoidance of tea or other iron-chelating substances (Thankachan et al. 2008). Numerous approaches are being tried to improve iron availability in the diet with a goal of low-cost and culturally acceptable implementation among the underprivileged (Hurrell et al. 2004).

In addition to the dietary components, the gut must be able to absorb iron to prevent or treat iron deficiency anemia. Inorganic iron absorption requires multiple mechanisms for entry and exit from duodenal and jejunal epithelial cells (Fig. 1). In cases of suspected malabsorption, a simple oral iron absorption test may prevent or direct more invasive studies (Alleyne et al. 2008). If the host's hepcidin expression is increased, inorganic iron from the diet will not be optimally absorbed into the blood from the intestine. Instead, that iron will be trapped in the intestinal epithelial cells, and then sloughed into the lumen and stool (Anderson et al. 2009). Therefore, infections, inflammation, or other hepcidin up-regulating mechanisms are likely to inhibit nonheme iron absorption despite a high bioavailability in the diet. Other diseases, including helicobacter infections (Vitale et al. 2011), bariatric surgery (Shankar et al. 2010), or decreased gastrin (Kovac et al. 2011) can inhibit iron absorption. Finally, celiac disease is a common cause of anemia owing to low iron absorption from the gut (Presutti et al. 2007). There is a high incidence (15%) of celiac sprue among the subjects who undergo endoscopy for evaluation of iron deficiency anemia (Oxentenko et al. 2002).

SOLVING THE GLOBAL PROBLEM

Iron Fortification

Iron fortification remains the mainstay of efforts aimed toward the treatment or prevention of iron deficiency anemia. As a general rule, menstruating and pregnant women along with their children clearly represent the largest at-risk population for this disease. In the underdeveloped world, iron may be provided with other micronutrients to reduce anemia in schoolchildren (Ahmed et al. 2010; Best et al. 2011; Lemaire et al. 2011). In-home food preparations with iron supplements present an alternate to industrial scale supplementation of grains or commercial food products (Lynch 2011). Numerous oral preparations and dosing regimens are available for menstruating and pregnant females (Fernández-Gaxiola and De-Regil 2011). Recently, intravenous iron preparations with improved toxicity profiles have been used for cases where rapid therapy was useful in reducing the need for transfusion (Gozzard 2011). Based on the large amount of iron that is provided with intravenous therapy, special care should be given toward dosing to prevent iatrogenic iron overload. As a general rule, therapy should be continued only until the anemia is resolved and iron stores are replenished. These goals are met with a normal hematocrit and serum ferritin level of 50–100 μg/L (corresponding to ~400–800 mg iron in stores for otherwise healthy adults) (Walters et al. 1973; Magnussen et al. 2008).

Delayed Clamping of the Umbilical Cord

Special attention must be given toward preventing iron deficiency in the newborn. In addition to maternal iron supplements, the simple maneuver of delayed umbilical cord clamping might help achieve this goal. In 1954, delayed umbilical cord clamping was found to be a relatively benign procedure (Colozzi 1954). The potential benefit of placental transfusion as an iron source was not initially considered. By the 1960s it was shown that placental transfusion causes a significant increase in the red cell volume when clamping is delayed for several minutes. A delay of 5 min results in 166 mL of added blood volume for a 3.5 kg infant (Usher et al. 1963). Placental transfusion is also achieved by placing the newborn on the mother's abdomen after vaginal delivery, and clamping the cord once it stops pulsating (Nelle et al. 1995). In cases of cesarean section or other clinical setting that may preclude delayed clamping, clamping near the placenta should increase the volume of blood delivered to the newborn (Daniel and Weerakkody 1996). In the 1990s, the benefit of delayed clamping for the purpose of preventing iron deficiency during the first 6 months of life was more clearly appreciated (Pisacane 1996). Widespread adaptation placental transfusion may have been prevented to date by unsupported concerns for polycythemia, hyperbilirubinemia, and increased blood viscosity (van Rheenen and Brabin 2004). Importantly, none of those concerns were realized by a 3 min delay in umbilical cord clamping at the time of delivery among 400 low-risk pregnancies (Andersson et al. 2011). Instead, neonatal anemia and iron parameters by 4 months of age were improved. None of the infants who had undergone delayed (>3 min) clamping had ferritin levels below 20 μg/L, versus decreased ferritin levels in 7.4% of subjects who had undergone clamping almost immediately after delivery. Overall, these studies suggest that some level of placental transfusion may be beneficial in most deliveries. Large, prospective multinational studies of this simple maneuver are needed to confirm these results. If those studies show a generic benefit toward preventing iron deficiency or anemia, then appropriate efforts should be made to widely institute placental transfusion via delayed umbilical cord clamping. This procedure may be particularly useful in underdeveloped communities, as it incurs little, if any, additional expense.

Understand Fundamental Iron Biology

It is predicted that advances in global therapy for iron deficiency anemia will be greatly assisted by basic research efforts. Perhaps the most significant advance in this regard is the discovery and development of hepcidin biology over the last decade (Ganz 2011). Hepcidin biology will undoubtedly evolve into applications for iron deficiency anemia among all world populations. For instance, the recognition that hepcidin expression is highly variable and influenced by a circadian rhythm should be advantageous in improving dosing regimens (Kemna et al. 2007). The kinetics of hepcidin expression in response to iron supplementation for iron deficiency remain largely unexplored as another research avenue aimed toward the optimization of therapy. Clinical comparisons of oral versus intravenous therapies will help determine if a rapid pulse of therapy can improve the chances of success for certain individuals or groups of patients.

Certain populations do not benefit from universal iron supplementation (Ghio 2011). With inherited hemochromatosis, the absorption of dietary iron increases. Some genetic variants are quite common, especially in northern Europeans (van Bokhoven et al. 2011). Although the clinical penetrance is quite variable in the most common forms of hereditary hemochromatosis, communal iron fortification for this group is generally not recommended. Another genetic disorder named thalassemia (see Fucharoen and Weatherall 2012; Gibbons 2012; Musallam et al. 2012; Nienhuis and Nathan 2012; Cao and Kan 2013; Higgs 2013; Thein 2013; Vichinsky 2013) has a more direct relationship with iron deficiency anemia, because both diseases are concentrated within malaria and hookworm-endemic regions of the world. Thalassemia is caused by mutations in the globin genes that lead to decreased production of hemoglobin. Interestingly, iron deficiency itself may affect the production of hemoglobin in α-thalassemia (O'Brien 1973). Thalassemia trait patients present with microcytosis similar to iron deficiency. Homozygous thalassemia mutations lead to iron overload in the host by mechanisms that are not fully understood. As such, it is essential to recognize thalassemia and avoid iron supplements in that population.

Recent basic discoveries of several important genes or mutations that modify iron metabolism should also be mentioned. Iron is regulated, in part, by a recently discovered gene named *TMPRSS6* (Du et al. 2008). Mutations in that gene cause iron deficiency that is refractory to iron supplements (Cau et al. 2010). If patients with malaria or hookworm benefit from iron deficiency, then some populations may have evolved by incorporating mutations into this gene. Population-based studies will help determine if mutations in the *TMPRSS6* gene could be confounding efforts to reverse iron deficiency in some global populations including those in sub-Saharan Africa.

As evidenced by the importance of iron for malarial pathogenesis, further research into the complex relationships between deprivation of iron for this pathogen and iron deficiency anemia are needed to determine the best course of therapy. Determination of hookworm effects on hepcidin expression should also be pursued. In populations afflicted with one or both of these parasites, efforts to supplement iron can be confounded by the host's inflammatory response. Relationships between hookworm and intestinal iron absorption should be studied further, understood, and incorporated into eradication efforts. Ideally, strategies will be tested that incorporate vaccination, sanitation, malarial treatment, deworming, and iron supplements into the same research plan. Although such strategies seem ambitious in a world of limited resources, it is crucial to remember that hundreds of millions stand to benefit worldwide.

CONCLUDING REMARKS

Unlike other prevalent anemias and hemoglobinopathies, the diagnosis and treatment of iron deficiency anemia is achievable in most, if not all individuals. However, consideration of iron deficiency anemia must include the possible convergence of several causative factors. If resources are adequate, care must be given toward individualized approaches to therapy (Alleyne et al. 2008). In the underdeveloped world, more communal approaches are being taken to overcome the overlapping causes of iron deficiency anemia that affect hundreds of millions worldwide. Ultimately, it is predicted that increased research and understanding of fundamental iron biology will assist in devising new strategies aimed toward the global elimination of this disease.

ACKNOWLEDGMENTS

J.L.M. thanks his mentors, colleagues, and all members of his section at National Institutes of Health for their inspiration over the last 20 years.

REFERENCES

Reference is also in this collection.

Ahlquist DA, McGill DB, Schwartz S, Taylor WF, Owen RA. 1985. Fecal blood levels in health and disease. A study using HemoQuant. *N Engl J Med* **312:** 1422–1428.

Ahmed F, Khan MR, Akhtaruzzaman M, Karim R, Williams G, Torlesse H, Darnton-Hill I, Dalmiya N, Banu CP, Nahar B. 2010. Long-term intermittent multiple micronutrient supplementation enhances hemoglobin and micronutrient status more than iron + folic acid supplementation in Bangladeshi rural adolescent girls with nutritional anemia. *J Nutr* **140:** 1879–1886.

Alleyne M, Horne MK, Miller JL. 2008. Individualized treatment for iron-deficiency anemia in adults. *Am J Med* **121:** 943–948.

Anderson BJ, Holford NH. 2008. Mechanism-based concepts of size and maturity in pharmacokinetics. *Annu Rev Pharmacol Toxicol* **48:** 303–332.

Anderson GJ, Frazer DM, McLaren GD. 2009. Iron absorption and metabolism. *Curr Opin Gastroenterol* **25:** 129–135.

Andersson O, Hellström-Westas L, Andersson D, Domellöf M. 2011. Effect of delayed versus early umbilical cord clamping on neonatal outcomes and iron status at 4 months: A randomised controlled trial. *BMJ* **343:** d7157.

Andrews NC. 2008. Forging a field: The golden age of iron biology. *Blood* **112:** 219–230.

Baker WF Jr. 2000. Iron deficiency in pregnancy, obstetrics, and gynecology. *Hematol Oncol Clin North Am* **14:** 1061–1077.

Best C, Neufingerl N, Del Rosso JM, Transler C, van den Briel T, Osendarp S. 2011. Can multi-micronutrient food fortification improve the micronutrient status, growth, health, and cognition of schoolchildren? A systematic review. *Nutr Rev* **69:** 186–204.

Black MM, Quigg AM, Hurley KM, Pepper MR. 2011. Iron deficiency and iron-deficiency anemia in the first two years of life: Strategies to prevent loss of developmental potential. *Nutr Rev* **69:** S64–S70.

Brooker S, Clements AC, Hotez PJ, Hay SI, Tatem AJ, Bundy DA, Snow RW. 2006. The co-distribution of *Plasmodium falciparum* and hookworm among African schoolchildren. *Malar J* **5:** 99.

Brugnara C, Zurakowski D, DiCanzio J, Boyd T, Platt O. 1999. Reticulocyte hemoglobin content to diagnose iron deficiency in children. *JAMA* **281:** 2225–2230.

Bungiro R, Cappello M. 2011. Twenty-first century progress toward the global control of human hookworm infection. *Curr Infect Dis Rep* **13:** 210–217.

Burgmann H, Looareesuwan S, Kapiotis S, Viravan C, Vanijanonta S, Hollenstein U, Wiesinger E, Presterl E, Winkler S, Graninger W. 1996. Serum levels of erythropoietin in acute *Plasmodium falciparum* malaria. *Am J Trop Med Hyg* **54:** 280–283.

Cable RG, Steele WR, Melmed RS, Johnson B, Mast AE, Carey PM, Kiss JE, Kleinman SH, Wright DJ, for the NHLBI Retrovirus Epidemiology Donor Study-II (REDS-II). 2011a. The difference between fingerstick and venous hemoglobin and hematocrit varies by sex and iron stores. *Transfusion* **52:** 1031–1040.

Cable RG, Glynn SA, Kiss JE, Mast AE, Steele WR, Murphy EL, Wright DJ, Sacher RA, Gottschall JL, Tobler LH, et al. 2011b. Iron deficiency in blood donors: The REDS-II Donor Iron Status Evaluation (RISE) study. *Transfusion* **52:** 702–711.

* Cao A, Kan YW. 2013. The prevention of thalassemia. *Cold Spring Harb Perspect Med* **3:** a011775.

Cau M, Melis MA, Congiu R, Galanello R. 2010. Iron-deficiency anemia secondary to mutations in genes controlling hepcidin. *Expert Rev Hematol* **3:** 205–216.

Cazzola M, Pootrakul P, Bergamaschi G, Huebers HA, Eng M, Finch CA. 1987. Adequacy of iron supply for erythropoiesis: In vivo observations in humans. *J Lab Clin Med* **110:** 734–739.

Clark SF. 2008. Iron deficiency anemia. *Nutr Clin Pract* **23:** 128–141.

Colozzi AE. 1954. Clamping of the umbilical cord; its effect on the placental transfusion. *N Engl J Med* **250:** 629–632.

Connolly RM. 1898. African Haemoglobinuric fever, commonly called Blackwater Fever. *Br Med J* **2:** 882–885.

Cook JD. 2005. Diagnosis and management of iron-deficiency anaemia. *Best Pract Res Clin Haematol* **18:** 319–332.

Craig WJ. 1994. Iron status of vegetarians. *Am J Clin Nutr* **59:** 1233S–1237S.

Crowell R, Ferris AM, Wood RJ, Joyce P, Slivka H. 2006. Comparative effectiveness of zinc protoporphyrin and hemoglobin concentrations in identifying iron deficiency in a group of low-income, preschool-aged children: Practical implications of recent illness. *Pediatrics* **118:** 224–232.

Daniel DG, Weerakkody AN. 1996. Neonatal prevention of iron deficiency. Blood can be transfused from cord clamped at placental end. *BMJ* **312:** 1102–1103.

Du X, She E, Gelbart T, Truksa J, Lee P, Xia Y, Khovananth K, Mudd S, Mann N, Moresco EM, et al. 2008. The serine protease TMPRSS6 is required to sense iron deficiency. *Science* **320:** 1088–1092.

Erdem A, Erdem M, Arslan M, Yazici G, Eskandari R, Himmetoglu O. 2002. The effect of maternal anemia and iron deficiency on fetal erythropoiesis: Comparison between serum erythropoietin, hemoglobin and ferritin levels in mothers and newborns. *J Matern Fetal Neonatal Med* **11:** 329–332.

Fairbanks VF, Klee GG, Wiseman GA, Hoyer JD, Tefferi A, Petitt RM, Silverstein MN. 1996. Measurement of blood volume and red cell mass: Re-examination of 51Cr and 125I methods. *Blood Cells Mol Dis* **22:** 169–186.

Fernández-Gaxiola AC, De-Regil LM. 2011. Intermittent iron supplementation for reducing anaemia and its associated impairments in menstruating women. *Cochrane Database Syst Rev* **12:** CD009218.

Finch CA, Huebers H. 1982. Perspectives in iron metabolism. *N Engl J Med* **306:** 1520–1528.

* Fucharoen S, Weatherall DJ. 2012. The hemoglobin E thalassemias. *Cold Spring Harb Perspect Med* **2:** a011734.

Ganz T. 2011. Hepcidin and iron regulation, 10 years later. *Blood* **117:** 4425–4433.

Ghio AJ. 2011. Anemia and global iron fortification and supplementation. *Ann Hematol* **91:** 957–958.

* Gibbons RJ. 2012. α-Thalassemia, mental retardation, and myelodysplastic syndrome. *Cold Spring Harb Perspect Med* **2:** a011759.

Goodman JD, Brodie C, Ayida GA. 1988. Restless leg syndrome in pregnancy. *BMJ* **297:** 1101–1102.

Gore FM, Bloem PJ, Patton GC, Ferguson J, Joseph V, Coffey C, Sawyer SM, Mathers CD. 2011. Global burden of disease in young people aged 10–24 years: A systematic analysis. *Lancet* **377:** 2093–2102.

Gozzard D. 2011. When is high-dose intravenous iron repletion needed? Assessing new treatment options. *Drug Des Devel Ther* **5:** 51–60.

Harrison PM, Arosio P. 1996. The ferritins: Molecular properties, iron storage function and cellular regulation. *Biochim Biophys Acta* **1275:** 161–203.

Hastka J, Lasserre JJ, Schwarzbeck A, Strauch M, Hehlmann R. 1992. Washing erythrocytes to remove interferents in measurements of zinc protoporphyrin by front-face hematofluorometry. *Clin Chem* **38:** 2184–2189.

* Higgs DR. 2013. The molecular basis of α-thalassemia. *Cold Spring Harb Perspect Med* **3:** a011718.

Hurrell RF, Lynch S, Bothwell T, Cori H, Glahn R, Hertrampf E, Kratky Z, Miller D, Rodenstein M, Streekstra H, et al. 2004. Enhancing the absorption of fortification iron. A SUSTAIN Task Force report. *Int J Vitam Nutr Res* **74:** 387–401.

Ingram CF, Lewis SM. 2000. Clinical use of WHO haemoglobin colour scale: Validation and critique. *J Clin Pathol* **53:** 933–937.

Kemna EH, Tjalsma H, Podust VN, Swinkels DW. 2007. Mass spectrometry-based hepcidin measurements in serum and urine: Analytical aspects and clinical implications. *Clin Chem* **53:** 620–628.

Kovac S, Anderson GJ, Alexander WS, Shulkes A, Baldwin GS. 2011. Gastrin-deficient mice have disturbed hematopoiesis in response to iron deficiency. *Endocrinology* **152:** 3062–3073.

Kroot JJ, Kemna EH, Bansal SS, Busbridge M, Campostrini N, Girelli D, Hider RC, Koliaraki V, Mamalaki A, Olbina G, et al. 2009. Results of the first international round robin for the quantification of urinary and plasma hepcidin assays: Need for standardization. *Haematologica* **94:** 1748–1752.

Lanas A, García-Rodríguez LA, Polo-Tomás M, Ponce M, Alonso-Abreu I, Perez-Aisa MA, Perez-Gisbert J, Bujanda L, Castro M, Muñoz M, et al. 2009. Time trends and impact of upper and lower gastrointestinal bleeding and perforation in clinical practice. *Am J Gastroenterol* **104:** 1633–1641.

Lee AI, Okam MM. 2011. Anemia in pregnancy. *Hematol Oncol Clin North Am* **25:** 241–259, vii.

Lemaire M, Islam QS, Shen H, Khan MA, Parveen M, Abedin F, Haseen F, Hyder Z, Cook RJ, Zlotkin SH. 2011. Iron-containing micronutrient powder provided to children with moderate-to-severe malnutrition increases hemoglobin concentrations but not the risk of infectious morbidity: A randomized, double-blind, placebo-controlled, noninferiority safety trial. *Am J Clin Nutr* **94:** 585–593.

Lewis SM, Emmanuel J. 2001. Validity of the haemoglobin colour scale in blood donor screening. *Vox Sang* **80:** 28–33.

Li ZS, Liao Z, Ye P, Wu RP. 2007. Dancing hookworm in the small bowel detected by capsule endoscopy: A synthesized video. *Endoscopy* **39:** E97.

López MA, Martos FC. 2004. Iron availability: An updated review. *Int J Food Sci Nutr* **55:** 597–606.

Lozoff B, Jimenez E, Wolf AW. 1991. Long-term developmental outcome of infants with iron deficiency. *N Engl J Med* **325:** 687–694.

Lutter CK. 2008. Iron deficiency in young children in low-income countries and new approaches for its prevention. *J Nutr* **138:** 2523–2528.

Lynch SR. 2011. Why nutritional iron deficiency persists as a worldwide problem. *J Nutr* **141:** 763S–768S.

Macdougall LG, Judisch JM, Mistry SB. 1970. Red cell metabolism in iron deficiency anemia. II. The relationship between red cell survival and alterations in red cell metabolism. *J Pediatr* **76:** 660–665.

Magnussen K, Bork N, Asmussen L. 2008. The effect of a standardized protocol for iron supplementation to blood donors low in hemoglobin concentration. *Transfusion* **48:** 749–754.

Mast AE, Blinder MA, Lu Q, Flax S, Dietzen DJ. 2002. Clinical utility of the reticulocyte hemoglobin content in the diagnosis of iron deficiency. *Blood* **99:** 1489–1491.

McLean E, Cogswell M, Egli I, Wojdyla D, de Benoist B. 2009. Worldwide prevalence of anaemia, WHO Vitamin and Mineral Nutrition Information System, 1993–2005. *Public Health Nutr* **12:** 444–454.

Mei Z, Cogswell ME, Parvanta I, Lynch S, Beard JL, Stoltzfus RJ, Grummer-Strawn LM. 2005. Hemoglobin and ferritin are currently the most efficient indicators of population response to iron interventions: An analysis of nine randomized controlled trials. *J Nutr* **135:** 1974–1980.

Milman N, Ibsen KK, Christensen JM. 1987. Serum ferritin and iron status in mothers and newborn infants. *Acta Obstet Gynecol Scand* **66:** 205–211.

Moser AM, Urkin J, Shalev H. 2011. Normal hemoglobin at the age of 1 year does not protect infants from developing iron deficiency anemia in the second year of life. *J Pediatr Hematol Oncol* **33:** 467–469.

* Musallam KM, Taher AT, Rachmilewitz EA. 2012. β-Thalassemia intermedia: A clinical perspective. *Cold Spring Harb Perspect Med* **2:** a013482.

Nelle M, Zilow EP, Bastert G, Linderkamp O. 1995. Effect of Leboyer childbirth on cardiac output, cerebral and gastrointestinal blood flow velocities in full-term neonates. *Am J Perinatol* **12:** 212–216.

Newman B. 2006. Iron depletion by whole-blood donation harms menstruating females: The current whole-blood-collection paradigm needs to be changed. *Transfusion* **46:** 1667–1681.

* Nienhuis AW, Nathan DG. 2012. Pathophysiology and clinical manifestations of the β-thalassemias. *Cold Spring Harb Perspect Med* **2:** a011726.

Njiru H, Elchalal U, Paltiel O. 2011. Geophagy during pregnancy in Africa: A literature review. *Obstet Gynecol Surv* **66:** 452–459.

Novacek G. 2006. Plummer-Vinson syndrome. *Orphanet J Rare Dis* **1:** 36.

O'Brien RT. 1973. The effect of iron deficiency on the expression of hemoglobin H. *Blood* **41:** 853–856.

Okebe JU, Yahav D, Shbita R, Paul M. 2011. Oral iron supplements for children in malaria-endemic areas. *Cochrane Database Syst Rev* CD006589.

O'Meara A, Infanti L, Stebler C, Ruesch M, Sigle JP, Stern M, Buser A. 2011. The value of routine ferritin measurement in blood donors. *Transfusion* **51:** 2183–2188.

Oski FA. 1993. Iron deficiency in infancy and childhood. *N Engl J Med* **329:** 190–193.

Oxentenko AS, Grisolano SW, Murray JA, Burgart LJ, Dierkhising RA, Alexander JA. 2002. The insensitivity of endoscopic markers in celiac disease. *Am J Gastroenterol* **97:** 933–938.

Pedersen LM, Milman N. 2003. Diagnostic significance of platelet count and other blood analyses in patients with lung cancer. *Oncol Rep* **10:** 213–216.

Perutz MF. 1982. Nature of the iron-oxygen bond and control of oxygen affinity of the haem by the structure of the globin in haemoglobin. *Adv Exp Med Biol* **148:** 31–48.

Phillips HM, Holland BM, Abdel-Moiz A, Fayed S, Jones JG, Turner TL, Wardrop CA, Cockburn F. 1986. Determination of red-cell mass in assessment and management of anaemia in babies needing blood transfusion. *Lancet* **1:** 882–884.

Pisacane A. 1996. Neonatal prevention of iron deficiency. *BMJ* **312:** 136–137.

Portugal S, Carret C, Recker M, Armitage AE, Gonçalves LA, Epiphanio S, Sullivan D, Roy C, Newbold CI, Drakesmith H, et al. 2011. Host-mediated regulation of superinfection in malaria. *Nat Med* **17:** 732–737.

Presutti RJ, Cangemi JR, Cassidy HD, Hill DA. 2007. Celiac disease. *Am Fam Physician* **76:** 1795–1802.

Radhika MS, Nair KM, Kumar RH, Rao MV, Ravinder P, Reddy CG, Brahmam GN. 2011. Micronized ferric pyrophosphate supplied through extruded rice kernels improves body iron stores in children: A double-blind, randomized, placebo-controlled midday meal feeding trial in Indian schoolchildren. *Am J Clin Nutr* **94:** 1202–1210.

Reinke S, Taylor WR, Duda GN, von Haehling S, Reinke P, Volk HD, Anker SD, Doehner W. 2010. Absolute and functional iron deficiency in professional athletes during training and recovery. *Int J Cardiol* **156:** 186–191.

Rocha LA, Barreto DV, Barreto FC, Dias CB, Moysés R, Silva MR, Moura LA, Draibe SA, Jorgetti V, Carvalho AB, et al. 2009. Serum ferritin level remains a reliable marker of bone marrow iron stores evaluated by histomorphometry in hemodialysis patients. *Clin J Am Soc Nephrol* **4:** 105–109.

Rockey DC. 1999. Occult gastrointestinal bleeding. *N Engl J Med* **341:** 38–46.

Rosenzweig PH, Volpe SL. 1999. Iron, thermoregulation, and metabolic rate. *Crit Rev Food Sci Nutr* **39:** 131–148.

Sazawal S, Black RE, Ramsan M, Chwaya HM, Stoltzfus RJ, Dutta A, Dhingra U, Kabole I, Deb S, Othman MK, et al. 2006. Effects of routine prophylactic supplementation with iron and folic acid on admission to hospital and mortality in preschool children in a high malaria transmission setting: Community-based, randomised, placebo-controlled trial. *Lancet* **367:** 133–143.

Shankar P, Boylan M, Sriram K 2010. Micronutrient deficiencies after bariatric surgery. *Nutrition* **26:** 1031–1037.

Sharp PA. 2010. Intestinal iron absorption: Regulation by dietary and systemic factors. *Int J Vitam Nutr Res* **80:** 231–242.

Skorokhod OA, Caione L, Marrocco T, Migliardi G, Barrera V, Arese P, Piacibello W, Schwarzer E. 2010. Inhibition of erythropoiesis in malaria anemia: Role of hemozoin and hemozoin-generated 4-hydroxynonenal. *Blood* **116:** 4328–4337.

Smith JL, Brooker S. 2010. Impact of hookworm infection and deworming on anaemia in non-pregnant populations: A systematic review. *Trop Med Int Health* **15:** 776–795.

Tanno T, Rabel A, Lee YT, Yau YY, Leitman SF, Miller JL. 2010. Expression of growth differentiation factor 15 is not elevated in individuals with iron deficiency secondary to volunteer blood donation. *Transfusion* **50:** 1532–1535.

Thankachan P, Walczyk T, Muthayya S, Kurpad AV, Hurrell RF. 2008. Iron absorption in young Indian women: The interaction of iron status with the influence of tea and ascorbic acid. *Am J Clin Nutr* **87:** 881–886.

* Thein SL. 2013. The molecular basis of β-thalassemia. *Cold Spring Harb Perspect Med* **3:** a011700.

Ullrich C, Wu A, Armsby C, Rieber S, Wingerter S, Brugnara C, Shapiro D, Bernstein H. 2005. Screening healthy infants for iron deficiency using reticulocyte hemoglobin content. *JAMA* **294:** 924–930.

Usher R, Shepard M, Lind J. 1963. The blood volume of the newborn infant and placental transfusion. *Acta Paediatr* **52:** 497–512.

van Bokhoven MA, van Deursen CT, Swinkels DW. 2011. Diagnosis and management of hereditary haemochromatosis. *BMJ* **342:** c7251.

van Rheenen P, Brabin BJ. 2004. Late umbilical cord-clamping as an intervention for reducing iron deficiency anaemia in term infants in developing and industrialised countries: A systematic review. *Ann Trop Paediatr* **24:** 3–16.

* Vichinsky EP. 2013. Clinical manifestations of α-thalassemia. *Cold Spring Harb Perspect Med* **3:** a011742.

Vitale G, Barbaro F, Ianiro G, Cesario V, Gasbarrini G, Franceschi F, Gasbarrini A. 2011. Nutritional aspects of *Helicobacter pylori* infection. *Minerva Gastroenterol Dietol* **57:** 369–377.

Viteri FE. 1998. A new concept in the control of iron deficiency: Community-based preventive supplementation of at-risk groups by the weekly intake of iron supplements. *Biomed Environ Sci* **11:** 46–60.

Vivarelli E, Siracusa G, Mangia F. 1976. A histochemical study of succinate dehydrogenase in mouse oocytes and early embryos. *J Reprod Fertil* **47:** 149–150.

Walters GO, Miller FM, Worwood M. 1973. Serum ferritin concentration and iron stores in normal subjects. *J Clin Pathol* **26:** 770–772.

World Health Organization. 2007. Conclusions and recommendations of the WHO consultation on prevention and control of iron deficiency in infants and young children in malaria-endemic areas. *Food Nutr Bull* **28:** S621–S627.

I

J

K

L

M

N

T

U

V

X